REDOX LIFE

A S Naidu

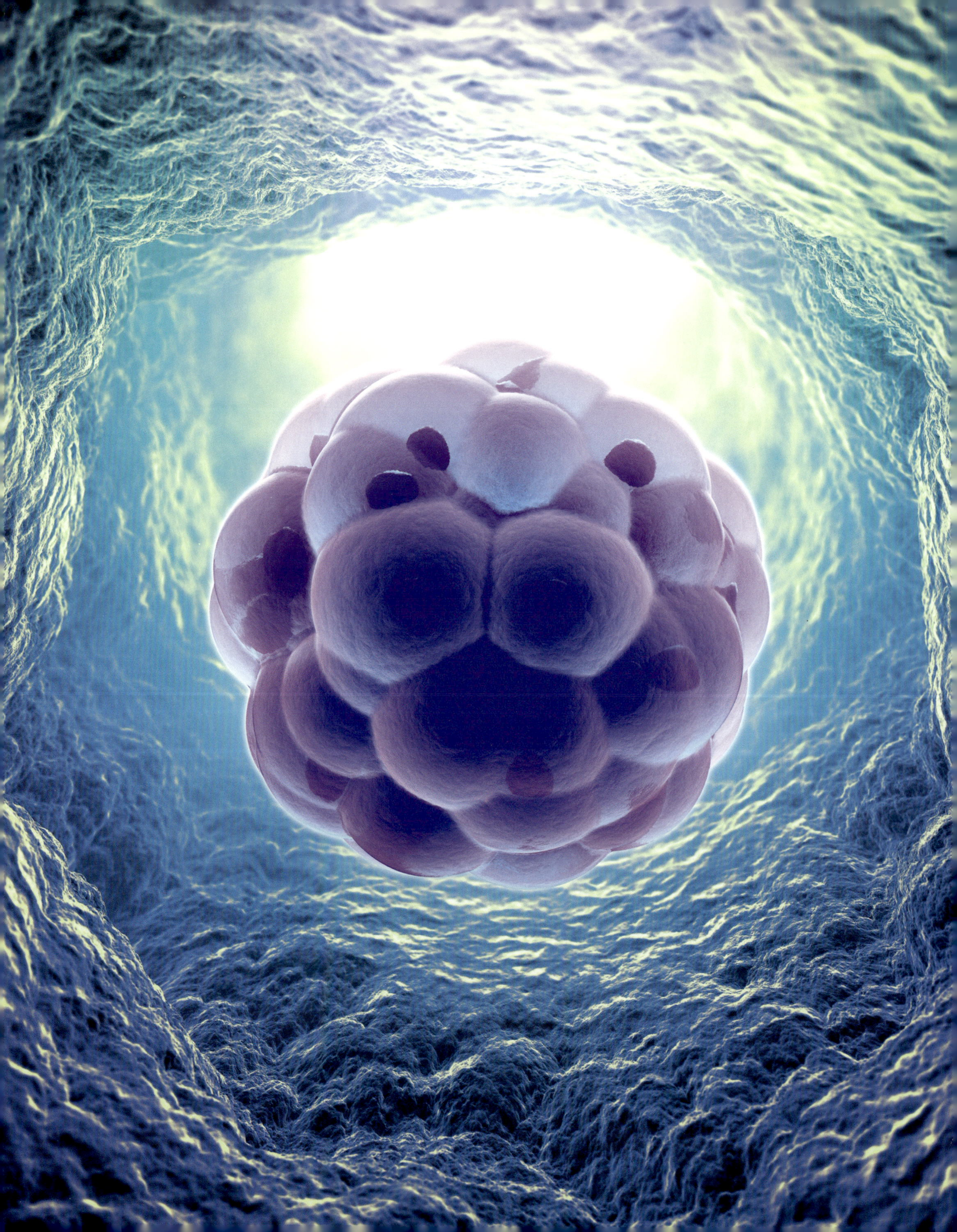

REDOX LIFE

A S Naidu

BIO-REP
NETWORK | MEDIA

Published by:
Bio-Rep Network | Media
Pomona, California

www.biorepnetwork.com

Library of Congress Control Number: 2013945485

DISCLAIMER: All reasonable precautions have been taken to verify the information contained in this publication. However, the published material is being distributed without warranty of any kind, either expressed or implied. The responsibility for the interpretation and use of the material lies with the reader. In no event shall the publishers be liable for damages arising from its use. Always consult your primary health care provider before starting any major change in your diet, health programs, or any course of supplementation or treatment, particularly if you are currently under medical care. If you have or suspect you may have a health problem, you should talk to your healthcare provider immediately. Only your physician should evaluate your health problems and prescribe treatment. Statements about theories and health conditions have not been evaluated by the U.S. Food and Drug Administration. The FDA has not reviewed or approved the Bio-Replenishments to diagnose, cure or prevent disease.

Design: Juan Carlos Torres

Book Layout: Bio-Rep Network | Media

First Printing: August 2013

10 9 8 7 6 5 4 3 2 **1**

ISBN 978-0-9824451-1-2

Printed in India at Pragati Offset (www.pragati.com)

BIO-REP

to

AISY

my own little free radical

Contents

2 REDOX: Chemistry

3 REDOX: Biology

4 REDOX: Medicine

5 REDOX: Nutrition

The Prologue

THE Origin of the Universe has been described by cultures world-wide. It was depicted as a manifestation of a giant egg – as in the Indian scripture *Brahmanda Purana*, the *Chinese Pangu*, and the *Finnish Kalevala*. In other ancient mythology, the creation idea is portrayed as a single entity emanated by Him or Her – as in the *Greek Gaia*, the *Tibetan Adi Buddha*, the ancient *Egyptian God Atum*, the *Aztec Goddess Coatlicue*, and the creation narrative of *Genesis*.

A number of alternate notions have also been concieved by other civilizations. In the *Maori folklore of Rangi and Papa*, the world was believed to be a result of the union of male and female energy. In other inscriptions, the Universe was crafted from pre-existing matter, such as the corpse of a dead God– as from *Tiamat in the Babylonian epic* or from the *Giant Ymir in Norse mythology*; or from chaotic materials, as in the *Izanagi-Izanami in Japanese mythology*. Still, in other civilizations, the Universe evolved from the cosmic resonance of *Brahman-Prakruti in the Hindu Manu Dharma*, the creation *myth of the Serers*, and the *yin-yang of the Tao*.

The two primordial forces that transverse through all of these extraordinary belief systems are – the 'Free Energy of Elements' and the 'Cosmic Power of Creation'. From a quantum mechanics stand point, both these forces collide and unify, deflect and divide, interact and equilibrate, in the same dimension of time, under the governing principles of REDOX. This book is a humble tribute to the ever expanding knowledge of REDOX Life, so – **Let there be light!**

A S 'Narain' Naidu
July 5, 2013
Anaheim Hills, California

'Y.S.' - The Miracle Man

"You've probably never heard of Dr. Yellapragada SubbaRow. Yet, because he lived, you may be alive and are well today. Because he lived you may live longer" – Doron Antrim

(1950, News Report from NY)

Discovery of adenosine triphosphate (ATP), the 'vital energy carrier' and an integral part of the genetic material (DNA/RNA), redefined life sciences; but never made it to the Nobel Prize. This tribute is to a dearest High School Alumni from my small hometown in India, who made the greatest discovery of the millennium and just faded away into the dusted layers of time.

Dr Yellapragada SubbaRow (Y.S.) accomplished several historic and extraordinary discoveries in biological chemistry and medicine between1924–1948. He was a scientist at Harvard Medical School and the director of research at Lederle Laboratories. In 1925, SubbaRow's discovery of phosphocreatine in muscle physiology has led him to the isolation and identification of adenosine triphosphate (ATP). Besides this landmark discovery of ATP (along with a battery of other nucleotides, i.e. ADP, AMP and other phospho-derivatives); SubbaRow is credited with several milestones in medicine that include: folic acid (role in anemia in 1946), methotroxate (cancer drug in 1947), diethylcarbamazine (anti-filarial drug in 1946) and tetracycline (antibiotic in 1948), SubbaRow's scientific contributions also span to other molecules such as pantothenic acid (vitamin B5), cobalamin (vitamin B12), steroids, therapeutic drugs against tuberculosis and blood pressure. Most of this work has been published in more than one hundred papers, but some of it has remained unpublished (due to 'Politics of Excellence'). In this context, George Hitchings (1988 Nobel Prize in Physiology or Medicine – a colleague of SubbaRow) has remarked that *"SubbaRow isolated several phosphorus compounds which were most likely nucleotides associated with RNA synthesis, and they had to be rediscovered by other scientists, because he was not allowed to publish that work".* This shows how SubbaRow's efforts were in vain, and this perhaps delayed considerably our understanding of RNA for decades! SubbaRow received little recognition during his lifetime, but is now hailed as one of the most eminent scientific minds of the century.

Dr SubbaRow's parting words (he died on August 8, 1948 at the age of 53), *"If God will spare me another couple of years, maybe we can cure another disease"*. A fitting epitaph for a beautiful mind!

Preface

THE government high school I attended as a teenager is located in Rajahmundry, a small river town in India. This brown granite high school has witnessed the metamorphosis of India from a British Raj into an independent nation through the 20th century and a transformation into Global India during the 21st century. Its corridors have survived the cheering noises of youngsters for more than a century; good enough to produce several luminaries, including two great presidents for the Indian subcontinent. In 1969, during my first walk though this school's historic hallways, an old sepia-toned photograph was brought to my attention by classmates. Upon closer examination, even I agreed that the image of the young student in the photograph looked almost like my identical twin! Later, our chemistry teacher would tell us that my look-alike was actually 'Y.S.', *Yellapragada Subbarow*, the discoverer of adenosine triphosphate (ATP), which is the fundamental energy carrier for every living organism. Though he was an alumni by almost half a century, learning about 'Y.S.' was my first encounter into the REDOX world!

REDOX Principles: During my college years, when *Peter Mitchell* won the 1978 Nobel Prize in Chemistry for putting forward the groundbreaking 'Chemiosmotic Theory' and elucidating the cellular generation of ATP, 'Bio-Energetics' became the buzz. Around this time, *Eugene Robinowitch* and *Govindjee* were also unraveling the energy transfer mechanisms of photosynthesis. The influence of membrane potentials on ATP generation became my graduate research focus track during the late 70s and eventually led to my doctoral thesis with *Kola Rajyalakshmi*, MD, on membrane-perturbing bacterial protein toxins. When I discovered the staphylococcal non-menstrual toxic shock syndrome several years later, in 1984, my academic interests were aligned to the field of molecular medicine. Though we were not aware of the REDOX mechanisms three decades ago, it is now known that toxic shock syndrome is a unique cell signaling phenomenon which is potent enough to inflict an army troop (the first identified susceptible population group) and manifest severe fatal outcomes within hours. Therefore, my initial foray into the protein chemistry of toxins, with respect to their structure-function and molecular interaction with lipid membranes, necessitated recalibrating my scientific curiosity into an intellectual obsession. This pursuit would eventually transport me to the Biomedical Center-Uppsala, which is known in the scientific world as the 'Vatican' of molecular separations. Here, Nobel Laureates like *Theodor Svedberg* (1926 Nobel Prize in Chemistry) invented ultra-centrifugation and *Arne Tiselius* (1948 Nobel Prize in Chemistry) developed electrophoresis to separate immunoglobulins. I was fortunate to work with Tiselius's successor – *Stellan Hjérten*, the inventor of agarose, free zone electrophoresis, SDS-PAGE and various modes of high-performance liquid chromatography (HPLC). The company of molecular separation chemists like *Stellan Hjérten*, *Jerker Porath* (who developed Sephadex columns), and *Per Åke Albertsson* (the discoverer of aqueous phase partition) opened my eyes to the amazing world of REDOX Chemistry!

REDOX Chemistry: REDOX (oxidation-reduction) reactions involve the escape of electrons (charged particles) from molecules or ions in one chemical substance and their capture by other substances. Free energy is the key to understanding this electron escape phenomenon. REDOX chemistry also refers to a change in the oxidation state, since the actual transfer of electrons may never occur. Accordingly, the REDOX potential reflects the physico-chemical attributes of molecules such as the net charge, surface hydrophobicity, van der waals forces, etc; all of which play a role in various biotic and abiotic interactions. Chromatography and electrophoretic methods are based on some of these REDOX attributes. For example, the binding (interaction) and release (elution) of molecules from a chromatography column is based on simple REDOX interference through gradients of salts or chaotropic agents. Interestingly, physiological systems (such as the alimentary canal, vascular lumen or respiratory tract) are coated with galactose-rich polysaccharides, such as heparan sulfate, and behave like chromatography columns.

In addition to the isolation of molecules, my early chromatography training also provided me with a powerful tool to distinguish between native (functional) and denatured (dysfunctional) proteins; the latter are also known as artifacts. I observed that a common laboratory protocol, such as the freezing/thawing of a biochemical sample or limiting oxygen exposure, could significantly affect the functional outcome of a protein. Furthermore, it fascinated me that a simple change in milieu conditions (such as ionic strength, pH, molarity of an eluting salt, or temperature) could significantly alter the protein's function. Our group would eventually report that a conformational change occurs in the structure of toxic shock syndrome toxin-1 (TSST-1) protein within few hours of exposure to oxygen; accordingly, a highly potent TSST-1 rapidly degrades into a less active or inactive form just by standing exposed on a lab bench. Interestingly, we also learned that the addition of certain metal chelators could prevent or delay such protein's oxidation. This snap shot of what oxidative damage could do to protein conformation led me to develop technologies that could reactivate (recoil) biological molecules to their native structural and functional states. It was an idea that resulted from sheer serendipity.

REDOX chemistry can influence the functional outcomes of bio-molecules by induction of conformational changes; which in turn, can alter surface charge distributions, structural rearrangement, and active domain exposure of the affected molecules. This finding led me to investigate specific chemical structures on biological surfaces that promote microbial adhesion and colonization to epithelial mucosa. These receptor-ligand interactions are prerequisite factors in the pathogenesis of several diseases. My work in the later part of 1980's was primarily in this exciting field of host-microbial interactions. I am indebted to *Arne Forsgren* (the discover or protein A) and *Torkel Wadström* (who pioneered the area of bacterial interactions with epithelial matrices) my two mentórs from Lund Univesity, Sweden.

In 1988, I was commissioned by the Swedish Dairies Association ('SMR')/ARLA to explore the possibility of human health applications for antimicrobial proteins from milk; this project brought lactoferrin, an iron-binding protein molecule, to my lab bench. As our research ensued, the REDOX chemistry of transition metals (i.e. iron, copper, manganese, cobalt, and vanadium), Haber-Weiss, and Fenton-type reactions suddenly marked their entries into our protocols. A research grant from the Swedish Biotechnology Board ('STU') and Semper AB (the Scandinavian Infant Food Formulary Giant) enabled me to assemble an elite team of scientists from 6 countries under one roof. Clinical management of gastrointestinal illnesses, especially with the Gram-negative bacteria, was the containment target. During the 1980's, technologies to interfere or block microbial attachment to epithelial surfaces were the focus of a newly emerging field of clinical research. Our discovery of the interaction of lactoferrin with endotoxin (LPS) enclosed porins (OmpC, OmpF and PhoE) on Gram-negative pathogens revealed the role of REDOX potentials (E_h) on bacterial outer membranes and colonization factors. We found that simple ionic/charge modifications in the milieu could significantly alter the expression and function of bacterial outer membrane proteins (Omps); surprisingly, such REDOX-based milieu manipulations could also potentiate the antibiotic susceptibility patterns of bacterial pathogens. In 1988, of the thousands of enteric pathogens tested in my laboratory, a newly discovered *Escherichia coli* O157:H7, the etiological agent of hemolytic uremic syndrome, made it to the test protocol. Four years later, this pathogen created public health havoc in the USA and was branded as a 'killer *E. coli*' that transmits through hamburgers. The massive media attention surrounding this outbreak prompted the Clinton administration to eventually rewrite USDA food safety guidelines. My tryst with REDOX Biology was about to begin.

REDOX Biology: In the early 1990s, the regulatory role of REDOX in cell signaling, protein activation/deactivation, cellular differentiation and apoptosis were just being explored in the scientific community. Among the various classes of ROS-induced macromolecular structural modifications, protein oxidation was identified with functional consequence; since proteins act as receptors, transporters, enzymes, transcription factors and cytoskeletal components, among others. Moreover, proteins are the largest constituents of the cell and probably the most frequent targets of free radical attacks.

In 1992, differential scanning calorimetric studies from my lab showed that the thermal folding and unfolding of milk proteins could result in immunogenicity (allergenicity) of dairy nutrients, especially in the UHT treated products (compared to pasteurized products). The vulnerability of cysteine and methionine residues and the resulting in thermal cleavage of disulfide bonds were identified as the underlying cause. Our laboratory also observed that negative charge distributions, which affect the REDOX potential of most proteins, are confined to the N-terminus end of a polypeptide chain. This finding continues to have seminal implications in REDOX biology, from the stand-point of receptor-mediated cell signaling and regulation of innate defense at mucosal transit. The protein immobilization method patented from my laboratory is now a globally recognized technology-transfer that has become a mainstay process in the field of biotechnology. This process turned out to be the crux in the protein engineering of activated lactoferrin, which was used for the control of *E.coli* O157:H7 in the beef industry world-wide.

The discovery of microbial blocking agents (MBAs) from my laboratory also brought a new class of natural antimicrobials to the forefront. These compounds can effectively attenuate and safely detach microbes from tissue and other biological surfaces instead of cidal effects. Lactoferrin, the first patented MBA from our laboratory, was used in the meat industry to decontaminate beef from *E.coli* O157:H7 and other food-borne pathogens. Two critical hurdles surfaced and needed a solution before we could begin industrial-scale application of this compound: (i) how to prevent oxidative damage to this 80-kD iron-binding protein, which needs to remain in aqueous solution (100s of gallons in volume); and (ii) how to deliver this basic protein (isoelectric point ~9.0) onto a beef carcass (>800 m^2 gigantic surface) with a uniform MBA distribution as low as ~8 ng/mm^2 of tissue surface within a time frame <2 seconds! This monumental challenge was met through REDOX technology; thus, an immobilized protein was engineered with a low threshold of electrostatic charge, which would behave like a charged particle colliding with an organic target in femto-seconds. To charge-transport the proteins to the organic targets (beef carcasses), we simply had to adapt the REDOX principles commonly used in the automobile industry (spray painting of cars) or electrostatic crop dusting of vineyards in Napa Valley. After testing this spray technique in a remote aircraft hangar in Dodge City, Kansas; the technology was transferred to the meat industry in 2001. A year later, 'activated lactoferrin', the world's first REDOX-amplified protein molecule, received GRAS (generally recognized as safe) status from the United States Food and Drug Administration (US-FDA) for industrial-scale food applications.

In 1988, when lactoferrin entered my resume, this protein was available for research purposes only, with selected chemical suppliers (i.e. Sigma or US Biochemicals Corp). In 2000, when the activated lactoferrin technology was announced from the power corridors of Washington DC, the news reached more than 300 TV channels, circulated through thousands of newspapers, and made it to the headlines of trade journals around the world. The impact could even be measured on Wall Street, where the stocks of beef and dairy corporates spiked within a day of our announcement. Over the next decade, over 500 tonnes of lactoferrin would be produced annually across 6 continents by more than two dozen giants in the dairy industry. The time had arrived where I needed to return my research priorities back into the field of medicine and, over the next few years, REDOX biology would play an integral role in the development of several human health applications.

REDOX Medicine: Nitroglycerin (NTG) is perhaps the first REDOX-based commercial treatment that originated from Sweden. In 1851, *Alfred Nobel* recognized the therapeutic potential of NTG and began its manufacture. Later in his life, Nobel suffered from angina pectoris; it is therefore ironic that, in 1890, physicians recommended NTG for his heart condition. In the late 1970s, the vasodilator effect of NTG was discovered to be mediated by nitric oxide (·NO). In 1998, approximately 130 years after the invention of dynamite by Alfred Nobel and the first observed clinical benefit of NTG, the Nobel Prize in Medicine or Physiology was awarded to *Robert Furchgott*, *Louis Ignarro* and *Ferid Murad* for discovering nitric oxide as a signaling molecule in the cardiovascular system.

REDOX is an integral mechanism of phagocytic respiratory burst, as well as host inflammatory responses. Lactoferrin is a major constituent of the secondary granules of neutrophils and its role in respiratory burst during phagocytosis has been well documented for more than three decades. Lactoferrin is also a key regulator of the inflammatory cell cascade, mediated through its specific receptors on neutrophils, T and B lymphocytes, monocytes, and macrophages. From 2005-2009, my scientific team's investigations demonstrated that a lactoferrin-ribonuclease (R-ELF) complex naturally found in milk could elicit specific regulatory effects on the balance of bone turnover (bone resorption versus bone formation) and optimize the bone inflammatory pathways (pro-inflammation versus anti-inflammation). Our subsequent understanding of REDOX switches in these two systems has the potential to significantly address the global issues of osteoporosis and arthritis. Several investigations are currently underway at our laboratory to define the role of specific REDOX switches that trigger oxidative stress in metabolic disorders such as obesity, diabetes and atherosclerosis.

REDOX Nutrition: Nutrition is the provision of chemicals that are necessary to support the metabolism of cells and organisms. Since metabolism/energy production requires oxygen (O_2), potential cell damaging free radical species are generated during this process. An optimum dietary intake of antioxidants and other REDOX supportive nutrients is critical to sustain good health. Accordingly, several diseases and dysfunctions can be prevented or alleviated with a proper diet that is enriched with antioxidant minerals, vitamins, free radical scavengers, and bio-replenishments. Moreover, modulation of optimum REDOX signaling through aqueous delivery systems could be the future of global healthcare.

In 2005, when a series of core patents on how to generate ATP from an extra-cellular stimulus was filed from my laboratory; it was a moment that brought my connection with 'Y.S.', the look-alike alumni from my high school who I had never had the opportunity to meet, to a full circle. It was also the marking of a new scientific hypothesis, the 'Bio-Replenishment Theory' – the summation of my lifetime of scientific observations (a topic for discussion for another day). As a teenager, while walking through the corridors of my high school, I was determined to foster science towards social relevance. In 1969, when I announced the name and logo of my company en-N-tech and its affiliate laboratory – the N-terminus; it must have been quite amusing to my classmates and teachers as I was only

13 years old then. But thirty years later, in 2000, when the company bearing the same name and logo made it into the corporate world, with its scope reaching 30+ countries, it was a feeling that could be compared to completing the first lap of a marathon. Despite the accomplishment, I realized even then that several more laps had to be sprinted before I could reach my finish line. In the true spirit of *Yellapragada Subbarow* – who wished, even with his last days looming, to cure just one more disease – for me there are still several more health issues that I am committed to help resolving. Just like those before me, the mantle shall be always be carried on. It always has been, and is, a privilege to embark upon such exciting journey in this beautiful REDOX Life.

Sigmund Freud often remarked that greatest revolutions in the history of science have but one common, and ironic, feature: they knock human arrogance off one pedestal after another of our previous convictions about our own self-importance. In Freud's three examples: Copernicus moved our home from center to periphery; Darwin then relegated us to descent from an animal world; and, finally (in one of the least modest statements of intellectual history), Freud himself discovered the unconscious and exploded the myth of a fully rational mind. REDOX is yet another anomaly to our conventional wisdom regarding our pre-casted scientific knowledge, in general, and the evolution of life, in particular.

This voluminous script has gone to print due to the relentless effort of several colleagues, friends and institutes. First and foremost, I am indebted to *Jagat Narula*, MD, PhD, MACC (Professor of Medicine-Cardiology, and Associate Dean for Global Health at Mount Sinai Hospital, New York) for being a true brother, for keeping my thoughts in check, and for providing me with multi-faceted insights into the deepest abyss of medical research. I am also proud to affiliated with *Narendra Paruchuri* (Pragati Offset), who has taken regular offset printing into an award winning global art. He is responsible for transforming my black and white script into a million shades of CMYK colors. I appreciate the collaborative effort of *Juan Carlos Torres* with this book's design. Working alongside Juan was a delight and he managed to fulfill this monumental task in time. Proof reading this voluminous technical subject was a strenuous effort, and I'd like to extend my sincere appreciation to *Smita Naidu*, MS; *Ashutosh Muchrikar*, MTech, *Tezus Naidu*, MBS, (PharmD); *Sreus Naidu*, MS, (PharmD); and *Aishwarya Naidu*, Esq., MS (Regulatory Science) for browsing the manuscript for edits and revisions. *Franklin Samuel Ramos* deserves special mention for his excellent help with literature search, tabulating the abbreviations, and arranging the glossary. Thanks to *Ken Carroll* and *Cameron St. George* for their administrative support. The talented Pragati Offset Team (Hyderabad, India) deserves a salute, especially, *Kamuni Ravindra Kumar, Devalla Laxmi Krishna Kishore, Krishnamurthy Shiva Shankar, Guduri Bhima Shankar, Jakka Vijaya Kumar,* and *Ganti Rama Krishna*, in creating the cover and preparing the final version to the press. Finally, I sincerely thank my publisher, Bio-Rep Media for supporting the Bio-Replenishment Book Series and launching 'REDOX Life' to the global audience.

As a concluding note, I believe that scientific research should prevail like child's innocence; with a virtue to admit ignorance and enthusiasm to raise questions. Scientific curiosity, therefore, is an instinctive evolution of three consequential phases leading to a belief system: (i) *Trust*, instituted by the body of information collected through academic pursuit (my reverence is due to my teachers, professors and mentors); (ii) *Experience*, attained through a calendar of events that imparts us with the patience to observe and face ignorance (salute to my students, lab technicians, fellows, research colleagues, and scientific collaborators for subjecting me through this process); and, (iii) *Understanding*, which needs a logical mind trained to process available data and interpret the unknown (thanks to my biotic and abiotic milieu, for keeping my inspiration afloat). I have learned about the phosphoric acid from my first Chem. lab in high school, to regression analysis of Scatchard plots in Sweden; from the tiny earthworm in my first biological dissection to every smile that my research brought to faces unknown to me around the world. It has been a privilege to review the original articles of over a hundred Nobel Laureates in Chemistry, Physiology or Medicine; thousands of REDOX-related original publications; and hundreds of reviews to generate a comprehensive manuscript with generally accepted principles and concepts. However, the daily increment of about 10 publications, which often conflicted with seemingly established views or opened up new perspectives, continuously forced me to revise this book. I, therefore, have to apologize for just delivering a snap shot of current opinions which is biased by medicinal chemistry, cellular biology and my own journey with REDOX Life.

A S 'Narain' Naidu
July 5, 2013
Anaheim Hills, California

CHAPTER

1

REDOX Principles

Life on Earth is an organized energy management system. Based on the source of energy utilized, living systems can be classified into two categories: i) autotrophs that derive energy from either sunlight or chemical reactions involving abiological compounds. Photosynthesis, for example, is an autotrophic life process that uses a narrow band of electro-magnetic radiation (visible light) emitted from the Sun. Thus, photosynthesis allows life to tap into practically unlimited source of energy. ii) Heterotrophic life, the other category uses high-energy organic molecules produced by autotrophs as a source of energy. Light, organic molecules, and reactive inorganic chemicals provide abundant sources of energy on Earth, so living organisms have evolved specialized adaptations to use the energy sources. Invariably, both living systems harness energy exclusively through REDOX reactions.

1

REDOX: Principles

The fundamental question of how the free radical chemistry evolved and life was sustained on Earth remains unanswered. This reflects our ignorance of basic REDOX Chemistry about the electron transfers that bring the ensemble of molecules in cells to 'LIFE'. Despite 150 years since the publication of 'The Origin of Species', the origin of life remains a black box.

Life is a highly ordered and intricately designed biochemical system. For instance, photons from the Sun do not directly incorporate into the chemical bonds of the cell; there has to be an intermediate 'high energy' carrier to store solar energy until the cell is ready to use it. The accrual of high-energy intermediates such as ATP (adenosine triphosphate) depends on the generation of proton gradients across membranes. Bio-informatics, the intricate genetic design and bio-energetics, the mechanics of the electron transfer system, interplay to accomplish three vital activities: (i) capture chemical energy to produce complex bio-molecules, (ii) expend energy to transport chemicals from the milieu into the cell, and (iii) generates mechanical energy to enable cells to change their physical location, mobilize and reorganize their interior structure. These activities require energy as well as a genetic blue print to direct the spatial arrangement of molecules.

REDOX is the driving force for life on Earth. The universe is always moving towards disorder (an increase in 'entropy' or 'randomness') and a living cell is a protest against this 2nd law of thermodynamics! Life is a constant battle with entropy; however, it ultimately loses and dies.

In a REDOX reaction, polar products are generated from non-polar reactants. REDOX chemicals (reactive species) also serve as cellular messengers. Changes in the REDOX status in the cell regulate several signal transduction pathways. A simple analogy that elucidates this mechanism is a regular lithium battery. Typically, a battery will discharge by itself. For a recharge, it needs to be plugged into an energy outlet, sufficient to 'carry' the electrons uphill (charge) back into the battery - so they can fall down (discharge) again. Organisms require a continuous flow of chemicals to refill the expended molecules that regulate cellular homeostasis, 'Bio-replenishment' fulfills this vital component. Thus, life is governed by the principles of REDOX Chemistry, supported by *Bio-Informatics*, *Bio-Energetics* and *Bio-Replenishment* as the operational triad.

1.1 REDOX: Laws and Theories

All living organisms require energy for survival, growth and multiplication. The ability to harness energy from a variety of metabolic pathways is a fundamental property of all living organisms. Life is dependent on energy transformations; living organisms survive because of exchange of energy within and without. Biological systems are composed of complex molecular configurations whose total bonding energy is less negative than that of their chemical precursors. Thus, the Gibbs free energy (ΔG) of living systems is quite high relative to the simple compounds from which they are formed.

In living plants, the energy flow is provided by solar radiation. Leaves (with chloroplasts) have relatively large surface area per unit volume; which allows them to capture essential solar energy. This solar power is then converted into biochemical function(s) that maintain the plants in their complex, high-energy configuration by a REDOX process called photosynthesis. Mass, such as water and carbondioxide, also flows through plants, providing necessary raw materials, but not energy.

For animals, energy flow is provided through consumption of high energy biomass (food, either plant or animal source). The assimilation (breaking down) of this energy-rich biomass with subsequent REDOX reactions supplies a continuous source of energy as well as raw materials.

If plants are deprived of sunlight or animals of food, dissipation within the system will manifest into death. However, a source of energy alone is not sufficient to sustain life. REDOX influxes are also vital in transforming energy into metabolism that maintains biological systems built from simple bio-monomers, the molecular building blocks. Therefore, without a REDOX mechanism to couple the available energy with the organism's metabolism, a mere availability of high-energy biomass is insufficient to sustain life. In living systems, the REDOX coupling mechanisms channel energy along specific chemical pathways to provide cellular vitality.

Life on earth is dependent on energy. Without a constant supply of energy, life on earth would cease to exist. One of the most common ways to define energy in living systems is the ability to do work (metabolism), which is the movement of molecules in and out of the cells. REDOX principles play a vital role in facilitating these metabolic pathways.

1.1.1. REDOX and Energy Demand

Energy is an essential attribute of all biological systems from the largest species to the smallest living organism in the biosphere. Survival of every life form is based on cellular development and organelle maintenance of its biological system. In order to accomplish these metabolic tasks, energy is often stored by cells in various chemical structures of compounds such as carbohydrates (including sugars), lipids, and proteins, which release energy when reacted with oxygen in respiration – a REDOX mechanism.

Living organisms rely on external sources of energy to survive, grow and multiply. These energy sources can be radiation from the Sun (in the case of green plants) or chemical energy in some form (in the case of animals.)

Energy can be divided into two broad categories: (i) kinetic energy (the energy of motion), and (ii) potential energy (stored energy). Kinetic energy is associated with movement, where as, potential energy is dependent upon fundamental forces of nature. Potential energy is stored in the arrangement of objects or particles that are under the influence of a particular type of force ('force field' or 'field'). These forces include gravitational energy (which is stored in a way that the masses are arranged in a gravitational field), several types of nuclear energy (which utilize potentials from the nuclear force and the weak force), electrical energy (from the electric field), and magnetic energy (from the magnetic field). Other familiar types of energy are a varying mix of both kinetic and potential energy. An example is mechanical energy, which is the sum of kinetic and potential energy in a system. Elastic energy in materials is also dependent upon electrical potential energy (which is between atoms), as is chemical energy, which is stored and released from a reservoir of electrical potential energy between electrons and the molecules or atomic nuclei that attract them.

Energy exists in different forms including thermal, mechanical, kinetic, chemical, electrical, magnetic, radiation, nuclear, elastic, sound, gravitational, and mass ($E = mc^2$). For example, a bicyclist riding (movement) up and down a hill exhibits kinetic energy. However, at the top of the hill he has potential energy which, based upon his position, imparts the ability to do work (ride down the hill). The daily energy requirement of 1500–2000 calories recommended for human adults is obtained through REDOX reactions (a combination of oxygen + food molecules). The dietary molecules are mostly carbohydrates and fats, of which glucose ($C_6H_{12}O_6$) and stearin ($C_{57}H_{110}O_6$) are prominent substrates. These molecules are further oxidized into carbon dioxide and water by the following REDOX process:

$$C_6H_{12}O_6 + 6\ O_2 \rightarrow 6\ CO_2 + 6\ H_2O$$

$$C_{57}H_{110}O_6 + 81.5\ O_2 \rightarrow 57\ CO_2 + 55\ H_2O$$

and some of the energy is used to convert ADP into ATP

$$ADP + HPO_4^{2-} \rightarrow ATP + H_2O$$

Mass Energy Equation ($E=mc^2$): The 1921 Nobel Prize in Physics was awarded to Albert Einstein for his contribution to Theoretical Physics, and especially for his discovery of the law of the photoelectric effect.

The recommended BMI in normal adults ranges between 19 and 24. A high BMI can potentially lead to obesity or health complications. In order to have an ideal BMI, an individual's dietary intake should not exceed the energy burned on a regular basis.

The rest of the chemical energy in the carbohydrate or fat molecules is converted into thermal energy (heat); the ATP is used as a 'cellular energy currency', and some of the chemical energy it contains is used for other metabolic functions such as conformational activation of proteins, transport of molecules, cell signaling, etc. The chemical energy that converted to heat serves a vital purpose – it keeps the molecular fabric of the body to remain active and highly functional.

The following points emphasize the basic relationship between REDOX, energy and life: (i) All biological processes require energy, from cell division, to secretion, to cell migration in an early embryo. Thus, living systems have strategies to harness and utilize energy. (ii) Energy does not arise from nothing. Since, it can neither be created nor destroyed; it transforms from one source to another. Biological processes depend on REDOX principles to control the transformation and use of energy. (iii) The biology of a cell is a complex and continuous web of REDOX couples that build new molecules and dismantle existing ones. Synthesis (anabolism) of new compounds requires energy; breakdown (catabolism) of the old structures releases energy – The REDOX coupling mechanism connects both these metabolic processes. (iv) ATP, through high-energy phosphate bonds, temporarily stores energy that a cell uses for a variety of activities. v) Without enzymes supported by REDOX, biochemical reactions cannot move fast enough for living organisms to survive; enzymes lower the energy demand (requirement) to start these reactions.

Food energy: The human body carries out its physiological functions by consuming and turning food into usable energy. Energy is derived from the diet through the process of cellular respiration, which is the REDOX chemistry of combining oxygen species with the molecules of food (aerobic respiration) or the reorganizing of atoms within the molecules (for anaerobic respiration). Food energy is typically derived from carbohydrates, fats, proteins, organic acids, polyols, and ethanol present in the diet. Some dietary components that provide little or no food energy, such as water, minerals, vitamins, and fiber, may still be necessary as REDOX co-factors.

1 kcal = 4.184 kJ

In the European Union, both the 'kcal' and 'kJ' appear on nutritional labels. In many other countries, only one of the units is displayed; in the US and Canada the unit is spelled out as "calorie".

In the International System of Units, energy is measured in joules (J) or its multiple – the kilo joule (kJ). An older metric system unit of energy, still widely used in diet-related contexts, is the calorie; more precisely, the 'food calorie', 'large calorie' or 'kilo calorie' (kcal or Cal). Fats and ethanol have the greatest amount of food energy per mass, 38 and 30 kJ/g (9 and 7 kcal/g), respectively. Proteins and most carbohydrates have about 17 kJ/g (4 kcal/g). Carbohydrates that are not easily absorbed, such as fiber or lactose in lactose-intolerant individuals, contribute less food energy. Polyols (including sugar alcohols) and organic acids have less than 17 kJ/g (4 kcal/g). Food energy can be measured in different ways, such as Gibbs free energy (ΔG) of combustion, or the amount of ATP generated by dietary metabolism. Conventional food energy is based on the heat of combustion in a bomb calorimeter and corrections that take into consideration the efficiency of digestion, absorption, and the production of urea and other substances in the urine.

The human equivalent (H-e or 'Human energy conversion') indicates, for a given amount of energy expenditure, the relative quantity of energy needed for human metabolism. The calculation assumes an average human energy expenditure of 12,500 kJ/day and a basal metabolic rate (BMR) of 80 watts. For example, if the body runs (on average) at 80 watts, then a light bulb running at 100 watts is running at 1.25 human equivalents (100 ÷ 80 = 1.25 H-e). For a difficult task lasting only a few seconds, a person can put out thousands of watts while climbing a flight of stairs may represent an energy expenditure of about 200 watts (note: 746 watts = 1 official horsepower). For tasks lasting a few minutes, a fit human can generate perhaps 1,000 watts or even more. For an activity that must be sustained for an hour, output drops to around 300; for an activity kept up all day, 150 watts is about the maximum. The H-e assists in understanding how energy flows in physical and biological systems by expressing energy units in human terms: it provides a 'feel' for the use of a given amount of energy.

An individual's energy requirement depends on the body's daily energy consumption and metabolic energy demand, which can be estimated by a simple using body weight and activity level. The following calculation represents daily energy consumption:

Daily Energy Consumption = Body weight (Kg) × 4 kJ × 24 hours/day / 4.18 kJ

Food intake that is not used as energy is stored in the body as fat. Excessive fat storage can lead to a higher body mass index (BMI). BMI indicates an individual's body fat and is determined by the subject's height and weight: BMI = Body Weight (Kg) / Height (m) × Height (m)

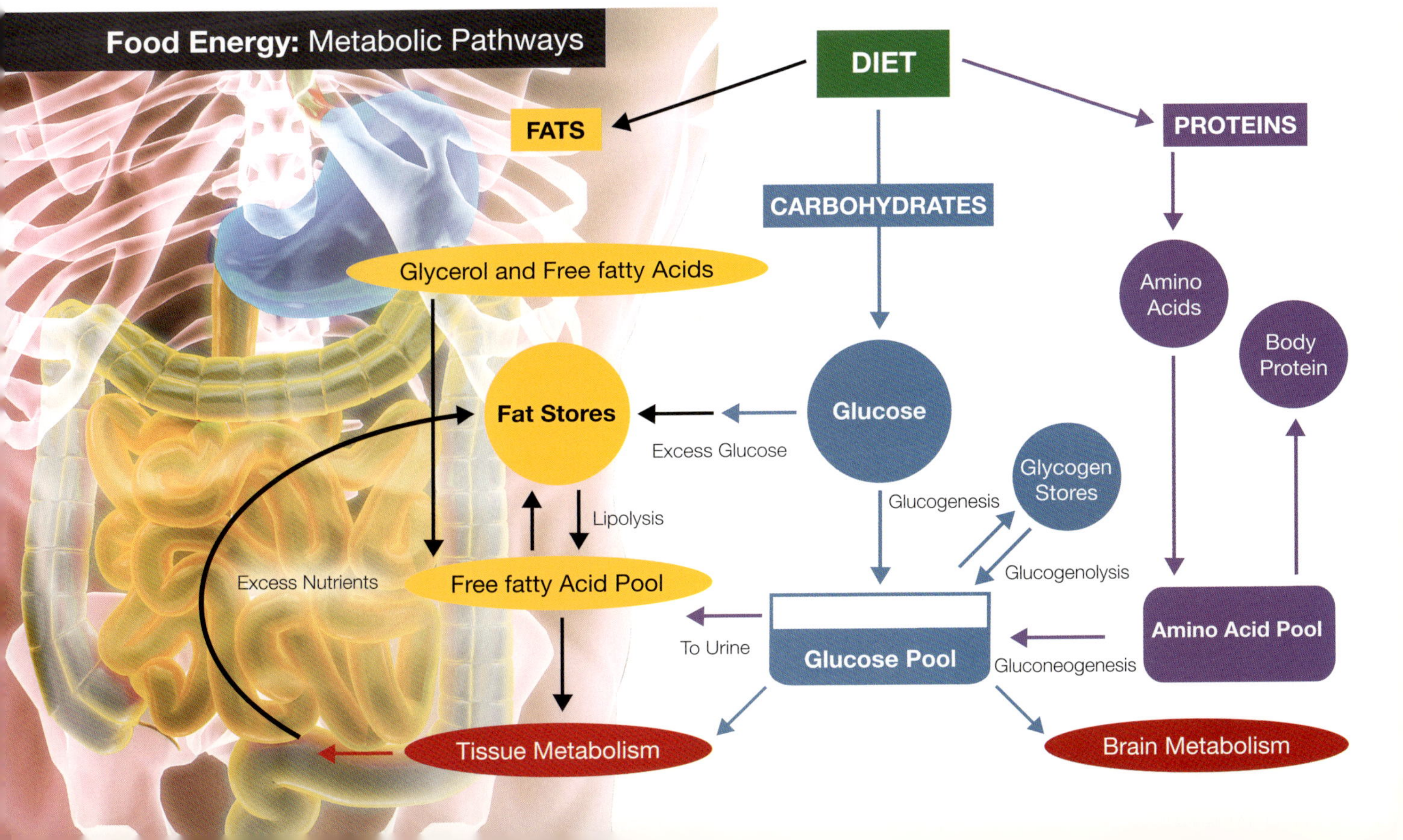

1.1.2. REDOX and Thermodynamics

"Life can be defined as regions of order which use energy to maintain their organization against the disruptive force of entropy" – Peter Molton.

Thermodynamics is an exact science that deals with energy; therefore, the laws of thermodynamics are fundamental for understanding biology. The laws of thermodynamics have made the evolution of higher life forms possible with the conjecture that the energy flows through biological systems are effective in organizing simple chemicals into complex arrangements associated with life. In existing living systems, the coupling of energy flow to the organizing "work" (metabolism) is regulated by REDOX principles. REDOX operates the metabolic motor of life that includes DNA, enzymes, and a number of biochemical pathways. Interestingly, the same REDOX principles also apply to automobiles that convert chemical (gasoline) energy into mechanical torque on wheels.

The laws of thermodynamics also govern the chemical processes (metabolism) in all biological organisms. Each atom and molecule that comprises a living cell totally obeys these energy laws and follows REDOX principles. Living systems use simple bio-monomers to construct complex polymers (such as DNA and proteins) through thermal, electrical, chemical, or solar energy flows regulated by REDOX. Accordingly, REDOX mechanisms provide a 'radical' explanation of the physio-chemical operation of life's metabolic motor and its underlying laws of thermodynamics.

The 1st Law of Thermodynamics

Also known as the law of conservation of energy, the first law states that energy can neither be created nor destroyed. It may change from one form to another, but the energy in a closed system remains constant. A major part of the science of thermodynamics is accounting, considering that the energy of a system has undergone some type of transformation. Thus, we derive from the first law of thermodynamics that the change in the energy of a system (ΔE) is equal to the work done on (or by) the system (ΔW) and the heat flow into (or out of) the system (ΔQ). In mathematical terms:

$$\Delta E = \Delta Q + \Delta W$$

The 2nd Law of Thermodynamics

The second law states that when energy is transferred, there will be less energy available at the end of the transfer process than at the beginning. Accordingly, the flow of energy in natural processes is irreversible. The energy flow in such processes always leans toward a more uniform energy distribution in the universe.

Negative Entropy: Schrodinger theorized that life, contrary to the second law of thermodynamics, runs on negative entropy. The 1933 Nobel Prize in Physics was awarded to Erwin Schrodinger (jointly with Paul Dirac) for discovery of new productive forms of atomic theory.

The concept of entropy (ΔS) gives a quantitative way to measure the tendency for energy to flow in a particular direction. The entropy change for a system is defined mathematically as the flow of energy divided by the temperature (T):

$$\Delta S \geq [\ \Delta Q / T]$$

where ΔS is the change in entropy, ΔQ is the heat flow into or out of a system, and T is the absolute temperature in degrees Kelvin (K). Due to entropy, which is the measure of disorder in a closed system, all available energy will not be useful to an organism. Entropy increases as energy is transferred. Life has sometimes been referred as 'reverse entropy (negative entropy)' because organisms maintain complex organized states compared to their surroundings. But they must obey the second law just as any other system of matter and energy; thus, all organisms must work continually to build and maintain non-random assemblages of matter and energy.

Entropy explains why it is easier to break something than to build it; and why the bio-machine called the human body will inevitably break down and die, or cease to function, someday.

From a REDOX stand point, the second law of thermodynamics can be summarized as follows: i) Entropy within the system be >0; ii) Sufficient energy flow is required to maintain negative entropy within the living system; iii) An efficient REDOX couple is needed to direct the negative entropy towards homeostasis of a living system and prevent disintegration. Therefore, REDOX coupling mechanisms are needed to bring specific kinds of energy-based molecular order to sustain and propagate life in biological systems.

The 3rd Law of Thermodynamics

The subject of entropy leads directly to the third law of thermodynamics, which states that, at the temperature of absolute zero, entropy also approaches zero (which is impossible to reach).

All matter is in motion at the molecular level, which helps define the three major phases of matter found on Earth. At one extreme phase is gas, wherein molecules exert little attraction toward one another and are therefore in constant motion at a high velocity rate. At the other end is the solid phase (with liquids in the middle) where molecular motion is minimized, and molecules merely vibrate in place. Absolute zero, or 0K on the Kelvin scale of temperature, is the point where all molecular motion stops entirely—or at least, it virtually comes to a halt.

Absolute zero is defined as the temperature at which the motion of the average atom or molecule is zero.

Scientists have attempted to reach the absolute zero, and indeed they have come very close. In 1993, physicists at the Helsinki University of Technology Low Temperature Laboratory in Finland used a nuclear demagnetization device to achieve a temperature of 2.8×10^{-10} K, or 0.00000000028 K. This means that a fragment equal to only 28 parts in 100 billion separated this temperature from absolute zero—but it was still above 0-K. Such extreme low-temperature research has a number of applications, most notably with superconductors or materials that exhibit virtually no resistance to electrical current at very low temperatures.

THE LAWS OF THERMODYNAMICS

1st LAW: The increase in internal energy of a closed system is equal to the difference of the heat supplied to the system and the work done by it.

2nd LAW: Heat cannot spontaneously flow from a colder location to a hotter location.

3rd LAW: As a system approaches absolute zero, the entropy of the system approaches a minimum value.

1.1.3. REDOX and the Quantum Theory

REDOX (oxidation-reduction) reactions involve the escape of electrons (charged particles) from molecules or ions in one chemical substance and their capture by other substances. Free energy is the key to understanding the electron escape phenomenon. REDOX reactions are important because they are the principal sources of energy on this planet, both natural (biological) and artificial. Oxidation of molecules by the removal of hydrogen or combination with oxygen normally liberates large quantities of energy. Current advances in free radical chemistry and REDOX principles are rooted in the 'Quantum Theory'.

A) The Quantum Theory

In 1900, Max Planck theorized that light beams were made of photons that are equivalent to particles of wave motion. Planck made the assumption that energy was made of individual units, or 'quanta'. In 1905, Albert Einstein theorized that not just energy, but radiation itself was quantized in the same manner. In 1924, Louis de Broglie proposed that there is no fundamental difference in the makeup and behavior of energy and matter; on the atomic and subatomic level either may behave as if it where made of either particles or waves. Broglie's theory became known as the 'Principle of Wave-Particle Duality', where elementary particles of both energy and matter behave, depending on the conditions, like either particles or waves. In 1927, Werner Heisenberg proposed that the precise, simultaneous measurement of two complementary values - such as the position and momentum of a sub-atomic particle - is impossible. Contrary to the principles of classical physics, simultaneous measurement is inescapably flawed; the more precisely one value is measured, the more flawed the measurement of the other value. This theory became known as the 'Uncertainty Principle'. When these concepts and discoveries were integrated, the 'Quantum Theory' emerged.

The Quantum Theory provided an accurate interpretation of the atomic and subatomic world. Quantum mechanics became the foundation of free radical chemistry, because it deals with subatomic particles, as well as atoms, molecules, elements, compounds, and much larger systems. The Quantum Theory has also provided each element with an atomic number, atomic weight (or atomic mass), atomic and ionic radii, most common and usual oxidation states, density, electron negativity, electron affinity, ionization potential, state at room temperature, crystal structure, metal, nonmetal, or metalloid character, melting point, boiling point, number of isotopes & abundance, electron configuration, thermal conductivity, molar volume, standard reduction potentials, etc.

The Quantum Theory: Nobel Prizes in Physics

1918 - Max Planck for his contributions to the advancement of Physics by his discovery of energy quanta.

1932 - Werner Heisenberg for the creation of quantum mechanics, the application of which led to the discovery of the allotropic forms of hydrogen.

How did matter transform into a self-propelled energy process called LIFE? Life is clearly a distinct state of highly organized matter made of atoms, where the behavior of sub-atomic particles is governed by Quantum Mechanics.

Furthermore, when coupled with the theory of relativity (developed by Einstein), there is no boundary between material (or mass) and energy. Energy and mass are equivalent, and they can convert into each other. The Quantum Theory does much more than explain the structure of the simplest atom, it rationalizes the existence of the chemical elements. The widely used periodic table of chemical elements today is based on the results of the Quantum Theory. For example, because of this theory, we know that the hydrogen atom is the simplest atom; it consists of a proton in the nucleus and an electron around it. This type of atom is also the most abundant atom in the solar system and the universe. Closer to home, hydrogen is also very abundant on the surface of planet Earth. Most hydrogen atoms combin with oxygen to form water molecules. In organic substances, however, hydrogen atoms are mostly attached to carbon, nitrogen, and oxygen atoms.
Hydrogen is also the most abundant atom in our body.

B) The Atomic Theory

Since the 20th century, chemists and physicists have studied the structure of atoms. The discovery of electrons in 1897 by JJ Thompson demonstrated the existence of fundamental particles present within atoms. Fourteen years later, Ernest Rutherford demonstrated experimentally that the atom is not solid (as previously assumed) but that it has an internal structure consisting of a small, dense nucleus about which electrons circle in orbits. Niels Bohr refined Rutherford's model by introducing different orbits in which electrons spin around the nucleus; this model is still used in chemistry. Today, elements are distinguished by their 'atomic number', which specifies the number of protons in the nucleus of the atom. Electrons are held in their orbits through the electrical attraction between the positive nucleus and the negative electron. Bohr argued that each electron has a certain amount of fixed energy, which corresponds to its orbit. Therefore, when an electron absorbs energy, it jumps to the next higher orbit rather than moving continuously between orbits. The characteristic of electrons with fixed energy quantities (quanta) is known as the quantum theory of the atom.

The Atomic Theory is the field of physics that describes the characteristics and properties of atoms that make up matter. The key point to note about atomic theory is the relationship between the macroscopic world (eg. human body) and the microscopic world of atoms. For example, the macroscopic world deals with concepts such as temperature and pressure to describe matter. In contrast, the microscopic world of the atomic theory deals with the kinetic motion of atoms that subsequently inform macroscopic quantities. Temperature is explained by the atomic theory as the motion of atoms (faster = hotter). Pressure is explained as the momentum transfer of those moving atoms on the walls of a container (faster atoms = higher temperature = more momentum/hits = higher pressure).

The Atomic Theory & The Exclusion Principle: The Nobel Prizes in Physics
1906 - Sir JJ Thomson for his theoretical and experimental investigations on the conduction of electricity by gases.
1922 - Niels Bohr for his investigation of the structure of atoms and of the radiation emanating from them.
1929 - Louis de Broglie for discovery of the wave nature of electrons.
1945 - Wolfgang Pauli for the discovery of the Pauli Exclusion Principle.

C) The Pauli Exclusion Principle

Pauli's Exclusion Principle is a basic energy relationship between electrons within the same atom. Thus there can be a maximum number of electrons in the outer most shell of an atom. This principle provided a theoretical basis for the placement of each element in the Period Table.

By the early 20th century, it became evident that atoms and molecules with even numbers of electrons are more chemically stable than those with odd numbers. Wolfgang Pauli (1900-1958) looked for an explanation for these numbers, which were at first only empirical. Pauli found that the complicated numbers of electrons in closed shells could be reduced to the simple rule of one electron per state, if the electron states were defined using four quantum numbers. For this purpose, he introduced a new two-valued quantum number, known as the 'electron spin'. The Pauli Exclusion Principle states that no two electrons can have the same four quantum numbers. We are now aware that a maximum number of two electrons can be found in one orbital and the two electrons must have opposing spins. That means one electron will spin up (+ ½) and the other will spin down (- ½).

TABLE: *REDOX and Electrochemistry*

REDOX	Assigning Oxidation Numbers
Rule 1	Elements in their elemental state have an oxidation number of zero (0). eg. Ag (Silver), Na (Sodium), H_2 (Hydrogen), O_2 (Oxygen)
Rule 2	The oxidation number of an atom ion is equal to the charge on that ion. eg. Na^{+1} = 1; Cl^{-1} = -1
Rule 3	Oxygen, in a compound, usually gets -2. Exceptions are: In peroxides: O = -1 With fluorine: O = +1 (the only time O is positive)
Rule 4	Hydrogen, in a compound, usually gets a +1. Exceptions are: In Hydrides (with metals, i.e. LiH): H = -1
Rule 5	Fluorine, in a compound, always gets a -1
Rule 6	In a neutral compound, the sum of all oxidation numbers must equal ZERO.

The Pauli Exclusion Principle: An important consequence of the principle is the elucidation of the elaborate electron shell structure of atoms and the way atoms share electrons. The principle explains the variety of chemical elements and their combinations. An electrically neutral atom contains bound electrons equal in number to the protons in the nucleus. Electrons, being fermions, cannot occupy the same quantum state, so electrons have to 'stack' within an atom – they have different spins while at the same place. An example is the neutral helium atom, which has two bound electrons, both of which can occupy the lowest-energy (1s) states by acquiring opposite spins. Since the spin is part of the quantum state of the electron, the two electrons are in different quantum states and do not violate the Pauli principle. Nonetheless, the spin can only take two different values. In contrast, in a lithium atom, which has three bound electrons, the third electron cannot reside in a 1s state; it must occupy one of the higher-energy 2s states instead. Similarly, successively larger elements must have shells of higher energy. The chemical properties of elements largely depend on the number of electrons in the outermost shell. Atoms with different numbers of shells but the same number of electrons in the outermost shell have similar properties, which gives rise to the Periodic Table of the Elements.

The gene expression and control in multi-cellular organisms is more complex than that of most single cell organisms. There are ~250 different cell types in humans, and each differentiated cell type uses only a subset of the 25,000 total human genes. For example, all cells in the body have the gene for hemoglobin, but only red blood cells make that protein. Along with this specialization, the cells must coordinate their activities by sending messages to each other. They also work cooperatively to develop their body, which is a 'stigmergy' structure. And they need apoptosis mechanisms to remove cells that have outlived their usefulness or become dangerous. Without all four of these organizing principles operating together in a coordinated manner, true multi-cellularity would not be possible.

1.2 REDOX: Bio-Informatics

The gene is a sequence of nucleotides in a DNA molecule that is translated by cells to produce a chain of amino acids (polypeptide chain). The sequence or order of amino acids in a protein corresponds to the order of nucleotides in the gene. This relationship between nucleotide sequence and amino acid sequence is known as the 'Genetic Code'. The folding of amino acids in a protein determines its 3-D shape; this structure is, in turn, responsible for the protein's function. Proteins carry out almost all the functions needed for cells to live. A change to the DNA in a gene can change a protein's amino acids, shape, and function; this can have a dramatic effect on the cell and the overall behavior of an organism.

In 1928, Frederick Griffith discovered the phenomenon of transformation, where-in a dead bacterium could transfer genetic material to other live bacteria. In 1944, Avery, McLeod and McCarty identified the transformation molecule as 'deoxyribonucleic acid' (also known as DNA). The Hershey-Chase experiment in 1952 confirmed that DNA is the genetic material of the viruses that infect bacteria, providing evidence that DNA is the molecule responsible for inheritance. In 1953, Watson and Crick determined the double-helical structure of DNA, based on the X-ray crystallography work of Rosalin Franklin and Maurice Wilkins. This structure elucidated that genetic information exists in the sequence of nucleotides on each strand of DNA. The structure also suggested a simple method for duplication: if the strands are separated, new strands can be reconstructed for each based on the sequence of the old strand. The discovery of genetic transcription processes [wherein cells use DNA as a template to create matching messenger RNA (with nucleotides similar to DNA) which, in turn, create amino acids in a polypeptide chain (or proteins)] has unraveled the Genetic Code. The development of DNA sequencing technology by Sanger (in 1977) allowed scientists to read the Genetic Code. In 1983, Mullis developed the polymerase chain reaction (PCR), a molecular tool to isolate and amplify a specific section of DNA from a mixture. The combined efforts of the 'Human Genome Project' and the parallel private effort by 'Celera Genomics' culminated in the sequencing of the human genome in 2003.

Although genetics define the appearance and behavior of organisms, the milieu plays an important role in determining the ultimate phenotype—a phenomenon often referred to as 'nature vs. nurture'. Nutrition affects gene expression, which is an emerging field of science known as 'nutri-genomics'. Gene expression is the process by which information from a specific fragment of DNA is used in the synthesis of a functional product – often proteins. The regulation of gene expression by oxidants, antioxidants, and the REDOX state has emerged as a novel sub-discipline in molecular biology with promising therapeutic implications.

1.2.1. The Genetic Code

In a human cell, there are about 6 billion nucleotide base pairs that stretch out to about 6 meters of DNA per cell,an amount 300,000 times greater than the diameter of the cell. In contrast, bacteria cells have only about 6 million base pairs.

Nucleic acids and proteins are 'aperiodic' polymers; this aperiodicity gives them the unique ability to carry a lot of biological information. By definition, a periodic structure has 'order' and an aperiodic structure has 'complexity'. In terms of information, periodic polymers (i.e. nylon) and crystals are analogous to a book in which the same sentence is repeated throughout. Also, a random arrangement of letters in a book can be periodic but still convey little or any useful information if it has no meaning (to cite an example, the word 'gift', which means a 'present' in English and 'poison' in German, is meaningless in French). Only specific sequences of letters correspond to sentences, and only certain sequences of sentences correspond to paragraphs, etc. In the same way only certain sequences of amino acids in polypeptides and bases along polynucleotide chains correspond to useful biological functions. The following three sets of letter arrangements illustrate the difference between order and complexity in relation to information:

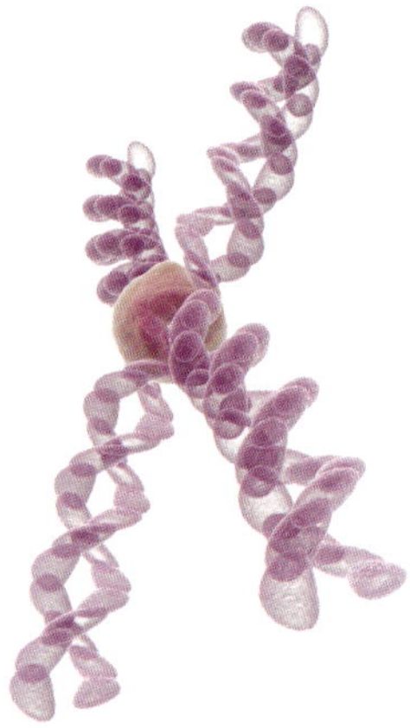

1) An ordered (periodic) and therefore specified arrangement (eg. nylon or crystal):
THE END THE END THE END THE END THE END

2) A complex (aperiodic) unspecified arrangement (eg. random polymers or peptides): *AGDCBFE GBCAFED ACEDFBG*

3) A complex (aperiodic) specified arrangement (eg. DNA or protein):
THIS SEQUENCE OF LETTERS CONTAINS A MESSAGE

It is widely agreed that both DNA and proteins are essential for living systems and indispensable components of every living cell. Both types of molecules are far more energy-rich and information heavy than the bio-monomers from which they are synthesized. The polymerization of amino acids to polypeptides (proteins) or nucleotides to polynucleotides (DNA) occurs through condensation reactions. These reactions provide a powerful thermodynamic process for the energy to flow through the system for synthesis of protein and DNA. Thus, there is a general relationship between information and entropy. The bio-informatics content in a given sequence of units, be it digits in a number, letters in a sentence, or amino acids in a polypeptide or protein, depends on the minimum number of instructions needed to specify or describe the molecular structure and its function. Many instructions are needed to specify a complex structure, such as DNA. In contrast, only a few instructions are needed to specify an ordered structure such as a crystal or nylon.

The genetic code is a set of rules defining how the 4-letter code of DNA (A-T-C-G) is translated into the 20-letter code of amino acids, which are the building blocks of proteins. The genetic code is a set of three-letter combinations of nucleotides called codons, each of which corresponds to a specific amino acid or stop signal.

The Genetic Code: The Nobel Prizes in Physiology or Medicine
1962 – Francis Crick, James Watson and Maurice Wilkins for discovering the molecular structure of nucleic acids and its significance for information transfer in living organisms.
1968 – Robert Holley, Har Gobind Khorana and Marshall Nirenberg for interpreting the Genetic Code and its function in protein synthesis.

There are 64 possible permutations, or combinations, of three-letter nucleotide sequences that can be made from the four nucleotides. Of these 64 codons, 61 represent amino acids and three are stop signals. Although each codon is specific for only one amino acid (or one stop signal), the Genetic Code is described as degenerate, or redundant, because a single amino acid may be coded by more than one codon. It is also important to note that the genetic code does not overlap, meaning that each nucleotide is part of only one codon – a single nucleotide cannot be part of two adjacent codons. Furthermore, the genetic code is nearly universal, with only rare variations reported. For instance, mitochondria have an alternative genetic code with slight variations.

The human body consists of approximately 100 trillion cells. Yet, if all of the DNA inside just one of these cells was arranged into a single straight line, it would stretch nearly two meters long!

Evolution is a mechanism which is associated with mutation, selection and reproduction. Mutation, the infidelity in self-replication, represents misinformation that occasionally alters the dynamics of a living organism to track and exploit sources of energy and nutrition. Thus, an innate ability to evolve through natural selection has emerged as a fundamental property of living organisms.

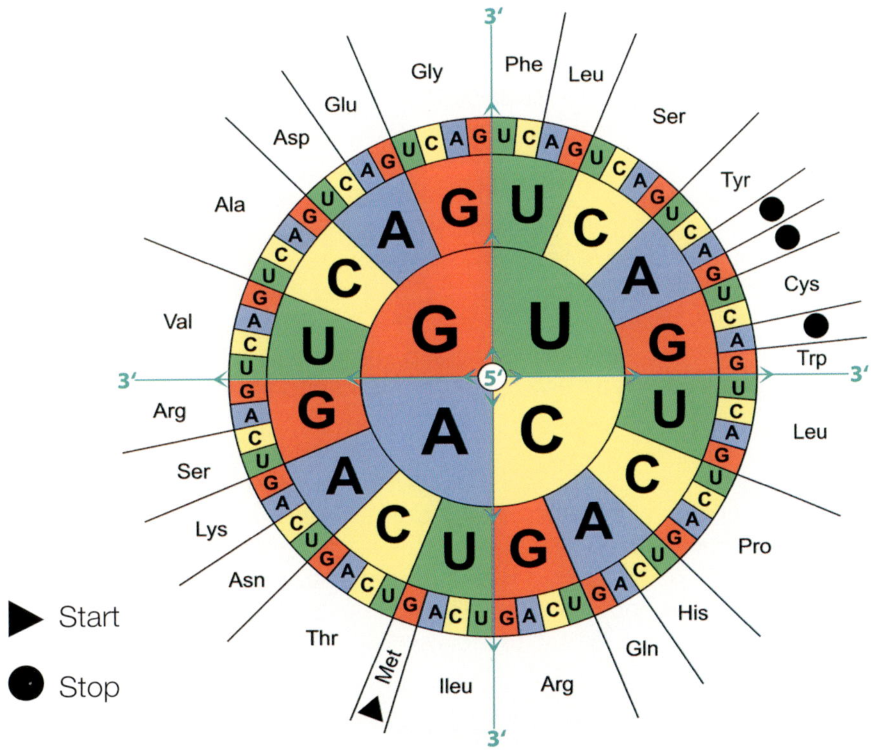

Polymerase Chain Reaction (PCR): The 1993 Nobel Prize in Chemistry was awarded to Kary B. Mullis for the invention of the polymerase chain reaction (PCR) method and to Michael Smith for his fundamental contributions to the establishment of oligonucleotide-based and site-directed mutagenesis.

1.2.2. Deoxyribonucleic Acid (DNA)

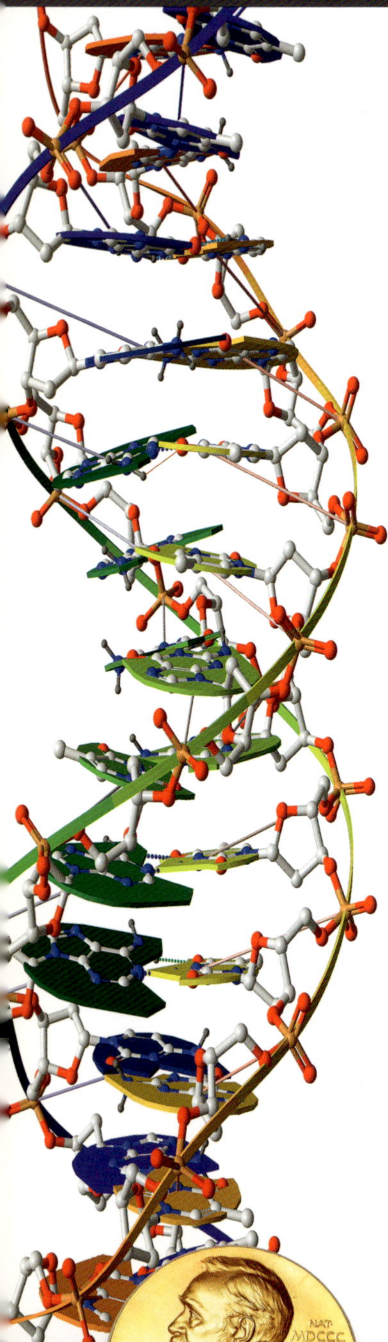

DNA is the genetic material found in nearly all living cells. In eukaryotes, DNA is housed within the nucleus, however, in prokaryotes, DNA is located directly in the cellular cytoplasm. DNA is a complex polymer with many repetitive chemical units, a portion of which are passed from the parent organisms to their offspring during the process of reproduction. Although each organism's DNA is unique, all DNA is composed of the same nitrogen-based molecules called 'nucleotides'.

What are Nucleotides? The nitrogenous base + a pentose sugar together is called a nucleoside, when combined with a phosphate group, the nucleoside becomes a nucleoside monophosphate, or commonly referred to as a nucleotide. In DNA, the pentose sugar is always deoxyribose and the phosphate group is always one phosphate molecule bonding with four oxygen molecules. It is the REDOX variance in the hydroxyl status of pentose sugar and the nitrogenous base that makes DNA highly functional in the development of life.

How does DNA differ from organism to organism? This genetic marvel is accomplished by a simple arrangement of nucleotides in a specific sequence within the DNA. In turn, this pattern of DNA arrangement ultimately determines each organism's unique characteristics. These specific patterns are known as genes, which are sections of DNA that code for polypeptide chains (proteins). A gene is the basic unit of heriditary information that creates an offspring.

How does DNA fit within a cell? This task is accomplished through a cellular process known as DNA packaging, wherein long pieces of double-stranded DNA are tightly looped, coiled, and folded to fit easily within the cell. Eukaryotes do this by wrapping their DNA around special proteins called histones and the compacted DNA coil is known as chromatin. DNA can be further compressed through a twisting process called super-coiling. In both eukaryotes and prokaryotes, this highly compacted DNA is arranged into structures called chromosomes.

Why Ribose (or Deoxyribose)? In nucleic acids, ribose sugar serves as the branch point between the polymer backbone and the side chains (i.e. the nucleo-bases). This role requires a chemical moiety that is, at least, tri-functional. Sugars with three or more carbons meet this minimum bonding requirement and among those known in the nature thus far, ribose appears to be perfect. The optimal functionality of ribose suggests that its inclusion in nucleic acids is more the result of evolutionary refinement.

Deoxyribonucleic Acid (DNA): The Nobel Prizes in Physiology or Medicine

1958 – George Beadle and Edward Tatum for discovering that genes regulate definite chemical events and to Joshua Lederberg for his discovery of genetic recombination and the organization of genetic material.

1959 – Severo Ochoa and Arthur Kornberg for discovering the mechanisms in the biological synthesis of RNA and DNA.

Why the DNA bases pair only with their complements

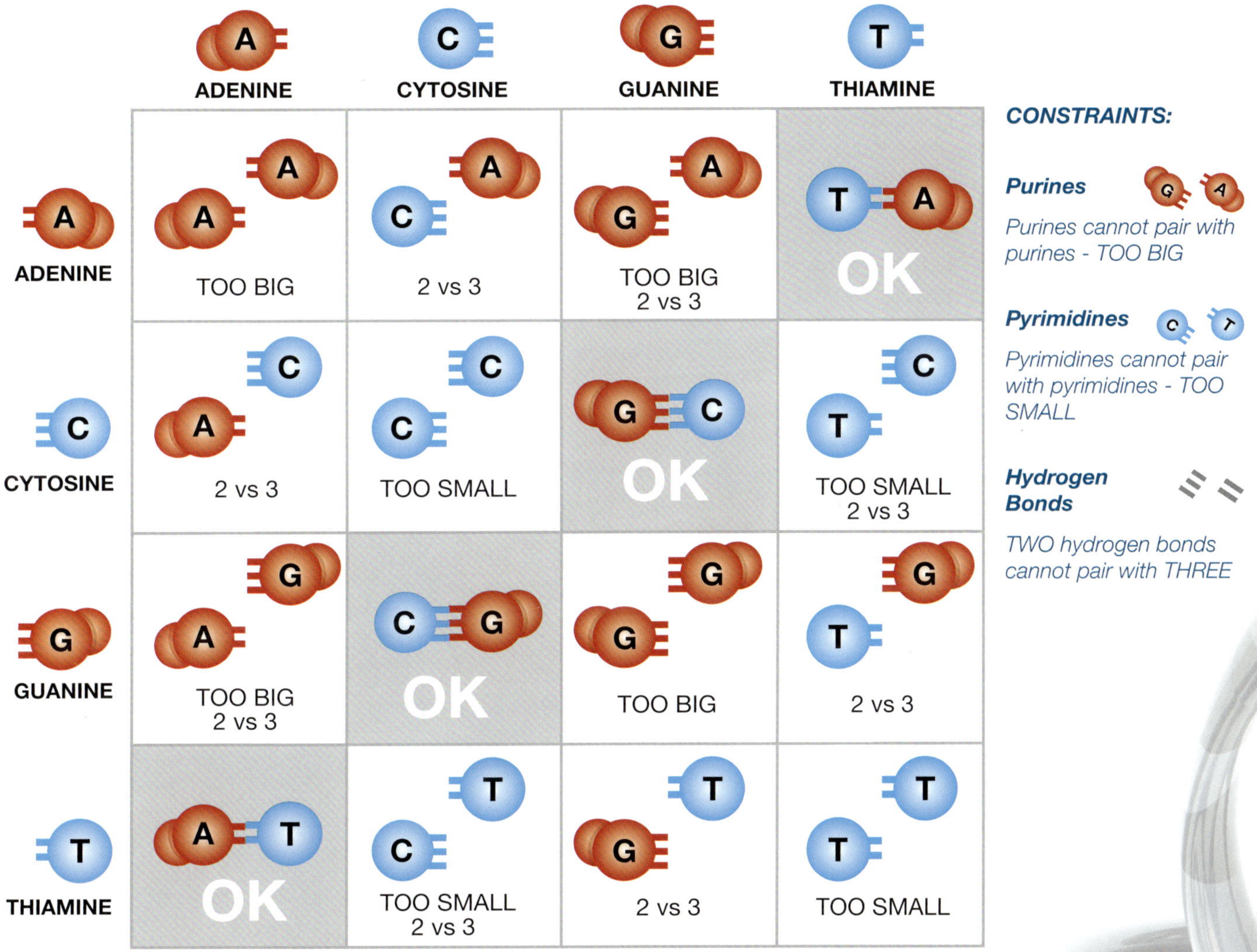

The four DNA nitrogenous bases - Adenine (A), Guanine (G), Cytosine (C) and Thymine (T) - fall into two sets of two categories based on their REDOX (hydrogen bonding). The pyrimidines (C and T) are smaller in molecular-size and have a single-ring of 6 carbon molecules. The larger molecular-sized purines (A and G) each have 2-rings of 6 carbon structures fused with a 5 carbon ring. Thus, the complementary bases on opposite strands of the double helix are always one pyrimidine and one purine. The REDOX potential causes adenine always to be paired with thymine rather than with cytosine, since A and T are prone to form two hydrogen bonds (A=T); whereas, guanine and cytosine form three bonds (G≡C). This REDOX phenomenon is known as 'complementary base pairing'. A pyrimidine dimer results from the formation of a covalent bond between adjacent pyrimidines on the same strand.

The Meselson-Stahl experiment (in 1958) supported the hypothesis that DNA replication was semi-conservative. It showed that the double stranded DNA helix replicated into 2 new double-stranded DNA helices consisting of one strand from the original helix and the other newly synthesized. It has been called 'the most beautiful experiment in biology'.

Why Phosphate? The phosphate backbone is a REDOX hallmark of both nucleic acids. As a tri-acid, phosphate provides nucleic acids with a means to connect two alcohols, in the form of sugars, via ester linkages, while maintaining a -ve charge. It is difficult to conceive of a chemical arrangement other than phosphate that would provide better REDOX potential for nucleic acid function-ality. The ester linkages provide connectivity that is sufficiently labile so that enzymes can hydrolyze the polymer, whereas the negative charge provides sufficient screening of the phosphorus center from nucleophilic attack to stabilize the polymer against hydrolysis. Additionally, the intra-strand electrostatic repulsion afforded by the regularly-spaced negative charges may be crucial to an extended helical conformation and for information transfer that is rigorously dependent on Watson-Crick base pairing. Phosphate is unique as a polyvalent acid with all acidic protons in close proximity; its -ve charge affords greater protection from nucleophilic attack and intra-residue repulsion than would any polyvalent acid that has ionizable moieties more distant from one another (e.g., citrate).

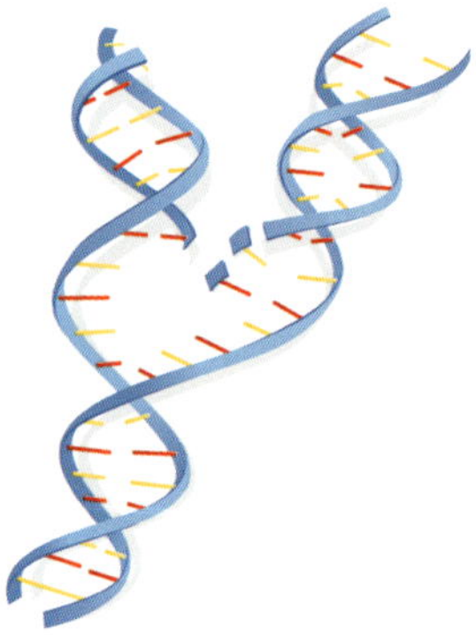

DNA replication and REDOX couples: Before cell division, DNA needs to be duplicated. DNA replication is regulated by various coupled REDOX reactions. It begins with the unzipping of the parent molecule as the hydrogen bonds between the base pairs are broken. Once exposed, the sequence of bases on each of the separated strands serves as a template to guide the insertion of a complementary set of bases on the strand under neo-synthesis. The new strands are assembled from deoxynucleoside triphosphates. Each incoming nucleotide (ATP, GTP, CTP and TTP) is covalently linked to the free 3' carbon atom on the pentose sugar. The second and third phosphates are removed as pyrophosphate (PPi) molecules. The nucleotides are assembled in the order that complements the order of bases on the strand serving as the template. Thus each cytosine (C) on the template guides the insertion of a guanine (G) on the new strand, forming a G≡C base pair.

Several enzyme catalytic processes act as REDOX couples to complete DNA duplication. A segment of the double helix is uncoiled by a helicase enzyme. A molecule of DNA polymerase binds to one strand of the DNA and begins moving along the 3' to 5' direction, using it as a template to assemble a leading strand of nucleotides and reform a double helix. Another enzyme, DNA ligase I, stitches these together into individual strands.

When the replication process is complete, 2 DNA molecules, identical to each other and identical to the original, are produced. Each strand of the original molecule remains intact, since it served as a template for the synthesis of a complementary strand. This mode of replication is described as semi-conservative, where one-half of each new molecule of DNA is old and the other-half totally new. Watson and Crick hypothesized this mode of DNA replication, which was later proven by the famous Meselson-Stahl experiment.

Reverse Transcriptase/Gene Splicing: Nobel Prizes in Physiology or Medicine
1975 – David Baltimore, Renato Dulbecco, and Howard Temin for discovering the reverse transcriptase that copies RNA into DNA (the opposite of the usual route of genetic transfer).
1993 – Philip Sharp and Richard Roberts for discovering introns and gene splicing.

1.2.3. Ribonucleic Acid (RNA)

Ribonucleic acid (RNA) belongs to a family of large bio-functional molecules with vital roles in the coding, decoding, regulation and expression of of genes. RNA, together with DNA and proteins, constitutes the 3 major macromolecules essential for all known forms of life. Cellular organisms use messenger RNA (mRNA) to convey genetic information to regulate protein synthesis. Several viruses also encode their genetic information using an RNA genome.

RNA molecules perform specific cellular roles including the catalysis of biological reactions, control of gene expression, and communication of cellular signals. One of these active processes is protein synthesis, a universal function whereby mRNA molecules direct the assembly of polypeptides on ribosomes. This process uses transfer RNA (tRNA) molecules to deliver amino acids to the ribosome, where ribosomal RNA (rRNA) links amino acids together to form the polypeptide chain (protein). Certain RNA molecules are chemical catalysts with enzymatic functions. For example, the active site of ribosomal structures is composed entirely of RNA, which catalyzes the peptide bond formation.

The 'RNA world hypothesis' places RNA at center-stage when life originated. The RNA world is proposed to have evolved into the DNA and protein world of today. DNA, through its greater chemical stability, took over the role of data storage – or 'bio-informatics'. Protein, on the other hand, is more flexible in catalysis through the great variety of amino acids, and has evolved into the specialized catalytic molecules – the 'enzymes'. The hypothesis suggests that RNA in modern cells, in particular ribosomal RNA (rRNA, which catalyzes protein production), is an evolutionary remnant of the RNA world. RNA is thought to have preceded DNA, because of the nature of its biosynthetic pathways. The deoxyribo- nucleotides used to make DNA are made from ribonucleotides, the building blocks of RNA, by removing the 2'-OH group. As a consequence, a cell must have the ability to make RNA before it can make DNA.

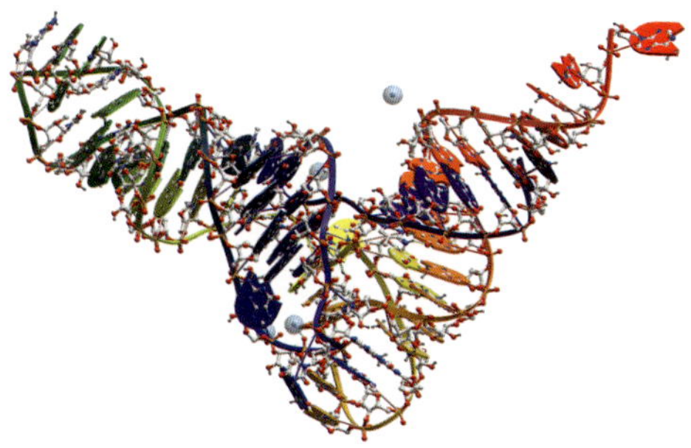

Transfer RNA

The following observations suggest that RNA world predates DNA and proteins: (i) RNA has the ability to store genetic data and pass down the hereditary information. (ii) RNA is the main component that links up DNA/gene formation to amino acids/protein synthesis via transcription/translation. (iii) RNA can duplicate itself, as well as the genetic information it carries, similar to DNA. (iv) RNA is less complex then DNA, and involves fewer types of molecules in order to self-replicate. (v) DNA requires RNA primer in order to replicate while RNA does not need any such primer. This shows that DNA has a functional dependence on

Ribonucleic Acid (RNA): The Nobel Prizes in Chemistry

1980 – Paul Berg for his fundamental studies on the biochemistry of nucleic acids, the other half jointly to Walter Gilbert and Frederick Sanger for the determination of base sequences in nucleic acids.

1989- Sidney Altman and Thomas Cech for discovering catalytic property of RNA.

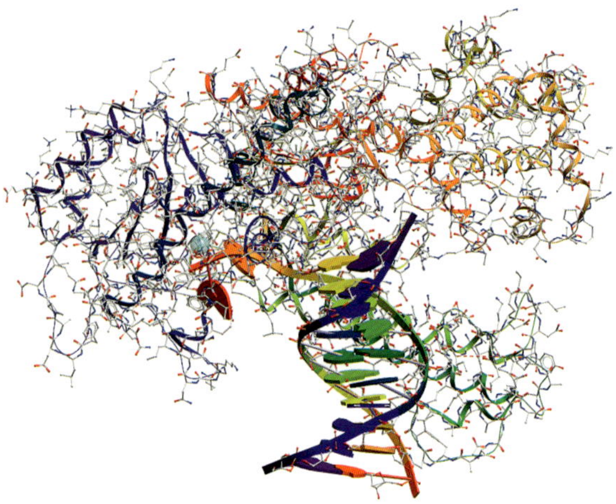

DNA Polymerase

RNA rather than the other way around. (vi) Like proteins, RNA is able to catalyze reactions. Furthermore, protein synthesis is regulated by RNA, which points heavily to its pre-existence over the proteins. (vii) RNA forms double helices similar to DNA, and tertiary structures similar to those of catalytic proteins. (viii) The structure of RNA consists of a ·OH group in the 2' position of the ribose sugar, which makes it less stable. This vulnerability (molecular breakdown) allows adaptation of RNA to different structural conformations which perhaps was beneficial to early phases of evolution. (ix) RNA contains uracil (a substitute base for 'thymine' in DNA), which makes it more prone to mutations and suitable for primitive life in early evolution.

The RNA world hypothesis is supported by the observation that many of the most critical components of cellular metabolism (those that evolve the slowest) are composed partially or totally of RNA. Also, many critical cofactors (ATP, Acetyl-CoA, NADH, etc.) are either nucleotides or substances clearly related to RNA. This would mean that the RNA and nucleotide cofactors in modern cells are an evolutionary remnant of an RNA-based system.

TABLE: *Differences between DNA and RNA*

Parameter	Deoxyribonucleic acid (DNA)	Ribonucleic acid (RNA)
Definition	A nucleic acid that contains genetic/heriditary information for growth and development of all living organisms.	A nucleic acid involved in protein synthesis and occassionally in the transmission of genetic information.
Function	Bio-informatics: Long-term storage and transmission of hereditary information.	Protein synthesis: Transfer of genetic code for the making of proteins in the ribosome.
Features	• The helix geometry of DNA is of B-Form. • DNA can be damaged by UV exposure. • DNA is relatively stable in the body.	• The helix geometry of RNA is of A-Form. • RNA is resistant to damage by UV rays. • RNA strands are continually made, broken down and reused.
Structure	Double- stranded molecule with a long chain of nucleotides.	A single-stranded molecule with shorter chain of nucleotides.
Bases & Sugars	• Deoxyribose sugar / phosphate backbone • Four bases: A/G/C and Thymine (T)	• Ribose sugar / phosphate backbone. • Four bases: A/G/C and Uracil (U)
Base Pairs	A=T (Adenine=Thymine) G≡C (Guanine≡Cytosine)	A=U (Adenine-Uracil) G≡C (Guanine≡Cytosine)
Stability	• Deoxyribose sugar is less reactive due to C-H (absence of hydroxyl) bonds. • Stable in alkaline conditions. • DNA has smaller grooves, which makes it resistant to enzyme hydrolysis.	• Ribose sugar is more reactive due to C-OH (hydroxyl) bonds. • Unstable in alkaline conditions. • RNA has larger grooves, which makes it susceptible to enzyme hydrolysis.
Propagation	DNA is self-replicating.	RNA is synthesized from DNA on demand.

1.2.4. DNA vs. RNA – The REDOX Variance

An obvious difference between DNA and RNA is the REDOX variance in their sugar moiety. While the sugar present in an RNA molecule is ribose, the sugar moiety in the DNA is deoxyribose – in deoxidized form. Deoxyribose is similar to ribose, except that the latter has an extra OH group. DNA does not usually exist as a single molecule, but as a tightly-associated pair of molecules. This pair of long strands entwines like vines in the shape of a double helix. This arrangement of DNA strands is known as anti-parallel. The asymmetric ends of DNA strands are referred to as the 5' (five prime) and 3' (three prime) ends. The four bases found in DNA are adenine (A), cytosine (C), guanine (G) and thymine (T). A fifth pyrimidine base, called uracil (U), usually takes the place of thymine in RNA. Chemically, U is similar to thymine, which differs only by a methyl group, but its production requires less energy. In terms of base pairing, uracil substitution has no effect, thus, A readily binds to U or T. Uracil, however, can damage cytosine bonding and make RNA particularly susceptible to mutations by replacing a GC base pair with a GU (wobble) or AU base pair.

Since DNA is a more stable molecule than RNA, it makes sense for DNA to adapt to the environment and take over the job of 'Bio-informatics'. And how is DNA more stable than RNA? Obviously, the REDOX variance between DNA and RNA is due to the deoxyribose sugar! The missing 2'-OH (hydroxyl) group on the deoxyribose sugar makes DNA more stable and less susceptible to reactive oxygen species (ROS) and other free radicals. Due to this oxidative stability, the chain of DNA nucleotides is not easily broken (or damaged by hydrolysis); thus, the DNA molecule remains intact in its double helical structure.

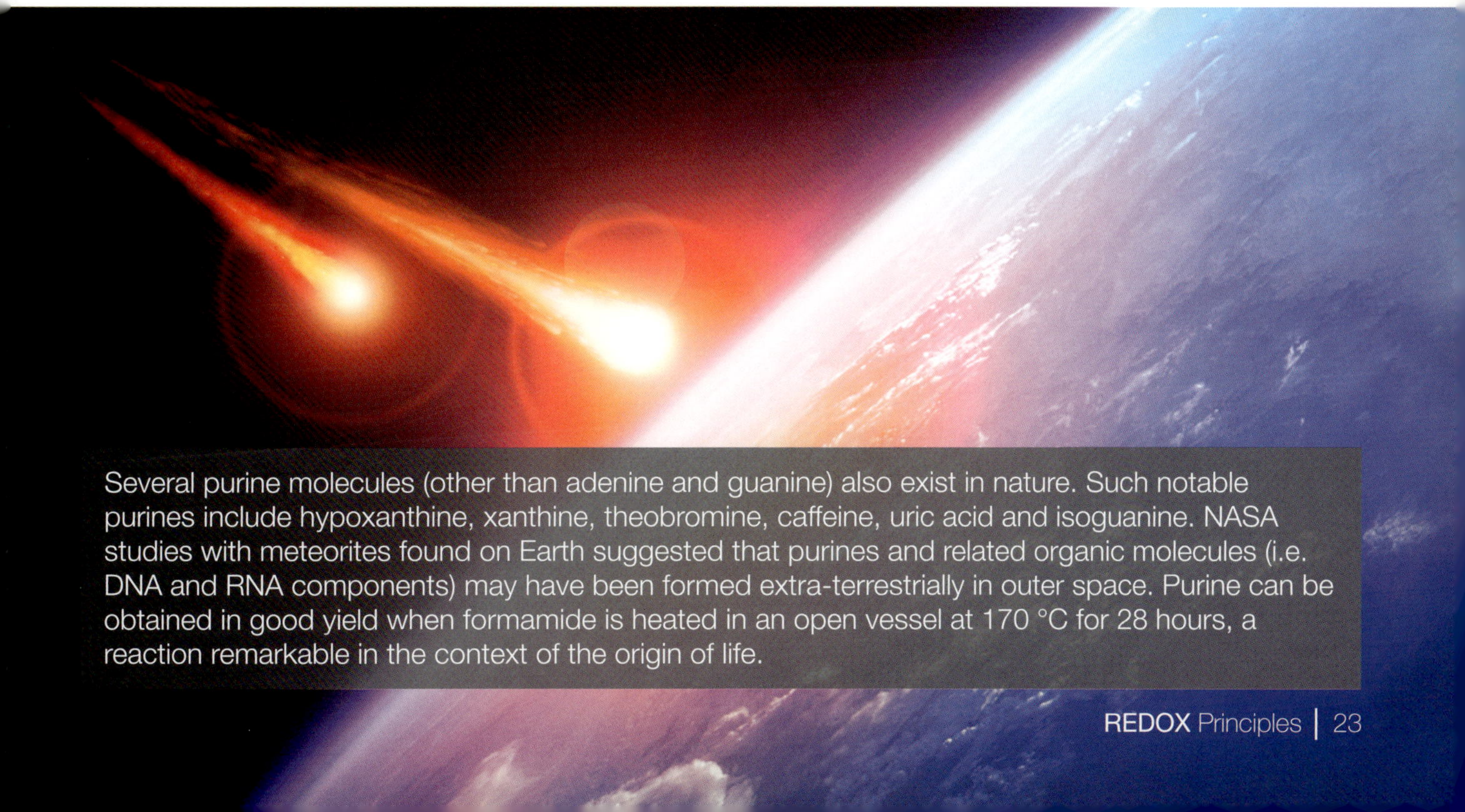

Several purine molecules (other than adenine and guanine) also exist in nature. Such notable purines include hypoxanthine, xanthine, theobromine, caffeine, uric acid and isoguanine. NASA studies with meteorites found on Earth suggested that purines and related organic molecules (i.e. DNA and RNA components) may have been formed extra-terrestrially in outer space. Purine can be obtained in good yield when formamide is heated in an open vessel at 170 °C for 28 hours, a reaction remarkable in the context of the origin of life.

Bio-Energetics involves the making and breaking of chemical bonds in living organisms. Vital cell processes like growth, development, and reproduction are central to the survival and propagation of biological species—the role of energy is fundamental to such processes. Life is dependent on energy transformations; organisms survive because of the exchange of energy within and without. Energy requirements vary based on factors like body size, activity level, and current environment.

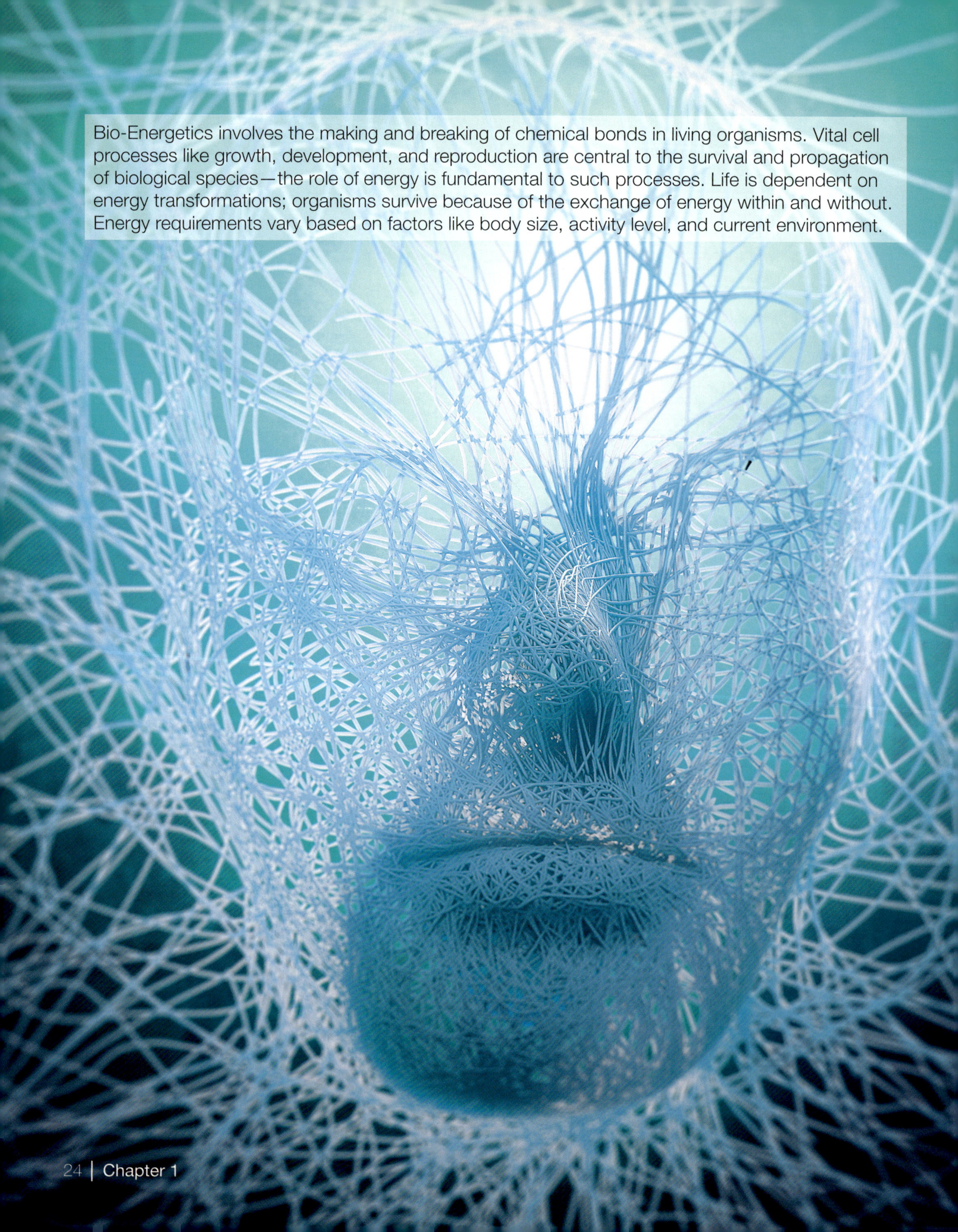

1.3 REDOX: Bio-Energetics

Survival, growth and multiplication are the three basic characteristics of a living organism. The role of energy is essential for maintaining the biological systems. Biological energy is harnessed from various chemical reactions and metabolic processes. The survival of a living organism exclusively depends on the exchange of this energy inside and out, by breaking chemical bonds from molecules found in biological systems and using this energy in forms such as adenosine triphosphate (ATP) molecules. This generated chemical energy is available for work (a run for survival, for example) or for other processes such as anabolic reactions that support growth and synthesis of hormones that facilitate reproduction. Bio-Energetics is the study of energy transfers regulated by REDOX principles within living systems. The laws of bio-energetics provide the rules upon which life works.

There are two laws of Bio-Energetics:

First Law: Energy can neither be created nor destroyed; however, can be transformed from one form to another. The main forms of energy in the human body are heat, light, chemical, mechanical, 'free energy' and entropy. A few examples of energy transfer inside the human body include: vision, which is transformation of light to electrical; muscle contraction, which is a transfer of chemical energy to mechanical; vitamin D synthesis, where light energy changes to chemical energy, etc.

Second Law: Energy transfer shall always proceed in the direction of increased entropy. The release of energy, also called 'free energy' (ΔG) is negative, since energy is lost from the bonds formed by the products. The more negative the ΔG, the greater the release of free energy during a reaction. Chemical reactions that have a $-\Delta G$ are known as 'exergonic' reactions. During these reactions, energy is invested to change the REDOX potential, in order to drive the reactants from a stable state to a highly energetic unstable configuration. These reactants are usually complex molecules that are broken into simple products. The entire reaction is usually catabolic in nature and potentially reversible. Conversely, 'endergonic' is an anabolic reaction that consumes energy. It has a positive ΔG because energy is required to break bonds. The free energy not used to do work is expressed as heat. Reactions that have no net change in substrate or product are termed equilibrium reactions, and have no change in free energy ($\Delta G = 0$). The direction and amount of free energy release of a chemical reaction can be modified by changing substrate and product concentrations. Thus, energy balance can be defined as the REDOX or biological homeostasis of energy in living systems. It is measured with the following equation:

$$\text{Energy Intake} = \text{Internal Heat Generated} + \text{External Work} + \text{Storage}$$

Entropy is a form of energy that cannot be re-used in chemical reactions, and is defined with increased randomness or disorder. 'Free energy' is referred to as Gibb's free energy, and is abbreviated 'G'. Typically, during energy transfers there is a change in energy forms, which is indicated by the 'Δ' symbol. Thus, a change in Gibb's free energy is expressed as a 'ΔG'.

1.3.1. ATP and other Nucleotides

If all the biochemical energy in a reaction was released at the same time, most of it would be lost as heat and burn the cells and little could be captured to do metabolic (or any other) work. Therefore, living organisms have evolved a multitude of mechanisms, such as enzymes that regulate and allow a strategic use of the released biochemical energy. Enzymes control the state of energy a molecule must attain before it can release energy and are the basic catalysts of biochemical reactions. They are neither consumed nor changed during the reactions. Basically, enzymes reduce the activation energy needed to start a reaction by temporary bonding with the reacting molecules and, in doing so, weaken the chemical bonds. Almost all of the over 2,000 known enzymes are proteins, nearly all of which operate with cofactors – metal ions or organic molecules (eg. Coenzyme Q10, NAD, FAD, cytochromes, etc.). Nucleotides are the subunits that are linked to form the nucleic acids ribonucleic acid (RNA) and deoxyribonucleic acid (DNA), which serve as the cell's storehouse of genetic information. Free nucleotides play important roles in cell signaling and metabolism, serving as convenient and universal carriers of biochemical energy and high-energy electrons.

Bio-Energetics and Nucleotides

Purine nucleotides (ATP and GTP) have a wide range of metabolic functions, while pyrimidines (CTP and UTP) act more as regulators. This is probably due to the more reactive nature of the purine rings, which makes ATP an especially ideal co-enzyme. Each nucleoside, phosphorylated to the triphosphate form, appears to have evolved to deal with a particular area of metabolism.

ATP: Adenosine is a multi-functional nucleotide with a vital role in almost every cellular pathway. Adenosine can be used as a source of energy, acting alone as ATP, or combined with other nucleotides (like niacin in NAD or riboflavin) in FAD. ATPases – enzymes that directly hydrolyze ATP into ADP and phosphate are found throughout the cell and perform a wide variety of functions from pumping ions across the membrane to running all of the cytoplasmic motors that shuttle material around the cell (i.e. movement of cilia, flagella, and muscles). Adenosine is also used extensively by the cell (as a source of phosphates) for modifying proteins - several proteins require phosphorylation for activation ('On') or inactivation ('Off'). REDOX signaling is used by cellular systems to control enzyme activities. Enzymes involved in phosphorylation of proteins are known as kinases and ATP is an essential co-factor for their function.

GTP: Guanosine is used similarly to Adenosine, but is limited to playing a role in protein synthesis. There are a few pathways in which GTP is used as a phosphate donor or an energy source (i.e. tubulin hydrolyzes GTP to GDP to form microtubules in the cytoskeleton). There are also other GTPases in the cell; however, most of these compounds are not used for their enzymatic properties, but rather to perform cell signaling. G-proteins are a specific class of GTPases that bind to GTP and interact with other enzymes in cellular activation. Many hormones and neurotransmitters have receptors that use G-proteins to transmit cell signals.

UTP: Uridine is used for a different purpose from the purine nucleotides. The most common utility is in glycogen synthesis. Within many cells in the body (especially in the liver), the sugars are stored in the form of glycogen, a complex starch polymer composed of long, branching chains of glucose molecules. UTP facilitates these polymer links by mobilizing free glucose molecules to produce UDP-glucose and free phosphate. This reaction makes the glucose molecules more reactive, since the glucose-phosphate bond in UDP-glucose is a high energy interaction. As the UDP-glucose is added to glycogen, the UDP is released, and the energy is used to attach the glucose to the glycogen molecule. In fact, uridine is used for UDP-glucose, UDP-galactose, UDP-mannose, etc., the building blocks of many carbohydrates that are essential for many cellular functions.

CTP: Cytidine functions are similar to uridine, however, unlike sugars, CTP is used in lipid metabolism. CDP-diacylglycerol, CDP-ethanolamine, and CDP-choline are the building blocks for phospholipids that constitute cell membranes. Since all cells require intact membranes to survive, this is an essential cellular function.

Thus, ATP is mainly involved in reactions concerned with electrical and mechanical work (ion pumps and contractile proteins), whereas UTP is involved in carbohydrate synthesis, GTP in protein synthesis and CTP in lipid synthesis.

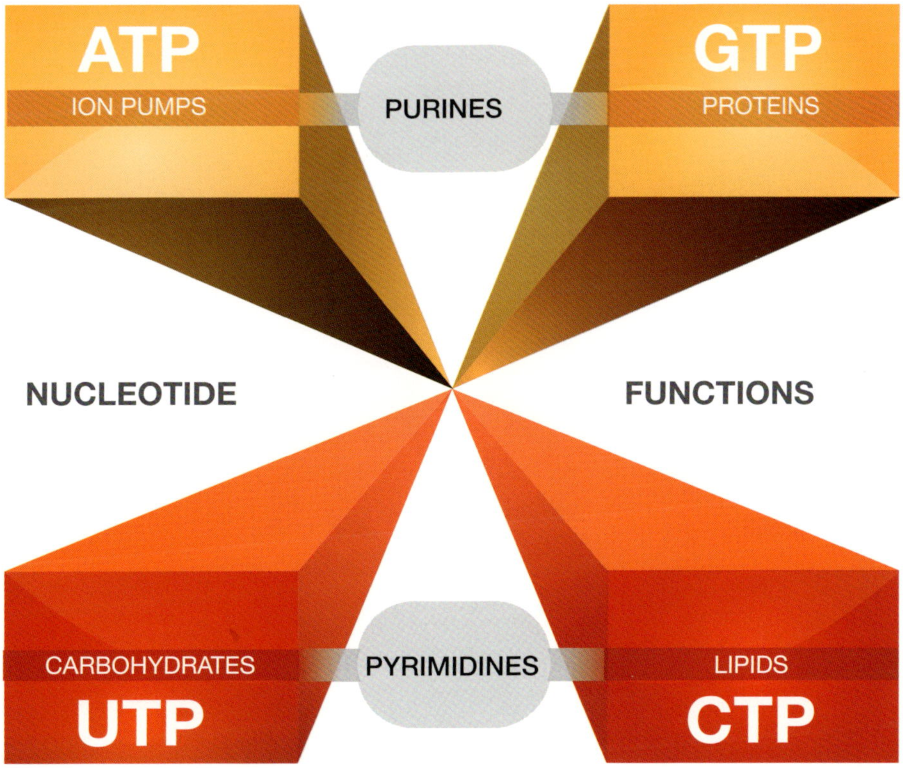

Bio-Energetics – Why ATP the cellular currency?

The evolution of nucleoside function seems to have its roots in the RNA world. The triphosphates used in metabolism contain ribose (not deoxyribose, which is found in DNA), and the base uracil (as UTP) rather than thiamine, (the base substituting for uracil in DNA). From a bioenergetics stand point, all four nucleoside triphosphates are used as energy currency. However, only ATP is replenishes itself by the oxidative phosphorylation cycle; the other three require mediation of NDP or NMP kinases in addition to the oxidative phosphorylation cycle, and they also use ATP as a phosphorylation source. Chemically, ATP is the easiest molecule to synthesize compared to other nucleotides. Therefore, ATP is likely to have been the first one to appear in biological use, and would thus have become the basis for further evolution.

The adenosine structure is part of the structure of many important biochemical compounds, including DNA, NADP, FAD, coenzyme A, and others. This suggests that ATP originated before the formation of nucleic acids and, consequently, the genetic code. In contrast, adenine structure has no oxygen atom. Therefore its synthesis in the reduced primitive atmosphere must have preceded synthesis of other nucleosides. Therefore, an anhydrous environment would be most favorable for ATP hydrolysis. The presence of water is, however, important in other respects, particularly for the mobility of reactants. Therefore, alteration of the dry/wet conditions is a favorable environment for ATP-based processes.

REDOX Potential / ATP Demand Stimulates Exercise Physiology

ATP Molecule and Functions

ATP hydrolysis is conjugated with reactions of biosynthesis. The participation of ATP can therefore supply energy to chemical reactions that cannot occur spontaneously. ATP hydrolysis is accompanied by the uptake of a water molecule, whereas the majority of reactions of synthesis proceed with the release of water.

$$ATP + H_2O \rightarrow ADP + Pi$$

A normal human body contains about 50 gm of ATP, which is recycled every day. The high-energy bonds of ATP are rather unstable; therefore, the energy is readily released when ATP is hydrolyzed in cellular reactions. It should be noted that ATP is an energy-coupling agent and not a fuel. ATP hydrolysis releases free energy $\Delta G = -31$ kJ/mol.

ATP is used for four major biological functions: (i) transportation of compounds across cell membranes; (ii) mechanical energy for muscle contraction; (iii) chemical energy for biosynthesis of various types of macromolecules; and (iv) also used as an on-off switch to control chemical reactions and to send cell signals.

ATP provides energy not only to heart muscle (for pumping blood) and skeletal muscle (for gross body movement), but also motion ability to chromosomes (spindles) and spermatozoa (flagella).

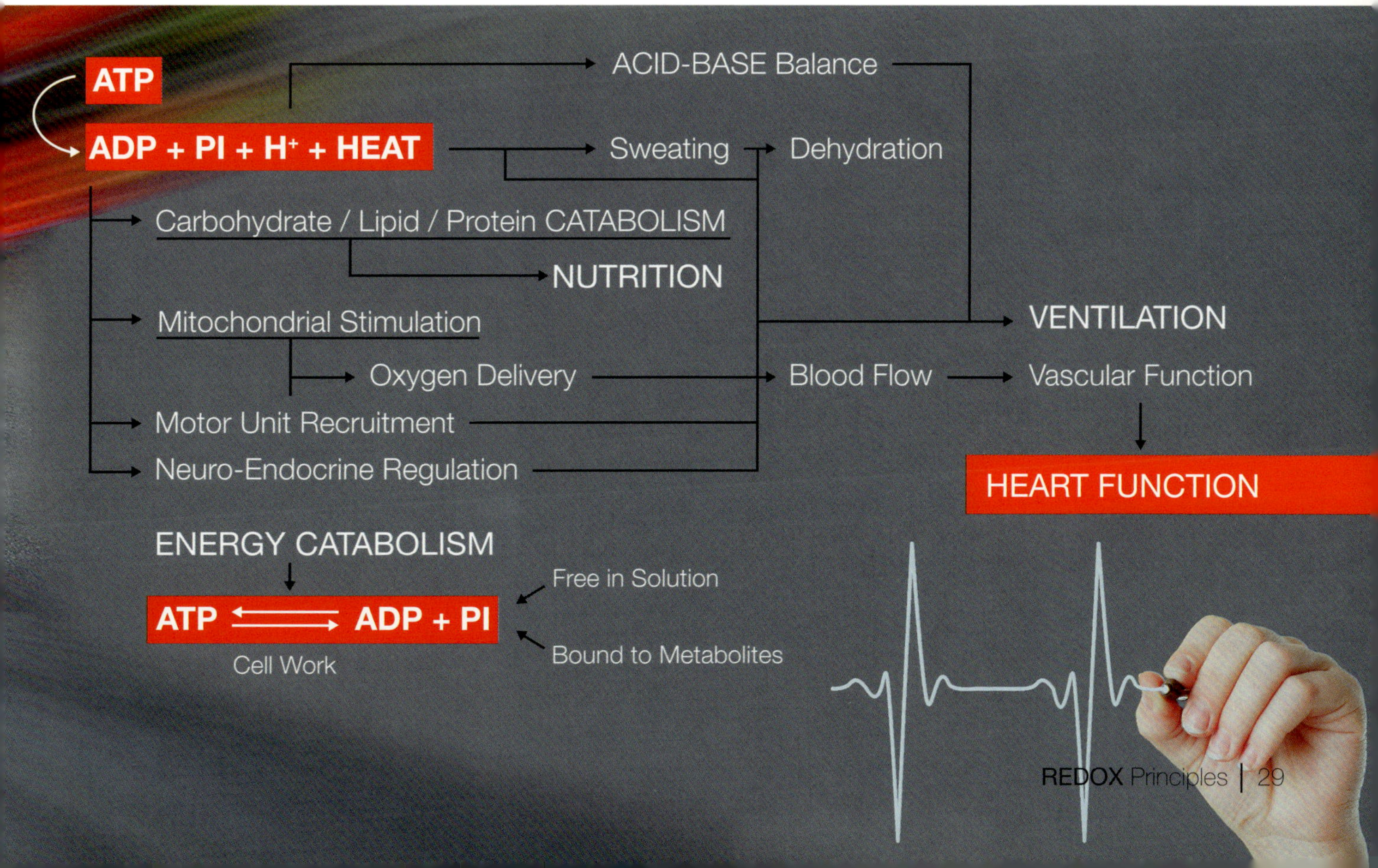

ATP Synthesis

All life forms exclusively produce ATP by three basic chemical methods: photo-phosphorylation, oxidative phosphorylation, and substrate-level phosphorylation. In prokaryotes (eg. bacteria or yeast), ATP is produced both in the cell wall and in the cytosol by glycolysis. In eukaryotes, most ATP is produced in chloroplasts (for plants) or in mitochondria (for both plants and animals). No means of producing ATP exists that is intermediate between these basic methods and no transitional forms have ever been found that bridge the gap between these different forms of ATP production. The bio-machinery required to manufacture ATP is so intricate that viruses are not able to make their own ATP. ATP synthase is a vital enzyme that drives ATP synthesis. This enzyme is located in the inner mitochondrial membrane, powered by H^+ ions flowing through it, like a 'molecular windmill'. ATP synthase is composed of four subunits (stator, rotor, rob, and knob), and the ionic flow causes it to spin and catalyze ATP synthesis. ATP synthase functions like an ion pump in reverse synthesis of ATP by using the energy created by different concentrations of H^+ on either side of the mitochondrial membrane. The rotor binds H^+ ions after passing through the stator. The binding causes a structural change in the rotor, inducing it to spin. After a complete turn, each H^+ ion leaves the rotor and enters the mitochondrial matrix via the stator. The spinning of the rotor causes an internal rod to rotate. The knob is attached to the rod but is held stationary by the stator. The rod's turning activates sites that catalyze the ATP synthesis in the knob. The H^+ gradient is created and maintained by the electron transport chain (ETC) The energy released by the ETC pumps H^+ ions against their concentration gradient into the inter-membrane space; the ions then diffuse back down their gradient through ATP synthase.

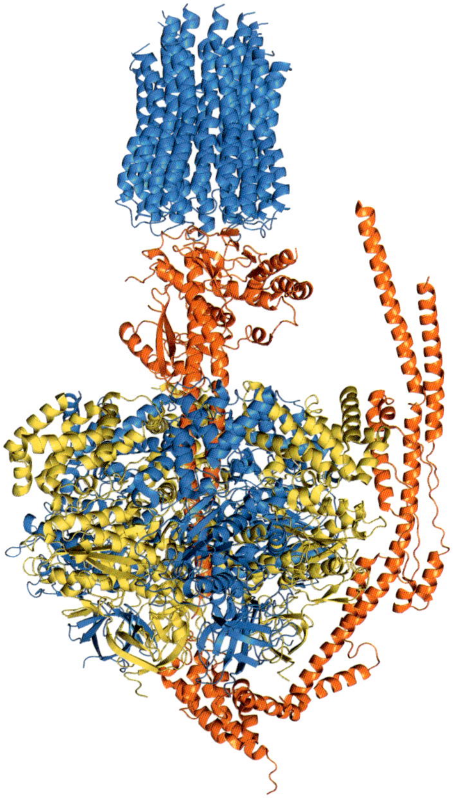

ATP-SYNTHASE

ATP is indispensable for short-term energy use, but not useful for either long-term energy storage or for processes requiring large amounts of energy. The former needs are met in plants chiefly by starch and lipids, the latter by sucrose. In humans, about 60% of the energy released from the hydrolysis of 1-mole of ATP produces metabolic heat, instead of any fuel that can support an actual function. Due to the acid-base properties of ATP, ADP, and inorganic phosphate, the hydrolysis of ATP has the effect of lowering the pH of the reaction medium. Under certain conditions, high levels of ATP hydrolysis can contribute to lactic acidosis.

ATP synthesis is coupled with a conformational change in the ATP synthase enzyme. For elucidating this mechanism, Paul Boyer and John E Walker shared half of the 1997 Nobel Prize in Chemistry. Jens Christian Skou received the other half of the Prize for the discovery of an ion-transporting enzyme Na^+, K^+-ATPase.

1.3.2. The Chemiosmotic Theory

The Chemiosmotic Theory explains the movement of ions across a selectively permeable membrane, down their electrochemical gradient. More specifically, it relates to the generation of ATP by the movement of protons (H^+ ions) across a membrane during cellular respiration. According to this theory, the transfer of electrons down an electron transport chain (ETC), through a series of REDOX reactions, releases energy. This energy allows certain carriers in the ETC to move hydrogen ions (H^+ or protons) across a membrane. As the H^+ ions accumulate on one side of a membrane, this proton concentration creates an electrochemical gradient, or REDOX potential (voltage), across the membrane. Thus, the fluid on the side of the membrane where protons accumulate acquires a positive charge and the fluid on the opposite side of the membrane is left with a negative charge. The energized REDOX state of the membrane with this charge separation is called 'Proton Motive Force' (PMF).This PMF provides the energy necessary for ATP synthase enzymes, also located in the membranes, to catalyze ATP synthesis from ADP and phosphate. ATP generation occurs as the protons cross the membrane through the ATP synthase complexes and re-enter either the bacterial cytoplasm or the matrix of the mitochondria. As the protons move down the concentration gradient through the ATP synthase, the energy released causes the rotor and stalk of the ATP synthase to rotate. The mechanical energy from this rotation converts into chemical energy as phosphate is added to ADP to form ATP.

The Chemiosmotic Theory explains how protons in aqueous solutions produce ATP in cellular membranes. Chemiosmotic coupling is the major energy producing process in mitochondria, chloroplasts, and several single-celled organisms.

In 1961, Peter Mitchell proposed that biological systems use H^+ ion batteries for ATP synthesis. The Chemiosmotic Theory consists of four postulates: (i) the inner membrane consists of electron transport enzymes that are oriented to facilitate the ejection of protons outward across the membrane. Energy from the substrate oxidation is converted and stored as a proton electrochemical gradient, the PMF; (ii) the inner membrane contains ATPase, a proton translocation enzyme, which is also oriented in a transportation-mode to facilitate the energy from ATP hydrolysis to drive protons outward across the inner membrane. The ATPase is reversible, so that protons driven inward through the enzyme, by the REDOX-generated PMF, will cause ATP synthesis; (iii) the inner membrane must have a low diffusive permeability to ions and to protons, in particular. Otherwise, ion leaks would short-circuit the PMF batteries, and ATP would not be synthesized; (iv) the inner membrane contains exchange carriers, which effectively couples anion entry with the proton entry. This provides a pathway for substrate anions to reach enzymes within the electro-negative matrix. The membrane was also postulated to contain exchange carriers in which cation exit is coupled to proton entry. This provides an effective process for removal of cations that enter the matrix by diffusion as a result of outward proton pumping.

Chemiosmotic Theory: A proposal that energy is stored as an electrochemical gradient across a vesicular membrane opened the door for understanding energy transformation by membrane systems. Peter Mitchell was awarded the Nobel Prize in Chemistry in 1978 for his theory of chemiosmotic energy transduction.

ETC consists of a spatially separated series of REDOX reactions in which electrons are transferred from a donor molecule to an acceptor molecule. The underlying force driving these reactions is the Gibbs free energy of the reactants and products.

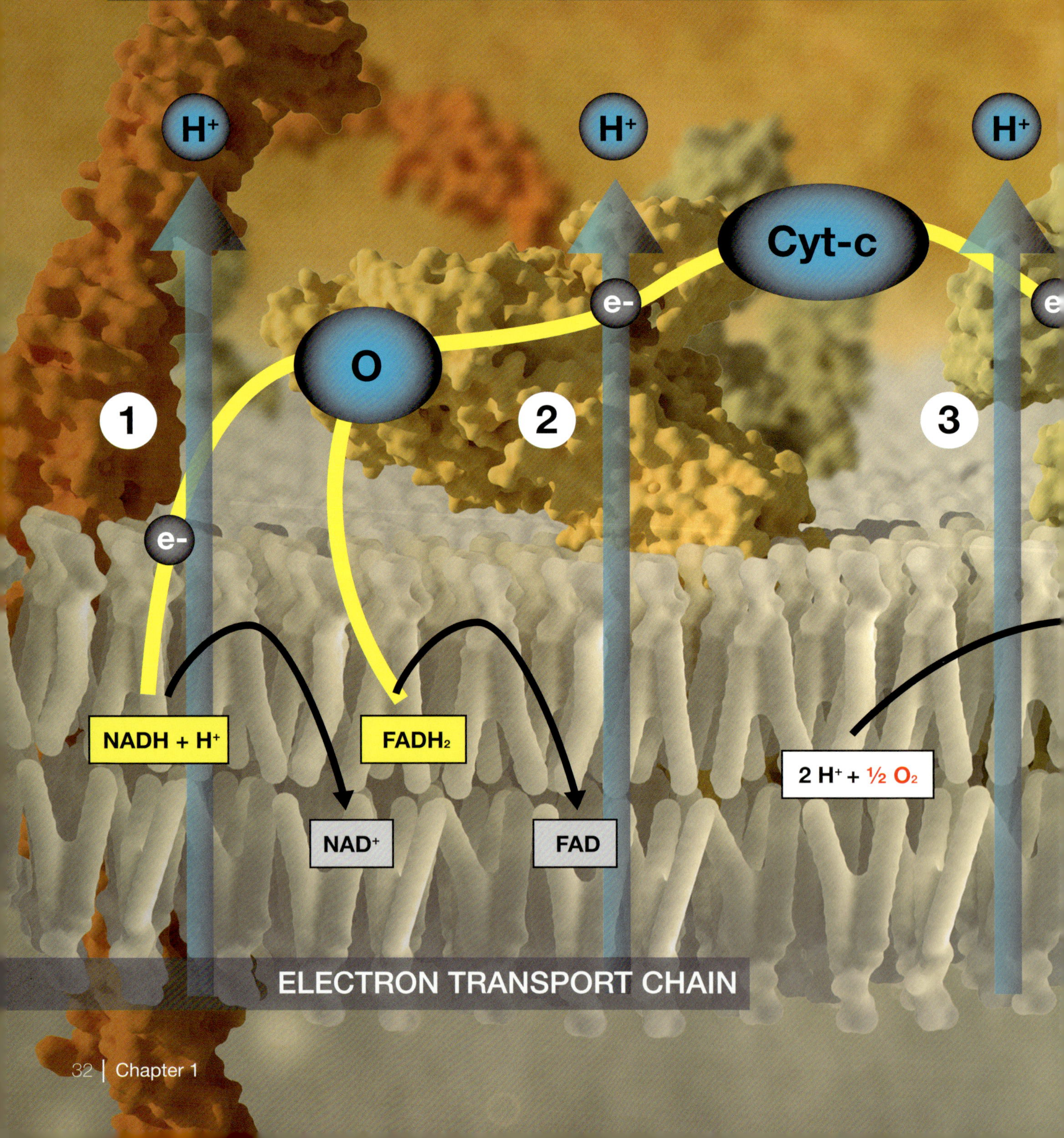

ELECTRON TRANSPORT CHAIN

ETC produces a trans-membrane proton (electro-chemical) gradient as a result of REDOX reactions. When protons flow back through the membrane, they enable mechanical work, such as movement of a cardiac muscle to support a heartbeat.

ATP Synthase

ATP synthase is an enzyme highly conserved among all species of life. It converts mechanical work into chemical energy by producing ATP, which powers most cellular reactions. A small amount of ATP is also produced from substrate-level phosphorylation (for example, glycolysis); however, biological systems generate most of their ATP through the ETC.

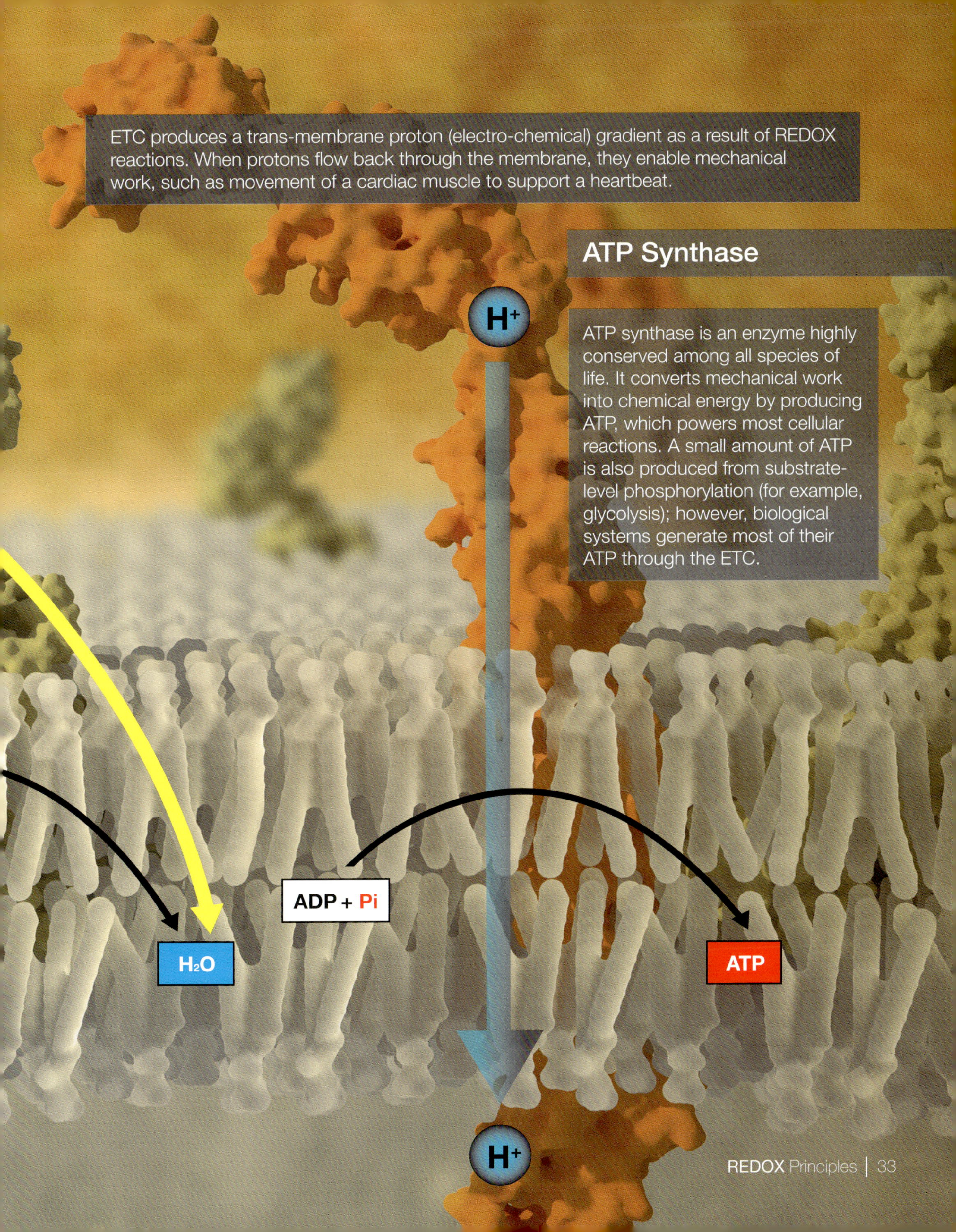

1.3.3. Photo-Phosphorylation (Photosynthesis)

Photosynthesis is the process of converting solar energy to chemical energy and storing it in the bonds of sugar. This process occurs in plants and some algae. Plants need only light energy, CO_2, and H_2O to make sugar. The process of photosynthesis takes place in the chloroplasts, using a green pigment called chlorophyll. The chemical equation for this process is:

$$6\ CO_2 + 12\ H_2O + light \rightarrow C_6H_{12}O_6 + 6\ O_2 + 6\ H_2O$$

Accordingly, 6 molecules of carbon dioxide (6 CO_2) and 12 molecules of water (12 H_2O) are consumed in the process, while glucose ($C_6H_{12}O_6$), six molecules of oxygen (6 O_2), and six molecules of water (6 H_2O) are produced.

Photosynthesis occurs mainly inside the leaves. Since photosynthesis requires CO_2, H_2O, and sunlight; these precursors must be procured or transported to the eaves. Carbon dioxide is obtained via tiny pores in plant leaves called stomata. Also, oxygen is released through the stomata. Water is obtained by the plant through the roots and delivered to the leaves through xylem/phloem, the vascular plant tissue systems. Sunlight is absorbed by chlorophyll, the site of photosynthesis in chloroplasts. Chloroplasts are located in specialized leaf cells, which often contain 50 or more chloroplasts per cell. Each chloroplast is defined by an inner and an outer envelope membrane and is shaped like a meniscus convex lens that is 5-10 microns in diameter. The inner envelope membrane acts as a barrier, controlling the flux of organic and charged molecules in and out of the chloroplast. Water passes freely through the envelope membranes, as do other small neutral molecules like CO_2 and O_2. The photosynthetic membrane is vesicular, defining a closed space with an outer water space (stromal phase) and an inner water space (lumen). The organization of the photosynthetic membrane can be described as groups of stacked membranes (like stacks of pita bread with the inner pocket representing the inner aqueous space), interconnected by non-stacked membranes that protrude from the edges of the stacks

Chloroplast

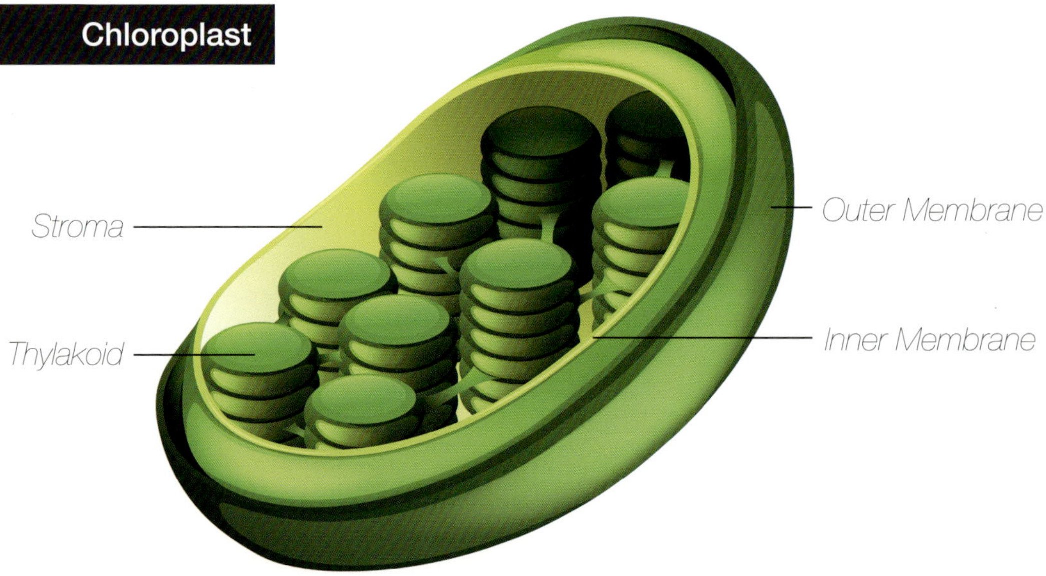

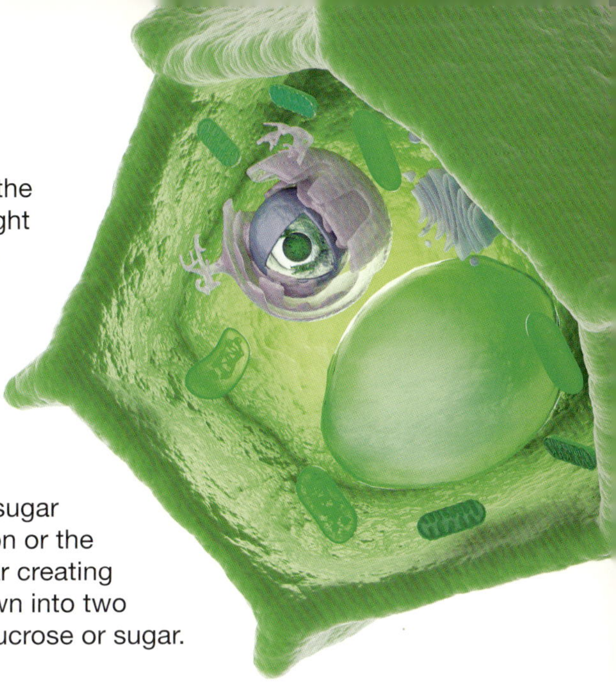

Photosynthesis occurs in two stages - the 'light reactions' and the 'dark reactions'. Light reactions take place in the presence of light and occur in the thylakoid stacks of the grana. Here, sunlight is converted to chemical energy in the form of ATP and NADPH. Chlorophyll absorbs light energy and starts a chain reaction that results in the production of ATP, NADPH, and oxygen (through the splitting of water). Oxygen is released through the stomata. Both ATP and NADPH are subsequently used in the dark reactions to produce sugar.

Dark reactions occur in the stroma, where CO_2 is converted to sugar using ATP and NADPH. This process is known as carbon fixation or the Calvin cycle. Carbon dioxide is combined with a 5-carbon sugar creating a 6-carbon sugar. This 6-carbon sugar is eventually broken-down into two molecules, glucose and fructose. These two molecules make sucrose or sugar.

The biochemical conversion of CO_2 to a carbohydrate is a reduction reaction that involves the rearrangement of covalent bonds between carbon, hydrogen and oxygen. The energy for the reduction of carbon is provided by energy rich molecules that are produced by the light driven electron transfer reactions. Carbon reduction can occur in the dark and involves a series of biochemical reactions that were elucidated in the late 1940s and 1950s.

Photosynthesis provides the energy and reduced carbon required for the survival of virtually all life on our planet. It also generates molecular oxygen necessary for the survival of aerobic organisms, including humans. Although photosynthesis occurs in cells or organelles that are microscopic, the process has a profound impact on the earth's atmosphere and climate. Each year more than 10% of the total atmospheric CO_2 is reduced to carbohydrate by photosynthesis.

Photosynthesis: Carbon reduction can occur in the dark and involves a series of biochemical reactions that were elucidated in the late 1940s and 1950s. Melvin Calvin was awarded the Nobel Prize for Chemistry in 1961 for this work.

PHOTOSYNTHESIS

Photosynthesis is driven by energy that originates in the center of the Sun, where mass is converted to heat by the fusion of hydrogen. Over time, the heat energy reaches the Sun's surface, where some of it is converted to light by black body radiation that reaches the Earth. A small fraction of the visible light incident on the Earth is absorbed by plants. A series of REDOX reactions enable photosynthetic organisms to transform light energy into chemical free energy in a stable form that can last for millions of years (e.g., fossil fuels).

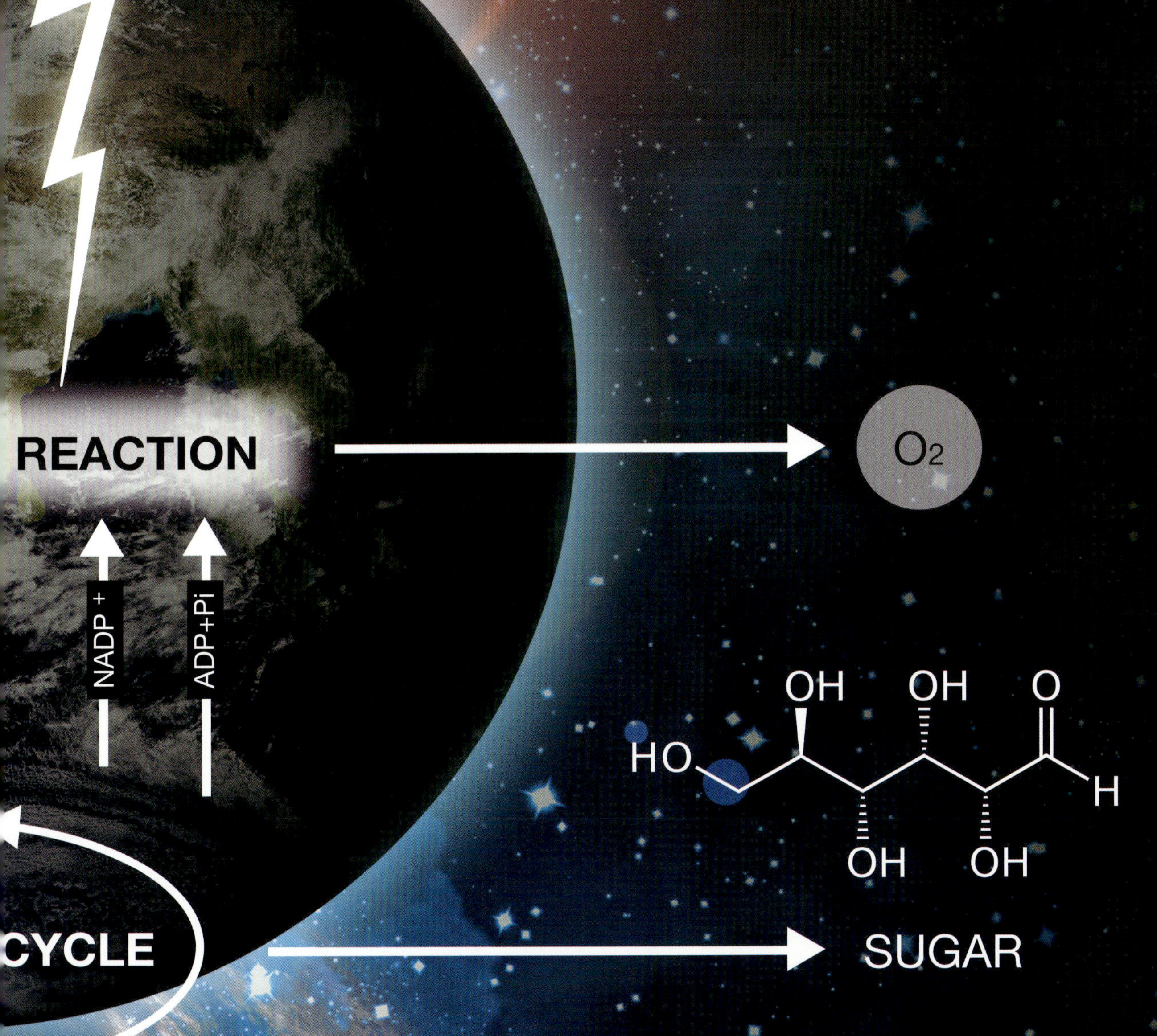

Bioenergetic pathways are equipped with multi-meric nano-machines, which contain numerous proteins, lipids, and several prosthetic groups. For example, oxygenic photosynthesis is itself the most complex metabolic machine known in nature. The minimal structure consists of two reaction centers. The electrons generated by the first water-splitting unit (PS-II) are coupled to a second reductant generating unit (PS-I) via a cytochrome complex. The core of the water-splitting complex is a hetero-dimer of two closely related proteins, namely D1 and D2. D1 is among the highly conserved proteins in nature, yet the protein is expended (turnover) about every 30 min; it works for approx. 10,000 electron transfers and then becomes irreversibly damaged. The damage leads to a loss of photosynthetic capacity, especially at high light under nutrient-limiting conditions. Yet, billions of years of evolution have not led to a 'better' reaction centre. So why has not nature developed a better oxygen evolving machine? The answer is: 'Because it can't'. Over 2.4 Ga of evolution, mutations of every amino acid in every combination of positions possible have occurred, yet, D1 is one of the most conserved proteins in the photosynthetic apparatus; the sequences are >80% identical between cyanobacteria and higher plants, and approx. 90% similar. The extraordinary conservation of this protein, which is irreversibly damaged by a product of its own catalysis, suggests that the restrictions imposed by protein–protein, protein–prosthetic group and protein–lipid interactions cannot be overcome.

According to the Kabelac and Drake analysis (1992), for one square meter on the earth's surface facing the sun, the energy received every second from a clear sky is 897.6 J, and the entropy received is 0.305 J/K. If these numbers are applied to a study of life on earth, these energy quantities should be spread over the entire earth's surface (of area $4\pi r^2$) rather than the cross-section of the earth (of area πr^2) that receives the rays perpendicular to the surface. Therefore, these numbers must be reduced by a factor of 4 to represent the energy and entropy that an average square meter of the earth receives every second, as 224.4 J and 0.076 J/K, respectively.

Simpler organisms can achieve higher energy efficiencies than complex life forms, but complex life forms can occupy ecological niches that are not accessible to their simpler counterparts. The conversion of a portion of the chemical energy to heat at each step in a metabolic pathway is the physical reason behind the pyramid of biomass observed in ecology: to take just the first step in the food chain, of the estimated 124.7 Pg/a of carbon that is fixed by photosynthesis, 64.3 Pg/a (52%) is used for the metabolism of green plants (i.e. reconverted into carbon dioxide and heat).

REDOX Principles of Energy Transfer: A key element in photosynthetic energy conversion is electron transfer within and between protein complexes and organic molecules by REDOX principles. Rudolph Marcus received the Nobel Prize in Chemistry in 1992 for his theory of electron transfer reaction in chemical systems.

1.3.4. Oxidative Phosphorylation (Respiration)

The 'Electron Transport Chain' creates a 'Proton Motive Force' which allows for the creation of ATP from ADP and Pi.

Mitochondria transform glucose, a simple sugar, into energy-rich ATP molecules through a multi-step process called cellular respiration. Mitochondria convert a single molecule of glucose into as many as 38 molecules of ATP. This conversion of a single large molecule into several smaller molecules allows cells to parcel out energy in little packages. A tiny biochemical reaction that requires only a little amount of energy can thus be powered by just a few molecules of ATP without wasting the energy of an entire glucose molecule! Indeed, ATP drives virtually every cellular process in the body.

The raw materials used to generate ATP are the foods that are consumed, or the tissues within the body that are broken down in a process called catabolism. The food is broken down to simpler basic units (such as carbohydrates, fats, and protein), which are then transferred into the mitochondria for further processing. The reactions within the mitochondria generate specific molecules with electrical charges that form a gradient within the inner mitochondrial membrane. These charged molecules are processed within the 5 electron transport chain complexes to finally combine with oxygen to make ATP. The reactive process of charged compounds with oxygen is called oxidation and the chemical process of ATP formation is called phosphorylation which is cumulatively known as oxidative phosphorylation.

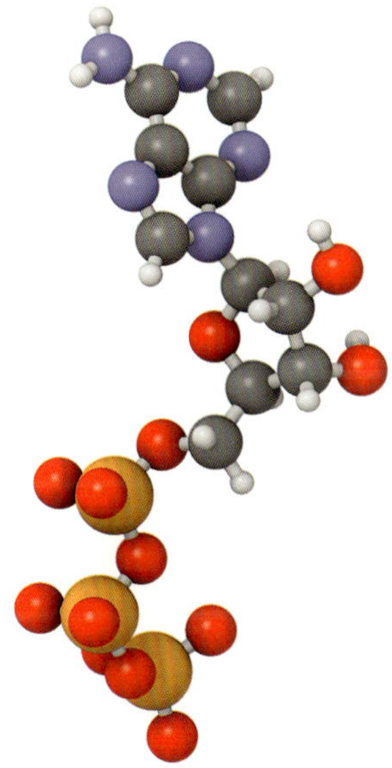

ATP

ATP Generation: Molecules such as glucose are metabolized to produce acetyl CoA (an energy-rich intermediate). The oxidation of acetyl CoA in the mitochondrial matrix is coupled to the reduction of a carrier molecule (such as NAD and FAD). The carriers pass electrons to the electron transport chain (ETC) in the inner mitochondrial membrane, which in turn passes them to other proteins in the ETC. The energy available in the electrons is used to pump protons from the matrix across the inner mitochondrial membrane, storing energy in the form of a transmembrane electrochemical gradient. The protons move back across the inner membrane through the enzyme ATP synthase. The flow of protons back into the matrix of the mitochondrion (via ATP synthase) provides enough energy for ADP to combine with inorganic Phosphate to form ATP. The electrons and protons at the last pump in the ETC are taken up by oxygen to form water. ATP synthase is the enzyme that makes ATP by chemiosmosis. It allows protons to pass through the membrane using kinetic energy to phosphorylate ADP and make ATP. The generation of ATP by chemiosmosis occurs in chloroplasts, mitochondria, and in most bacteria.

Citric Acid Cycle and Metabolism: Hans A Krebs identified two vital metabolic cycles: the urea cycle and the citric acid cycle. The latter, the key sequence of metabolic chemical reactions that produces energy (ATP) in cells, is also known as the Krebs cycle and earned him the Nobel Prize for Medicine in 1953.

1.3.5. Substrate-level Phosphorylation (Fermentation)

Substrate-level phosphorylation is a type of metabolic reaction that results in the formation of ATP [or guanosine triphosphate (GTP)] by the direct transfer and donation of a phosphoryl (PO_3) group to ADP or guanosine diphosphate (GDP) from a phosphorylated reactive intermediate. Unlike oxidative phosphorylation, oxidation and phosphorylation are not coupled in the process of substrate-level phosphorylation. However, both types of phosphorylation result in the formation of ATP and reactive intermediates. Substrate-level phosphorylation serves as a fast source of ATP, independent of external electron acceptors and respiration. This is the case, for example, in human erythrocytes, which have no mitochondria, and in the muscle during oxygen deprevation. The main part of substrate-level phosphorylation occurs in the cytoplasm of cells as part of ‘glycolysis’ and in mitochondria as part of the ‘Kreb’s Cycle’ under both aerobic and anaerobic conditions.

Fermentation is a form of anaerobic digestion that generates ATP by the process of substrate-level phosphorylation. The energy for generating ATP comes from the oxidation of organic compounds, such as carbohydrates. Fermentation is important in anaerobic conditions when there is no oxidative phosphorylation to maintain ATP synthesis. During respiration (oxidative phosphorylation), electrons are donated to an exogenous electron acceptor, such as oxygen, via ETC. In contrast, fermentation uses an endogenous, organic acceptor, such as pyruvate. Fermentation occurs in mammalian muscles during periods of intense exercise where oxygen supply becomes limited, resulting in the creation of lactic acid.

Glycolysis

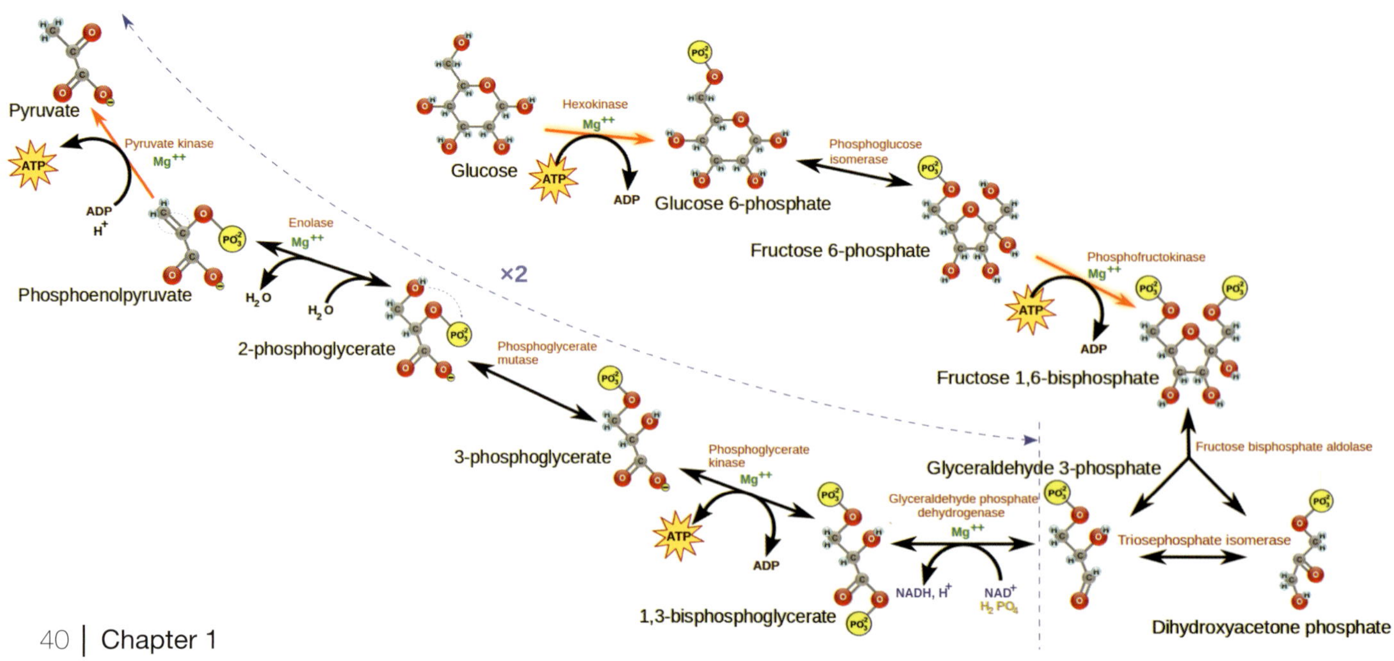

Lactic acid fermentation is the simplest type of REDOX reaction. In anaerobic conditions, the primary mechanism of cellular ATP production is glycolysis. Glycolysis reduces NAD^+ and forms NADH. Due to a limited supply of NAD^+ in any given cell, NADH must be oxidized to regenerate the NAD^+ for glycolysis to continue. Therefore, NADH donates its extra electrons to the pyruvate molecules formed during glycolysis and regenerates NAD^+ for another round of glycolysis. Lactic acid, which this process is named after, is formed by the reduction of pyruvate. Before lactic acid fermentation can occur, a single molecule of glucose is broken down into two molecules of pyruvate. This process is called glycolysis.

Glycolysis: The most common type of Glycolysis is the Embden–Meyerhof–Parnas (EMP) Pathway. To harness chemical energy from glucose, the sugar molecule needs to be split into two molecules of pyruvate. This splitting process generates two molecules of NADH and four molecules of ATP; there is only net gain of two ATP molecules considering the two initially consumed.

$$\text{Glucose} + 2\ \text{ADP} + 2\ \text{Pi} + 2\ \text{NAD}^+ \rightarrow 2\ \text{Pyruvate} + 2\ \text{ATP} + 2\ \text{NADH} + 2\ H_2O + 2\ H^+$$

As shown by the reaction equation, glycolysis causes causes the reduction of two molecules of NAD^+ to NADH. Two ADP molecules are also converted to two ATP and two water molecules via substrate-level phosphorylation.

Aerobic respiration completely oxidizes the pyruvate produced by glycolysis and generates additional ATP and NADH by entering enzymatic pathways of the Kreb's (also known as 'citric acid') cycle and by oxidative phosphorylation. This ATP synthesis can occur only in the presence of O_2. ATP can also be generated by substrate-level phosphorylation in the mitochondrial matrix, a pathway that is independent from the proton motive force (PMF). In the mitochondrial matrix, there are two reactions capable of substrate-level phosphorylation: the mitochondrial PEP-carboxykinase (PEPCK), and the succinate-CoA ligase. Mitochondrial PEPCK participates in the transfer of the phosphorylation REDOX potential from the matrix to the cytosol and vice versa. The ADP-forming succinate-CoA ligase is the only matrix enzyme that generates ATP in the absence of a PMF and is capable of maintaining matrix ATP levels under energy-limited conditions (i.e. hypoxia). Another form of substrate-level phosphorylation is also reported in the skeletal muscles and the brain. Phosphocreatine is stored as a readily available high-energy phosphate supply; the enzyme creatine phosphokinase transfers a phosphate from phosphocreatine to ADP to generate ATP.

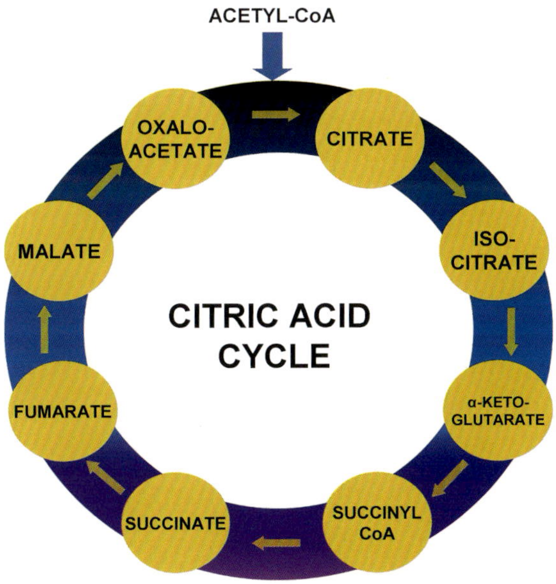

Glycolysis: The 1922 Nobel Prize in Physiology or Medicine was shared between Archibald Vivian Hill for his discovery relating to the production of heat in the muscle and Otto Fritz Meyerhof for his discovery of the fixed relationship between the consumption of oxygen and the metabolism of lactic acid in the muscle.

1.4 | REDOX Principles: **Bibliography**

1.1 | REDOX: Laws and Theories

1.1.1. REDOX and Energy Demand

1. Abbott D, Davies PCW, Pati AK (2008) *Quantum Aspects of Life*. London: Imperial College Press. ISBN 978-1848162679.
2. Grandy WT Jr (2008) *Entropy and the Time Evolution of Macroscopic Systems*, Oxford, UK: Oxford University Press. ISBN 978-0-19-954617-6.

1.1.2. REDOX and Thermodynamics

1. Haynie DT (2008). *Biological Thermodynamics* (2nd ed.). Cambridge, UK: Cambridge University Press.
2. Kondepudi D (2008) *Introduction to Modern Thermodynamics*, Chichester: Wiley. ISBN 978-0-470-01598-8.
3. Molton PM (1978) On the likelihood of a human interstellar civilization. *Journal of the British Interplanetary Society* 31:203-208.
4. Perrot P (1998). *A to Z of Thermodynamics*. Oxford, UK: Oxford University Press. ISBN 0-19-856552-6.
5. Schrodinger E, Penrose R (1992) *What is Life?* Cambridge, UK: Cambridge University Press. ASIN: B000SEKM3I.
6. Thaxton CB, Bradley WL, Olsen RL, Kenyon DH (1992) *The Mystery of Life's Origin: Reassessing Current Theories*, (2nd ed) Lewis & Stanley ASIN: B000XEKK1W

1.1.3. REDOX and the Quantum Theory

1. Bohr N (2011) *Atomic Theory and the Description of Nature: Four Essays with an Introductory Survey* (Reissue Edition). Cambridge, UK: Cambridge University Press. ISBN: 978-1107628052.
2. Massimi M (2012) *Pauli's Exclusion Principle: The Origin and Validation of a Scientific Principle.* Cambridge, UK: Cambridge University Press. ISBN: 978-1107410732.
3. Peskin ME, Schroeder DV (1995) *An Introduction to Quantum Field Theory.* Westview Press. ASIN: B0052TUEM8
4. Rae A (2011) *Quantum Physics: A Beginner's Guide.* Oneworld Publications. ASIN: B0052TNZA6
5. Thornton ST, Marion JB (2003) *Classical Dynamics of Particles and Systems* (5th ed). Cengage Learning. ISBN: 978-0534408961

1.2 | REDOX: Bio-Informatics

1. Astley SB (2007) An introduction to nutrigenomics developments and trends. *Genes and Nutrition* 2(1):11-13.
2. Bauer CE, Elsen S, Bird TH (1999) Mechanisms for redox control of gene expression. *Annual Review of Microbiology* 53:495-523.
3. Gregory SG, Barlow KF, Mclay KE, Kaul R, et al. (2006) The DNA sequence and biological annotation of human chromosome 1. *Nature* 441 (7091):315-321.
4. Griffiths WM, Miller JH, Suzuki DT, Lewontin RC, et al (eds.) (2000) *An Introduction to Genetic Analysis* (7th ed.). New York: WH Freeman. ISBN 0-7167-3520-2.
5. Hartl D, Jones E (2005) *Genetics: Analysis of Genes and Genomes* (6th ed.). Jones & Bartlett. ISBN 0-7637-1511-5.
6. Pearson H (2006) Genetics: what is a gene? *Nature* 441 (7092):398-401.
7. Sanger F, Nicklen S, Coulson AR (1977) DNA sequencing with chain-terminating inhibitors. *Proceedings of the National Academy of Sciences of the USA* 74(12): 5463-5467.
8. Sen CK, Packer L (1996) Antioxidant and redox regulation of gene transcription. *FASEB Journal* 10(7):709-720.
9. Watson JD, Crick FH (1953) Genetical implications of the structure of deoxyribonucleic acid. *Nature* 171 (4361): 964-967.

1.2.1. The Genetic Code

1. Crick FHC (1968) The origin of the genetic code. *Journal of Molecular Biology* 38:367-379.
   Knight RD, Freeland SJ, Landweber LF (1999) Selection, history and chemistry: the three faces of the genetic code. *Trends in Biochemical Sciences* 24(6):241-247.
2. Lewis R (2005) *Human Genetics: Concepts and Applications* (6th ed.). Boston, Mass:McGraw Hill. ISBN 0-07-111156-5.
3. Nirenberg M, Leder P, Bernfield M, Brimacombe R, et al (1965) RNA codewords and protein synthesis, VII. On the general nature of the RNA code. *Proceeding of the National Academy of Sciences USA* 53(5):1161-1168.
4. Szathmary E (1999) The origin of the genetic code: amino acids as cofactors in an RNA world. *Trends in Genetics* 15(6):223-229.
5. Wang Q, Parrish AR, Wang L (2009) Expanding the genetic code for biological studies. *Chemical Biology* 16(3):323-336.
6. Westheimer FH (1987) Why Nature chose phosphates. *Science* 235:1173-1178.
7. Yarus M (2010) Life from an RNA World: The Ancestor Within. Cambridge:Harvard University Press. ISBN 0-674-05075-4.

1.2.2. Deoxyribonucleic Acid (DNA)

1. Alberts B, Johnson A, Lewis J, Raff M, Roberts K, Walter P (2002) *Molecular Biology of the Cell* (4th ed) New York:Garland Science. ISBN 0-8153-3218-1.
2. Caspersson T, Schultz J (1939) Pentose nucleotides in the cytoplasm of growing tissues. *Nature* 143(3623):602-6033
3. Dahm R (2005) Friedrich Miescher and the discovery of DNA. *Developmental Biology* 278(2):274-288.
4. De Grey ADNJ (1999) *The Mitochondrial Free Radical Theory of Aging.* Austin, Texas:RG Landes Co. ISBN: 1-57059-564-X.
5. Griffiths AJF, Miller JH, Suzuki DT, Lewontin RC, Gelbart WM (1999) *An Introduction to Genetic Analysis.* (7th ed.). San Francisco: WH Freeman. ISBN 0-7167-3771-X.
6. Lodish HF, Berk A, Zipursky SL, Matsudaira P, Baltimore D, Darnell JE (2000) *Molecular Cell Biology* (4th ed). San Francisco:WH Freeman. ISBN 0-7167-3706-X.
7. Meselson M, Stahl FW (1958) The replication of DNA in *Escherichia coli. Proceedings of the National Academy of Sciences* 44:671-682.

1.2.3. Ribonucleic Acid (RNA)

1. Anet FAL (2004) The place of metabolism in the origin of life. *Current Opinion in Chemical Biology* 8:654-659.
2. Benner SA, Ricardo A, Carrigan MA (2004) Is there a common chemical model for life in the universe? *Current Opinion in Chemical Biology* 8:672-689.
3. Cech TR (1993) The efficiency and versatility of catalytic RNA: implications for an RNA world. *Gene* 135(1-2):33-36.
4. Cech TR (2000) Structural biology. The ribosome is a ribozyme. *Science* 289(5481):878-879.
5. Cech TR (2009) Evolution of biological catalysis: ribozyme to RNP enzyme. *Cold Spring Harbor Symposium Quantitave Biology* 74:11-16.
6. Eigen M, Gardiner W, Schuster P, Winkler-Oswatitsch R (1981) The origin of genetic information. *Scientific American*. 244(4):88-92.
7. Eschenmoser A (2007) The search for the chemistry of life's origin. *Tetrahedron* 63:12821-12844.
8. Gilbert W (1986) The RNA World. *Nature* 319 (6055):618.
9. Orgel LE (2004) Prebiotic chemistry and the origin of the RNA world. *Critical Reviews in Biochemistry & Molecular Biology* 39:99-123.
10. Ruvkun G (2001) Glimpses of a tiny RNA world. *Science* 294(5543):797-799.
11. Unrau PJ, Bartel DP (1998) RNA-catalysed nucleotide synthesis. *Nature* 395(6699):260-263.

1.2.4. DNA vs RNA – The REDOX Variance

1. Amaral PP, Mattick JS (2008) Non-coding RNA in development. Mammalian Genome: *Official Journal of the International Mammalian Genome Society* 19(7–8):454-492.
2. Callahan MP, Smith KE, Cleaves HJ 2nd, Ruzicka J, et al (2011) Carbonaceous meteorites contain a wide range of extra-terrestrial nucleobases. *Proceedings of the National Academy of Sciences USA* 108(34):13995-13998.

3. Herd CD, Blinova A, Simkus DN, Huang Y, et al (2011) Origin and evolution of prebiotic organic matter as inferred from the Tagish Lake meteorite. *Science* 332(6035):1304-1307.
4. Parker ET, Cleaves HJ, Dworkin JP, Glavin DP, et al (2011) Primordial synthesis of amines and amino acids in a 1958 Miller H2S-rich spark discharge experiment. *Proceedings of the National Academy of Sciences USA* 108(14):5526-5531.
5. Sanchez RA, Ferris JP, Orgel LE (1967) Studies in prebiotic synthesis. II. Synthesis of purine precur sors and amino acids from aqueous hydrogen cyanide. *Journal of Molecular Biology* 30(2): 223–53.
6. Yamada H, Okamoto T (1972) A one-step synthesis of purine ring from formamide. *Chemical & Pharmaceutical Bulletin* 20(3):623

1.3 | REDOX: Bio-Energetics

1. Klingenberg M (2010) Wanderings in bioenergetics and biomembranes. *Biochimica et Biophysica Acta* 1797(6-7):579-594.
2. Kushnareva Y, Newmeyer DD (2010) Bioenergetics and cell death. *Annals of the New York Academy of Sciences* 1201:50-57.
3. Nath S (2010) Beyond the chemiosmotic theory: analysis of key fundamental aspects of energy coupling in oxidative phosphorylation in the light of a torsional mechanism of energy transduction and ATP synthesis--invited review part 2. *Journal of Bioenergetics & Biomembranes* 42(4):301-309.
4. Nath S (2010) Beyond the chemiosmotic theory: analysis of key fundamental aspects of energy coupling in oxidative phosphorylation in the light of a torsional mechanism of energy transduction and ATP synthesis--invited review part 1. *Journal of Bioenergetics & Biomembranes* 42(4):293-300.
5. Schoepp-Cothenet B, van Lis R, Atteia A, Baymann F, et al (2013) On the universal core of bioenergetics. *Biochimica et Biophysica Acta* 1827(2):79-93.

1.3.1. ATP and other Nucleotides

1. Bonora M, Patergnani S, Rimessi A, De Marchi E, et al (2012) ATP synthesis and storage. *Purinergic Signal* 8(3):343-357.
2. Calhoun KA, Swartz JR (2007) Energy systems for ATP regeneration in cell-free protein synthesis reactions. *Methods in Molecular Biology* 375:3-17.
3. Klingenberg M (2008) The ADP and ATP transport in mitochondria and its carrier. *Biochimica et Biophysica Acta* 1778(10):1978-2021.
4. Ricquier D (2005) Respiration uncoupling and metabolism in the control of energy expenditure. 64:47-52.

1.3.2. The Chemiosmotic Theory

1. Boyer PD (1997) The ATP synthase – a splendid molecular machine. *Annual Review of Biochemistry* 66:717-749.
2. Graber P, Milazzo G (1997) *Bioelectrochemistry: Principles and Practice.* Basel: Birkhauser. ISBN-10:3764352957.
3. Mitchell P (1961) Coupling of phosphorylation to electron and hydrogen transfer by a chemiosmotic type of mechanism. *Nature* 191 (4784):144-148.
4. Palmeira CM, Moreno AJ (2012) *Mitochondrial Bioenergetics: Methods and Protocols* (Methods in Molecular Biology). Humana Press. ISBN-10:1617793817

1.3.3. Photo-Phosphorylation (Photosynthesis)

1. Blankenship R (1992) Origin and early evolution of photosynthesis. *Photosynthesis Research* 33:91-111.
2. Calvin M (1989) Forty years of photosynthesis and related activities. *Photosynthesis Research* 21:3-16.
3. Duysens LNM (1989) The discovery of the two photosynthetic systems: a personal account. *Photosynthesis Research* 21:61-79.
4. Frenkel AW (1995) Photosynthetic phosphorylation. *Photosynthesis Research* 46:73-77.
5. Govindjee (1975) *Bioenergetics of Photosynthesis.* Academic Press:NY. ISBN 0-12-2943503
6. Govindjee, Coleman WJ (1990) How plants make oxygen. *Scientific American* 262:50-58.
7. Kabelac S, Drake FD (1992) The entropy of terrestrial solar radiation. *Journal of Solar Energy Science and Engineering* 48 (4):239-248
8. Marcus RA, Sutin N (1985) Electron transfers in chemistry and biology. *Biochimica et Biophysica Acta* 811:265-322.

9. Melis A (1991) Dynamics of photosynthetic membrane composition and function. *Biochimica et Biophysica Acta* 1058:87-106.
10. Moser CC, Keske JM, Warncke K, Farid RS, Dutton PL (1992) Nature of biological electron transfer. *Nature* 355:796-802.
11. Prasil O, Adir N, Ohad I (1992) Dynamics of photosystem II: mechanism of photoinhibtion and recovery processes. In: *The Photosystems: Stucture, Function and Molecular Biology,* Vol. 11 (ed. JR Barber), pp. 295-348. New York, NY: Elsevier. ISBN-10: 0444894403
12. Shi T, Bibby TS, Jiang L, Irwin AJ, Falkowski PG (2005) Protein interactions limit the rate of evolution of photosynthetic genes in cyanobacteria. *Molecular Biology of Evolution* 22:2179-2189.
13. Witt HT (1991) Functional mechanism of water splitting photosynthesis. *Photosynthesis Research* 29:55-77.

1.3.4. Oxidative Phosphorylation (Respiration)

1. Alberts B, Johnson A, Lewis J, Raff M, et al (2002) *The importance of H+ driven transport in bacteria. Molecular Biology of the Cell*. Garland. ISBN 0-8153-4072-9.
2. Ankel-Simons F, Cummins JM (1996) Misconceptions about mitochondria and mammalian fertiliza tion: implications for theories on human evolution. *Proceedings of the National Academy of Sciences USA* 93:13859-13863.
3. Cooper GM (2000) *Electron Transport and ATP Synthesis during Photosynthesis. The Cell: A Molecular Approach* (2nd ed.). Sinauer Associates, Inc. ISBN 0-87893-119-8.
4. Elston T, Wang H, Oster G (1998) Energy transduction in ATP synthase. *Nature* 391:510-513.
5. Kaneda H, Hayashi J, Takahama S, Taya C, Lindahl KF, Yonekawa H (1995) Elimination of paternal mitochondrial DNA in intra-specific crosses during early mouse embryogenesis. *Proceedings of the National Academy of Sciences USA* 92:4542-4546.
6. Krebs HA (1937) The intermediate metabolism of carbohydrates. *Lancet* 2:736-738.
7. Martin W, Muller M (1998) The hydrogen hypothesis for the first eukaryote. *Nature* 392:37-41.
8. Nicholls D. G., Ferguson S, J. (1992). *Bioenergetics 2* (2nd ed.). San Diego: Academic Press. ISBN 978-0125181242.
9. Pfanner N (1998) Mitochondrial import: Crossing the aqueous intermembrane space. *Current Biology* 8:R262-R265.
10. Stryer L (1995). *Biochemistry* (4th ed.). New York - Basingstoke: W. H. Freeman & Company. ISBN 978-0716720096.

1.3.5. Substrate-level Phosphorylation (Fermentation)

1. Johnson JD, Mehus JG, Tews K, Milavetz BI, Lambeth DO (1998) Genetic evidence for the expression of ATP- & GTP-specific succinyl-CoA synthetases in multicellular eucaryotes. *The Journal of Biological Chemistry* 273(42):27580-27586.
2. Lambeth DO (2002) What is the function of GTP produced in the Krebs citric acid cycle?. *IUBMB Life* 54(3):143-144.
3. Lambeth DO, Tews KN, Adkins S, Frohlich D, Milavetz BI (2004) Expression of two succinyl-CoA synthetases with different nucleotide specificities in mammalian tissues. *The Journal of Biological Chemistry* 279 (35):36621-36624.
4. Ottaway JH, McClellan JA, Saunderson CL (1981) Succinic thiokinase and metabolic control. *The International Journal of Biochemistry* 13 (4):401-410.
5. Romano AH, Conway T (1996) Evolution of carbohydrate metabolic pathways. *Research in Microbiology* 147(6–7):448–55
6. Voet D, Voet JG (1995) *Biochemistry* (2nd ed.). New York, NY: John Wiley & Sons. ISBN 978-0-471 -58651-7.
7. Wilson DF, Ereciska M, Schramm VL (1983) Evaluation of the relationship between the intra- and extra-mitochondrial ATP/ADP ratios using phosphoenolpyruvate carboxykinase. *The Journal of Biological Chemistry* 258(17):10464-10473.

CHAPTER 2

REDOX Chemistry

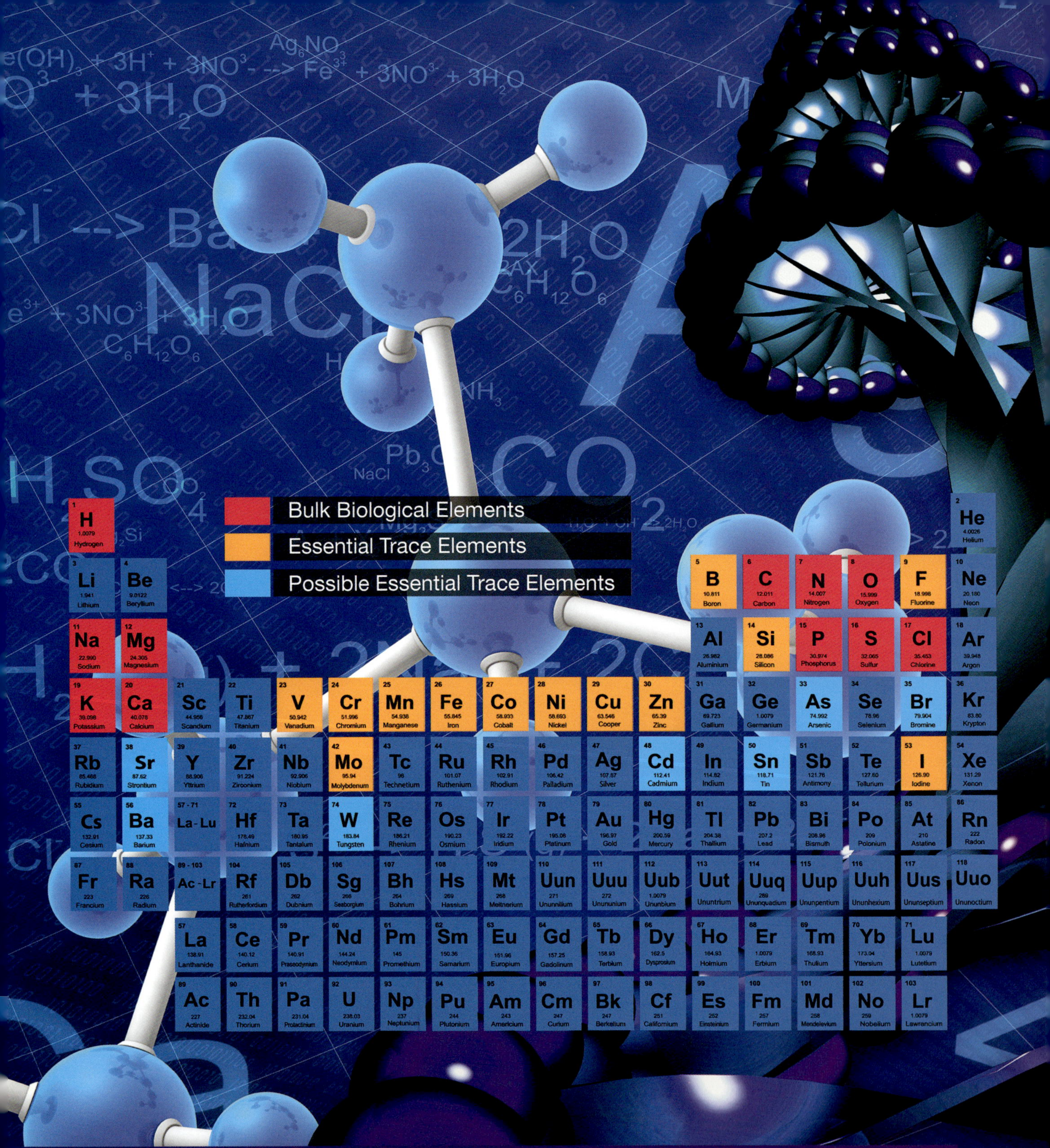

The Distribution of Elements Essential for Life in the Periodic Table

2

REDOX: Chemistry

Life evolved on Earth about 3.8 billion years ago using six major elements, namely: hydrogen (H), carbon (C), nitrogen (N), oxygen (O), sulfur (S) and phosphorus (P). These elements are the fundamental building blocks of all the major biological macromolecules including proteins, nucleic acids, lipids and carbohydrates. Synthesis of these macromolecules requires input of energy and rearrangement of chemical bonds. Bioenergetics, the process of energy transduction- governed by the principles of REDOX chemistry - is the basis for evolution of life on our Blue Planet.

The average adult human body contains at least detectable traces of 60 chemical elements, and 20 of these are known to play an active positive role in life and health (refer to TABLE). The relative amounts of each element vary by individual, based on fat/muscle/bone body composition. The adult human body averages ~53% water. This varies substantially by age, sex, and adiposity. In a large sample of adults of all ages and both sexes, the figure for water fraction by weight was found ~49% for females and 59% for males. Water is ~11% hydrogen (H) by mass but ~67% by atomic per cent, and these numbers along with the complementary % numbers for oxygen (O) in water, are the largest contributors to overall mass and atomic composition of the human body. Due to water content, the human body contains more oxygen by mass than any other element, and more hydrogen by atom-fraction than any element. The other major 10 elemental constituents include: carbon (C), nitrogen (N), calcium, phosphorus, sulfur, potassium, chlorine, sodium, magnesium and iron. The amounts of HOCN contained in the adult human body represent the integration of the day to day accretions from the time of conception to the termination of growth.

Element	Mass (kg)
Oxygen	43
Carbon	16
Hydrogen	7
Nitrogen	1.8
Calcium	1.0
Phosphorus	0.78
Potassium	0.14
Sulfur	0.14
Sodium	0.10
Chlorine	0.095
Magnesium	0.019
Iron	0.0042
Zinc	0.0023
Silicon	0.0010
Copper	0.000072
Iodine	0.000020
Selenium	0.000015
Chromium	0.000014
Manganese	0.000012
Cobalt	0.000003

TABLE: *Elemental mass in the adult human body (70 kg)*

2.1. **REDOX** and the Biosphere

Biosphere represents the role of living organisms and their metabolic by-products in controlling and interacting with the other spheres in the global bio-geo-chemical cycles and energy budgets. The biosphere plays a central role in the bio-geo-chemical processing of carbon, nitrogen, phosphorus, sulfur and other elements. As a result, bio-geo-chemical processes such as photosynthesis and nitrogen fixation are critical for the REDOX balance of the earth systems as a whole. The REDOX properties of the biosphere in terms of its surface reflectance (light), thermal transfers (heat exchange) and moisture with the atmosphere (humidity) are critical for understanding the global climate. Accordingly, the REDOX principles play a significant role in our understanding the current issues of global warming.

The term 'biosphere' was coined by geologist Eduard Seuss in 1875, defined as – the place on Earth's surface where life dwells.

The biosphere includes every living organism on our Blue Planet, the global sum of all ecosystems. It can also be called as the zone of life on Earth, a closed (besides solar and cosmic radiation) and self-regulating system. A key component of earth systems, the biosphere interacts with and exchanges matter and energy with the other spheres (lithosphere, hydrosphere, and atmosphere), helping to drive the global biogeochemical cycling of carbon, nitrogen, phosphorus, sulfur and other elements. From a thermodynamic standpoint, the biosphere integrates all living organisms and their relationships, biological functions, including photosynthesis, respiration, decomposition, nitrogen fixation and denitrification – every pathway regulated under the exclusive governance of REDOX principles.

The biosphere has evolved since the first single-celled organisms originated 3.5 billion years ago under atmospheric conditions resembling those of our neighboring planets Mars and Venus, which have atmospheres composed primarily of CO_2. Billions of years of primary production by plants released oxygen from this CO_2 and deposited the carbon in sediments, eventually producing the oxygen-rich atmosphere that exists today. The biosphere is a dynamic system influenced by seasonal cycles with its primary productivity driven by the energy captured by photosynthesis. Free oxygen (O_2) for aerobic respiration and ozone (O_3) in the stratosphere to protect the biosphere from harmful UV radiation, has made life possible (as we know it), while transforming the chemical systems of this Blue Planet forever.

The Level of REDOX Organization

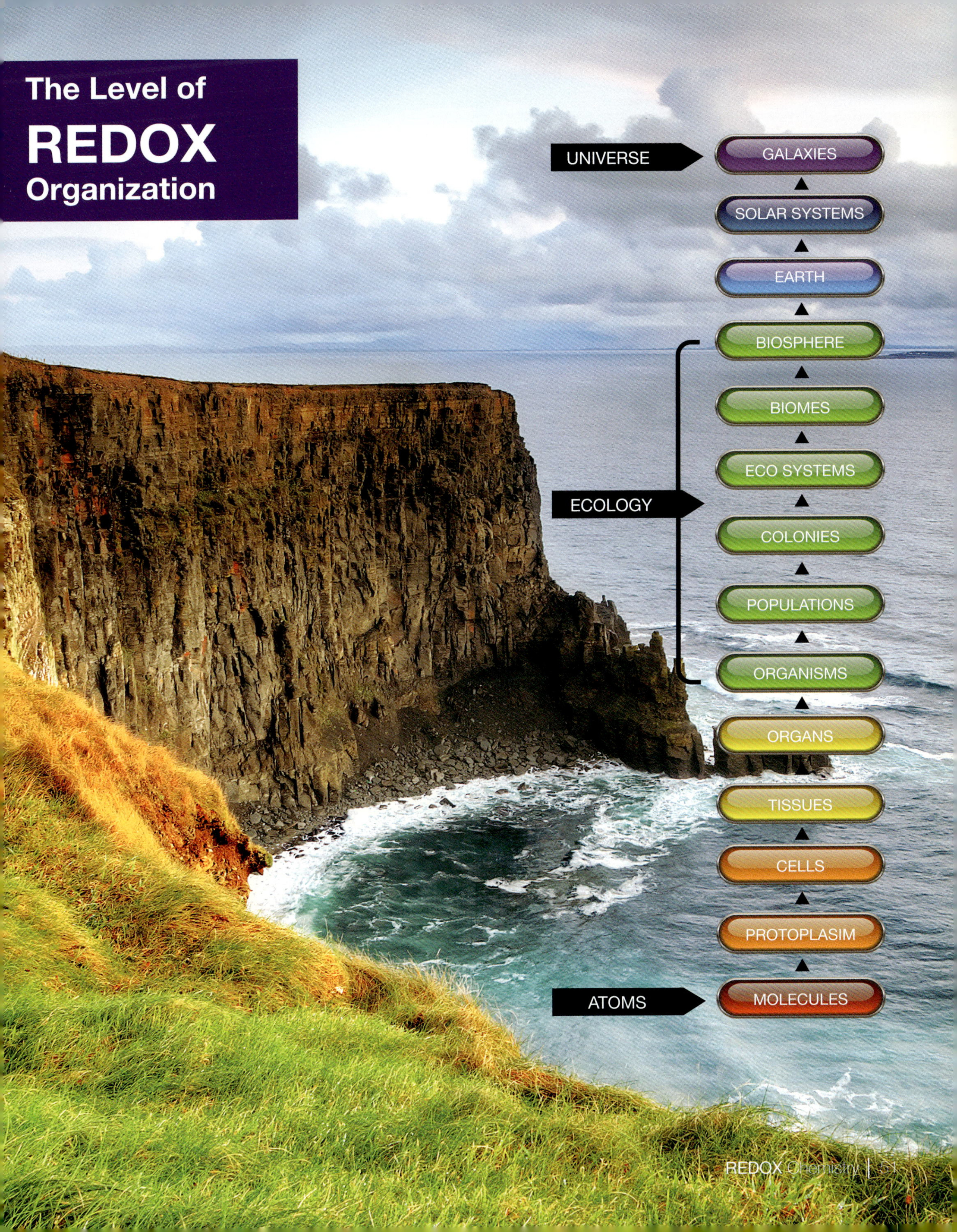

2.1.1. The Carbon Cycle

The carbon cycle is one of the Earth's fastest recycling processes where each atom of carbon has been recycled numerous times. Therefore, the carbon cycle has no specific beginning or ending point. Carbon-based molecules are crucial for life on earth, because carbon is the main component of biological compounds. Carbon passes through different cycles on the Blue Planet, via three main stages: reservoirs, assimilation, and release.

Atmospheric Carbon Cycle: Carbon in the earth's atmosphere exists in two main forms: carbon dioxide (CO_2) and methane (CH_4). Both of these gases absorb and retain heat in the atmosphere and are partially responsible for the greenhouse effect. Atmospheric CO_2 has two major sinks: terrestrial ecosystems and marine ecosystems, both of which deal with photosynthesis (as a part of assimilation) and respiration (as a part of release). Terrestrial ecosystems draw CO_2 from the atmosphere and use it in photosynthesis.

The equation: $CO_2 + H_2O + \text{light} \rightarrow C_6H_{12}O_6 + O_2 + \text{energy}$ – shows how CO_2 is broken down and used to produce glucose for the plants and oxygen as a byproduct. All plants act as a sink for CO_2 because it is a necessary gas for photosynthesis. Of the terrestrial ecosystems, forests have the highest rates of productivity, thus utilizing carbon at a higher rate compared to oceans.

Terrestrial Carbon Cycle: The terrestrial biosphere includes the organic carbon in all land-living organisms, both alive and dead, as well as carbon stored in soils. About 500 gigatons of carbon are stored above ground in plants and other living organisms, while soil holds approximately 1,500 gigatons of carbon. Most carbon in the terrestrial biosphere is organic carbon, while about a third of soil carbon is stored in inorganic forms, such as calcium carbonate. Organic carbon is a major component of all organisms living on earth. Autotrophs extract it from the air in the form of carbon dioxide, converting it into organic carbon, while heterotrophs receive carbon by consuming other organisms. Thus, CO_2 is captured by plants throughout their lives and heterotrophic organisms in turn obtain a part of this carbon. The element is transferred from organism to organism when plants are consumed by herbivores, which are in turn eaten by carnivores along the food chain. All these organisms go through respiration, excrete organic waste, and eventually die and decompose which releases carbon into the soil either as carbonates or fossil fuels. The CO_2 drawn into marine ecosystems is eventually released through oceanic respiration. Through time and pressure, the organic material buried in soil and sediments may eventually become fossil fuels such as coal and oil, which then become additional sources of carbon. When these are burned, they emit tremendous amounts of carbon back into the atmosphere. Thus, the atmospheric carbon sequestered through weathering, terrestrial and marine ecosystem sinks is eventually recycled through the processes of decomposition, respiration, and tectonic forces, releasing the carbon again into the atmosphere.

Carbon leaves the terrestrial biosphere in several ways and on different time scales. The combustion or respiration of organic carbon releases it rapidly into the atmosphere. It can also be exported into the oceans through rivers or remain sequestered in soils in the form of inert carbon. Carbon stored in soil can remain there for up to thousands of years before being washed into rivers by erosion or released into the atmosphere through soil respiration. The length of carbon sequestering in soil is dependent on local climatic conditions and thus changes in the course of climate change.

Oceanic Carbon Cycle: Oceans contain the greatest quantity of actively cycled carbon, which enters the ocean mainly through the dissolution of atmospheric CO_2, which is converted into carbonate. It can also enter the oceans through rivers as dissolved organic carbon. Marine ecosystems are separated into two areas: coastal ecosystems and the open ocean. Both have the capacity to store significant amounts of carbon in sediments and are also able to sequester carbon in photosynthesis or chemosynthesis through phytoplankton, seaweeds, and other marine algae. Most storage of carbon is in marine sediments and rocks, although some carbon is used by marine life in the formation of calcium carbonate. CO_2 absorption also makes water more acidic, which affects ocean biosystems. The projected rate of increasing oceanic acidity could slow the biological precipitation of calcium carbonates, thus decreasing the ocean's capacity to absorb CO_2.

Geological Carbon Cycle: Geology is one of the most important determinants of the amount of carbon in the atmosphere, and thus impacts global temperatures. Of the carbon stored in the geosphere, about 80% is limestone and its derivatives, which form from the sedimentation of calcium carbonate stored in the shells of marine organisms. The remaining 20% is stored as kerogens formed through the sedimentation and burial of terrestrial organisms under high heat and pressure. Organic carbon stored in the geosphere can remain there for millions of years. Another carbon sink is the weathering of mountains and other rock formations formed by plate tectonics, mainly silicate weathering. Carbon dioxide is consumed from silicate weathering as seen in the following equation:

$$CaSiO_3 + 2\ CO_2 + 2\ H_2O \rightarrow CaCO_3 + SiO_2 + CO_2 + 2\ H_2O$$

Carbon can also leave the geosphere in several ways. Carbon dioxide is released during the metamorphosis of carbonate rocks when they are subducted into the earth's mantle. This carbon dioxide can be released into the atmosphere and ocean through volcanoes and hotspots. It can also be removed by humans through the direct extraction of kerogens in the form of fossil fuels. After extraction, fossil fuels are burned to release energy, thus emitting the carbon they store into the atmosphere.

Human Factor: The largest and most direct human influence on the carbon cycle is through direct emissions from burning fossil fuels, which transfers carbon from the geosphere into the atmosphere. Over the past several centuries, human activities including land use and land cover change, deforestation for agricultural purposes, has led to the loss of biodiversity and reduced the resilience of ecosystems to handle environmental stresses and remove carbon from the atmosphere. The net result of these activity is that more carbon stays in the atmosphere. Also, acid rain and polluted runoff from agriculture and industry change the ocean's chemical composition. Such changes can have dramatic effects on highly sensitive ecosystems such as coral reefs, thus limiting the ocean's ability to absorb carbon from the atmosphere on a regional scale and reducing oceanic biodiversity globally.

The carbon cycle has major effects on global climate. The burning of fossil fuels at the present alarming rate increases global warming. This is due to the increasing amount of greenhouse gases, specifically CO_2, which capture heat easily thus increasing global temperature. The excess CO_2 from the burning of fossil fuels is responsible for the gradual yet constant rise in temperature over the past few decades. Certain wave lengths coming from the sun, mainly infrared radiation, are trapped by greenhouse gases and kept in the earth's atmosphere. These gases block the re-radiation of infrared waves back out into space, causing the global temperature to increase.

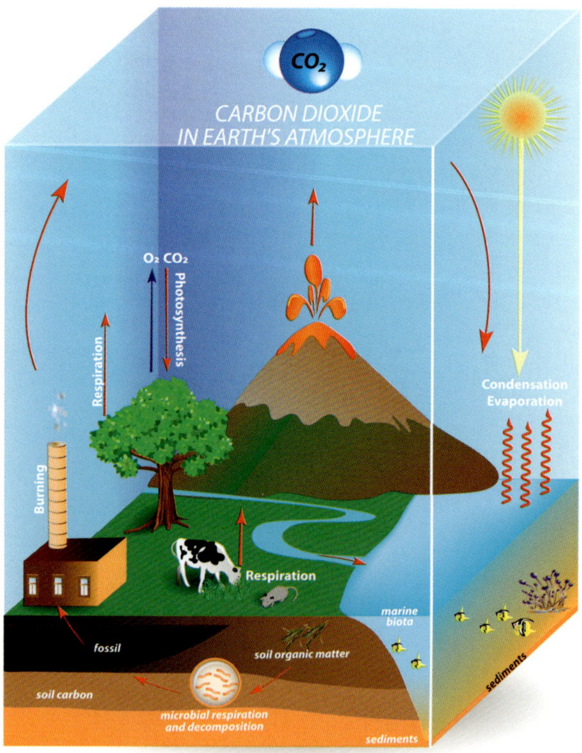

THE CARBON CYCLE

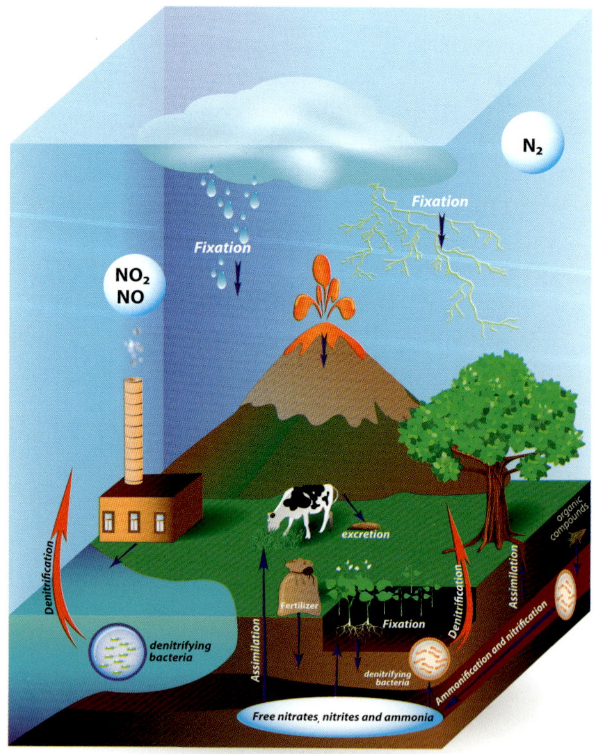

THE NITROGEN CYCLE

2.1.2. The Nitrogen Cycle

Nitrogen is essential for many processes and is crucial for any life on Earth. It is a component in all amino acids that make peptide chains (proteins), and is present in the bases that make up nucleic acids (DNA, RNA). In plants, much of the nitrogen is used in chlorophyll molecules, essential for photosynthesis. Although earth's atmosphere is an abundant source of nitrogen (N), most is relatively unusable by plants. Nitrogen is present in the environment in a wide variety of chemical forms including organic nitrogen, ammonium (NH_4^+), nitrite (NO_2^-), nitrate (NO_3^-), nitrous oxide (N_2O), nitric oxide (NO) or inorganic nitrogen gas (N_2). Organic nitrogen may be in the form of a living organism, humus or in the intermediate products of organic matter decomposition. Many of those processes are carried out by microbes, either in their effort to harvest energy or to accumulate nitrogen in a form needed for their growth. These biological processes put together forms the nitrogen cycle.

All living systems can convert ammonia (NH_3) to organic nitrogen (molecules containing C—N bonds) but few can synthesize ammonia from atmospheric nitrogen (N_2) or from nitrate (NO_3^-), a common inorganic form of nitrogen in soil. The fixation and utilization of nitrogen by organisms involves a series of REDOX reactions which form an essential component of the energetics of the biosphere. The reduction of N_2 to NH_3 is termed 'nitrogen fixation'. Nitrogen fixation is one of the most energy-expensive reactions in biology (compare the four electrons and three ATP molecules required to fix each CO_2 in the Calvin cycle of photosynthesis (described in Chapter 3). Although some nitrogen fixation occurs in atmospheric lightning discharges and by burning in air at high temperature, around 85% of nitrogen fixation is of biological origin. The ability to carry out the reaction is limited to certain bacteria and cyanobacteria. No eukaryotic organisms can fix nitrogen. The abundance or scarcity of this 'fixed' form of nitrogen, (also known as reactive nitrogen), dictates how much food can be grown on a piece of land. The nitrogen and carbon cycles work together to maintain an ecosystem. These processes of the nitrogen cycle transform nitrogen from one form to another.

Human Factor: Human alterations to the global nitrogen cycle are most intense in developed countries and in Asia, where vehicle emissions and industrial agriculture (with fertilizer) are highest. Nitrous oxide (N_2O) has risen in the atmosphere as a result of agricultural fertilization, biomass burning, cattle and feedlots, and industrial sources. N_2O has deleterious effects in the stratosphere, where it destroys the atmospheric ozone. In the atmosphere nitrous oxide is a greenhouse gas, and is currently the third largest contributor to global warming, after carbon dioxide and methane. While not as abundant in the atmosphere as carbon dioxide, it is for an equivalent mass, nearly 300 times more potent in its ability to warm the planet. Ammonia (NH_3) in the atmosphere has tripled as the result of human activities. Atmospheric ammonia and nitric acid damage respiratory systems.

2.1.3 The Hydrogen (Water) Cycle

The global water cycle, also known as the hydrologic cycle or the H_2O cycle, describes the circulation of water as a vital and dynamic substance in its liquid, solid, and vapor phases as it moves through the atmosphere, oceans and land. Water moves from one reservoir to another, such as from river to ocean, or from the ocean to the atmosphere, by the physical processes of evaporation, condensation, precipitation, infiltration, runoff, and subsurface flow. Each day, about 4 trillion gallons of this water falls to earth as precipitation, such as rain, snow, or hail; some of the water soaks into the ground and provides runoff to rivers, lakes, and oceans. The remainder—more than 2.5 trillion gallons—returns to the atmosphere through evaporation, and the process begins again. This continuous process has been going on ever since oceans were formed on this planet 3.8 billion years ago. Life in its many forms has evolved due to this extraordinary water cycle.

The water cycle involves the exchange of energy, which leads to temperature changes. When water evaporates, it takes up energy from its surroundings and cools the environment. When it condenses, it releases energy and warms the environment. These heat exchanges influence climate. During its transfer cycle water is purified and replenishes the land with freshwater, and transports minerals all over the planet. It also plays a major role in reshaping the geological features of the Earth, through processes such as erosion and sedimentation. Finally, the water cycle is vital for the maintenance of life and ecosystems.

Oceans cover about 70% of the Earth's surface withholding roughly 97% of water resource on this Blue Planet. The water cycle is powered by solar energy, with 86% of the global evaporation occurs from the oceans. The oceans, then distribute heat around the planet, keeping heat and cold circulating by way of surface currents. Aquifer drawdown or over drafting and the pumping of fossil water increases the total amount of water in the hydrosphere, and has been postulated to be a contributor to sea-level rise.

Why are the oceans salty? *As water flows through rivers, small amounts of minerals are picked from the rocks and soil while pouring into the ocean. The water in the oceans only leaves by evaporation; however, the salt remains dissolved in the ocean and does not evaporate. So the sea water gets saltier as time goes.*

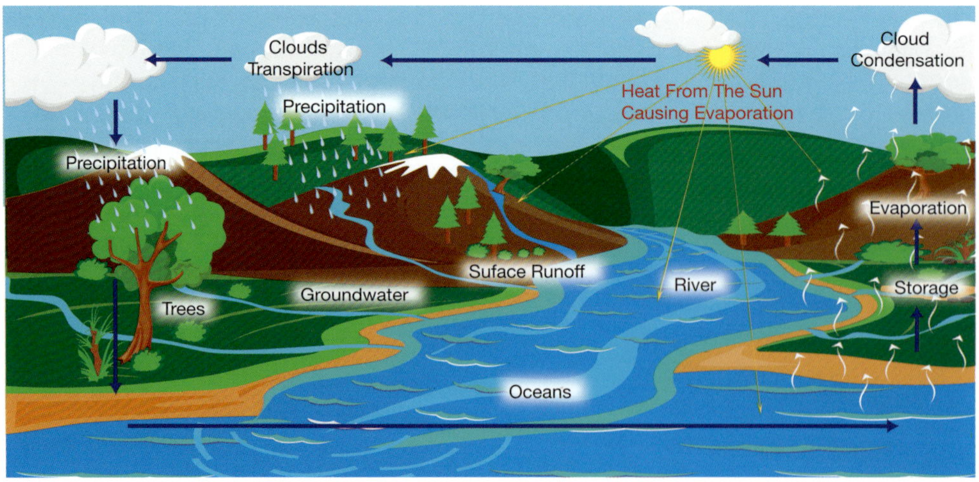

2.1.4 The Sulfur Cycle

Sulfur cycle is comprised of processes where this element is mobilized to and from minerals (including the waterways) and living systems. Such biogeochemical cycles are important for life, since sulfur is one of the elements that constitutes several proteins, vitamins and co-factors. Proteins consist of amino acids that contain sulfur atoms. Plants absorb sulfur when it is dissolved in water. Animals consume these plants, so that they take up enough sulfur to maintain their health.

Most of the earth's sulfur is tied up in rocks and salts or buried deep in the ocean in oceanic sediments. Sulfur can also be found in the atmosphere. It enters the atmosphere through both natural and human sources. Natural sources can be for instance volcanic eruptions, bacterial processes, evaporation from water, or decaying organisms. When sulfur enters the atmosphere via human activity, this is mainly a consequence of industrial processes where sulfur dioxide (SO_2) and hydrogen sulfide (H_2S) gases are emitted. When SO_2 enters the atmosphere it will react with oxygen to produce sulfur trioxide gas (SO_3), or with other chemicals in the atmosphere, to produce sulfur salts. Sulfur dioxide may also react with water to produce sulfuric acid (H_2SO_4). Sulfuric acid may also be produced from demethylsulfide, which is emitted into the atmosphere by plankton species. All these particles will settle back onto earth, or react with rain and fall back onto earth as acid deposition. The particles will then be absorbed by plants again and are released back into the atmosphere, so that the sulfur cycle will restart.

Human activities in the recent years has had a major impact on the global sulfur cycle. The burning of coal, natural gas, and other fossil fuels has significantly increased the amount of sulfur in the atmosphere and ocean, while depleting sulfur in the sedimentary rock sink. Also, H_2S plays a central role in the biogeochemical cycle of sulfur on Earth. H_2S has been implicated in several mass extinctions that have occurred in the Earth's past. In particular, a buildup of hydrogen sulfide in the atmosphere may have caused the Permian-Triassic extinction event 252 million years ago.

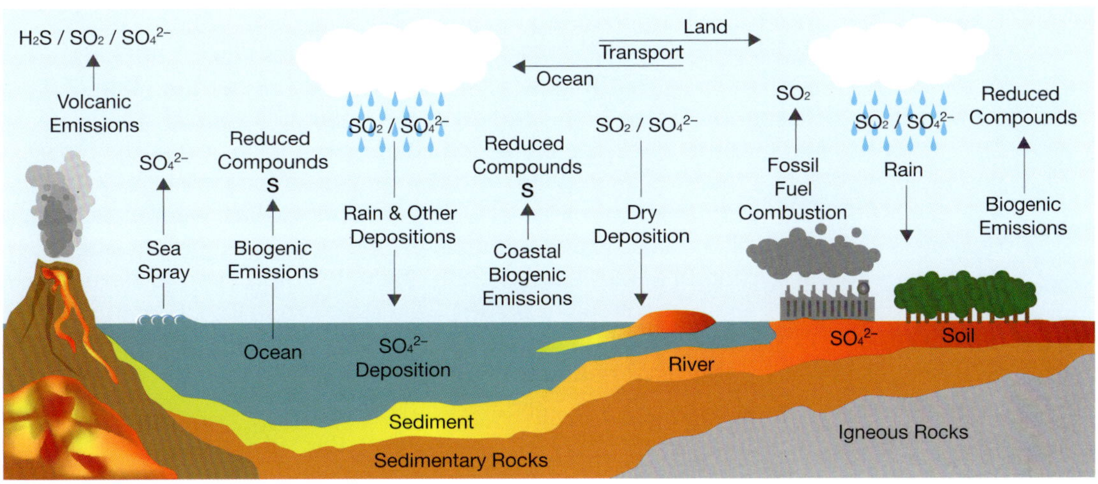

2.2 **REDOX** Chemistry: Oxygen

The two most abundant gases are nitrogen (N_2) and oxygen (O_2). Nitrogen behaves as a noble gas and it is virtually non-reactive. The atmospheric levels of N_2 remained constant since the planet was first formed about 4.6 Ga ago. In contrast, O_2, the second most abundant gas in Earth's atmosphere, is highly reactive, and without a continuous source would become rapidly depleted. This gas exists far from thermodynamic equilibrium with a virtually infinite source of reductant in Earth's mantle. Indeed, high concentrations of gaseous diatomic oxygen are unique to this planet in our Solar System and this feature of our planetary atmosphere has not yet been found on any other planet within approximately 20 parsecs of our Solar System.

Oxygen constitutes a majority of the mass of living organisms, because water is the primary constituent (about two-thirds of human body mass). Many major classes of organic molecules in living organisms, such as proteins, nucleic acids, carbohydrates and fats contain oxygen. Elemental oxygen is used in cellular respiration for all complex life. Oxygen plays contradictory roles, one essential for life and the other as a toxic substance. In 1775, Priestly described the toxicity of the O_2 molecule to organisms and compared its effect on the body as similar to that of burning a candle. As a candle burns out much faster in O_2 than in air, the body becomes quickly exhausted in this "pure kind of air." The toxicity of the atmospheric oxygen molecule had already been used by our ancestors for therapeutic purposes, such as treatment of infected sites. The beneficial use of the toxic effect of oxygen has been utilized in hyperbaric and irradiation therapy. Widespread cases of blindness in young infants born prematurely in the 1940s were associated with the high oxygen concentration in the newly invented incubators. This pathology, called retinopathy of prematurity (ROP), was easily controlled by modulating the O_2 concentration in the incubators.

Oxygen in the Universe is formed along the so-called 'main line' sequence from the high-temperature fusion of four 4 helium (He) atoms in hot stars. The mass of the resulting element is approximately equal to or slightly higher than that of carbon in the solar atmospheres in this region of our galaxy. Molecular orbital calculations reveal that the atom has 6 valence electrons, and the ground state of this molecular O_2 is a triplet. This unusual electron configuration prevents O_2 from reacting readily with atoms or molecules in a singlet configuration without forming radicals. Through reactions catalyzed by metals or photochemical processes, O_2 often leads to the formation of oxides of group I, II, III, IV, V and even VI elements such as H_2O (water), MgO (magnesium oxide), CaO (calcium oxide), CO_2 (carbon dioxide), SiO_2 (silicon dioxide), NO_x (nitrites, nitrates, etc), PO_4 (phosphates) and SO_x (sulfites, sulfates, etc). The reactivity of O_2 is driven by electron transfer

An adult human in rest inhales 1.8 to 2.4 grams of oxygen per minute. This amounts to more than 6 billion tonnes of oxygen inhaled per year.

Oxygen was discovered independently by Carl Wilhem Scheele in 1773 (called it the 'fire air'), and Joseph Priestley in 1774 (named it the 'dephlogisticated air') but Priestley is often credited because his work was published first. In 1777, Antoine Lavoisier showed that air is a mixture of two gases; one is the 'vital air', which is essential to combustion and respiration processes; the other is 'azote' (meaning 'lifeless' in Greek), which did not support either processes. 'Azote' later became Nitrogen in English, although it has kept the name in French and several other European languages. Lavoisier renamed 'vital air' to Oxygène (which has Greek roots); oxys means 'sharp' (from the taste of acids) and gene means 'produce' – in literal sense, gas that produces a sharp acid taste.

The free oxygen partial pressure in the human body is highest in the respiratory system, and decreases along any arterial system, peripheral tissues and venous system, respectively.

(REDOX) reactions, leading to highly stable products, such as H_2O, CO_2, HNO_3 (nitric acid), H_2SO_4 (sulfuric acid) and H_3PO_4 (phosphoric acid). The abiotic reactions of O_2 often involve unstable reactive intermediates such as H_2O_2 (hydrogen peroxide), NO (nitric oxide), NO_2 (nitric dioxide), CO (carbon monoxide) and SO_2 (sulfur dioxide). The reactions of O_2 with other elements are always exergonic (energy-releasing). Therefore, without a continuous source, free molecular oxygen would be depleted from earth's atmosphere within a few million years. Also, free molecular oxygen allowed a thermodynamic efficiency in the oxidation of organic matter via aerobic respiration and, simultaneously, the presence of the gas in the atmosphere allowed Earth to form stratospheric ozone (O_3) layer that absorbed UV radiation. Oxygen also greatly influenced the evolution of lipids, and other secondary metabolites that are critical for maintaining electron traffic within organisms.

Oxygen, by itself qualifies as a radical species due to its two unpaired electrons, each of which is located in a different anti-bonding orbital. These 2 electrons also have parallel spins and share the same spin quantum number. Consequently, ground state O_2 is sparingly reactive. To oxidize another molecule, O_2 should accept an electron pair; both electrons would have to possess anti-parallel spins relative to the unpaired electrons in O_2, according to Pauli's exclusion principle. This criterion is seldom met in a typical electron pair, so O_2 tends to accept electrons one at a time. *In vivo*, multi-electron reduction of O_2 is carried out by a coordinated series of enzymes that reduce O_2 with 1-electron at a time. If O_2 accepts a single electron, it must enter an anti-bonding orbital producing the $O_2^{\bullet-}$ radical. Two-electron reduction of O_2, with addition of $2H^+$, generates H_2O_2. Most $O_2^{\bullet-}$ is metabolized by SOD to H_2O_2 and oxygen. In reactions catalyzed by free transition metals such as Fe^{2+}, $O_2^{\bullet-}$ and H_2O_2 can generate the extremely reactive ·OH radical, the major cause of damage to proteins, lipids, and DNA.

In essence, O_2 is required by prokaryotic and eukaryotic cells for energy production, often via the mitochondrial electron transport chain. In most cases oxygen is consumed as O_2 in the form of a diatomic molecule, its common configuration in the atmosphere. Its derivative ozone (O_3) has allowed the absorbance of harmful solar UV radiation, thereby, enabled organisms to survive on dry land. The atmospheric O_2 level (21%) is a dynamic parameter that is constantly changing. Massive deforestation of the rain forests and the gradual decline in O_2 levels could lead to changes in the biochemical response of the living cell.

Human activities, including the burning of 7 billion tones of fossil fuels each year, have had very little effect on the amount of free oxygen in the atmosphere. At the current rate of photosynthesis it would take about 2,000 years to regenerate the entire O_2 in the present atmosphere.

Ozone in the Atmosphere: The Nobel Prize in Chemistry 1995 was awarded to Paul Crutzen, Mario Molina and F. Sherwood Rowland for their work in atmospheric chemistry, concerning the formation and decomposition of ozone.

Although O_2 is the substrate that cells use, the tissues have no major storage system for O_2. Therefore, living organisms rely on a continuous supply of O_2 at a rate that precisely matches their metabolic demand. Any shortage in O_2 delivery (hypoxia) and failure of cellular use of O_2 could lead to metabolic dysfunction. O_2 transport from environmental air to the mitochondria of individual cells occurs in a series of steps. The heart, lungs, and circulation extract O_2 from the atmosphere and generate a flow of oxygenated blood to the tissues to maintain aerobic metabolism. The system must be energy efficient, match O_2 distribution with metabolic demand, and allow efficient O_2 transport across the extra-vascular tissue matrix. At the tissue level, cells must extract O_2 from the extra-cellular environment and use it efficiently in cellular metabolism.

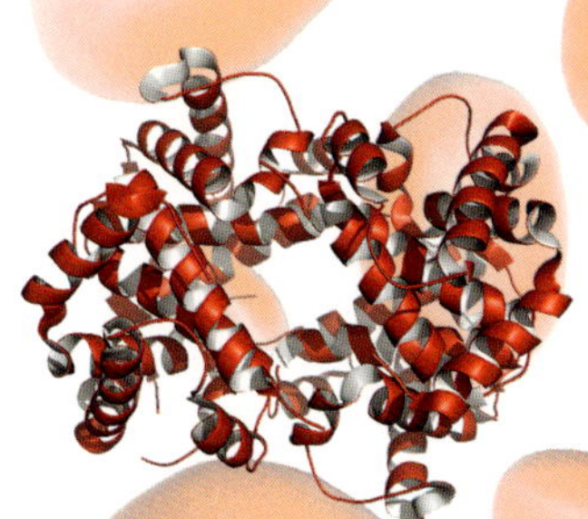

Hemoglobin is the heme-containing O_2-transporter in the red blood cells of all vertebrates (with the exception of certain fish) as well as the tissues of some invertebrates. Hemoglobin in the blood carries O_2 from the respiratory organs to the rest of the body (i.e. the tissues) where it releases the O_2 to burn nutrients to provide energy to power the functions of the organism, and collects the resultant CO_2 to bring it back to the respiratory organs to be dispensed from the organism.

Different O_2-transport Molecules found in Aerobic Life Forms.
Myoglobin (Iron-based): Found in the muscle tissue of many vertebrates, including humans. It gives muscle tissue a distinct red or dark gray color. Stores O_2 rather than transporting it.
Hemocyanin (Copper-based): The 2nd most common O_2-transporting protein in nature. Found in the blood of many arthropods/mollusks. Has copper prosthetic groups and is blue in color when oxygenated.
Hemerythin (Iron-based): Some marine invertebrates and annelids use this iron-containing non-heme protein to carry O_2 in their blood. Appears pink/violet when oxygenated, clear when not bound to O_2.
Chlorocruorin (Iron-based): Found in many annelids, it is very similar to erythrocruorin, but the heme group is significantly different in structure. Appears green when deoxygenated and red when oxygenated.
Vanabins (Vanadium-based): Found in the blood of sea squirts. They use the rare metal vanadium as an oxygen binding prosthetic group and bind little oxygen. They may act as toxins.
Erythrocruorin (Iron-based): Found in many annelids, including earthworms. It is a big free-floating blood protein containing several iron- and heme-bearing subunits bound together into a single protein complex.
Pinnaglobin (Manganese-based): Only seen in the mollusc Pinna squamosa. Brown manganese-based porphyrin protein.
Leghemoglobin (Iron-based): Found in N_2-fixing bacteria in the leguminous roots. Protects nitrogenase enzyme, which is unable to reduce nitrogen gas in the presence of free O_2.
Coboglobin (Cobalt-based): A synthetic cobalt-based porphyrin. Coboprotein appears colorless when oxygenated, but yellow when in veins.

Hemin: The Nobel Prize in Chemistry 1930 was awarded to Hans Fischer for his researches into the constitution of hemin and chlorophyll and especially for his synthesis of hemin.

2.3 REDOX Reactions

The body produces more free radicals when stressed. An imbalance in the levels of free radicals in the body could lead to diseases such as cancer, arthritis, stroke, diabetes, genetic damage, and rapid aging. On the other hand, free radicals are essential for life as they support several vital functions in the body. Free radicals serve as messengers in a process known as REDOX cell signaling. Radical reactions are the basis for bioenergetics – the vital mechanism for survival, growth and multiplication of every life form.

Free radicals are defined as "any transient (chemically unstable) species (atom, molecule, or ion)" – Herzberg

At a basic level, atoms form molecules by sharing electrons, and electrons pair up with one another. Each pair of electrons constitutes a chemical bond. Most molecules contain an even number of electrons. A free radical is an atom or a group of atoms with an odd number of electrons. The odd, unpaired electron in a free radical seeks to pair (form a bond) with another electron. Most free radicals are so reactive they exist only for fraction of a moment.

A free radical can react with another free radical, but more often it reacts with a stable, evenly paired molecule. It does so by (i) donating an electron to the molecule, (ii) removing an electron from the molecule, (iii) removing a group of atoms from the molecule, or (iv) adding itself to the molecule. When a free radical reacts with a stable molecule in these ways, it turns the stable molecule into a free radical. This can set in motion a chain reaction of free radical formation. A notable example of free radical is ·OH, a molecule that is one hydrogen atom short of a water molecule and thus has one bond 'dangling' from the oxygen.

Although radicals are generally short-lived due to their reactivity, there are long-lived radicals – divided into three categories as follows: (i) Stable radicals are less reactive than other radicals. Examples of stable radicals include molecular dioxygen (O_2) and nitric oxide (NO). (ii) Persistent radicals are those with longevity due to steric crowding around the radical center, which makes it physically difficult for the radical to react with another molecule. Examples of these include Gomberg's triphenylmethyl radical, nitroxides, etc. Persistent radicals are generated in large quantity during combustion. (iii) Diradicals are molecules containing two radical centers. Multiple radical centers can exist in a molecule. Atmospheric oxygen naturally exists as a diradical in its ground state as triplet oxygen (ozone). The low reactivity of atmospheric oxygen is due to its diradical state. Non-radical states of dioxygen are actually less stable than the diradical.

Free Radical Chemistry: The Nobel Prize for Chemistry in 1971 was awarded to Gerhard Herzberg for his contributions to the knowledge of electronic structure and geometry of molecules, particularly free radicals. He used atomic and molecular spectroscopy to determine the structures of free radicals

Over time, paints and plastics peel, crack and degrade due to free radical damage. We can witness such damage on works of art, even on our own house and vehicles. Art preservation is a technology that limits canvas exposure to oxygen and direct light, which could generate free radicals and damage pigments. The smile on daVinci's Mona Lisa, as of now, is taken care by the principles of REDOX chemistry.

2.3.1. REDOX Mechanisms

Oxidation *is the loss of electrons or an increase in oxidation state by a molecule, atom, or ion.*

Reduction *is the gain of electrons or a decrease in oxidation state by a molecule, atom, or ion.*

The term REDOX originates from two opposite chemical reactions, REDuction and OXidation. REDOX reactions are a matched pair, accordingly, there cannot be an oxidation without having a reduction simultaneously. Oxidation and reduction, each by itself is known as a half-reaction and the two half-reactions always occur together to make a whole reaction.

REDOX can be defined in two ways. A simpler one refer to reactions involving some form of oxygen. As an example, pure iron can be produced from iron oxide in a blast furnace by the following reaction:

$$3\ C + 2\ Fe_2O_3 \rightarrow 4\ Fe + 3\ CO_2$$

In this reaction, iron oxide (Fe_2O_3) gives away its oxygen to carbon. In chemical terms, the carbon is oxidized, since it has gained oxygen. At the same time, the iron oxide is reduced because it has lost oxygen. Based on its ability to give away oxygen, iron oxide is called an oxidizing agent. Similarly, the ability to take on oxygen makes the carbon, a reducing agent. For years, oxidation and reduction were considered as reactions that involve oxygen in some way or another. However, several other elements were found to behave chemically in much the same way as oxygen. Therefore, REDOX was redefined to more general term to include reactions other than oxygen-based ones.

REDOX and Electron Exchanges: The 2nd definition for REDOX is based on electron exchange where a reductant transfers electrons to an oxidant. Oxidants, or oxidative agents can remove electrons from another substance. Since they receive electrons, are also known as electron acceptors. Oxidants usually have high oxidation states (i.e. H_2O_2), or else highly electronegative (-) elements (O_2, F_2, Cl_2, Br_2) that can gain extra electrons by oxidizing another substance. Reductants or reductive agents are able to transfer electrons to another substance. Since, they give away electrons; they are also called as electron donors. Reductants in chemistry are very diverse. Electropositive (+) metals, such as lithium (Li), sodium (Na), magnesium (Mg), iron (Fe), zinc (Zn), and aluminium (Al), are good reducing agents. These metals donate or give away electrons readily. A REDOX couple is a pair consisting of a reducing species and its corresponding oxidized form. In a REDOX reaction, the element that is oxidized always loses (or donates) electrons; and the element that is reduced always gains (or accepts) electrons. In more general definition, REDOX reactions refer to a change in the oxidation state, since the actual transfer of electrons may never occur. Thus, oxidation is an increase in oxidation state, and reduction as a decrease in oxidation state. For example, sodium (Na) metal reacts with chlorine (Cl_2) gas in such a way that sodium atoms lose one electron each to chlorine atoms:

$$2\ Na + Cl_2 \rightarrow 2\ NaCl$$

In the above reaction, sodium is oxidized by losing electrons; whereas chlorine is reduced by gaining electrons in the reaction. The transfer of electrons always cause a change in oxidation state; however, there are many reactions classified as REDOX even though no electron transfer occurs.

REDOX reactions are the most common chemical reactions that occur in our day-to-day life. REDOX reactions are classified based on speed (reaction kinetics). Combustion is an example of a rapid REDOX reaction that occurs fast with noticeable production of heat and light. Corrosion and decay are examples of a slow REDOX reaction that happens at a delayed pace without any noticeable heat and light generation.

Key terms used in the REDOX Chemistry are often confusing – an element when oxidized, loses electrons and that element is referred to as a reducing agent. Whereas, an element when reduced gains electrons and is referred to as the oxidizing agent.

The following mnemonics could help remember the REDOX terminology: ***"OIL RIG"*** *–* ***O****xidation* ***I****s* ***L****oss of electrons,* ***R****eduction* ***I****s* ***G****ain of electrons*

Combustion means burning, a rapid REDOX reaction. The following 2 equations elucidates the chemical reaction that occurs during the burning of coal (which is nearly pure carbon) or gasoline (C_8H_{18} – hydrocarbon). The fuel (carbon or hydrocarbon) is oxidized in each case:

$$C + O_2 \rightarrow CO_2$$

$$2\ C_8H_{18} + 25\ O_2 \rightarrow 16\ CO_2 + 18\ H_2O$$

During combustion, the oxidation reaction proceeds very fast and release energy by breaking the chemical bonds. Such generated energy is put to use for heating homes and buildings; to run automobiles, trucks, ships, airplanes, and trains; to operate industrial processes; and for numerous other purposes.

Corrosion: Most metals react with oxygen to form compounds known as oxides. Rust is the name given to the oxide of iron and, sometimes, the oxides of other metals. This rusting process is also known as corrosion. Corrosion is similar to combustion, except that it occurs at an extremely slow pace. The equation below represents the most common form of corrosion, the rusting of iron:

$$4\ Fe + {}^{3}O_2 \rightarrow 2\ Fe_2O_3$$

Decay: Biological compounds, such as carbohydrates are complex organic molecules essential for vital metabolic functions. These organic chemicals are composed primarily of carbon (C), oxygen (O), and hydrogen (H). A simple way to represent such compounds is by using letters x, y, and z to show the number of C, H, and O atoms present in the compound. When a plant or animal dies, its organic mass starts reacting with oxygen. The process is known as decay, which is yet another example of a common REDOX reaction. The equation below represents the decay (oxidation) of a biomass that might be present in a dead organism:

$$C_xH_yO_z + O_2 \rightarrow CO_2 + H_2O$$

Many changes that take place within living organisms are also REDOX reactions. For example, the digestion of food is an oxidation process. Food molecules react with O_2 in the body to form CO_2 and H_2O. Energy is also released in the process. The CO_2 and H_2O are eliminated from the body as waste products, but the energy is used in metabolism to help keep an organism survive, grow and multiply.

2.3.2. REDOX Potential (E_h)

REDOX potential, also known as standard electrode potential, is a measure of the affinity of a substance for electrons (its electro-negativity). Every REDOX reaction is divided into two half reactions: one in which a chemical species undergoes oxidation and one in which another chemical species undergoes reduction. If a half-reaction is a reduction, then the driving force is the reduction potential. If the half-reaction is an oxidation, then the driving force is the oxidation potential related to the reduction potential by a sign change. Accordingly, the REDOX potential is the reduction/oxidation potential of a compound measured under standard conditions against a standard reference half-cell. In a REDOX reaction, when electrons flow is'downhill', free energy is released (denoted with the symbol ΔG preceded by a minus sign). Conversely, a REDOX reaction would require an input of free energy to force electrons to move 'uphill' (indicated with ΔG preceded by a plus sign).

Oxidants and reductants occur as couples, with a strong reducing agent coupled with a weak oxidizing agent and vice versa. Since these processes involve the transfer of electrons, the measurement of the resulting charge separation can be quantified in voltage measured between the couple and a standard hydrogen half-cell. In biological systems the standard REDOX potential is defined at pH 7.0 versus the hydrogen electrode and partial pressure of hydrogen = 1 bar. The REDOX potential is denoted by E_h, and measured in volts (V) or milli-volts (mV). The standard is always hydrogen with its E_h value set at zero (0).

The E_h value of REDOX couples that commonly occur in biological reactions can give insight into their roles in bioenergetic processes. A negative number for E_h indicates a reducing agent whereas a strong oxidizer will have a positive (+) E_h. Since, harness of useful energy in a bioenergetic pathway involves oxidation of some food compound, it requires a large negative (-) E_h, which about -0.32 V for NAD^+ suggests a useful role in producing a reduced product, like NADH. A reduced co-enzyme NADH can be oxidized to provide energy for biological processes.

If two substances with varying E_h values are put together with a potential path for electron flow between them – that creates a battery. Although it may be in a mitochondrion, it is just as much a battery as the lead-acid storage battery in an automobile. Greater the voltage (E_h) variance between the two components of a battery, greater the energy available during an electron flow. The free energy (ΔG) yields of the two common bioenergetic pathways, cellular respiration and photosynthesis, are described below.

In photosynthesis, ΔG is positive because electrons have moved against the gradient (from positive to negative) in contrast to cellular respiration. Thus, energy (from light) could be put into the system.

Cellular Respiration: For every molecule of glucose respired, 24 electrons travel down the respiratory chain to the final acceptor, the oxygen molecule. Carbon reduced in carbohydrates (eg. glucose) has an E_h value of ~ –0.42 V. Oxygen, the most electronegative (-) substance in the system, has the largest E_h (+0.82 V).The E_h difference is 1.24 volts. *Thus, allowing 24 moles of electrons to pass through this potential gives a free energy yield: $\Delta G = -686$ kcal.*

Photosynthesis: To synthesize a molecule of glucose by photosynthesis, 24 electrons must be removed from water molecules and held by the E_h of oxygen (+0.82 V) and pumped 'uphill'. *Carbon atoms are reduced to carbohydrate with an E_h of −0.42 V. Again, the E_h difference is 1.24 V, so the $\Delta G = +686$ kcal.*

To establish standard values for elements, compounds and ions; E_h is measured against a hydrogen 'standard' (which is taken to have an E_h of zero, so all such values are relative to hydrogen). Elements and ions differ in the ease with which they are reduced or oxidized. Reduction half-reactions for several ions and elements are listed in the adjacent TABLE. The reduction potentials listed were measured at 25°C and 1 atm pressure, using solutions in which one mole of reactant was dissolved in 1 L solution. In each equation, the E_h value compares the tendency for the listed reaction of the compound to occur with the tendency of aqueous hydrogen ion to be reduced:

$2\,H^+ + 2\,e^- \rightarrow H_2$ ($E_h = 0.0$ V)

The E_h for the reduction of hydrogen ion to hydrogen has been arbitrarily assigned a value of 0.0 volts (V). A substance whose reduction potential is less than that of H^+ ion (has a negative (-) value) is reduced less easily than the H^+ ion. A substance whose reduction potential is positive (+) (>H^+ ion) is more easily reduced than hydrogen.

HIGHLY REDUCING

REDOX Couple	E_h (mV)
CO_2 / $CO_2^{\bullet -}$	−1800
CO_2 / CO	−520
Acetyl-Co / A-pyruvate	−500
Succinyl-CoA / 2-Oxoglutarate	−491
CO_2 / $HCOO^-$	−421
H^+ / H_2	−414
NAD^+ / NADH	−316
$NADP^+$ / NADPH	−315
CO_2 / Acetate	−291
TrxC [TRXSS /TRX(SH_2)]	−269
TrxB [TRXSS /TRX(SH_2)]	−262
TrxA [TRXSS /TRX(SH_2)]	−248
$2H^+$ / 2Cys-SH (Cystine)	−230
FAD^+ / $FADH_2$	−219
FMN^+ / $FMNH_2$	−219
Pyruvate, H^+ / Lactate	−183
Oxaloacetate, $2H^+$ / Malate	−166
Manaquinone	−74
CoQ / $CoQ^{\bullet -}$	−36
Fumarate / Succinate	+32
Ubiquinone / Ubiquinol	+45
Fe^{3+} / Fe^{2+} (aq)	+110
$Ascorbate^{\bullet -}$ / $Ascorbate^-$	+282
O_2 / H_2O_2	+295
Cytochrome a_3 (Fe^{3+}) / Cyt-a_3 (Fe^{2+})	+350
NO_3^- / NO_2^-	+421
α-$Tocopheroxyl^\bullet$ / α-Tocopherol	+500
O_2 / H_2O	+818
$RS^\bullet$ / RS^- (Cystine)	+920
$GS^\bullet$ / GS^- (Glutathione)	+920
$NO_2^\bullet$ / NO_2^-	+990
$ROO^\bullet$, H^+ / ROOH (Alkyl Peroxyl)	+1000
$HO_2^\bullet$, H^+ / H_2O_2	+1060
ONOO– / $NO_2^\bullet$ (aq)	+1400
$RO^\bullet$, H^+ / ROH (Aliphatic Alkoxyl)	+1600
NO_2^+ / $NO_2^\bullet$	+1600
$CO_3^{\bullet -}$, H^+ / HCO_3^-	+1780
$HO^\bullet$, H^+ / H_2O	+2310

HIGHLY OXIDIZING

TABLE: *Reduction Potential (E_h) Values of selected REDOX couples measured at 25°C (in comparison to hydrogen E_h=0)*

2.3.3. REDOX: Metal Catalysts

Most of the transition metals - those in the first row of the D block in the periodic table - contain unpaired electrons (with the exception of zinc). They can participate in the chemistry of radicals and convert relatively stable oxidants into powerful radicals. Indeed, most transition-metal ions are radicals, since their biological effects (whether beneficial or detrimental) involve the ability to accept or donate single electrons. The reduction potential of transition-metal ions depends on the ligands to the metal, and thus can be altered in different enzymes to allow the same metal to catalyze different reactions. Unless their availability is carefully controlled, transition metals will catalyze unwanted free radical reactions such as auto-oxidations and ·OH formation.

Transition elements play a major role in the production of ·OH radicals or endo genous oxidizing agents, via Fenton-like reactions, and probably in the catalysis of the Haber-Weiss reaction of O_2 with H_2O_2. Transition metal complexes react preferentially with secondary (aliphatic C-centered) and tertiary (alkyl-peroxyl) radicals formed in the biological systems. These reactions direct the radical-initiated processes to biological sites where the transition metal complexes are located. They provide an alternative to the explanation offered by the site-specific effect. The products of radical reactions with transition metal complexes, especially those resulting from the decomposition of the transient complexes, are probably the cause of many biologically deleterious effects initiated by radicals.

In 1890, Henry JH Fenton described the interaction between Fe^{2+} and H_2O_2. Forty years later, this reaction was found to generate ·OH radicals, which provided an explanation to its oxidizing power. In Fenton reaction, ferrous iron(II) is oxidized by H_2O_2 to ferric iron(III), a ·OH radical, and a OH- anion. Iron(III) is then reduced back to iron(II), a peroxide radical, and a proton (H+) by the same hydrogen peroxide. The net effect is a disproportionation of hydrogen peroxide to create two different oxygen-radical species, with water (H+ + OH–) as a byproduct.

(Reaction-1) $Fe^{2+} + H_2O_2 \rightarrow Fe^{3+} + \cdot OH + OH^-$

(Reaction-2) $Fe^{3+} + H_2O_2 \rightarrow Fe^{2+} + HOO\cdot + H^+$

The free radicals generated by this process then engage in secondary reactions. For example, the ·OH is a powerful, non-selective oxidant. Oxidation of an organic compound by Fenton's reagent is rapid and exothermic (heat-producing) and results in the oxidation of contaminants to primarily CO_2 and water. Reaction-1 was suggested by Haber and Weiss in the 1930s. Broadly, the Fenton oxidation encompasses three types of oxidation routes for the production of ·OH:
(i) through the use of Fenton reagents alone, termed classical Fenton oxidation, which involves the catalytic decomposition of H_2O_2 by suitable transition metals to generate ·OH radical, (ii) assisted by means of UV (photo-Fenton), or (iii) electron beams, the electro-Fenton process, where H_2O_2 is produced in the required amount from the electrochemical reduction of oxygen.

Since $O_2^{\bullet -}$ ions and transition metals act in a synergistic manner in the creation of free radical damage, iron supplementation must not be done in patients with any active infections or in general any diseases.

The Fenton reaction has importance in biology because it involves the creation of free radicals by chemicals that are present in vivo. Transition-metal ions such as iron and copper donate or accept free electrons via intra-cellular reactions and help in creating free radicals. Although most intra-cellular iron is in Fe(II) form, the $O_2^{\bullet-}$ ions can convert it to the Fe(III) form to take part in Fenton reaction.

Among the various transition metals, copper and iron are most abundant (in relatively high concentrations) and are major players in the Fenton-type reaction and the metal-mediated Haber-Weiss reaction. The metal ions participating in this reaction are those bound to the surface of proteins, DNA, and other macromolecules or chelates. These particular ions can still undergo REDOX process, interact with O_2 -derivatives, and are often called 'loosely bound metals' or 'removable metals'. Metals that are hidden in proteins (as in catalytic sites and cytochromes) or storage complexes; are not exposed to oxygen radicals; or are kept under 1 oxidation state that cannot participate in this REDOX chemistry.

TABLE: *Biological importance of some d-block elements*

Metal	Orbital (Mass/mg)*	Biological Significance
Vanadium (V^{2+})	$3d^3$ (0.11)	Involved in regulation of glucose metabolism. Inhibits the ATPase enzyme, affects protein kinase and phosphatase enzymes.
Chromium (Cr^{2+})	$3d^4$ (14)	Involved in regulation of glucose metabolism. Chromates can damage DNA.
Manganese (Mn^{2+})	$3d^5$ (12)	Needed for Mn-SOD, and arginase. Activates a number of hydrolase and carboxylase enzymes. Toxic in excess. In brain can cause a Parkinsonian-type syndrome. (Plays a key role in photosynthesis).
Iron (Fe^{2+})	$3d^6$ (4,200)	Needed for hemoglobin, myoglobin, cyclo-oxygenases, cytochromes, many hydroxylase/oxidase enzymes, ribonucleotide reductase, aconitase, succinate dehydrogenase, catalase and many others. Both iron excess and iron deficiency can cause oxidative stress.
Cobalt (Co^{2+})	$3d^7$ (3)	Essential in vitamin B12. Cobalt compounds can cause DNA damage.
Nickel (Ni^{2+})	$3d^8$ (15)	Found in urease in plant cells and bacteria (including *H. pylori*) and in several bacterial enzymes, such as *E. coli* glyoxalase I. Some bacterial hydrogenases and carbon monoxide dehydrogenase. Some nickel-containing SOD enzymes exist.
Copper (Cu^{2+})	$3d^9$ (72)	Essential for enzymes such as EC-SOD, Cu-(Zn)-SOD, cytochrome oxidase, lysine oxidase, dopamine-α-hydroxylase and caeruloplasmin. Toxic in excess; both excess and deficiency can cause oxidative stress.
Zinc (Zn^{2+})	$3d^{10}$ (-)	Acts as an antioxidant by displacing iron ions from their binding sites and inhibiting iron-dependent oxidative damage. Found in RNA polymerase, carbonic anhydrase, Cu-(Zn)-SOD, glyoxalase I, EC-SOD, 'zinc fingers'. Important to the immune system. Toxic in excess, especially to neurons.

** Mass of the first row transition metal ions present in a 70 kg human*

Iron (Fe^{2+}) at physiological pH, is mostly oxidized and attached to biological chelate in the form of Fe^{3+}. In order to participate in the Fenton-type reaction, iron must be converted to its reduced form, Fe^{2+}; in some cases, metals in a higher oxidation state can also participate. Reducing equivalents with a suitable oxidation potential can easily be responsible for this reaction. For example, ascorbic acid can reduce ferric ions to ferrous ions, enabling it to undergo the Fenton-type reaction and produce ·OH radicals for hydroxylating aromatic compounds. Many of the deleterious effects of oxygen radicals are attributed to the continuous production of highly toxic ·OH radicals via the metal-mediated Haber-Weiss reaction in which ferric ions are reduced by $O_2^{\bullet-}$ radicals to ferrous ions. The latter interact with H_2O_2 (produced in *vivo* from the spontaneous or enzymatic dismutation of $O_2^{\bullet-}$ radicals)to yield ·OH via the Fenton-type reaction. The interaction between superoxide radicals and H_2O_2 leads to the production of ·OH radicals *in vivo*.

$$O_2^{\bullet-} + H_2O_2 \rightarrow O2 + \cdot OH + OH^-$$

Copper (Cu^{2+}) has two common oxidation numbers, Cu(I) and Cu(II). The one electron difference between these two oxidation states allows copper to promote radical reactions. Copper ions interact rapidly with $O_2^{\bullet-}$

$$Cu^{2+} + O_2^{\bullet-} \rightarrow Cu^+ + O_2$$

$$H^+ + Cu^+ + HO_2^- \rightarrow Cu^{2+} + H_2O_2$$

$$O_2^{\bullet-} + Cu^+ + H_2O \rightarrow Cu^{2+} + OH^- + HO_2^-$$

$$\text{Net: } O_2^{\bullet-} + O_2^{\bullet-} + 2\,H^+ \rightarrow H_2O_2 + O_2$$

The copper, by changing its oxidation number, is catalyzing the conversion of two $O_2^{\bullet-}$ radicals and two H+ ions to H_2O_2 and O_2; that is, it is catalyzing the dismutation of $O_2^{\bullet-}$.

Manganese (Mn^{2+}) is most stable in aqueous solution as Mn^{2+}, oxidizing species is such as Mn(III), Mn (IV) and Mn(VII) also exist. Mn^{2+} can interact with $O_2^{\bullet-}$.

$$Mn^{2+} + O_2^{\bullet-} \rightarrow Mn(O_2)+$$

$$Mn(O_2)^+ + O_2^{\bullet-} + 2H^+ \rightarrow Mn^{2+} + H_2O_2 + O_2$$

$$Mn(O_2)^+ + 2\,H^+ \rightarrow H_2O_2 + Mn(III)$$

Like copper, Mn^{2+} is capable of catalyzing the dismutation of $O_2^{\bullet-}$ to H_2O_2, to an extent considerably affected by the ligands bound to it. This is physiologically important in some bacteria. In contrast to Fe^{2+} and Cu^+, unchelated Mn^{2+} appears not react with H_2O_2 in Tris buffer (if both CO_2 and HCO_3^- are present) and in HCO_3^- buffer (Mn^{2+} and H_2O_2 can oxidize LDL and amino acids).

Commonly, 'transition metal' refers to the 40 elements that belong to groups 3–12 of the periodic table. In the atomic ground state, the highest-energy electron of those elements is in the 'd-orbital'. Most of these elements have an incomplete 'd' sub-shell, or can give rise to cations with an incomplete 'd' sub-shell.

Zinc (Zn^{2+}) is a catalytic agent in hydroxylation and other enzymatic reactions. The metal also has a flexible coordination geometry, which allows proteins using it to rapidly shift conformations to perform biological reactions. Two examples of zinc-containing enzymes are carbonic anhydrase and carboxypeptidase, which are vital to the processes of CO_2 regulation and digestion of proteins, respectively. In the blood, carbonic anhydrase converts CO_2 into bicarbonate and the same enzyme transforms the bicarbonate back into CO_2 for exhalation through the lungs. Without this enzyme, this conversion would occur about one million times slower in normal blood (pH 7.0). The non-related α-carbonic anhydrase is required in plants for leaf formation, the synthesis of indole acetic acid (auxin) and alcohol fermentation. Also, carboxypeptidases cleave peptide linkages during digestion of proteins. A coordinate covalent bond is formed between the terminal peptide and a C=O group attached to zinc, which gives the carbon a positive charge. This helps to create a hydrophobic pocket on the enzyme near the zinc, which attracts the non-polar part of the protein being digested.

Thus, metal ions play essential roles in about one third of enzymes. These ions can modify electron flow in a substrate or enzyme, and effectively control enzyme-catalyzed reactions. They can bind and orient substrate with respect to functional groups in the active site, and provide a site for REDOX activity if the metal has several valence states. Without the appropriate metal ion, a biochemical reaction catalyzed by a particular metallo-enzyme would proceed at a very slow pace, if at all. No set of general rules describe how a given metal ion will behave in an enzyme. Generally, metal ions are positively charged and act as electrophiles, seeking the possibility of sharing electron pairs with other atoms so that a bond or charge-charge interaction can be formed. They behave like protons (H^+). Metal ions often have positive charges greater than one, and have a larger ionic volume so that they can accommodate many ligands around them at the same time. In addition, metal ion concentrations can be high at neutral pH values, while H^+ ion concentrations are, by the definition of pH, low at these values.

Metals act as a prosthetic group in biologically active metallo-proteins. Their inter-changeable prevalent ionic forms and affinity for functional groups occurring in proteins are unique properties of transition metals that make them useful in biochemical REDOX reactions. However, the same chemical properties that make metal ions useful cofactors, can also be responsible for undesired reactions in a cellular environment. Since they are able to form and sustain several stable bonds with donor atoms of organic ligands, transition metal ions can become toxic when their interaction with substrates are not fully controlled. Metal-induced uncontrolled REDOX reactions or displacement of endogenous metal cofactors from their cell binding sites can lead to poisoning and endanger the organism's survival. Thus, to maintain levels of essential metal ions within physiological limits and to cope with the toxic effects of metal overload, photosynthetic organisms have developed a tightly controlled and sophisticated metal homeostasis network that ensures the balance between metal uptake, chelation, distribution and storage processes.

2.3.4. REDOX: Type of Radicals

Free Radicals: Most free radicals are short-lived species and quickly react with other molecules. Some of the ROS are highly reactive with a short half-life. The milieu factors such as pH and the presence of other ionic species influence the half-life and activity of ROS. Generally, a longer half-life of a species might imply a higher toxicity of the compound, since it allows adequate time to diffuse and reach a sensitive location where it can interact and cause damage a long distance from its site of production. For example, the relatively long half-life of $O_2^{\bullet -}$ radicals allows movement to locations and interact with other molecules; these radicals can be produced in the mitochondrial membrane, diffuse towards the mitochondrial genome, and reduce transition metals bound to the genome.

On the other hand, a highly reactive species with an extremely short life span, like ·OH, is produced in locations where it can cause damage by interacting with its immediate surroundings. For example, ·OH survives for 10^{-10} (ten billionth) of a second in biological systems; its reaction rate constants ($m^{-1}s^{-1}$) for biological components are extremely high (10^7-10^9 $m^{-1}s^{-1}$) and in many cases diffusion-controlled. The life span of other radicals is also short but depends on the environmental milieu. For instance, the half-life of ·NO in an air-saturated solution might be a few minutes. RO· can survive about 10^{-6} (one millionth) of a second, while the half-life of ROO· is about 17 seconds. If there is no essential biological target adjacent their production site, radicals will not cause oxidative damage. The high reactivity of radicals and their short life span illustrate the potential toxic effect and difficulties in preventing oxidative damage. To prevent the interaction between radicals and biological targets, the antioxidant should be present at the location where the radicals are being produced in order to compete with the radical for the biological substrate. This is an important guideline in the design of appropriate antioxidant therapy.

Free Radical Type	Formula	Reactive Properties
Carbon-centered radical	C·	The unpaired electron residing on carbon that rapidly reacts with O_2 to make peroxyl radicals.
Superoxide anion Hydroperoxyl radical	$O_2^{\bullet -}$ $\cdot HO_2$	The 'primary' oxygen-centered radical in its anionic and protonated forms.
Peroxyl radical Alkoxyl radical	RO_2 RO·	Oxygen-centered radicals that form from the reaction of carbon-centered radicals with O_2 (RO_2), or from the breakdown of organic peroxides, such as lipid hydroperoxides (RO_2, RO·).
Hydroxyl radical	·OH	Highly reactive, oxygen-centered radical that reacts with most of the bio-molecules.
Nitric oxide Nitrogen dioxide	·NO $\cdot NO_2$	Formed from L-arginine. Formed from reaction of ·NO with O_2.
Thyl radical Perthyl radical	RS· RSS·	Radicals with unpaired electrons residing on sulfur atom.
Transition-metal ions	Fe^{2+}, Cu^{2+}, Mn^{2+}, Co^{2+}	Ability to change REDOX numbers by 1, allows them to accept/donate single electrons; they can catalyze REDOX reactions.

Although radicals are generally short-lived due to their reactivity, there are long-lived radicals. These are categorized as follows: (i) Stable radicals: The prime examples are molecular dioxygen (O_2) and NO. Organic radicals can be long lived if they occur in a conjugated system, like those derived from α-tocopherol. (ii) Persistent radicals are compounds whose longevity is due to steric crowding around the radical center, which makes it physically difficult for the radical to react with another molecule. Examples of these include nitroxides, nitronyl nitroxides, and azephenylenyls and radicals derived from PTM (perchlorophenylmethyl radical). Persistent radicals are generated in large quantity during combustion, and may be responsible for the oxidative stress resulting in cardiopulmonary disease and probably cancer that has been attributed to exposure to airborne fine particles. (iii) Diradicals are molecules containing two radical centers. Atmospheric O_2 naturally exists as a diradical in its ground state as triplet oxygen. The low reactivity of atmospheric O_2 is due to its diradical state. Non-radical states of dioxygen are actually less stable than the diradical. The relative stability of the oxygen diradical is primarily due to the spin-forbidden nature of the triplet-singlet transition required for it to grab electrons (i.e. oxidation).

Non-free radical derivatives of molecular oxygen (O_2), such as hydrogen peroxide (H_2O_2), hypochlorous acid (HOCl), singlet oxygen ($^1O_2$) and peroxynitrite ($ONOO^-$) are important molecules in REDOX chemical spectrum. Hydrogen peroxide (H_2O_2) is considered to be one of the most powerful REDOX signaling ROS. Although peroxynitrite ($ONOO^-$) is an important ROS in cell signaling, it can also be categorized as the RNS as it is derived from the reaction of a $O_2^{\bullet-}$ radical (an ROS) with ·NO (an RNS); it will therefore be discussed in the RNS section.

Non-Free Radical Type	Formula	Reactive Properties
Hydrogen peroxide	H_2O_2	A diffusible weak oxidizing and poor reactive agent. Participates in cellular signaling. In the presence of transition metals, can generate ·OH radicals.
Hypochlorite Hypochlorous acid	OCl^- HOCl	Weak acid but strong oxidant. Can give rise to secondary ROS, chloramines and amino acid-derived aldehydes.
Peroxynitrite Peroxynitrous acid	$ONOO^-$ ONOOH	Forms via reaction of O_2 with ·NO and comparable to $O_2^{\bullet-}$ that reacts with SOD. The protonated form is highly reactive.
Alkylperoxynitrites Dinitrogen trioxide Nitryl chloride Nitronium (nitryl) ion	ROONO N_2O_3 NO_2Cl NO_2	Additional, Reactive Nitrogen Species (RNS).
Nitrosothiols	RSNO	Forms via reaction of RS· with ·NO, or thiols with higher oxides of nitrogen. Nitrosothiols are weak oxidants.
Singlet oxygen Ozone	$^1O_2$ O_3	Non-radical oxidants produced from H_2O_2.

2.4 **REDOX**: Reactive Oxygen Species (ROS)

In 1818, Thenard isolated the prominent reactive oxygen-derivative – the hydrogen peroxide (H_2O_2). Identification and isolation of several other reactive oxygen species (ROS) and their REDOX Chemistry soon followed in the early 19th Century and continued with a strong research hold into the 21st Century. In 1954, Commoner and co-workers described the biological ROS production. Nearly a decade later, during 1968-69, the discovery of superoxide dismutases (SODs) by McCord and Fridovich has completely revolutionized the field of free radical chemistry and related oxidants in biology. They elucidated that superoxide can be produced by enzymes and that free radicals can serve as substrates for an enzyme.

This triggered a total paradigm shift from the old school of thought that free radicals bring only harmful oxidations. Initial research focused on the chemistry of the individual ROS and the damage they could inflict on biological systems. In those early years much energy was invested in determining the kinetics and mechanisms of these detrimental oxidative reactions. Further identification of two major sources of ROS; the mitochondrial superoxide generation and the phagocytic respiratory burst, has opened new frontiers in free radical research. The first record of mitochondrial superoxide generation was reported by Boveris et al (1972). During oxidative phosphorylation and transfer of energy along the electron transport chain, 1–5% of electrons escape from complexes I and III to form superoxide ($O_2^{\bullet-}$) in the presence of oxygen. This process of $O_2^{\bullet-}$ generation, known as 'electron leak', has been implicated in the pathophysiology of numerous diseases. Babior et al (1973) suggested a role for ROS in innate immunity. He hypothesized that the elevated neutrophil oxygen consumption and the concurrent increase in cellular $O_2^{\bullet-}$ that accompanied bacterial phagocytosis was vital to the bactericidal activity of granulocytes. The rapid increase in the REDOX status of neutrophils and macrophages while engulfing pathogens became known as the 'respiratory burst'. During the 1970s, few studies reported that exogenously added H_2O_2 could mimic the signaling activity of insulin. This finding remained within the circles of a small free radical research community. In the 1980s, several laboratories reported that exogenous H_2O_2 could also stimulate cell proliferation at lower concentrations. Subsequent research into the ROS generating NOx enzymes revealed the true physiological roles of endogenous H_2O_2. In the 1980's the thinking in this new field was broadened with the introduction of a new, and then controversial, idea that at lower concentrations, $O_2^{\bullet-}$ or H_2O_2 could induce cell division. The following section is a brief review about the chemistry of such ROS, a family of molecules that includes highly reactive free oxygen radicals such as: superoxide, hydroxyl, peroxyl, and alkoxyl; or the non-radicals including: hypochlorous acid, hypochlorite, ozone, and singlet oxygen.

Free radicals are molecules with unpaired electrons. They are everywhere -in the air, our bodies, and the materials around us. In search of finding another electron, free radicals are very reactive and can cause damage to surrounding molecules.

2.4.1. Superoxide ($O_2^{\bullet-}$) Anion

SUPEROXIDE

Superoxide (or hyperoxide) consists of the anion with the chemical formula $O_2^{\bullet-}$. It is generated as a product of the 1-electron reduction of dioxygen (O_2), which occurs widely in nature. With a single unpaired electron, the $O_2^{\bullet-}$ ion is a free radical, and, like dioxygen, it is paramagnetic in nature.

Properties: The chemical nature of superoxide depends on pH and milieu conditions. Due to its pKa (acid dissociation constant) of 4.8, superoxide can exist in the form of either $O_2^{\bullet-}$ or, at low pH, hydroperoxyl ($HO_2^{\bullet-}$). The latter can more easily penetrate biological membranes than the charged form. Hydroperoxyl can therefore be considered an important species, although under physiological pH most of the superoxide is in the charged $O_2^{\bullet-}$ form. In a hydrophilic environment both the $O_2^{\bullet-}$ and $HO_2^{\bullet-}$ can act as reducing agents, for example, they can reduce ferric (Fe3+) ions to ferrous (Fe2+) ions. In organic solvents the solubility of $O_2^{\bullet-}$ is higher, and its ability to act as a reducing agent is increased. It also acts as a powerful nucleophile capable of attacking positively charged centers, and as an oxidizing agent it can react with compounds capable of donating protons (H+) (i.e. ascorbate and tocopherol). The most important reaction of $O_2^{\bullet-}$ radicals is dismutation; in this reaction, superoxide radical reacts with another $O_2^{\bullet-}$ / $HO_2^{\bullet-}$ species. One is oxidized to O_2, and the other is reduced to H_2O_2.

$$O_2^{\bullet-} / HO_2^{\bullet-} + O_2^{\bullet-} / HO_2^{\bullet-} \rightarrow H_2O_2 + O_2$$

Although the rate for this spontaneous reaction is low, it can become much higher in acidic pH where the $HO_2^{\bullet-}$ radical is formation is favored.

Synthesis: Biological sources of $O_2^{\bullet-}$ include phagocytotic cells and mitochondria (described in Chapter 3). In phagocytes, $O_2^{\bullet-}$ is produced in large quantities by the enzyme NADPH oxidase for use in oxygen-dependent killing mechanisms of invading pathogens. Phagocytic cells experience a 'respiratory burst' during the immune response and their oxygen consumption is greatly increased. During this process, the formation of $O_2^{\bullet-}$ is catalyzed by the enzyme, NADPH oxidase, where NADPH is utilized as a source of electrons.

$$2\ O_2 + NADPH \rightarrow H+ \rightarrow NADP+ + H+ + 2\ O_2^{\bullet-}$$

The principal source of $O_2^{\bullet-}$ in biological systems is probably the mitochondrial respiration. As electrons are passed from one complex to another along the electron transport chain, an occasional electron is leaked from one of the complexes and is able to react with O_2 leading to the formation of $O_2^{\bullet-}$. Additional enzymes also catalyze the formation of $O_2^{\bullet-}$, especially xanthine oxidase, an enzyme that catalyzes the metabolism of purines to uric acid. In addition, $O_2^{\bullet-}$ is formed when certain molecules undergo auto-oxidation in the presence of O_2. Some examples of these compounds include reduced flavin mononucleotide, reduced flavin adenine dinucleotide, L-DOPA, adrenalin, noradrenalin, dopamine, etc. Heme containing proteins, such as hemoglobin and myoglobin, are also responsible for the release of $O_2^{\bullet-}$. It is believed that during the process of O_2 binding to the coordinated Fe^{2+} atom of a heme group, an occasional $O_2^{\bullet-}$ escapes from the transition state.

Activity: Despite its name, the 'superoxide' anion, $O_2^{\bullet-}$ itself is not very lethal, since $O_2^{\bullet-}$ is metabolized by superoxide dismutase (SOD) to hydrogen peroxide (H_2O_2). However, it can further undergo the Fenton reaction that ultimately leads to formation of the highly lethal hydroxyl (·OH) radical. Superoxide has been implicated in diseases ranging from Alzheimer's to diabetes. The wide range of pathologies associated with $O_2^{\bullet-}$ is the result of its ability to react with a variety of cellular targets, including enzyme active sites, nucleic acids and lipids. Cells protect themselves against $O_2^{\bullet-}$ damage by producing superoxide dismutase enzymes (SODs). These enzymes accept an electron from a $O_2^{\bullet-}$ molecule, reducing it to produce H_2O_2 and regenerating the enzyme. The high reactivity of $O_2^{\bullet-}$ in aqueous solutions almost precludes its being added to a biological test system in a controlled manner. As a result, much of what is known of the involvement of $O_2^{\bullet-}$ in disease states comes from studies in which SOD has been inactivated, leaving the cells vulnerable to endogenous $O_2^{\bullet-}$ damage.

Potassium superoxide is used as an oxygen source in chemical oxygen generators, such as those used on the space shuttle and on submarines. Superoxides are also used in diver's oxygen tanks in order to provide a readily available source of oxygen.

2.4.2. Hydroxyl (·OH) Radical

IUPAC name:
Hydroxyl radical

Systematic names:
Oxidanyl
Hydridooxygen

Other names:
Hydroxy
Hydroxyl

Hydroxyl radical (·OH) is the neutral form of the hydroxide (HO^-). Hydroxyl radicals are highly reactive and consequently short-lived; however, they form an important part of free radical chemistry.

Properties: Hydroxyl radical (·OH) is considered the most deleterious radical encountered in biological systems due to its high reactivity. Its reaction with a substrate is only limited by its ability to maneuver (diffusion-controlled) and it is considered to be a non-selective radical. In contrast to $O_2^{\bullet-}$ radicals (that are considered relatively stable and have constant, relatively low reaction rates with biological components), ·OH radicals are short-lived species possessing high affinity toward other molecules. Hydroxyl radical is a powerful oxidizing agent that can react at a high rate with most organic and inorganic molecules in the cell, including DNA, proteins, lipids, amino acids, sugars, and metals. The three main chemical reactions of ·OH radicals include hydrogen abstraction, addition, and electron transfer.

Synthesis: Hydroxyl radicals are produced from the decomposition of hydroperoxides (ROHO) or, in atmospheric chemistry, by the reaction of excited atomic oxygen with water. The two most common routes for the formation of ·OH consist of the Fenton reaction or homolytic bond scission by UV irradiation. The production of ·OH by the Fenton reaction relies on the presence of Fe^{2+} salt and H_2O_2.

i) $Fe^{2+} + H_2O_2 \rightarrow Fe^{3+} + HO- + \cdot OH$

This well-known reaction that was first observed by Fenton in the late 1800s. In the 1930s, Haber and Weiss elucidated the subsequent mechanisms associated with the original Fenton reaction:

ii) $\cdot OH + H_2O_2 \rightarrow H_2O + O_2^{\bullet-} + H^+$
iii) $O_2^{\bullet-} + H^+ + H_2O_2 \rightarrow H_2O + O_2^{\bullet-} + H^+$

Another source of ·HO is due to absorption of UV irradiation by H_2O_2 that results in a homolytic scission reaction yielding two hydroxyl radicals.

iv) $H–O–O–H \rightarrow hV \rightarrow 2\ \cdot OH$

In general, the ·OH is able to react with various species by three different pathways. First, ·OH can extract a H^+ from the substrate, usually through a C–H bond.

v) $\cdot OH + R–H \rightarrow H_2O + R\cdot$

Global Warming: The Nobel Peace Prize 2007 was awarded jointly to the Intergovernmental Panel on Climate Change (IPCC) and Albert (Al) Gore Jr for their efforts to build up and disseminate greater knowledge about man-made climate change, and to lay the foundations for the measures to counteract such change.

Secondly, ·HO can also oxidize metals as a result of the transfer of an electron from the metal to the radical. As shown below, this reaction yields water and a higher valence state of the metal.

vi) $Metal^{+} + \cdot OH + (e^-) \rightarrow Metal^{+1} + H_2O$

Activity: Hydroxyl radicals can be produced as a byproduct of immune host defense. Macrophages and microglia frequently generate this ROS when exposed to bacterial pathogens. The destructive action of ·OH radicals has been implicated in several neurological autoimmune diseases. Hydroxyl radical can damage virtually all types of macromolecules: carbohydrates, nucleic acids (cause mutations), lipids (lipid peroxidation) and amino acids. The ·OH radical has a very short *in vivo* half-life of approximately 10^{-9} seconds and a high reactivity. Unlike $O_2^{\bullet -}$, which can be detoxified by SOD enzyme, the ·OH radical cannot be eliminated by an enzymatic reaction. Therefore, mechanisms for OH radical scavenging includes endogenous antioxidants such as melatonin and glutathione, and dietary antioxidants such as mannitol and vitamin E.

Role of Hydroxyl Radical in Global Warming: The ·OH radical is one of the critical ROS that controls the oxidizing capacity of the global Earth atmosphere. This ROS has a major impact on the levels of greenhouse gases and pollutants in the Earth atmosphere. It is the most widespread oxidizer in the troposphere (the lowest part of the atmosphere); therefore, understanding the ·OH variability is important in evaluating human impacts on the atmosphere and climate. Hydroxyl radical has a lifetime of less than one second in the Earth atmosphere; thus, the rate of reaction with the hydroxyl radical often determines how long many pollutants last in the atmosphere, if they do not undergo photolysis or are rained out. For instance methane, which reacts relatively slowly with hydroxyl radical, has an average lifetime of >5 years and many chloroflurocarbons (CFCs) have lifetimes of 50+ years. Pollutants, such as larger hydrocarbons, can have very short average lifetimes of less than a few hours. The role of ·OH in the oxidation process of methane present in the atmosphere to first carbon monoxide (CO) and then CO2 is critical for assessing the residence time of this greenhouse gas, the overall carbon budget of the troposphere, and its influence on the process of global warming.

2.4.3. Hydrogen Peroxide (H_2O_2)

IUPAC name:
Dihydrogen dioxide

Other names:
Dioxidane
Oxidanyl

Hydrogen peroxide (H_2O_2) is the simplest peroxide (a compound with an oxygen-oxygen single bond). Hydrogen peroxide is a clear liquid, slightly more viscous than water. In dilute solution, it appears colorless. Pure H_2O_2 has a pH of 6.2, therefore, considered to be a weak acid. When diluted at approximately 60%, the pH of H_2O_2 can be as low as 4.5. In acidic solutions, H_2O_2 acts as a powerful pro-oxidant, stronger than chlorine or chlorine dioxide.

Properties: Although H_2O_2 molecules are considered reactive oxygen metabolites, they are not radical by definition; they can, however, cause damage to the cell at a relatively low concentration. They are freely dissolved in aqueous solution and can easily penetrate biological membranes. Their deleterious chemical effects can be divided into the categories of direct activity, originating from their oxidizing properties, and indirect activity in which they serve as a source for more deleterious species, such as ·OH or hypochlorite (HOCl).

Synthesis: Living organisms naturally produce H_2O_2 as a by-product of oxidative metabolism. It is formed by oxido-reductases by two-electron reduction of O_2 or by dismutation of $O_2^{\bullet-}$, in respiratory or photosynthetic electron transport chains or simply by auto-oxidation of biomolecules. Hydrogen peroxide (unlike $O_2^{\bullet-}$) is an uncharged species that penetrates membranes; therefore, H_2O_2 stress arises inside cells whenever H_2O_2 is present in their extra-cellular environment. Such sources of H_2O_2 probably drove the evolution of H_2O_2 scavengers and other defensive enzymes. Accordingly, enzymes known as catalases and peroxidases, safely and catalytically decompose low concentrations of H_2O_2 to water and oxygen, through following reactions:

$$RH2 + H_2O_2 \rightarrow R + 2\ H_2O \text{ (Peroxidase-mediated)}$$

$$H_2O_2 + H_2O_2 \rightarrow O_2 + 2\ H_2O \text{ (Catalase-mediated)}$$

Activity: Direct activities of H_2O_2 include degradation of heme proteins; release of iron; inactivation of enzymes; and oxidation of DNA, lipids, thiol (-SH) groups, and keto acids. Hydrogen peroxide also plays an important role in aging (graying of hair is one of the evident effects). Endogenous H_2O_2 have long been recognized as destructive molecules. The well-established roles they have in the phagosome and genomic instability has led to the characterization of these molecules as non-specific agents of destruction. Recent studies, however, demonstrated that at lower physiological levels, H_2O_2 can act as a classical intra-cellular signaling molecule and regulate kinase-driven pathways. The newly discovered biological functions attributed to H_2O_2 in several regulatory cellular events including activation of transcription factors, cell proliferation, and apoptosis. H_2O_2 produced from the mitochondrial electron transport chain has been shown to play a role inhemato poietic cell differentiation and cell division. NADPH oxidase (NOx) generated H_2O_2 can affect cardiac differentiation, vascularization, and angiogenesis. In targeting cysteine and methionine residues of protein kinases and phosphatases, H_2O_2 is also capable of modulating a number of cell signaling cascades.

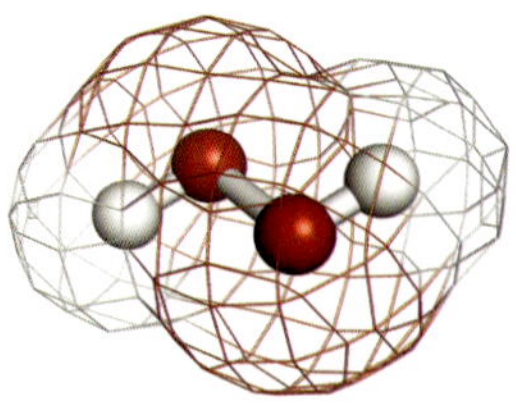

HYDROGEN PEROXIDE

Broad-spectrum Disinfectant: Hydrogen peroxide is generally recognized as safe (GRAS) as a disinfectant, an oxidizing agent and for other purposes by the US FDA. For example, 35% hydrogen peroxide is used to prevent infection transmission in the hospital environment. Also, H_2O_2 is registered with the US EPA as a sporicidal agent and dilute solutions of H_2O_2 has long been used for cleaning small surface wounds. Hydrogen peroxide is used, in sufficient concentrations, to disinfect inanimate objects. It can be used as a toothpaste, or oral debriding agent, when mixed with correct quantities of baking soda and salt. H_2O_2 and benzoyl peroxide mixtures are sometimes used to treat acne.

2.4.4. Hypochlorous Acid (HOCl)

IUPAC name:
Hypochlorous acid
Chloric(I) acid
Chloranol
Hydroxidochlorine

Other names:
Hydrogen hypochlorite
Chlorine hydroxide

Hypochlorous acid is a weak acid with the chemical formula HOCl. It is formed when chlorine dissolves in water, and cannot be isolated in pure form due to rapid equilibration with its precursor. HOCl is an a strong oxidant. In aqueous solution, HOCl partially dissociates into the anion – hypochloride (OCl^-). Salts of hypochlorous acid are called hypochlorites. One of the best-known hypochlorites is NaClO, the active ingredient in bleach. HClO is a stronger oxidant than chlorine under standard conditions.

Synthesis: Immune host defense has an innate ability to wade off any microbial attack by deploying a group of highly reactive chemicals, including oxidized halogens, oxidizing radicals, and singlet oxygen collectively known as 'neutrophil-released extra-cellular traps (NET)'. The precursor of these ROS is the $O_2^{\bullet-}$ radical generated by specialized immune cells, mostly neutrophils, eosinophils, mononuclear phagocytes, and B lymphocytes. Production of ROS in these cells is accompanied by a significant rise in oxygen consumption, a series of events referred to as the respiratory burst. The primary enzyme responsible for ROS production is a mitochondrial-membrane–bound NADPH oxidase (NOx) enzyme.

During respiratory burst, neutrophils-produced myeloperoxidase (MPO) catalyzes the oxidation of chloride by H_2O_2 and forms a strong oxidant – the hypochlorous acid (HOCl) in the following reaction:

$$H_2O_2 + Cl^- + H^+ \rightarrow HOCl + H_2O$$

MPO is present on NETs and has the potential (given a supply of H_2O_2) to generate HOCl in close proximity to trapped bacteria, thus providing a prospective mechanism for oxidative NET-mediated killing.

Activity: Hypochlorous acid is known to be the major strong oxidant produced by neutrophils, and is a potent microbicidal agent within these cells. It has been estimated that 10^6 (million) neutrophils stimulated in vitro can produce 0.1 μM HOCl. This quantity of HOCl can kill 1.5×10^7 (10 million) *Escherichia coli* in less than 5 minutes. HOCl reacts readily with a range of biological molecules, particularly those with thiol, thiolether, heme proteins, and amino groups, and may lead to tissue injury. Taurine, a non-essential amino acid naturally found at aboutr 15 mM within neutrophils acts as a scavenger molecule for HOCl via the following mechanism, and evades any possible collateral damage to cellular macromolecules from HOCl:

$$HOCl + NH_2\text{-} CH_2\text{-} CH_2\text{-} SO_3H \rightarrow Cl_{(1 \text{ or } 2)} N\text{-}CH_2\text{-}CH_2\text{-}SO_3H + H_2O$$

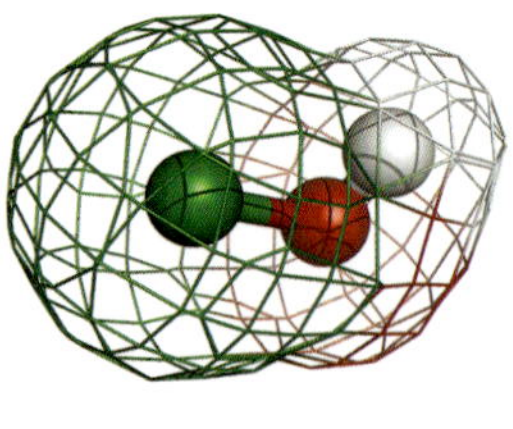

HYPOCHLOROUS ACID

HOCl elicits broad-spectrum antimicrobial effects. Inhibition of glucose oxidation by HOCl has been implicated as one of the modes of its antimicrobial mechanisms. HOCl has been shown to diffuse across the cytoplasmic membranes and inactivate key SH-containing enzymes in the glycolytic pathway. Inhibition of

ATP hydrolysis and proton co-transport by HOCl has also been suggested as an alternative antimicrobial mechanism. Hypochlorous acid has also been shown to inhibit DNA replication. Thus, when bacteria are exposed to HOCl, a significant decline in DNA synthesis, followed by a drop in protein synthesis is observed. Ultimately, these cellular events could lead to inhibition of microbial growth. HOCl is also known to induce post-translational modifications to proteins, especially the oxidation of cysteine and methionine residues.

Furthermore, HOCl reacts with a wide variety of biomolecules, including nucleic acids (DNA, RNA), fatty acid groups, cholesterol and proteins. Hypochlorus acid is known to oxidize sulfhydryl groups, leading to the formation of disulfide bonds (crosslinking) and inactivation of proteins. A single -SH-containing amino acid can scavenge up to 4 molecules of HOCl. Hypochlorous acid reacts readily with amino acids that have amino group side-chains, with the chlorine from HOCl displacing hydrogen, resulting in an organic chloramine. Chlorinated amino acids rapidly decompose. A concentration of 10 mM or greater HOCl is necessary to fragment proteins *in vivo*. Hypochlourous acid reacts slowly with DNA and RNA (as well as all nucleotides), which could interfere with DNA base pairing. Hypochlorous acid reacts with unsaturated (but not saturated) bonds in lipids. This reaction occurs by hydrolysis, resulting in the formation of chlorhydrin. The polar chlorine disrupts lipid bilayers and could increase permeability. Chlorhydrin formation in the lipid bilayers of red blood cells lead to significant alterations in membrane permeability.

HOCl - the commonly used Disinfectant in the world

2.4.5. Hypochlorite (ClO^-) Ion

IUPAC name:
Hypochlorite
Chlorate(I)

Other names:
Sodium hypochlorite
Calcium hypochlorite

The hypochlorite (ClO^-), also known as chlorate anion is a chemical compound containing chlorine in oxidation state +1. Hypochlorites are the salts of hypochlorous acid (HOCl). Chlorine is a naturally occurring gas and hypochlorite is a chlorine-containing compound in an oxidated state (meaning that it has lost electrons). The two most common forms are sodium hypochlorite [NaClO] and calcium hypochlorite [$Ca(ClO)_2$].

Properties: Hypochlorites are unstable compounds — for example, sodium hypochlorite is not available as a solid, since removal of the water from NaClO solution converts it to a mixture of sodium chloride and sodium chlorate. Heating of NaClO solution also causes this reaction. Hypochlorites decompose in sunlight, giving chlorides and oxygen. Due to their low stability, hypochlorites are very strong oxidizing agents and can react with many organic and inorganic compounds. Reaction with organic compounds is exothermic (generates heat) and may cause ignition, therefore, hypochlorites should be handled with care. They can oxidize manganese compounds and convert them to permanganates.

Synthesis: Hypochlorite is created in vivo when the human immune system destroys pathogens. Viruses and bacteria are engulfed into an intra-cellular vacuole, the phagosome, where they are digested. For this purpose, several proteases (e.g. elastase) are released into the phagosome from intra-cellular granules. In addition, there is also an enzyme-mediated 'respiratory burst', which results in the production of a set of ROS compounds. Contact with a foreign particle causes the activation of membrane bound NADPH oxidase (NOx), which produces $O_2^{\bullet -}$ in the phagosome. Superoxide dismutates to oxygen and H_2O_2: the H_2O_2 formed is used as a substrate for the myeloperoxidase (MPO)-catalysed reaction, which converts chloride to ClO^-.

Activity: Hypochlorite is created in vivo from the downstream of HOCl-mediated reactions. The dissociation of HOCl to the less microbicidal ClO^- ionic form is dependent on pH. As the pH increases, more ClO^- ion is formed and the microbicidal activity decreases. Disinfection by ClO^- is optimal at around pH 6.0 because the concentration of HOCl is optimal and its dissociation is minimal. Other useful functions of ClO^- include its ability to breakdown phenolic compounds and remove phenolic tastes and odors. Its lack of reaction with ammonia, and its reduced tendency to form trihalomethanes or chlorophenols makes it an ideal disinfectant. It is widely used in the chlorination of drinking water and for elimination of cyanides, sulfides, aldehydes, and mercaptans.

HYPOCHLORITE

The production of hypochlorite dates back to 1789; synthesized when chlorine gases passed through a sodium carbonate solution. The resulting sodium hypochlorite was called 'Javel water', named after the place of manufacture, in France. This chemical compound is a commonly used disinfectant in a solution.

Chlorinated lime (a white or grayish-white powder, also available as tablets) is a strong oxidizer and is considered more stable than sodium hypochlorite. Chlorinated lime is believed to provide more chlorine and widely used to purify drinking water.

Hypochlorite has been used as a disinfectant for more than 100 years. It has many properties of an ideal disinfectant, including a broad antimicrobial activity, rapid bactericidal action, reasonable persistence in treated potable water, ease of use, solubility in water, relative stability, relative non-toxicity at use concentrations, no poisonous residuals, no color, no staining, and low cost. Hypochlorites are lethal to most microbes, although viruses and vegetative bacteria are more susceptible than endospore-forming bacteria, fungi, and protozoa. Activity is reduced by the presence of heavy metal ions, a biofilm, organic material, low temperature, low pH, or UV radiation. Clinical uses in health-care facilities include hyperchlorination of potable water to prevent Legionella colonization, chlorination of water distribution systems used in hemodialysis centers, cleaning of environmental surfaces, disinfection of laundry, local use to decontaminate blood spills, disinfection of equipment, decontamination of medical waste prior to disposal, and dental therapy. Despite the increasing availability of other disinfectants, hypochlorites continue to find wide use in hospitals.

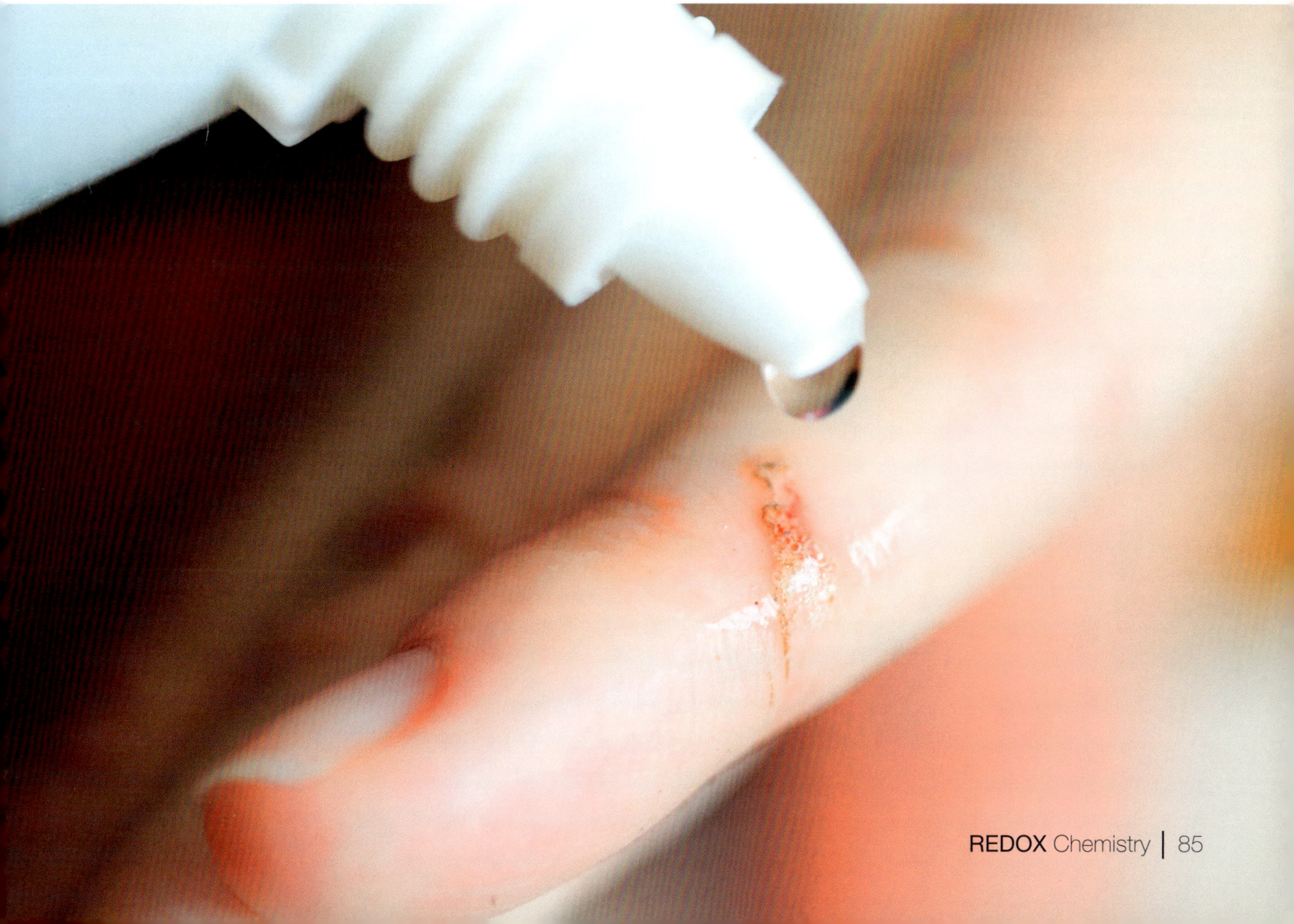

2.5 REDOX: Reactive Nitrogen Species (RNS)

Reactive Nitrogen Species (RNS) are various ·NO-derived compounds, that includes nitroxyl (·HNO) anion, nitrosonium (NO^+) cation, higher oxides of N_2, S-nitrosothiols (RSNO), and dinitrosyl iron complexes. RNS are known to play a crucial role in the physiological regulation of several cellular functions, such as smooth muscle cells, cardiomyocytes, platelets, and nerve cells. They possess pleiotropic properties on cellular targets after both post-translational modifications and interactions with ROS. Elevated levels of RNS have been implicated in cell injury and death by inducing nitrosative stress.

NITROXYL

RNS are produced in starting with the reaction of nitric oxide with superoxide to form peroxynitrite, as shown in the following reaction:

$$(ONO_2^-): \cdot NO + O2^{\bullet -} \rightarrow ONO_2^-$$

Vascular generation of ONO_2^- depletes the bioactivity of ·NO, which is important because NO is a key mediator in many important vascular functions including regulation of smooth muscle tone and blood pressure, platelet activation, and vascular cell signaling.

Nitroxyl (HNO) is highly reactive towards nucleophiles (especially thiols) and dimerizes to hyponitrous acid ($H_2N_2O_2$), which is then dehydrated to nitrous oxide (N_2O). Nitroxyl has a potential in the treatment of heart failure and ongoing research is focused on finding new ·HNO donors.

S-Nitrosothiols, also known as thionitrites, serve as donors of the nitrosonium ion. NO^+ and ·NO and some organic nitroso derivatives serve as signaling molecules, especially related to vasodilation. Red blood cells release S-nitrosothiols into the bloodstream under low-O_2 conditions, causing the blood vessels to dilate. The addition of a nitroso group to a sulfur atom of an amino acid residue of certain proteins is known as S-nitrosation or S-nitrosylation. This is a reversible process and a major form of post-translational modification of proteins. S-Nitrosated proteins (SNOs) facilitate ·NO bioactivity and regulate protein function through mechanisms analogus to phosphorylation. Thiol S-nitrosylation and ·NO transfer reactions are involved in virtually all classes of cell signaling.

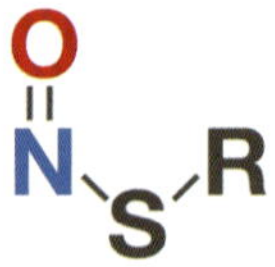

S-NITROSOTHIOL

Formation of dinitrosyl iron complexes (DNICs) are observed in many kinds of living organisms, in a wide spectrum of physiological conditions associated with inflammation, ischemia/reperfusion and cancer. Accumulation of DNICs coincides with intensified production of ·NO in macrophages, neurons, endothelial cells, Langerhans' cells and hepatocytes. Low-molecular thiol-containing DNICs show vasodilatory activity and play a role as NO transducers and stabilizers. DNICs also modulate cellular REDOX potential via inhibition of GSH-dependent enzymes, such as glutathione reductase, S-transferase, and peroxidase.

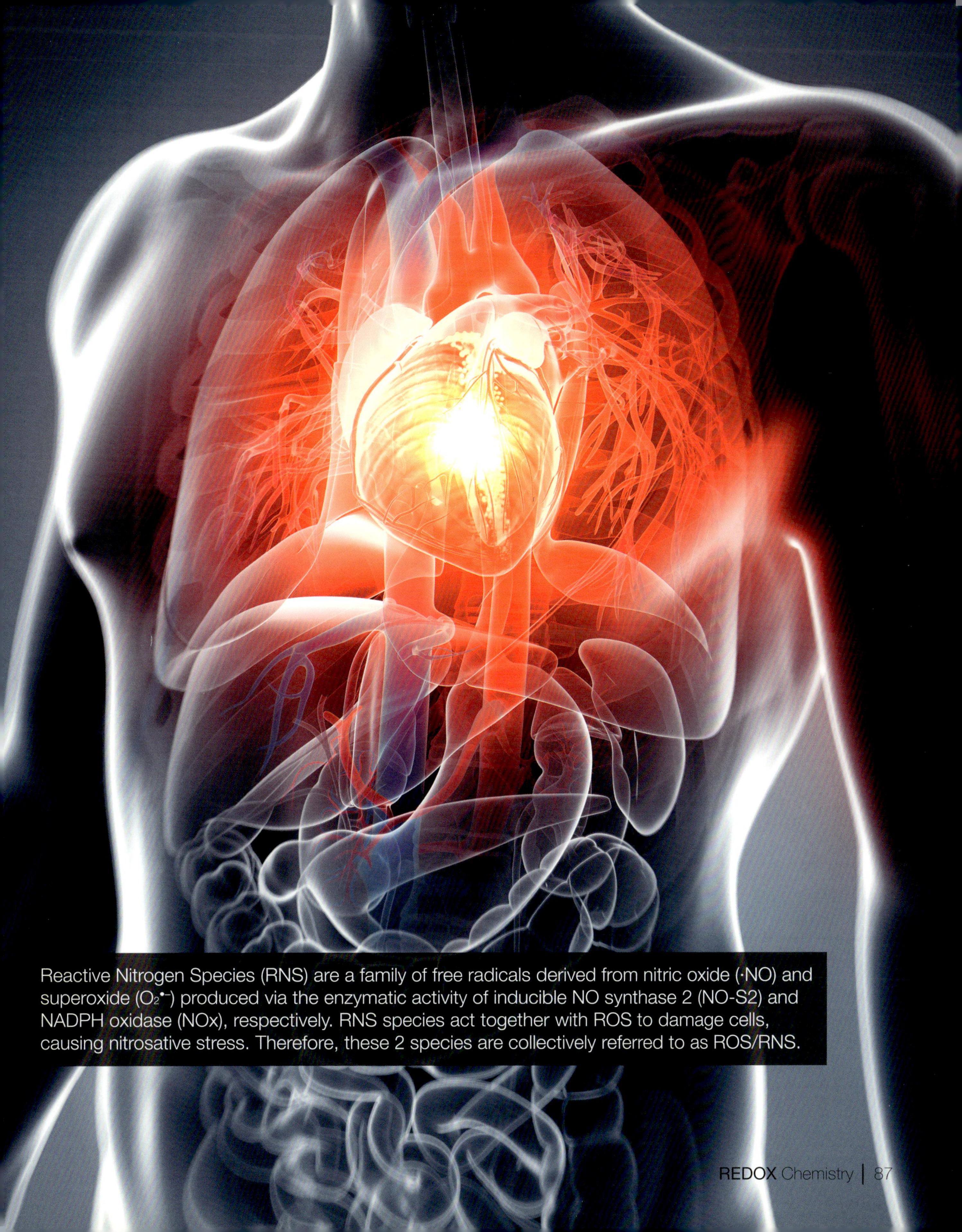

Reactive Nitrogen Species (RNS) are a family of free radicals derived from nitric oxide (·NO) and superoxide ($O_2^{\bullet-}$) produced via the enzymatic activity of inducible NO synthase 2 (NO-S2) and NADPH oxidase (NOx), respectively. RNS species act together with ROS to damage cells, causing nitrosative stress. Therefore, these 2 species are collectively referred to as ROS/RNS.

2.5.1. Nitric Oxide (·NO)

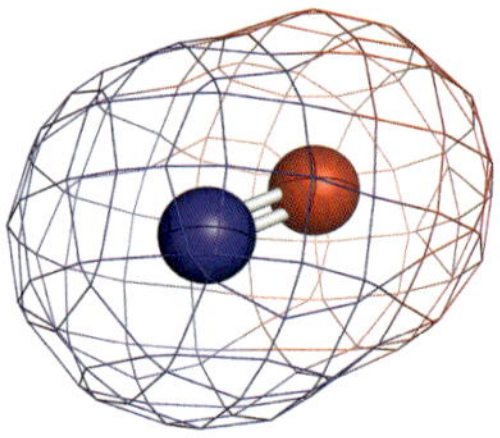

NITRIC OXIDE

Nitric oxide (·NO; also known as nitrogen monoxide) is a colorless gas, moderately soluble in water, more soluble in organic solvents. Hence ·NO can easily cross membranes and diffuse readily between and within cells. Nitric oxide has an unpaired electron in anti-bonding orbital: thus it is a free radical. If the unpaired electron is removed, nitrosonium (NO^+) cation is produced. One-electron reduction gives nitroxyl (NO^-) anion, which can also exist as a protonated HNO form. Nitroxyl is a reactive short-lived species, it reacts with O_2 to give peroxynitrite ($ONOO^-$).

Nitric oxide from vascular endothelium was originally identified as an endothelium-derived relaxing factor (EDRF), a 'factor' that relaxes blood vessels. Nitric oxide is synthesized from the amino acid L-arginine using different isoforms of nitric oxide synthase (NOS). Three NOS have been described: an endothelial isoform (eNOS), a neuronal isoform (nNOS) in 'nitrergic' nerves and a macrophage or inducible isoform (iNOS). NOS enzymes cleave a nitrogen from the semi-essential aminoacid L-arginine, and combine it with oxygen from molecular oxygen to form ·NO. The byproduct of this reaction is the amino acid L-citrulline. The eNOS and nNOS enzymes require Ca^{2+} and the calmodulin proteins for their action; consequently, low intra-cellular 'free' Ca^{2+}' within cells restricts NOS activity. At sites of inflammation, an 'extra' NOS is often present, called inducible NOS (iNOS, NOS-2). This enzyme was first identified in macrophages and liver cells after treatment with endotoxin or certain cytokines.

Much of the ·NO generated in the vascular system is eventually removed by interaction with hemoglobin. Nitric oxide forms complexes with Fe^{2+} and ferrous compounds (a process called nitrosylation) including deoxyhaemoglobin. Nitric oxide also reacts with oxyhemoglobin, oxidizing it to methemoglobin and forming nitrate. As a result, once ·NO has entered an erythrocyte, its half-life is <1 μsec. The rate of entry of ·NO into erythrocytes limits its reaction with oxyhemoglobin and it has been estimated that the half-life of ·NO in whole blood is about 1.8 milli-seconds. Free oxyhemoglobin scavenges ·NO much faster and as a result can cause problems, for example in the brain. Prolonged inhibition can increase $O_2^{\bullet-}$ formation from the electron transport chain. However, the simultaneous activation of $O_2^{\bullet-}$ synthesis along with ·NO will completely transform the biological actions of nitric oxide by forming peroxynitrite ($ONOO^-$).

Nitric Oxide Physiology: Nitric oxide is an omnipresent inter-cellular messenger in all vertebrates, modulating blood flow, thrombosis, and neural activity. It readily binds to certain transition metal ions, and many of its physiological effects arise from its binding to Fe^{2+}-heme groups in the enzyme guanylate cyclase. For example, ·NO synthesized by the vascular endothelial cells that line the interior of blood vessels presumably diffuses in all directions, but some of it will reach the underlying smooth muscle, bind to guanylate cyclase and activate it. As a result, more cyclic GMP is made, which lowers intra-cellular free Ca^{2+} and relaxes the muscle, dilating the vessel and lowering the blood pressure.

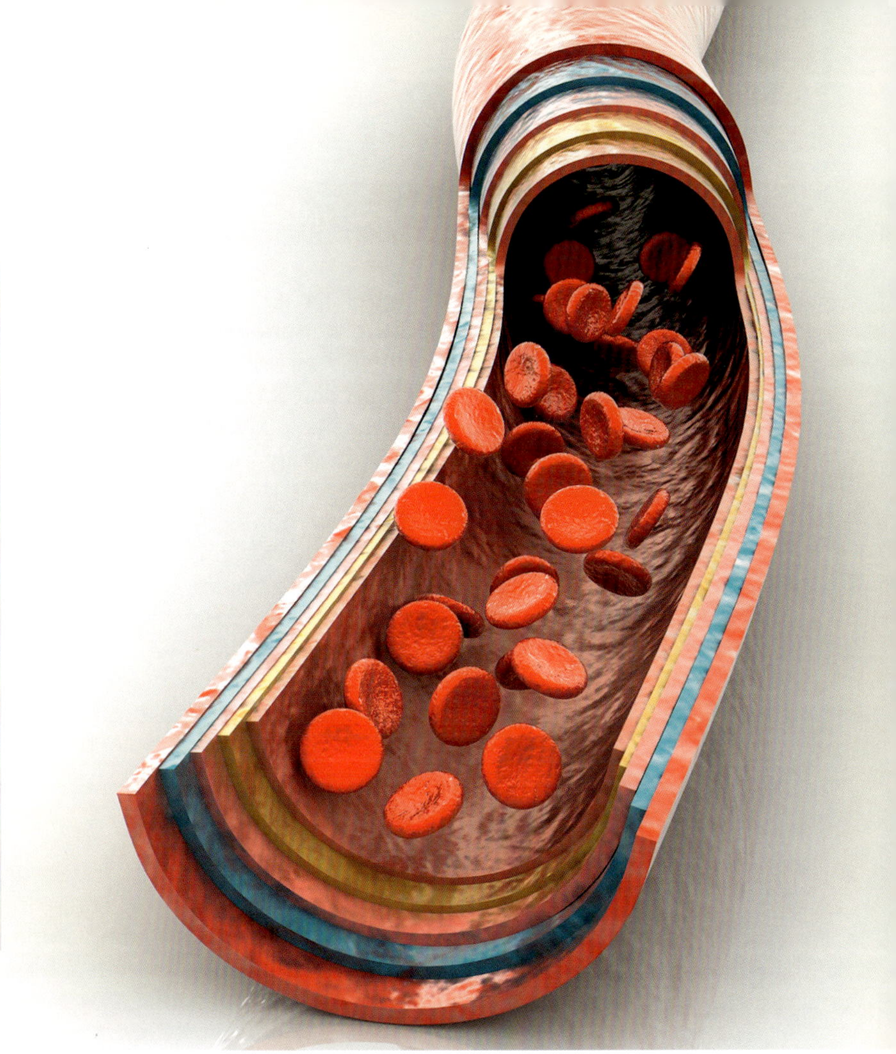

What happens when superoxide and nitric oxide are produced simultaneously in close proximity?

Any rise in $O_2^{\bullet-}$ and ·NO by 10-fold will increase peroxynitrite formation, the most potent oxidative agent, by 100-fold. During pro-inflammatory conditions, the production of $O_2^{\bullet-}$ and ·NO strongly rise by 1,000-fold, which simultaneously enhance peroxynitrite formation by 1,000,000-fold. ·NO and $O_2^{\bullet-}$ do not even have to be produced within the same cell to form peroxynitrite, because ·NO can so readily move through membranes and between cells.

Nitric oxide is known to elicit several signaling functions. It can readily react with $O_2^{\bullet-}$ to form peroxynitrite ($ONOO^-$) a potent oxidative agent. Both $O_2^{\bullet-}$ and NO are synthesized within a few cell diameters from each other; and combine spontaneously to form $ONOO^-$ by a diffusion-limited reaction. In essence, when NO and $O_2^{\bullet-}$ collide, an $ONOO^-$ radical is formed. No enzyme is needed for $ONOO^-$ formation, since no enzyme can possibly catalyze any reaction this fast. NO is the only known biological molecule that reacts faster with $O_2^{\bullet-}$ and is produced in high enough concentrations to outcompete endogenous levels of superoxide dismutase (SOD). Peroxynitrite formation also reduces the bioavailability of both $O_2^{\bullet-}$ and ·NO, which could also influence cell signalling events.

Inhaling low concentrations of gaseous ·NO is approved by the US Food and Drug Administration (FDA) for the treatment of persistent pulmonary hypertension of the newborn. In addition, ·NO can be produced for 80 years by neurons in human brain without overt toxicity. Paradoxically, the production of the same molecule can become highly damaging to the same neurons within a few minutes during pathological challenges as occur after cerebral ischemia.

2.5.2. Peroxynitrite ($ONOO^-$)

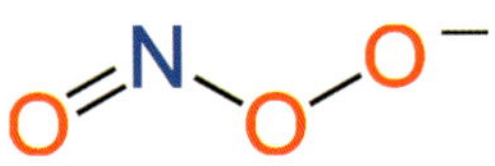

PEROXYNITRITE

Peroxynitrite ($ONOO^-$) is a strong oxidant that reacts with various biological targets and cellular components including lipids, thiols, amino acid residues, DNA bases, and low-molecular weight antioxidants. However, these reactions occur at a relatively slow rate, which allows $ONOO^-$ to react more selectively throughout the cell. Peroxynitrite is able to cross the cell membranes through anion channels. Peroxynitrite directly interacts with proteins that contain transition metal centers; thus, $ONOO^-$ can modify proteins such as hemoglobin, myoglobin, and cytochrone c by oxidizing ferrous heme into its corresponding ferric forms. Peroxynitrite may also be able to change protein structure through the reaction with various amino acids in the peptide chain. The most common reaction with amino acids is cysteine oxidation. Another reaction is tyrosine nitration; however peroxynitrite does not react directly with tyrosine. All of these reactions affect protein structure and function and thus have the potential to cause changes in the catalytic activity of enzymes, alter cytoskeletal organization, and impair cellular signals. Additionally $ONOO^-$ can react with other molecules to form different types of RNS including nitrogen dioxide ($\cdot NO_2$) and dinitrogen trioxide (N_2O3) as well as other types of chemically reactive free radicals.

Although the reaction is at a slow rate with most biological molecules, $ONOO^-$ can diffuse quite far on a cellular scale. It is able to traverse cell membranes through anion channels. For example, the $\cdot OH$ radical is formed by a slow reaction between Fe^{2+} ion and H_2O_2, but this ion can only diffuse about the diameter of a typical protein. In contrast, $ONOO^-$ radical is formed every time when a superoxide ($O_2^{\bullet -}$) collides with $\cdot NO$; however, it reacts slowly and more selectively throughout the cell. That makes the biological and pathological implications of $ONOO^-$ radical far more interesting because it can have more subtle and specific effects on cells.

Peroxynitrite is not a free radical in nature, however, it is highly reactive compared to its parent molecules $\cdot NO$ and $O_2^{\bullet -}$. The half-life of $ONOO^-$ is short (about 10-20 milli-sec), but sufficient to cross biological membranes. It can diffuse about 1 to 2 cell diameters, and allow significant interactions with most critical biomolecules. Kinetic studies have indicated that $ONOO^-$ oxidizes target molecules through two distinct mechanisms. First, $ONOO^-$ and its protonated form peroxynitrous acid (ONO_2H) can exert direct oxidative modifications through one- or two-electron oxidation processes. Only a few chemical groups directly react with $ONOO^-$, which favors selective reactions with key moieties in proteins, such as thiols, iron/sulfur centers, and zinc fingers. The second mechanism involves $ONOO^-$, which indirectly mediates oxidation by decomposing into highly reactive radicals. Peroxynitrite can yield $\cdot OH$ radical and nitrogen dioxide ($\cdot NO_2$) during the homolytic decomposition of peroxynitrous acid. The direct reaction of $ONOO^-$ with CO_2 gives rise to $CO_3^{\bullet -}$ (carbonate radical) and $\cdot NO_2$. Carbonate radical is likely to be more toxic than $\cdot OH$ radical and yields many of the same types of oxidation commonly attributed to $\cdot OH$ radical. Thus CO_2 redirects much of the $ONOO^-$ mediated radical effects. The limited reactivity of $ONOO^-$ with most molecules makes it unusually selective as an oxidant, which increases its influence over biological processes.

2.6 **REDOX**: Reactive Sulfur Species (RSS)

Sulfur is an essential and vital element for living organisms. It is a constituent of many organic molecules, amino acids (such as cysteine and methionine), and the small tripeptide glutathione (GSH). Sulfur is also important in the formation of Fe-S clusters for many enzymes, particularly, those that are involved in REDOX reactions. In the same way that oxygen and nitrogen have reactive species (ROS and RNS), sulfur too can form reactive molecular species (RSS), for example when a -SH group is oxidized. Indeed, several REDOX reactions occur via RSS intermediates. Sulfur in the form of thiol groups is central to the REDOX control of cellular metabolism. Since thiol groups are oxidized and reduced easily and reversibly, the REDOX control of cellular metabolism has become an increasing focus of research. Therefore, high cellular concentrations of thiols and disulfides make sulfur a prime target for oxidation under conditions of oxidative stress. Under these conditions, the oxidation of thiols can lead to sulfur species of different oxidation states and various REDOX behaviors. Whereas disulfides are generally only mild oxidizing species, the reaction of disulfides with ROS (i.e. H_2O_2) also generates reactive sulfur species (RSS) *in vitro*.

Sulfur is a multi-valent REDOX molecule able to appear in a wide range of different oxidation states and 'chemotypes', each with its own formation pathways, as well asphysical and chemical properties. It also reacts via a range of different REDOX mechanisms. In some cases, the same sulfur chemotype may participate in various reactions, with different mechanisms. This mechanistic diversity enables a wide range of sulfur species to connect with diverse biological REDOX processes, which may include various reaction partners and mechanisms. Then again, not all sulfur-containing biomolecules are REDOX active under normal physiological conditions, and not all of them participate in REDOX signaling.

Among the sulfur chemotypes: thiols, disulfides, thiyl radicals, perthiols, sulfenic acids, sulfinic acids, sulfonic acids, sulfenyl-amides, S-nitrosothiols, isothiocyanates, thiosulfinates, thiosulfonates, sulfoxides, sulfones and polysulfides play a significant role as part of various cellular REDOX processes. Furthermore, inorganic sulfur species, such as hydrogen sulfide (H_2S) and the ·HS anion have recently been implicated in the control of cellular processes, such as vasodilation and inflammation.

Thiol groups are found in several small biological molecules, peptides, proteins and enzymes. These groups are important for a range of sulfur modifications, so-called sulfur chemotypes. The thiol/disulfide (RSH/RSSR) REDOX couple

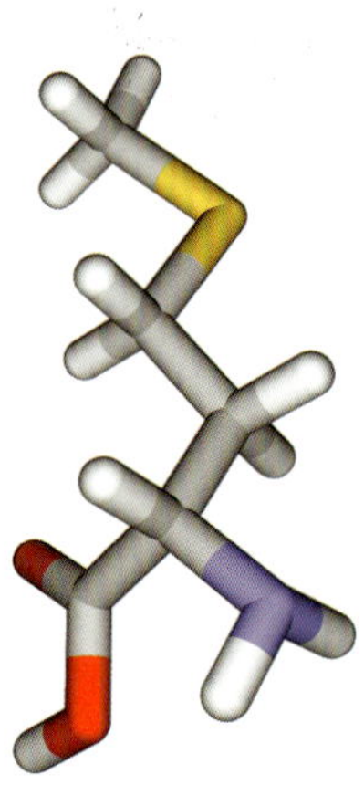

METHIONINE

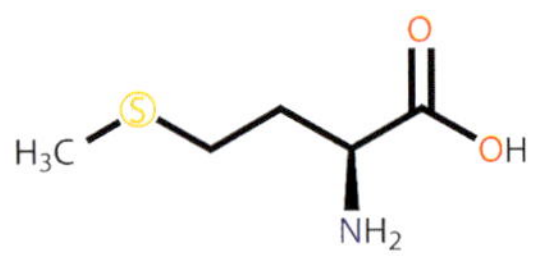

represents the classical biological REDOX system associated with sulfur. Similar to the GSH/GSSG system, this sulfur-based couple forms the major intra-cellular REDOX buffer. It links the intra-cellular processes via the enzyme glutathione disulfide reductase and maintains the GSH levels to NADPH consumption and energy metabolism. The controlled formation of intra- and inter-molecular disulfides also plays a role in protein structure and function. Oxidation of thiol groups results in a range of sulfur-containing products, not just disulfides.

Disulfide bond formation is a two-electron chemical oxidation, wherein two cysteines (RSH) are covalently bonded (RS=SR) with the formation of two protons and two electrons, as shown in the following reaction:

$$RSH + RSH \rightarrow RS{=}SR + 2\ H+ + 2\ e^-$$

The two interacting cysteines may reside in the same polypeptide chain (intra-molecular disulfide) or in different polypeptide chains within the same complex (inter-molecular disulfide). The two cysteines in an intra-molecular disulfide may be close to each other in primary sequence or far away; however, to form a disulfide, the two cysteine residues must be able to come in close contact to form a covalent bond and an electron acceptor must be able to remove electrons, a process that is additive with other interactions in the protein folding.

Methionine residues could act as 'integrated antioxidants' in protein structures. A protective mechanism for methionine in the form of a 'sacrificial' antioxidant residing within (REDOX-sensitive) proteins and peptides – such as for the enzyme glutamine synthase has been suggested. This methionine oxidation to the sulfoxide is reversible. On the other hand, irreversible 'over-oxidation' to the sulfone may fulfill the function of a REDOX sensor for oxidative stress in the form of a 'REDOX Switch', similar to the sulfenic/sulfinic acid REDOX switch found in PRx enzymes.

Nonetheless, there are also reports against such beneficial role of methionine as a protective antioxidant. For example, a diet rich in methionine promotes aging and decreases the life span of an organism, while a diet with low methionine content extends the life span. This effect may be partly due to the fact that bio-molecules (i.e. proteins) that contain high amounts of methionine are more susceptible to REDOX modification compared to molecules that contain no or little methionine.

Another couple of inorganic reactive sulfur species, such as thiocyanate (SCN^-), thiocyanogen [$(SCN)_2$], trithiocyanate [$(SCN)_3^-$] and hypothiocyanite ($OSCN^-$), also occur in vivo. Indeed, the environment of the cell provides the ingredients and reaction conditions for the formation of a whole range of organic and inorganic sulfur species, many of which have not yet been conclusively identified *in vivo*.

REDOX mechanisms without the flow of 'free' electrons are widespread in biology and avoid certain complications associated with direct electron transfer. Not surprisingly, the fourth pathway, that is, the thiol/disulfide exchange reaction, which is perhaps the most important avenue leading to disulfides, is a substitution reaction which does not require the transfer of electrons between molecules. Among the various reactive molecules of the intra-cellular REDOX systems, sulfur-containing species occupy a prominent role as targets of oxidative stressors, REDOX sensors, response elements and participants in numerous cellular signaling pathways. In many ways, reactive sulfur species (RSS) are unique, since they are targets and oxidative stressors at the same time.

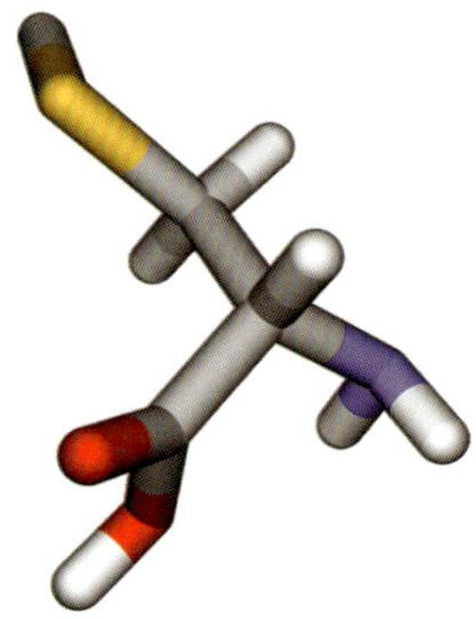

CYSTEINE

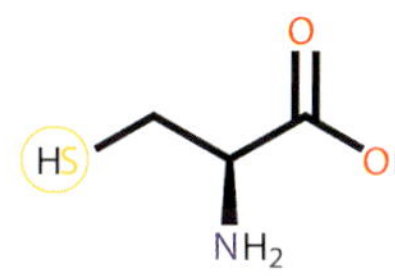

The Cysteine Proteome: Proteins are unique among biological macromolecules with elements (i.e. sulfur-containing amino acids: cysteine, methionine, selenocysteine) that undergo reversible oxidation-reduction (REDOX) reactions as part of their normal function. Cysteine (*Cys*) residues of proteins, in particular, play a vital role in several catalytic mechanisms, as well as in stabilizing proteins by formation of disulfide bridges. The interaction of *Cys* residues with Zn^{2+} could protect cells from oxidants and reactive electrophiles. Also, the reversible covalent modifications of specific *Cys* residues in a protein structure could contribute to other functions such as the signal transduction.

Cysteine (*Cys*) is derived from methionine via the trans-sulfuration pathway or can be obtained via import of cysteine or cystine (followed by reduction). In contrast to the 1–10 mM steady-state concentrations of GSH found inside cells, *Cys* levels are considerably low and limit GSH synthesis. The first of two steps in the GSH synthesis pathway is catalyzed by γ-glutamylcysteine ligase, the rate-limiting enzyme. The second enzyme, GSH synthetase, catalyzes the ATP-dependent addition of glycine to γ-glutamylcysteine to generate GSH. In addition to antioxidant function, GSH also serves as a repository of cysteine. GSH is rapidly turned over in many cell types, with half-lives varying from 2 to 6 hours, which suggests a high export rate. Multiple transporters with tissue-specific distribution exist for the import of cystine, the predominant form of the amino acid in circulation, as well as for *Cys*, which is present at substantially lower concentrations. Broadly, the *Cys* transporters can be categorized into Na^+-dependent and Na^+-independent types.

The proteome encoded by the human genome contains about 214,000 *Cys* residues. According to the analysis, about 92% of mammalian proteins contain at least 1 *Cys*; however, only 50% of Archaea proteins have at least 1 *Cys*. Such an increase in *Cys* content could reflect the evolution of signaling and the complexity of control functions. A large number of these functions are associated with covalent and non-covalent modifications of *Cys* residues. Reversible REDOX reactions can occur with single *Cys* or with vicinal *Cys*. Thus, a single *Cys* residue could form disulfide bond with GSH, *Cys* or other proteins while vicinal *Cys* form disulfide bridges or metal clusters.

Oxidation of thiol/disulfide couples in human plasma is a hallmark of aging and a common feature of age-related disease.

Interestingly, there is selection against random use of *Cys* in animal protein structures. In other words, *Cys* residues are used sparingly but are retained within structures during evolution of complexity, a characteristic of functionally important amino acid residues. REDOX proteomics helps in identification of proteins that

REDOX-sensing thiols could optimize structure-function relationships and account for a large number of reversible oxidation, nitrosylation, acylation, sulhydration or metal binding reactions affecting macromolecular structure, activity, interactions and trafficking.

undergo REDOX changes under relevant biological conditions. Targeted analyses using REDOX western blotting show that selected proteins in the mitochondria, cytoplasm, nuclei, endoplasmic reticulum and plasma membrane are partially oxidized under normal physiological conditions. H_2O_2 production rates in cells are sufficient to oxidize about 0.5% of all cell thiols each minute, and computational modeling indicates that protein oxidation contributes in a quantitatively important way to elimination of low levels of H_2O_2. The specific *Cys* residues that oxidize under normal physiological conditions are present in a broad spectrum of structural and functional classes. For example, cytoskeletal proteins (i.e. actin, tubulins, cofilin and dyneins) change in fractional oxidation in aortic endothelial cells exposed to a relatively oxidized extra-cellular cysteine/cystine REDOX potential. Other diverse group of proteins with *Cys* residues that are partially oxidized under physiologic conditions include: signal transduction molecules (i.e. Ras, phosphatases), cell stress proteins such as heat shock proteins, detoxification proteins (i.e. PRx and glutathione S-transferases), mitochondrial proteins (i.e. ATP synthase and Complex I subunits), metabolic enzymes including several from fatty acid metabolism, and pro-inflammatory signaling proteins (i.e. IL-1β).

The cytoplasmic and extra-cellular REDOX potentials of the primary thiol buffers are vastly different and influence the structure, stability, and function of macromolecules that reside within each compartment. The REDOX status of exofacial membrane proteins, receptors, and transporters should be sensitive to the *Cys*/cystine REDOX couple and influence, in turn, their stability and/or function. A number of studies have revealed systemic changes in the plasma or extra-cellular GSH/GSSG and *Cys*/cystine REDOX couples. High oxidized potentials are correlated with aging, smoking, cardiovascular disease, and diabetes.

A comparison of postprandial plasma cysteine/cystine REDOX potential showed that a meal with sulfur amino acids causes a 10-mV reductive shift compared with one missing these amino acids. This study elucidated both the dynamic nature of the systemic extra-cellular *Cys*/cystine REDOX potential and its modulation by diet. Thus, dietary sulfur amino acid intake can affect plasma REDOX status in the short term and that Zn^{2+} supplementation and a combination of vitamin C and E can protect against an oxidation over a 5-year period in older individuals. Also, systemic oxidation of thiol/disulfide couples could impact many systems because of the presence of REDOX-sensing thiols. However, cell proliferation and repair mechanisms may decline with age and oxidative stress conditions. Cells would respond to oxidized E_h through REDOX-sensing thiols with increased expression of pro-apoptotic machinery, shifting the balance of cell populations. Processes such as platelet activation, fibrotic signaling, and inflammatory signaling would be sensitized by REDOX-sensing thiols in the signal transduction pathways. Application of new technologies to define REDOX-sensing thiols will provide the basis for development of relevant REDOX interventions.

Bioactive Sulfur Compounds: The Nobel Prize in Chemistry 1955 was awarded to Vincent du Vigneaud for discovering biochemically relevant sulfur compounds, especially for the first synthesis of a polypeptide hormone.

2.7 REDOX: Other Reactive Species

During their highly charged motion path, ROS and RNS randomly hit various biological molecules, which sometimes trigger chain reactions generating other reactive species. For example, when ROS and RNS react with the unsaturated fatty acids of phospholipids, a cascade of lipid peroxidation reactions is initiated, which leads to the formation of a range of secondary breakdown products. While such lipid peroxidation contributes extensively to oxidative damage by perturbing biological membranes, there is also growing evidence that of these reactive breakdown species of lipid peroxidation could act as REDOX signals. Several other reactive species, including gaso-transmitters such as hydrogen sulfide (H_2S) and carbon monoxide (CO) are described below.

A) Hydrogen sulfide (H_2S)

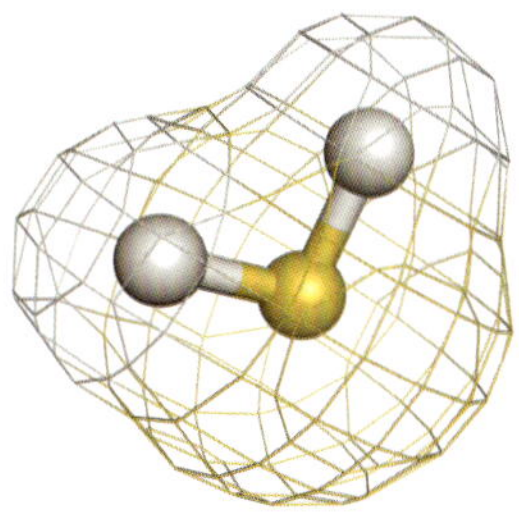

HYDROGEN SULPHIDE

Hydrogen sulfide (H_2S), also known as hydrosulfuric acid or sulfhydric acid, is a weak acid with a number of REDOX signaling functions. Hydrogen sulfide, is in gaseous form, and is produced in small amounts by some cells of the mammalian body. The gas is produced from *Cys* by the enzymes cystathionine β-synthase and cystathionine γ-lyase. It acts as a relaxant of smooth muscle, as a vasodilator, and is active in the brain (where it is involved in the formation of memory). Eventually the gas is converted to sulfite in the mitochondria by thiosulfate reductase, and the sulfite is further oxidized to thiosulfate and sulfate by sulfite oxidase. The sulfates are excreted in the urine.

Both NO and H_2S demonstrate strong vasodilatory effects, however, their mechanisms of action are different. While NO activates the enzyme guanylyl cyclase, H_2S activates ATP-sensitive potassium channels in smooth muscle cells. NO does most of the vessel-relaxing work in large vessels and H_2S is responsible for similar action in smaller blood vessels. The vasodilatory effects of NO and H_2S are inter- dependent, with a potential protective role in cardiovascular system.

Similar to NO, the involvement of H_2S in the relaxation of smooth muscle could induce erection of the penis, thus presents a possible new therapy opportunity for erectile dysfunction. In Alzheimer's disease the brain's H_2S concentration is severely decreased. In Parkinson's disease, the brain's H_2S concentration shows a decline, and H_2S administration alleviates the condition in animal models. Hydrogen sulfide is also involved in type 1 diabetes. The pancreatic beta cells produce excess of H_2S in type 1 diabetes which leads to cellular apoptosis (death); while reduced production of insulin by the residual active cells compounds the problem.

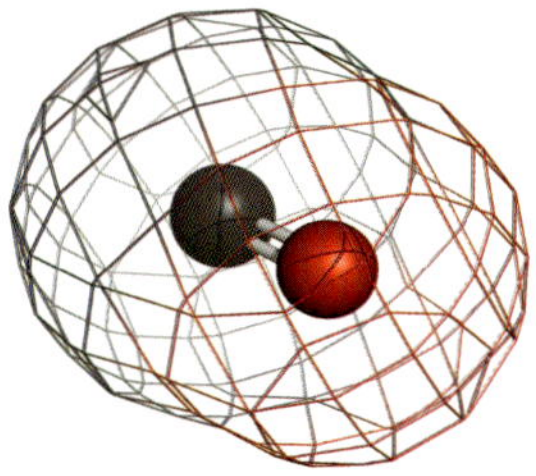

CARBON MONOXIDE

B) Carbon monoxide (CO)

Carbon monoxide consists of one carbon atom and one oxygen atom, connected by a triple bond. It is the simplest oxo-carbon, and isoelectronic with the cyanide ion and molecular nitrogen. In coordination complexes the carbon monoxide ligand is called carbonyl. Carbon monoxide is absorbed through breathing and enters the blood stream through gas exchange in the lungs. Normal circulating CO levels in the blood are 0% to 3%, and are higher in smokers.

Carbon monoxide is produced naturally by the human body by the action of heme oxygenase 1 and 2 on the heme from hemoglobin breakdown. This process produces a certain amount of carboxy-hemoglobin in normal individuals, even if they do not breathe any CO. Abnormal CO metabolism has been linked to neuro-degenerative disorders, hypertension, heart failure, and inflammation. Carbon monoxide is a REDOX signaling molecule, with a physiological role as a blood vessel relaxant.

Following the first report that CO is a neurotransmitter in 1993, as well as being one of three gases that naturally modulate inflammatory responses in the body (the other two being NO and H_2S), CO has received a great deal of clinical attention as a biological regulator. In many tissues, all three gases are known to act as anti-inflammatory agents, vasodilators, and promoters of neo-vascular growth. These functional properties have potential in the prevention of a series of pathological conditions including ischemia reperfusion injury, transplant rejection, atherosclerosis, severe sepsis, severe malaria, or autoimmunity. Clinical trials of small amounts of CO as a drug are ongoing.

C) Carbonate radical ($CO_3^{\bullet-}$) anion

The carbonate radical ($CO_3^{\bullet-}$) is believed to be an important intermediate oxidant derived from the oxidation of bicarbonate anions and nitroso-peroxo-carboxylate anions (formed in the reaction of CO_2 with ONO_2^-) in cellular environments as follows:

$$ONO_2^- + CO_2 \rightarrow NO_2\cdot + CO_3^{\bullet-}$$

Carbonate radical is also produced when ONO_2^- reacts with CO_2 and when Mn^{2+} reacts with H_2O_2 in the presence of HCO_3^-, and CO_2. It can be made in the laboratory by the decomposition of cobalt(III) carbonato complex, $[Co(NH_3)5CO_3]+$, induced by UV light. Also, $CO_3^{\bullet-}$ radicals are formed when ·OH reacts with bicarbonate ions, and it has only recently dawned on most of the free radical community that the CO_2/HCO_3 content of reaction systems is an important parameter in determining the molecular nature of oxidative damage. $CO_3^{\bullet-}$ radical is a potent one-electron oxidizing agent and oxidizes a range of biomolecules, including hyaluronic acid. For example, it can abstract H^+ from cysteine or NADPH and it can oxidize ascorbate, methionine, and guanine. $CO_3^{\bullet-}$ radical is also involved in the 'peroxidase' reactions of Cu/Zn-SOD.

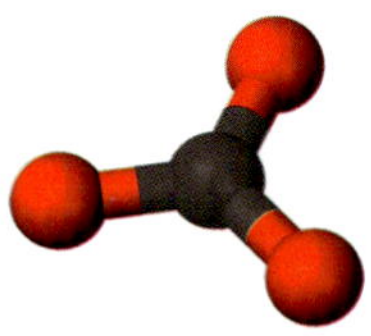

CARBONATE

D) Ozone (O_3)

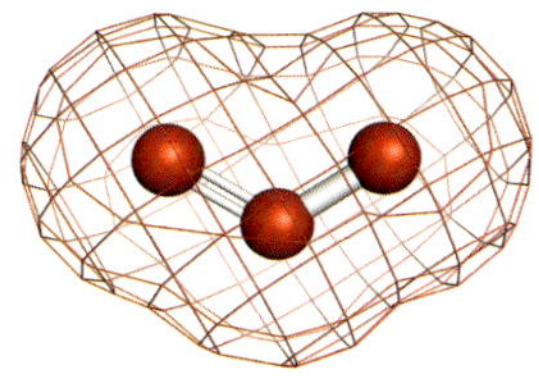

OZONE

Ozone (O_3), or trioxygen, is a tri-atomic molecule, consisting of three O_2 atoms. Ozone performs an essential 'antioxidant' function in the atmosphere by screening out UV light. Ozone is produced in the atmosphere by the photo-dissociation of O_2 into oxygen atoms, which then combine with O_2. In total, ozone makes up to 0.6 ppm of the atmosphere.

Ozone, along with ROS such as $O_2^{\bullet-}$, $^1O_2$, H_2O_2, and $HOCl^-$ ions, is naturally produced by white blood cells and other biological systems during host defense. Endogenously generated O_3 could cause ozonolysis of cholesterol in human atherosclerotic lesions. Ozone is also produced when phagocytes are stimulated. Also, when ozone breaks down to dioxygen it gives rise to oxygen free radicals, which are highly reactive and capable of damaging several organic molecules. During lipid interaction, O_3 directly attack fatty acids to generate ozonides, which gradually decompose to cytotoxic aldehydes. Ozone also oxidizes proteins, by attacking thiol (-SH) groups in tyrosine, tryptophan, histidine and methionine residues. Both lipids and proteins may be targets for O_3 attack in the respiratory tract. Ozone, as low as 0.5 ppm, can cause lung damage in a short time. Ozone induces inflammation, activates pulmonary macrophages, recruits neutrophils to the lung; ROS produced by these cells are an additional source of oxidative stress after O_3 exposure. Damage to lung macrophages can also decrease resistance to infections. Ozone irritates the eyes and can oxidize proteins (e.g. lysozyme) and lipids in tear fluids. Also, the powerful oxidizing properties of O_3 could induce inflammation and cause skin damage.

E) Singlet oxygen ($^1O_2$)

Singlet oxygen ($^1O_2$) is an electronically excited state of molecular oxygen (O_2), which is less stable than the normal triplet oxygen. Due to its unusual properties, $^1O_2$ can stay for over an hour at room temperature. Because of differences in their electron shells, singlet and triplet oxygen differ in their chemical properties. Singlet oxygen is highly reactive and one of the most active intermediates involved in chemical and biochemical reactions. Singlet oxygen is often generated by photo-sensitization reactions. If certain molecules are exposed to light at specific wavelength, the absorbed photo-energy activates the molecule to an excited state. The excitation energy transfers to an adjacent O_2 molecule, converting it to a single state whilst the photo-sensitizer returns to its ground state. Several bio-molecules act are sensitizers, including vitamin B2 (riboflavin) and its derivatives FMN and FAD, chlorophylls a and b, the bile pigment bilirubin, and various porphyrins. In mammalian biology, $^1O_2$ is one of the ROS, which is linked to oxidation of LDL cholesterol and cardiovascular effects. Polyphenol antioxidants can scavenge ROS and may prevent such deleterious oxidative effects. Ingestion of pigments capable of producing $^1O_2$ with activation by light can produce severe photo-sensitivity of skin. Singlet oxygen is the active species in photodynamic therapy, which is generated to kill cancer cells.

2.8 **REDOX**: Free Radical Formation

Molecular oxygen is present in all compartments of the human body. Activation of molecular oxygen to generate ROS may occur either by electron transfer (which is reduction of oxygen) or by energy transfer reactions (such as those that occur under the influence of light in the presence of appropriate photosensitizers). Singlet oxygen ($^1O_2$), a charge excited and non-radical form of molecular oxygen, is generated in the latter type of reaction; whereas the superoxide anion ($O_2^{\bullet-}$) results from monovalent reduction of oxygen. Molecular oxygen can also be chemically activated to turn into a ROS both as a by-product of metabolism (endogenous formation) and as a result of biochemical processes triggered by the interaction of human cells and tissues with exogenous stimuli.

Xenobiotic is a chemical which is not normally produced in the body, especially drugs such as antibiotics. A natural compound could turn into a xenobiotic if taken up by another organism; such as the uptake of natural human hormones by fish found downstream of human habitat. The term xenobiotic is also often used in the context of pollutants.

The interaction of oxygen with REDOX chains in the cellular compartments are characterized by intense electron flow, which may yield as a $O_2^{\bullet-}$ by-product. The respiratory (ETC) chain of the inner mitochondrial membrane is the most prominent REDOX chain of this category. Electron flow along the respiratory chain with reducing equivalents (such as NADH, $FADH_2$ and cytochromes) fuels the molecular oxygen to drive ATP synthesis. Electrons, instead of serving as substrates for the complete reduction of molecular oxygen to water, sometimes may reduce oxygen to $O_2^{\bullet-}$. An estimated 1% of the oxygen is reduced to $O_2^{\bullet-}$ and contributes to endogenous free radical formation. Other endogenous sources of $O_2^{\bullet-}$ production include enzymatic reactions such as the reduction of oxygen at the expense of NADPH as catalyzed by NADPH oxidase (NOx) enzymes or by formation of uric acid by xanthine oxidoreductase (XOR) catalysis of hypoxanthine and xanthine. Furthermore, several 'xenobiotics', if taken up by cells, may trigger REDOX cycling processes and generate $O_2^{\bullet-}$.

The interaction of ROS with biological molecules may initiate chain reactions involving the formation of secondary ROS. For example, ·OH radicals easily abstract H^+ atoms from hydroxyl groups and generate alkoxyl (RO·) radicals. Similarly, RO· radicals resulting from H^+ abstraction may combine with molecular oxygen to produce peroxyl (ROO·) radicals, which may react further to form hydroperoxides (ROOH). Peroxidation of poly unsaturated fatty acids will also result in the generation of reactive lipid peroxidation products that interact with other biomolecules such as proteins. For example, malondialdehyde may cross link proteins, and 4-hydroxynonenal (4-HNE) is an alkylating agent that could modify nucleophiles such as thiols of low molecular mass (eg. glutathione, GSH) or proteins.

In 1900, Moses Gomberg discovered an organic free radical and affirmed what had been thought impossible. A century later, free radical researchers look back to Gomberg as the founder of REDOX chemistry.

The Origin of 'Reactive' Species: There is no absolute answer to origin of free radicals in the Universe. Hypothetically, it all started with the 'Big Bang Theory', where the Universe expanded from an extremely hot, dense phase called the Planck epoch, in which all the matter and energy of the observable universe was concentrated. Since the Planck epoch theory, the Universe has been expanding. Recent observations indicate that this expansion is accelerating because of dark energy. Following the 'Big Bang', the origin of reactive species started with the formation of hydrogen and to a lesser extent, helium. Eventually, these gases coalesced under the force of their own gravity into stars. The more massive ones created, through fusion, the elements of carbon, nitrogen, and oxygen. With time, increased density due to the formation of the heavier elements caused the core of these stars to collapse. The release of this gravitational energy blew off the outer regions, erupting in a supernova, during which heavier atoms, such as iron, were produced in the process. Debris from these explosions sent vibrations and cosmic dust (free radicals) through the Universe, enriching inter-stellar space with matter for the next generation of stars. This cascade led, about 5 billion years ago, to the formation of our Solar system, leading to the creation of, and, with time, the diversity of many life forms - from the prebiotic assembly of proteins to single cell organisms to the evolution of humans. On a lighter note, the free radical chemistry in the cerebral hemispheres and the resulting REDOX potential has empowered humans to satisfy their trail of blazing questions with complex hypotheses.

2.8.1. Free Radical Formation: General Pathways

Free radicals are chemically generated in three following pathways:

i) Photon-induced Formation: This is the major and more significant way of free radical generation. Free radicals are highly charged chemical species in general, and require energy for their formation. The rich energy source for the free radical formation is sunlight – photons, in particular. The plant photosynthesis generates the most abundant free radicals – oxygen molecules via a variety of ROS intermediates. This way of free radical generation is crucial for the existence of life on earth. The sun light also causes diverse free radical formation in the atmosphere. It can split oxygen molecule (diatomic molecule) into 2 oxygen atoms (or 2 radicals) in the upper atmosphere. Each oxygen free radical atom in the atmosphere may combine with an oxygen molecule to form ozone (another highly energetic chemical species). Ozone gas derived from ROS forms a layer in the upper atmosphere, protecting life on the earth from the damages of UV light emitted from the Sun.

ii) Free Radical Chain Reactions: In the human body, free radicals are generated from other existing free radicals (the oxygen molecule). The exception is cancer radiation therapy, which also generates free radicals in the body. The oxygen molecule is converted to H_2O_2 in the electron transfer processes during ATP synthesis. The resulting H_2O_2 contains a highly energetic oxygen-oxygen bond, and may be further converted to free radicals by homolytic cleavage of the oxygen-oxygen bond, or by another electron transfer. Oxygen free radicals and H_2O_2 in the human body are called reactive oxygen species (ROS) in biology.

iii) Thermal Chemical Reactions: Free radicals can be generated at high temperature as it provides high energy. For example, when saturated alkanes are heated at high temperatures in an isolated vessel without oxygen, the carbon-carbon and/or carbon-hydrogen chemical bonds are broken and free radicals are generated. Free radicals are also rich as the reactive intermediates in combustion; however, mainly free radical chain reactions involve oxygen. Free radicals may be generated at low temperature or room temperature, only from chemicals that have a highly energetic chemical bond.

2.8.2. Free Radical Formation: Biological Systems

Free radicals are produced in biological systems in a number of ways:

i) Exposure to ionizing radiation is a major cause of free radical production. When irradiated water is ionized, and an electron is removed, what is left behind is an ionized water molecule. The damaging species resulting from the radiolysis of water are the free radicals ·H and ·OH and (hydrated electrons). They are highly reactive and have a lifetime of about 10^{-9} to 10^{-11} seconds. The hydroxyl radical is extremely reactive. Since water presents the largest number of target molecules in a cell, most of the energy transfer goes on in water when a cell is irradiated, rather than the solute consisting of protein, carbohydrate, nucleic acid, and bioinorganic molecules. Oxygen is an excellent electron acceptor and can combine with the hydrogen radical (·H) to form a peroxyl radical ($\cdot H + O_2 \rightarrow \cdot HO_2$). Hydrogen peroxide is toxic, and when present in sufficient quantities, can interfere with normal cellular metabolism.

ii) Enzymes and transport molecules generate free radicals as a normal consequence of their catalytic function. Examples of two enzymes which have been extensively studied in biological systems are xanthine oxidase (XO) and aldehyde oxidase. Both of these enzymes generate the superoxide anion radical ($O_2^{\bullet-}$) by adding a single electron to molecular oxygen. Other enzymes may use $O_2^{\bullet-}$ for their normal catalytic activity. The mitochondria of cells are the major source of endogenous free radical generation and are utilized in the synthesis of adenosine triphosphate (ATP) from adenosine diphosphate (ADP), the primary energy currency of the body. Thus, the mitochondrion serves as the powerhouse of the cell and contains most of the respiratory enzymes of the citric acid cycle.

iii) Auto-oxidation reactions produce free radicals from the spontaneous oxidation of biological molecules involved in non-enzymatic electron transfers. Although these reactions are a normal part of cellular metabolism, these free radicals may, under certain adverse conditions, achieve serious clinical significance. Examples of compounds that may be auto-oxidized in the body include thiols, hydroquines, catecholamines, flavins, ferredoxins, and hemoglobin. In all these auto-oxidation reactions, $O_2^{\bullet-}$ is the main free radical species that is produced initially. The processes involved in REDOX reactions are of immense biochemical importance since the transfer of electrons is the means by which the body derives most of its free energy. In oxidation, electrons are lost; in reduction, electrons are gained.

iv) Toxic metals may produce free radicals in the body. The metals (copper, iron, cadmium, arsenic, mercury, chromium, antimony, beryllium, thallium, silver, and nickel) are believed to derive their toxic effects from their inherent ability to transfer electrons, which is also an expression of their capability to generate free radicals. Transition metals (scandium, titanium, vanadium, chromium, manganese, iron, cobalt, nickel, copper, and zinc) usually promote free radical reactions. These free radicals can adversely affect cellular health by producing lipid peroxidation of intra-cellular membranes and cross linkages of membrane macromolecules.

Heavy-metal free radicals have a tendency to form covalent bonds with sulfhydral groups. In this manner they are able to modify the functions of many enzymes, not to mention non-enzymatic antioxidant compounds (such as glutathione) which depend on these groups for their biological activity. Toxic metals are known to affect cell membrane permeability, sub-cellular organelles, and the structure and function of proteins and nucleic acid. Toxic metals may affect the biosynthetic formation of hormones and depress enzymatic and other metabolic processes. These metals may also stimulate metabolic functions that lead to free radical production and carcinogenesis. Heavy metals have been found to have a deleterious effect on the immune system and contribute to atherogenesis.

2.8.3. Free Radical Formation: The Human Body

In the human body, free radicals are generated by two principle sources: both endogenous and exogenous.

A) Endogenous sources: free radicals are generated intra-cellularly (acting within the cell) and may also be released into the surrounding area. These intra-cellular free radicals result from auto-oxidation and consequent inactivation of small molecules such as reduced thiols and flavins. Paradoxically, antioxidants may also produce free radicals. A wide range of free radical molecular species are endogenous. The singlet oxygen is not a free radical, but is nevertheless a reactive oxygen species and capable of causing tissue damage.

i) The primary endogenous source of free radicals is during the energy production process within the cells. During respiration, about 98% of the oxygen is used within our cells in the production of energy, however, a very small amount (2%) of O_2 that was not used, ends up as loss of electrons and creates free radicals.

ii) Another important endogenous source that creates lots of free radicals is the presence of toxic heavy metals in the body. Electron transfer from metals such as iron to oxygen-containing molecules can also initiate free radical reactions. All transition metals with the exception of copper, contains one electron in their outermost shell and can be considered a free radical. Copper, although having a full outer shell of 2 electrons, loses its electron easily and can also be con sidered a free radical. The following metals are major free radicals creating agents in the body: mercury (Hg), chromium (Cr); aluminium (Al), lead (Pb), iron (Fe), and cadmium (Cd).

iii) Certain foods that we consume may generate free radicals. An example would be fried foods in high temperature, which causes the bonds in the food molecules become unstable.

iv) During normal metabolism of fat, carbohydrates, and proteins, free radicals are produced.

v) Oxidative stress: REDOX enzymes and certain endocrine hormones create free radicals. Oxidative stress may also occur as a result of the activity of certain oxidases, lipoxygenases, cyclo-oxygenases, dehydrogenases and peroxidases.

vi) Low blood supply, which generally occurs during heart attacks and strokes is known to form free radicals.

B) Exogenous sources: free radicals production is stimulated by various external factors including irradiation, chemical pollutants, and some medications such cancer chemotherapeutic agents. External stimulants of free radical formation comprise environmental exposure to carcinogens. Other exogenous sources for free radicals include:

i) Environmental contaminants such as air and water pollution. Cigarette smoke tremendously increases free radical load.

ii) The exogenous sources of free radicals resulting from ionizing and ultraviolet radiation such as from overexposure to sunlight and prolonged exposure to x-rays ionizing radiation play a major role in free radical production. The energy transferred into water from ionizing particles charges the water molecule. The water ions thus produced dissociate and yield free radicals.

iii) Excessive or over-exercising increases metabolic energy demand and generates free radicals.

iv) Toxic chemicals such as pesticides, herbicides, asbestos are known free radical generators.

v) Petroleum based products and chemicals found in furniture polish and paints such toluene, benzene and formaldehyde are strong pro-oxidants.

vi) Prescription drugs, especially, anti-inflammatory agents and diuretics produce free radicals.

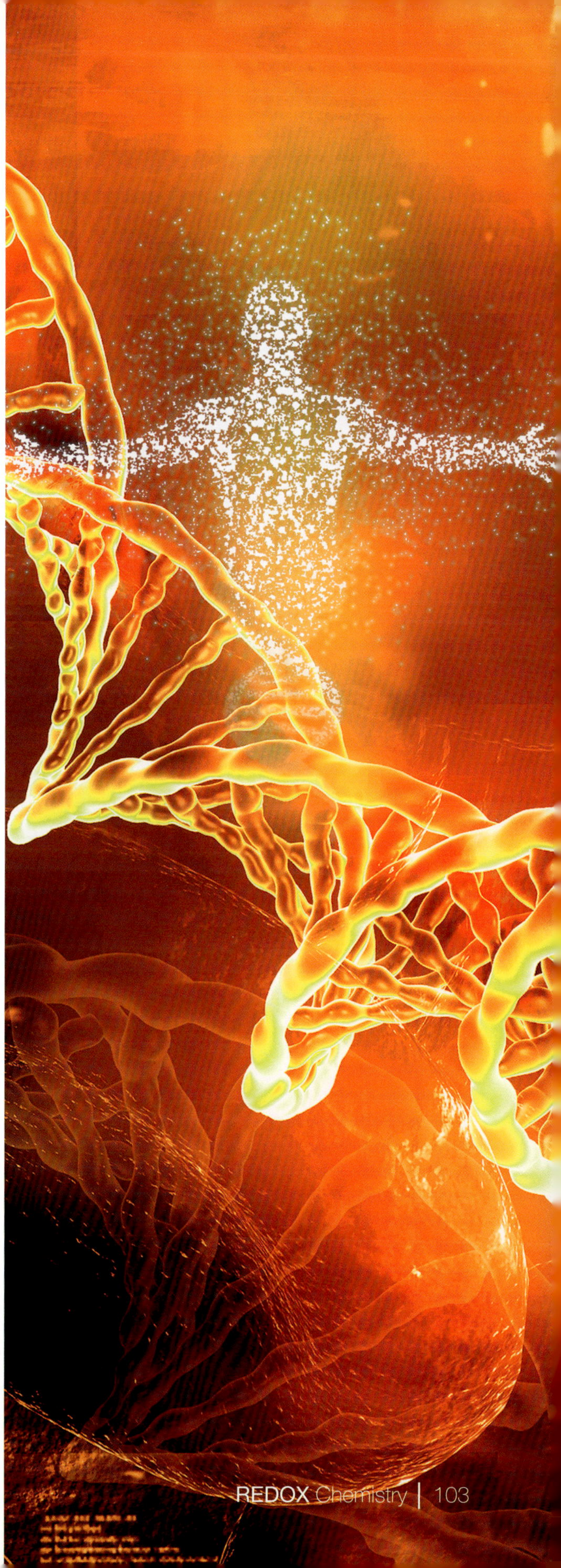

2.9 Free Radicals: **Bio-Molecular Targets**

Free radicals are charged ions with high motility, extreme agility, and instant reactivity. These reactive species hit molecules that come in their motion path and result in two types of physico-chemical outcomes: they either form new chemical bonds or breakdown an existing bond. Such outcomes result either from donating an electron (reduction) or pulling out an electron (oxidation) from the molecular target. This REDOX reaction can put an inert molecule into motion (to induce a cell signal); allow a chemical to form complex with another complementary molecule to perform a function such as transport or enzyme hydrolysis; activate a compound to capture energy for storage or release energy to perform work; or destroy a susceptible molecular target and induce chemolysis (i.e. phagocytic respiratory burst during innate host defense). The free radical interaction with biochemicals is undoubtedly, an extraordinary process that set the pace for evolution of life on this Blue Planet.

In the human body, cellular DNA is constantly attacked by free radicals (both ROS and RNS). In mammalian cells, ~1.5×10^5 oxidative adducts in DNA per cell are found. Such hits may induce mutations and play a role in the evolution process. Moderate levels of DNA damage triggers cell cycle arrest and initiates DNA-repair processes that ensure DNA integrity. In contrast, excessive damage or failure in DNA repair can induce apoptosis. Oxidative modifications of lipids, proteins, or DNA play a crucial role in physiologic processes such as differentiation, maturation, and trafficking of intra-cellular vesicles. However, when the ROS/RNS levels are in excess, the biologic consequences are often deleterious. Therefore, regulation of ROS/RNS levels is critical in maintaining cellular homeostasis.

The continuous efflux of free radicals from endogenous and exogenous sources results in both non-stop and accumulative oxidative damage to cell components. Such damage may alter several cellular functions and trigger cell signals. Among the susceptible biological targets are proteins (enzymes, in particular), lipids (most commonly the cell membranes), and nucleic acids (i.e. DNA). In general, free radicals interact with molecular targets in the cellular locations or milieu where they are produced. For example, hydroxyl (·OH) radicals, produced during cellular respiration in mitochondrial compartments, are responsible for damaging that very mitochondrion but not the nucleus. Due to high reactivity with bio-molecules, free radicals have an extremely short life span. This limitation restricts their ability to damage long distance cellular targets. On the other hand, oxygen metabolites that are less reactive, such as $O_2^{\bullet -}$ anion, may exist in the intra-cellular environment for longer periods of time and reach locations far from their site of generation. For example, H_2O_2 produced in mitochondria may interact elsewhere in the cytoplasm or in the nucleolus.

Free radicals can elicit varying functional outcomes based on their site of generation, pathways of chain reaction and proximity to susceptible target. These effects include: cell signaling, cellular apoptosis, or a simple denaturation and inactivation of a biochemical target.

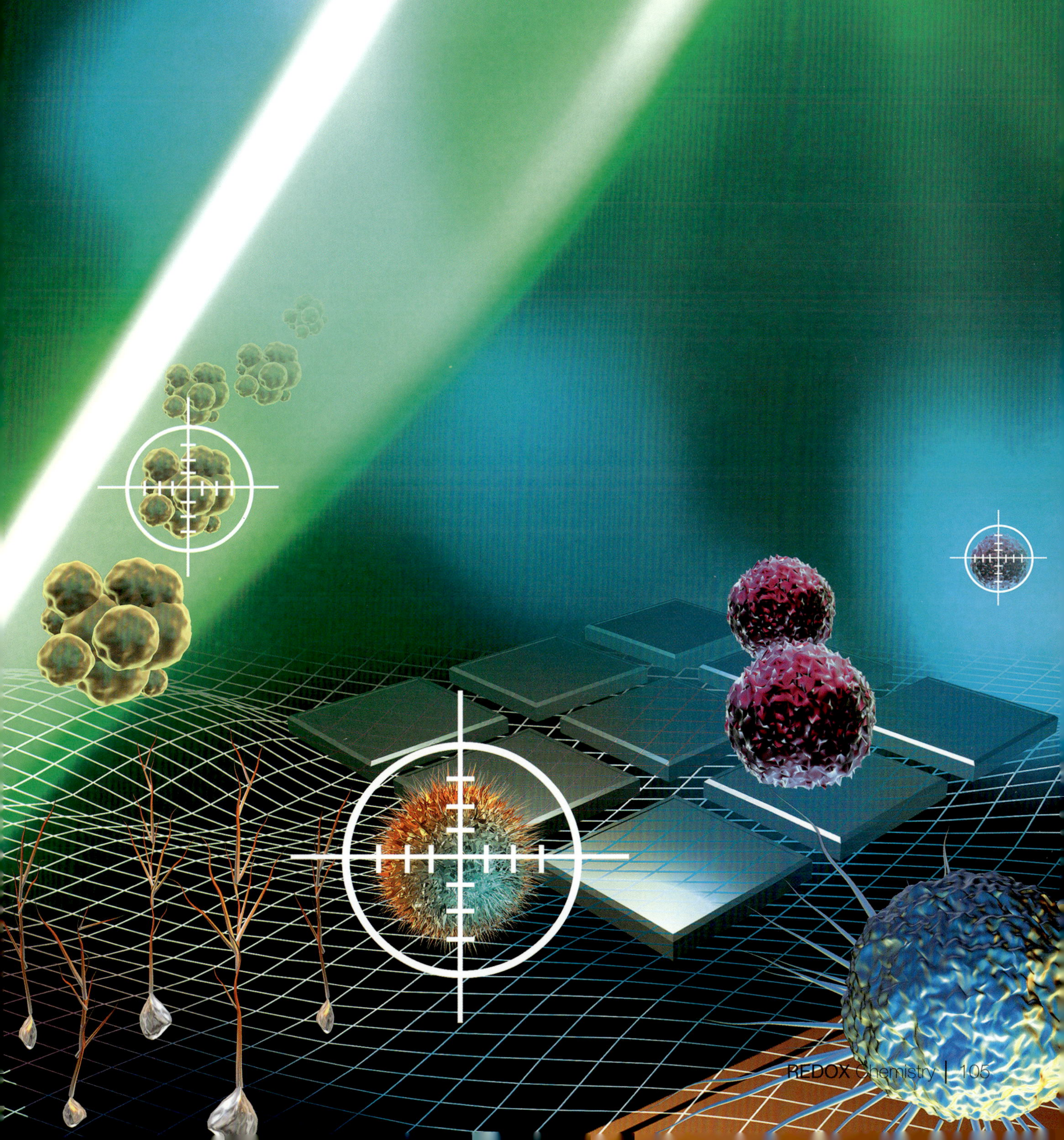

2.9.1. Free Radical Target: Lipids

Lipids are most susceptible to oxidative modification. Lipid peroxidation generates lipid radicals, which can further attack the subsequent lipid molecules and propagate as a chain reaction. Cellular membranes are especially vulnerable to oxidation due to their high concentrations of unsaturated fatty acids. The damage to lipids (lipid peroxidation) occurs in three stages:

i) Initiation stage: involves ROS attack that pulls out a hydrogen atom from a methylene group in the lipid. The double bond adjacent the methylene group weakens and can easily be removed from the molecule. The resulting fatty acid radical retains 1 electron and is stabilized by structural rearrangement forming a conjugated diene.

ii) Propagation stage: when oxygen is sufficient in the milieu, the fatty acid radical reacts with O_2 to form ROO·. These radicals are capable of pulling out another hydrogen atom from a neighboring fatty acid molecule, which leads again to the production of fatty acid radicals that undergo the same reactions – rearrangement and interaction with oxygen. The ROO· becomes a lipid-H_2O_2 that can further decompose to an aldehyde or form cyclic endoperoxide and hydrocarbons. The propagation stage allows the reaction to continue; a single initiation can lead to a chain reaction resulting in peroxidation of all unsaturated lipid in the membrane. An antioxidant that can stop this process is therefore defined as a chain-breaking antioxidant. Fatty acids with no double bonds or with a single double bond can undergo oxidation but not a chain lipid peroxidation process; for example, oleic acid with 18 carbon atoms and 1 double bond (18:1) cannot undergo the lipid peroxidation process.

iii) Chain termination stage: this final step occurs following interaction of one ROO· with another radical or antioxidant. Polyunsaturated fatty acid residues of phospholipids are attacked by a radical either at an internal position or near the end of the conjugated system, generating a peroxyl radical. Attack at an internal position allows the peroxyl radical to further undergo either a cyclization or metal-catalyzed reaction and produce reactive alkoxyl radicals.

After cyclization, the fatty acid may form a hydroperoxide or undergo another cyclization, which produces aldehydes, including malondialdehyde (MDA) and 4-hydroxy-2-nonenal (HNE). While MDA can further react with DNA bases, resulting in gene mutations, HNE reacts mostly with proteins, leading to significant functional alterations affecting signaling pathways. Oxidative damage of phospholipids can lead to cell death, not only through membrane damage, but also through the lipid peroxidation product HNE. Attack of various proteins where found to be a mechanism of cell death induced by lipid peroxidation.

2.9.2. Free Radical Target: Proteins

Proteins are major constituents of enzymes and membrane transport systems. These compounds serve as vulnerable targets for free radical attack, especially radicals from reactive oxygen species (ROS), reactive nitrogen species (RNS), and RO·. Hydrogen peroxide itself and $O_2^{\bullet-}$ radicals (in physiological concentrations) exert weak effects on proteins; however, protein targets containing -SH groups (cysteine and methionine residues), can undergo oxidation when encountered with oxidative radicals such as H_2O_2, hypochlorous acid and hypochlorite ions. Sulfur-containing amino acids such as cysteine and methionine are also susceptible to either reversible or irreversible oxidation. Reversible oxidation of the sulfhydryl (-SH) group includes intra-molecular or inter-molecular protein cross-linkages and glutathionylation. Irreversible protein oxidation includes nitrosylation of cysteine -SH groups, tyrosine, methionine, and tryptophan by ONO_2^-. Nitration of tyrosine residues may inhibit phosphorylation or adenylation, important for protein function.

Proteins can undergo direct and indirect damage following interaction with ROS, including peroxidation, damage to specific amino-acid residues, alterations to their tertiary structure, degradation, and fragmentation. The consequences of protein damage as a response mechanism to stress are loss of enzymatic activity, altered cellular functions such as energy production, interference with the creation of membrane potentials, and changes in the type and level of cellular proteins.

Protein oxidation products are usually aldehydes, ketones, and carbonyls. Since several mechanisms of protein oxidation can lead to formation of protein carbonyls (which are easily detectable), the level of protein carbonyls has been used as a quantitative marker of protein oxidation and oxidative stress. Also, 3-nitro-tyrosine is another major adducts that can easily be detected; therefore, it can serve as a marker for protein oxidative damage. This adduct is produced following the interaction between perdoxynitrate (ONO_2^-) and other RNS with the amino acid tyrosine. Following ·OH attack, a series of compounds can be formed, including hydroxyproline, glutamyl semialdehyde, and others.

Protein oxidation can be reversible or irreversible, depending on the target and the form of oxidative damage. The highly reactive ·OH, generated through ionizing radiation or the Fenton reaction, and ONO_2^- are common reactive species that target proteins. Although all amino acid residues could be oxidized by ROS/RNS, certain side chains are particularly susceptible to oxidation. For instance, lysine, arginine, histidine, proline, and threonine are highly sensitive to metal-catalyzed oxidation. Oxidation of these side chains results in carbonyl derivatives, which can also be generated through glycation/glycoxidation pathways, lipid peroxidation, α-amidation, and glutamate pathways. Severe oxidative stress can induce disulfide bond–mediated protein cross-linkage or secondary oxidative modifications such as adduct formation between oxidized proteins and lipid peroxides or glycation products. These products can generate aggregation of bulky protein complexes, leading to accumulation of damaged proteins and cell death.

2.9.3. Free Radical Target: Nucleic Acids (DNA)

Although DNA is a stable and well-protected molecule, ROS can interact with nucleotides and cause several types of damage including: (i) modification of DNA bases, (ii) single- and double-DNA breaks, (iii) loss of purines, (iv) damage to the deoxyribose sugar, (v) DNA-protein cross-linkages, and (vi) damage to the DNA repair system. Not all ROS can cause damage, but many have the potential to, and the most destructive radical is the hydroxyl (·OH) species. For example, an ·OH exposure to DNA, similar to induction by ionizing irradiation, triggers formation of a variety of adducts. The ·OH can attack guanine at its C-8 position to yield an oxidation product. Other positions are also vulnerable for ·OH attack and other damaged DNA fragments could be formed. Hydroxyl radicals can also attack other bases like adenine to yield 8-(OH)-adenine. Other products are the result of interactions between pyrimidines and ·OH radicals leading to the formation of thymine peroxide, thymine glycols, 5- (hydroxyl-methyl) uracyl (in case of ribonucleic acid), and other such degraded products. The direct interaction of DNA with other less reactive ROS, such as $O_2^{\bullet-}$ and H_2O_2, does not lead to damage at their physiological concentrations; however, these oxidative species serve as sources for other reactive intermediates that can easily attack nucleic acids and cause damage. For example, H_2O_2 and $O_2^{\bullet-}$could lead to the production of the ·OH via the Haber-Weiss reaction, and NO and $O_2^{\bullet-}$ might lead to the formation of peroxynitrite (ONO_2^-), which can easily cause DNA damage similar to ·OH radical attack. Transition metals such as iron with high-binding affinity to DNA sites can catalyze the production of ·OH in close proximity to the DNA molecule, thus indulging in repeated attacks on the DNA through an effux of highly powerful oxidative radicals.

Compared with lipids and proteins, nuclear DNA may seem less susceptible to oxidative modifications because of its double helix structure and the protective shield from histone and other coating proteins. However, oxidative nuclear DNA damage is detectable under various conditions. Thus, oxidized DNA base intermediates such as 8-OHdG are used as markers for measuring oxidative stress-induced damage. A correlation between oxidative DNA damage and various stages of carcinogenesis has also been reported. DNA can be damaged in nearly all of its components. Both purine and pyrimidine bases and the sugar backbone contain nitrogen (N) and oxygen (O) as nucleophilic centers; both of which are highly susceptible to electrophlic ·OH attack. Furthermore, the double bonds between the bases are prime targets for ·OH radicals. Reactions are primarily centered at the C-5 and C-6 of pyrimidines and C-4 and C-8 of purines. Nitric oxide and ONO_2^- could react with DNA bases and induce single-strand breaks. Oxidative damage to the sugar backbone, through H abstraction, has been known to cause single-strand breaks and double-strand breaks. Unlike nuclear DNA, mitochondrial DNA (mt-DNA) is more susceptible to oxidative damage, not only because of its close proximity to the major site of ROS generation (electron transport chain), but also due to the limited capacity of mtDNA repair.

2.10 | REDOX Chemistry: **Bibliography**

2.1 | REDOX and the Biosphere

1. Vernadsky VI (1998) *Biosphere: Complete Annotated Edition.* New York: Springer Verlag. ISBN: 038798268X.

2.1.1. The Carbon Cycle

1. Falkowski P, Scholes RJ, Boyle E, Canadell J, et al. (2000) The global carbon cycle: A test of our knowledge of Earth as a system. *Science* 290 (5490):291-296.
2. Crowley TJ (2000) Causes of climate change over the past 1000 years. *Science* 289 (5477):270-277.
3. Lal R (2008) Sequestration of atmospheric CO2 in global carbon pools. *Energy & Environmental Science* 1:86-100.
4. Kleypas JA, Buddemeier RW, Archer D, Gattuso JP, et al (1999) Geochemical consequences of increased atmospheric carbon dioxide on coral Reefs. *Science* 284 (5411):118-120.

2.1.2. The Nitrogen Cycle

1. Schuur EAG (2011) Ecology: Nitrogen from the deep. *Nature* 477 (7362):39-40.
2. Morford SL, Houlton BZ., Dahlgren RA (2011) Increased forest ecosystem carbon and nitrogen storage from nitrogen rich bedrock. *Nature* 477 (7362):78-81.
3. Moir JWB (Editor) (2011) Nitrogen Cycling in Bacteria: Molecular Analysis.New York:Acaemic Press. ISBN: 978-1-904455-86-8.
4. Vitousek PM, Aber J, Howarth RW, Likens GE, et al (1997) Human alteration of the global nitrogen cycle: Sources and consequences. *Issues in Ecology* 1:1-17.

2.1.3. The Water (Hydrogen) Cycle

1. Bamford VA, Bruno S, Rasmussen T, Appi-Ayme C, et al (2002) Structural basis for the oxidation of thiosulfate by a sulfur cycle enzyme: *European Molecular Biology Organization* 21(21):5599-5610
2. Canfield DE, Raiswell R (1999) The evolution of the sulfur cycle. *American Journal of Science* 299:697-723
3. Machel HG (1998) Bacterial and thermochemical sulfate reduction in diagenetic settings – old and new insights. *Sedimentary Geology* 140:143-175.
4. Schlesinger WH (1997) Biogeochemistry an analysis of global change (2nd ed). New York: Academic Press. ISBN:978-0126251555

2.1.4. The Sulfur Cycle

1. Jouzel J, Masson-Delmotte V, Cattani O, Dreyfus G, et al (2007) Orbital and millennial Antarctic climate variability over the past 800,000 Years. *Science* 317(5839):793-796.
2. Shiklomanov I (1993) World fresh water resources, In: *Water in Crisis: A Guide to the World's Fresh Water Resources.* PH Gleick (ed). New York: Oxford University Press. ISBN: 978-0195076288
3. Harman R (2005) *The Water Cycle: Evaporation, Condensation & Erosion (Earth's Processes).* New York:Heineman. ISBN: 978-1403470669.

2.2 | REDOX Chemistry: Oxygen

1. Balentine JD (1982) *Pathology of Oxygen Toxicity.* New York: Academic Press. ISBN-10: 0120770806
2. Falkowski PG, Godfrey LV (2008) Electrons, life and the evolution of Earth's oxygen cycle. *Philosophical Transactions of the Royal Society* B 363:2705-2716.
3. Farquhar J, Bao H, Thiemens M (2000) Atmospheric influence of Earth's earliest sulfur cycle. *Science* 289:756-758.
4. Gitto E, Reiter RJ, Karbownik M, Xian-Tan D, Barbereri I (2001) Respiratory distress syndrome in the newborn: Role of oxidative stress. *Intensive Care Medicine* 27:1116-1123.

5. Lane N (2003) *Oxygen: The Molecule that made the World.* Oxford: Oxford University Press. ISBN: 0198-607830.
6. Raymond J, Segre D (2006) The effect of oxygen on biochemical networks and the evolution of com plex life. *Science* 311:1764–1767.
7. Sawyer DT, Williams RJP (1991) *Oxygen Chemistry.* New York:Oxford University Press. ISBN: 978-0195057980.
8. Valentine JS, Foote CS, Greenberg A, Liebman JF (1995) *Active Oxygen in Biochemistry.* London: Chapman & Hall. ISBN-10: 0751402931.

2.3 | REDOX Reactions

2.3.1. REDOX Mechanisms

1. Phillips J, Strozak V, Wistrom C (2000) *Chemistry: Concepts and Applications.* Glencoe McGraw-Hill. ISBN:978-0-02-828210-7.
2. Rodgers G (2012) *Descriptive Inorganic, Coordination, and Solid-state Chemistry.* Brooks/ Cole:Cengage Learning. ISBN: 978-0-8400-6846-0.
3. Schüring J, Schulz HD, Fischer WR, Böttcher J, Duijnisveld WH, eds. (1999) *Redox: Fundamentals, Processes and Applications.* Heidelberg: Springer-Verlag, Heidelberg. ISBN: 978-3-540-66528-1.
4. Tratnyek PG, Grundl TJ, Haderlein SB, eds. (2011) *Aquatic Redox Chemistry.* ACS Symposium Series 1071. ISBN: 978-0-841-22652-4.
5. Zumdahl S, Zumdahl S (2009) *Chemistry.* Houghton Mifflin. ISBN: 978-0-547-05405-6.

2.3.2. REDOX Potential

1. Benitez-Alfonso Y, Jackson D, Maule A (2011) Redox regulation of intercellular transport. *Protoplasma* 248(1):131-140.
2. Chiu J, Dawes IW (2012) Redox control of cell proliferation. *Trends in Cellular Biology* 22(11):592-601.
3. Droge W (2002) Free radicals in the physiological control of cell function. *Physiological Reviews* 82:47-95.
4. Genestra M (2007) Oxyl radicals, redox-sensitive signalling cascades and antioxidants. *Cell Signaling* 19(9):1807-1819.
5. Go YM, Jones DP (2008) Redox compartmentalization in eukaryotic cells. *Biochimica et Biophysica Acta* 1780(11):1273-1290.
6. López-Mirabal HR, Winther JR (2007) Redox characteristics of the eukaryotic cytosol. *Biochimica et Biophysica Acta* 1783(4):629-640.
7. Ottaviano FG, Handy DE, Loscalzo J (2008) Redox regulation in the extracellular environment. *Circulation Journal* 72(1):1-16.
8. Sarsour EH, Kumar MG, Chaudhuri L, Kalen AL, Goswami PC (2009) Redox control of the cell cycle in health and disease. *Antioxidants & Redox Signaling* 11(12):2985-3011.
9. Song MY, Makino A, Yuan JX (2011) Role of reactive oxygen species and redox in regulating the function of transient receptor potential channels. *Antioxidants & Redox Signaling* 15(6):1549-1565.

2.3.3. REDOX: Metal Catalysts

1. Fenton HJH (1894) Oxidation of tartaric acid in presence of iron. *Journal of Chemical Society Transactions* 65 (65):899-911.
2. Finney LA, O'Halloran TV (2003) Transition metal speciation in the cell: Insights from the chemistry of metal ion receptors. *Science* 300:931-936.
3. Frausto da Silva JJR, Williams RJP (2009) *The Biological Chemistry of the Elements – The Inorganic Chemistry of Life.* Oxford: Clarendon Press. ISBN 978-0-19-850848-9.
4. Glusker JP (1991) Structural aspects of metal liganding to functional groups in proteins. *Advances in Protein Chemistry* 42:1-73.
5. Goldstein S, Meyerstein D, Czapski G (1993) The Fenton reagents. *Free Radical Biology & Medicine* 15(4):435-445.
6. Goyer RA (1997). Toxic and essential metal interactions. *Annual Reviews in Nutrition* 17:37-50.
7. Katz AR, Glusker JP (1998) Roles of zinc and magnesium ions in enzymes. *Advances in Molecular Structure Research* 4:227-279.
8. Mildvan AS (1970) Metals in enzymes catalysis, In: *The Enzymes,* Vol. 2, PD Boyer (ed.), pp. 446-536. New York: Academic Press.

2.3.4. REDOX: Type of Radicals

1. Pauling L (1979) The discovery of the superoxide radical. *Trends in Biochemical Sciences* 4:N270–N271.
2. Sawyer DT, Valentine JS (1981) How super is superoxide? *Accounts of Chemical Research* 14:393-400.

2.4 | REDOX: Reactive Oxygen Species (ROS)

1. Babior BM, Kipnes RS, Curnutte JT (1973) Biological defense mechanisms. The production by leuko cytes of superoxide, a potential bactericidal agent. *Journal of Clinical Investigation* 52:741-744.
2. Babior BM, Lambeth JD, Nauseef W (2002) The neutrophil NADPH oxidase. *Archives in Biochemistry and Biophysics* 397:342-344.
3. Bielski BHJ, Cabelli DE, Arudi RL, Ross AB (1985) Reactivity of hydroperoxyl/superoxide radicals in aqueous solution. *Journal of Physical & Chemical Reference Data* 14:1041-1100.
4. Boveris A, Oshino N, Chance B (1972) The cellular production of hydrogen peroxide. *Biochemistry Journal* 128:617-630.
5. Burdon RH, Rice-Evans C (1989) Free radicals and the regulation of mammalian cell proliferation. *Free Radical Research Communications* 6:345-358.
6. Buxton GV, Greenstock CL, Helman WP, Ross AB (1988) Critical review of rate constants for reactions hydrated electrons, hydrogen atoms and hydroxyl radicals in aqueous solution. *Journal of Physical & Chemical Reference Data* 7:513–886.
7. Commoner B, Townsend J, Pake GE (1954) Free radicals in biological materials. *Nature* 174:689-691.
8. Czech MP, Lawrence JC, Lynn W (1974) Evidence for the involvement of sulfhydryl oxidation in the regulation of fat cell hexose transport by insulin. *Proceedings of the National Academy of Sciences USA* 71:4173-4177.
9. Fridovich I (1995) Superoxide radical and superoxide dismutases. *Annual Review of Biochemistry* 64:97-112.
10. Hoffman MZ, Bolletta F, Moggi L, Hug GL (1989) Rate constants for the quenching of excited states of metal complexes in fluid solution. *Journal of Physical & Chemical Reference Data* 18:219–543.
11. Jeffree GM (1958) Hydrogen peroxide and cancer. *Nature* 182:892.
12. McCord JM, Fridovich I (1968) The reduction of cytochrome c by milk xanthine oxidase. *Journal of Biological Chemistry* 243:5753–5760.
13. McCord JM, Fridovich I (1969) Superoxide dismutase: An enzymatic function for erythrocuprein (hemocuprein). *Journal of Biological Chemistry* 244:6049-6055.
14. Oberley LW, Buettner GR (1979) The role of superoxide dismutase in cancer: A review. *Cancer Research* 39:1141-1149.
15. Plaine HL (1955) The effect of oxygen and hydrogen peroxide on the action of a specific gene and on tumour induction in drosophila melanogaster. *Genetics* 40:268-280.
16. St-Pierre J, Buckingham JA, Roebuck SJ, Brand MD (2002) Topology of superoxide production from different sites in the mitochondrial electron transport chain. *Journal of Biological Chemistry* 277:44784–44790.
17. Wardman P (1989) Reduction potentials of one-electron couples involving free radicals in aqueous solution. *Journal of Physical & Chemical Reference Data* 18:1637-1755.

2.4.1. Superoxide ($O_2^{\bullet-}$)

1. Afanasev IB (1989) *Superoxide Ion Chemistry and Biological Implications.* Boca Raton:CRC Press. ISBN: 978-0849354519.
2. Bielski BHJ, Cabelli BH, Arudi RL, Ross AB (1985) Reactivity of RO2/O2 radicals in aqueous solution. *Journal of Physical & Chemical Reference Data* 14:1041-1100.
3. Bielski BHJ, Cabelli DE (1995) Superoxide and hydroxyl radical chemistry in aqueous solution. In: *Active Oxygen in Chemistry,* Foote CS, Valentine JS, Greenberg A, Liebman JF (eds). Pp. 66-104. London: Chapman and Hall.
4. Schafer FQ, BuettnerGR(2001). Redox environments of the cell as viewed through the redox state of the glutathione disulfide/glutathione couple. *Free Radical Biology & Medicine* 30:1191-1212.

2.4.2. Hydroxyl Radical (·OH)

1. HalliwellB, KaurH (1997) Hydroxylation of salicylate and phenylalanin as assays for hydroxyl radicals: A cautionary note visited for the third time. *Free Radical Research* 27:239-244.
2. Heal MR, Heard DE, Pilling MJ, Whitaker BJ (1995) On the development and validation of FAGE for local measurement of tropospheric HO and HO2. *Journal of Atmospheric Sciences* 52(19):3428-3448.
3. Isaksen ISA, Dalsøren SB (2011) Getting a better estimate of an atmospheric radical. *Science* 331(6013):38-39.
4. Montzka SA, Krol M, Dlugokencky E, Hall B, et al (2011) Small inter-annual variability of global atmo spheric hydroxyl. *Science* 331(6013):67-69.
5. Reiter RJ, Carneiro RC, Oh CS (1997) Melatonin in relation to cellular antioxidative defense mechanisms. *Hormone & Metabolic Research* 29(8):363-372.

2.4.3. Hydrogen Peroxide (H_2O_2)

1. Giorgio M, Trinei M, Migliaccio E, Pelicci PG (2007) Hydrogen peroxide: a metabolic by-product or a common mediator of ageing signals?". *Nature Reviews of Molecular & Cell Biology* 8(9):722-728.
2. López-Lázaro M (2007) Dual role of hydrogen peroxide in cancer: possible relevance to cancer chemo prevention and therapy. *Cancer Letters* 252:1-8.
3. Milani M, Bigardi A, Zavattarelli M (2003) Efficacy and safety of stabilized hydrogen peroxide cream (Crystacide) in mild-to-moderate acne vulgaris: A randomized, controlled trial versus benzoyl peroxide gel.*Current Medical Research & Opinion* 19(2):135-138.
4. Niethammer P, Clemens GA,. Look T, Mitchison TJ (2009) A tissue-scale gradient of hydrogen peroxide mediates rapid wound detection in zebra fish. *Nature* 459 (7249):996-999.
5. Rojkind M, Dominguez-Rosales JA, Nieto N, Greenwel P (2002) Role of hydrogen peroxide and oxidative stress in healing responses. *Cellular & Molecular Life Sciences* 59 (11):1872-1891.
6. Veal EA, Day AM, Morgan BA (2007) Hydrogen peroxide sensing and signaling. *Molecular Cell* 26:1-14.
7. Wilgus TA, Bergdall VK, Dipietro LA, Oberyszyn TM (2005). Hydrogen peroxide disrupts scarless fetal wound repair. *Wound Repair & Regeneration* 13 (5):513-519.

2.4.4. Hypochlorous Acid (HOCl)

1. Barrette WC, Hannum DM, Wheeler WD, Hurst, JK (1989) General mechanism for the bacterial toxicity of hypochlorous acid: abolition of ATP production. *Biochemistry* 28(23):9172-9178.
2. Brinkmann V, Reichard U, Goosmann C, et al (2004) Neutrophil extracellular traps kill bacteria. *Science* 303(5663):1532-1535.
3. Hurst JK, Barrette WC, Michel BR, Rosen H (1991) Hypochlorous acid and myeloperoxidase-catalyzed oxidation of iron-sulfur clusters in bacterial respiratory dehydrogenases. *European Journal of Biochemistry / FEBS 202(3):1275-1282.*
4. Klebanoff SJ (1970) Myeloperoxidase: contribution to the microbicidal activity of intact leukocytes. *Science* 169(950):1095-1097.
5. Knox WE, Stumpf PK, Green DE, Auerbach VH (1948). The inhibition of sulfhydryl enzymes as the basis of the bactericidal action of chlorine. *Journal of Bacteriology* 55(4):451-458.
6. Lapenna D, Cuccurullo F (1996) Hypochlorous acid and its pharmacological antagonism: an update picture. *General Pharmacology* 27:1145-1147.
7. McKenna SM, Davies KJ (1988) The inhibition of bacterial growth by hypochlorous acid: Possible role in the bactericidal activity of phagocytes. *The Biochemical Journal* 254(3):685-692.
8. Prütz WA (1996) Hypochlorous acid interactions with thiols, nucleotides, DNA, and other biological substrates. *Archives of Biochemistry & Biophysics* 332(1):110-120.
9. Vissers MC, Winterbourn CC (1995) Oxidation of intra-cellular glutathione after exposure of human red blood cells to hypochlorous acid. *Biochemistry Journal* 307(Pt 1):57-62.
10. Weiss SJ, LampertMB, Test ST (1983) Long-lived oxidants generated by human neutrophils: Charac terization and bioactivity. *Science* 222:625-628
11. Winter J, Ilbert M, Graf PCF, Özcelik D, Jakob U (2008) Bleach activates a Redox-regulated chaperone by oxidative protein unfolding. *Cell* 135(4):691-701.

2.4.5. Hypochlorite (OCl^-)

1. Block SS (1991) Historical review. In SS Block (ed.), *Disinfection, Sterilization, and Preservation,* (4th ed) pp. 3-17. Philadelphia: Lea & Febiger

2. Cotter JL, Fader RC, LilleyC, Herndon DN (1985) Chemical parameters, antimicrobial activities, and tissue toxicity of 0.1 and 0.5% sodium hypochlorite solutions. *Antimicrobial Agents and Chemotherapy* 28(1):118-122.
3. Dakin HD (1915) On the use of certain antiseptic substances in the treatment of infected wounds. *British Medical Journal* 2:318-320.
4. Ellis KV (1991) Water disinfection: a review with some consideration of the requirements of the third world. *Critical Reviews in Environment Control* 20:341-407.
5. Heggers JP, Sazy JA, Stenberg BD, et al (1991) Bactericidal and wound-healing properties of sodium hypochlorite solutions: the 1991 Lindberg Award. *Journal of Burn Care Rehabilitation* 12(5):420-424.
6. Rutala WA, Weber DJ (1997) Uses of inorganic hypochlorite (bleach) in health-care facilities. *Clinical Microbiology Reviews* 10(4):597-610
7. Young SB, Setlow P (2003) Mechanisms of killing of Bacillus subtilis spores by hypochlorite and chlorine dioxide. *Journal of Applied Microbiology* 95:54-67.

2.5 | REDOX: Reactive Nitrogen Species (RNS)

1. Borodulin RR, Kubrina LN, Mikoyan VD, Poltorakov AP, et al (2013) Dinitrosyl iron complexes with glutathione as NO and NO_2^- donors. *Nitric Oxide* 29:4-16.
2. Diesen DL, Hess DT, Stamler JS (2008) Hypoxic vasodilation by red blood cells: evidence for an S-nitrosothiol-based signal. *Circulation Research* 103(5):545-553.
3. Guzik TJ, West NE, Pillai R, Taggart DP, Channon KM (2002) Nitric oxide modulates superoxide release and peroxynitrite formation in human blood vessels. *Hypertension* 39(6):1088-1094.
4. Martínez MC, Andriantsitohaina R (2009) Reactive nitrogen species: molecular mechanisms and poten tial significance in health and disease. *Antioxidant & Redox Signal* 11(3):669-702.
5. O'Donnell VB, Eiserich JP, Chumley PH, Jablonsky MJ, et al (1999) Nitration of unsaturated fatty acids by nitric oxide-derived reactive nitrogen species peroxynitrite, nitrous acid, nitrogen dioxide, and nitro nium ion. *Chemical Research in Toxicolology* 12(1):83-92.
6. Pacher P, Beckman JS, Liaudet L (2007). Nitric oxide and peroxynitrite in health and disease. *Physiological Reviews* 87(1):315-424.
7. Patel RP, McAndrew J, Sellak H, White CR, et al (1999) Biological aspects of reactive nitrogen species. *Biochimica Biophysica Acta* 1411:385-400.
8. Patel RP, Moellering D, Murphy-Ullrich J, Jo H, et al (2000) Cell signaling by reactive nitrogen and oxygen species in atherosclerosis. *Free Radicals in Biology & Medicine* 28:1780-1794.
9. Squadritoa GL, Pryor WA (1998) Oxidative chemistry of nitric oxide: the roles of superoxide, peroxyni trite, and carbon dioxide. *Free Radical Biology & Medicine* 25(4–5):392-403.
10. van Faassen E, Vanin AF(2007). *Radicals for Life: the various forms of Nitric Oxide.* Amsterdam:Elsevier. ISBN 978-0-444-52236-8.

2.5.1. Nitric oxide (·NO)

1. Beckman JS, Tsai JHM (1994) Reactions and diffusion of nitric oxide and peroxynitrite. *Biochemist* 16:8-10.
2. Boveris A, Alvarez S, Navarro A (2002) The role of mitochondrial nitric oxide synthase in inflammation and septic shock. *Free Radical Biology & Medicine* 33:1186-1193.
3. Brookes PS, Bolanos JP, Heales SJ (1999) The assumption that nitric oxide inhibits mitochondrial ATP synthesis is correct. *FEBS Lett* 446:261-263.
4. Burney S, Caulfield JL, Niles JC, Wishnok JS, Tannenbaum SR (1999) The chemistry of DNA damage from nitric oxide and peroxynitrite. *Mutation Research* 424:37-49.
5. Furchgott RF, Zawadzki JV (1980) The obligatory role of endothelial cells in the relaxation of arterial smooth muscle by acetylcholine. *Nature* 288:373-376.
6. Gow AJ, Farkouh CR, Munson DA, Posencheg MA, Ischiropoulos H (2004) Biological significance of nitric oxide-mediated protein modifications. *American Journal of Physiology: Lung Cellular & Molecular Physiology* 287:L262-L268.
7. Hogg N, Kalyanaraman B (1999) Nitric oxide and lipid peroxidation. *Biochimca Biophysica Acta* 1411:378-384.
8. Ignarro LJ (1999) Nitric oxide: a unique endogenous signaling molecule in vascular biology. *Bioscience Reports* 19:51-71.
9. Massion PB, Feron O, Dessy C, Balligand JL (2003) Nitric oxide and cardiac function: ten years after, continuing. *Circulation Research* 93:388-398.

10. McCall TB, Boughton-Smith NK, Palmer RMJ, Whittle BJR, Moncada S (1989) Synthesis of nitric oxide from L-arginine by neutrophils. *Biochemistry Journal* 261:293-296.
11. Moncada S, Higgs EA (2006) The discovery of nitric oxide and its role in vascular biology. *British Journal of Pharmacology* 147(Suppl 1):S193-S201.
12. Nathan C, Xie QW (1994) Regulation of biosynthesis of nitric oxide. *Journal of Biological Chemistry* 269:13725-13728.
13. Thatcher GR, Nicolescu AC, Bennett BM, Toader V (2004) Nitrates and NO release: contemporary aspects in biological and medicinal chemistry. *Free Radical Biology & Medicine* 37:1122-1143.

2.5.2. Peroxynitrite (ONO_2^-)

1. Alvarez B, Radi R (2003) Peroxynitrite reactivity with amino acids and proteins. *Amino Acids* 25:295-311.
2. Beckman JS, Koppenol WH (1996) Nitric oxide, superoxide, and peroxynitrite: the good, the bad, and ugly. *American Journal of Physiology: Cell Physiology* 271:C1424-C1437.
3. Boccini F, Herold S. Mechanistic studies of the oxidation of oxyhemoglobin by peroxynitrite. *Biochemistry* 2004;43:16393–16404.
4. Denicola A, Souza JM, Radi R (1998) Diffusion of peroxynitrite across erythrocyte membranes. *Proceedings of the National Academy of Sciences USA* 95:3566-3571.
5. Ebadi M, Sharma SK, Ghafourifar P, Brown-Borg H, El Refaey H (2005) Peroxynitrite in the pathogenesis of Parkinson's disease and the neuroprotective role of metallothioneins. *Methods in Enzymology* 396:276–298.
6. Hogg N, Darley-Usmar VM, Graham A, Moncada S (1993) Peroxynitrite and atherosclerosis. *Biochemical Society Transactions* 21:358–362.
7. Koppenol WH (2001) 100 years of peroxynitrite chemistry and 11 years of peroxynitrite biochemistry. *Redox Reports* 6:339–341.
8. Levrand S, Vannay-Bouchiche C, Pesse B, Pacher P, et al (2006) Peroxynitrite is a major trigger of cardiomyocyte apoptosis in vitro and in vivo. *Free Radical Biology & Medicine* 41:886-895.
9. Pacher P, Beckman JS, Liaudet L (2007) Nitric oxide and peroxynitrite in health and disease. *Physiological Reviews* 87(1):315-424.
10. Plumb RC, Tantayanon R, Libby M, Xu WW (1989) Chemical model for Viking biology experiments: implications for the composition of the martian regolith. *Nature* 338:633-635.
11. Pryor WA, Squadrito GL (1995) The chemistry of peroxynitrite: a product from the reaction of nitric oxide with superoxide. *Americal Journal of Physiology: Lung Cellular & Molecular Physiology* 268:L699–L722
12. Radi R (1998) Peroxynitrite reactions and diffusion in biology. *Chemical Research in Toxicology* 11:720-721.
13. Radi R, Cassina A, Hodara R, Quijano C, Castro L (2002) Peroxynitrite reactions and formation in mitochondria. *Free Radical Biology & Medicine* 33:1451-1464.
14. Radi R, Peluffo G, Alvarez MN, Naviliat M, Cayota A (2001) Unraveling peroxynitrite formation in biological systems *Free Radical Biology & Medicine* 30:463-488.
15. Rubbo H, Trostchansky A, Botti H, Batthyany C (2002) Interactions of nitric oxide and peroxynitrite with low-density lipoprotein. *Biological Chemistry* 383:547-552.

2.6 | REDOX: Reactive Sulfur Species

1. Fry FH, Holme AL, Giles NM, Giles GI, et al (2005) Multifunctional redox catalysts as selective enhancers of oxidative stress. *Organic Biomolecular Chemistry* 3(14):2579-2587.
2. Giles GI, Jacob C (2002) Reactive sulfur species: an emerging concept in oxidative stress. *Biological Chemistry* 383(3-4):375-388.
3. Giles GI, Tasker KM, Collins C, Giles NM, et al (2002) Reactive sulphur species: an in vitro investigation of the oxidation properties of disulphide S-oxides. *Biochemistry Journal* 364(2):579-585.
4. Giles GI, Tasker KM, Jacob C (2001) Hypothesis: the role of reactive sulfur species in oxidative stress. *Free Radical Biology & Medicine* 31(10):1279-1283.
5. Giles GI, Tasker KM, Jacob C (2002) Oxidation of biological thiols by highly reactive disulfide-S-oxides. *General Physiology & Biophysics* 21(1):65-72.
6. Giles NM, Watts AB, Giles GI, Fry FH, et al (2003) Metal and redox modulation of cysteine protein function. *Chemical Biology* 10(8):677-693.

7. Gruhlke MC, Slusarenko AJ (2012) The biology of reactive sulfur species (RSS). *Plant Physiology & Biochemistry* 59:98-107.
8. Jacob C, Lancaster JR, Giles GI (2004) Reactive sulphur species in oxidative signal transduction. *Biochemical Society Transactions* 32(6):1015-1017.
9. Jones DP (2006) Extracellular redox state: refining the definition of oxidative stress in aging. *Rejuvenation Research* 9:169-181.
10. Jones DP (2008) Radical-free biology of oxidative stress. *American Journal of Physiology: Cell Physiology* 295:C849-868.
11. Meister A, Anderson ME (1983) Glutathione. *Annual Review of Biochemistry* 52:711-760.
12. Miseta A, Csutora P (2000) Relationship between the occurrence of cysteine in proteins and the complexity of organisms. *Molecular Biology of Evolution* 17:1232-1239.
13. Moriarty-Craige SE, Ha KN, Sternberg P Jr, Lynn M, et al (2007) Effects of long-term zinc supple mentation on plasma thiol metabolites and redox status in patients with agerelated macular degeneration. *American Journal of Ophthalmology* 143:206-211.
14. Park Y, Ziegler TR, Gletsu-Miller N, Liang Y, et al (2010) Postprandial Cysteine/Cystine redox potential in human plasma varies with meal content of sulfur amino acids. *Journal of Nutrition* 140:1-6.

2.7 | REDOX: Other Reactive Species

1. Canonica S, Kohn T, Mac M, Real FJ, et al (2005) Photo-sensitizer method to determine rate constants for the reaction of carbonate radical with organic compounds. *Environmental Science & Technology* 39(23):9182-9188.
2. Clennan EL, Pace A (2005) Advances in singlet oxygen chemistry. *Tetrahedron 61*(28):6665-6691.
3. Coletta C, Papapetropoulos A, Erdelyi K, Olah G, et al (2012) Hydrogen sulfide and nitric oxide are mutually dependent in the regulation of angiogenesis and endothelium-dependent vasorelaxation. *Proceedings of the National Academy of Sciences USA* 109(23):9161-9166.
4. Horvath M, Bilitzky L, Huttner J (1985) *Ozone.* Amsterdam: Elsevier. ISBN: 0-444-99625-7.
5. Kimura H (2002) Hydrogen sulfide as a neuromodulator. *Molecular Neurobiology* 26 1):13-19.
6. Benavides GA, Giuseppe L, Mills SRW, Patel HD, et al (2007) Hydrogen sulfide mediates the vaso-activity of garlic. *Proceedings of the National Academy of Sciences USA* 104 (46):17977-17982.
7. Lefer DJ (2007) A new gaseous signaling molecule emerges: Cardioprotective role of hydrogen sulfide. *Proceedings of the National Academy of Sciences* 104(46):17907-17908.
8. Ogilby PR (2010) Singlet oxygen: there is indeed something new under the sun. *Chemical Society Reviews* 39(8):3181-3209.
9. Prockop LD, Chichkova RI (2007) Carbon monoxide intoxication: an updated review. *Journal of Neurological Sciences* 262(1–2): 122–130.
10. Ramasamy S, Singh S, Taniere P, Langman MJS, Eggo MC (2006) Sulfide-detoxifying enzymes in the human colon are decreased in cancer and upregulated in differentiation. *American Journal of Physiology Gastrointest. Liver Physiology* 291(2):G288-G296.
11. Schweitzer C, Schmidt R (2003) Physical mechanisms of generation and deactivation of singlet oxygen. *Chemical Reviews* 103(5): 1685-1757.
12. Smith LL (2004) Oxygen, oxysterols, ouabain, and ozone: a cautionary tale. *Free Radical Biology & Medicine* 37(3):318-324.
13. Tikuisis P, Kane DM, McLellan TM, Buick F, Fairburn SM (1992) Rate of formation of carboxyhemo-globin in exercising humans exposed to carbon monoxide. *Journal of Applied Physiology* 72(4):1311-1319.
14. Verma A, Hirsch D, Glatt C, Ronnett G, Snyder S (1993) Carbon monoxide: A putative neural messenger. *Science* 259(5093):381-3844.
15. Wentworth P, Nieva J, Takeuchi C, Galve R, et al (2003) Evidence for ozone formation in human athero sclerotic arteries. *Science* 302 (5647):1053-1056.
16. Wu L, Wang R (2005). Carbon monoxide: Endogenous production, physiological functions, and pharmacological applications. *Pharmacology Reviews* 57(4):585-630.
17. Yadav A, Mishra PC (2012) Carbonate radical anion as an efficient reactive oxygen species: Its reaction with guanyl radical and formation of 8-oxoguanine. *Chemical Physics* 405(11):76-88.

2.8 | Free Radical Formation

1. Stephen Hawking (1988). *A Brief History of Time.* Bantam Books. ISBN 0-553-05340-X.
2. Einstein A, de Sitter W (1932) On the relation between the expansion and the mean density of the Universe. *Proceedings of the National Academy of Sciences* 18(3):213-214.
3. Ellis GFR, Kirchner U, Stoeger WR (2004) Multiverses and physical cosmology. *Monthly Notices of the Royal Astronomical Society* 347(3):921-936.

2.8.1. Free Radical Formation: General Pathways

1. Bano T, Kumar N, Dudhe R (2012) Free radical scavenging properties of pyrimidine derivatives. *Organic Medicinal Chemistry Letters* 2(1):34.
2. Borm PJ, Kelly F, Künzli N, Schins RP, Donaldson K (2007) Oxidant generation by particulate matter: from biologically effective dose to a promising, novel metric. *Occupational & Environmental Medicine* 64(2):73-74.
3. Castranova V (1994) Generation of oxygen radicals and mechanisms of injury prevention. *Environmental Health Perspective* 102 (Suppl 10):65-68.
4. Chignell CF, Motten AG, Buettner GR (1985) Photo-induced free radicals from chlorpromazine and related phenothiazines: relationship to phenothiazine-induced photosensitization. *Environmental Health Perspective* 64:103-110.
5. Park SG, Kim JH, Xia Y, Sung JH (2011) Generation of reactive oxygen species in adipose-derived stem cells: friend or foe? *Expert Opinion on Therepeutic Targets* 15(11):1297-306.
6. Yu H, Xia Q, Yan J, Herreno-Saenz D, et al (2006) Photo-irradiation of polycyclic aromatic hydrocarbons with UVA light - a pathway leading to the generation of reactive oxygen species, lipid peroxidation, and dna damage. *International Journal of Environmental Reseach & Public Health* 3(4):348-354.
7. Zhu S, Manuel M, Tanaka S, Choe N, et al (1998) Contribution of reactive oxygen and nitrogen species to particulate-induced lung injury. *Environmental Health Perspective* 106 (Suppl 5):1157-1163.

2.8.2. Free Radical Formation: Biological Systems

1. Adam-Vizi V, Starkov AA (2010) Calcium and mitochondrial reactive oxygen species generation: how to read the facts. *Journal of Alzheimers Disease* 20(Suppl 2):S413-426.
2. Leto TL, Morand S, Hurt D, Ueyama T (2009) Targeting and regulation of reactive oxygen species generation by Nox family NADPH oxidases. *Antioxidants & Redox Signaling* 11(10):2607-2619.
3. Murphy MP (2009) How mitochondria produce reactive oxygen species. *Biochemistry Journal* 417(1):1-13.
4. Romero FJ, Bosch-Morell F, Romero MJ, Jareño EJ, et al (1998) Lipid peroxidation products and antioxidants in human disease. *Environmental Health Perspective* 106(Suppl 5):1229-1234.
5. Starkov AA (2008) The role of mitochondria in reactive oxygen species metabolism and signaling. *Annals of NY Academy of Science*s 1147:37-52.

2.8.3. Free Radical Formation: The Human Body

1. Jackson MJ (2011) Control of reactive oxygen species production in contracting skeletal muscle. *Antioxidants & Redox Signaling* 15(9):2477-86.
2. Jackson MJ, McArdle A (2011) Age-related changes in skeletal muscle reactive oxygen species generation and adaptive responses to reactive oxygen species. *Journal of Physiology* 589:2139-2145.
3. Tostes RC, Carneiro FS, Lee AJ, Giachini FR, et al (2008) Cigarette smoking and erectile dysfunction: focus on NO bioavailability and ROS generation. *Journal of Sex Medicine* 5(6):1284-1295.
4. Zinkevich NS, Gutterman DD (2011) ROS-induced ROS release in vascular biology: redox-redox signaling. *American Journal of Physiology: Heart & Circulation Physiology* 301(3):H647-653.

2.9 | Free Radicals: Bio-Molecular Targets

1. Comporti M (1989) Three models of free radical-induced cell injury. *Chemical & Biological Interactions* 72:1-56.
2. Davies KJ (1999) The broad spectrum of responses to oxidants in proliferating cells: A new paradigm for oxidative stress. *Life* 48:41-47.

3. Gracy RW, Talent JM, Kong Y, Conrad CC (1999) Reactive oxygen species: The unavoidable environmental insult? *Mutation Research* 428:17-22.
4. Trachootham D, Lu W, Ogasawara MA, Valle NS, Huang P (2008) Redox regulation of cell survival. *Antioxidants & Redox Signaling* 10(8):1343-1374.

2.9.1. Free Radical Target: Lipids

1. Awasthi YC, Sharma R, Cheng JZ, Yang Y, et al (2003) Role of 4-hydroxynonenal in stress-mediated apoptosis signaling. *Molecular Aspects of Medicine* 24:219-230.
2. Girotti AW (1998) Lipid hydroperoxide generation, turnover, and effector action in biological systems. *Journal of Lipid Research* 39:1529-1542.
3. Halliwell B, Gutteridge JM (2007) *Free Radicals in Biology and Medicine,* 4th edition. Avon, England: Oxford University Press. ISBN-10: 019856869X
4. Meral A, Tuncel P, Surmen-Gur E, Ozbek R, et al (2000) Lipid peroxidation and antioxidant status in beta-thalassemia. *Pediatric Hematology & Oncology* 17:687–693.

2.9.2. Free Radical Target: Proteins

1. Davis KJ (1987) Protein damage and degradation by oxygen radicals. I. general aspects. *Journal of Biological Chemistry* 262:9895-9901.
2. Dean RT, Fu S, Stocker R, Davies MJ (1997) Biochemistry and pathology of radical-mediated protein oxidation. *Biochemistry Journal* 324:1-18.
3. Grune T, Reinheckei T, Davies KJA (1997) Degradation of oxidized proteins in mammalian cells. *FASEB Journal* 11:526-534.
4. Requena JR, Fu MX, Ahmed MU, Jenkins AJ, Lyons TJ, Thorpe SR (1996) Lipoxidation products as biomarkers of oxidative damage to proteins during lipid peroxidation reactions. *Nephrology & Dialysis Transplant* 11(suppl 5):48-53.
5. Stadtman ER (1986) Oxidation of proteins by mixed-function oxidation systems, implication in protein turnover, aging and neutrophil function.*Trends in Biochemical Sciences* 11:11–12.

2.9.3. Free Radical Target: Nucleic acid (DNA)

1. Beckman KB, Ames BN (1997) Oxidative decay of DNA. *Journal of Biological Chemistry* 272:19633-19636.
2. Breen AP, Murphy JA (1995) Reactions of oxyl radicals with DNA. *Free Radical Biology & Medicine* 18:1033-1077.
3. DizdarogluM, Jaruga P, Birincioglu M, Rodriguez H (2002) Free radical induced damage to DNA: Mechanisms and measurement. *Free Radical Biology & Medicine* 32:1102-1115.
4. Halliwell B (2000) Why and how should we measure oxidative DNA damage in nutritional studies? How far have we come? *Americal Journal of Clinical Nutrition* 72:1082-1087.
5. Kasprzak KS (2002) OxidativeDNAand protein damage in metal-induced toxicity and carcinogenesis. *Free Radical Biology & Medicine* 32:958-967.

CHAPTER

3

REDOX Biology

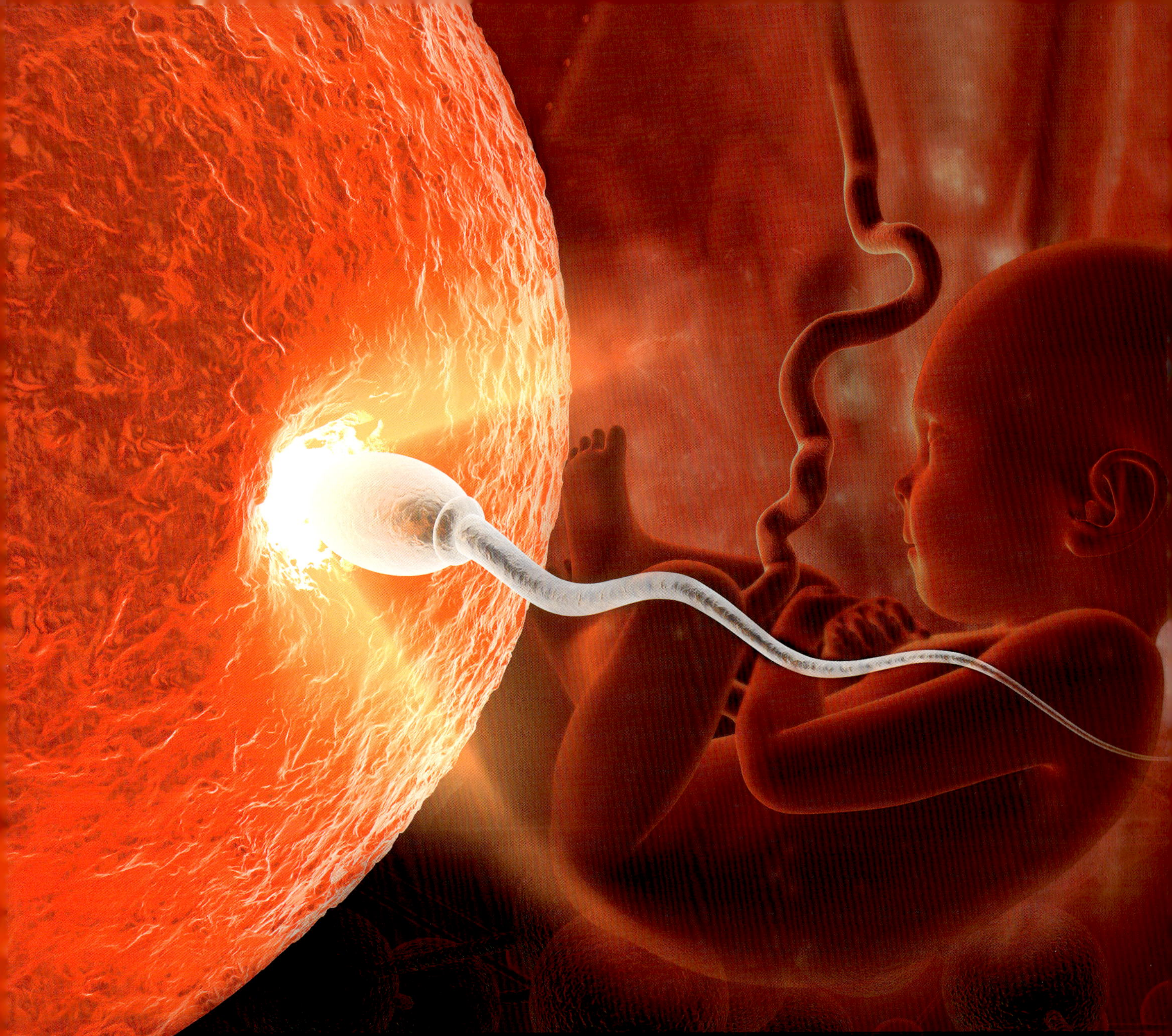

Power of Life: A REDOX Inheritence

Mitochondria – the generators of cellular energy, are passed on from one generation to the next through egg cells (from the mother), and almost none contributed by sperm (from the father). This is largely because sperm are tiny with less than 100 mitochondria (located in the tail); and any paternal mitochondrial DNA that gets into the fertilized egg is actively destroyed. This is in total contrast to the way that the rest of our DNA is inherited. We receive one copy of each chromosome from each parent (with the exception of the sex chromosomes). The exclusive inheritance of maternal mt-DNA has helped us to discover the patterns of migration of early man.

3

REDOX: Biology

REDOX biology takes a pivotal role in various metabolic pathways of living organisms. These biochemical processes include mechanisms of energy production (ATP generation) and host defense (phagocytic killing of pathogens). Several other vital processes in biological systems are dependent on the smooth interplay of several REDOX active organic molecules and inorganic ions and complexes. The cellular REDOX mechanisms are tightly controlled by various regulatory proteins and enzymes.

Reactive oxygen species (ROS) are produced by living organisms as a result of normal cellular metabolism. ROS are a double-edge sword; at low to moderate concentrations, they promote cell proliferation and survival. However, when levels reach certain threshold, ROS may overwhelm the cellular antioxidant capacity and lead to adverse modifications to cell components, such as lipids, proteins, DNA and may trigger cellular death (apoptosis). A shift in balance between the oxidant and antioxidant ratio in favor of oxidants is termed 'oxidative stress'. Oxidative stress contributes to many pathological conditions. Human body has antioxidant mechanisms, which include enzymatic and non-enzymatic compounds that are usually effective in blocking the harmful effects of ROS.

Accordingly, normal cells maintain REDOX homeostasis through proper balance between ROS generation and elimination. Cells with higher ROS generation (as in the case of cancer cells) would be more dependent on the antioxidant system and more vulnerable to oxidative agents. Increasing the oxidative stress with exogenous ROS-generating agents or drugs could disable the endogenous antioxidant system and may preferentially kill cancer cells. In contrast, normal cells could tolerate such induced oxidative stress due to their low ROS outputs and normal metabolic regulation. However, reducing ROS levels with proper antioxidants may be a useful strategy to prevent or delay aging and age-related diseases.

3.1 REDOX: Prooxidants

Prooxidant is a generic collective term for a number of active and reactive, partially reduced O_2 metabolites. Some of them, such as $O_2^{\bullet-}$ and ·OH, can be defined as true prooxidants, which are reactive molecular species with an impaired electron in their outer orbital. Prooxidants are paramagnetic and reactive chemical entities that can undergo REDOX reactions by interacting with surrounding molecules in order to regain the more stable non-radical condition. Other prooxidants, such as H_2O_2, are more properly prooxidant non-radical agents. Indeed, $O_2^{\bullet-}$, ·OH and H_2O_2 are by far the most relevant in physiological or pathological conditions. Prooxidants primarily produced by two metabolic sources inside the cell: the mitochondrial electron-transport chain (ETC) and O_2-metabolizing enzymatic reactions. Prooxidant levels also are dependent on O_2 concentrations in the milieu. Major prooxidants, their sources and reactions are briefly summarized below.

3.1.1. ROS Reservoirs

Both exogenous and endogenous sources contribute to the formation of intracellular ROS/RNS. Exogenous sources include irradiation (i.e., UV, X-ray, gamma-ray), atmospheric pollutants, and chemicals. A major endogenous source of cellular ROS is from the mitochondria, where $O_2^{\bullet-}$ is generated by electron leakage from complex I and III of the electron-transport chain. Microsomes and peroxisomes are also sources of ROS, primarily H_2O_2, whereas immune cells such as neutrophils and macrophages possess oxygen-dependent mechanisms to fight against invading microorganisms. Other endogenous sources of ROS include the membrane-associated NAD(P)H oxidase, cytochrome c oxidase, and xanthine oxidase. The presence of REDOX-active metals such as iron (Fe) and copper (Cu) also contributes to ROS generation. In the presence of Fe(II) and Fe(III), hydroxyl (·OH) radicals can be generated through the Fenton reaction or Haber–Weiss reaction. Similarly, nitric oxide (·NO) generation occurs through specific nitric oxide synthase isozymes, including mitochondrial nitric oxide synthase (mtNOS), neuronal NOS (nNOS), endothelial NOS (eNOS), and inducible NOS (iNOS). NO can react with $O_2^{\bullet-}$ to generate $ONOO^-$.

Daily, humans use about 250 g of O_2 out of which 2–5% is converted to reactive oxygen species (ROS). ROS means all oxygen associated species that have higher oxidative potential (higher reactivity) than molecular oxygen (O_2): singlet oxygen ($^1O_2$), superoxide anion ($O_2^{\bullet-}$), hydrogen peroxide (H_2O_2) and hydroxyl radical (·OH).

A) Mitochondria

Mitochondria is the 'cellular power house', the active site for most ATP generation though oxidative phosphorylation (OXPHOS), which accounts for 90% to 95% of cellular oxygen consumption. The major bulk of mitochondrial ROS generation occurs at the electron transport chain, as a byproduct of respiration. The first mitochondrial site producing ROS was identified at the Complex III located inside the inner mitochondrial membrane. The primary ROS produced at this site is $O_2^{\bullet-}$. ROS production is dependent on the mitochondrial membrane potential. Mitochondria possess at least 10 known sites that are capable of generating ROS, but they also feature a sophisticated multilayered ROS defense system.

Mitochondria has a double-membrane architecture, with an outer membrane of eukaryotic origin and an inner membrane characterized by the absence of cholesterol and the presence of cardiolipin, which is typical to bacterial membrane. The inner membrane is organized in characteristic folds, termed cristae, which protrude into the matrix and accommodate the respiratory chain complexes. In healthy cells, the inner membrane is impermeable to ions, which allow the electrons transport chain (ETC) to actively build up the proton gradient. The mitochondrial membrane potential results from the difference in electrical potential generated by the electrochemical gradient across the inner membrane. Mitochondria are responsible for 90% of the cellular energy production that fuels various tissues and physiological systems.

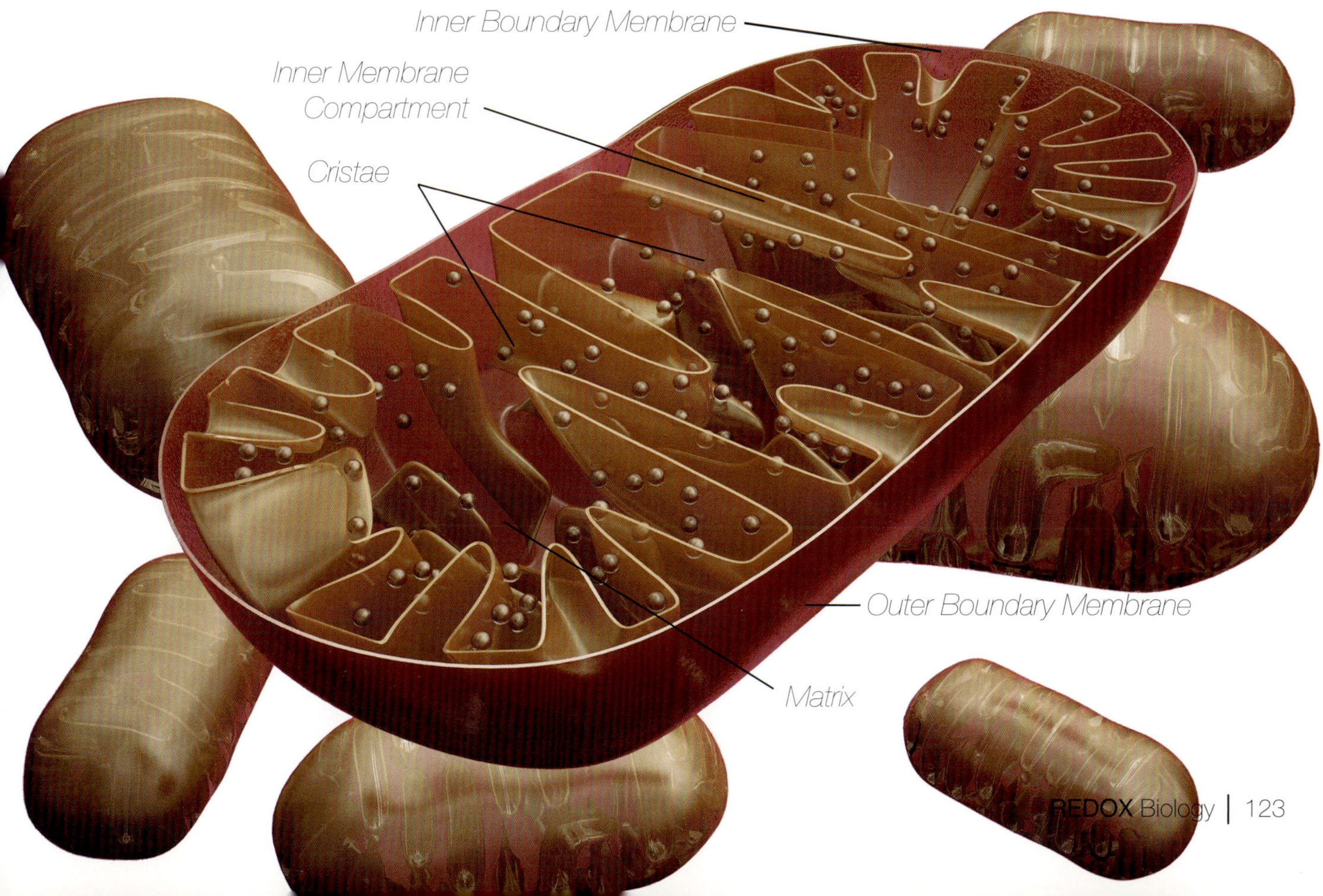

Mitochondria enable us to utilize oxygen for energy production. Quantity and quality of this cellular organelle determines our physiological performance and endurance.

Mitochondria are sub-cellular organelles. Different cell types contain varying numbers; on one extreme, red blood cells (erythrocytes) contain none and on the other, egg cells (ova) contain about 100,000. Most human cells contain between a few hundred and a few thousand. Most cells (except platelets and, red blood cells) have a nucleus, which is in the chromosomal compartment. Mitochondria are never found in the nucleus, only in the rest of the cell (the cytoplasm). Mitochondria are distributed throughout the cytoplasm. Even in very long and thin cells, such as nerves, there are mitochondria all the way to the end—sometimes several feet away from the nucleus. They occupy a substantial fraction of the cell volume: about 1/5 in liver cells, for example.

A mitochondrion, like all organelles, is essentially a tiny drop of water encased in a double membrane. The surface area of its inner membrane is intricately folded, known as cristae. The outer membrane is not folded and has large holes (called porins) that allow small molecules to easily pass through. The inner membrane, on the other hand, is impermeable to most molecules (especially charged ones). The space enclosed by the inner membrane is called the mitochondrial matrix. The space between the inner and outer membranes is known as the mitochondrial inter-membrane space. The membranes are both composed of the same type of molecules as all other cells, except for the 'cardiolipin', a diphospholipid exclusive to mitochondrial inner membrane. The two membranes are not completely separate and come together at a few places called contact sites. The inner membrane and the matrix, each contain many molecules whose roles are relevant to the mitochondrial free radical theory of aging. It has been estimated that half the mass of the matrix is protein, mostly enzymes: catalysts of chemical reactions that are involved in mitochondrial respiration. The matrix also contains smaller molecules, many of which are substrates or cofactors of these enzymes. Finally, it also contains a small amount of DNA, and the machinery (ribosomes) for decoding that DNA. The matrix is one of the most alkaline environments in the cell. It has a pH of between 8 and 8.5, roughly half a unit higher than the cytosol.

Superoxide anion ($O_2^{\bullet-}$) is the primary ROS generated by mitochondria, as a result of mono-electronic reduction of oxygen. Superoxide anion is the precursor for several types of ROS and a mediator in many oxidative chain reactions. *In vivo*, $O_2^{\bullet-}$ is produced both enzymatically (i.e. NADPH oxidase, cytochrome P450-dependent oxygenases, and xanthine oxidase) and non-enzymatically when a single electron is directly transferred to O_2. Also, $O_2^{\bullet-}$ may react with other radicals including nitric oxide (NO·) and generate reactive nitrogen species (RNS). Dismutation of $O_2^{\bullet-}$, either spontaneously, or through a reaction catalyzed by superoxide dismutases (SODs), produces hydrogen peroxide (H_2O_2). H_2O_2 generated in mitochondria has many possible fates. Being relatively stable and membrane permeable, H_2O_2 can diffuse into the cell and be eliminated by mitochondrial antioxidant systems. Mitochondria H_2O_2 can also act as a signaling molecule in the cytosol and regulate several pathways such as cell cycles, stress responses, energy metabolism, and REDOX balance. When not metabolized, H_2O_2 can be transformed to hydroxyl radical (·OH) in the presence of metal ions, by Fenton reaction. Hydroxyl radical is highly reactive, potentially damaging and one of the strongest oxidant molecules in nature. Accordingly, mitochondria have developed efficient H_2O_2 removal systems, as well as metal-chelating mechanisms to prevent the formation of ·OH radical. At least, 10 mammalian mitochondrial enzymes contribute to ROS production and their ROS generating abilities vary in a tissue-specific manner.

Mitochondrial ROS signals regulate homeostasis through cellular apoptosis and cell survival. Mitochondria ROS may be considered as "balance of power," that takes cell towards life or death.

The mitochondrial respiratory chain, due to its natural imperfections, can be considered a peculiar superoxide generator. In such a case, complexes of the respiratory chain are in the firsthand exposition to oxidative modification because they are present at the sites of ROS production. A wide range of mitochondrial ROS-induced damages have been described, including protein carbonylation, lipid peroxidation, or mitochondrial-DNA damage. These damages, either individually or collectively, can lead to a cellular energetic catastrophe.

Under certain conditions, protons can re-enter the mitochondrial matrix without contributing to ATP synthesis. This process is known as proton leak or mitochondrial uncoupling and is due to the facilitated diffusion of protons into the matrix. The process results in the unharnessed potential energy of the proton electrochemical gradient being released as heat. The process is mediated by a proton channel called thermogenin, primarily found in brown adipose (fat) tissue, which is responsible for non-shivering thermogenesis. In humans, brown adipose tissue is present at birth and decreases with age.

The discovery of mitochondria came in 1886 when Richard Altman, a cytologist, identified the organelles using a dye technique, and called them 'bioblasts'. He postulated that these structures were the basic units of cellular activity. Carl Benda, in 1898, coined the term mitochondria, derived from the Greek words thread (mitos) and granule (chondros). In 1912, a German biochemist named Otto Heinrich Warburg hypothesized that an enzyme within cells enabled the processing of oxygen. He showed that cyanide had an effect on respiration at the cellular level. Later in 1923 David Keilin demonstrated that the oxidation state within cytochromes (hemoproteins responsible for electron transport) was changed during respiration. He later identified the existence of cytochrome c, part of the inner membrane of mitochondria.

Studies by Kalckar and Belitser showed that the addition of a phosphate to protein increases cellular respiration in a process called oxidative phosphorylation. Isolation of mitochondria as an organelle harvested from the liver, gave scientists an access to discover its REDOX functions. In 1950, Kennedy and Lehninger showed that oxidation occurs within mitochondria. In1961, Mitchell postulated the chemiosmotic theory to describe the diffusion of protons (H^+ ions) across membranes, and the role of REDOX in ATP synthesis during respiration. These studies established the overall purpose of mitochondria, to work as the organelle that converts the potential energy of food molecules into ATP.

B) Phagocytes

The respiratory burst was discovered by Baldridge and Gerard in 1933. The purpose of this oxidative burst is not to provide energy for phagocytosis, but to furnish lethal oxidants for microbial killing. Phagocytes are estimated to produce relatively high amounts of $O_2^{\bullet-}$, on the order of 0.5 mM/sec. Taking into account the rate of spontaneous dismutation and approximate volume of the phagosome, this is estimated to yield 50 μM $O_2^{\bullet-}$ at pH 7.4 or 2 μM at pH 4.5 (the approximate pH of the acidified phagosome). Because the resulting H_2O_2 can diffuse across membranes, including out of the phagosome, the steady state concentration of the H_2O_2 would be approximately 1–4 μM. The H_2O_2 that diffuses into the bacterial cytoplasm could potentially cause damage. Indeed a dogma is that phagocytic ROS kill bacteria by damaging DNA

During phagocytosis the phagocytic cells (neutrophils, eosinophils, and mononuclear phagocytes) undergo an increase in glucose and oxygen consumption termed the respiratory burst. The 'respiration' associated with this burst, however, has nothing to do with energy production; rather, its purpose is to generate powerful microbicidal agents by the partial reduction of oxygen. Their rates of oxygen uptake increase greatly, sometimes more than 50-fold; they begin to produce large amounts of superoxide ($O_2^{\bullet-}$) and hydrogen peroxide (H_2O_2); and they start metabolizing large quantities of glucose. The respiratory burst generates several oxygen-containing compounds that kill the microbes undergoing phagocytosis – oxygen-dependent intra-cellular killing. Reactive oxygen species (ROS) are critical weapons in the phagocyte arsenal.

The respiratory burst is mediated by the NADPH-oxidase complex, a multicomponent system that is rapidly assembled following activation of neutrophils. The NADPH oxidase catalyzes the production of superoxide anion:

$$O_2 + NADPH \rightarrow O_2 + NADP^+ + H^+$$

A dismutation reaction occurs with $O_2^{\bullet-}$ to produce peroxide:

$$2\,O_2^{\bullet-} + 2H^+ \rightarrow H_2O_2 + O_2$$

Both H_2O_2 and $O_2^{\bullet-}$ are components of the oxygen-dependent, antimicrobial system of phagocytic leukocytes. Another phagocytic enzyme, myeloperoxidase, catalyses a reaction between chloride ions and hydrogen peroxide to yield hypochlorous acid (HOCl), which reacts further with hydrogen peroxide to yield more singlet oxygen. Nitric oxide (·NO), and its peroxynitrite derivatives, are also produced. The net effect of this toxic cocktail is to disable the target cell by oxidizing crucial components of its cellular apparatus.

Phagocytosis: The 1908 Nobel Prize in Physiology or Medicine was awarded to Ilya Mechnikov for the discovery of phagocytes/phagocytosis. His 'Phagocytic Theory' is the basis of immunological concept "self and non-self' as the prerequisite for inflammatory response and host defense.

Hypochlorite, the prototype of the oxidized halogens, is itself an exceedingly powerful microbicidal agent. It is also the precursor of the chloramines, a group of microbicidal oxidized halogens that are formed by the reaction between hypochlorite and ammonia or amines:

$$OCl^- + RNH_3 \rightarrow RNH_2Cl + \cdot OH$$

Oxidizing radicals, the other class of microbicidal oxidants, are formed in reactions involving $O_2^{\bullet -}$. The hydroxyl radical ($OH^{\bullet}$), the only member of this class to have been identified unequivocally as a product of stimulated phagocytes, is thought to be made by a metal-catalyzed reaction between $O_2^{\bullet -}$ and H_2O_2 (the well-known Haber-Weiss reaction):

$$O_2 + H_2O_2 \rightarrow \textit{(Fe or Cu)} \rightarrow \cdot OH + OH^- + O_2$$

Other microbicidal oxidizing radicals are undoubtedly produced as well, but (apart from the likelihood that $O_2^{\bullet -}$ serves as a precursor) their nature and origin are presently obscure.

Neutrophils and macrophages are able to protect themselves by detoxifying the toxic ROS intermediates that they generate. Granulocyte self-protection is achieved in reactions employing the dismutation of superoxide anion to hydrogen peroxide by superoxide dismutase and the conversion of hydrogen peroxide to water by catalase.

Phagocytes are estimated to produce relatively high amounts of $O_2^{\bullet -}$, on the order of 0.5 mM/sec. Taking into account the rate of spontaneous dismutation and approximate volume of the phagosome, this is estimated to yield 50 µM $O_2^{\bullet -}$ at pH 7.4 or 2 µM at pH 4.5 (the approx. pH of the acidified phagosome). Because the resulting H_2O_2 can diffuse across membranes, including out of the phagosome, the steady state concentration of the H_2O_2 would be approximately 1–4 µM. The H_2O_2 that diffuses into the bacterial cytoplasm could potentially cause damage. Indeed, a dogma is that the phagocytic ROS kill bacteria by damaging DNA.

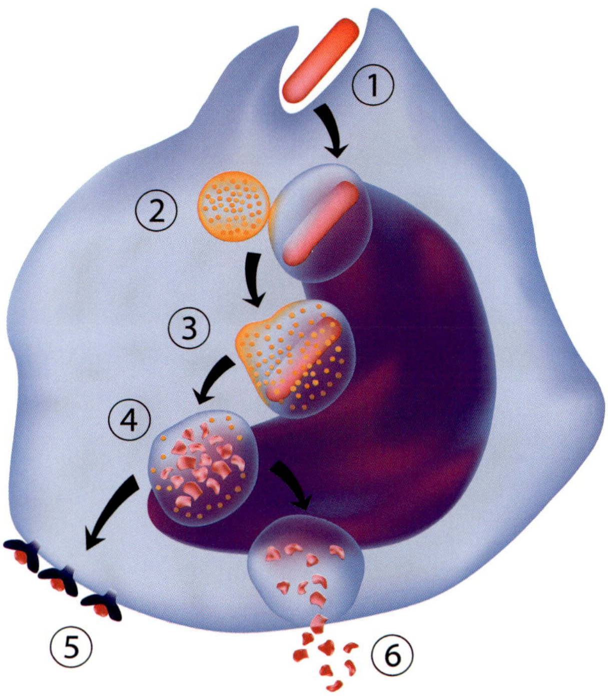

ROS and Phagocytosis

1) Phagocytosis of enemy cell (antigen)

2) Fusion of lysosome and phagosome

3) ROS start to degrade enemy cell

4) Enemy cell broken into fragments

5) Antigen fragments presented on APC surface

6) Leftover fragments released by exocytosis.

3.1.2. ROS Generators

Specific oxidative enzymes are widely distributed in biological systems for ROS generation. Such enzyme catalyzed, kinetic regulation of free radical formation is important for several vital functions including digestive process in the gastro-intestinal lumen, microbial killing of phagocytes, bone turnover of the skeletal system, vascular tone of the circulatory system and synaptic transmission of the central nervous system (CNS). This highly controlled REDOX generation provides molecular movement, cell signaling capacity and metabolic grid for the body. Three major ROS generating enzyme systems, xanthine oxidoreductase (XOR), NADPH oxidase (NOx) and nitric oxide synthase (NOS), are fundamental for the REDOX-based function of the human body. Different homologs or isotypes of these enzymes exist in the human body based on tissue location and system function. Any regulatory disruption (qualitative or quantitative) of these enzymes results in severe pathological conditions ranging from inflammatory diseases (Crohn's disease, Atherosclerosis, arthritis, asthma) to chronic neurodegenerative disorders such as Alzheimers and Parkinson's. These ROS generators are vital of REDOX homeostasis.

A) Xanthine Oxido-Reductase (XOR)

XOR exists as xanthine dehydrogenase (XDH) and xanthine oxidase (XO), both of which metabolize purines to form uric acid. The enzymes are molybdopterin-containing flavoproteins that consist of two identical subunits. The enzyme from mammalian sources, including man, is synthesized as the XDH form; however, it can be readily converted to the XO form by oxidation of sulfhydryl residues or by proteolysis. In its oxidase form, XOR produces $O_2^{\bullet-}$ and H_2O_2 when O_2 is used as an electron acceptor during purine metabolism. Superoxide and hydrogen peroxide can negatively impact multiple metabolic processes in the body either independently or after reaction with NO.

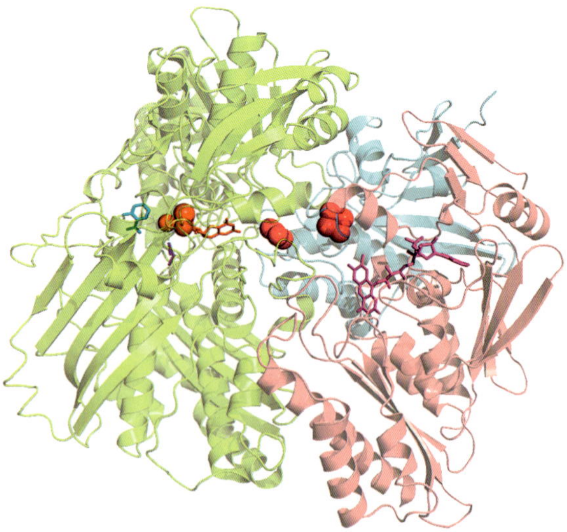

XANTHINE OXIDO-REDUCTASE (XOR)

Wide tissue distribution of XOR may implicate several physiological functions. Expression of XOR is high in intestinal epithelial cells, where a microbicidal role for this enzyme has been proposed. The enzyme is particularly abundant in the first part of the digestive tract, and it is found in goblet cells and enterocytes of the small intestine, especially in the basal and apical layers. It is also present in epithelial Paneth cells, which are recognized to play an antimicrobial defensive role. The enzyme was also present in the apical cell layers of esophagus and tongue epithelia. XOR has an antimicrobial function in the gut, which results from the generation of ROS, $O_2^{\bullet-}$, and H_2O_2. High XOR activity was also reported in the epithelial cells from skin, vagina, uterus, penis, liver, oral and and nasal cavities. In addition, activity was also observed in the sinusoidal cells of liver and adrenal cortex, endothelial cells of various organs, and connective tissue fibroblasts.

XOR Physiology: XO and XDH participate in several biochemical reactions including the hydroxylation of various purines, pterins, and aromatic heterocycles, as well as aliphatic and aromatic aldehydes, thereby contributing to the detoxification or activation of endogenous compounds and xenobiotics. One of XOR's primary roles is the conversion of hypoxanthine to xanthine and xanthine to uric acid. XOR is a generator of oxygen radicals in polymorphonuclear, endothelial, epithelial, and connective tissue cells; thus, plays an important role in cellular oxidative status. During free radical metabolism and oxidative stress, XO is a major cellular source of $O_2^{\bullet-}$, and therefore, a causative factor in ischemia and reperfusion damage. On the other hand, uric acid (a product of the XDH catalytic reaction) is a strong protective agent against oxidative damage of lipids, proteins, and purines. Also, XOR serves as a messenger or mediator in the activation of neutrophils, T-cells, cytokines, or transcription in defense mechanisms. XOR plays an important role in intestinal iron absorption and iron mobilization in the liver. Dietary iron exists in ferrous state, which is oxidized and converted to the ferric state by XO in the mucosal cell. XO is also capable of oxidizing a wide variety of substrates such as pteridenes and aldehydes. XOR serves as a rate-limiting enzyme to catalyze the nucleic acids.

XOR Pathophysiology: XOR is an essential component of innate immunity. XOR regulates leukocyte adhesion and contributes to cytokine-induced acute toxicity. XOR plays a fundamental role in inflammation and inflammatory diseases. XOR is involved in a wide range of pathophysiological conditions such as Crohn disease, acute pancreatitis, lens induced uveitis, atherosclerosis, chronic heart failure, essential hypertension, and diabetic vascular injury. In the lung, inhibition of XOR is protective in chronic obstructive pulmonary disease airways, ischemia reperfusion injury, acute lung injury, and other respiratory disorders that exhibit an inflammatory component. Xanthinuria is a rare genetic disorder where the lack of XO leads to high concentration of xanthine in blood and can cause health problems such as renal failure. Inhibition of XO has been proposed as a mechanism for improving cardiovascular health.

REDOX enzymes transfer electrons to any good acceptor, which is encountered in their path. Electron transfer to oxygen creates $O_2^{\bullet-}$ and H_2O_2; both are partially reduced species that can oxidize bio-molecules.

XOR-inhibitors: Allopurinol, a xanthine oxidase inhibitor was discovered and developed for the treatment of gout. Gertrude B Elion and George H Hitchings were jointly awarded the 1988 Nobel Prize in Physiology or Medicine for this new REDOX therapeutic agent.

B) NADPH Oxidase (NOx)

NOx is a superoxide ($O_2^{\bullet-}$) generating enzyme first identified in phagocytes, and was determined to be cellularly essential for their antimicrobial activities. Such NADPH-dependant $O_2^{\bullet-}$ generation was also identified in several non-phagocytic cells. The biological function of NOx enzymes is therefore the generation of ROS. The catalytic subunit of these oxidases, NOx, constitutes the NOx family. The NOx family NADPH oxidases are proteins that transfer electrons across biological membranes. In general, the electron acceptor is oxygen and the product of the electron transfer reaction is $O_2^{\bullet-}$. All NOx family members are trans-membrane proteins that transport electrons across biological membranes to reduce oxygen to $O_2^{\bullet-}$. In accordance with this preserved function, there are conserved structural properties of NOx enzymes that are common to all family members.There are five homolog enzymes in the family, NOx1 to NOx5, and two related enzymes, DUOx1 and DUOx2.

NOx1 is most highly expressed in colon epithelium; however, it is also expressed in a variety of other cell types, including vascular smooth muscle cells, endothelial cells, uterus, placenta, prostate, osteoclasts, retinal pericytes, as well as in several cell lines, such as the colon tumor cell lines and the pulmonary epithelial cell lines. In addition to its constitutive expression in a variety of tissues, NOx1 is induced under many circumstances. For example, vascular smooth muscle, platelet-derived growth factor (PDGF), prostaglandins, and angiotensin II; all induce NOx1 expression. NOx1 is upregulated in restenosing carotid artery after injury and in prostate as a response to castration.

NOx2 was first described in neutrophils and macrophages and is often referred to as the phagocyte NADPH oxidase (NOx). NOx2 is found in a large number of tissues, including thymus, small intestine, colon, spleen, pancreas, ovary, placenta, prostate, and testis. This wide tissue distribution is due to the presence of phagocytes and/or blood in these sites. NOx2 is also expressed in non-phagocytic cells including neurons, cardiomyocytes, skeletal muscle myocytes, hepatocytes, endothelial cells and hematopoietic stem cells. In phagocytes, NOx2 is intra-cellular, localized in close association with the membrane proteins. In resting neutrophils, most of the NOx2 is found in the intra-cellular compartments, in particular, the secondary granules (i.e. lactoferrin) and tertiary (i.e., gelatinase-containing) granules. During phagocyte stimulation, NOx2 is translocated to the surface as the granules fuse with the phagosomal or the plasma membrane. This fusion is the key event for the microbicidal activity of NOx2. In cells other than phagocytes, the sub-cellular distribution varies depending on the specific cell type. In smooth muscle cells, NOx2 is found to co-localize with the perinuclear cytoskeleton. In hippocampal neurons, NOx2 is suggested to be localized in the membranes of synaptic sites, possibly playing a role in $O_2^{\bullet-}$-dependent long-term potentiation and memory function.

TABLE: *Tissue Distribution/Localization an Pathophysiology of NOx Homologs*

Type	Tissue Localization	Pathophysiology
NOx1	Colon, smooth muscle, endothelium, uterus, placenta, prostate, osteoclasts, retinal pericytes	Hypertension, aneurysm, neointima formation, inflammatory pain, cerebral ischemia, neuroinflammation, hyperoxia-induced acute lung injury, colorectal cancer
NOx2	Phagocytes, B lymphocytes, neurons, cardiomyocytes, skeletal muscle, hepatocytes, endothelium, smooth muscle	Cardiac hypertrophy, fibrosis, heart failure, myocardial infarction, Alzheimer's disease, Parkinson's disease, schizophrenia, liver fibrosis, liver ischemia and reperfusion injury, amyotrophic lateral sclerosis (ALS)
NOx3	Inner ear, fetal kidney, fetal spleen, skull bone, brain	Development of the otoconia, hearing loss, insulin resistance (diabetes)
NOx4	Kidney, blood vessels, osteoclasts, endothelium, fibroblasts, keratinocytes, melanoma cells, neurons	Mitochondrial dysfunction, intestinal fibrosis, sympathetic nerve activity, pulmonary fibrosis, pulmonary hypertension, diabetic nephropathy, renal cancer
NOx5	Lymphoid tissue, testis, pancreas, placenta, ovary, uterus, stomach	Barrett's esophagus, prostate cancer
DUOx1	Thyroid, airway epithelia, tongue epithelium, cerebellum, testis	Host defense
DUOx2	Thyroid, salivary and rectal glands, GI, uterus, gall bladder, pancreas	Hypothyroidism, host defense

NOx3 expression is very high in the inner ear, including the cochlear and vestibular sensory epithelia and the spiral ganglion. Low levels of NOx3 can also be detected in other tissues, including fetal spleen, fetal kidney, skull bone, and brain.

NOx4 was originally identified as an NADPH oxidase homolog highly expressed in the kidney. In addition to its strong expression in the kidney, NOx4 is also found in osteoclasts, endothelial cells, smooth muscle cells, hematopoietic stem cells, fibroblasts, keratinocytes, melanoma cells, and neurons. Induction of NOx4 expression is observed under conditions in response to endoplasmic reticulum (ER) stress, shear stress, carotid artery injury, hypoxia and ischemia, and transforming growth factor (TGF)-α1 and tumor necrosis factor (TNF)-α stimulation of smooth muscle.

NOx5 expression is described in testis, spleen, lymph nodes, vascular smooth muscle, bone marrow, pancreas, placenta, ovary, uterus, stomach, and in various fetal tissues. Within the testis, NOx5 is localized in spermatocytes. NOx5 shows a distinct localization in mature B cells and T lymphocytes. Interestingly, NOx5 could not be detected within circulating lymphocytes.

DUOx1 and DUOx2 (originally called thyroid oxidase) are highly expressed in the thyroid. In addition, DUOx1 has been described in airway epithelia as well as in the prostate. DUOx2 is found in the ducts of the salivary gland; in rectal mucosa; all along the gastrointestinal tract including duodenum, colon, and cecum; in airway epithelia; and in prostate. DUOx1 could be induced with interleukin response in the respiratory tract epithelium. DUOx2 expression could be induced in response to interferon-γ in respiratory tract epithelium and in response to insulin in thyroid cell lines.

NOx Physiology

i) *Host defense and inflammation:* Host defense is an important role for members of the NOx enzyme family, mediated through several oxygen-dependent killing mechanisms. The mechanism of NOx2 in microbial killing is mediated by directly via ROS and indirectly through moduation of pH and ion homeostasis. There is mounting evidence that NOx enzymes have a role in limiting the inflammatory response. This anti-inflammatory activity of NOx represents a regulatory function. Bacterial killing by H_2O_2 derived from $O_2^{\bullet-}$ has been suggested as a mechanism for NOx activity. DUOx2-derived H_2O_2 in combination with lactoperoxidase in the digestive tract is yet another antimicrobial mechanism. NOx enzymes may also inactivate microbial virulence factors through the H_2O_2/myeloperoxidase system. It is also important to note that phagocytes have a number of oxygen-independent killing mechanisms, in particular, the release of microbicidal proteins from the granules, such as lactoferrin.

ii) *Effects on cell signaling:* NOx-derived ROS can regulate protein tyrosine phosphorylation in several different cell types.

iii) *Effects on gene expression:* NOx-dependent ROS generation can induce the expression of TNF-α, TGF-α1, angiotensin II, monocyte chemo-attractant protein-1, and plasminogen activator inhibitor-1. NOx-derived ROS can affect gene expression through NF-κB and AP-1. NOx-derived ROS may also alter gene expression through the alternation of mRNA stability.

iv) *Effects on cellular death and cellular senescence:* NOx-derived ROS have a pro-survival effect. It may act as anti-apoptotic signal through activation of the NF-κB pathway. NOx activation is most commonly associated with cell death; however, under certain circumstances it may be anti-apoptotic.

v) *Regulation of cell growth:* NOx-derived ROS may lead to either cellular senescence or to enhanced cell growth.

vi) *Role in oxygen sensing:* Traditionally NOx enzymes were thought to be involved in oxygen sensing through a decreased ROS generation in hypoxic tissues. ROS generation by NOx enzymes depends on the concentration of its electron acceptor, i.e., oxygen. Indeed, when NOx2-dependent respiratory burst is measured at O_2 levels below 1%, there is a steep drop in ROS generation.

vii) *Role in biosynthesis and protein cross-linking:* Peroxidation reactions are important in physiology. One of the best-documented roles for NOx enzymes is the iodination of thyroid hormones, a reaction catalyzed by the thyroid peroxidase using DUOX-derived H_2O_2.

viii) *Regulation of cellular REDOX potential:* Disulfide bond formation is determined by the REDOX potential. NOx-dependent ROS generation has a broad influence on protein disulfide bridges in the entire cell or in restricted cellular compartments.

ix) *Reduction of metal ions:* Iron starvation of microorganisms is a potent host defense mechanism. Removal of iron from the phagosome is achieved in two steps: (A) reduction of Fe^{3+} through an oxido-reductase and (B) translocation of divalent iron across the phagosome membrane by lactoferrin. An important role for NOx2- derived $O_2^{\bullet-}$ to reduce iron has also been suggested.

x) *Regulation of matrix metalloproteinases:* NOx-derived ROS are involved in the regulation of expression and/or activation of matrix metallo-proteinases.

xi) *Role in angiogenesis:* Various pathological conditions with neo-vascularization are associated with ROS generation. Angiogenesis-inducing ROS were found to be generated by NOx2 in endothelial cells. NOx-derived ROS have also been implicated in angiogenesis during differentiation of embryonic stem cells.

xii) *Cross-talk with the NO system:* One of the key activities of NOx-derived ROS is posttranslational modification of proteins (in particular phosphatases, transcription factors, and other signaling molecules). In many instances, NOx-derived ROS interact as reversible signaling molecules. Other important activities of NOx-derived ROS include the interaction with other macromolecules (lipids, carbohydrates, nucleic acids) or with small molecules (i.e. ·NO).

NOx Pathophysiology: Mutations in the NADPH oxidase subunit genes cause 'Chronic Granulomatous Disease (CGD)', a condition in which there is a defect in NADPH oxidase; therefore, the phagocyte is unable to make the ROS required for bacterial killing, resulting in bacteria being able to survive inside the phagocyte.

NADPH oxidase is a major cause of atherosclerosis. Atherosclerosis is caused by the accumulation of macrophages containing cholesterol (foam cells) in artery walls (in the intima). NADPH oxidase produces ROS, which activates an enzyme that makes the macrophages adhere to the artery wall (by polymerizing actin fibers). This process is counter-balanced by NADPH oxidase inhibitors and by antioxidants. An imbalance in favor of ROS produces atherosclerosis. *In vitro* studies have found that NADPH oxidase inhibitors in combination with antioxidants such as N-acetyl-cystine and resveratrol can break the adhesive bonds, and allow foam cells to migrate out of the intima.

Currently available anti-hypertensive agents, angiotensin-converting enyzme inhibitors and angiotensin receptor blockers also effectively inhibit NADPH oxidase activation. Similarly, the cholesterol-lowering agents, statins, have been shown to attenuate NADPH oxidase activation. Although, antioxidants act to scavenge the ROS produced by these enzymes, their effectiveness is limited. Targeting NADPH homologues may have a distinct advantage over current therapies because it would specifically prevent the patho-physiological formation of ROS that contributes to hypertension.

C) Nitric Oxide Synthase (NOS)

Nitric Oxide Synthases are a family of enzymes that catalyze ·NO production from L-arginine. ·NO is an important cellular signaling molecule. It helps modulate vascular tone, insulin secretion, airway tone, and peristalsis, and is involved in angiogenesis and neural development. NOS catalyzes the reaction:

$$\text{L-arginine} + \tfrac{3}{2}\,\text{NADPH} + \text{H}^+ + 2\,\text{O}_2 \rightarrow \text{citrulline} + \cdot\text{NO} + \tfrac{3}{2}\,\text{NADP}^+$$

NOSs require five cofactors and the electron flow in the NO synthase reaction is:

$$\text{NADPH} \rightarrow \text{FAD} \rightarrow \text{FMN} \rightarrow \text{heme} \rightarrow \text{O}_2$$

Tetrahydrobiopterin provides an additional electron during the catalytic cycle which is replaced during turnover. NOS is the only known enzyme that binds flavin adenine dinucleotide (FAD), flavin mononucleotide (FMN), heme, tetrahydropterin, (BH4) and calmodulin.

Arginine-derived ·NO synthesis has been identified in mammals, where 3 distinct genes encode NOS isozymes, neuronal (nNOS or NOS-1), cytokine-inducible (iNOS or NOS-2) and endothelial (eNOS or NOS-3). iNOS and nNOS are soluble and found predominantly in the cytosol, while eNOS is membrane associated.

Neuronal nitric oxide synthase (nNOS) is expressed in specific neurons of the central nervous system (CNS), and is regulated by Ca^{2+} and calmodulin. nNOS exists in particulate or soluble forms, and also found in the spinal cord, sympathetic ganglia and adrenal glands, peripheral nerves, epithelial cells of various organs, kidney cells, pancreatic islet cells, and in the vascular smooth muscle. The largest source of nNOS in terms of tissue mass is the skeletal muscle.

nNOS Physiology: nNOS modulates physiological functions such as learning, memory, and neurogenesis. In the central nervous system (CNS), nNOS regulates synaptic transmission. nNOS-mediated formation of ·NO is involved in the central regulation of blood pressure. In the peripheral nervous system (PNS), nNOS-derived ·NO acts as an atypical neurotransmitter, which mediates relaxing components of gut peristalsis, vasodilation, and penile erection.

nNOS Pathophysiology: Abnormal ·NO signaling may contribute to a variety of neurodegenerative conditions (i.e. excitotoxicity following stroke, multiple sclerosis, Alzheimer's, and Parkinson's diseases. Hyperactive nNOS has been implicated in neuronal death in cerebrovascular stroke. Elevated level of ·NO inhibits mitochondrial respiration and glycolysis, which may lead to energy depletion. Increased ·NO generation by nNOS in peripheral nitrergic nerves disturb smooth muscle tone of the GI tract.

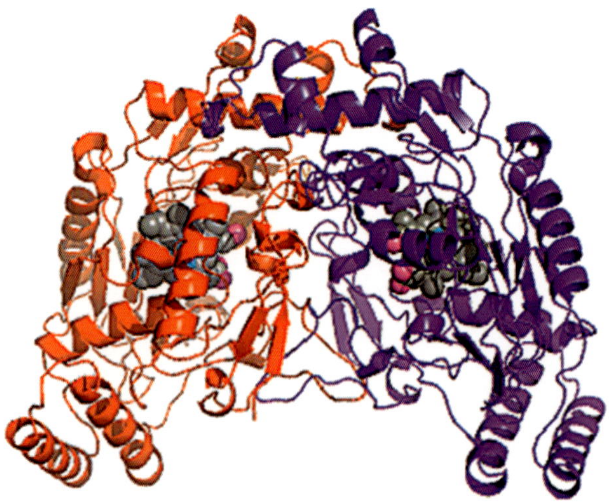

NITRIC OXIDE SYNTHASES (NOS)

Inducible nitric oxide synthase (iNOS) is not usually expressed in cells, but induced by bacterial lipopolysaccharide (endotoxin), cytokines, and other agents. Although primarily identified in macrophages, iNOS synthesis can be stimulated in virtually any cell or tissue, in the presence of appropriate inducing agent. Once expressed, iNOS is constantly active and is not regulated by intra-cellular Ca^{2+} levels.

nNOS effects pro-erectile function of phosphodiesterase 5 (PD5)inhibitors such as sildenafil (Viagra®), vardenafil (Levitra®), and tadalafil (Cialis®). PD5 is also found in pulmonary arteries, therefore, sildenafil (Revatio®) and tadalafil (Adcirca®) are used in the treatment of pulmonary arterial hypertension.

iNOS Physiology: iNOS produces large amounts of ·NO; a major cytotoxic radical of macrophages. Higher concentrations of ·NO, produced by iNOS induced macrophages, can directly interfere with the DNA of target cells and cause strand breaks and fragmentation. A combination of these reactions is likely to induce both cytostatic and cytotoxic effects of ·NO on pathogens and certain tumor cells. Interestingly, non-immune cells can also be induced with cytokines to release ·NO in large amounts can effect the adjacent cells. Cytokine-activated endothelial cells, for example, have been shown to kill tumor cells. Induced hepatocytes can use ·NO to kill malarial parasites. Inducible NOS activity is likely to be responsible for all of these effects.

iNOS Pathophysiology: High levels of ·NO produced by activated macrophages (and probably neutrophils and other cells) are not only toxic to undesired microbes, parasites, or tumor cells, but—when released at the wrong site—may also harm healthy cells. *In vivo*, cell and tissue damage can be related to the ·NO radical itself or an interaction of ·NO with O_2 leading to the formation of peroxynitrite ($ONOO^-$). iNOS-derived ·NO is also involved in non-specific allograft rejection. Inducible NOS-derived ·NO, in combination with hypoxia can induce neuronal death and results in excitotoxicity. Excessive ·NO production by iNOS also plays a crucial role in septic shock. This condition is characterized by massive arteriolar vasodilatation, hypotension, and microvascular damage. Bacterial endotoxins usually initiate these symptoms. The fall in blood pressure is predominantly due to excess ·NO production by iNOS induced in the vascular wall.

Endothelial nitric oxide synthase (eNOS) is mostly expressed in endothelial cells. It is also detected in cardiac myocytes, platelets, certain neurons of the brain, kidney tubular epithelial cells and in the human placenta. Similar to nNOS, Ca^{2+}-activated calmodulin is important for the regulation of eNOS activity. eNOS synthesizes ·NO in a pulsatile manner with its activity markedly increasing when intra-cellular Ca^{2+} rises. Ca^{2+} induces the binding of calmodulin to the enzyme. However, eNOS can also be activated by stimuli that do not produce sustained increases in intra-cellular Ca^{2+}, but still induce a long-lasting release of ·NO. The best established example of such stimulus is fluid shear stress.

Nitric Oxide (NO): The 1998 Nobel Prize in Physiology or Medicine was awarded to Robert F Furchgott, Louis J Ignarro and Ferid Murad for their discoveries concerning the nitric oxide as a signaling molecule in the cardiovascular system.

Enhanced eNOS activity could reverse impaired bone marrow cells in ischemic cardiomypathy patients. Simvastatin (Zocor®), the stimulator of eNOS, increases active endothelial progenitor cells in patients with myocardial infarction.

eNOS Physiology: (i) *Vasodilation and inhibition of platelet aggregation/adhesion* – eNOS is a homeostatic regulator of several cardiovascular functions. eNOS-derived ·NO dilates all types of blood vessels by stimulating soluble guanylyl cyclase and increasing cyclic GMP in smooth muscle cells. Deletion of the eNOS gene leads to elevated blood pressure. ·NO is a potent inhibitor of platelet aggregation and adhesion to the vascular wall, which protects from thrombosis. eNOS is also critical for adaptive vascular remodeling to chronic changes in blood flow. (ii) *Inhibition of leucocyte adhesion and vascular inflammation* – eNOS controls the expression of genes involved in atherogenesis. ·NO could inhibit leucocyte adhesion to the vessel wall, an early event in the development of atherosclerosis, which may protect against the onset of atherogenesis. A disturbed integrity of the endothelial monolayer barrier can initiate proinflammatory events. eNOS-derived ·NO prevents endothelial cell apoptosis induced by pro-inflammatory cytokines and pro-atherosclerotic factors including ROS and angiotensin II. The suppression of apoptosis may also contribute to the anti-inflammatory and anti-atherosclerotic effects of endothelium-derived ·NO. (iii) *Control of vascular smooth muscle proliferation* – furthermore, eNOS-derived NO could inhibit DNA synthesis, mitogenesis, and proliferation of vascular smooth muscle cells. ·NO also prevents a later step in atherogenesis, fibrous plaque formation. Based on theose combination effects, ·NO produced in endothelial cells can be considered an anti-atherosclerotic principle. (iv) *Stimulation of angiogenesis* – eNOS-derived ·NO plays a critical role in post-natal angiogenesis, mediating signals downstream of angiogenic factors. (v) *Activation of endothelial progenitor cells* – eNOS expressed by bone marrow stromal cells influences recruitment of stem and progenitor cells.

Stages of Atherosclerosis

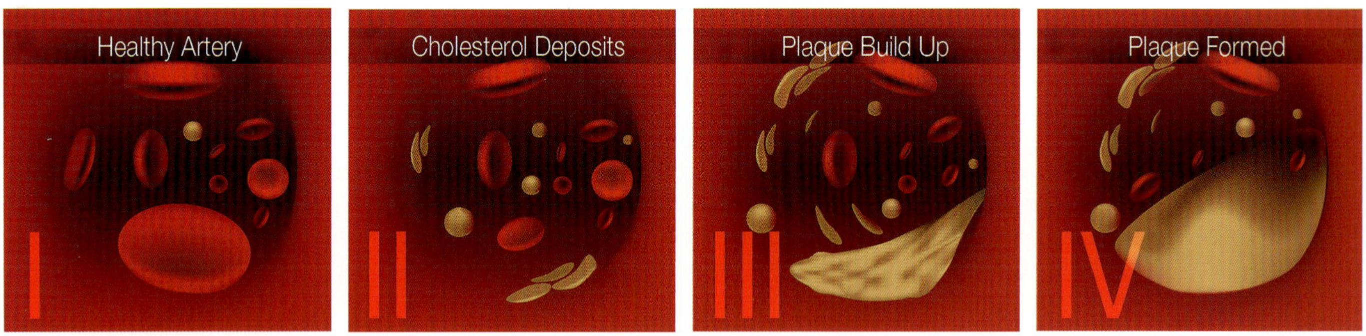

Statins: Michael Brown and Joseph Goldstein were awarded the 1985 Nobel Prize in Physiology or Medicine for discovering the LDL-receptor and its role in regulating cholesterol. This led to the development of cholesterol-lowering HMG CoA reductase inhibitors, or statins.

eNOS Pathophysiology: Patients with cardiovascular risk factors (such as hypertension, hypercholesterolemia, diabetes mellitus, etc.) and with vascular disease show endothelial dysfunction, i.e. inability of the endothelium to generate adequate amounts of bioactive ·NO (and ·NO-mediated vasodilation). Cardiovascular risk factors and vascular disease are also associated with an increased production of ROS. There are several enzyme systems that generate ROS in the vessel wall. These include the NOx, XOR and uncoupled eNOS. Of these, NOx are of primary importance for ROS generation. Due to enhanced oxidative stress in vascular disease, an increased degradation of NO will occur. However, oxidative stress converts eNOS from an ·NO-producing enzyme to an enzyme that generates O_2.

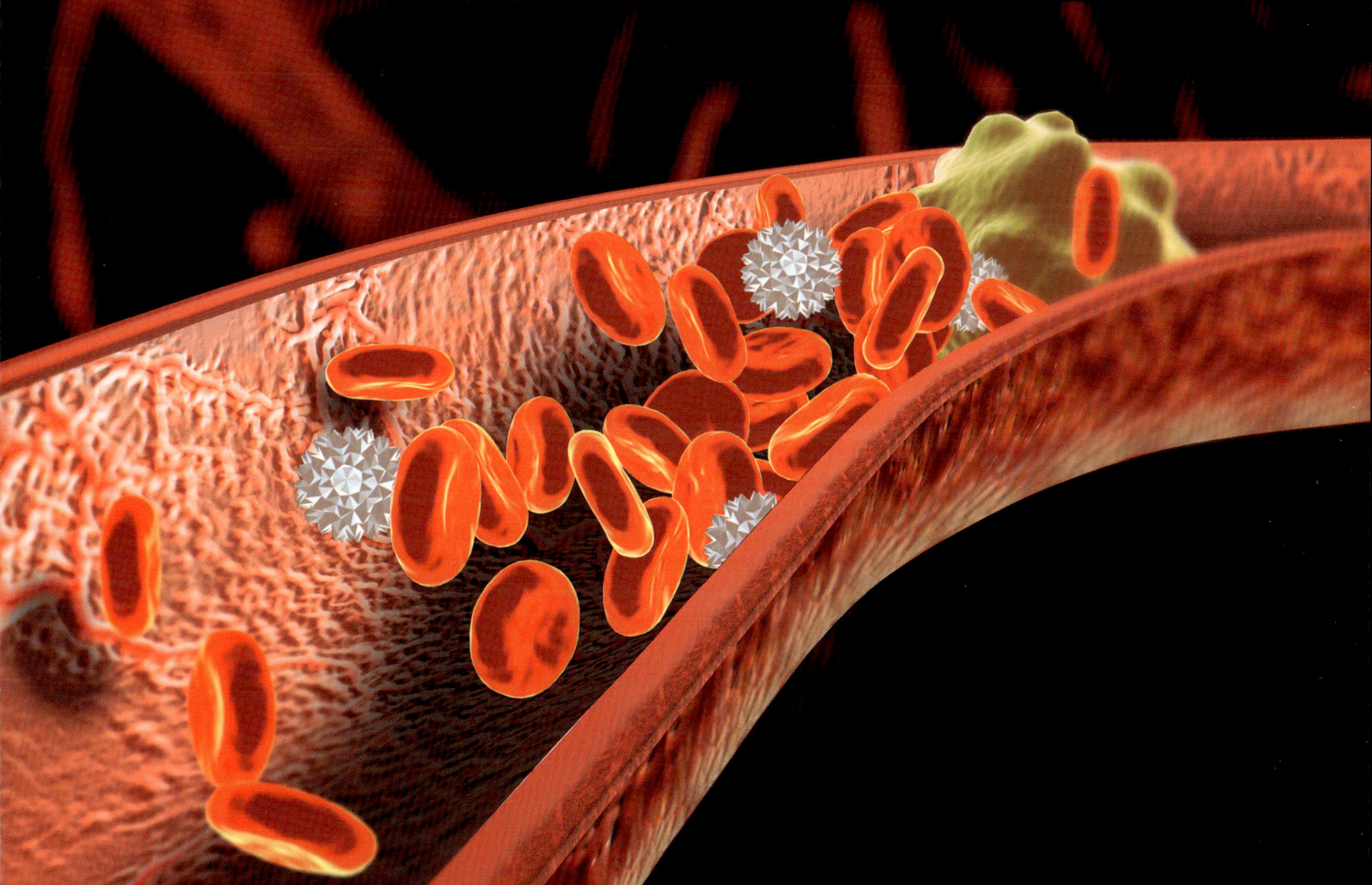

Reduced systemic ·NO bioactivity in ischemic heart disease may contribute to impaired neo-vascularization. During such condition, bone marrow mononuclear cells show reduced neo-vascularization. Accordingly, ·NO plays an important role in neo-vascularization.

3.2 REDOX: Antioxidants

Living organisms are equipped with enzymatic and non-enzymatic antioxidant systems to eliminate free radicals as part of cellular defense and to maintain REDOX homeostasis. Broadly, antioxidant systems can be divided into different categories based on their functions (enzymatic or non-enzymatic), their physical properties (water-soluble or lipid-soluble), and, for humans, their sources (endogenous or exogenous). Some antioxidants are enzymes or serve as co-factors for enzymes. The major non-enzymatic cellular REDOX buffer systems rely on water-soluble antioxidants such as glutathione (GSH) and ascorbic acid. However, several other less abundant antioxidants also serve important functions in REDOX homeostasis and are important for cellular function and disease prevention. These antioxidant molecules function not only in quenching chain reactions, scavenging ROS, and preventing lipid peroxidation, but also influence cellular functions by binding to specific receptors to initiate signal transduction cascades, inhibiting or activating enzymes, and regulating gene expression.

3.2.1. Antioxidant Enzymes

A major class of enzymatic antioxidants, which catalyze the dismutation of $O_2^{\bullet-}$ to H_2O_2, is known as superoxide dismutase (SOD). Multiple isoforms of SOD exist in different cellular compartments. Further conversion of H_2O_2 to $H_2O + O_2$ occurs through the action of catalase, a heme-based enzyme that is normally localized in the peroxisome. Catalase has high substrate-turnover rates, scavenging ~6 million molecules of H_2O_2 per minute. H_2O_2 also can be converted to O_2 through coupled reactions with the conversion of reduced glutathione (GSH) to oxidized glutathione (GSSG), catalyzed by glutathione peroxidase (GPx). Five isoforms of selenium (Se)-dependent GPxs are found in humans.

NADPH, thioredoxin (Trx) and Trx-reductase enzyme form the TRX system, which is present in all living organisms. The TRX system is responsible for maintaining a reducing environment in the cytosol with a low REDOX potential. Similar function is also provided by GSH, glutaredoxins, and NADPH-dependant GSH-reductase. In humans, Trx is a stress-inducible antioxidant factor. Glutaredoxin, also known as thioltransferase, is an important component of the GSH system that contributes to defense against oxidative stress. Oxidative processes may result in structural and functional changes in proteins by affecting methionine residues. However, in contrast to other forms of oxidative damage, oxidation of methionines is reversible due to antioxidant enzyme, methionine sulfoxide reductase. This enzyme activity is considered to be an important process that protects cells against oxidative stress, regulates protein function and delays the aging process.

A) Superoxide dismutase (SOD)

SODs are oxidoreductases, a group of antioxidant enzymes found in all living cells. The discovery of SOD by McCord and Fridovich (1968); the concomitant finding that superoxide can be produced by enzymes (1969); and that free radicals can be substrates for an enzyme completely revolutionized the field of REDOX biology. SOD not only blunts the cascade of oxidations initiated by its substrate, $O_2^{\bullet-}$ free radical; but also initiates a plethora of cell signaling pathways. The mechanism of dismutation of $O_2^{\bullet-}$ to H_2O_2 by SOD involves alternate reduction and reoxidation of a REDOX active transition metal, such as copper (Cu) and manganese (Mn) at the active site of the enzyme as depicted in the following half-reactions. This indicates that SOD activity requires a catalytic metal [where M= copper (Cu), manganese (Mn), iron (Fe) or nickel (Ni)]:

$$M^{(n+1)+}\text{-SOD} + O_2^{\bullet-} \rightarrow M^{n+}\text{-SOD} + O_2$$
$$M^{n+}\text{-SOD} + O_2^{\bullet-} + 2H^+ \rightarrow M^{(n+1)+}\text{-SOD} + H_2O_2$$

Three forms of SOD are present in humans: SOD1 (Cu-Zn) located in the cytoplasm, SOD2 (Mn) in the mitochondria, and the extra-cellular SOD3 (Cu-Zn). The distinct subcellular location of these SOD isoforms is particularly important for compartmentalized REDOX signaling.

TABLE: *SOD Isoforms, Metal Co-factors and Tissue Localization*

Type	Isoform	Metal co-factor	Location
SOD1	Cytosolic Cu/Zn-SOD	Cu^{2+} (catalytic) Zn^{2+} (stability)	Cytoplasm, mitochondrial IMS, nucleus, lysosomes, peroxisomes
SOD2	Mitochondrial Mn-SOD	Mn^{2+} (catalytic)	Mitochondrial matrix
SOD3	Extra-cellular Cu/Zn-SOD	Cu^{2+} (catalytic) Zn^{2+} (stability)	Extra-cellular matrix and fluids, cell surface, extracelluar fluids

SOD1 (cytosolic Cu/Zn-SOD): SOD1 is the major intra-cellular SOD, and is mainly localized in the cytosol with a smaller fraction in the inter-membrane space (IMS) of mitochondria. SOD1 is sensitive to cyanide that distinguishes it from SOD2 (which is relatively resistant). Enzymatic activity of SOD1 depends on the presence of the Cu and Zn. Further, Cu is not replaceable with another metal, whereas Zn is replaceable with cobalt and Cu, and is not necessary for enzyme activity at low pH. More importantly, the mechanism of scavenging $O_2^{\bullet-}$ by SOD involves alternate reduction and reoxidation of the Cu at the active site of the enzyme. Thus, SOD1 activity requires a catalytic Cu to scavenge $O_2^{\bullet-}$. Under physiological conditions, the level of intra-cellular free Cu is extraordinarily restricted. Thus, soluble Cu carrier proteins termed 'Copper chaperones' are required to directly transfer Cu to specific cellular targets. SOD1 has a half-life of 6–10 min in the sera.

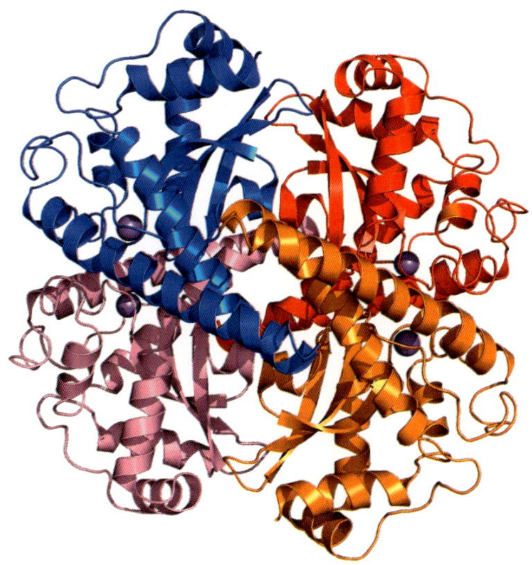

SUPEROXIDE DISMUTASE (SOD)

SOD2 (mitochondrial Mn-SOD): SOD2 is Mn-containing enzyme localized in the mitochondrial matrix. Manganese at the active site of SOD2 serves to catalyze the disproportionation of $O_2^{\bullet -}$ to oxygen and H_2O_2 in a similar fashion as SOD1 and SOD3 (Cu/Zn-SODs). It is synthesized in the cytoplasm and directed to the mitochondria by a signal peptide, where it is involved in dismutating $O_2^{\bullet -}$ generated by the respiratory chain of enzymes. Also, SOD2 has a half-life in sera of 5–6 hr. The essential role of SOD2 in maintaining mitochondrial function is demonstrated by the neonatal lethality of mice with targeted disruption of the gene for SOD2. Such complete ablation of SOD2 causes dilated cardiomyopathy and neurodegeneration leading to early postnatal death.

SOD3 (extra-cellular Cu/Zn-SOD and ec-SOD): The primary location of SOD3 in tissues is in the extra-cellular matrix and on cell surfaces with a smaller fraction in the plasma and extra-cellular fluids. Tissue SOD3 is thought to account for 90%–99% of the SOD3 in the body. SOD3 is expressed highly in selected tissues, such as blood vessels, the lung, kidney, uterus, and, to a lesser extent, in heart. SOD3 activity is extremely sensitive to cyanide, like SOD1. In vascular tissue, SOD3 is mainly synthesized by vascular smooth muscle cells and fibroblasts. In injured tissue and atherosclerosis, SOD3 is also found in inflammatory cells. SOD3 plays an important role in various oxidative stress-dependent diseases, including hypertension, heart failure, ischemia-reperfusion injury, and lung injury.

The end product of dismutation reaction is hydrogen peroxide (H_2O_2), a harmful byproduct of many normal metabolic processes. In order to prevent damage to cells and tissues, H_2O_2 must be quickly converted into other, less dangerous substances. Enzymes such as catalases are frequently used by cells to quickly decompose hydrogen peroxide into less-reactive gaseous oxygen and water molecules. However, the predominant scavengers of H_2O_2 in normal mammalian cells are likely peroxiredoxins rather than catalase.

B) Catalase

Catalase is a vital enzyme found in nearly all living organisms exposed to oxygen. It catalyzes the decomposition of H_2O_2 to water and oxygen. A single catalase molecule can convert millions of molecules of H_2O_2 to water and oxygen each second. It contains four porphyrin heme (iron) groups that allow the enzyme to react with the H_2O_2. The optimum pH for human catalase is ~7.0 and works at an optimum temperature of 37°C, which is approximately the temperature of the human body. Catalase is usually located in a cellular, bipolar environment organelle called the peroxisome.

Oxidative Enzymes: The Nobel Prize in Physiology or Medicine 1955 was awarded to Hugo Theorell for his discoveries concerning the nature and mode of action of oxidation enzymes.

The enzyme consists of 4 protein subunits, each of which contains ferric (Fe^{3+}) ions of the heme group that undergo oxidation following interaction with the first molecule of H_2O_2 to produce a structure called Compound 1. A second molecule of H_2O_2 serves as an electron donor and results in the destruction of the two H_2O_2 molecules involved to produce an oxygen molecule as follows:

CATALASE

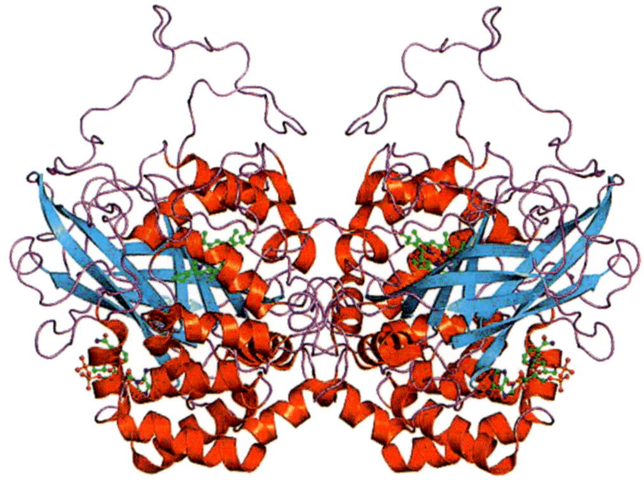

$$H_2O_2 \rightarrow (\textit{catalase}) \rightarrow \{\text{Compound 1}\}$$

$$\{\text{Compound 1}\} + H_2O_2 \rightarrow O_2 + 2\,H_2O$$

$$2H_2O_2 \rightarrow (\textit{catalase}) \rightarrow O_2 + 2\,H_2O$$

Catalase contributes to ethanol metabolism in the body after ingestion of alcohol, but it only breaks down a small fraction of the alcohol in the body. A catalase deficiency may increase the risk of developing type 2 diabetes. Also, low levels of catalase may play a role in the graying process of human hair. Hydrogen peroxide is naturally produced by the body and catalase breaks it down. If catalase levels decline, hydrogen peroxide cannot be broken down as well. This allows the H_2O_2 to bleach the hair from the inside out.

C) Peroxidase

In contrast to catalase, peroxidase possesses high affinity for and can remove H_2O_2 even when it is present in low concentration; however, the electron donors in these reactions are small molecules, such as glutathione (GSH) or ascorbate. Thus, the removal of H_2O_2 is an 'expensive' reaction from the cell's point of view, as it consumes valuable molecules in the cellular environment; two molecules of glutathione are consumed for the removal of 1 molecule of H_2O_2. That no oxygen is produced in the latter reaction distinguishes the activity of peroxidase from that of catalase, as shown:

$$2\,GSH + H_2O_2 \rightarrow (\textit{peroxidase}) \rightarrow GSSG + 2\,H_2O$$

For many of these enzymes the optimal substrate is hydrogen peroxide, but others are more active with organic hydroperoxides such as lipid peroxides. Peroxidases can contain a heme co-factor in their active sites, or alternately REDOX-active cysteine or selenocysteine residues.The glutathione peroxidase (GPx) family consists of 8 known human isoforms. Glutathione peroxidases (GPx) use glutathione (GSH) as an electron donor and are active with both hydrogen peroxide and organic hydroperoxide substrates.

Glutathione peroxidase (GPx) is the general name of an enzyme family with peroxidase activity whose main biological role is to protect the organism from oxidative damage. The biochemical function of glutathione peroxidase is to reduce lipid hydroperoxides to their corresponding alcohols and to reduce free hydrogen peroxide to water.

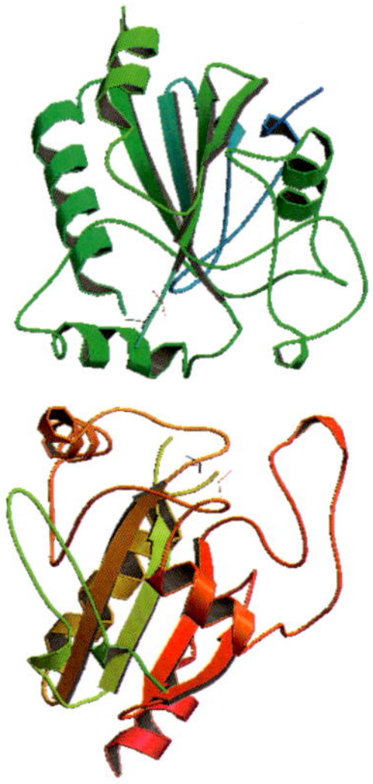

GPx

Several isozymes are encoded by different genes, which vary in cellular location and substrate specificity. Glutathione peroxidase-1 (GPx1) is the most abundant version, found in the cytoplasm of nearly all mammalian tissues, whose preferred substrate is hydrogen peroxide. Glutathione peroxidase 4 (GPx4) has a high preference for lipid hydroperoxides; it is expressed in nearly every mammalian cell, though at much lower levels. Glutathione peroxidase-2 is an intestinal and extra-cellular enzyme, while glutathione peroxidase-3 is an extra-cellular enzyme found abundantly in the plasma. So far, eight different isoforms of glutathione peroxidase (GPx1-8) have been identified in humans.

Haloperoxidases are peroxidases that are able to mediate the oxidation of halides by H_2O_2. The following table shows the classification of haloperoxidases according to the halides whose oxidation they are able to catalyze:

Haloperoxidase	Oxidizable halide	Origin
Chloroperoxidase (CPO)	Chlorine Bromine Iodine	Neutrophiils (myeloperoxidase) Eosinophils (eosinophil peroxidase)
Bromoperoxidase (BPO)	Bromine Iodine	Milk, saliva, tears (lactoperoxidase)
Iodoperoxidase (IPO)	Iodine	Thyroid (thryroid peroxidase)

Myeloperoxidase (MPO) is a chloroperoxidase, most abundantly expressed in neutrophil granulocytes. It is a lysosomal protein stored in azurophilic granules of the neutrophil. MPO has a heme pigment, which results in the green color found in secretions rich in neutrophils, such as pus and some forms of mucus. MPO produces hypochlorous acid (HOCl) from hydrogen peroxide (H_2O_2) and chloride anion (Cl^-) (or the equivalent from a non-chlorine halide) during the respiratory burst of neutrophils. It requires heme as a cofactor. Furthermore, it oxidizes tyrosine to tyrosyl radical using hydrogen peroxide as an oxidizing agent. Hypochlorous acid and tyrosyl radical are cytotoxic, so they are used by the neutrophil to kill bacteria and other pathogens. MPO deficiency is hereditary, a predisposing factor for immune-compromised condition. MPO plays a significant role in the development of the atherosclerosis.

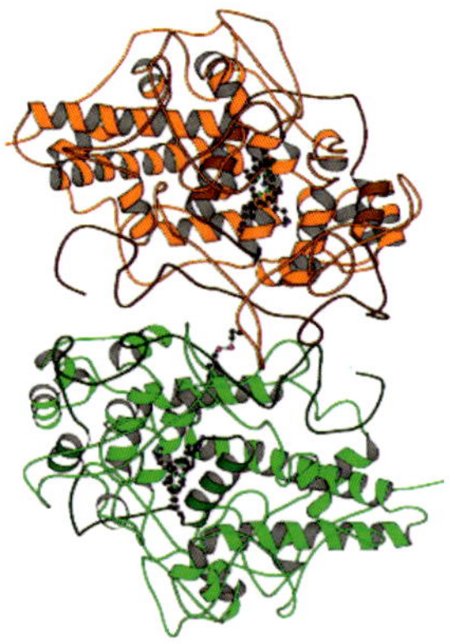

MYELOPEROXIDASE (MPO)

Lactoperoxidase (LPO) is secreted from mammary, salivary, and other mucosal glands with potent antimicrobial activity. LPO catalyzes the oxidation of a number of inorganic and organic substrates by H_2O_2. These substrates include bromide and iodide, and therefore, LPO is categorized as a haloperoxidase. Also, another important substrate for LPO is thiocyanate. The oxidized products produced through the action of this enzyme have potent bactericidal effects. LPO together with its inorganic ion substrates, hydrogen peroxide, and oxidized products constitutes the lactoperoxidase system.

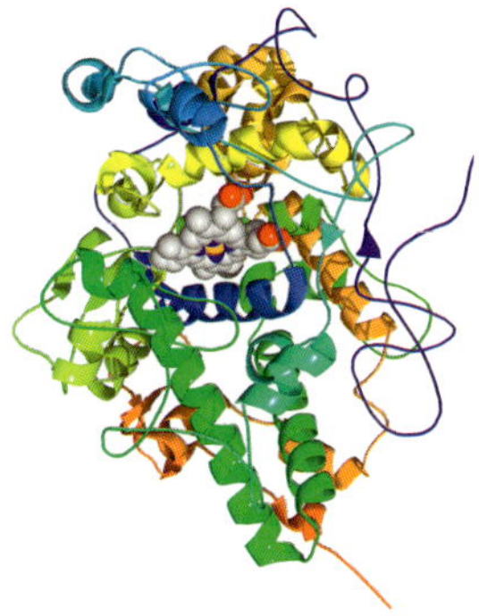

LACTOPEROXIDASE (LPO)

Lactoperoxidase catalyzes the H_2O_2 oxidation of several acceptor molecules, specific examples include:

> thiocyanate (SCN^-) → hypothiocyanite ($OSCN^-$)
> bromide (Br^-) → hypobromite (BrO^-)
> iodide (I^-) → hypoiodite (IO^-)

Source of the H_2O_2 is usually from the reaction of glucose with O_2 in the presence of the enzyme glucose oxidase, and takes place in saliva. In its antimicrobial capacity, LPO appears to acts synergistically with lactoferrin and lysozyme.

Thyroid peroxidase or thyroperoxidase (TPO) is expressed mainly in the thyroid, which liberates iodine for addition onto tyrosine residues on thyroglobulin for the production of thyroid hormones - thyroxine (T_4) or triiodothyronine (T_3). Inorganic iodine enters the body primarily as iodide, (I^-); after entering the thyroid follicle, TPO oxidizes iodide to atomic iodine (I) or iodinium (I^+).

Other enzymes also exist in the cellular environment to support the activity of the above three antioxidant enzymes – SOD, catalase and peroxidase. For example, glucose-6-phosphate dehydrogenase provides reducing equivalents (NADPH) necessary for cellular function and important for the regeneration of oxidized antioxidants. Also, enzymes such as xanthine dehydrogenase can produce uric acid, which serves as an effective endogenous antioxidant.

TPO and Thyroid Hormones

TPO enzyme plays a vital role in the function of the thyroid gland, a butterfly-shaped tissue in the lower neck. TPO assists the chemical reaction that adds iodine to a protein called thyroglobulin, a critical step in generating thyroid hormones – thyroxine (T_4) or tri-iodo-thronine (T_3). Thyroid hormones regulate growth, and development of the brain; and also control the rate of chemical reactions in the body (metabolism). To function properly, TPO must be located in the cell membrane of certain thyroid cells, called follicular cells. TPO has several different versions (isoforms), which vary by size and location within the cell. Some versions are inactive because they are not located in the cell membrane.

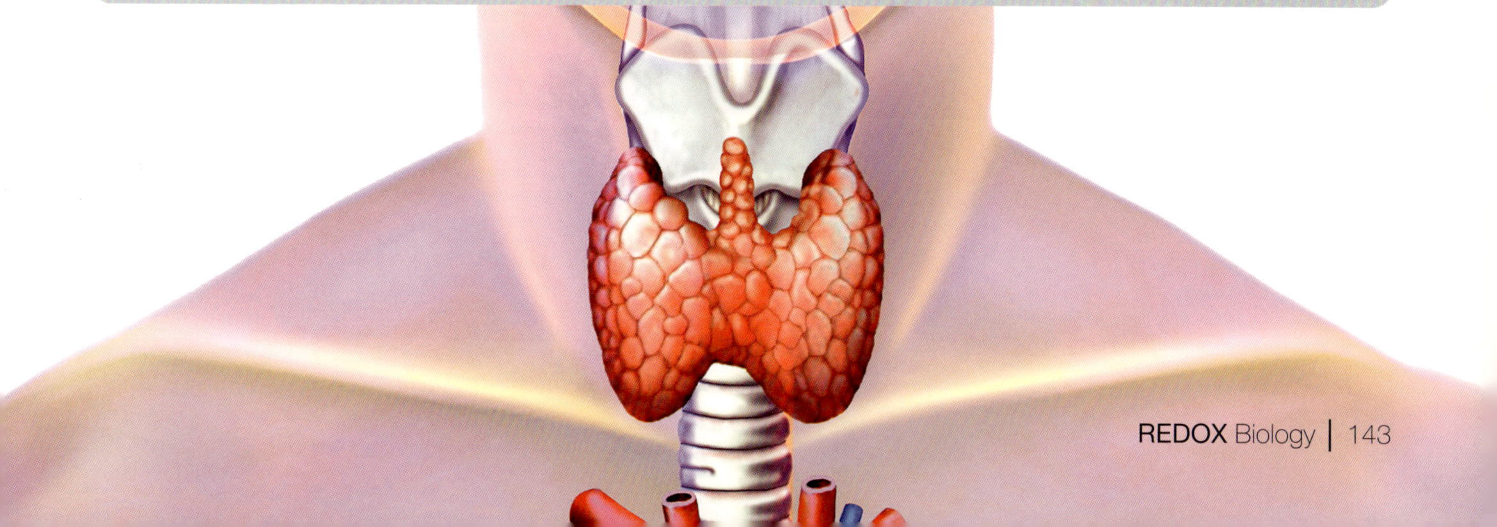

3.2.2. Antioxidant Scavengers (Non-enzymes)

Non-enzymatic antioxidants also play a major role in maintaining cellular REDOX balance. In addition to being a cofactor of various antioxidant enzymes, GSH, which is the most abundant peptide in cells, possesses a plethora of functions. These include direct scavenging of ·OH, singlet oxygen, and regeneration of other antioxidants such as vitamin C and E to their active forms. The TRX system, in collaboration with the GSH system, plays an important role in reducing oxidized thiol-containing proteins.

A) Glutathione (GSH)

Glutathione (GSH) is the smallest intra-cellular thiol molecule synthesized by the addition of cysteine to glutamate and followed by the addition of glycine. The sulfhydryl (-SH) group of cysteine is involved in reduction and conjugation reactions are considered as the most important functions of GSH. Its high electron-donating capacity (high negative REDOX potential) combined with high intra-cellular concentration (milli-molar levels) generate great reducing power.

GSH concentration in human tissues normally range from 0.1-10 mM, mostly concentrated in the liver (~10 mM) as well as in the spleen, kidney, lens, erythrocytes, and leukocytes. Plasma concentration of GSH is approx. 4.5 μM. In the liver, GSH is carried through the bile into the intestinal lumen. Epithelial linings of kidney tubules, intestinal mucosa and lungs also have modest capacity to export GSH.

GSH levels in the body are readily depleted by oxidative stress. Other contributing factors for GSH deficiency include: UV; viral infections, toxins, regular household chemicals, heavy metals; surgery, inflammation, burns, septic shock; and dietary deficiencies of GSH precursors. The homeostatic GSH-REDOX cycle replenish GSH as it is being consumed. Amounts available from foods are limited (less that 150 mg/day), and oxidative depletion could outpace GSH synthesis. Fasting, protein-energy malnutrition, or other dietary amino acid deficiencies limit GSH biosynthesis in the body. Low GST activity may increase risk for disease such as cystic fibrosis, multiple sclerosis and stroke.

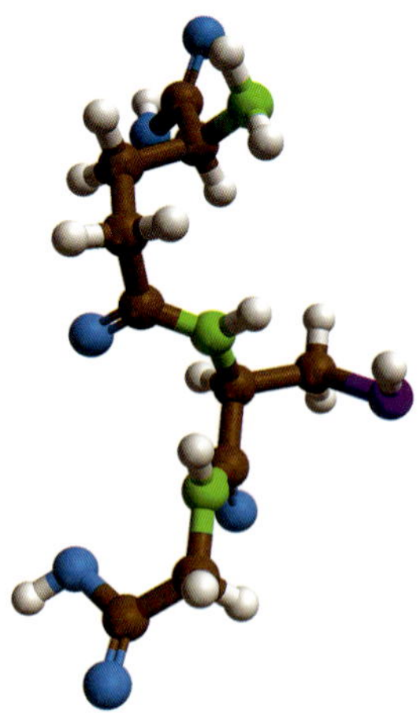

GLUTATHIONE (GSH)

As a vital cellular antioxidant, GSH directly quenches reactive hydroxyl free radicals, other oxygen-centered free radicals, and radical centers on DNA and other bio-molecules. GSH is a primary protectant of skin, lens, cornea, and retina against radiation damage; as well as other biochemical pathways of P450 detoxification in the liver, kidneys, lungs, intestinal, epithelia and other organs. GSH also plays a major role in different cellular compartments. In mitochondria it plays a key role in regulating apoptosis versus necrosis. In the nucleus, GSH is a key regulator of cellular division. GSH is used as a cofactor by (i) multiple peroxidase enzymes, to detoxify peroxides generated from $O_2^{\bullet-}$ attack on biological molecules; (ii) transhydrogenases, to reduce oxidized centers on DNA, proteins, and other molecules; and (iii) GSH S-transferases (GST) to conjugate GSH with endogenous substances (e.g., estrogens), exogenous electrophiles, and diverse xenobiotics. Glutathione is the essential cofactor for many enzymes that require thiol-reducing equivalents, and helps keep REDOX-sensitive active sites on enzyme in the necessary reduced state. GSH and the GSH/GSSG ratios are critical in the regulation of several thiol-based cellular systems including metallothioneins, thioredoxins and other REDOX regulator proteins. Accordingly, the GSH/GSSG balance is vital for metabolism and homeostasis. GSH and its metabolites also interface with the bio-energetics and the neurotransmitter syntheses through several prominent metabolic pathways.

B) Uric acid

Uric acid provides an excellent example of the adaptation made by organisms to oxidative stress. It is a cellular waste product originating from the oxidation of hypoxanthine and xanthine by xanthine oxidase and dehydrogenase. Humans lack urate oxidase, an enzyme that converts uric acid to allantoin; therefore, ureate accumulates in the human blood plasma to a concentration of about 300 μM. In humans, over half the antioxidant capacity of blood plasma comes from uric acid. Uric acid is a marker of oxidative stress, and has a potential therapeutic role as an antioxidant. An excess of ureate might lead to its precipitation and crystallization in the joint leading to joint inflammation known as gout. High uric acid levels may, however, also provide efficient antioxidant activity for the organism, and suggested to increase life span. Urate, the physiological state of uric acid, reacts with ·OH radicals producing a stable urate radical that can be regenerated by ascorbate to its prior state, urate. This compound can act with peroxyl radicals, $^1O_2$, O_3, ·NO and other nitrogen-oxygen radicals. Urate also protects protein from nitration; it can chelate metal ions [i.e. Cu(II) and Fe(III)] and prevent them from participating in REDOX cycling. The degradation product of urate, allantoin, was found *in vivo*, supporting claims that urate indeed is an efficient antioxidant.

The ability of urate to scavenge oxygen radicals and protect the erythrocyte membrane from lipid oxidation was described by Kellogg and Fridovich in 1977. Urate is relevant to a variety of physiological conditions, such as autoxidation of hemoglobin or peroxide production by macrophages. One of the major sites where the antioxidant effects of uric acid have been proposed is in the central nervous system, particularly in conditions such as multiple sclerosis, Parkinson's disease, and acute stroke. While chronic elevation of uric acid levels are associated with increased stroke risk, acute elevations in uric acid may provide some antioxidant protection. Urate was found to block peroxynitrite ($ONOO^-$) mediated nitrosylation of neuronal proteins and to block the increase in the blood brain barrier resulting in less infiltration of leukocytes. Uric acid is a unique scavenger of peroxynitrite in the extra-cellular space. However, uric acid cannot scavenge superoxide ($O_2^{\bullet -}$) without the presence of ascorbic acid and thiols. Uric acid is an antioxidant only in the hydrophilic environment, which is probably a major limitation of the antioxidant function of uric acid.

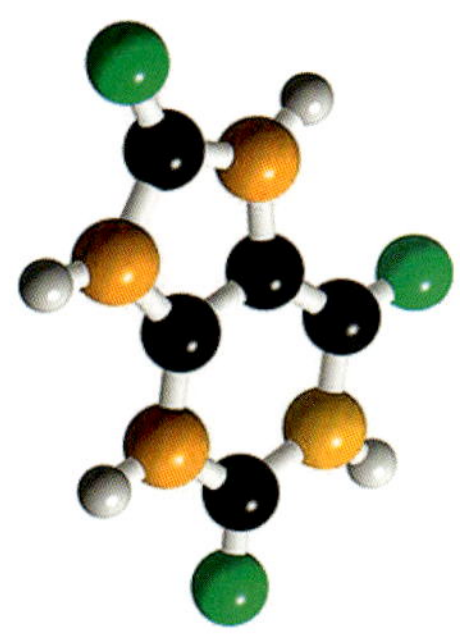

URIC ACID

C) Melatonin

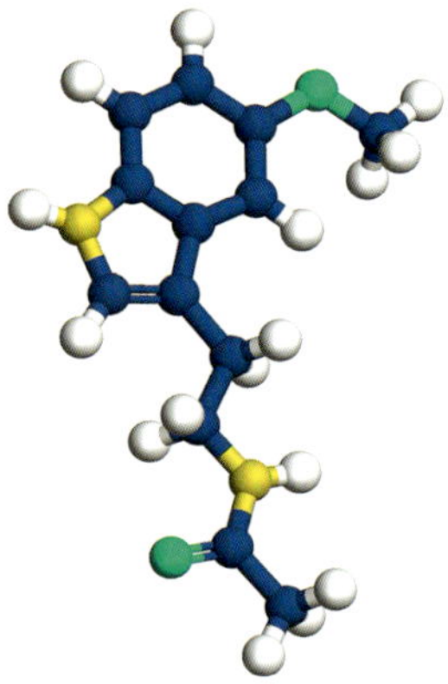

MELATONIN

Melatonin (N-acetyl-5-methoxytryptamine) is secreted during the dark hours at night by pineal gland. Melatonin regulates a variety of important central and peripheral actions related to circadian rhythms and reproduction. Melatonin is a powerful free radical scavenger and a broad-spectrum antioxidant. Due to its small size and highly lipophilic properties, melatonin crosses all cell membranes and easily reaches subcellular compartments, including mitochondria and nuclei, where it seems to accumulate in high concentrations. Melatonin can prevent lipid peroxidation, protein, and DNA damage. In particular, melatonin has been found to preserve optimal mitochondrial function and homeostasis by reducing and preventing mitochondrial oxidative stress, thereby curtailing subsequent apoptotic events and cell death.

The ability of melatonin to scavenge ·OH radicals is much higher than other antioxidants including mannitol, glutathione (GSH) and vitamin E. Melatonin is a powerful and broad antioxidant, since it has been shown to scavenge different types of free radicals including $O_2^{\bullet -}$, ·OH, singlet oxygen ($^1O_2$), H_2O_2, hypochlorous acid ($HOCl^-$), nitric oxide (·NO), and the peroxynitrite anion ($ONOO^-$). Not only is melatonin itself a direct free radical scavenger, but metabolites that are formed during these interactions are excellent antioxidants. Furthermore, melatonin plays an important role in activating antioxidant defenses such as SOD, catalase (CAT), GPx, glutathione reductase (GSH-Rd) and glucose-6-phosphate dehydrogenase (G6PD).

Beneficial antioxidant effects of melatonin have been reported for several chronic diseases, including rheumatoid arthritis and hypertension. Melatonin improves mitochondrial respiration and increases ATP synthesis under physiological and toxic conditions. Melatonin has also been shown to ameliorate inflammation by blocking transcriptional factors and TNF-α. Melatonin treatment increases intra-follicular melatonin concentrations in infertile women, reduces intra-follicular oxidative damage, and elevates fertilization and pregnancy rates.

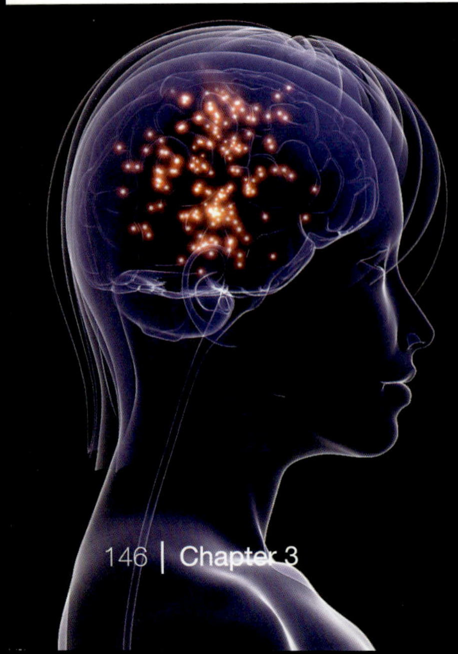

Melatonin forms part of the signal system that regulates the sleep-wake cycle by chemically causing drowsiness and lowering the body temperature. Production of melatonin by the pineal gland is inhibited by light to the retina and permitted by darkness. Its onset each evening is called the dim-light melatonin onset (DLMO). Melatonin levels become regular in about the third month after birth, with the highest levels measured between midnight and 8 AM. It is believed that as children become teenagers, the nightly schedule of melatonin release is delayed, leading to late sleeping and waking times. Human melatonin production decreases with aging.

3.2.3. Antioxidant Vitamins

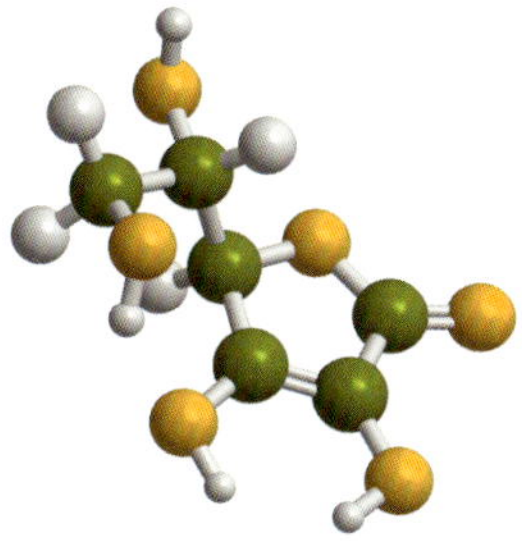
ASCORBATE

Vitamins are organic chemical compounds that cannot be synthesized in sufficient quantities in our body, therefore, must be obtained from the diet. Vitamins are classified by their biological and chemical activity, not by their structure. Therefore, each 'vitamin' refers to a number of vitamer compounds that all show the biological activity associated with a particular vitamin. Such a set of chemicals is grouped under an alphabetized vitamin 'generic descriptor' title, such as 'Vitamin A', which includes the compounds retinal, retinol, and four known carotenoids. Vitamers by definition are convertible to the active form of the vitamin in the body, and are sometimes inter-convertible to one another. Vitamins A, C, and E can carry chemical groups or electrons between molecules and act as antioxidants or prooxidants. Accordingly, antioxidant vitamins play an important protective role in the body by quenching harmful free radicals. Besides being antioxidants, A, C, and E vitamins also work as precursors to several enzyme cofactors in cellular respiration (in mitochondria) or during respiratory burst (in phagocytes).

A) Ascorbic acid

L-ascorbic acid ($C_6H_8O_6$) is the trivial name for vitamin C, first isolated from natural sources by Nobel Laureate Szent-Gyorgyi. The biological role of ascorbate is to act as a reducing agent, donating electrons to various enzymatic and a few non-enzymatic reactions. The 1- and 2-electron oxidized forms of ascorbic acid semi-dehydroascorbic acid and dehydroascorbic acid (DHA), respectively, can be reduced in the body by GSH and NADPH-dependent enzymatic mechanisms. The presence of GSH in cells and extra-cellular fluids helps maintain ascorbate in a reduced state.

Ascorbic acid is a powerful free radical scavenger that neutralizes an array of ROS and RNS, at very low concentrations. In addition, ascorbate can regenerate other antioxidants such as alpha-tocopherol (vitamin E) and beta-carotene (vitamin A). Ascorbate has special functions in the cellular REDOX inter-relationship, as an anti-oxidant and enzyme cofactor. Ascorbate can reduce REDOX-active metals such as copper and iron, thereby increasing their pro-oxidant activity. Thus, ascorbate can act as both a prooxidant and an antioxidant. In general, at low concentrations, ascorbate behaves as a pro-oxidant, and at high concentrations, it tends to be an antioxidant. As a water-soluble antioxidant, ascorbate serves as an electron donor for 8 different enzymes in the human body. Ascorbic acid is a co-factor for hydro-xylases and monooxygenase enzymes involved in the biosynthesis of collagen, carnitine and neurotransmitters. Ascorbic acid plays an important role in several biological pathways, including: (i) maintenance of collagen, the matrix polymer that represents about one third of the total body protein; (ii) synthesis of muscle carnitine, the chemical required for lipid transport into mitochondria for ATP production; (iii) serves as a co-factor for the enzyme dopamine-α-hydroxylase that catalyzes the conversion of neurotransmitter dopamine to norepinephrine; (iv) enzyme catalysis (amidation) for optimal function of hormones oxytocin, vasopressin, cholecystokinin and α-melanotripin; and, (v) conversion of chole-sterol to bile acids, a rate limiting step for cholesterol catabolism in liver.

Ascorbate also takes part in several non-enzymatic reactions. It is a powerful REDOX agent in the physiological milieu. It protects low density lipoproteins (LDL) from oxidation, reduces harmful oxidants in the gut and promotes iron absorption. Ascorbic acid is easily oxidized to an unstable DHA, which is absent in normal plasma but may occur transiently during oxidant stress. Another essential role of ascorbate is attributed to REDOX homeostasis. Since oxidative respiration in mitochondria is a major source of intra-cellular ROS production, the presence of mitochondrial ascorbate is of particular importance. Furthermore, the role of ascorbate in oxidative protein folding and in the maintenance of intra-lumenal oxidative environment makes it an important molecule in endoplasmic reticulum (ER)-related processes. The REDOX state of a cellular milieu determines the regulatory role for ascorbate in cell signaling.

Ascorbate is important for oral health. This antioxidant helps strengthen gums and fights against gingivitis

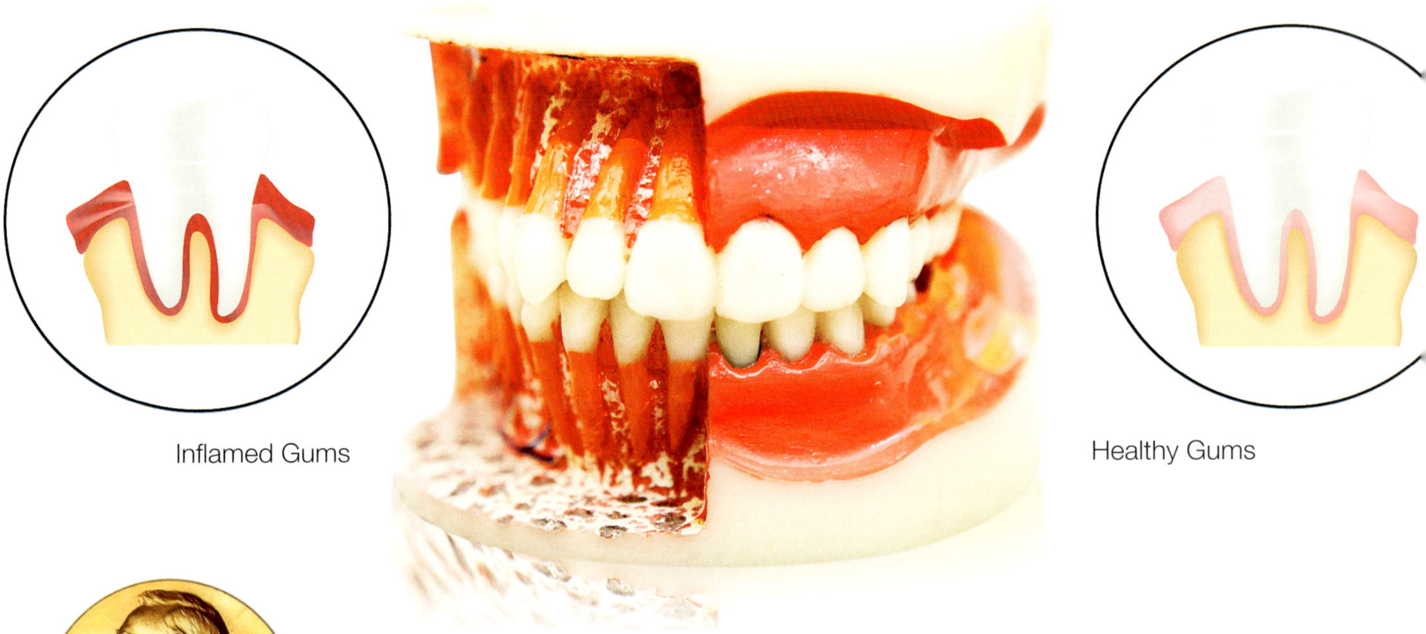

Vitamin C: Albert Szent-Györgyi won the 1937 Nobel Prize in Physiology or Medicine for the discovery of vitamin C and its role in biological combustion processes with special reference to catalysis of the citric acid cycle.

B) α-Tocopherol

α-Tocopherol (α-TP), also known as vitamin E, is an important lipid-soluble antioxidant with a vital role in the glutathione (GSH) peroxidase pathway. The localization of α-TP within the highly unsaturated phospholipid bilayer of cell membranes provides a mechanism to control lipid oxidation. It protects cell membranes from oxidation by reacting with lipid radicals produced during lipid peroxidation chain reaction. α-TP removes free radical intermediates and blocks further continuation of the oxidation reaction. As an antioxidant, α-TP prevents the propagation of proxyl radicals in tissues, by reacting with these free radicals and forming a toco-pheroxyl radical. This reaction is then reduced by a hydrogen donor (such as vitamin C) and returns α-TP to its reduced state.

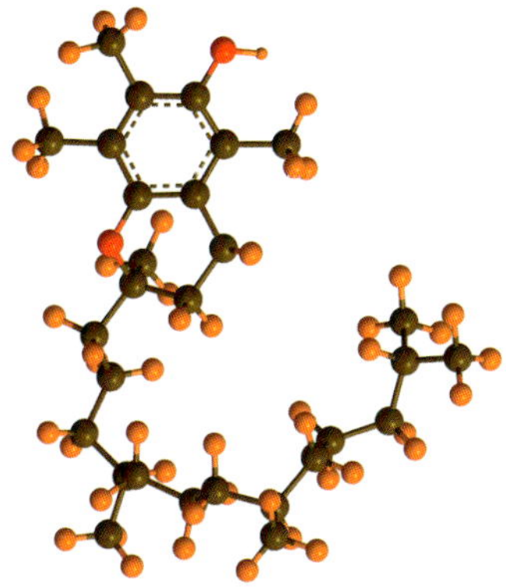

TOCOPHEROL

Lipids associated with subcellular organelles (e.g., mitochondria and microsomes) are especially susceptible to oxidation by virtue of their high contents of phospholipids containing relatively large amounts of polyunsaturated fatty acids PUFAs). Mitochondria are the site for major oxidative processes and are important in fat oxidation and energy production, but a side effect is leakage of ROS. It has been estimated that approx. 1–2% of all O_2 consumed may result in the formation of ROS, with most ROS being generated in the mitochondria. Oxidative stress is characterized by increased ROS generation and subsequent mitochondrial damage to mitochondria, has been proposed to cause a wide range of metabolic disorders. Thus, incorporation of α-TP into mitochondria and other cellular compartments is important in order to maintain oxidative stability of the mebrane-bound lipids and prevent damage from the ROS. α-TP prevents the propagation of free radicals in membranes, and when peroxyl radicals are formed, these react 1,000 times faster with vitamin E than with polyunsaturated fatty acids. Interest in preventing oxidant-induced pathologies and reduced stabilization of PUFAs in the mitochondria and other subcellular fractions has therefore led to studies on α-TP antioxidant therapy.

α-TP is a multifunctional molecule and its spectrum includes enzymatic activities, gene expression, and neurological function(s). The most important function of α-TP has been suggested to be in cell signaling. Among other signaling pathways that have been shown to be affected by α-TP, PKC promises to be an important area of interest. Beside its role in retinal vaso-regulation, the PKC pathway boasts a decisive factor in the pathogenesis of glaucoma, one of the neurodegenerative conditions; and α-TP regulates this enzyme pathway. α-TP also affects the gene expression; it down-regulates the scavenger receptor gene and modulates the connective tissue growth factor (CTGF) gene. The CTGF gene, when expressed, is responsible for the repair of wounds and regeneration of the extra-cellular tissue lost or damaged during atherosclerosis. α-TP also plays a role in neurological functions and in the inhibition of platelet aggregation.

Sterols and Vitamins: The Nobel Prize in Chemistry 1928 was awarded to Adolf Windaus for the services rendered through his research into the constitution of the sterols and their connection with the vitamins.

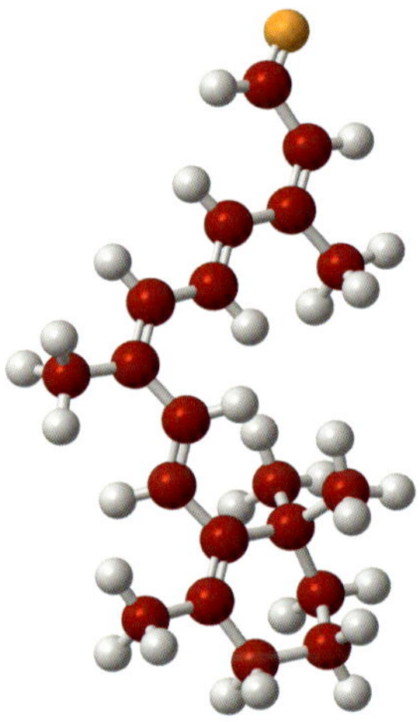

VITAMIN-A

C) Carotenoids

Carotenoids (vitamin A) are tetraterpenoids containing up to 15 conjugated double bonds. Based on chemical properties these are divided into 2 classes; carotenes (i.e. β-carotene, lycopene) and xanthophylls (zeaxanthin, lutein).

In the human body, carotenoids demonstrate various functional outcomes: (i) some are converted to retinoids (i.e. have provitamin A activity), (ii) certain types modulate the enzymatic activities of lipoxygenases (proinflammatory and immuno-modulatory molecules), (iii) most have antioxidants properties, and iv) few could activate the expression of genes that encode Connexin-43 biosynthesis, which is an integral protein component of the gap junctions required for inter-cellular communication. Such gene activation is not associated with antioxidant capacity and is independent of pro-vitamin A activity.

The antioxidant activity of carotenoids is based on the ability to quench singlet oxygen and trap peroxyl radicals. An excited carotenoid could dissipate newly acquired energy through a series of rotational and vibrational interactions with the solvent. This results in the regeneration of original unexcited carotenoid, which can be reused for further cycles of singlet oxygen quenching. The antioxidant quenching activity of a carotenoid depends mainly on the number of conjugated double bonds in its structure. Lycopene is one of the most efficient singlet oxygen quenchers among the natural carotenoids. The prevention of lipid peroxidation by carotenoids has been suggested to be mainly via singlet oxygen quenching.

A number of studies have shown that β-carotene and others carotenoids have lipid-soluble antioxidant activity. Carotenoids partially or completely protect intact cells (e.g. human liver cell line HepG2) against oxidant-induced lipid peroxidation, and the protective effect is independent on their provitamin A activity. Primarily, β-carotene can react with peroxyl (ROO·), hydroxyl (·OH), and superoxide ($O_2^{\bullet-}$) radicals. Carotenoids show their antioxidant effects in low O_2 partial pressure but may have pro-oxidant effects at higher oxygen concentrations. Both carotenoids and retinoic acids are capable of regulating transcription factors. β-Carotene inhibits the oxidant-induced NF-kB activation and interleukin (IL)-6 and TNF-α production. Carotenoids also affect apoptosis of cells.

Human tissues retain considerable amounts of carotenoids that may play a role in preventing oxidative damage. Additionally, the central part of the human retina, the macula lutea, owes its yellow color to high levels of the xanthophylls, lutein and zeaxanthin. These macular pigments have been suggested to filter short-wavelength visible light, to lessen chromatic aberration and to prevent the onset of age-related macular degeneration

Carotenoids and Vitamins: The Nobel Prize in Chemistry 1938 was awarded to Richard Kuhn for his work on carotenoids and vitamins.

Vitamin A was recognized as an essential factor in food a century ago. Early research suggested that certain yellow plant pigments had the same activity as vitamin A. In 1930, Moore explained this phenomenon with the discovery of an intestinal pathway that converts β-carotene into vitamin A; thus provided the first evidence that a plant-derived carotenoid is the direct precursor of retinoids. β-Carotene is the major source for retinoids in the human diet. The formal first step in retinoid metabolism is the conversion of β-carotene to retinaldehyde, from which all other retinoids can be synthesized by endogenous enzymes. Over 600 carotenoids found in nature, about 40 are present in a typical human diet. Of these carotenoids, only 14 and some of their metabolites have been identified in blood and tissues.

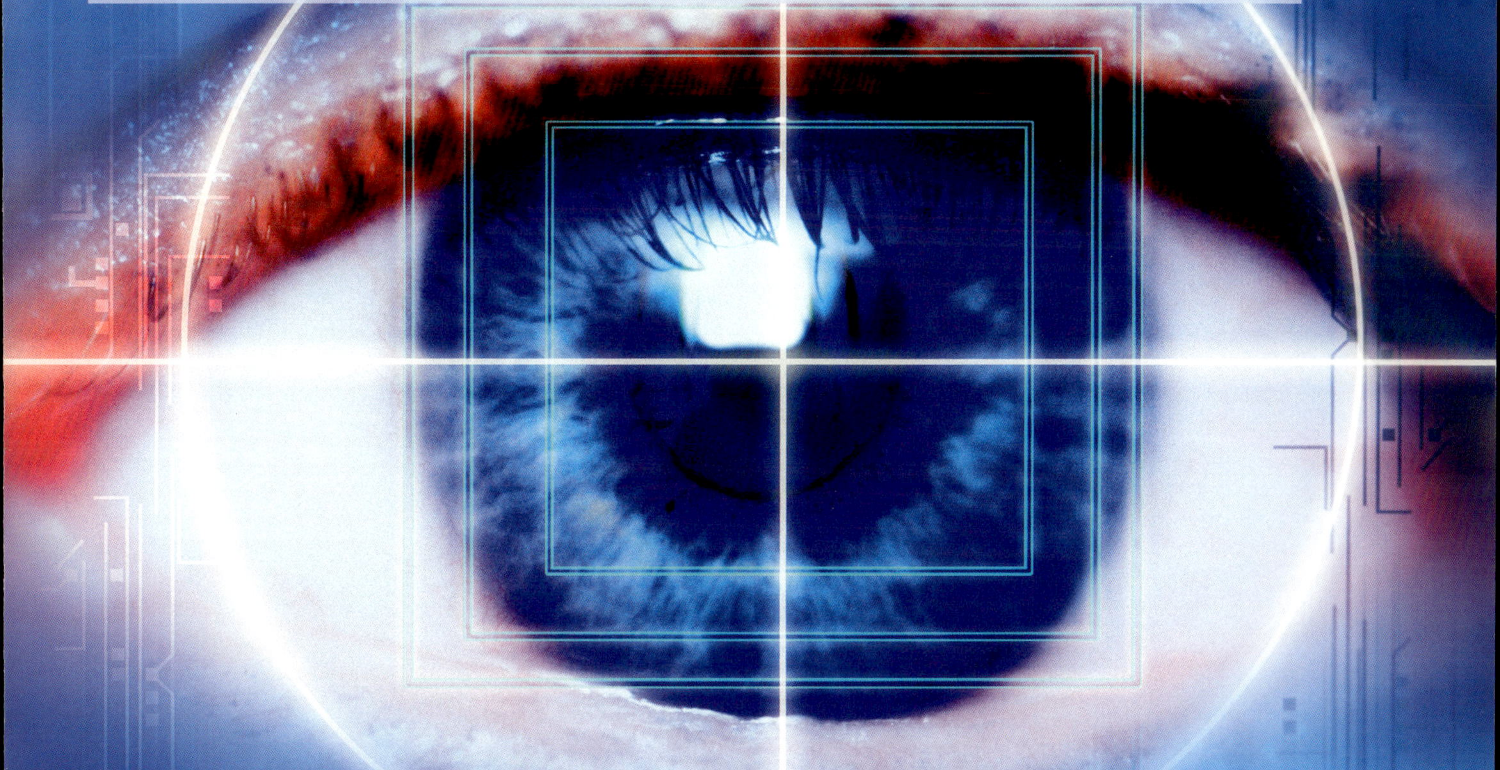

Carotenoids are vital for eye health

Vitamin A (retinol): Rhodopsin (eye pigment) on photo-oxidation with light yields 'retinaldehyde', which converts to vitamin A. The Nobel Prize in Physiology or Medicine 1967 was awarded to George Wald for his discovery of vitamin A and its role in the physiological and chemical visual processes in the eye.

3.3 REDOX: Coupling Systems

An oxidizing and reducing agent which appear on opposite sides of a half-equation constitute a REDOX couple. Human body consists of three major REDOX coupling systems: (i) the glutathione/glutathione disulfide (GSH/GSSG) system, (ii) the Trx/Trx-SS system and the cysteine/cystine (Cys/CySS) system. Each REDOX couple functions as unique rheostat 'on/off switches' in the REDOX regulation of cellular pathways. Any disruption of these distinct 'REDOX control switches or circuitry' would dramatically affect several vital body functions. Accordingly, these 'on/off switches' represent a generalized REDOX signaling mechanism that control various REDOX-sensitive biological processes in normal cell physiology, and redefines their role in the cellular oxidative stress during the disruption of physiological REDOX signals.

The 'REDOX state' of a specific REDOX couple such as glutathione (GSH), thioredoxin (Trx) or cysteine (*Cys*) refers to the ratio of their inter-convertible reduced and oxidized forms, i.e., GSH/GSSG, Trx/TrxSS, or Cys/CySS. The term, 'REDOX environment' represents the sum total of the product of the reducing potential and reducing capacity of these linked REDOX couples.

3.3.1. GSH/GSSG REDOX System

Glutathione (GSH) is the primary determinant of the cellular REDOX environment. It is present in milli-molar concentrations (2–10 mM) inside the cell, mainly in a biologically reduced-thiol form. The oxidation of GSH to GSSG and subsequent decrease in the GSH-to-GSSG ratio is often associated with oxidative stress. Thus, the ratio of GSH to GSSG is a measure of oxidative stress. GSH synthesis, GSSG reduction, and exogenous GSH uptake are critical for the maintenance of cellular REDOX homeostasis. REDOX modulation of GSH/GSSG system is an important regulatory mechanism in the removal of reactive oxygen or nitrogen species (ROS/RNS), carcinogens, drugs, and xenobiotics from the cellular milieu.

GSH/GSSG REDOX couple processes most of the ROS and RNS in the body to maintain REDOX homeostasis.

Cytosolic GSH exists in a highly reduced form [REDOX potential (E_h) at -260 mV for GSH/GSSG], and is the source for distinct GSH REDOX pools inside the cellular compartments of mitochondria, endoplasmic reticulum and nucleus. Mitochondrial GSH, at levels similar to those in the cytosol, can account for 15% to 30% of total GSH content. The inner mitochondrial membrane transporters and carriers maintain the matrix GSH/GSSG at –300 mV REDOX potential. Also, the GSH/GSSG REDOX state of the inter-membrane space (IMS) is more oxidized despite free access of cytosolic GSH through porin channels. A REDOX potential (E_h) of –255 mV in the IMS is an ideal REDOX environment. Similarly, the highly

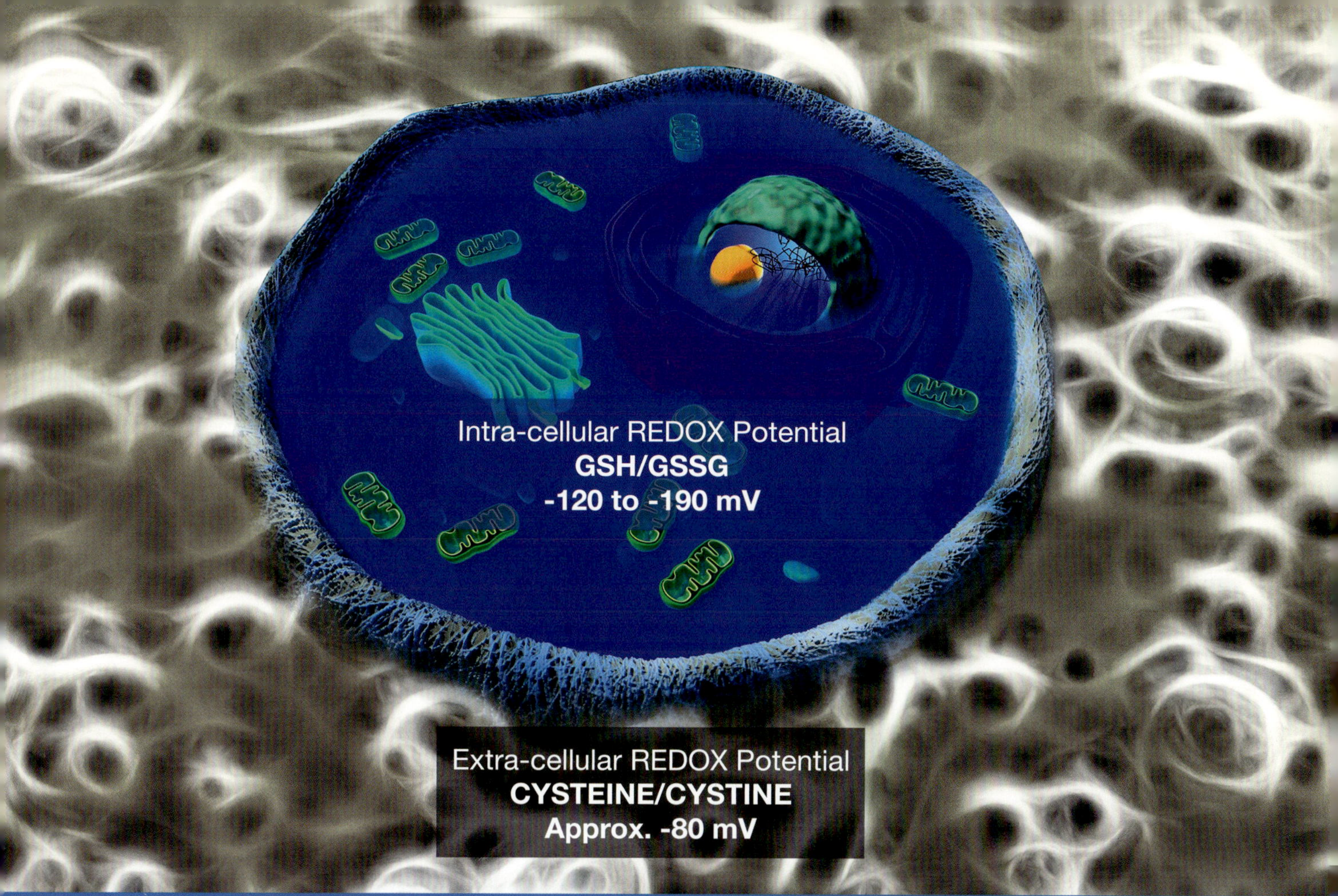

Cellular REDOX Potential

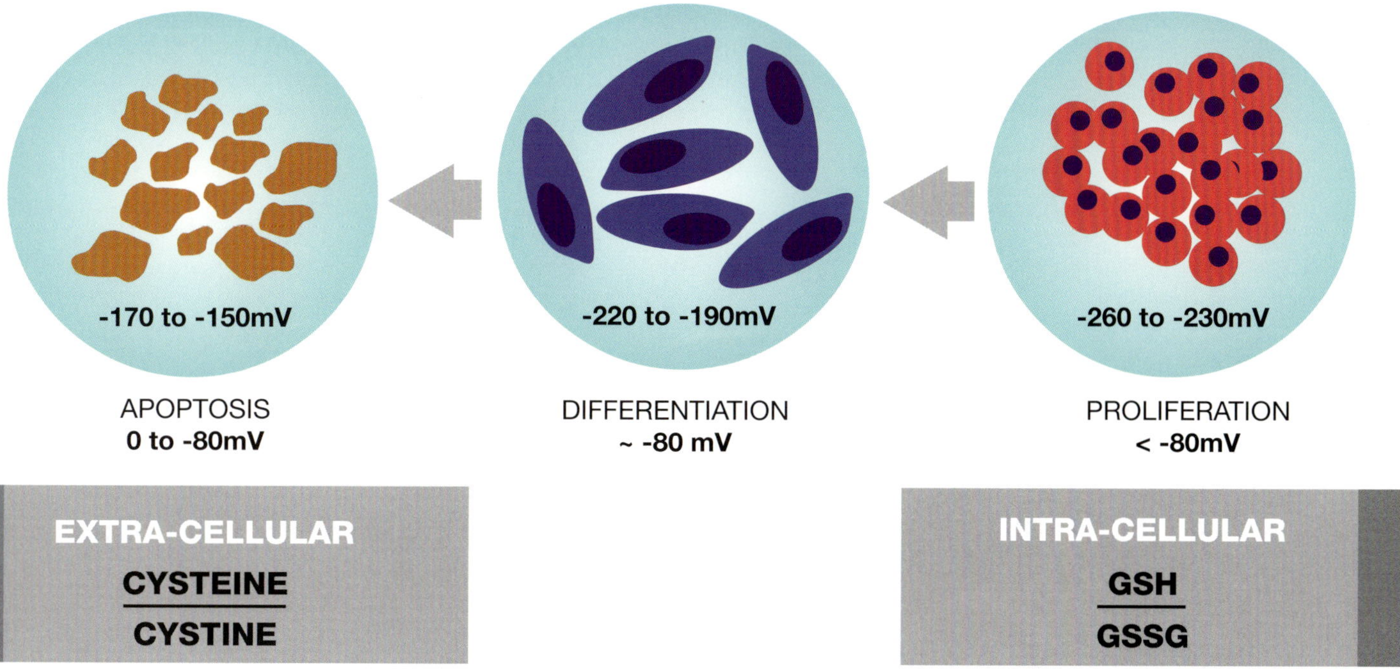

oxidized GSH-REDOX milieu of the endoplasmic reticulum (GSH:GSSG between an E_h of −170 mV and −185 mV maintains the REDOX buffering capacity. The interaction of nuclear and cytosolic GSH is dynamic. During cell cycle, cytosolic GSH is distributed to the nucleus (4 times higher) while confluent (matured) cells show equal ratios of nuclear-to-cytosol GSH distribution.

3.3.2. Trx/Trx-SS REDOX System

TRx retention throughout the evolution indicates its vital role in life. Animals that over express TRx show reduced oxidative stress, more resistance to inflammation and live 35% longer.

Trx is an oxido-reductase enzyme, a class of small REDOX protein, found nearly in all known organisms and is essential for mammalian survival. Trx proteins have a characteristic tertiary structure termed the 'thioredoxin fold'; maintained in a reduced state by the flavoenzyme 'thioredoxin reductase' in a NADPH-dependent reaction. In conjunction with GSH and glutaredoxin, the TRX system maintains the reducing environment of the cell and detoxifies ROS. The antioxidant activity of the TRX system is exerted by peroxiredoxins, which are reduced by Trx to enable scavenging of ROS. Trx acts as electron donor to peroxidases and ribonucleotide reductase. Trx plays a vital role as a cellular antioxidant in a multitude of biological processes, including the maintenance of intra-cellular REDOX homeostasis and intercellular signaling. Any deficiency in the TRX-reductase gene expression could be lethal at the four-cell stage of embryonic development.

The TRX system exists within distinct cellular compartments in the cytosol and nucleus, which is traditionally described as the Trx-1 isoform. Separately, the TRX system in the mitochondria is known as the thioredoxin-2 (Trx2) isoform. Trx1 and 2 are regulated independently and both play an essential role in hematopoiesis, angiogenesis, heart development, and heart function.

Peroxiredoxins (PRxs) are a family of small (22-27 kDa) non-seleno thiol-specific peroxidases. PRxs are known to exist in 6 mammalian isoforms. Although their individual roles in cellular REDOX regulation and antioxidant protection are quite distinct, they all catalyze peroxide reduction of H_2O_2, organic hydroperoxides and peroxynitrite. They are found to be expressed ubiquitously and in high levels, suggesting that they are both an ancient and important enzyme family. PRxs play a major role not only in peroxide defense, but also in the regulation of peroxide-mediated cellular signaling. PRxs can be divided into 3 major subclasses: typical 2-cysteine (2-Cys) PRxs (PRx1-4), atypical 2-Cys PRx (PRx 5) and 1-Cys PRx (PRx 6). The cellular localization of these isoforms is shown in the TABLE to the left.

Isoform	Localization
PRx 1	Cytosol
PRx 2	Cytosol
PRx 3	Mitochondria
PRx 4	Extra-cellular space
PRx 5	Cytosol, Mitochondria, Nucleus, Peroxisomes
PRx 6	Cytosol

Elevated Prx expression (up to 1% of the cellular proteins) and its high catalytic activity suggest that Prx may be responsible for the bulk of intra-cellular H_2O_2 reduction under physiological conditions. Localized H_2O_2 accumulation and the inactivation of Prx have been implicated in cell signaling.

3.3.3. Cys/CySS REDOX System

Cysteine (Cys) and cystine (CySS) constitute the most abundant low-molecular thiols in extra-cellular fluids with concentrations reaching 40 μM and 8–10 μM, respectively. The status of Cys and CySS contributes to maintaining the extra-cellular REDOX environment. The measured Cys/CySS-REDOX potential (E_h) in cell culture media or in plasma of healthy human subjects is around –80mV, a value that is more oxidized than that of GSH/GSSG and Trx/TrxSS-REDOX systems. Plasma Cys/CySS homeostasis is highly regulated by dietary Cys/CySS, hydrolysis of exported GSH, thiol-disulfide exchange between CySS and plasma GSH or homocysteine, and intra-cellular/extra-cellular Cys/CySS shuttle. An oxidized Cys/CySS-REDOX state has been associated with age-related diseases like diabetes, cardiovascular disease, and atherosclerosis. Since the plasma Cys/CySS and GSH/GSSG-REDOX states serve as quantitative measures of oxidative stress, the REDOX states of these couples in the plasma has been suggested as predictive markers of health and disease.

The REDOX potential (E_h) values for thiol-disulfide couples range from -360 mV for Trx-2 in mitochondria to a mean value of -60 mV for Cys/CySS in plasma. This means that a E_h of 60 mV between two couples is sufficient to drive a 100-fold change in dithiol-to-disulfide ratio. The 300-mV span between mitochondrial Trx-2 and Cys/CySS is therefore sufficient to control different REDOX functions through relatively shallow REDOX gradients provided that spatial or catalytic mechanisms are available to control rates. Because H_2O_2 and O_2 always exist in aerobic conditions, coupling of electron transfer to peroxidase or oxidase reactions can further provide an energetic driving force to maintain function of low-current REDOX control pathways.

If a hypothetical protein with a dithiol-disulfide motif with E_h value of -210 mV were equilibrated with cytoplasmic GSH/GSSG at -210 mV, the dithiol to disulfide ratio would be 1. If the same protein were equilibrated with cytoplasmic Trx-1 at -270 mV, the ratio would be 100:1. If the protein were equilibrated with mitochondrial Trx-2 at -360 mV, the ratio would be 100,000:1. If the protein were equilibrated with cytoplasmic Cys/CySS at -150 mV, the ratio would be 1:100. Equilibration with plasma Cys/CySS at -60 mV would yield 1:100,000. Thus, if catalysts or autocatalytic structures are present to allow selective interaction of thiols with the central REDOX control nodes, the range of REDOX potentials is sufficient to provide a considerable organization of protein structures and functions.

3.4 **REDOX:** Aging

Survival, growth and multiplication are an integral part of life cycle. During this cycle, the physiological fitness reaches its peak performance at reproductive phase and gradually declines to a dysfunctional endpoint - the death of an organism. The time sequential deterioration of an organism (commonly the post-reproductive phase) with gradual loss of ability to maintain homeostasis increases the probability of death. These cumulative events are conventionally referred to as the "aging process". Two fundamental questions needed to be addressed, to explain any mechanism of the aging process: (i) How individual organisms lose physiological fitness during the latter part of life? (ii) What is the biochemical basis of the wide variations in longevity among different (or similar) genotypes?

The aging of an organism is mainly dependent upon three biological characteristics: (i) the genetic efficiency of the physiological systems to maintain homeostasis, (ii) the magnitude of metabolic events that diminish the ability of the organism to maintain homeostasis, and (iii) the synthetic ability of an organism to refill its innate molecules to maintain homeostasis – the bio-replenishment.

Oxidative damage of bio-molecules increases with age and is postulated to be a major causal factor of cellular biochemical senescence. Aging results in an accelerated functional decline. Several theories have been postulated to explain some of the processes that occur with aging and the degenerative diseases of aging. Theories based on free radical-based oxidative stress propose that elevated oxygen radical production with age increases the risk of cellular damage. Other observations point out a strong inverse correlation of aging with the rate of mitochondrial ROS generation. Evidently, animals with high mitochondrial metabolism have a short life span since they have higher oxygen consumption and elevated ROS production.

Aging has a strong genetic component, but it can be viewed partly as a degenerative process dictated by free radical reactions, possibly increasing the incidence of several degenerative disease states. The exact interrelationship is unclear, but there is a significant amount of evidence that free radicals are involved. Free radicals increase with age, alter physiological function, increase disease states and accelerate patho-physiological conditions.

'Aging is the result of the progressive accumulation of deleterious changes that reduce an organism's ability to resist stress causing a decrease in survival possibilities [Denham Harman]'. There are three possible ways to increase longevity: (i) decrease damage accumulation, (ii) increase defenses and/or (iii) increase repair. From an energy conservation point, the most effective way to increase longevity is to decrease endogenous damage generation and this is the pathway which evolution has favored.

3.4.1. Free Radical (Oxidative Stress) Theories

In 1908, German physiologist Max Rubner postulated that the rate of metabolic activity may be a determinant of aging process. He suggested that organisms are born with fixed energy resource and limiting factors such as the amount of oxygen inhaled or number of heartbeats spent define longevity.This rate of living theory was advanced in 1928 by biostatistician Raymond Pearl, who observed the effects of caloric restriction on domestic animals. Pearl suggested that duration of life is dependent upon two variable factors: (i) the genetically determined vitality or metabolic potential of the individual, and (ii) the average rate of energy expenditure during life. The phrase 'live fast, die young' emerged from this aging theory.

In 1956, Denham Harman proposed the free radical theory of aging, when prevailing scientific opinion held that free radicals were highly unstable to exist in biological systems. This causal mechanism of aging, implicates the endogenously generated oxygen radicals cause physiological damage resulting in aging and death. Harman's hypothesis was based on that: (i) free radicals are produced *in vivo*, (ii) free radicals cause extensive damage to macromolecules *in vitro*, especially lipids, and (iii) the harmful effects of exposure to irradiation and hyperoxia are ameliorated by antioxidant administration.

In 1985, the free radical theory was modified and eventually merged with the oxidative stress hypothesis advanced by Helmut Sies. Oxidative stress was defined as a disturbance in the prooxidant/antioxidant balance in favor of the former, leading to potential damage. In context of the aging process, oxidative stress represents a cellular state in which the antioxidant defenses are insufficient to completely remove the ROS, thereby resulting in the age-dependent accrual of macromolecular damage. Although there are several cellular sites of ROS production, the largest source for the generation of $O_2^{\bullet-}$ and H_2O_2 are mitochondria, which could play a crucial role in the modulation of oxidative stress and the aging process. The rates of mitochondrial ROS production tend to increase during the post-reproductive phase of life. Age-related elevation in mitochondrial ROS production has been hypothesized to be a consequence of a 'vicious cycle' of self-inflicted damage to mitochondrial DNA, encoding the components of the electron transport chain (ETC). This results in a progressively defective ETC, increased production of ROS and decreased respiratory efficiency.

Nevertheless, there are relatively few substantive differences between the basic propositions of the original "free radical" and the later 'oxidative stress' hypothesis. Both postulate that the rate of aging, i.e. the progression of the age-related deleterious alterations, is a function of the imbalance between ROS fluxes and the antioxidant defenses; also, both predict that the shortening of this gap should reduce the amount of structural damage and prolong the life span. Therefore, from this historical perspective, the postulated mechanism, implicating ROS in the aging process, can be aptly characterized as the structural damage-based oxidative stress hypothesis.

According to the above hypotheses, the lowering of ROS levels and the enhancement of antioxidant defenses was predicted to retard the physiological damage, slow the progression of age-related alterations and prolong life span. Despite numerous efforts, however, augmentations of non-enzymatic as well as enzymatic antioxidant defenses have, in general, not only failed to extend longevity beyond the species-specific maximum life spans, but in some cases also resulted in shortening of the life span.

3.4.2. Mitochondrial Mutation Theory

Several studies have demonstrated an age-related increase in ROS generation of mitochondria. Animal experiments showed that $O_2^{\bullet-}$ and H_2O_2 production increase with age, especially in isolated mitochondria and sub-mitochondrial particles from the aged heart. The enhanced ROS generation by older mitochondria may itself cause oxidative damage to mitochondrial membranes and proteins. These findings have widely promulgated the mitochondrial mutation theory of aging. In this theory, senescence is linked to mutations of mitochondrial DNA (mtDNA) in differentiated cells. Isolated mitochondria from aged animals have an increased production of ROS compared to young animals. This observation may be of key importance because it is likely to increase the rate at which proteins and DNA are oxidized. Since these mutations occur in post-mitotic cells and mtDNA lacks excision and recombination repair mechanisms, it has been postulated that these mutations lead to problems in replication, leading to a decline in physiological performance and the pathogenesis of many age-related diseases. In addition, mtDNA is not protected by histones or DNA-binding proteins; therefore, it is directly exposed to a high steady-state level of ROS and RNS. Thus, oxidative modification and mutation of mtDNA may occur with great ease. Mutations in mtDNA can lead to the production of less functional respiratory chain proteins, resulting in increased free radical production and more mtDNA mutations. This vicious cycle operates in different tissues at variable rates and leads to differential accumulation of oxidatively modified mtDNA. This may ultimately reduce energy output and contribute to the common signs of normal aging.

Mitochondria have two main characteristics: i) Indispensability – as intra-cellular machines they enable us to use oxygen, and maintain vital aspects of cellular survival; ii) Vulnerability – they constantly inflict self damage as a side-effect of aerobic respiration, which may be permanently harmful to the cell, especially the DNA. Mitochondria have their own genome; the mutation of mitochondrial genes cause several diseases.This theory predicts that free radicals damage mt-DNA and cause mutations, which then accumulate during aging. These mutations would in turn alter mitochondrial function, and expedite the aging process. Thus, oxidative damage to mt-DNA negatively correlates with maximum life span.

How much $O_2^{\bullet-}$ and H_2O_2 is generated during mitochondrial respiration?
Mitochondrial electron transport is imperfect and results in the production of $O_2^{\bullet-}$ from the 1 electron reduction of O_2. Respiration is characterized by a high degree of reduction of electron carriers and a limiting supply of ADP, while H_2O_2 production at its maximum. Under such mitochondrial respiratory milieu the release of reactive intermediates accounts for 2% of the oxygen consumed.

3.4.3. Cell Signaling (REDOX Stress) Hypothesis

During the past two decades, there has been a significant progress in elucidating the role of ROS in cell physiology. Contrary to previous characterization of ROS as being solely toxic, certain oxidants, particularly H_2O_2, have now been recognized to be essential for cell survival. The regulatory role of ROS in a wide range of vital functions, including gene regulation, cell signaling, protein activation/deactivation, cellular differentiation and apoptosis has been established. Furthermore, among the various classes of ROS-induced macromolecular structural modifications, protein oxidation has a functional consequence, since proteins act as receptors, transporters, enzymes, transcription factors and cytoskeletal components, among others. Moreover, proteins are the largest constituents of the cell and probably the most frequent targets of ROS attacks. Thus, based on the more contemporary understanding of the multifunctional role of ROS and the nature of oxidative stress, Sohal and Orr (2012) have proposed an alternative notion, the REDOX stress hypothesis of aging.

Oxidation/reduction reactions of specific protein moieties (i.e. cysteinyl thiols) constitute a fundamental mechanism for the modulation of basic cellular activity. Also the cellular REDOX state becomes increasingly more pro-oxidizing in the latter part of life, especially: (i) an age-related increase in the rate of mitochondrial production of H_2O_2, (ii) an increase in the steady-state amounts of GSSG with age, (iii) a progressive decline in the GSH:GSSG ratios and REDOX potential, and (iv) an elevation in the level of protein mixed disulfides. Indeed, an increase in the rate of mitochondrial H_2O_2 production in the post-reproductive phase of life is one of the most consistent features of the aging process. Interestingly, the rates of mitochondrial $O_2^{\bullet -}/H_2O_2$ production, are inversely correlated with maximum life spans of the species.

ROS and the REDOX state are also involved in cellular transitions during the earlier phases of life cycle [i.e. cellular proliferation (growth) and differentiation]. ROS, particularly H_2O_2, and GSH, in concert, play a key role in cellular proliferation and differentiation. Studies have shown that incubation of embryonic cells in low concentrations of H_2O_2 stimulates cell proliferation, whereas exposure to higher concentrations leads to differentiation or apoptosis. In contrast, antioxidants impede these transitions. One of the most commonly observed correlates of aging is the increase in the mitochondrial production of H_2O_2, which implies that this particular alteration should be considered as a possible triggering factor in the initiation of the senescent decline. Increased production of H_2O_2 would quite likely decrease the GSH/GSSG-REDOX potential and elevate the levels of protein modification. Thus, the process of aging may constitute the terminal phase of differentiation, during which the REDOX state becomes less benign due to the over-production of oxidants.

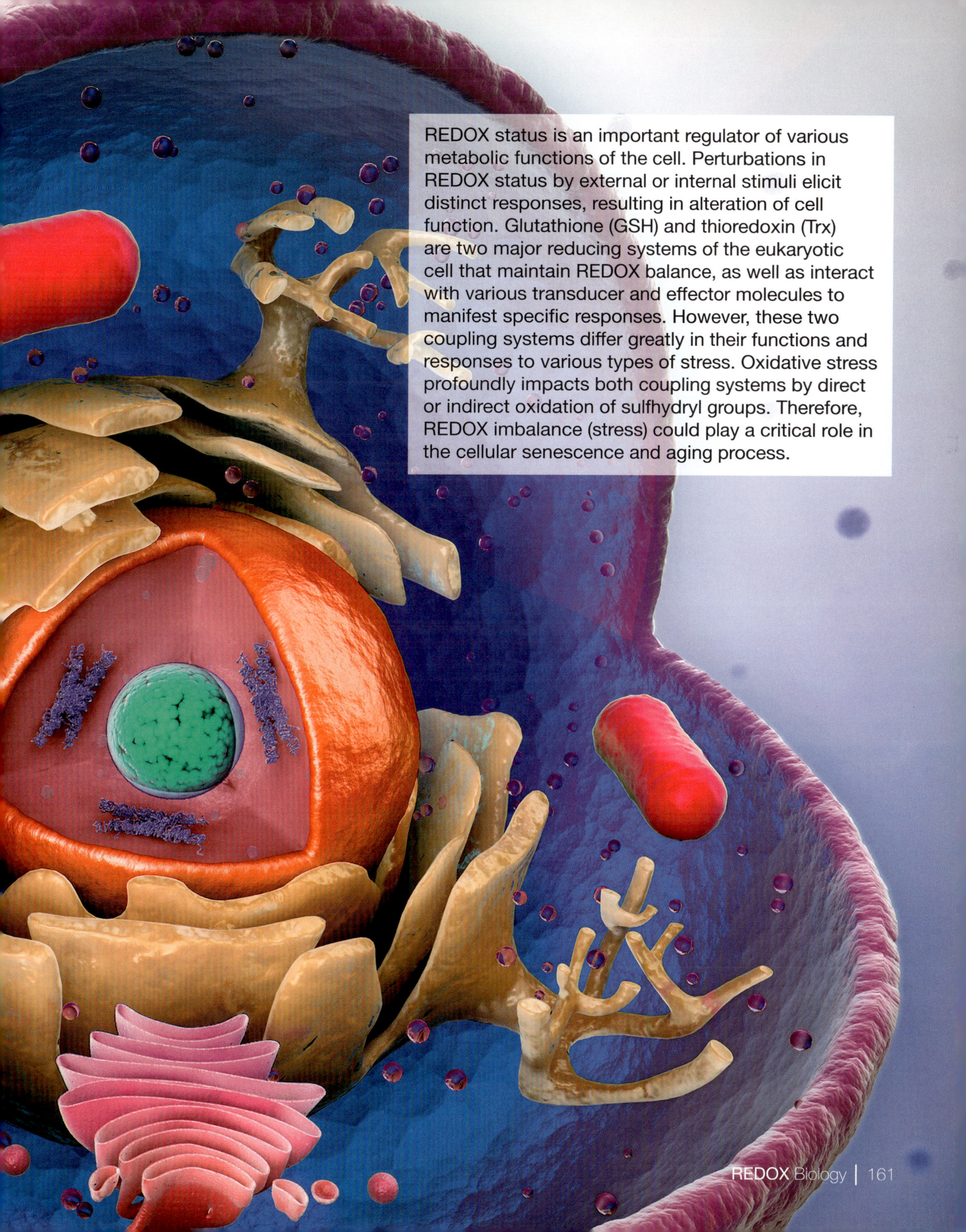

REDOX status is an important regulator of various metabolic functions of the cell. Perturbations in REDOX status by external or internal stimuli elicit distinct responses, resulting in alteration of cell function. Glutathione (GSH) and thioredoxin (Trx) are two major reducing systems of the eukaryotic cell that maintain REDOX balance, as well as interact with various transducer and effector molecules to manifest specific responses. However, these two coupling systems differ greatly in their functions and responses to various types of stress. Oxidative stress profoundly impacts both coupling systems by direct or indirect oxidation of sulfhydryl groups. Therefore, REDOX imbalance (stress) could play a critical role in the cellular senescence and aging process.

3.5 REDOX: Signaling

REDOX signaling is cellular communication using short-lived ROS, or RNS, or through changes in the electrochemical potential of REDOX couples within biological systems. REDOX signaling is a vital feature of normal physiology, as well as of disease and aging. The cellular operation of REDOX signaling requires ROS or RNS that either act directly on a target, or converts to another signal, or changes the REDOX potential of the biological milieu with orchestration of one or several REDOX couples – thereby altering functions of a cellular target. Once a REDOX signal is completed, biological processes operate to reverse the effect of the signal and return the system to its basal (ground) state. REDOX signaling is an important mechanism that regulates physiological processes.

Life based on oxidative metabolism has necessitated the evolution of antioxidative systems to protect against over-oxidation, and to combat ROS generated as by-products of a leaky electron transport chain (ETC) of mitochondrial respiration. ROS are also produced in a multitude of other oxygen-metabolizing reactions, which at low levels serve in biological signaling pathways. Accordingly, biological systems must achieve a delicate balance between using these reactive species for cell signaling functions, as well as in circumventing any collateral damage. Excessive production and accumulation of ROS and/or reactive nitrogen species are detrimental to cells and tissues, resulting in injury and ultimately loss of viability and death by oxidative damage to cellular macromolecules. However, low levels of ROS modulate protein phosphorylation mediated by protein kinases and phosphatases, alter intra-cellular calcium levels, stimulate phospholipases, and regulate transcription factors and growth factors/growth factor receptors. ROS/RNS can play a variety of roles in signaling pathways that lead to changes in expression of proteins important to health and disease. Thus, ROS/RNS serve as direct injury-causing agents, as exogenous stimulants that cause production of inflammatory mediators, as intra-cellular signaling molecules, and as modulators of the REDOX state in cellular systems. In these capacities, ROS/RNS appear to function as modulators of the expression of epithelial genes whose products are pro-inflammatory or act in other ways to prolong the presence of inflammatory cells in diseased conditions. Intra-cellular ROS/RNS can serve as signaling molecules in signal transduction cascades that ultimately trigger changes in expression of cellular genes. As the cellular REDOX state change during the production of such reactive species, transcription factors become activated. In this manner, the transcriptionally controlled expression of gene products modulate biological outcome. An effective antioxidative defensive mechanism or an oxidative pathologic state results in the survival of an organism.

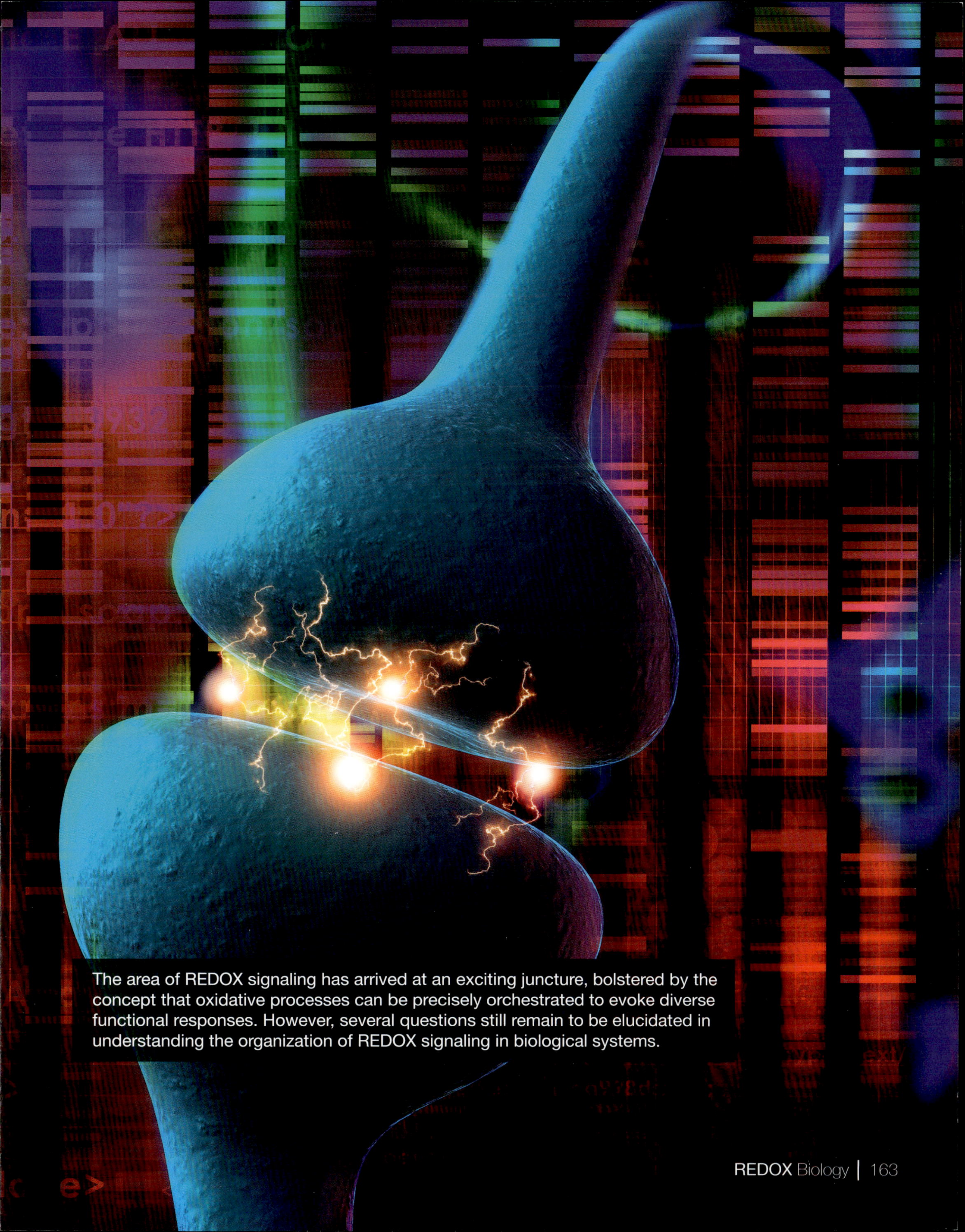

The area of REDOX signaling has arrived at an exciting juncture, bolstered by the concept that oxidative processes can be precisely orchestrated to evoke diverse functional responses. However, several questions still remain to be elucidated in understanding the organization of REDOX signaling in biological systems.

Alternations in REDOX homeostasis of intra- and extra-cellular compartments could trigger REDOX signals leading to vital processes such as proliferation, differentiation, and death of a cell.

ROS, RNS, and lipid peroxidation products are generated and utilized in physiological settings as signaling molecules to regulate cellular and tissue homeostasis, cell division, migration, contraction, and mediator function. These signaling oxidants, in particular ·NO, S-nitrosothiols (SNO), and hydrogen peroxide (H_2O_2) are produced mainly by NADPH-dependent enzymes; whose expression is tightly controlled, compartmentalized and tissue-specific. Signaling by these reactive molecules is mainly carried out by targeted modifications of cysteine residues in proteins, including S-nitrosylation and S-oxidations.

The specificity and selectivity of oxidant (ROS/RNS) molecules is dictated by their chemical reactivity. Signal transduction cascades are rapid, reversible and activated by specific agonists and stimuli, that transmit precisely regulated, compartmentalized signals, to produce a distinct molecular or biochemical outcome. Oxidants employed in signal transduction in physiological settings must exhibit substrate specificity and produce reversible oxidations. Highly reactive oxidant species such as ozone (O_3), hypochlorous acid (HOCl), peroxynitrite ($ONOO^-$), nitrogen dioxide (NO_2) and the hydroxyl (·OH) radical, oxidize biological molecules without preference or specificity. Most oxidative modifications caused by these highly reactive oxidants, such as the 3-nitrotyrosine and protein carbonyls are also not easily reduced, and thus represent poor signaling events. Protein modifications by the aforementioned species may lead to the aberrent activation of signal transduction cascades, often resulting in pathophysiology.

In contrast, physiological oxidants such as ·NO/SNO, and $O_2^{\bullet-}/H_2O_2$, have been implicated in reversible cysteine (*Cys*)-based modifications that establish homeostatic control and diverse biological responses. For example, the ·NO radical in smooth muscle biology, activates guanylate cyclase via binding to heme iron and generates cGMP. Hydrogen peroxide (H_2O_2) a small and non-charged molecule that is readily produced and removed following the physiological stimuli, exhibits properties that are ideal for signal transduction. Although it was initially believed that H_2O_2 can freely diffuse across membranes, recent genetic evidence suggests that some membranes are poorly permeable to H_2O_2, and that its transport may be regulated by aquaporin channel proteins, which also regulates transport of NO. H_2O_2 exerts its functions in signaling by causing reversible amino-acid oxidations. H_2O_2 also stimulates growth factor receptors, toll like receptors (TLR), cytokine receptors, and also leads to the activation of other signaling cascades [i.e. activation of non-phagocytic oxidases (NOX), or dual oxidase (DUOX)]. The expression of multiple isoforms of these enzymes in a variety of tissues provides evidence that the deliberate production of low levels of oxidants is a feature of many cells.

From a pharmacological stand point, it would be meaningful to explore on what extent that locally generated ROS/RNS act within a cellular milieu, and to what length do such reactive species diffuse to reach distant targets and exert their signaling effects? Such knowledge could tremendously brighten our chances to design effective prophylactics and therapeutics to combat several chronic diseases such as CVD, diabetes, neurodegenerative disorders and cancer.

Cellular signaling pathways regulated by ROS

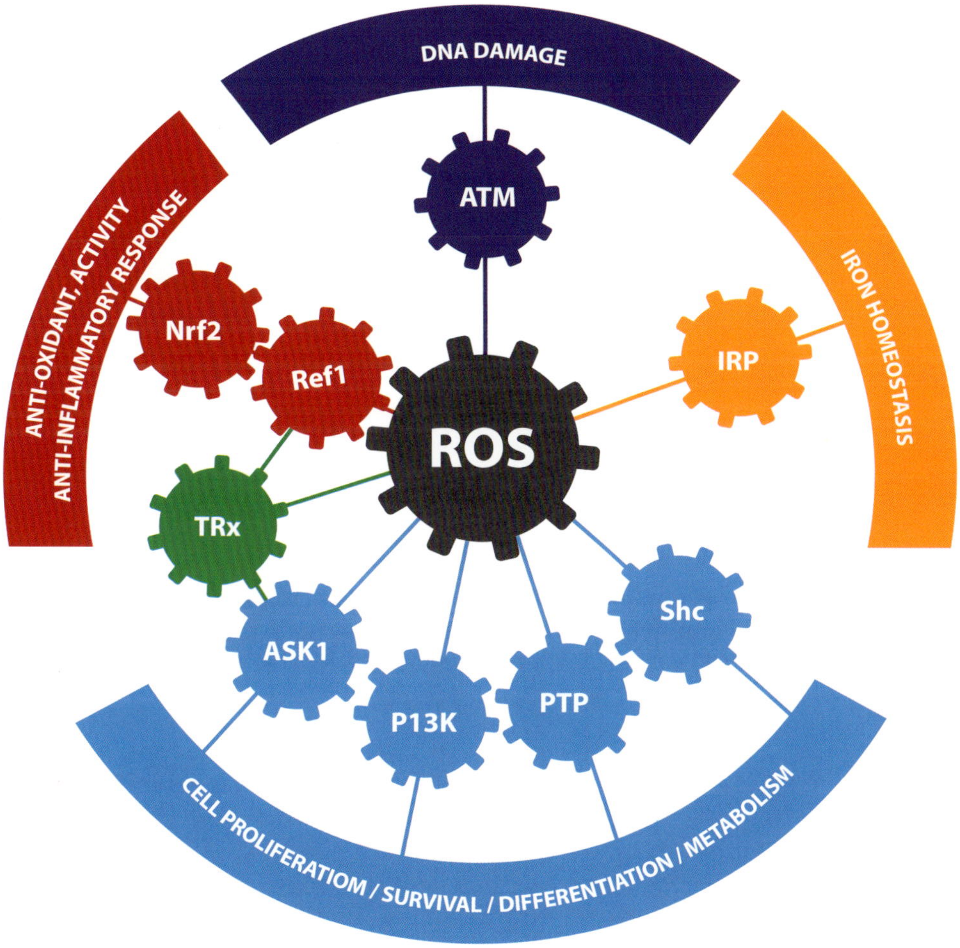

Abbreviations: Apoptosis signal-regulated kinase 1 (ASK1), PI3 kinase (PI3K), Protein tyrosine phosphatase (PTP), and Src homology 2 domain-containing (Shc)); *Antioxidant/Antiinflammatory Response*: Thioredoxin (TRX), REDOX-factor 1 (Ref-1), and NFE2-like 2 (Nrf-2)); *Iron homeostasis:* Iron regulatory protein (IRP)); and *DNA damage response:* Ataxia–telangiectasia mutated (ATM).

Reactive oxygen species (ROS) regulate several signaling pathways through interaction with critical signaling molecules, affecting a variety of cellular processes, such as proliferation, metabolism, differentiation, and survival.

3.5.1. REDOX Signaling: Molecular Mechanisms

The REDOX regulatory circuit needs a minimum set of elements to adapt the metabolic system to special requirements: a signal and a sensor to 'switch-on' the adaptive process, a transducer, a modulator of sensitivity, an effector, and a 'switch-off' device. With increasing complexity of the organism, the regulatory systems have to cross-talk with different compartments of the cell, with the entire organism, and its environment. They have no chance to operate in splendid isolation; moreover, the resting position is not equilibrium but a snap shot of steady states of competing reactions within the metastable and open system that defines life. The complexity, thus created, has two important implications: (i) each regulatory step under consideration has to be kinetically competitive with a realm of competing reactions and (ii) the signal has to be specifically transduced to the effector despite possible side reactions and cross-talks between signaling cascades. These two aspects are particularly relevant in the molecular mechanism of REDOX regulation.

REDOX Signaling: Reactive Nitrogen Species (RNS): RNS could elicit REDOX signals similar to H_2O_2. Most of the reactions between RNS and proteins lead to the similar end products as with H_2O_2. Thus, RNS can modify key catalytic residues to inhibit protein function or can modify non-catalytic residues to allosterically activate or inhibit protein function in similar ways to other ROS. Some different modifications, however, also arise from the reaction of RNS with proteins. S-Nitrosation is one example, which forms key catalytic residues such as in caspases and phosphatases in cells. Nitric Oxide (·NO) is the principal RNS with a central role in signaling. Nitric oxide metabolism and reactivity leads to the formation of several other RNS, primarily peroxynitrite ($ONOO^-$) but also other reactive species such as nitrogen dioxide ($\cdot NO_2$), dinitrogen trioxide (N_2O_3), and dinitrogen tetroxide (N_2O_4), which may be involved in REDOX signaling.

Statins regulate ·NO availability. The commonly used drug family, the HMG-CoA reductase inhibitors (statins), treatment leads to more ·NO bioavailability and a reversal of endothelial dysfunction.

Nitric Oxide (·NO) Signaling: ·NO regulates several vital physiological pathways by at least two major arms of REDOX regulation. Activation of cGMP-dependent signaling is the classical arm, which is achieved through reversible binding of ·NO to certain transition metal ions, such as Fe(II) heme prosthetic groups of proteins.The interaction induces a conformational change in soluble guanylate cyclase, results in elevated cGMP levels, ultimately leading to smooth muscle relaxation. The ·NO-induced vasodilation is associated with the rise in cGMP levels within the underlying vascular smooth muscle cells. Several studies have shown that ·NO is an important regulator of the number of mitochondria within cells, a program called 'mitochondrial biogenesis'. Treatment of cells with ·NO could increase mitochondrial number through a cGMP-dependent pathway. Nitric oxide also reversibly binds other heme-containing proteins such as complex IV in mitochondria and decreases respiration at low levels of oxygen. Alternatively, ·NO can directly modulate protein function through S-nitrosylation (S-NO). Generally, ·NO does not react with most non-radicals at physiological concentrations;

however, at certain conditions it is able to nitrosate thiols to form S-nitrosothiols. This post-translational modification results in the reversible, covalent modification of target proteins when ·NO interacts with a reactive cysteine thiol to form an S-NO protein Many proteins appear capable of forming S-NO intermediaries including caspases. phosphatases, signaling molecules and channels. Certain S-nitrosylated proteins demonstrate an increased enzymatic activity; however, under other circumstances the enzyme activity is down-regulated.

Peroxynitrite ($ONOO^-$) Signaling: Peroxynitrite is a potent oxidant and nitrating species formed from the reaction between the free radicals ·NO and $O_2^{\bullet-}$. An excessive formation of $ONOO^-$ represents an important mechanism contributing to cell death and dysfunction in multiple cardiovascular pathologies, such as myocardial infarction, heart failure and atherosclerosis. Peroxynitrite is a potent modulator of various cell signal transduction pathways due to its ability to nitrate tyrosine residues. Thus, $ONOO^-$ affects cellular processes dependent on tyrosine phosphorylation. Peroxynitrite also exerts complex effects on the activity of various kinases and phosphatases, resulting in the up- or down-regulation of signaling cascades, in a dose- and cell type-dependent manner. Such roles of $ONOO^-$ in the REDOX regulation of key signal pathways for cardiovascular homeostasis, including protein kinase B and C, the MAP kinases, NF-κB, as well as signaling dependent on insulin and the sympatho-adrenergic system.

REDOX Signaling: Reactive Oxygen Species (ROS): Until recently, ROS have been studied for their harmful effects on proteins, lipids, and DNA that lead to cell damage and death; however, there is a growing evidence in the emerging role of ROS, especially $O_2^{\bullet-}$ and H_2O_2, in modulating cellular processes through REDOX signaling. The fate of most of the $O_2^{\bullet-}$

Transcriptional Regulation

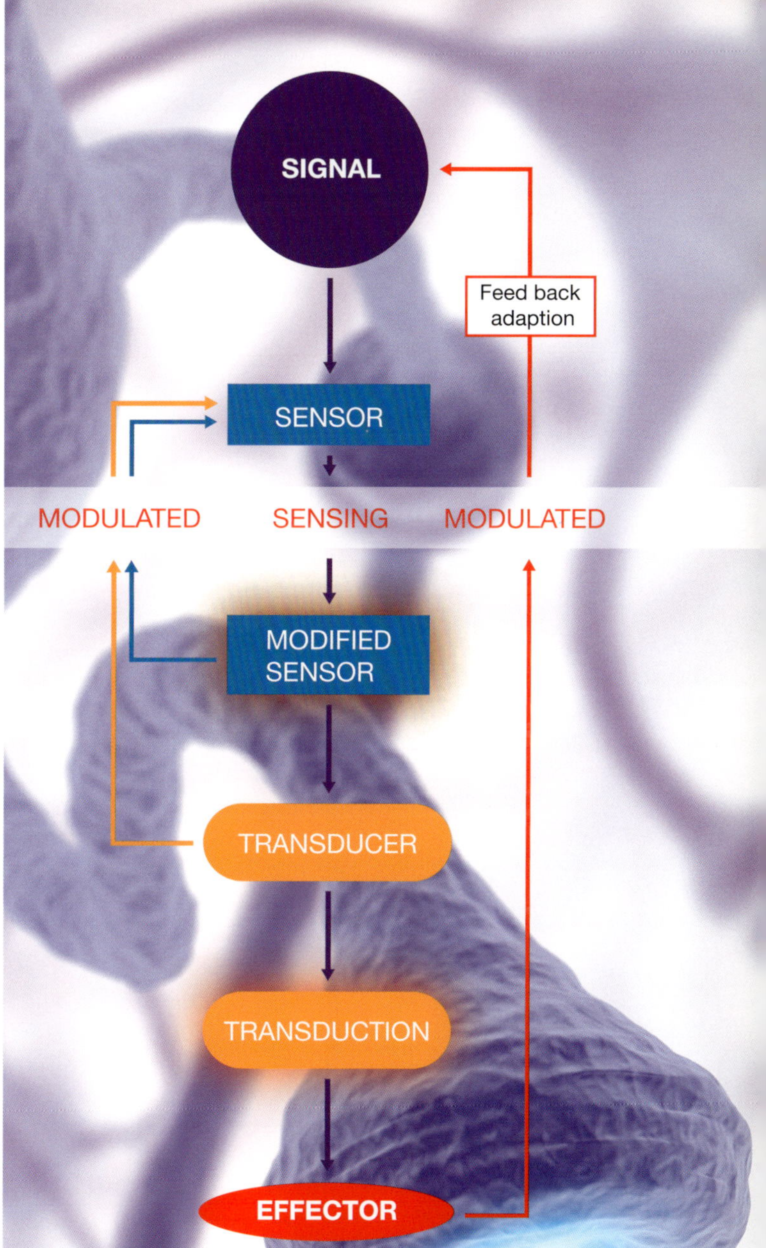

The molecules that initially act as REDOX signals can be divided into three broad categories: ROS, RNS and lipid peroxidation products.

generated is dismutated to H_2O_2 by both enzymatic (i.e. SOD) and non-enzymatic mechanisms. The $O_2^{\bullet-}$ that escapes degradation by the cellular antioxidant system is not particularly reactive with most biomolecules (proteins, nucleic acids, lipids and other small molecules); instead reacts mainly with ·NO and transition metals, such as Fe(II) found in Fe/S center-containing proteins. Therefore, most of the proteins that react with $O_2^{\bullet-}$, such as aconitase, guanylate cyclase, ribonucleotide reductase, phosphatase calcineurin and SoxR contain iron.

Hydrogen Peroxide (H_2O_2) Signaling: Hydrogen peroxide is the most characterized ROS involved in REDOX signaling. Physiological levels of H_2O_2 are regulated to some extent by $O_2^{\bullet-}$ formation, and its metabolic fate as a REDOX messenger, by a range of degradation pathways. These characteristics make H_2O_2 ideal for REDOX signaling. H_2O_2 acts as a REDOX signal through specific, reversible modification of key amino acids on proteins. Thiol modification by H_2O_2 is one of the most prominent mechanisms that influence structure and functional properties of protein molecules. H_2O_2 reacts with the thiolate moieties on proteins. Access of H_2O_2 to the thiolate group and the extent to which the thiol moieties are deprotonated are two critical factors that govern the REDOX signaling. Cysteine residues are the most readily susceptible protein residues; however, other sites on proteins are also vulnerable to H_2O_2. Hydrogen peroxide acts as a REDOX signal either by reacting directly with protein residues or indirectly through biological REDOX couples, such as the TRx and GSH systems. In certain conditions, products of these reactions inhibit protein function by blocking important catalytic active sites, such as cysteine residues on tyrosine phosphatase enzymes. These modifications could alter protein conformation or interactions with other biological molecules and thereby activate or inhibit protein function.

The prominent REDOX signaling mechanism related to H_2O_2 is the regulation of protein function by glutathiolation. GSH is present in millimolar concentrations in cells and provides an essential role in maintaining a reducing environment within the cell. It represents an important H_2O_2 scavenging system, where two molecule of GSH can interact with H_2O_2 to produce two molecules of water and one molecule of GSSG. The GSSG can then be reduced back to GSH by the action of glutathione reductase with NADPH being consumed in the reaction. GSSG can directly react with the thiolate anion of reactive cysteine residues in proteins. Also, GSH can directly react with cysteine sulfenic intermediates (S-OH). A number of proteins undergo such modifications during physiological or pathological stress. Such glutathiolated proteins includes transcription factors, metabolic enzymes, and structural proteins. Taken together, these observations suggest that cells purposely produce $O_2^{\bullet-}/H_2O_2$ to fight infection in the case of phagocytes and to transmit signals in other cell types.

Carbon Monoxide (CO) Signaling: Carbon monoxide production and heme-oxygenase expression are strongly linked to oxidative stress. Two different heme-oxygenase (HO) enzymes are known to stimulate the degradation of heme to produce biliverdin, Fe(II), and CO. This reaction also requires the presence of molecular O_2 and NADPH. While HO-2 is expressed in the brain and vascular tissues, HO-I is generally inducible and only seen after exposure to certain stress conditions. Activation of heme targets (i.e. sGC) induces vasorelaxation, while ·NO regulates the vascular tone. On a mole per mole basis, CO is a much weaker inducer of sGC, compared to ·NO. Thus, exposure of CO and increased expression of HO-I in the vessel wall results in increased vaso-relaxation.

Oxygen sensing is a CO-dependent (HO-2) vital signal transduction mechanism. The carotid body, a small specialized locus of cells in the neck, contains the main arterial chemo-receptors that sense changes in arterial O_2, CO_2, and pH. The carotid body, when activated, releases a number of transmitters that regulate the central respiratory centers in the brain. In particular, a fall in arterial O_2 saturation triggers the carotid body, which in turn stimulates an increase in the rate of breathing. Carbon monoxide is an important regulator of hypoxic sensing of the carotid body. Glomus cells in the carotid body respond to low O_2 levels by altering the opening of specialized channels on their cell surface. Channel activity regulates potassium entry and subsequent neurotransmitter release. The O_2-dependent production of CO by HO-2 is integral part of the overall hypoxic sensing system.

Another intriguing aspect of CO signaling involves the transcription factor, NPAS2. This protein binds to DNA with another protein co-factor, BMALI. Together, these transcription factors have been implicated in circadian rhythms. Low doses of CO may have anti-inflammatory, anti-proliferative, and anti-apoptotic effects. Some of these effects may be mediated through cGMP-dependent pathways. In addition, there is considerable evidence that CO may affect certain arms of the mitogen activated protein kinase (MAPK) family.

Hydrogen Sulfide (H_2S) Signaling: Hydrogen sulfide formation results from the activity of two different enzymes, cystathionine-synthase or cystathionine-lyase. H_2S is a colorless gas that readily diffuses across plasma membranes. Similar to both ·NO and CO, H_2S is produced abundantly in the brain, highly expressed especially in the hippocampus and the cerebellum. The neuronal actions of H_2S have been reported to include regulation of memory and long term potentiation (LTP). This effect is thought to occur through regulation of NMDA receptor activity. Interestingly, both NO and carbon monoxide also regulate LTP although H_2S does not appear to regulate cGMP levels like these two other gases. H_2S may function as a signaling molecule in processes such as neuromodulation in the brain and smooth muscle relaxation in the vascular system. It also has a cyto-protective effect, since it can protect neurons and cardiac muscle from oxidative stress and ischemia reperfusion injury, respectively. H_2S can also modulate inflammation, insulin release, and angiogenesis.

Cell Signaling: The Nobel Prize in Physiology or Medicine in 1999 was awarded to Günter Blobel for the discovery that proteins have intrinsic signals that govern their transport and localization in the cell.

Lipid Peroxidation Products and REDOX Signaling: ROS and RNS react with polyunsaturated fatty acids (PUFAs) containing phospholipids and initiate a cascade of lipid peroxidation reactions. Lipid peroxidation causes extensive oxidative damage to cells by disrupting biological membranes; however, some of the breakdown products of lipid peroxidation may act as REDOX signals. PUFAs are oxidized *in vivo* by two distinct mechanisms: (i) enzyme-catalyzed oxidation and (ii) ROS- or RNS-mediated oxidation. The latter oxidation of PUFAs can generate products such as isoprostanes. Alkoxyl radicals (RO·), peroxyl radicals (RO_2^-)), hydroperoxyl radicals (HO_2^-), several iron-oxygen complexes and ·OH radicals, are all capable of abstracting hydrogen atoms and initiating lipid peroxidation. At high concentrations, exposure to lipid peroxidation products results in a range of responses from acute toxic effects to inhibition of cell proliferation. At low concentrations, lipid peroxidation products may stimulate several processes, such as enzymatic activities, including adenylate cyclase, guanylate cyclase and phospholipase C, and the transcriptional regulation of antioxidant genes, such as those involved in GSH metabolism and heme oxygenase (HO)-I. Lipid peroxidation products can activate signaling pathways in several ways, such as by forming covalent adducts with proteins or non-covalently binding to protein receptors.

Phosphorylation/Dephosphorylation Signaling: Most REDOX signaling systems are based on post-translational chemical modifications of certain proteins and enzymes, which upon modification change their structure and/or function and activity. Such modifications may differ widely. One of the most significant signaling systems is built around phosphorylation (i.e. the attachment of a phosphate group) and its counterpart, dephosphorylation (i.e. the removal of a phosphate group). The phosphorylation/dephosphorylation signaling pair is universally present in every cell. The key chemical component of phosphorylation/dephosphorylation is the phosphate group (PO_4^{3-}). This group may be attached to a range of nucleophiles in the form of phosphate esters. The most common target molecules for phosphorylation are alcohols and phenols, which form fairly stable esters with phosphate (other esters, such as the rare sulfinic phosphoryl ester in PRx, are less stable). As far as biological molecules are concerned, the amino acids serine, threonine and tyrosine contain hydroxyl groups which are preferred sites for phosphorylation (note: serine and threonine are alcohols, tyrosine is a phenol). Phosphorylation and dephosphorylation are under the tight control of specific enzymes. Phosphorylation is usually catalyzed by protein kinases, which formally transfer the PO_4^{3-} moiety from the triphosphate ATP to the alcohol/phenol, resulting in the phosphorylated amino acid and ADP. The dephosphorylation reaction is catalyzed by protein phosphatases, which hydrolyze the phosphate ester. This reaction yields inorganic phosphate (PO_4^{3-}) and a free hydroxyl group of the corresponding amino acid. Phosphorylation and dephosphorylation have a significant impact on proteins and enzymes.

In summary, REDOX signaling is involved in many biological processes from the sensing of ROS, RNS and lipid peroxidation products to protection against damage caused by oxidative stress. REDOX signals transmit information in cellular processes, such as signal transduction pathways. REDOX signaling mechanisms require a source of ROS or RNS, proteins that can sense the REDOX signal and alter their function. The affected systems ultimately reverse the effects of a REDOX signal on the target protein.

Antioxidant enzymes are extremely fast in eliminating cellular ROS that most of these reactions take place in less than one millionth of a second. For example, SODs dismutate superoxide ($O_2^{\bullet-}$) at a rate constant of ~10^9 $M^{-1}s^{-1}$. The reaction kinetics for heme enzymes are at ~10^7 $M^{-1}s^{-1}$; perioredoxins (Prxs) range from 10^4-10^7 $M^{-1}s^{-1}$; glutathione peroxidases (GPxs) that work with sulfur catalysis are at ~10^8 $M^{-1}s^{-1}$, whereas for the GPxs that work with selenium catalysis the rate constants are at 10^7 $M^{-1}s^{-1}$

The signal/sensor interaction in a cytokine signaling, for example, involves specific binding of a peptide to its receptor. In case of REDOX biology, however, it is enigmatic how signaling by a promiscuously reactive radical (ROS or RNS) complies with the specificity requirement of a meaningful cellular regulation. Furthermore, the abundance of antioxidant enzymes (i.e. SODs, GPxs, and Prxs) that readily eliminate most of the ROS at rates complicates the enigma.

Micrograph of Blood Vessel (Artery & Vein)

3.5.2. REDOX Signaling: Gene Regulation

REDOX reactions are important chemical processes that regulate signal transduction. The response of a cell to a ROS/RNS-rich environment often involves the activation of numerous intra-cellular signaling pathways, which can cause transcriptional changes and allow the cells to respond appropriately to the perceived oxidative stress. For example, at least two well-defined transcription factors, NF-kB and activation protein-1 (AP1), are regulated and influenced by REDOX signal and are implicated in the transcriptional regulation of a wide range of genes involved in oxidant stress and cellular response. In addition to the activation of transcription factors, signaling pathways such as the family of mitogen-activated protein kinases (MAPKs) are directly or indirectly altered by REDOX changes. In the nucleus, REDOX affects histone acetylation and deacetylation status, which partly regulates inflammatory gene expression by activation of the REDOX sensitive transcription factors.

Activation of gene transcription is regulated by protein phosphorylation and de-phosphorylation. According to Brigelius-Flohe and Flohe (2011) the mammalian gene transcription factors are regulated by the following REDOX mechanism(s):

- Direct signaling by free radicals is restricted to $O_2^{\bullet -}$ and ·NO and excluded for fast reacting radicals such as ·OH, ·OR, or Cl^-.
- Oxidant signals are elicited by H_2O_2, enzymatically generated lipid hydro-peroxides, and peroxynitrite ($ONOO^-$).
- Free radical damage is sensed via generation of Michael acceptors.
- Protein thiol oxidation/alkylation is a prominent mechanism that modulates gene function.
- REDOX sensors must be thiol peroxidases or proteins with reactive cysteine or seleno-cysteine (*Sec*) residues to kinetically compete with GSH-peroxidase (GPx) and peroxiredoxin (Prx)-type peroxidases or GSH-S-transferases.

The Michael addition is mild formation of C–C bonds, where, the R and R' substituents on the nucleophile (a 'Michael donor') are electron-withdrawing groups that make methylene hydrogen acidic and forms carbanion. The substituent on the activated alkene, is a 'Michael acceptor', a ketone or a nitro group.

REDOX Signaling and the KEAP1 Sensing System: Prokaryotic cells utilize transcription factors OxyR and SoxRS to sense the REDOX state of the cell, and during oxidative stress these factors induce the expression of nearly 80 defensive genes. Eukaryotic cells have similar mechanisms to protect against oxidative stress. Initial effect of oxidative/electrophilic stress leads to activation of a battery of defensive gene expression that leads to detoxification of chemicals and ROS and prevention of free radical generation and cell survival. Some of these genes have end products that regulate a wide variety of cellular activities including signal transduction, proliferation, and immunologic defense reactions. There is a wide variety of factors associated with the cellular response to oxidative stress.

Nrf2/INrf2 (KEAP1) are cellular sensors of chemical and radiation induced oxidative and electrophilic stress. Nrf2 is a nuclear transcription factor that controls the expression and coordinated induction of a battery of defensive genes encoding detoxifying enzymes and antioxidant proteins. This is a mechanism of critical importance for cellular protection and cell survival. Nrf2 is retained in the cytoplasm by an inhibitor INrf2, which functions as an adapter for degradation of Nrf2. In response to oxidative/electrophilic stress, Nrf2 is switched on and then off by distinct early and delayed mechanisms. Nrf2 is stabilized and translocates to the nucleus, forms heterodimers and binds antioxidant response element (ARE) that leads to coordinated activation of gene expression. It takes less than 15 minutes from the time of exposure to switch on nuclear import of Nrf2. This is followed by activation of a delayed mechanism that controls 'switching off' of Nrf2 activation of gene expression. The switching 'on and off' of Nrf2 protect cells against ROS damage, prevents apoptosis and promotes cell survival. Activation of the KEAP1 signaling pathway is an adaptive response to environmental and endogenous stresses, which provides resistance to oxidative stress and other forms of toxicity. Any disruption of this pathway exacerbates pathobiological outcomes. KEAP1 signaling can be activated by sulfhydryl-reactive agents, and responds to H_2O_2, organic hydroperoxides, peroxynitrite ($ONOO^-$), and electrophiles that are generated by oxidative tissue damage. KEAP1 regulates the inducible expression of an extended battery of cytoprotective genes. Some of the protective effects may be mediated indirectly via cross talk with additional pathways affecting cell survival and other aspects of cell fate. These interactions provide a multi-tiered, integrated REDOX response to oxidative stresses.

REDOX Signaling: Nrf2 Regulation

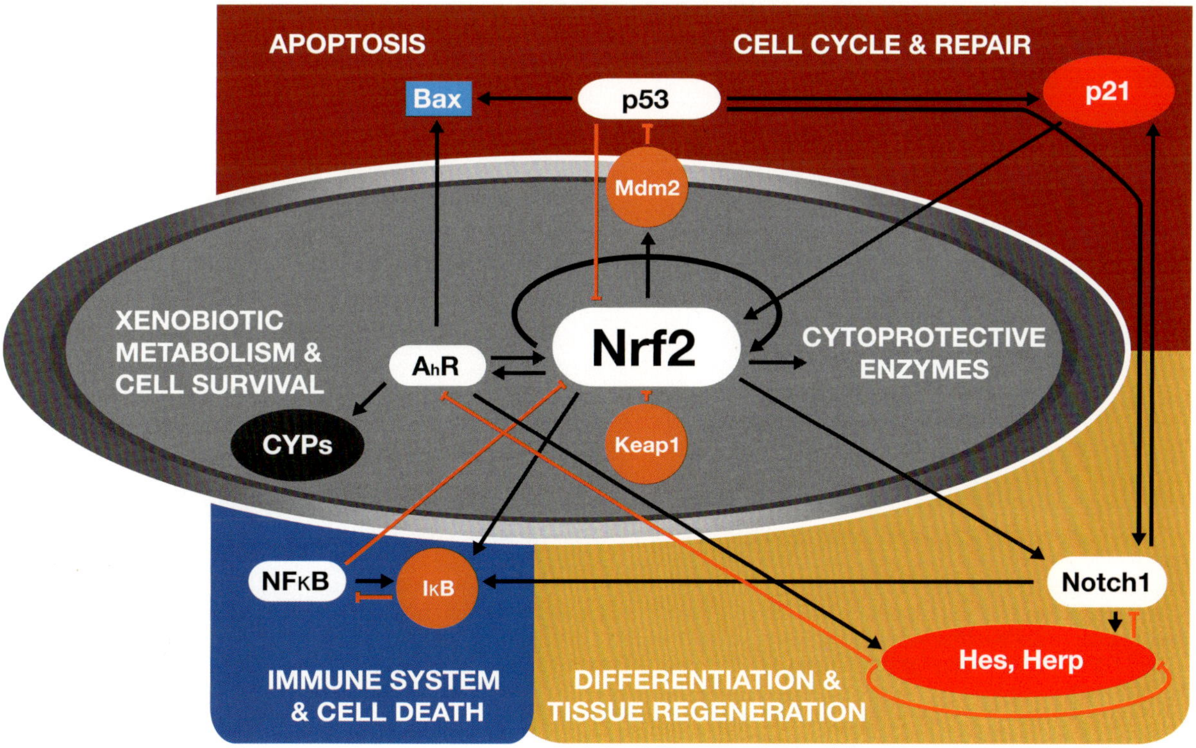

Table: *REDOX Regulation and Mammalian Transcription Factors (from Brigelius-Flohe and Flohe, 2011)*

Factor	REDOX Regulation	Physiological Effects
AhR	Hypoxia; competition of HIF and AhR for ARNT	Inhibition of DNA binding
AP-1	Phosphorylation of c-Jun/c-Fos by activation of kinases	Enhanced association of c-Jun and c-Fos with CBP/p300 and transcriptional activation
	Oxidation of Trx (or Grx)	Activation of ASK-1 and JNK
	Oxidant-induced dissociation of GST from ASK-1 & JNK	Activation of JNK
	Oxidation of *Cys*(154) in c- Fos and *Cys*(272) in c-Jun	Inhibition of DNA binding
	Reversal of *Cys* oxidation by Trx and APE1/Ref-1	Prerequisite for DNA binding
β-catenin	Thiol oxidation in nucleoredoxin	Release of nucleoredoxin and β-catenin activation
Egr-1	Cysteine oxidation of in 3 Zn-fingers by APE1/Ref-1	Inhibition of DNA binding
FOXO	Cysteine oxidation of PTP1b	Activation of Akt, phosphorylation and export of FOXO
	Disulfide formation between *Cys*(297) & *Cys*(311) in Akt	Affinity increase of Akt for PP2A, de-phosphorylation and inactivation of Akt. FOXO remains in the nucleus
	Oxidative activation of JNK	Phosphorylation of FOXO
GR	Zn-finger oxidation, reversible by Trx	Inhibition of DNA binding
HIF-1α	Oxidation of Fe(II) to Fe(III) in HPH	Inactivation of HPH; prevention of hydroxylation and degradation of HIF, activation of HIF
	Reduction of *Cys*(800) by APE1/Ref-1 and Trx/TrxR	Increase in trans-activation capacity of HIF
P53	Low oxidative activation of MAPKs	Low phosphorylation/activation of p53, induction of antioxidant and defense systems, cell growth
	Enhanced oxidative activation of MAPKs	Enhanced phosphorylation and activation of p53, elimination of cells with mutations by apoptosis
	Oxidation of *Cys* in DNA-binding domain and Zn-fingers	Inhibition of DNA binding
Pax-5, - 8	Oxidation of *Cys* in Pax-8; reversible by APE1/Ref-1	Inhibition of DNA binding
	Disulfide formation in Pax-5, reversible by APE1/Ref-1	Inhibition of DNA binding
NF-κB	Formation of disulfide bond in LC8, reversible by TRP14	Phosphorylation of IκBα by IKKβ followed by IκB degradation, release and activation of NF-κB
	Reduction of *Cys*(62) in p50 by APE1/Ref-1 & Trx/TrxR	Enhancement of DNA binding
Nrf2	Modulation of *Cys*(272) and *Cys*(288) in Keap1	Conformational change of Keap1, prevention of Nrf2 ubiquitination, and activation of Nrf2
	Modulation of *Cys*(151)	Prevents ubiquitination & activation/degradation of Nrf2
Sp1	Oxidation of cysteines in 3 Zn-fingers	Inhibition of DNA binding
TTF	Oxidation of cysteines in Zn-finger domain	Inhibition of DNA binding
USF	*Cys*(229) and *Cys*(248) cross-linking by oxidants	Inhibition of DNA binding
	Interaction of HMG-1 with reduced USF	Enhancement of DNA binding

Abbreviations: *ARNT*, Aryl hydrocarbon receptor nuclear translocator; *ASK-1*, Apoptosis signal-regulating kinase-1; *CBP*, CREB-binding protein; *Egr*, Early growth response; *FOXO*, Forkhead box O; *GR*, Glucocorticoid receptor; *Grx*, Glutaredoxin; *GSK3β*, Glycogen synthase kinase-3β; *GST*, Glutathione-S-transferase; *HIF-1*, Hypoxia-inducible factor 1; *HPH*, HIF prolyl hydroxylase; *IKK*, IκB kinase; *JNK*, c-jun N-terminal kinase; *Keap1*, Kelch-like ECH-associated protein-1; *LC8*, Dynein light chain 8; *MAPK*, Mitogen-activated protein kinase; *NF-κB*, Nuclear transcription factor of bone marrow-derived lymphocytes; *PKC*, Protein kinase C; *PTP*, Protein tyrosine phosphatase; *RTK*, Receptor tyrosine kinase; *TRP14*, Thioredoxin-related protein-14; *Trx/TrxR*, Thioredoxin/Thioredoxin reductase; *TTF*, Thyroid transcription factor; *USF*,Upstream stimulatory factor.

REDOX Control of Gene Expression: REDOX-regulated transcription factors are among the possible targets of cell signaling pathway. REDOX-regulated changes to transcription factor activity can occur via (i) oxidative modification of the DNA binding motif of the transcription factor by ROS, or (ii) a phosphorylation/dephosphorylation switch as a result of REDOX-regulated signaling. The DNA binding activity of a number of transcription factors is abolished by oxidizing conditions *in vitro*. Zinc finger motifs, a common DNA binding domain, are particularly sensitive to REDOX changes; the thiol group of cysteine residues in the zinc finger protein confers REDOX sensitivity. Oxidative burst, induced by stress or alteration of antioxidant defenses, is associated with specific changes in gene expression. Gene/protein expression is influenced by cellular REDOX status. Modification of specific amino acid residues of proteins is fundamental for modulation of gene expression. In particular, modulation of the REDOX states of *Cys*-residues has emerged as a widespread mechanism in cell regulation and signaling pathways. *Cys*-residues play ubiquitous roles in mediating cellular responses to REDOX status by detecting changes in the REDOX milieu and inflict changes to protein structure/function. Several transcription factors are modulated by REDOX-regulated DNA binding.

REDOX Signaling and MAP Kinase/Cell Cycle: A number of important signal pathways are activated when cells are treated with oxidants. The association between MAPK signaling and oxidative stress is well documented. The MAPKs are modules of three successive protein kinases that are central for regulation of various cellular processes. These enzymes are modulated, in part, by REDOX regulation. Depleting cellular GSH causes transient increases in ROS, leading to ERK2 activation, a classic MAPK. This activation can be inhibited by N-acetyl cysteine (NAC) treatment. Stress-activated protein MAP kinases (SAPK) are also sensitive to REDOX modulation. Activation of p38 SAPK by tumor growth factor β requires free radical production. Similarly, protein phosphatases, particularly protein tyrosine phosphatases, need tight control to maintain mitogenic signaling. Normally, ROS levels are low when protein tyrosine phosphatase activity is high; however, after growth factor stimulation ROS increases leading to transient inactivation of protein tyrosine phosphatases. This condition is followed by a concomitant burst of kinase activity, until ROS returns to basal threshold and protein tyrosine phosphatase activity is restored. Both protein tyrosine phosphatases and serine/threonine phosphatase are REDOX sensitive. The active sites of protein tyrosine phosphatases include *Cys* and arginine residues separated by five amino acids, creating a low acid dissociation milieu (pKa); which renders *Cys*-residues more susceptible to ROS. Oxidation of the *Cys*-residue results in protein tyrosine phosphatase inactivation. Several oxidative stressors, including $O_2^{\bullet -}$, H_2O_2 and peroxynitrite ($ONOO^-$) activate receptor tyrosine kinases following downstream signaling including MAPK. The multifunctional p53 tumor suppressor gene encodes a transcription factor involved in cell cycle arrest. Reduced *Cys* are essential for p53 DNA binding; mutations involving these residues are associated with many cancers. Several studies have shown that the cell cycle can be arrested in response to ROS and/or RNS. Alteration of cellular REDOX state by GSH depletion leads to elevated ROS levels, which may induce programmed cell death.

3.5.3. REDOX Signaling: Protein/Enzyme Regulation

The fact that proteins become oxidized and lose activity during tissue fractionation has often been considered 'artifact' without questioning why evolution produced such reactive and unstable proteins. The rates of auto-oxidation of cysteine (*Cys*)-thiols in proteins are similar to the kinetics from oxidative turnover of H_2O_2 in cells. Because the function of REDOX-sensing thiols is dependent upon being oxidizable, the *Cys*- residues are programmed to undergo oxidation during extraction and separation from REDOX-control systems. With this interpretation, inactivation of proteins by auto-oxidation provides a valuable insight to elucidate the role of REDOX-sensing thiols in cell cycle regulation and apoptosis.

REDOX Signaling via Protein Oxidation: Proteins with oxidizable thiols are essential to many cellular functions including transcription, chromatin stability, nuclear protein import and export, as well as DNA replication and repair. REDOX states of proteins involve elimination of ROS/RNS to prevent oxidation and reduction of oxidized thiols to restore function. These processes depend on the common thiol reductants, glutathione (GSH) and thioredoxin-1 (Trx1). This oxidative interface mainly regulates the REDOX-reactive *Cys*-residues on proteins. Oxidation of *Cys*-residues generates reactive sulfenic acid (–SOH) that can form disulfide bonds with nearby cysteines (–S–S–) or undergo further oxidation to sulfinic (–SO_2H) or sulfonic (–SO_3H) acid; if nearby nitrogen is available, sulfenic acid may also form a sulfenamide. These oxidative modifications result in changes in structure and/or function of a protein. With the exception of sulfonic acid, and to a lesser degree sulfinic acid, these REDOX modifications are reversible by reducing systems such as thioredoxin (TRx) and peroxiredoxin (PRx).

S-nitrosylation of specific cysteine residues has been detected in well over 100 proteins of all classes and is arguably the principal mechanism by which ·NO signals. S-nitrosylation influences protein activity, protein-protein interactions and protein location. S-nitrosylation serves as the fundamental REDOX-based signal. S-nitrosylation has been implicated in downstream signal transmission of all classes of receptors, acting locally within subcellular signaling domains as well conveying signals from the cell surface to intra-cellular compartments, including the mitochondria and the nucleus. S-nitrosylation exhibits a number of essential features that are basic to REDOX signaling, useful when considering the possibility that additional ROS signaling roles: (i) temporal regulation on physiological timescales, (ii) the existence of motifs within proteins that facilitate S-nitrosylation and/or provide specificity, (iii) co-localization of target proteins with a source of ·NO, (iv) reversibility, and (v) enzymatic controls. In addition, dysregulation of protein S-nitrosylation is associated with a growing list of diseases, including cystic fibrosis, asthma, heart disease, and neurodegenerative diseases. Therefore, understanding the regulatory events that control S-nitrosothiol biology has the potential to lead to novel therapeutic opportunities.

Table: *REDOX Signaling through Protein Modification(s)*

Protein type	Protein Name	Group Modified	REDOX (Modification Type)	Effect
Transcription factor	SoxR	Fe/S clusters	ROS ($O_2^{\bullet-}$)	Activation
			RNS	Activation
	OxyR	Thiol	H_2O_2 (intraprotein disulfide)	Activation
			RNS (S-nitrosation)	Activation
	OhrR	Thiol	H_2O_2 (mixed disulfide, sulfonamide)	Inhibits repressor
	Keap-1	Thiol	Electrophilic lipids (Michael adduct)	Nrf2 release
	PerR	Fe(II)-His	H_2O_2 (histidine oxidation)	Gene expression
Kinase	JNK	Thiol	RNS (S-nitrosation)	Inhibition
	Styl (Spcl, Phhl)	Thiol	H_2O_2	Activation
	IκB kinase	Thiol	Electrophilic lipids	Inactivated
Phosphatase	PTEN	Thiol	H_2O_2	Inactivation
	PTP1B	Thiol	H_2O_2	Inactivation
Ion Channels	Ryanodine receptors	Thiol	ROS and RNS	Activation
Other	Guanylate cyclase	Heme	Nitric oxide	Activation
	Mt-Complex IV	Heme	Nitric oxide	Inhibition

Oxidizable Amino Acids as REDOX Targets: Amino acids that are targets for reversible oxidation such as cysteine residues have unique susceptibility to oxidation. In addition, methionine, tryptophan, and tyrosine residues are also prone to oxidative modification, although the functional impact of those events in physiological signal transduction remains to be established. The cellular REDOX status governs the balance between oxidized versus reduced states of cysteines and methionines. During an elevated oxidative environment, ER and extra-cellular compartment have a higher prevalence of oxidized cysteines. Also, the extent of steady-state thiol modifications (e.g. glutathionylation and cysteinylation) is modulated by the ambient REDOX potential of the cellular compartment. A shift in the steady-state REDOX potential by any metabolic changes or environmental insults could impact the balance between reduced versus oxidized protein thiols. Thus, for every 30-mV change in the REDOX potential, an estimated 10-fold shift in the equilibrium between protein dithiols and disulfides is expected.

REDOX Signaling via Cysteine *(Cys)* Oxidation: Various types of proteins contain free cysteine residues that are highly conserved across species, suggesting regulatory possibilities, beyond structural roles and metal ion coordination. In addition, cysteine residues can be modified through alternative REDOX-based modifications that may enable differential effects on protein function. In other words, distinct reversible modifications of cysteines may lead to unique functional. For example, sulfenic acid modification (SOH) (hydroxylation) of a single allosteric cysteine has been shown to elicit effects on structure and function that are distinct from those elicited by mixed disulfide (glutathionylation).

Cysteines are thus believed to serve as molecular switches, capable of processing different REDOX-based signals into distinct functional responses. A number of cysteine targets in proteins have been identified and cysteine oxidation events can have important functional consequences for an array of signal transduction cascades. It has become apparent that multiple classes of regulatory proteins are reversibly oxidized by H_2O_2 or ·NO/SNO, and that these oxidation events also have broad implications.

Cysteines are an integral part of protein structure and these residues bear a thiol (sulfhydryl) group that represents the most reduced state of sulfur in proteins. Conversion of these thiol groups to oxidized or alkylated species occurs during post-translation stage and can act as a switch, changing the catalytic properties of an enzyme. A number of enzymes involved in metabolic, cell signaling, and antioxidant defense pathways utilize activated cysteinyl residues at their active sites. Since cysteine could typically change its REDOX state during the course of reaction and/or may be inherently reactive toward oxidants and electrophiles given the protein microenvironment that supports their catalytic reactivity; these enzymes are also subject to regulation through thiol modification.

REDOX Signaling via Enzyme Modification: Enzymes are the primary catalysts that promote chemical reactions fundamental to biological processes. Without regulatory mechanisms to ensure the proper expression and activation state at the appropriate time and site for such enzymes, cells would be unable to function properly or respond to their environment. Enzymes involved in metabolism, cell signaling, and oxidative protection exert their activity by modulation of their REDOX state either at the catalytic site or at distinct regulatory sites. Post-translational modification of *Cys*-residues in the enzyme structure is the predominant targets of REDOX-based regulation. Thus, enzymes involved in metabolism and signaling are regulated by post-translational modifications that influence their catalytic activity, rates of turnover, and targeting to subcellular locations. Most prominent among these has been phosphorylation/dephosphorylation.

Proteases such as caspases and matrix metalloproteinases (MMPs) are major examples of enzyme regulation by oxidative events. Caspases contain *Cys* in its active site. In response to a pro-apoptotic stimulus (such as Fas ligand of TNFα), caspases are activated and execute apoptotic pathway. H_2O_2 has been shown to inhibit the activation of caspases. Caspase-3 activation during TNFα-induced apoptosis is negatively regulated by S-glutathionylation, and could be reversed through the thiol transferase activity of GSH-reductase (Grx). Caspase activation can also be negatively regulated by direct cysteine alkylation with biologically generated electrophiles, such as acrolein or hydroxynonenal. In contrast to the oxidant sensitivity of critical cysteines in the family of caspases, the metzincin MMP super family contains an invariant *Cys*-residue in their prodomain region, which maintains latency by chelating the active site Zn(II) center. Activation of these proteases requires disruption of this *Cys*-Zn interaction (the cysteine switch) and can be activated by oxidants or by ·NO/SNO.

3.5.4. REDOX Signaling: Mitochondrial Regulation

REDOX-dependent processes influence most cellular functions, such as differentiation, proliferation, and apoptosis. Mitochondria are at the center of these processes, as these organelles generate reactive oxygen species (ROS) that drive REDOX-sensitive events as well as respond to ROS-mediated changes in the cellular REDOX state. Mitochondria produce ROS as a consequence of electron leak during respiration. Often this leak is referred to as a by-product of electron transport chain (ETC); however, lack of mitochondrial oxidant production leads to diminished cell signaling in response to growth factors, such as insulin, and decreases the formation of necessary protein disulfides. Although some increases in ROS can stimulate these essential pathways, there is a threshold range of concentrations beyond which ROS become harmful. Additionally, ROS-damaged mitochondria tend to produce more ROS, thereby activating mitochondrial-mediated apoptosis or necrotic pathways. Adaptive antioxidant systems are essential for maintaining the critical REDOX balance in cells. REDOX signaling allows for the rapid adaptation of cells to a variety of environmental stimuli, such as hormones, nutrients, and O_2 tension. The mitochondrion plays a central role in such cellular adaptation, which includes: modulation of metabolic, energetic, and ROS responses to promote cell growth and survival or cell death. The overall REDOX status is kept in balance by a host of antioxidant enzyme systems.

Mitochondrial REDOX Switches: Proteins offer unique REDOX switches that are reversibly modified by ROS to alter their function. Protein localization, thiol accessibility, pKa, and protein–protein interactions may all contribute to the specificity of these responses. Mitochondrial function can be modulated directly by protein thiol oxidation or by lipid peroxidation, and indirectly following the REDOX-dependent activation of phosphatases or other REDOX-sensitive proteins. Other proteins, such as mitochondrial STAT3, may protect ETC (complex I) from oxidative inactivation by decreasing ROS production after oxidative stress. In contrast, oxidative mechanisms that activate *p66shc* to promote its mitochondrial localization can augment ROS generation, contributing to REDOX modifications to proteins and phospholipids. Downstream of ROS, other modifications of mitochondrial proteins can lead to mitophagy. Mitophagy and biogenesis are seen as two complementary processes that help to maintain metabolic homeostasis and reduce ROS accumulation. Brief openings of mitochondrial pore may serve a protective function, but more prolonged pore exposure can lead to cell death. Excess mitochondrial ROS can also lead to detrimental phospholipid oxidation, which leads to mitochondrial-dependent apoptosis.

REDOX Regulation of Mitochondria: Mitochondria are essential regulators of cellular bioenergetics that respond to changes in environmental stimuli, such as hormones, nutrients, and O_2 tension. REDOX-sensitive cellular targets play a role in these processes, providing sensitive and rapid responses to changes in metabolism and ROS fluxes. Mitochondrial function is influenced by

Table: *REDOX Modifications and Effects on Mitochondrial Functions (from Handy & Loscalzo, 2012)*

Molecular Targets	REDOX Modification	Functional Outcome(s)
Energetics		
Complex I	S-SG	Either increases activity/decreases $O_2^{\bullet-}$, or decreases activity/increases $O_2^{\bullet-}$
	S-NO; Acetylation	Decreases activity/increases or decreases $O_2^{\bullet-}$
Complex II	Acetylation	Inhibition
Complex IV	*Cys*-oxidation/S-NO	Inhibition
ATPase	S-NO	Inhibition
Tricarboxylic Acid (TCA) Enzymes		
Aconitase/ α-KGDH	S-SG/*Cys*-oxidation	Inhibition
Isocitrate dehydrogenase (ICDH)	Acetylation	Inhibition
Malate dehydrogenase (MDH)	Acetylation	Activatiion
Mitochondrial Transporters		
Adenine Nucleotide Transporter (ANT)	S-S; S-NO	Modulates ADP transport and apoptosis
Voltage-dependent Anion Channel	*Cys* oxidation	Possible role in Apoptosis
Uncoupling Proteins (UCP)	S-SG	Maintains inactive state
Antioxidant Enzymes		
Glutathione peroxidase-1 (GPx-1)	Selenium-oxidation Selenium-NO	Loss of Selenium, Inhibition Inhibition
Glutaredoxin-1 (Grx1)	S-S; *Cys*-NO	Inhibition
Glutaredoxin-2 (Grx2)	Oxidation	Loss of [Fe-S] group
Glutathione reductase (GR)	S-NO; S-OH; S-O_2H	Inhibition
Thioredoxin (Trx)	S-S	ASK1 release and JNK-mediated apoptosis
Peroxiredoxin (Prx)	*Cys*-oxidation	Inhibition
Mitochondrial Networks		
Dynamin-related protein-1 (Drp1)	S-NO	Stimulates mitochondrial fission
Mitochondrial Biogenesis		
Peroxisome proliferator-activated receptor-γ coactivator-1α (PGC-1α)	Acetylation	Inhibition
Apoptosis/Autophagy		
Caspase 3	S-NO	Inhibition; attenuates apoptotic signaling
BCL-2	S-NO	Decreases proteasomal degradation
ATG4	*Cys*-oxidation	Protease inhibition to promote autophagy
L-type Ca(II) channel	*Cys*-oxidation	Inhibition; Ca(II) -mediated apoptosis

ASK1, apoptosis signal-regulated kinase 1; *Cys*, cysteine; *Fe-S*, iron–sulfur; *JNK,* c-jun N-terminal kinase; α-*KGDH*, α-ketoglutarate dehydrogenase.

growth factors, nutrients, Ca(II) fluxes, ROS and its generators in a number of ways. These effects include modulation of respiration and mitochondrial ROS production, or regulation of cellular mitochondrial content through stimulation of biogenic or mitophagy mediators. Cellular antioxidants and 'uncoupling proteins' modulate cellular responses based on essential and excess ROS in the milieu. Mitochondria both generate and respond to these ROS. In mitochondria, essential ROS are generated during normal respiration or in response to growth factor signaling. Lack of this essential signaling could affect cellular REDOX balance, protein disulfide formation, and cell signaling (via reversible thiol oxidation) or other mechanisms. Antioxidants like GPx-1 may attenuate these responses. Essential ROS may also play a role in modulating mitochondrial biogenesis. AII receptor activation is one stimulus that may lead to the generation of excess ROS. Thus, angiotensin II-mediated stimulation of NOx-dependent ROS could enhance mitochondrial production of ROS, leading to mitochondrial dysfunction. Excess ROS could cause oxidative damage to components of the ETC (i.e. complex I) or TCA cycle (i.e. aconitase or a-ketoglutarate), or other mitochondrial proteins. Excess ROS can also promote the mitochondrial recruitment of proteins (i.e. *p66shc*) that further amplify ROS production, which may contribute to cell death. Other mitochondrial proteins, such as *mitochondrial STAT3*, could preserve ETC activity through blocking of ROS leak at complex I. Excess ROS leads to REDOX imbalance, over-oxidation of protein thiols, oxidation of mitochondrial lipids, and 'mitochondrial permeability transition'. At some threshold of ROS, these changes result in necrosis or apoptosis. Cells are equipped with several mechanisms to prevent excess ROS, including the biogenic program that generates new mitochondria and increases the expression of certain antioxidant genes; as well as mitophagy that may target and remove damaged mitochondria. ROS up-regulation of antioxidants or 'uncoupling proteins' may also modulate the responses of excess ROS, and, under some circumstances, brief 'mitochondrial permeability transition' could alleviate mitochondrial stress and prevent mitochondrial dysfunction.

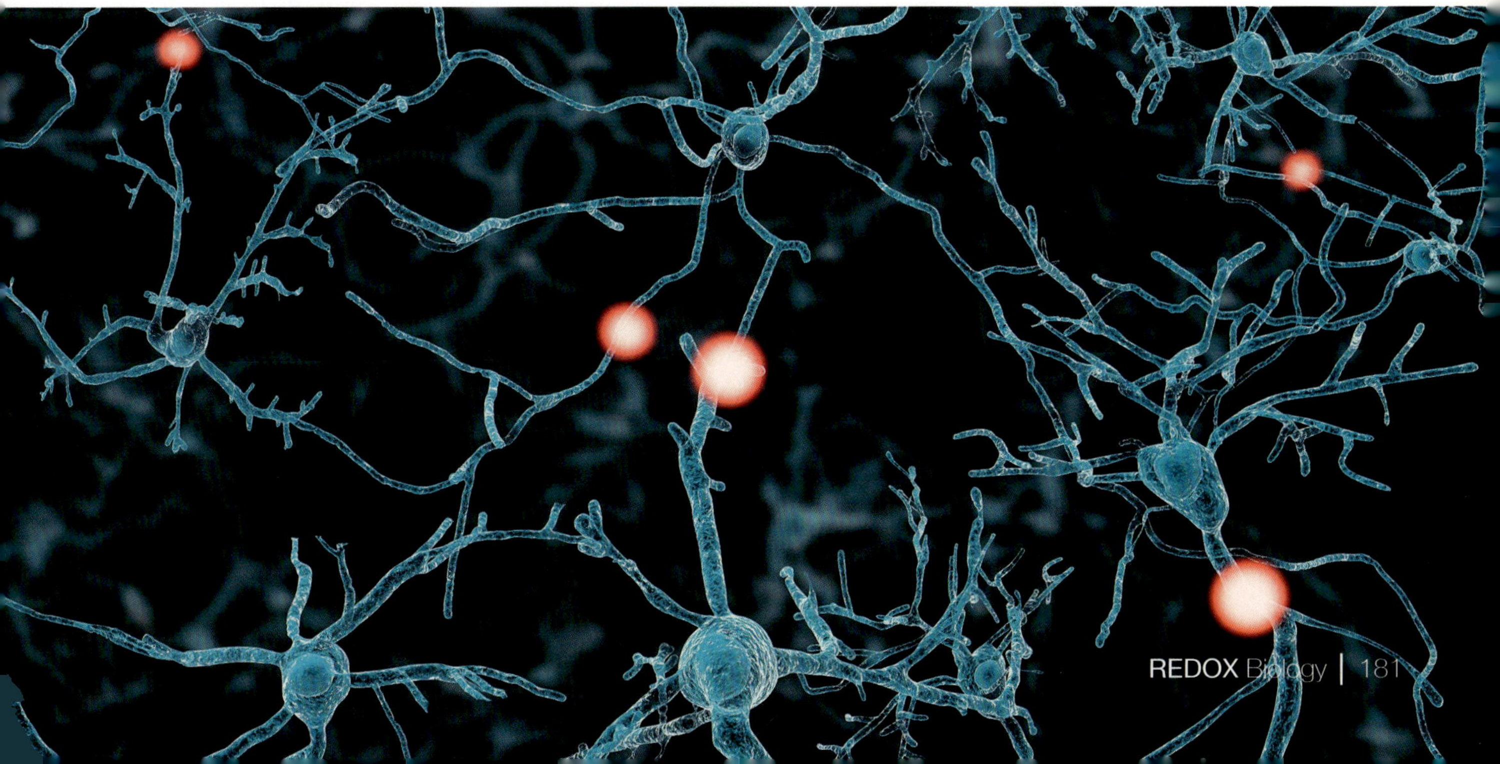

3.5.5. REDOX Signaling: Vascular Regulation

The heart has complex mechanisms that facilitate the maintenance of an oxygen supply–demand balance necessary for its contractile function in response to physiological fluctuations in workload as well as in response to chronic stresses such as hypoxia, ischemia, and overload. REDOX-sensitive signaling pathways are centrally involved in many of these homeostatic and stress-response mechanisms. REDOX signaling plays a key role in signal transduction related to contraction and relaxation of blood vessels, growth, death and migration of vascular cells, extra-cellular matrix (ECM) modifications and angiogenesis.

The nitric oxide (·NO) pathway with endothelium-derived relaxing factor is widespread and plays a variety of physiological roles in the cardiovascular system. These include the maintenance of vascular tone, neurotransmitter function in both the central as well as peripheral nervous systems, and mediation of cellular defense. In addition, ·NO interacts with mitochondrial systems to regulate cell respiration and to augment the generation of ROS, thus triggering mechanisms of cell survival or death.

REDOX Signaling in the Cardiomyocyte: Electrons channel through REDOX couples (i.e. $NADP^+$/NADPH; FAD/$FADH_2$) is fundamental for O_2-based cellular metabolism such as oxidative phosphorylation, protein folding in the endoplasmic reticulum (ER), the enzymatic activities of many oxygenases, and the generation of reductive power. Oxygen or O_2-derived reactive species (ROS), are also involved in post-translational protein modifications that transmit and amplify REDOX signals. In cardiomyocytes, ROS are generated by various sources among which the most important are mitochondria, NADPH oxidases of the Nox family, and nitric oxide synthases (NOSs). The quantity of ROS generation by these sources may increase multi-fold under different cellular conditions (i.e. disease) or upon specific stimulation (i.e. NADPH oxidases). These ROS include both 1-electron oxidants (i.e. $O_2^{\bullet-}$ and ·NO) and 2-electron oxidants (i.e. H_2O_2 and $ONOO^-$). These reactive species are influenced not only by the site and level of production, but also by the cell compartment-specific antioxidant pools and by the interplay between 1- and 2-electron species to form more powerful oxidants such as ·OH radicals. Also, the inactivation of ·NO by $O_2^{\bullet-}$ may lead to modulation of signaling pathways both as a consequence of $ONOO^-$ generation due to the decrease in ·NO bioavailability. At low threshold, the highly reactive ·OH radicals may cause severe oxidation of macromolecules and organelles, which may lead to maladaptive cardiac dysfunction. On the other hand, less oxidative ROS (such as $O_2^{\bullet-}$, ·NO, and H_2O_2) participate in cellular signaling that may have an impact on both adaptive and maladaptive cardiac responses. Therefore, oxidative stress resulting from an imbalance between increased ROS generation and inadequate endogenous antioxidant pools contributes to heart failure. This condition may also contribute to the activation of maladaptive signaling cascades, such as impaired Ca(II) handling —a rudimentary feature of most forms of advanced heart disease.

REDOX Status in Cardiac Metabolism

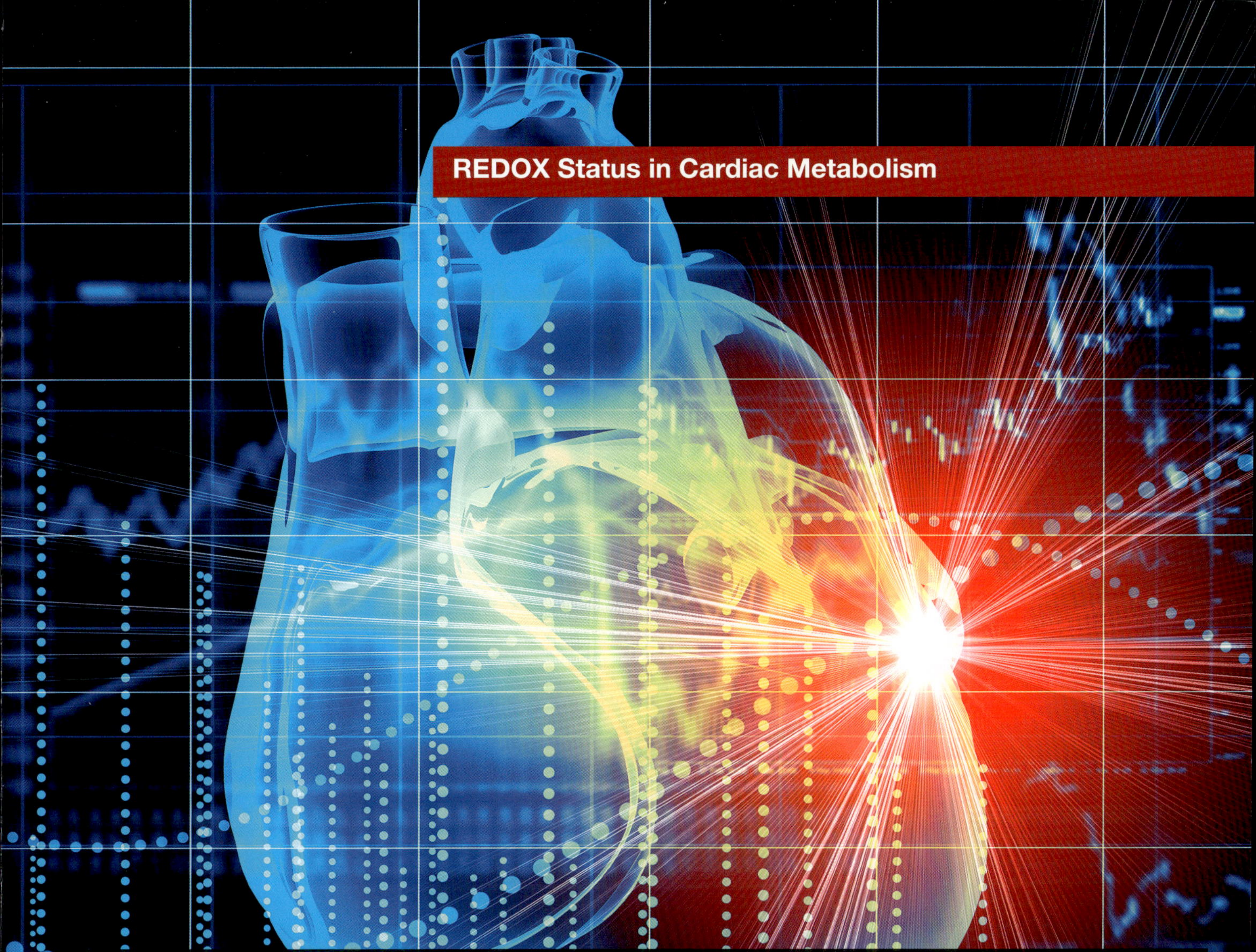

Mitochondria occupy 30% of cardiomyocyte volume, organized in rows, spaced between myofilaments, closely associated with the sarcoplasmic reticulum (SR), and clustered around the nucleus. This cellular organization supports the functional interplay between mitochondrial ATP production, SR-regulated Ca(II) homeostasis and myofilament-dependent contraction. This cyto-architecture also favors efficient high ATP generation and localized energy consumption. In contrast to organs such as brain that depend exclusively on glucose metabolism, the adult mammalian heart normally uses lipids as the major fuel. The mitochondria supply over 90% of the total ATP to the heart through β-oxidation of plasma fatty acids. During hypoxia, under ischemia or settings of increased cardiac workload, however, there is a substantial increase in glycolytic ATP generation in cardiac metabolism. This phenomenon may be cardio-protective during ischemia–reperfusion by ensuring an adequate ATP supply for membrane and SR ion pumps. This metabolic shift is accompanied by changes in levels and activities of numerous proteins involved in glucose metabolism. During this phase, a major increase in the expression and activity of glucose-6-phosphate dehydrogenase (G6PDH) contributes to the generation of NADPH. The metabolic shift to glycolysis has important implications for REDOX status of the heart. NADPH is critical for the maintenance of antioxidant defense through the regeneration of reduced pools of glutathione (GSH), and G6PDH activity is critical for the maintenance of REDOX status, Ca(II) homeostasis, and contractile function in cardiomyocytes subjected to oxidative stress.

An imbalance between ROS and antioxidants in the opposite direction leading to 'reductive stress', could also be detrimental in certain cardiac conditions. Such varying effects on cell signaling can be considered within an overall scheme in which O_2 and local compartment-specific generation of ROS modulate signaling pathways that can potentially drive adaptive or maladaptive stress responses. Ultimately, the balance between such pathways determines whether the heart adapts or fails in different pathological settings (i.e. cardiac overload or ischemia).

Induction of Cardiovascular REDOX Signaling

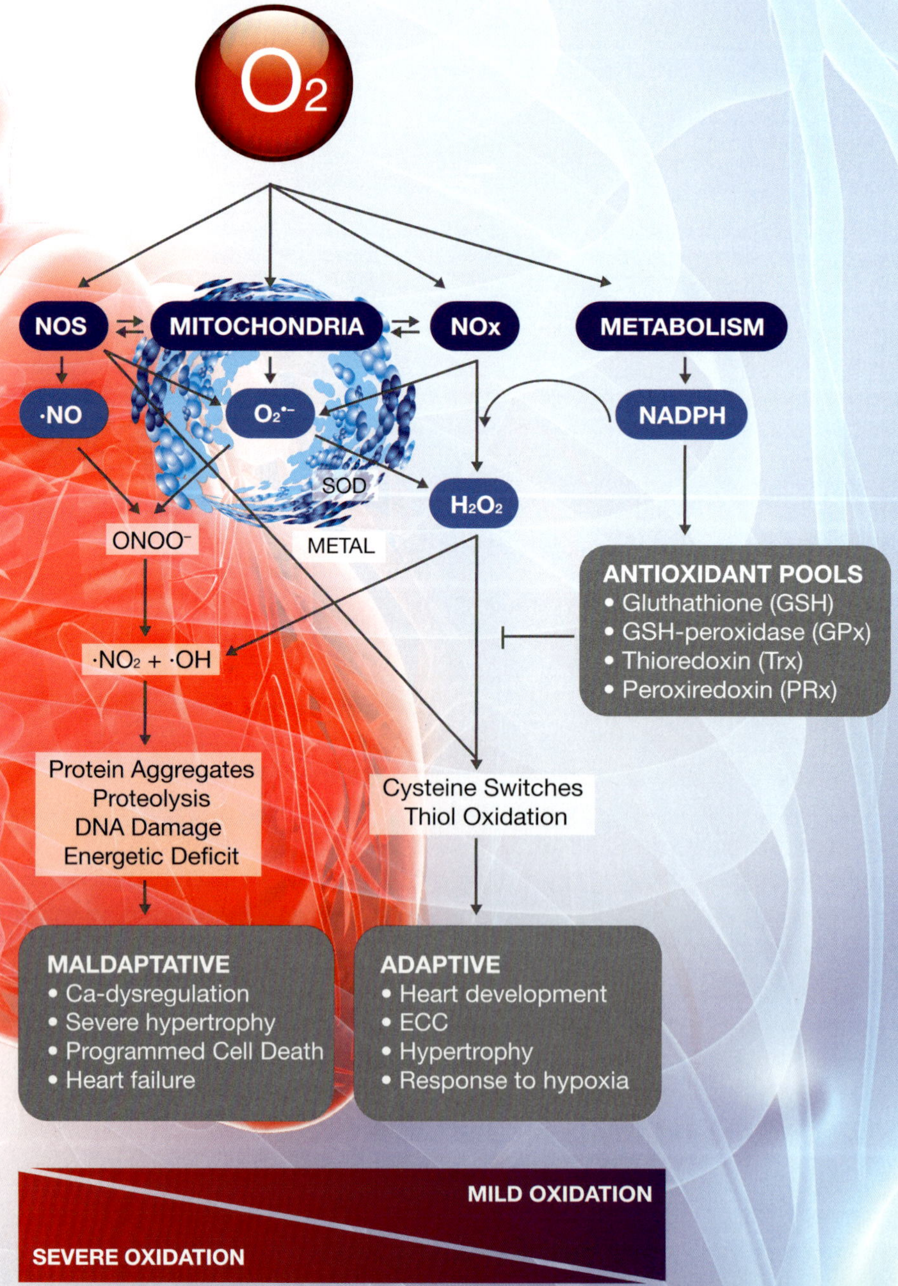

Post-translational REDOX modifications to myocardial proteins (such as *Cys* and *Met* thiol oxidation, proline and arginine hydroxylation, and tyrosine nitration) may potentially affect the conformation, stability, and activity of various receptors, ion transporters (pumps/ exchangers/channels), kinases, phosphatases, caspases, translocators (GTPases), transcription factors, and structural/ contractile proteins. Protein cysteine thiols are the most susceptible REDOX targets of ROS (i.e. H_2O_2) and their oxidation may result in reversible intra- or inter-molecular disulfide formation or other thiol modifications such as nitrosylation and glutathiolation. The balance between oxidized and reduced forms of signaling proteins is influenced by both local ROS generation and reductants such as glutathione (GSH) can reduce a wide range of oxidized proteins. Thioredoxin 1 (Trx 1) is a more specific modulator of REDOX switching of key signaling proteins in the heart, which catalyzes the reduction of cysteine disulfides and nitrosothiols. Irreversible acute protein oxidation may also modulate signaling. Both aged and failing hearts show increased levels of irreversibly oxidized protein amino acids (i.e. 3-nitrotyrosine). In summary, a fine balance between REDOX state and metabolism is more important than oxidative stress *per se* and that an imbalance in either the oxidative or the reductive direction could be detrimental for the heart.

REDOX Regulation in Cardiomyocyte Differentiation, Proliferation and Apoptosis: Adult cardiomyocytes are terminally differentiated cells with absent or extremely limited proliferative capacity, and therefore, differentiation and proliferation are processes that are pertinent mainly to the developing heart. REDOX balance is a critical regulator for differentiation and proliferation of cardiomyocytes. Both mechanical strain and electrical stimulation that enhance intra-cellular ROS levels, also increase the proportion of beating cardiomyocytes population. By contrast, agents that scavenge or reduce ROS levels (i.e. Trolox, catalase, and N-acetyl-cysteine) impair cardiomyocyte formation. An important source of the ROS involved in cardiomyogenesis is NADPH oxidase (NOx), which is abundantly expressed in embryonic stem cells. Cardiomyocyte apoptosis is recognized to be important in both ischemia–reperfusion injury and the transition from compensated cardiac hypertrophy to heart failure. Oxidative stress activates the intrinsic apoptotic pathway in cardiomyocytes (e.g., during ischemia–reperfusion) through multiple REDOX signaling mechanisms.

REDOX Signaling and Angiogenesis: Angiogenesis, a process of new blood vessel formation, is a key process involved in normal development and wound repair as well as in the various pathological conditions such as ischemic heart and limb diseases and atherosclerosis. ROS such as $O_2^{\bullet-}$ and H_2O_2 function as signaling molecules in many aspects of growth factor-mediated responses including angiogenesis. Vascular endothelial growth factor (VEGF) is a key angiogenic growth factor that stimulates proliferation, migration, and tube formation of endothelial cells primarily through the VEGF receptor type-2. The major source of ROS in endothelial cells is a NADPH oxidase (Nox) which is activated by angiogenic factors including VEGF and angiopoietin-1. ROS derived from this enzyme stimulate diverse REDOX signaling pathways leading to angiogenesis-related gene induction as well as endothelial cell migration and proliferation, which may contribute to angiogenesis *in vivo*.

3.5.6. REDOX Signaling: Immune Regulation

The mechanism to differentiate and select peptides by the major histocompatibility complex (MHC) Class I molecules during antigen processing is facilitated by thiol-based REDOX regulation. This REDOX switch provides a checkpoint for the MHC assembly of proteins (antibodies) which elicits rapid and sustained responses to pathogens and evades any unwanted immune response. It is evident that cells employ REDOX signals to regulate a wide variety of physiological functions. The susceptibility of reactive cysteine residues to oxidation is the basis for housing REDOX switches into proteins, which can respond to changes in cellular REDOX status. The thiolate anion is stabilized by proximity to hydrogen bond donors, basic residues, and metal ions. Thus, reactive cysteine residues affect REDOX status and provide a simple and rapid mechanism to regulate biological responses.

REDOX Signaling and NF-κB-mediated Immune Regulation: Nuclear factor kappa B (NF-κB) is a pleiotropic transcription factor that regulates expression of more than 100 genes, which collectively dominate the innate and acquired immune response and associated inflammation. NF-κB is activated via different receptor families that signal to NF-κB by distinct phosphorylation cascades, the common denominator being release from an inhibitory cytosolic complex and nuclear import of the active transcription factor. NF-κB was the first mammalian transcription factor shown to be REDOX regulated and directly activated by a variety of ROS. However, the NF-κB system is primarily activated by cytokines and non-oxidant foreign stressors via TNF, Toll-like, and other receptors (TLR). REDOX regulation of NF-kB activation can be summarized as follows: (i) most agents that activate the NF-κB trigger formation of $O_2^{\bullet-}/H_2O_2$ via NOx or are oxidants by themselves such as $O_2^{\bullet-}$, H_2O_2 or LOx products; (ii) NF-κB activation can be triggered by H_2O_2 or organic hydroperoxides in the absence of any physiological stimulus; (iii) NF-κB activation is inhibited by a broad range of chemically unrelated antioxidants.

The pivotal role of NF-κB system is the activation of an innate immune response upon challenge by microorganisms or mimicking ligands. The main receptors of the system are TLRs, of which so far 13 distinct ones have been identified in mammals, four residing in endosomal vesicles, the other ones at the cell surface. Most of the TLRs respond to microbial structures known as pathogen-associated molecular pattern (PAMP), the most prominent example is the lipopolysaccharide (LPS), which are components of the outer membrane of Gram-negative bacteria. NF-κB is also activated by inflammatory mediators via their respective receptors as well as by G-protein-coupled receptors and antigen receptors.

The overall outcome of the NF-κB activation is an inflammatory response characterized by enhanced expression of pro-inflammatory cytokines (i.e. TNFα, IL-1, and IL-6), chemokines, adhesion molecules, growth factors, and enzymes that produce secondary inflammatory mediators such as COX-2 and induced NO synthase (iNOS). NF-κB links inflammation to carcinogenesis by at least three distinct roles: (i) enforce and sustain an inflammatory response and bring out the mutagenic potential of $O_2^{\bullet-}$, H_2O_2 and lipid hydroperoxides; (ii) possibly kill tumor cells via massive oxidative attack through TNFα induction; and (iii) activate gene expression for anti-apoptotic factors, survivin proteins, and Mn-SOD; collectively support tumor development. The consequences of NF-κB activation, therefore, depend on type and stage of tumor.

REDOX signaling and T-cell Regulation: REDOX is vital for intercellular signaling of the adaptive immune system. An early step in immune processing is the engagement of naive T-cells with antigen-presenting cells in the immune synapse. This triggers a signaling cascade that leads to activation and subsequent clonal expansion of mature T-cells. Low expression of the X_C^- (system X_C^- is an amino acid antiporter that typically mediates the exchange of extra-cellular L-cystine and intra-cellular L-glutamate across plasma membrane makes T-cells inefficient in importing cystine and depend on antigen-presenting cells for cysteine, which is needed for activation. In contrast to naive T-cells, antigen-presenting cells (i.e. dendritic cells), express the X_C^- antiporter, are proficient in cystine uptake, and further up-regulate expression of X_C^- when co-cultured with T-cells. The net outcome of these changes is a shift in the extra-cellular milieu to a more reducing REDOX potential.

Intra-cellular transport and conversion of cystine to cysteine (*Cys*) by dendritic cells involves the transmembrane GSH/*Cys* cycle. The increased *Cys* consumption and *Cys* release by dendritic cells result in an approximately -30-mV decrease in the extra-cellular *Cys*/cystine REDOX potential to -110 mV. In contrast, when dendritic cells and naive T-cells are cultured separately, the extra-cellular REDOX potentials are approximately -80 and -50 mV, respectively, consistent with dendritic cells being differentiated and T-cells being fated for apoptosis in the absence of activation. The reductive remodeling of the extra-cellular environment is conducive for T-cell proliferation and also causes a significant shift in the membrane thiol status.

Regulatory T-cells prevent activation of effector T-cells as part of natural surveillance mechanism for keeping autoimmunity in check. One strategy for immune-suppression by regulatory T-cells is to interfere with the extra-cellular REDOX remodeling during T-cell activation. Cysteine supplementation to the culture medium seems to alleviate immune-suppression by regulatory T-cells. Thus, a change in intra-cellular REDOX metabolism in dendritic cells that leads to a shift in the extra-cellular REDOX potential is a critical determinant of the outcome of T-cell activation and hence immunity.

G-Proteins: The Nobel Prize in Physiology or Medicine in1994 was awarded to Alfred Gilman and Martin Rodbell for their discovery of G-proteins and the role of these proteins in signal transduction in cells

3.5.7. REDOX Signaling: Apoptosis

Apoptosis, or programmed cell death, represents a physiologically conserved mechanism of cell regulation by REDOX signaling. This cellular process is critical for embryonic development and tissue homeostasis. Apoptosis can be triggered by a variety of stimuli, including death receptor ligands, environmental agents, and cytotoxic drugs. Apoptosis has also been implicated in the etiology of several diseases including cancer, neurodegenerative, lung, heart, and auto-immune disorders. Apoptosis involves assembly of specific ligand-receptor pairs leading to activation of caspases, systematic degradation of cellular proteins, plasma membrane disruption, chromatin condensation, and cell fragmentation. Multiple pathways of apoptosis, including both extrinsic and intrinsic pathways, comprised of distinct receptors and organelles have been unraveled.

The morphological features of apoptosis are generated during the activation of cellular cysteine proteases (caspases) that can occur through two main apoptotic pathways: (i) *Death receptor-mediated apoptosis* is triggered by extrinsic signals, such as extra-cellular hormones or components of the tumor necrosis factor (TNF) family that bind to death receptors on the plasma membrane. Pro-apoptotic proteins interact through their death effector domains and form death-inducing signalling complex (DISC). Initiator caspases, such as caspase-8 or caspase-10 are recruited to the DISC that triggers a downstream activation of executioner caspases-3 and -7, leading to apoptosis. (ii) *An intrinsic mitochondrial apoptotic pathway* that involves mitochondria-to-cytosol release of pro-apoptotic proteins. Different apoptotic stimuli, such as reactive oxygen and nitrogen species (ROS/RNS) cause mitochondrial DNA damage, alter mitochondrial outer membrane permeability and release of mitochondrial pro-apoptotic factors, such as cytochrome c (cyt c). Within the cytosol, these apoptogenic factors trigger caspase-dependent or caspase-independent signaling events. For example, cyt c binds to the apoptotic protease-activating factor-1 (Apaf-1) and forms the apoptosome that recruits procaspase-9. The ATP-dependent cleavage of procaspase-9 signals downstream cleavage/activation of effector caspases-3 and -6/7. Additionally, Smac/Diablo, which antagonizes inhibitors of caspases, enhances caspase activation. The mitochondria-to-nuclear translocation induces chromatin condensation and DNA fragmentation leading to apoptosis.

REDOX Signaling and Apoptosis: REDOX-based changes are required for efficient activation of apoptosis. Among these REDOX changes, alterations to glutathione (GSH) and its oxidative post-translational modification to protein S-glutathionylation (PSSG) are considered as critical regulators of apoptosis. GSH acts as a reducing agent and antioxidant in the cells, constituting a major component in the overall cellular REDOX homeostasis. Cellular conditions that create a REDOX imbalance (also referred to as oxidative stress) are known to decrease the ratio of reduced GSH to oxidized glutathione disulfide (GSSG). During oxidative stress, GSH is conjugated to reactive cysteine moieties in protein

Cellular Apoptosis: Death Receptor Pathway

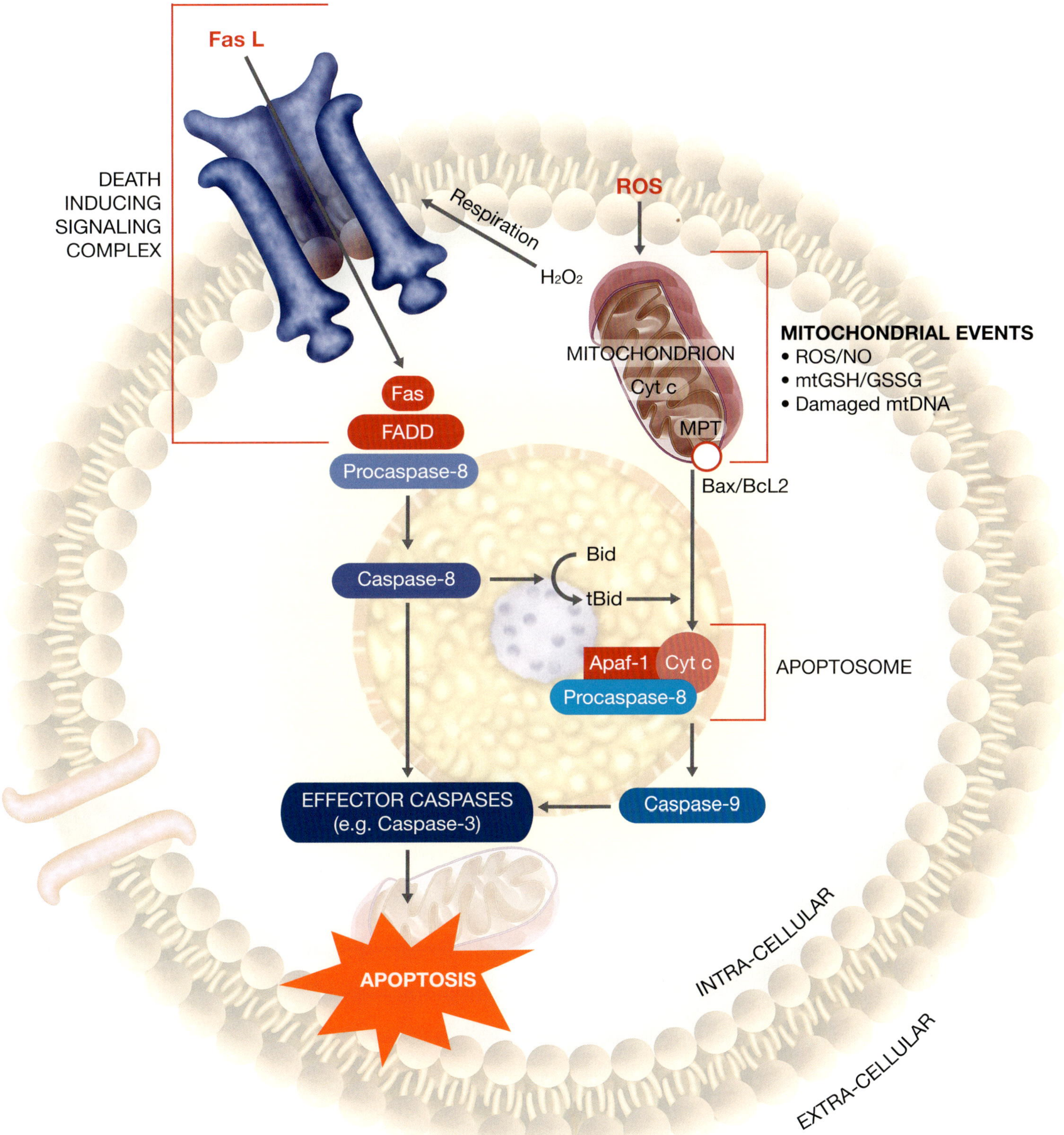

Replenishment of cellular GSH pools by N-acetyl-cysteine (NAC) is known to protect against apoptosis.

structures to form mixed disulfides; a posttranslational modification known as protein S-glutathionylation (PSSG). This oxidative post-translational modification could affect structure and function of cells. Biological 'on' and 'off' switches such as phosphorylation–dephosphorylation, acetylation–deacetylation, and ubiquitination–deubiquitination regulate the cellular response to extra-cellular and intra-cellular stimuli. Similar to these post-translational modifications, S-glutathionylation and de-glutathionylation of proteins also regulate the strength of cellular signals. GSH is critical for survival, based on findings that GSH synthesis-deficient animals die from massive apoptosis. GSH depletion in response to oxidants has been widely reported, and linked to cell death. GSH depletion occurs due to physiological death ligands during early stages, resulting in accumulation of ROS that can either be stimulatory or inhibitory to apoptosis dependent on the cell type and stimulus.

REDOX regulation of death inducing ligands: Apoptosis is triggered by death receptors such as Fas, TNF-receptor 1, and death receptors 4 (DR4) and 5 (DR5). These receptors are activated by *FasL*, TNF-α, and TRAIL, respectively. Multimers of death receptor ligands such as TNF-α and *FasL* are efficient in triggering oligomerization of death receptors *in vitro* and *in vivo*, which is important for effective induction of apoptosis. Chronic disease conditions are known to be accompanied by an increased oxidative environment, due to enhanced production of oxidants by inflammatory cells, or activated or damaged structural cells. Oxidant-induced signaling cascades can lead to transcriptional activation of death receptor ligands (i.e. *FasL*), which has been shown to require activation of *c-Jun-N-terminal kinase*, and the transcription factor, activator protein-1 (AP-1). It is also possible that direct oxidative modification of death-inducing ligands increases their potency as inducers of cell death.

Activation of *Fas*, DR4, and DR5 leads to the formation of the death-inducing signaling complex (DISC), consisting of Fas-associated death domain (FADD) and procaspase-8/10. Activation of initiator caspase-8/10 further amplifies the apoptotic cascade by activation of executioner caspases-3/6/ 7. Cleavage and activation of caspase-3 are known to play central roles in apoptosis. Substantial evidence exists that the activity of caspase-3 is regulated via REDOX-dependent posttranslational modifications. Thus, apoptosis is regulated at multiple layers via REDOX-controlled mechanisms that include S-glutathionylation of multiple proteins in ligand and cell dependent manner.

Oxidation of thiol systems induce cell death due to dysregulation of Ca(II) homeostasis, activation of the mitochondrial permeability transition and activation of apoptosis signal-regulating kinase-1. During apoptosis, GSH efflux is activated so that regardless of how apoptosis is signaled, a common feature is a REDOX potential (E_h) of about −170 mV. Oxidation of GSSG to E_h values more positive than −150 mV cause necrosis.

3.6 | REDOX Biology: **Bibliography**

1. Banarjee R (ed.) (2008) *Redox Biochemistry.* New Jersey: Wiley-Interscience. ISBN 978-0-471786245.
2. Droge W (2002) Free Radicals in the physiological control of cell function *Physiological Reviews* 82:47-95.
3. Imlay JA (2003) Pathways of oxidative damage. *Annual Review of Microbiology* 57:395-418.
4. Trachootham D, et al (2007) Redox Regulation of Cell Survival. *Antioxidants & Redox Signaling* 10(8):1343-1374.

3.1 | REDOX: Pro-oxidants

3.1.1. ROS Reservoirs

A) Mitochondria

1. Cadenas E, Davies KJ (2000) Mitochondrial free radical generation, oxidative stress, and aging. *Free Radical Biology & Medicine* 29(3-4):222-230.
2. Hamanaka RB, Chandel NS (2010) Mitochondrial reactive oxygen species regulate cellular signal ing and dictate biological outcomes. *Trends in Biochemical Sciences* 35(9):505-513.
3. Kowaltowski AJ, de Souza-Pinto NC, Castilho RF, Vercesi AE (2009) Mitochondria and reactive oxygenspecies. *Free Radical Biology & Medicine* 47(4):333-343.
4. Lenaz G (2002) The mitochondrial production of reactive oxygen species: mechanisms and implications in human pathology. *IUBMB Life* 52(3–5):159-164.
5. Liu Y, Fiskum G, Schubert D (2002) Generation of reactive oxygen species by the mitochondrial electron transport chain. *Journal of Neurochemistry* 80(5):780-787.
6. Muller FL, Liu Y, van Remmen H (2004) Complex III releases superoxide to both sides of the inner mitochondrial membrane. *Journal of Biological Chemistry* 279(47):49064-49073.
7. Murphy MP (2009) How mitochondria produce reactive oxygen species. *Biochemistry Journal* 417:1-13
8. Palade GE (1953) An electron microscope study of the mitochondrial structure. *Journal of Histo-chemistry & Cytochemistry* 1(4):188-211.
9. Starkov AA (2008) The role of mitochondria in reactive oxygen species metabolism and signaling. *Annuals of the New York Academy of Sciences* 1147:37-52.
10. Turrens JF (2003) Mitochondrial formation of reactive oxygen species. *Journal of Physiology* 552(2):335-344.

B) Phagocyte

1. Baldridge CW, Gerard RW (1933) The extra respiration of phagocytosis. *American Journal of Physiology* 103:235-236.
2. Babior BM (1984) The respiratory burst of phagocytes. *Journal of Clinical Investigation* 73:599-601.
3. Slauch JM (2011) How does the oxidative burst of macrophages kill bacteria? Still an open question. *Molecular Microbiology* 80(3):580-583
4. Fang FC (2004) Antimicrobial reactive oxygen and nitrogen species: concepts and controversies. *Nature Reviews Microbiology* 2:820-832.
5. Robinson JM (2008) Reactive oxygen species in phagocytic leukocytes. *Histochemisty & Cell Biology* 130:281-297.

3.1.2. ROS Generators

A) Xanthine Oxidoreductase (XOR)

1. Berry CE, Hare JM (2004) Xanthine oxidoreductase and cardiovascular disease: molecular mechanisms and pathophysiological implications. *Journal of Physiology* 555(Pt 3):589-606.
2. Chung HY, Baek BS, Song SH, Kim MS, et al (1997) Xanthine dehydrogenase/ xanthine oxidase and oxidative stress. *Age (Omaha)* 20(3):127-140.

3. Gibbings S, Elkins ND, Fitzgerald H, Tiao J, et al (2011) Xanthine oxidoreductase promotes the inflammatory state of mononuclear phagocytes through effects on chemokine expression, peroxisome proliferator-activated receptor-{gamma} sumoylation, and HIF-1{alpha}. *Journal of Biological Chemistry* 286(2):961-975.
4. Houston M, Chumley P, Radi R, Rubbo H, Freeman BA (1998) Xanthine oxidase reaction with nitric oxide and peroxynitrite. *Archives in Biochemistry & Biophysics* 355:1-8.
5. Martin HM, Hancock JT, Salisbury V, Harrison R (2004) Role of xanthine oxidoreductase as an anti microbial agent. *Infection & Immunity* 72(9):4933-4939.
6. Riaz AA, Schramm R, Sato T, Menger MD, et al (2003) Oxygen radicaldependent expression of CXC chemokines regulate ischemia/reperfusion-induced leukocyte adhesion in the mouse colon. *Free Radical Biology & Medicine* 35(7):782-789.
7. Topham RW, Walker MC, Calisch MP, Williams RW (1982) Evidence for the participation of intestinal xanthine oxidase in the mucosal processing of iron. *Biochemistry* 21:4529-4535.
8. Vorbach C, Harrison R, Capecchi MR (2003) Xanthine oxidoreductase is central to the evolution and function of the innate immune system. *Trends in Immunology* 24(9):512-517.
9. Xiang L, Klintman D, Thorlacius H (2003) Allopurinol inhibits CXC chemokine expression and leukocyte adhesion in endotoxemic liver injury. *Inflammation Research* 52(9):353-358.

B) NADPH Oxidases (NOX)

1. Banfi B, Malgrange B, Knisz J, Steger K, et al (2004) NOX3: a superoxide-generating NADPH oxidase of the inner ear. *Journal of Biological Chemistry* 279:46065-46072.
2. Banfi B, Molnar G, Maturana A, Steger K, et al (2001) A Ca(2+)- activated NADPH oxidase in testis, spleen, lymph nodes. *Journal of Biological Chemistry* 276:37594-37601.
3. Bedard K, Krause K-H (2007) The NOX family of ROS-generating NADPH oxidases: *Physiology and Pathophysiology. Physiological Reviews* 87:245-313.
5. Castier Y, Brandes RP, Leseche G, Tedgui A, Lehoux S (2005) p47phox-dependent NADPH oxidase regulates flow-induced vascular remodeling. *Circulation Research* 97:533-540.
6. Cheng G, Cao Z, Xu X, Meir EG, Lambeth JD (2001) Homologs of gp91phox: cloning and tissue expression of Nox3, Nox4, Nox5. *Gene* 269:131-140.
7. De Deken X, Wang D, Many MC, Costagliola S, et al (2000) Cloning of two human thyroid cDNAs encoding new members of the NADPH oxidase family. *Journal of Biological Chemistry* 275:23227-23233.
8. Geiszt M, Kopp JB, Varnai P, Leto TL (2000) Identification of renox, an NAD(P)H oxidase in kidney. *Proceedings of the National Academy of Sciences USA* 97:8010-8014.
9. Geiszt M, Leto TL (2004) The Nox family of NAD(P)H oxidases: host defense and beyond. *Journal of Biological Chemistry* 279:51715-51718.
10. Goldblatt D, Thrasher AJ (2000) Chronic granulomatous disease. *Clinical Experimental Immunology* 122:1-9.
11. Morand S, Dos Santos OF, Ohayon R, Kaniewski J, et al (2003) Identification of a truncated dual oxidase 2 (DUOX2) messenger ribonucleic acid (mRNA) in two rat thyroid cell lines. Insulin and forskolin regulation of DUOX2 mRNA levels in FRTL-5 cells and porcine thyrocytes. *Endocrinology* 144:567-574.
12. Morawietz H, Bornstein SR (2006) Leptin, endothelin, NADPH oxidase, heart failure. *Hypertension* 47:20-21.
13. Moreno JC, Bikker H, Kempers MJ, van Trotsenburg AS, et al (2002) Inactivating mutations in the gene for thyroid oxidase 2 (THOX2) and congenital hypothyroidism. *New England Journal of Medicine* 347:95-102.
14. Rude MK, Duhaney TA, Kuster GM, Judge S, et al (2005) Aldosterone stimulates matrix metallo proteinases and reactive oxygen species in adult rat ventricular cardiomyocytes. *Hypertension* 46:555-561.
15. Sauer H, Bekhite MM, Hescheler J, Wartenberg M (2005) Redox control of angiogenic factors and CD31-positive vessel-like structures in mouse embryonic stem cells after direct current electrical field stimulation. *Experimental Cellular Research* 304:380-390.
16. Szocs K, Lassegue B, Sorescu D, Hilenski LL, et al (2002) Upregulation of Nox-based NAD(P)H oxidases in restenosis after carotid injury. *Arteriosclerosis Thrombosis & Vascular Biology* 22:21-27.

C) Nitric Oxide Synthases (NOS)

1. Forstermann U, Sessa WC (2012) Nitric oxide synthases: regulation and function. *European Heart Journal* 33:829-837.
2. Hemmens B, Mayer B (1998) Enzymology of nitric oxide synthases. *Methods in Molecular Biology* 100:1-32.
3. Huang PL, Huang Z, Mashimo H, Bloch KD, et al (1995) Hypertension in mice lacking the gene for endothelial nitric oxide synthase. *Nature* 377:239-242.
4. Lange M, Enkhbaatar P, Nakano Y, Traber DL (2009) Role of nitric oxide in shock: the large animal perspective. *Frontiers of Bioscience* 14:1979-1989.
5. Li H, Forstermann U (2000) Nitric oxide in the pathogenesis of vascular disease. *Journal of Pathology* 190:244-254.
6. MacMicking J, Xie QW, Nathan C (1997) Nitric oxide and macrophage function. *Annual Review of Immunology* 15:323–350.
7. Nicholson S, Bonecini-Almeida Mda G, Lapa e Silva JR, Nathan C, et al (1996) Inducible nitric oxide synthase in pulmonary alveolar macrophages from patients with tuberculosis. *Journal of Experimental Medicine* 183:2293-2302.
8. Rapoport RM, Draznin MB, Murad F (1983) Endothelium-dependent relaxation in rat aorta may be mediated through cyclic GMP-dependent protein phosphorylation. *Nature* 306:174-176.
9. Rudic RD, Shesely EG, Maeda N, Smithies O, et al (1998) Direct evidence for the importance of endothelium-derived nitric oxide in vascular remodeling. *Journal of Clinical Investigation* 101:731-736.
10. Stenger S, Thuring H, Rollinghoff M, Bogdan C (1994) Tissue expression of inducible nitric oxide synthase is closely associated with resistance to Leishmania major. *Journal of Experimental Medicine* 180:783-793.
11. Wei XQ, Charles IG, Smith A, Ure J, et al (1995) Altered immune responses in mice lacking inducible nitric oxide synthase. *Nature* 375:408-411.
12. Wong JM, Billiar TR (1996) Regulation and function of inducible nitric oxide synthase during sepsis and acute inflammation. *Advances in Pharmacology* 34:155-170.

3.2 | REDOX: Antioxidants

3.2.1. Antioxidant Enzymes

A) Superoxide dismutase (SOD)

1. Buettner GR (2011) Superoxide dismutase in redox biology: the roles of superoxide and hydrogen peroxide. *Anticancer Agents in Medicinal Chemistry* 11(4):341-346.
2. Fridovich I (1995) Superoxide radical and superoxide dismutases. *Annual Reviews in Biochemistry* 64:97–112.
3. Fukai T, Ushio-Fukai M (2011) Superoxide dismutases: Role in redox signaling, vascular function, and diseases. *Antioxidants & Redox Signaling* 15(6):1583-1606.
4. Halliwell B, Gutteridge JM (2007) *Free Radicals in Biology and Medicine* (4th ed.) England:Oxford University Press. ISBN-10:019856869X
5. Holley AK, Dhar SK, Xu Y, St Clair DK (2012) Manganese superoxide dismutase: beyond life and death. *Amino Acids* 42(1):139-158.
6. Kohen R, Nyska A (2002) Oxidation of biological systems: oxidative stress phenomena, antioxidants, redox reactions, and methods for their quantification. *Toxicology Pathology* 30(6):620-50.
7. McCord JM, Fridovich I (1968) The reduction of cytochrome c by milk xanthine oxidase. *Journal of Biological Chemistry* 243:5753–5760.
8. McCord JM, Fridovich I (1969) Superoxide dismutase: An enzymatic function for erythrocuprein (hemocuprein). *Journal of Biological Chemistry* 244:6049–6055.
9. Miriyala S, Holley AK, St Clair DK (2011) Mitochondrial superoxide dismutase – signals of distinction. *Anticancer Agents in Medicinal Chemistry* 11(2):181-190.

B) Catalases

1. Alberts B, Johnson A, Lewis J, Raff M, (2002) *Perioxisomes. Molecular Biology of the Cell* (4th ed). New York:Garland Science. ISBN 0-8153-3218-1
2. Chance B, Sies H, Boveris A (1979) Hydroperoxide metabolism in mammalian organs. *Physiological Reviews* 59:527-605

3. Gaetani G, Ferraris A, Rolfo M, Mangerini R, et al (1996) Predominant role of catalase in the disposal of hydrogen peroxide within human erythrocytes. *Blood* 87(4):1595-1599.
4. Lardinois OM (1995) Reactions of bovine liver catalase with superoxide radicals and hydrogen peroxide. *Free Radical Research* 22:251-274.
5. Wood JM, Decker H, Hartmann H, Chavan B, et al (2009) Senile hair graying: H_2O_2-mediated oxidative stress affects human hair color by blunting methionine sulfoxide repair. *FASEB Journal* 23(7):2065–2075.

C) Peroxidases

1. de Wit JN, van Hooydonk ACM (2000) Lactoperoxidase: physico-chemical properties, occurrence, mechanism of action and applications. *British Journal of Nutrition* 84:S19-S25.
2. Nagy K, Urban E, Fazwkas O, Thurzo L, Nagy E (2007) Controlled study of lactoperoxidase gel on oral flora and saliva in irradiated patients with oral cancer. *Journal of Craniofacial Surgery* 18(5):1157-1164.
3. Naidu AS (2000) Lactoperoxidase, In: *Natural Food Antimicrobial Systems* (Ed: AS Naidu). Boca Raton:CRC Press. ISBN-10: 084932047X
4. Neidleman SL, Geigert J (1986) *Biohalogenation - Principles, Basic Roles and Applications.* Chichester:Ellis Horwood Ltd Publishers. ISBN 0-85312-984-3.
5. Ruf J, Carayon P (2006) Structural and functional aspects of thyroid peroxidase. *Archives in Biochemistry & Biophysics* 445(2):269-277.
6. Tenovuo JO, Pruitt KM, ed. (1985). *The Lactoperoxidase System: Chemistry and Biological Significance*. New York:Dekker. ISBN 0-8247-7298-9.
7. Zhang R, Brennan ML, Fu X, Aviles RJ, et al (2001) Association between myeloperoxidase levels and risk of coronary artery disease. *JAMA* 286(17):2136-2142.

3.2.2. Antioxidant Scavengers

A) Glutathione (GSH)

1. Forman HJ, Zhang H, Rinna A (2009) Glutathione: overview of its protective roles, measurement, and biosynthesis. *Molecular Aspects of Medicine* 30(1-2):1-12.
2. Franco R, Schoneveld OJ, Pappa A, Panayiotidis MI (2007) The central role of glutathione in the pathophysiology of human diseases. *Archives in Physiological Biochemistry* 113(4-5):234-258.
3. Marí M, Morales A, Colell A, García-Ruiz C, (2013) Mitochondrial glutathione: Features, regulation and role in disease. *Biochimica et Biophysica Acta* 1830:3317-3328
4. Meister A (1974) Glutathione, metabolism and function via the gamma-glutamyl cycle. *Life Sciences* 15(2):177–190.
5. Meister A (1994) Glutathione, ascorbate, and cellular protection. *Cancer Research* 54:1969S-1975S.
6. Sen CK (1997) Nutritional biochemistry of cellular glutathione. *Nutritional Biochemistry* 8:660-672.
7. Sies H (1999) Glutathione and its role in cellular functions. *Free Radical Biology & Medicine* 27:916-921.
8. Wu G, Fang YZ, Yang S, Lupton JR, Turner ND (2004) Glutathione metabolism and its implications for health. *Journal of Nutrition* 134(3):489-492.
9. Zhang H, Forman HJ (2012) Glutathione synthesis and its role in redox signaling. *Seminars in Cell & Developmental Biology* 23(7):722-728.

B) Uric acid

1. Ames BN, Cathcart R, Schwiers E, Hochstein P (1981) Uric acid provides an antioxidant defense in humans against oxidant- and radical-caused aging and cancer: a hypothesis. *Proceedings of the National Academy of Sciences USA* 78:6858-6862.
2. Becker BF (1993) Towards the physiologica l function of uric acid. *Free Radical Biology & Medicine* 14:615-631.
3. Glantzounis GK, Tsimoyiannis EC, Kappas AM, Galaris DA (2005) Uric acid and oxidative stress. *Current Pharmaceutical Design* 11(32):4145-4151.
4. Hooper DC, Spitsin S, Kean RB, Champion JM, et al (1998) Uric acid, a natural scavenger of peroxyni trite, in experimental allergic encephalomyelitis and multiple sclerosis. *Proceedings of the National Academy of Sciences USA* 95:675-680.
5. Kellogg EW III, Fridovich I (1977) Liposome oxidation and erythrocyte lysis by enzymically generated superoxide and hydrogen peroxide. *Journal of Biological Chemistry* 252:6721-6728.

6. Kuzkaya N, Weissmann N, Harrison DG, Dikalov S (2005) Interactions of peroxynitrite with uric acid in the presence of ascorbate and thiols: implications for uncoupling endothelial nitric oxide synthase. *Biochemical Pharmacology* 70:343-354.
7. Yu ZF, Bruce-Keller AJ, Goodman Y, Mattson MP (1998) Uric acid protects neurons against excitotoxic and metabolic insults in cell culture, and against focal ischemic brain injury in vivo. *Journal of Neuroscience Research* 53:613-625.

C) Melatonin

1. Forrest CM, Mackay GM, Stoy N, Stone TW, et al (2007) Inflammatory status and kynurenine metabolism in rheumatoid arthritis treated with melatonin. *British Journal of Clinical Pharmacology* 64:517-526.
2. Kedziora-Kornatowska K, Szewczyk-Golec K, Czuczejko J, Pawluk H, et al (2008) Antioxidative effects of melatonin administration in elderly primary essential hypertension patients. *Journal of Pineal Research* 45:312–317.
3. Peyrot F, Houee-Levin C, Ducrocq C (2006) Melatonin nitrosation promoted by NO_2; comparison with the peroxynitrite reaction. *Free Radical Research* 40:910-920.
4. Reiter RJ (2003) Melatonin: clinical relevance. *Best Practice & Research: Clinical Endocrinology & Metabolism* 17:273-285.
5. Reiter RJ, Paredes SD, Korkmaz A, Jou MJ, Tan DX (2008) Melatonin combats molecular terrorism at the mitochondrial level. *Interdisciplinary Toxicology* 1:137-149.
6. Reiter RJ, Tan DX, Manchester LC, Lopez-Burillo S, et al (2003) Melatonin: detoxification of oxygen and nitrogen-based toxic reactants. *Advances in Experimental Medicine & Biology* 527:539-548.
7. Tamura H, Takasaki A, Taketani T, Tanabe M, et al (2013) Melatonin as a free radical scavenger in the ovarian follicle. *Endocrine Journal* 60(1):1-13.
8. Tomas-Zapico C, Coto-Montes A (2005) A proposed mechanism to explain the stimulatory effect of melatonin on antioxidative enzymes. *Journal of Pineal Research* 39:99-104.

3.2.3. Antioxidant Vitamins

A) Ascorbic acid

1. Campbell JD, Cole M, Bunditrutavorn B, Vell AT (1999) Ascorbic acid is a potent inhibitor of various forms of T cell apoptosis. *Cellular Immunology* 194:1-5.
2. Frei B, England L, Ames BN (1989) Ascorbate is an outstanding antioxidant in human blood plasma. *Proceedings of the National Academy of Sciences USA* 86:6377-6381.
3. Kc S, Cárcamo JM, Golde DW (2005) Vitamin C enters mitochondria via facilitative glucose transporter 1 (Glut1) and confers mitochondrial protection against oxidative injury. *FASEB Journal* 19:1657-1667.
4. Levin M (1986) New concepts in the biology and biochemistry of ascorbic acid. *New England Journal of Medicine* 31:892-902.
5. Mandl J, Szarka A, Banhegyi G (2009) Vitamin C: Update on physiology and pharmacology. *British Journal of Pharmacology* 157:1097-1110.
6. Meister A (1994) Glutathione-ascorbic acid antioxidant system in animals. *Journal of Biological Chemistry* 269(13):9397-9400.
7. Ness AR, Chee D, Elliott P (1997) Vitamin C and blood pressure – an overview. *Journal of Human Hypertension* 11:343-350.
8. Rumsey SC, Levine M (1998) Absorption, transport, and disposition of ascorbic acid in humans. *Journal of Nutritional Biochemistry* 9:116-30.
9. Sharma P, Mongan PD (2001) Ascorbate reduces superoxide production and improves mitochondrial respiratory chain function in human fibroblasts with electron transport chain deficiencies. *Mitochondrion* 1:191-198.
10. Svirbely JL, Szent-Gyorgyi A (1932) The chemical nature of vitamin C. *Biochemistry Journal* 26:865-870.

B) Tocopherols

1. Atkinson J, Epand RF, Epand RM (2008) Tocopherols and tocotrienols in membranes: a critical review. *Free Radical Biology & Medicine* 44(5):739-764.
2. Azzi A (2007) Molecular mechanism of alpha-tocopherol action. *Free Radical Biology & Medicine* 43 (1):16-21.

3. Brigelius-Flohe R, Traber MG (1999) Vitamin E: function and metabolism. *FASEB Journal* 13:1145–1155.
4. Engin KN, Engin G, Kücüksahin H, Oncu M, et al (2007) Clinical evaluation of the neuro-protective effect of α-tocopherol on retına against glaucomatous damage. *European Journal of Ophthalmology* 17:528-533.
5. Herrera; Barbas C (2001) Vitamin E: action, metabolism and perspectives. *Journal of Physiology & Biochemistry* 57 (2): 43-56.
6. Lauridsen C, Jensen SK (2012) α-Tocopherol incorporation in mitochondria and microsomes upon supranutritional vitamin E supplementation. *Genes & Nutrition* 7:475-482.
7. Nell S, Bahtz R, Bobecker A, Kipp A, et al (2007) PCR-verified microarray analysis and functional in vitro studies indicate a role of α-tocopherol in vesicular transport. *Free Radical Research* 41:930–942.
8. Schneider C (2005) Chemistry and biology of vitamin E. *Molecular Nutrition & Food Research* 49(1):7-30.
9. Sies H, Murphy ME (1991) Role of tocopherols in the protection of biological systems against oxidative damage. *Journal of Photochemistry & Photobiology* 8:211-218.

C) Carotenoids

1. Burton GW, Ingold KU (1984) Beta-Carotene: an unusual type of lipid antioxidant. *Science* 224:569–573.
2. El-Agamey A, Lowe GM, McGarvey DJ, Mortensen A, et al (2004) Carotenoid radical chemistry and antioxidant/pro-oxidant properties. *Archives in Biochemistry & Biophysics* 430:37-48.
3. Krinsky NI (1998) Overview of lycopene, carotenoids, and disease prevention. *Proceedings of the Society for Experimental Biology & Medicine* 218:95-97.
4. Moore T (1930) Vitamin A and carotene. VI. The conversion of carotene to vitamin A in vivo. *Biochemistry Journal* 24:692-702.
5. Niles RM (2004) Signaling pathways in retinoid chemoprevention and treatment of cancer. *Mutation Research* 555:81-96.
6. Rice-Evans CA, Sampson J, Bramley PM, Holloway DE (1997) Why do we expect carotenoids to be antioxidants in vivo? *Free Radical Research* 26:381-398.
7. Sies H, Stahl W (1997) Carotenoids and intercellular communication via gap junctions. *International Journal for Vitamin & Nutrition Research* 67:364-367.
8. Snodderly DM (1995) Evidence for protection against age-related macular degeneration by carotenoids and antioxidant vitamins. *American Journal of Clinical Nutrition* 62:1448S-1461S.
9. Woodall AA, Britton G, Jackson MJ (1997) Carotenoids and protection of phospholipids in solution or in liposomes against oxidation by peroxyl radicals: Relationship between carotenoid structure and protective ability. *Biochimica et Biophysica Acta* 1336:575-586.

3.3 | REDOX: Coupling systems

1. Go YM, Jones DP (2008) Redox compartmentalization in eukaryotic cells. *Biochimica et Biophysica Acta* 1780:1273-1290.
2. Jones DP, Go YM (2010) Redox compartmentalization and cellular stress. *Diabetes Obesity & Metabolism* 12(2):116-125.
3. Jones DP (2006) Redefining oxidative stress. *Antioxidants & Redox Signaling* 8:1865-1879.

3.3.1. GSH/GSSG Redox System

1. Bass R, Ruddock LW, Klappa P, Freedman RB (2004) A major fraction of endoplasmic reticulum located glutathione is present as mixed disulfides with protein. *Journal of Biological Chemistry* 279:5257-5262.
2. Circu ML, Aw TY (2010) Reactive oxygen species, cellular redox systems, and apoptosis. *Free Radical Biology & Medicine* 48:749-762.
3. Herrmann JM, Riemer J (2010) The intermembrane space of mitochondria. *Antioxidants & Redox Signaling* 13:1341-1358.
4. Kemp M, Go YM, Jones DP (2008) Nonequilibrium thermodynamics of thiol/disulfide redox systems: A perspective on redox systems biology. *Free Radical Biology & Medicine* 44:921-937.
5. Koehler CM, Beverly KN, Leverich EP (2006) Redox pathways of the mitochondrion. *Antioxidants & Redox Signaling* 8:813-822.

6. Markovic J, Borras C, Ortega A, Sastre J, et al (2007) Glutathione is recruited into the nucleus in early phases of cell proliferation. *Journal of Biological Chemistry* 282:20416-20424.
7. Schafer FQ, Buettner GR (2001) Redox environment of the cell as viewed through the redox state of the glutathione disulfide/glutathione couple. *Free Radical Biology & Medicine* 30:1191-1212.
8. Schnellmann RG (1991) Renal mitochondrial glutathione transport. *Life Sciences* 49:393-398.

3.3.2. Trx-S/Trx-SS Redox System

1. Ahsan MK, Lekli I, Ray D, Yodoi J, Das DK (2009) Redox regulation of cell survival by the thioredoxin superfamily: an implication of redox gene therapy in the heart. *Antioxidants & Redox Signaling* 11:2741-2758.
2. Conrad M, Jakupoglu C, Moreno SG, Lippl S, et al (2004) Essential role for mitochondrial thioredoxin reductase in hematopoiesis, heart development, and heart function. *Molecular Cell Biology* 24:9414-9423.
3. Dunn LL, Buckle AM, Cooke JP, Ng MK (2010) The emerging role of the thioredoxin system in angiogenesis. *Arteriosclerosis Thrombosis & Vascular Biology* 30(11):2089-2098.
4. Hansen JM, Go YM, Jones DP (2006) Nuclear and mitochondrial compartmentation of oxidative stress and redox signaling. *Annual Reviews in Pharmacology & Toxicology* 46:215-234.
5. Yoshida T, Nakamura H, Masutani H, Yodoi J (2005) The involvement of thioredoxin and thioredoxin binding protein-2 on cellular proliferation and aging process. *Annals of the NY Academy of Sciences* 1055:1-12.

Peroxiredoxins (PRXs)

1. Cox AG, Pearson AG, Pullar JM, Jonsson TJ, et al (2009) Mitochondrial peroxiredoxin 3 is more resilient to hyperoxidation than cytoplasmic peroxiredoxins. *Biochemistry Journal* 421:51-58.
2. Knoops B, Goemaere J, Van der Eecken V, Declercq JP (2011) Peroxiredoxin 5: Structure, mechanism and function of the mammalian atypical 2-Cys peroxiredoxin. *Antioxidants & Redox Signaling* 15:817-829.
3. Neumann CA, Cao J, Manevich Y (2009) Peroxiredoxin I and its role in cell signaling. *Cell Cycle* 8(24):4072-4078.
4. Woo HA, Yim SH, Shin DH, Kang D, et al (2010) Inactivation of peroxiredoxin I by phosphorylation allows localized H2O2 accumulation for cell signaling. *Cell* 140:517-528.

3.3.3. Cys/CySS Redox System

1. Blanco RA, Ziegler TR, Carlson BA, Cheng PY, et al (2007) Diurnal variation in glutathione and cysteine redox states in human plasma. *American Journal of Clinical Nutrition* 86:1016-1023.
2. Park Y, Ziegler TR, Gletsu-Miller N, Liang Y, et al (2010) Postprandial cysteine/cystine redox potential in human plasma varies with meal content of sulfur amino acids. *Journal of Nutrition* 140:760-765.
3. Stipanuk MH (2004) Sulfur amino acid metabolism: pathways for production and removal of homo cysteine and cysteine. *Annual Reviews of Nutrition* 24:539-577.
4. Moriarty-Craige SE, Jones DP (2004) Extracellular thiols and thiol/disulfide redox in metabolism. *Annual Reviews of Nutrition* 24:481-509.

3.4 | REDOX: Aging

3.4.1. Free Radical (Oxidative Stress) Theories

1. Gruber J, Schaffer S, Halliwell B (2008) The mitochondrial free radical theory of ageing--where do we stand? *Frontiers of Bioscience* 13:6554-6579
2. Harman D (1956) Aging: a theory based on free radical and radiation chemistry. *Journal of Gerontology* 11(3):298-300.
3. Pearl, R (1928) The rate of living. New York:Alfred A. Knopf, Inc
4. Sies H, Cadenas E (1985) Oxidative stress: damage to intact cells and organs. *Philosophical Transactions of Royal Society of London (B) Biological Sciences* 311(1152):617-631.
5. Sohal RS, Mockett RJ, Orr WC (2002) Mechanisms of aging: an appraisal of the oxidative stress hypothesis. *Free Radical Biology & Medicine* 33(5):575-586.

3.4.2. Mitochondrial Mutation Theory

1. Esser K, Martin GM (1995) *Molecular aspect of aging.* Chichester:J.Wiley and Sons. ISBN-10:0471956899.
2. Miquel J (1998) An update on the oxygen stress-mitochondrial mutation theory of aging: genetic and evolutionary implications. *Experimental Gerontology* 33(1-2):113-126.
3. Miquel J, Fleming JE (1984) A two-step hypothesis on the mechanisms of in vitro cell aging: cell differentiation followed by intrinsic mitochondrial mutagenesis. *Experimental Gerontology*19:31-36.
4. Yakes FM, Van Houten B (1997) Mitochondrial DNA damage is more extensive and persists longer than nuclear DNA damage in human cells following oxidative stress. *Proceedings of the National Academy of Sciences USA* 94(2):514-519.

3.4.3. Cell signaling (Redox Stress) Hypothesis

1. Bandy B, Davison AJ (1990) Mitochondrial mutations may increase oxidative stress: implications for carcinogenesis and aging? *Free Radical Biology & Medicine* 8(6):523-539.
2. Brandes N, Schmitt S, Jakob U (2009) Thiol-based redox switches in eukaryotic proteins. *Antioxidants & Redox Signaling* 11(5):997-1014.
3. Halliwell B, Gutteridge JMC (2007) *Free Radicals in Biology and Medicine* (4th Ed) New York: Oxford University Press. ISBN 10:019856869X
4. Sohal RS, Orr WC (2012) The redox stress hypothesis of aging. *Free Radical Biology & Medicine* 52(3):539-555.
5. Sohal RS, Weindruch R (1996) Oxidative stress, caloric restriction, and aging. *Science* 273:59-63.

3.5 | REDOX: Signaling

1. Afanasev IB (2009) *Signaling Mechanisms of Oxygen and Nitrogen Free Radicals.* Boca Raton:CRC. Press. ISBN:978-1420073744.
2. Banerjee R (2012) Redox outside the box: linking extracellular redox remodeling with intracellular redox metabolism. *Journal of Biological Chemistry* 287(7):4397-4402.
3. Bienert GP, Moller AL, Kristiansen KA, Schulz A, et al (2007) Specific aquaporins facilitate the diffusion of hydrogen peroxide across membranes. *Journal of Biological Chemistry* 282:1183-1192.
4. Irani K, Xia Y, Zweier JL, Sollott SJ, et al (1997) Mitogenic signaling mediated by oxidants in Ras-trans formed fibroblasts. *Science* 275:1649-1652.
5. Jacob C, Winyard PG (2009) *Redox Signaling and Regulation in Biology and Medicine.*Weinheim: Wiley-VCH. ISBN:978-3-527-31925-1.
6. Packer L, Yodoi J (eds.) (1999) *Redox Regulation of Cell Signaling and its Clinical Applications.* New York: Marcel Dekker. ISBN:0-8247-1961-1.
7. Ray PD, Huang B-W, Tsuji Y (2012) Reactive oxygen species (ROS) homeostasis and redox regulation in cellular signaling. *Cell Signaling* 24(5):981-990.
8. Rhee SG (2006) Cell signaling. H2O2, a necessary evil for cell signaling. *Science* 312:1882-1883.

3.5.1. REDOX Signaling: Molecular Mechanisms

1. Banerjee R, Becker D, Dickman M, Gladyshev V, Ragsdale S (ed.) (2008) *Redox Biochemistry.* New Jersey: John Wiley. ISBN:978-0-471-78624-5.
2. Barrett DM, Black SM, Todor H, Schmidt-Ullrich RK et al (2005) Inhibition of protein-tyrosine phos phatases by mild oxidative stresses is dependent on S-nitrosylation. *The Journal of Biological Chemistry* 280(15):14453-14461.
3. Beckman JS, Koppenol WH (1996) Nitric oxide, superoxide, and peroxynitrite: the good, the bad, and ugly. *American Journal of Physiology* 271(5 Pt 1):C1424-C1437.
4. Brigelius-Flohe R, Leopold Flohe (2011) Basic principles and emerging concepts in the redox control of transcription factors. *Antioxidants & Redox Signaling* 15(8): 2335–2381.
5. Ceaser EK, Moellering DR, Shiva S, Ramachandran A, et al (2004) Mechanisms of signal transduction mediated by oxidized lipids: the role of the electrophile-responsive proteome. *Biochemical Society Transactions* 32(Pt 1):151-155.
6. Hess DT, Matsumoto A, Kim SO, Marshall HE, et al (2005) Protein S-nitrosylation: purview and param eters. *Nature Reviews Molecular Cell Biology* 6(2):150-166.
7. Kim HP, Ryter SW, Choi AM (2006) CO as a cellular signaling molecule. *Annual Review of Pharmacology & Toxicology* 46:411-449.

8. Kimura H, Nagai Y, Umamura K, Kimura Y (2005) Physiological roles of hydrogen sulfide: synaptic modulation, neuroprotection and smooth muscle relaxation. *Antioxidants & Redox Signaling* 7:795-803.
9. Kimura H, Shibuya N, Kimura Y (2012) Hydrogen sulfide is a signaling molecule and a cytoprotectant. *Antioxidants & Redox Signaling* 17(1):45-57.
10. Pacher P, Beckman JS, Liaudet L (2007) Nitric oxide and peroxynitrite in health and disease. *Physiological Reviews* 87(1):315-424.
11. Ray PD, Huang BW, Tsuji Y (2012) Reactive oxygen species (ROS) homeostasis and redox regulation in cellular signaling. *Cell Signal* 24(5):981-990.
12. Salmeen A, Andersen JN, Myers MP, Meng TC et al (2003) Redox regulation of protein tyrosine phos phatase 1B involves a sulphenyl-amide intermediate. *Nature* 423 (6941):769-773.
13. Stamler JS, Hausladen A (1998) Oxidative modifications in nitrosative stress. *Nature Structural Biology* 5(4):247-249.
14. Sundaresan M, Yu ZX, Ferraris VJ, Irani K, Finkel T (1995) Requirement for generation of H2O2 for platelet-derived growth factor signal transduction. *Science* 270:296-299.
15. Szabo C, Ischiropoulos H, Radi R (2007) Peroxynitrite: biochemistry, pathophysiology and development of therapeutics. Nature Reviews: *Drug Discovery* 6(8):662-680.
16. van Montfort RL, Congreve M, Tisi D, Carr R, et al (2003) Oxidation state of the active-site cysteine in protein tyrosine phosphatase 1B. *Nature* 423(6941):773-777.
17. Wink DA and Ford PC (1995) Nitric oxide reactions important to biological systems: A survey of some kinetics investigations. *Methods* 7(1):14 –20.
18. Wood ZA, Poole LB, Karplus PA (2003) Peroxiredoxin evolution and the regulation of hydrogen peroxide signaling. *Science* 300(5619):650-653.

3.5.2. REDOX Signaling: Gene Regulation

1. Brigelius-Flohé R, Flohé L (2011) Basic principles and emerging concepts in the redox control of tran scription factors. *Antioxidants & Redox Signaling* 15(8):2335-2381.
2. Chen PR, Bae T, Williams WA, Duguid EM, et al (2006) An oxidation-sensing mechanism is used by the global regulator MgrA in Staphylococcus aureus. *Nature Chemistry & Biology* 2:591-595.
3. Finkel T (2003) Oxidant signals and oxidative stress. *Current Cell Biology* 15:247-254.
4. Green J, Pager MS (2004) Bacterial redox sensors. *Nature* 2:954-966.
5. Guyton KZ, Lin Y, Gorospe M, Xu O, Holbrook NJ (1996) Activation of mitogen-activated protein kinase by H_2O_2. *Journal of Biological Chemistry* 271:4138-4142.
6. Kaspar JW, Niture SK, Jaiswal AK (2009) Nrf2:INrf2 (Keap1) signaling in oxidative stress. *Free Radical Biology & Medicine* 47(9):1304-1309.
7. Niture SK, Kaspar JW, Shen J, Jaiswal AK (2009) Nrf2 signaling and cell survival. *Toxicology & Applied Pharmacology* 244(1):37-42.
8. Park B, Lee S, Kim E, Cho K, et al (2006) Redox regulation facilitates optimal peptide selection by MHC Class 1 during antigen processing. *Cell* 127:369-382.

3.5.3. REDOX Signaling: Protein & Enzyme Regulation

1. Bae YS, Kang SW, Seo MS, Baines IC, et al (1997) Epidermal growth factor (EGF)-induced generation of hydrogen peroxide. Role in EGF receptor-mediated tyrosine phosphorylation. *Journal of Biological Chemistry* 272:217-221.
2. Finkel T (2003) Oxidant signals and oxidative stress. *Current Opinion in Cell Biology* 15:247-254.
3. Fu X, Kassim SY, Parks WC, Heinecke JW (2001) Hypochlorous acid oxygenates the cysteine switch domain of pro-matrilysin (MMP-7). A mechanism for matrix metalloproteinase activation and athero-sclerotic plaque rupture by myeloperoxidase. *Journal of Biological Chemistry* 276:41279-41287.
4. Fu X, Kassim SY, Parks WC, Heinecke JW (2003) Hypochlorous acid generated by myeloperoxidase modifies adjacent tryptophan and glycine residues in the catalytic domain of matrix metalloproteinase-7 (matrilysin): an oxidative mechanism for restraining proteolytic activity during inflammation. *Journal of Biological Chemistry* 278:28403-28409.
5. Georgiou G (2002) How to flip the (redox) switch. *Cell* 111:607-610.
6. Hess DT, Matsumoto A, Kim SO, Marshall HE, Stamler JS (2005) Protein S-nitrosylation: purview and parameters. Nature Reviews: *Molecular & Cellular Biology* 6:150-166.
7. Irani K, Xia Y, Zweier JL, Sollott SJ, et al (1997) Mitogenic signaling mediated by oxidants in Ras-transformed fibroblasts. *Science* 275:1649-1652.

8. Iwakiri Y, Satoh A, Chatterjee S, Toomre DK, et al (2006) Nitric oxide synthase generates nitric oxide locally to regulate compartmentalized protein S-nitrosylation and protein trafficking. *Proceedings of the National Academy of Sciences USA* 103:19777-19782.
9. Janssen-Heininger YM, Mossman BT, Heintz NH, Forman HJ, et al (2008) Redox-based regulation of signal transduction: principles, pitfalls, and promises. *Free Radical Biology & Medicine* 45(1):1-17.
10. Klomsiri C, Karplus PA, Poole LB (2011) Cysteine-based redox switches in enzymes. *Antioxidants & Redox Signaling* 14(6):1065-1077.
11. Mannick JB, Hausladen A, Liu L, Hess DT, et al (1999) Fas-induced caspase denitrosylation. *Science* 284:651-654.
12. Paget MS, Buttner MJ (2003) Thiol-based regulatory switches. *Annual Review of Genetics* 37:91-121.
13. Shao MX, Nadel JA (2005) Dual oxidase 1-dependent MUC5AC mucin expression in cultured human airway epithelial cells. *Proceedings of the National Academy of Sciences USA* 102:767-772.
14. Xu D, Rovira II, Finkel T (2002) Oxidants painting the cysteine chapel: redox regulation of PTPs. *Developmental Cell* 2:251-252.

3.5.5. REDOX Signaling: Cardiac/Vascular Regulation

1. Brutsaert DL (2003) Cardiac endothelial–myocardial signaling: its role in cardiac growth, contractile performance, and rhythmicity. *Physiological Review* 83:59-115.
2. Massion PB, Balligand JL (2007) Relevance of nitric oxide for myocardial remodeling. *Current Heart Failure Reports* 4:18-25.
3. Seddon M, Shah AM, Casadei B (2007) Cardiomyocytes as effectors of nitric oxide signaling. *Cardiovascular Research* 75:315-326.
4. Cave AC, Brewer AC, Narayana Panicker A, Ray R, et al (2006) NADPH oxidases in cardiovascular health and disease. *Antioxidants & Redox Signaling* 8:691-728.
5. Opie LH, Sack MN (2002) Metabolic plasticity and the promotion of cardiac protection in ischemia and ischemic preconditioning. *Journal of Molecular & Cellular Cardiology* 34:1077-1089.
6. Jain M, Brenner DA, Cui L, Lim CC, et al (2003) Glucose-6-phosphate dehydrogenase modulates cytosolic redox status and contractile phenotype in adult cardiomyocytes. *Circulation Research* 93:9-16.
7. Gupte SA, Levine RJ, Gupte RS, Young ME, et al (2006) Glucose-6- phosphate dehydrogenase-derived NADPH fuels superoxide production in the failing heart. *Journal of Molecular & Cellular Cardiology* 41:340–349.
8. van Empel VP, Bertrand AT, Hofstra L, Crijns HJ, et al (2005) Myocyte apoptosis in heart failure. *Cardiovascular Research* 67:21-29.
9. Foo RS, Mani K, Kitsis RN (2005) Death begets failure in the heart. *Journal of Clinical Investigation* 115:565-571.
10. Sauer H, Rahimi G, Hescheler J, Wartenberg M (2000) Role of reactive oxygen species and phosphatidylinositol 3-kinase in cardiomyocyte differentiation of embryonic stem cells. *FEBS Letters* 476:218-223.
11. Buggisch M, Ateghang B, Ruhe C, Strobel C, et al (2007) Stimulation of ES-cell-derived cardiomyo genesis and neonatal cardiac cell proliferation by reactive oxygen species and NADPH oxidase. *Journal of Cellular Science* 120:885-894.
12. Gurusamy N, Mukherjee S, Lekli I, Bearzi C, et al (2009) Inhibition of ref-1 stimulates the production of reactive oxygen species and induces differentiation in adult cardiac stem cells. *Antioxidants & Redox Signaling* 11:589-600.
13. Santos CXC, Anilkumar N, Zhang M, Brewer AC, Shah AM (2011) Redox signaling in cardiac myocytes. *Free Radical Biology & Medicine* 50:777-793.
14. Ushio-Fukai M (2006) Redox signaling in angiogenesis: role of NADPH oxidase. *Cardiovascular Research* 71(2):226-235.

3.5.4. REDOX Signaling: Mitochondrial Regulation

1. Cheng Z, Tseng Y, White MF (2010) Insulin signaling meets mitochondria in metabolism. *Trends in Endocrinology & Metabolism* 21: 589-598.
2. Doughan AK, Harrison DG, Dikalov SI (2008) Molecular mechanisms of angiotensin II-mediated mito chondrial dysfunction: linking mitochondrial oxidative damage and vascular endothelial dysfunction. *Circulation Research* 102:488-496.
3. Handy DE, Loscalzo J (2012) Redox regulation of mitochondrial function. *Antioxidants & Redox Signaling* 16(11):1323-1367.

4. Handy DE, Lubos E, Yang Y, Galbraith JD, et al (2009) Glutathione peroxidase-1 regulates mitochon drial function to modulate redox-dependent cellular responses. *Journal of Biological Chemistry* 284:11913-11921.
5. Koopman WJ, Nijtmans LG, Dieteren CE, Roestenberg P, et al (2010) Mammalian mitochondrial complex I: biogenesis, regulation, and reactive oxygen species generation. *Antioxidants & Redox Signaling* 12:1431-1470.
6. Migliaccio E, Giorgio M, Mele S, Pelicci G, et al (1999) The p66shc adaptor protein controls oxidative stress response and life span in mammals. *Nature* 402:309-313.
7. Mookerjee SA, Divakaruni AS, Jastroch M, Brand MD (2010) Mitochondrial uncoupling and lifespan. *Mechanisms of Ageing & Development* 131:463-472.
8. Rigoulet M, Yoboue ED, Devin A (2011) Mitochondrial ROS generation and its regulation: mechanisms involved in H(2)O(2) signaling. *Antioxidants & Redox Signaling* 14:459-468.
9. Ristow M, Zarse K (2010) How increased oxidative stress promotes longevity and metabolic health: the concept of mitochondrial hormesis (mitohormesis). *Experimental Gerontology* 45:410-418.
10. Semenza GL (2007) Oxygen-dependent regulation of mitochondrial respiration by hypoxia-inducible factor 1. *Biochemistry Journal* 405:1-9.
11. Taylor ER, Hurrell F, Shannon RJ, Lin TK, et al (2003) Reversible glutathionylation of complex I in creases mitochondrial superoxide formation. *Journal of Biological Chemistry* 278:19603-19610.
12. Yang Y, Song Y, Loscalzo J (2007) Regulation of the protein disulfide proteome by mitochondria in mammalian cells. *Proceedings of the National Academy of Sciences USA* 104:10813-10817.
13. Youle RJ, Narendra DP (2011) Mechanisms of mitophagy. *Nature Reviews – Molecular & Cellular Biology* 12:9-14.

3.5.6. REDOX Signaling: Immune Regulation

1. Angelini G, Gardella S, Ardy M, Ciriolo MR, et al (2002) Antigen-presenting dendritic cells provide the reducing extracellular microenvironment required for T lymphocyte activation. *Proceedings of the National Academy of Sciences USA* 99:1491-1496.
2. Banerjee R (2012) Redox outside the box: Linking extracellular redox remodeling with intra-cellular redox metabolism. *Journal of Biological Chemistry* 287:4397-4402.
3. Beutler B and Rietschel ET (2003) Innate immune sensing and its roots: the story of endotoxin. Nature Reviews - *Immunology* 3:169-176.
4. Gloire G, Piette J (2009) Redox regulation of nuclear posttranslational modifications during NF-kappaB activation. *Antioxidants & Redox Signaling* 11:2209-2222.
5. Kumar H, Kawai T, Akira S (2009) Toll-like receptors and innate immunity. *Biochemical Biophysics Research Communications* 388:621-625.
6. Rodbell M (1995) Nobel Lecture. Signal transduction: evolution of an idea. *Bioscience Reports* 15(3:117-133.
7. Schreck R, Rieber P, Baeuerle PA (1991) Reactive oxygen intermediates as apparently widely used messengers in the activation of the NF-kappa B transcription factor and HIV-1. *EMBO Journal* 10:2247-2258.
8. Shi Z, Cai Z, Wen S, Chen C, et al (2009) Transcriptional regulation of the novel Toll-like receptor Tlr13. *Journal of Biological Chemistry* 284:20540-20547.
9. Staal FJ, Roederer M, Herzenberg LA (1990) Intra-cellular thiols regulate activation of nuclear factor kappa B and transcription of human immunodeficiency virus. *Proceedings of the National Academy of Sciences USA* 87:9943-9947.
10. Yan Z, Garg SK, Banerjee R (2010) Regulatory T cells interfere with glutathione metabolism in dendritic cells and T cells. *Journal of Biological Chemistry* 285:41525-41532.
11. Yan Z, Garg SK, Kipnis J, Banerjee R (2009) Extracellular redox modulation by regulatory T cells. *Nature Chemical Biology* 5:721-723.

3.5.8. REDOX Signaling: Apoptosis

1. Anathy V, Aesif SW, Guala AS, Havermans M, et al (2009) Redox amplification of apoptosis by caspase-dependent cleavage of glutaredoxin 1 and S-glutathionylation of Fas. *Journal of Cell Biology* 184:241-252.
2. Anathy V, Roberson EC, Guala AS, Godburn KE, et al (2012) Redox-based regulation of apoptosis: S-glutathionylation as a regulatory mechanism to control cell death. *Antioxidant & Redox Signaling* 16(6):496-505.

3. Circu ML, Aw TY (2008) Glutathione and apoptosis. *Free Radical Research* 42(8):689-706.
4. Dalton TP, Chen Y, Schneider SN, Nebert DW, Shertzer HG (2004) Genetically altered mice to evaluate glutathione homeostasis in health and disease. *Free Radical Biology & Medicine* 37:1511-1526.
5. Edinger AL, Thompson CB (2004) Death by design: apoptosis, necrosis and autophagy. *Current Opinion in Cell Biology* 16:663-669.
6. Franco R, Cidlowski JA (2006) SLCO/OATP-like transport of glutathione in FasL-induced apoptosis: Glutathione efflux is coupled to an organic anion exchange and is necessary for the progression of the execution phase of apoptosis. *Journal of Biological Chemistry* 281:29542-29557.
7. Franco R, Cidlowski JA (2009) Apoptosis and glutathione: Beyond an antioxidant. *Cell Death and Differentiation* 16:1303-1314.
8. Hengartner MO (2000) The biochemistry of apoptosis. *Nature* 407:770-776.
9. Holler N, Tardivel A, Kovacsovics-Bankowski M, Hertig S, et al (2003) Two adjacent trimeric Fas ligands are required for Fas signaling and formation of a death-inducing signaling complex. *Molecular & Cellular Biology* 23:1428-1440.
10. Janssen-Heininger YM, Mossman BT, Heintz NH, Forman HJ, et al (2008) Redox-based regulation of signal transduction: Principles, pitfalls, and promises. *Free Radical Biology & Medicine* 45:1-17.
11. Okouchi M, Ekshyyan O, Maracine M, Aw TY (2007) Neuronal apoptosis in neurodegeneration. *Antioxidants & Redox Signaling* 9:1059-1096.
12. Qanungo S, Starke DW, Pai HV, Mieyal JJ, Nieminen AL (2007) Glutathione supplementation potentiates hypoxic apoptosis by S-glutathionylation of p65-NFkappaB. *Journal of Biological Chemistry* 282:18427-18436.
13. Samraj AK, Keil E, Ueffing N, Schulze-Osthoff K, Schmitz I (2006) Loss of caspase-9 provides genetic evidence for the type I/II concept of CD95-mediated apoptosis. *Journal of Biological Chemistry* 281:29652–29659.
14. Susin SA, Lorenzo HK, Zamzami N, Marzo I, et al (1999) Molecular characterization of mitochondrial apoptosis-inducing factor. *Nature* 397:441-446.
15. Xiong Y, Uys JD, Tew KD, Townsend DM (2011) S-glutathionylation: From molecular mechanisms to health outcomes. *Antioxidants & Redox Signaling* 15:233-270.

REDOX Life in the Universe

Two of the most prominent bio-energetic pathways in our biosphere are oxygenic photosynthesis and aerobic respiration, which together constitute the proton (H^+) cycle. It is estimated that ~75% of the oxygen produced by photosynthesis is reduced to water, thus coupling two life-sustaining high energy pathways in the biosphere. The most important property of a life-sustaining solvent is that it has to occur in large quantities on the planetary surface or wherever the life-sustaining reactions are to occur. With oxygen, the most common element on Earth, and hydrogen, the most abundant element in the Universe, water would be expected to fulfill this vital criterion on Earth. Water is also known, from the spectral analysis, to be a common molecule in the Universe and, thus, may be an obvious choice as a solvent elsewhere, as well. There is, however, no compelling reason to limit the solvent criterion for life to water alone, although terrain life is constrained to use carbon as the structural element for most of its bio-molecules. Several candidates could potentially replace water as a polar solvent on other worlds; thus, sulfuric acid could be a possible solvent for Venus; and ammonia for Jupiter. Most of the potential solvent candidates are liquids at lower temperatures than water. However, the key to developing a suitable bio-chemistry at a given temperature depends on selecting REDOX reactions suited to that temperature. For example, those reactions involving unstable and highly reactive free radicals have very low activation energies and would be suited for low temperature organisms.

CHAPTER

4

REDOX Medicine

Periodic oscillations in the REDOX state could function as cellular signals that regulate vital biological processes, such as cell proliferation, differentiation and apoptosis. A shift in the REDOX balance could lead to aberrant cellular functions, a hallmark of several proliferative disorders such as cancer, wound healing, fibrosis, CVD, diabetes, and neurodegenerative diseases. Restoration of REDOX homeostasis may reverse such human pathologies and improve health.

4

REDOX: Medicine

Human body and its metabolic pathways are critically dependent on oxygen (O_2). The physiological impact of O_2 in the human body is reflected by the involvement of three major organ systems that have exclusively evolved to transport O_2 from the surrounding environment to the cells systems – namely, the cardiovascular, respiratory, and hematopoietic systems. Cells do not have the ability to store O_2 and are therefore rely upon its continuous supply to meet the body's metabolic demands. If the supply of O_2 is not aligned with the normal metabolic demand, tissue hypoxia will ensue, eventually resulting in cellular injury and/or death. Ironically, O_2 is inherently toxic to the same cells that are critically dependent upon a steady supply of O_2 for normal cellular function – the so-called oxygen paradox.

Hypoxia is defined as a lower than normal partial pressure of oxygen (PO_2) in the tissues. Whereas, hypoxemia is defined as a lower than normal partial pressure of oxygen in the blood (PaO_2). There are different types of hypoxia: (i) Hypoxemic hypoxia (due to lower than normal PaO_2); (ii) Anemic hypoxia (due to decreased O_2 carrying capacity of the blood, due to either a low RBC count or presence of abnormal hemoglobin); (iii) Circulatory hypoxia (also known as stagnant hypoxia, due to lower than normal cardiac output); and (iv) Histo-toxic hypoxia (due to cyanide poisoning or toxicity from sepsis).

The toxicity of O_2 rests in its unique atomic structure. Each oxygen atom contains one unpaired electron in its outer orbit. Elemental oxygen is therefore considered a free radical. However, oxygen exists in the atmosphere as O_2, and contains two unpaired electrons with parallel spins. During normal cellular respiration, O_2 is reduced to H_2O through the sequential addition of four electrons at the inner membrane of the mitochondria, thereby generating the necessary chemical energy

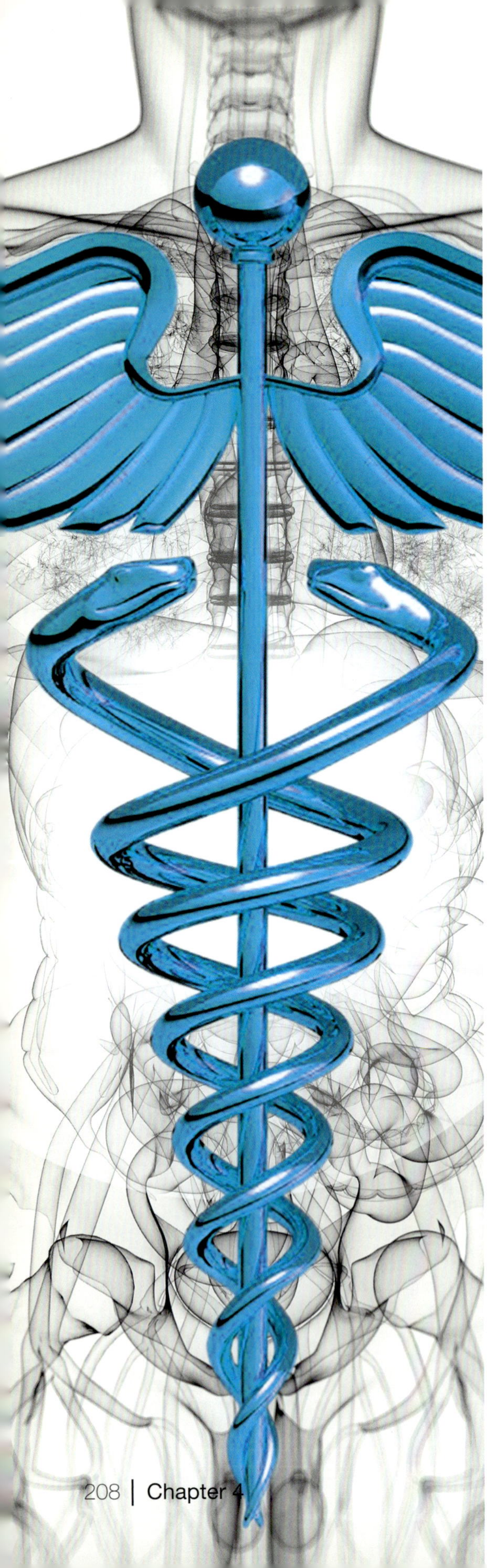

to generate 36 molecules of adenosine triphosphate (ATP). Nearly 98% of oxygen is converted to water in this reaction. The remaining 2% of O_2 undergoes partial reduction due to the leakage of electrons across the respiratory chain, resulting in the formation of intermediate ROS such as $O_2^{\bullet-}$, H_2O_2 and ·OH radicals.

REDOX Medicine deals with the metabolic states of health and disease, based on the movement of O_2 and its reactive oxygen species (ROS) through cellular pathways. The movement of ROS species can be controlled and equilibrated by innate systems to establish a physiological homeostasis – a sign of healthy status. REDOX homeostasis can be defined as a state of equilibrium between different but inter-dependent elements in an organism. REDOX homeostasis is important to effectively harness reducing power produced through the catabolism of various substrates and to utilize this power in the anabolism of cellular components such as DNA, lipids and proteins. REDOX optimization is also vital for various signaling pathways that control cellular proliferation, maturation and apoptosis.

On the other hand, an uncontrolled or random movement of pro-oxidant species can lead to a toxic condition, known as oxidative stress. Oxidative stress results from an imbalance between ROS/RNS generation and the body's antioxidant defenses, due to (i) increased generation of ROS/RNS; (ii) decreased synthesis or consumption of the body's natural antioxidants; or (iii) a combination of both processes. Both ROS and RNS contribute to the host inflammatory response either via direct cellular toxicity or through the activation of pro-inflammatory gene expression pathways. ROS and RNS have direct effects (via oxidation or nitration reactions, respectively) on the structure and function of many enzymes, proteins, lipids, and DNA. These effects disrupt the function of the cellular machinery. Several critical enzymes, transcription factors, and scaffolding proteins become non-functional through such oxidative modifications; while lipid peroxidation results in disruption of the cell membrane and release highly cytotoxic products (such as malondialdehyde). The release of these contents, which may include endogenous danger signals such as heat shock proteins may further activate the inflammatory response in adjacent cells. Thus, physiological shifts in REDOX homeostasis contribute to the aging of cells, tissues, and organs. Any imbalance in REDOX homeostasis impairs cellular pathways, which may lead to dysfunction, disease or death.

4.1 REDOX: Homeostasis

REDOX homeostasis is defined by John Allen (2010) as 'the maintenance of constant electro-chemical potential and ionic-concentration gradient across a cellular boundary, despite fluctuations in the electro-chemical potential of the external environment and changing identities and activities of electron donors and acceptors'. The REDOX homeostasis ensures that the cells respond properly to endogenous and exogenous stimuli. However, when the REDOX homeostasis is disturbed, the electro-chemical imbalance may led to aberrant cellular dysfunction and contribute to disease development or death. Therefore, REDOX chemistry and homeostasis are considered as the two most fundamental properties of life.

REDOX homeostasis is controlled by two opposite chemical pathways in a living organism. From one end, a regulated production of pro-oxidative reactants (i.e. ROS, RNS and RSS) that ensure a continuous flow of charged particles (electrons and protons) for breaking or forming chemical bonds. The influx and efflux of this free energy flow through chemical bonds imparts reactivity, in other words, an ionic instability due to loss or gain of charged particles to molecules. This highly reactive nature sets these molecules in a kinetic motion to find a suitable chemical partner to relieve their acquired charge either by covalent interaction or through an ionic discharge. This basic principle of energy transfer in small packets ('quanta') is the driving force for all chemical reactions and every metabolic pathway of life.

On the other end, free flow of charged particles obey the laws of energy (or thermodynamics) and transfer their 'quanta' to any random encounter in their motion path. An organized capture of such 'quanta' is vital for life force; therefore, organisms have evolved specialized REDOX couples to avoid randomness in the quantum energy transfer – even for a millionth fraction of a second! Thus, living organisms have evolved anti-oxidants and radical quenchers, alongside the REDOX couples, with an array of genetic controls, based on the cellular mass and complexity of an organism. No system can ever hold an organization against the universal nature of randomness (the entropy) for too long – accordingly, free radical damage, aging and death are absolute consequences of the 2nd Law of Thermodynamics. In an amazing twist, however, life has also evolved pathways to utilize certain low thresholds of oxidative stress to its own advantage; for example, cell signaling, programmed cell death (apoptosis) and gene mutations to enhance the chance of survival. In essence, REDOX homeostasis is a monumental effort of a living system to maintain the narrow margin of its survival through its genetic design between two variable effectors – the milieu (environment) and the time.

Any disturbance in REDOX homeostasis may affect cell death and contribute to the development of diseases such as cancer and degenerative disorders. On the other hand, REDOX system controls key transcription factors, signal-transduction pathways, and cell-death regulators that affect cell survival. Basic knowledge of REDOX regulation of cell survival can be applied in the development of strategies for the treatment or prevention of diseases.

Genetic transcription factors and REDOX homeostasis: 'Klotho', a suppressor gene for aging can increase resistance to oxidative stress at the cellular and organismal level in mammals. Klotho activates the FoxO forkhead transcription factors, negatively regulated by insulin/insulin-like growth factor (IGF)-1 signaling, which induce expression of manganese (Mn)-SOD. This in turn facilitates removal of ROS and confers oxidative stress resistance. Thus, Klotho-induced inhibition of insulin/IGF-1 signaling is associated with increased resistance to oxidative stress, which potentially contributes to the anti-aging properties of Klotho.

FOX (Forkhead box) proteins are a family of transcription factors that play critical roles in regulating the expression of genes involved in cell growth, proliferation, differentiation, and longevity. Many FOX proteins are involved in the embryonic development. The defining feature of FOX proteins is the forkhead box, which is a sequence of 80 to 100 amino acids forming a motif that binds to DNA. This fork-head motif is also known as the winged helix due to the butterfly-like appearance of the loops in the protein structure of the domain. Forkhead genes are a subgroup of the helix-turn-helix class of proteins.

FoxOs, a subclass of the 'Forkhead box', represent another major cell defense mechanism against oxidative insults. In humans, FoxO comprises of four members: FoxO1, FoxO3, FoxO4, and FoxO6. FoxOs shuttle between the cyto-plasm and the nucleus depending on the phosphorylation of specific sites by distinct sets of kinases. In the setting of stimulation by growth factors, such as insulin and IGFs, FoxO1, 3, and 4 are subject to phosphorylation, resulting in their nuclear export. Oxidative stress promotes translocation of FoxOs into the nucleus through a mechanism that activates jun-N-terminal kinase (JNK) by ROS. Oxidative stress also promotes other post-translational modifications of FoxOs, including ubiquitylation and acetylation. These changes in turn lead to the activation of FoxOs and thereby to FoxO-mediated oxidative stress responses by regulating the transcription of antioxidant enzymes such as Mn-SOD and catalase, as well as, genes involved in cell cycle, DNA repair, and lifespan. Through these effects, FoxOs promote mammalian cell survival by inducing cell cycle arrest and quiescence in response to oxidative stress, and they also regulate longevity in model organisms.

Importantly, stress stimuli over ride the sequestration of FoxO in the cytoplasm by growth factors. FoxOs can also induce apoptosis through FoxO-mediated regulation of several pro-apoptotic genes. The role of FoxOs in cell death seems counter-intuitive considering their role in the protection against oxidative stress.

However, elimination of damaged or abnormal cells by FoxO-induced apoptosis evidently plays a major role in the ability of these transcription factors to promote tumor suppression and longevity. The importance of FoxOs in mammalian biology has been recently highlighted by the profound effects of the deletion of individual FoxOs or the combined deletion of FoxO1, -3, and -4 in animal models. FoxO1 is critical for development because FoxO1-null animals die at embryonic stage (at day: 10.5) due to defects in angiogenesis. The antioxidant defense provided by FoxOs is ultimately compromised by increased oxidative stress or ROS-activated pathways that interfere with the activity of FoxOs, or both.

β-Catenin plays an important role in the regulation of oxidative stress-induced transcription pathways. In the recent years, 'Wnts', a large family of secreted glycoproteins, and their signaling pathway was found to be important for REDOX homeostasis. Wnt/β-catenin signaling is required for cellular protection against oxidative stress-induced apoptosis. The inhibition of FoxO through its phosphorylation by β-catenin-induced gene expression may reduce the apoptotic function of FoxO3, which may result in increased cell survival. These findings have relevance for future therapies directed at cellular protection, regeneration, based on REDOX homeostasis.

Normal cells maintain REDOX homeostasis by controlling the proper balance between ROS generation and elimination. The REDOX dynamics may fluctuate within a tolerable range. The pro-oxidant activity depends on the regulated cellular production of ROS via enzymatic sources such as the xanthine oxido-reductase (XOR), NADPH oxidase (NOx) complex, nitric oxide synthase (NOS). Alterations in REDOX homeostasis by exogenous stimuli or endogenous stress or both can result in increased oxidative stress with elevated cellular ROS. The ability to adapt to ROS stress determines the overall fate of the cells. A successful adaptation to increased survival signals (in combination with ROS-mediated mutations and loss of critical regulatory mechanisms) lead to defective cell death and aberrant cell proliferation. These cellular conditions may contribute to development of cancer. A failure in adaptation to the ROS stress (while the cells accumulate oxidative DNA, protein, and lipid damage) may result in excessive cell death, leading to degenerative disorders and aging.

Low molecular weight antioxidants, such as ascorbate, GSH, and tocopherol, are information-rich REDOX buffers that interact with several cellular components. In addition to crucial roles in defense and as enzyme cofactors, cellular antioxidants influence growth and development by modulating processes from metabolism to senescence and death of a cell. Most importantly, antioxidants provide essential information on cellular REDOX state, and influence gene expression associated with biotic and abiotic stress responses to maximize defense. Growing evidence suggests a model for REDOX homeostasis, where the ROS–antioxidant interaction acts as a metabolic interface for signals derived from metabolism and from the milieu. This interface modulates appropriate induction of acclimation processes or, alternatively, execution of cell death (apoptosis) programs. An increase of ROS may promote cell proliferation and survival, as in the case of many cancer cells. However, when the increase of ROS reaches a critical level (threshold), it may overwhelm the cellular antioxidant capacity and trigger cellular death (apoptosis). Chronic ROS stress may cause accumulation of damage to a level that induces cell death. This mechanism may contribute to neural degenerative diseases and aging. For therapeutic purposes, it is possible to use agents that promote ROS generation or inhibit the cellular antioxidant system to trigger cancer cell death by pushing the ROS above the threshold level. In contrast, antioxidants may be used to prevent cells from oxidative damage and delay aging and neurodegenerative processes.

Table: *Generation of various Reactive Species in Human Cellular Systems*

Reactive Oxygen Species (ROS) Generation		
NADPH Oxidase (NOx)	$2\ O_2$ + NADPH	→ $2\ O_2^{\bullet-}$ (Superoxide) + $NADPH^+$ + H^+
	$2\ O_2^{\bullet-} + 2\ H^+$ (Spontaneous conversion)	→ [$2\ HO_2\cdot$ (Hydroperoxyl radical)] → $O_2 + H_2O_2$
Superoxide dismutase (SOD)	$2\ O_2^{\bullet-} + 2\ H^+$	→ $O_2 + H_2O_2$
Myeloperoxidase (MPO)	$Cl^- + H_2O_2$	→ OCl^- (Oxidized Halide) + H_2O
Secondary Reactions of ROS		
	$H_2O_2 + Fe^{2+}$	→ $OH^- + \cdot OH$ (Hydroxyl radical) + Fe^{3+}
	$Fe^{3+} + O_2^{\bullet-}$	→ $O_2 + Fe^{2+}$
	$O_2^{\bullet-}$ + HOCl	→ $O_2^{\bullet-} + \cdot OH + Cl^-$
	$H_2O_2 + OCl^-$	→ $^1O_2$ (Singlet Oxygen) + $H_2O + Cl^-$
	NH_3 + HOCl	→ NH_2Cl (Chloramine) +H_2O
	R-CHNH$_2$-COOH + HOCl (Amino acids)	→ R-CHNHCl–COOH +H_2O → R-CHO + CO_2 + NH_4+ + Cl^- (Chloramines) (Aldehydes)
Reactive Nitrogen Species (RNS) Generation and Secondary Reactions of RNS		
Nitric Oxide Synthase (NOS)	Arginine + O_2 + NADPH	→ ·NO + Citrulline + $NADP^+$
	$\cdot NO + O_2^{\bullet-}$	→ $ONOO^-$ (Peroxynitrite)
	$ONOO^- + H^+$	→ ONOOH (Peroxynitrous acid)
	ONOOH (Peroxynitrous acid)	→ NO_3^- (Nitrate ion)
	ONOOH	→ NO_2^+ (Nitronium ion)
	ONOOH	→ $NO_2\cdot$ (Nitrogen Dioxide radical)
	ONOOH	→ OH^-
	ONOOH	→ ·OH (Hydroxyl radical)
Lipid peroxidation		
	LH + Radical·	→ L· + RH
	L· + O_2	→ LOO· (Lipid Peroxyl radical)
	LH + LOO·	→ L· + LOOH (leading to lipid propagation)
	LOO· + TocOH (α-tocopherol)	→ LOOH + TocO· (chain termination)

Table: *Localization and Activities of the Major Non-enzymatic Antioxidants.*

Antioxidants	Localization	Activity
DIRECT ANTIOXIDANTS		
Vitamin E	Lipids / Cells Mitochondrial membranes	Lipid peroxidation inhibition Membrane stabilization
Vitamin A	Lipids / Cell membranes	Lipid peroxidation reduction
Vitamin C	Aqueous middle / Cytosol Extra-cellular liquids	Vitamin E regeneration LDL protection
Glutathione (GSH)	Aqueous middle	Substrate for GPx Vitamin E and C regeneration
Cystein (*Cys*)	Intra-cellular middle	Glutathione precursor
Lipoic acid	Aqueous middle	Lipid peroxidation inhibition Vitamin C and E and cystein regeneration
Thioredoxin (TRx)	Aqueous middle	Mn-SOD synthesis Vitamin C regeneration
Glutaredoxin (GRx)	Aqueous middle	Electron carriers for nucleotides
Flavonoids	Linked with glucids	Prooxidant enzymes inhibition Prooxidant ions (Fe^{2+}, Fe^{3+}, Cu^{2+}) trapping LDL protection
Uric acid	Aqueous middle	Prooxidant ions (Fe^{2+}, Fe^{3+}, Cu^{2+}) trapping Erythocytes, hemoglobin, DNA, lipids protection
INDIRECT ANTIOXIDANTS		
HSP	Aqueous middle	Protection of proteins (cells)
Iron (Fe)	Aqueous middle	Catalase cofactor
Ferritin	Aqueous middle	Free iron trapping
Zinc (Zn)	Aqueous middle	SOD cofactor (Cu-Zn-SOD) LDL and thiol protection FR production inhibition
Copper (Cu)	Aqueous middle	SOD cofactor (Cu-Zn-SOD)
Selenium (Se)	Aqueous middle	GPX cofactor
Manganese (Mn)	Aqueous middle	SOD cofactor (Mn-SOD)
Albumin	Aqueous middle	Donate electron to FR Cu^{2+} trapping
Ceruloplasmin	Aqueous middle	Donate electron to FR Cu^{2+}, Fe^{2+} and Fe^{3+} trapping
Bilirubin	Aqueous middle	Donate electron to FR Lipid peroxidation inhibition Erythrocyte protection

4.2 **REDOX:** Oxidative Stress

Oxidative stress is caused by an imbalance between the production of reactive species (ROS and RNS) and the ability of a biological system to readily detoxify the reactive intermediates or easily repair the resulting damage. All forms of life maintain a reducing environment within their cells. This reducing environment is sustained by enzymes that maintain the reduced state through a constant input of metabolic energy. Imbalance between REDOX status causes oxidative or reductive stress, which could trigger cellular damage or aberrant signaling, leading to cellular dysfunction. Oxidative stress causes detrimental effects through production of reactive species such as ROS/RNS in various molecular compartments of a cell. ROS/RNS also act as cellular messengers in REDOX signaling. Reductive stress, which is the down-regulation of ROS/RNS by antioxidants below certain levels, can disrupt normal cellular functions and cause adverse effects. Oxidative stress, defined as the excessive accumulation of reactive oxygen species (ROS) and the inability of an organism to neutralize the reactive intermediates, can also result in damage. In contrast, reductive stress is an ionic condition where the antioxidant capacity exceeds the ROS production and interferes in regular oxidative physiology, such as the mitochondrial respiration.

Enhanced oxidative stress does not always emanate from a single cellular source; it may represent a complex interplay among pro-oxidant enzymes, antioxidants, nuclear transcription, and disruption of normal electron transfer in key regulatory enzymes. The complexity of this REDOX system has several advantages to the cell: (i) amplification of ROS generation which enhances efficiency of the REDOX process, (ii) inter-enzymatic stimulation allows stereo-specific intra-cellular targeting of ROS generation where small amounts of ROS from one site can activate larger amounts from a separate intra-cellular site, (iii) one ROS can stimulate the production of another distinct ROS within the same cell, and (iv) better temporal control of ROS generation in response to diverse stimuli.

Cellular response to oxidative stress involves elimination of, protection against, and repair of damage caused by ROS. Antioxidant enzymes including SOD, peroxiredoxin (PRX), and catalase are responsible for the direct elimination of ROS; whereas systems that reconstitute antioxidants [e.g., glutathione (GSH) and GSH peroxidase (GPX)] can indirectly reduce ROS. Molecules sensitive to oxidative stress can be surrounded by decoys that are preferentially oxidized, thereby protecting key cellular molecules (e.g., GSH, α-tocopherol, bilirubin, ascorbate, urate, albumin plasmogens, and cytosolic amino acids). Direct physical quenching of $O_2^{\bullet-}$ or other free radicals (e.g. carotenoids) is another biologically deployed mechanism. Each of these processes contributes to local and systemic cellular REDOX balance with implications in cell signaling, growth, and apoptosis.

Oxidative Stress and Genotoxic Effects: ROS/RNS directly damage DNA and impair DNA repair mechanisms. This damage can occur in the form of DNA strand breakage or individual nucleotide base damage. ·NO, especially in high concentrations, causes the deamination of deoxynucleotides and DNA modification, all of which can lead to somatic mutations. Mitochondrial DNA (mtDNA) is particularly susceptible to oxidative stress without effective repair, it can cause loss of mitochondrial function, cell growth arrest, and apoptosis. The increased susceptibility relates to the proximity of mtDNA to ROS and lipid peroxidation products that generate from inner mitochondrial membrane. Mitochondrial inner membrane contains components of the respiratory electron transport chain (ETC) and lack protective histones or potentially efficient repair mechanisms.

Sources of Oxidative Stress

Effects at Molecular & Cellular Level

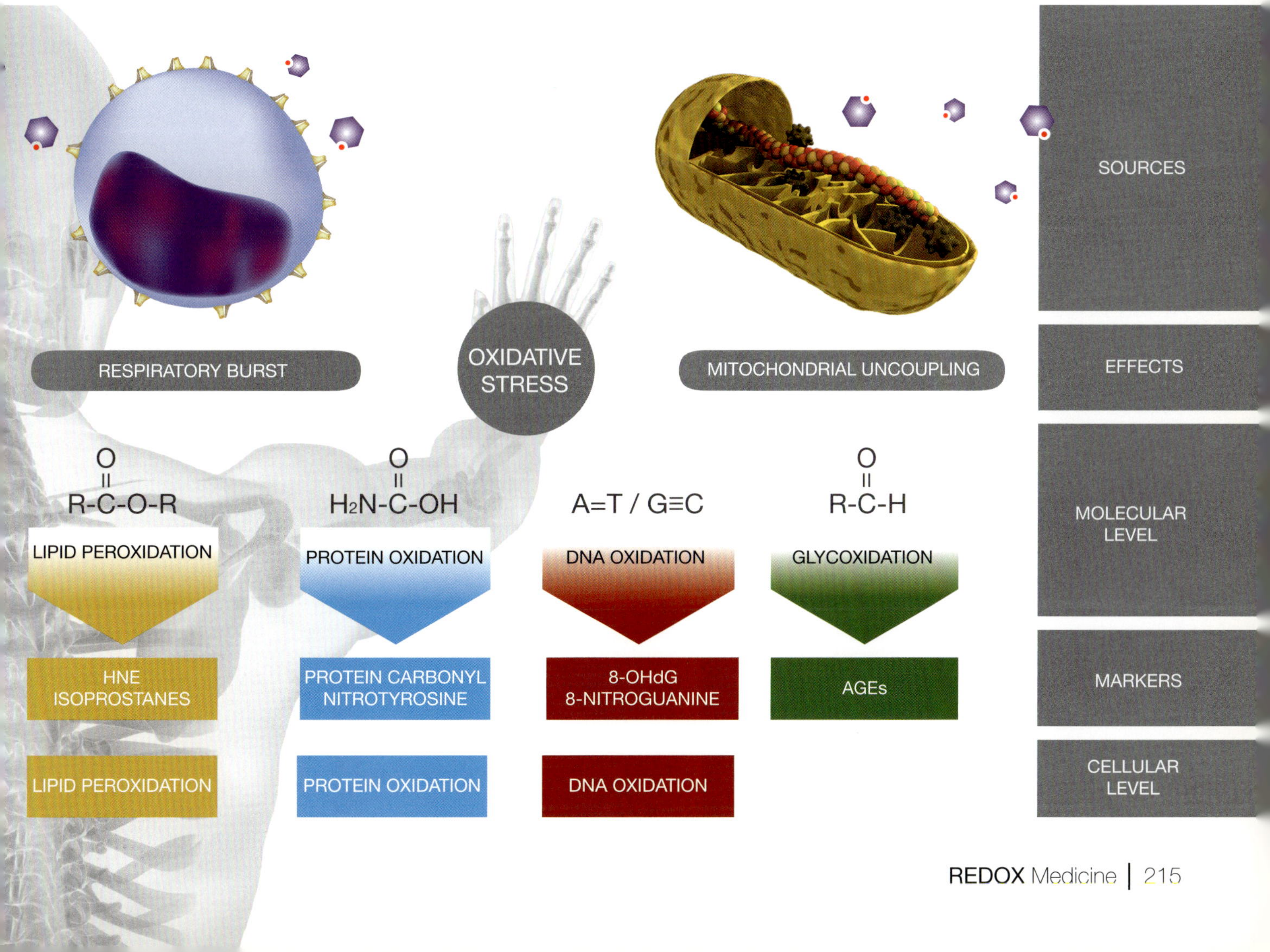

Oxidative stress is implicated in cellular processes of senescence, angiogenesis, apoptosis, and a myriad of diseases including atherosclerosis, hypertension, hypercholesterolemia, obesity, cancer, diabetes mellitus, and neurodegenerative disorders. Most pathology is the result of excessive ROS formation that promotes inflammation in surrounding tissues and accelerates cell death or senescence. It could also cause necrosis, ATP depletion, and may interfere with controlled apoptotic death. Oxidative stress is strongly linked to cardiovascular disease, since oxidation of LDL in the vascular endothelium is a precursor to plaque formation. Oxidative stress also plays a role in the ischemic cascade due to oxygen reperfusion injury following hypoxia. This cascade includes both stroke and myocardial infarction. Oxidative stress has also been associated with chronic fatigue syndrome, as well as, in diabetes. Oxidative stress is likely to be involved in age-related development of cancer. The reactive species produced in oxidative stress can cause direct damage to the DNA (therefore is mutagenic). It may also suppress apoptosis, promote proliferation, invasiveness and metastasis. Gastritis, the infection by *Helicobacter pylori,* increases the production of ROS and RNS in human stomach and is also thought to be important in the development of gastric cancer. During gastritis, levels of ROS can reach high concentrations in sites of inflammation or injury. Interestingly, excessively low levels of ROS can also invoke pathological changes by interrupting the physiological role of oxidants in proliferation, vasodilation, host defense and may promote carcinogenesis.

Oxidative stress could be beneficial, when used by the immune system as defense mechanism to combat pathogens. At optimum physiological doses, ROS can function as REDOX signaling molecules and support cell growth, protein phosphorylation, and cell migration. This REDOX dose dependency has a significant pathophysiological relevance.

TABLE: Oxidative Stress - *The Interplay of Prooxidant and Antioxidant Systems*

INTRA-CELLULAR Oxidant Sources	EXTRA-CELLULAR Oxidant Sources	ANTIOXIDANT DEFENSE	
		Enzymatic Antioxidants	**Small Molecules**
• NADPH Oxidase (NOx) • Mitochondrial ETC • Nitric oxide synthase (NOS) • Xanthine oxidase (XOR) • Arachidonic acid metabolism *[Lipoxygenase (LOX) & Cyclooxygenase (COX)]*	• Cigarette smoke • Radiation *(UV, gamma)* • Neutrophil activation • Ischemia-reperfusion • Hyperoxia • Electrical current • Elevated glucose levels • Lipid peroxidation • Metal-catalyzed oxidation • Catecholamine oxidation • Neuromelanin	• Superoxide dismutase (SOD) *(Cu-Zn-SOD) & (Mn-SOD)* • Catalase • Glutathione (GSH) system *[(GSH/GSSG) & GPx]* • Peroxiredoxins (PRx) • Thioredoxin reductase (TRx) • Heme oxygenase • Biliverdin reductase	• Salicylic acid • Carotenoids • Vitamins A, C, E • α-tocopherol • Ascorbic acid • Uric acid • Thioredoxins • Coenzyme Q • Melatonin • Flavonoids • Polyphenols

Involvement of Reactive Species in various Clinical Conditions

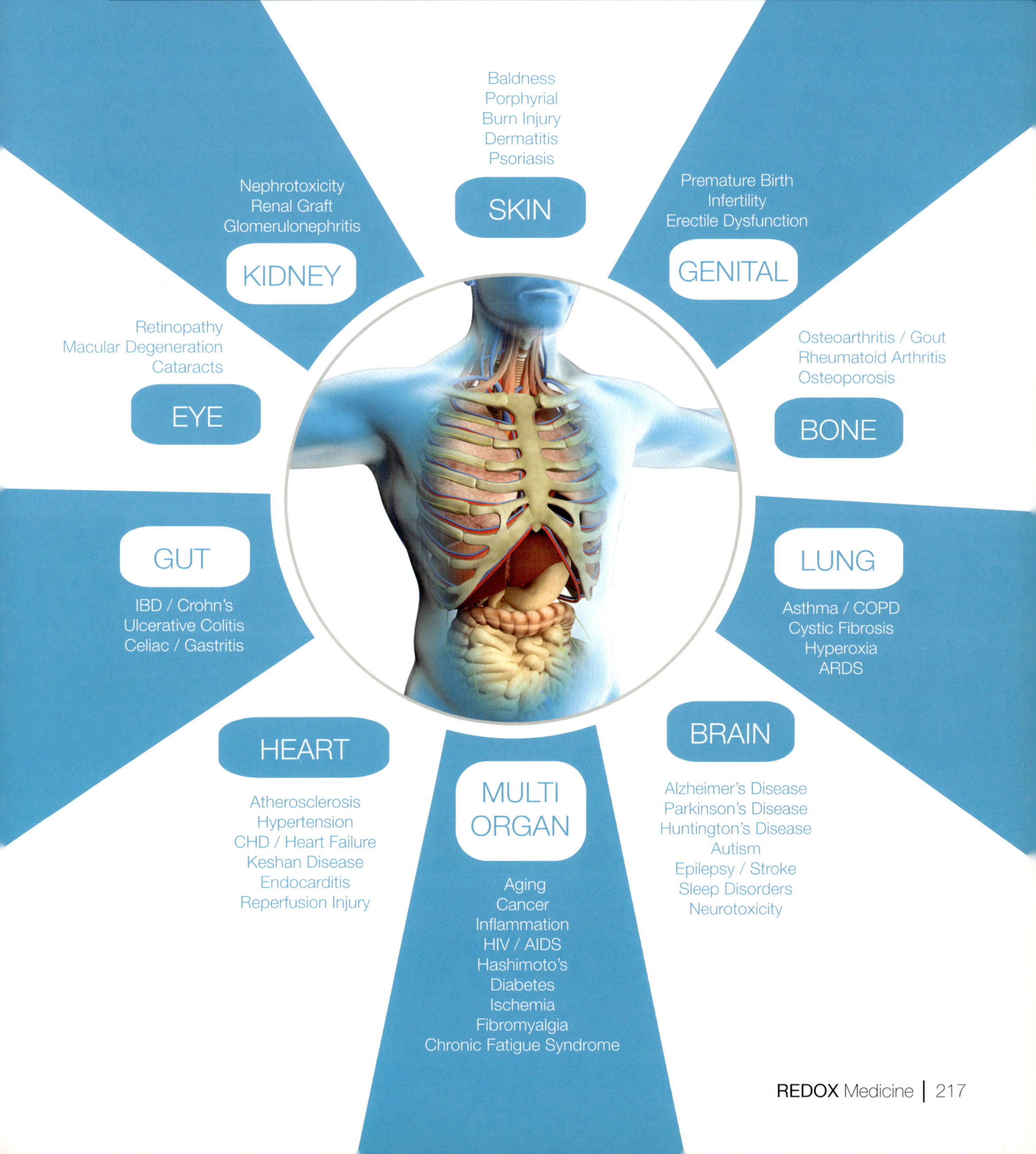

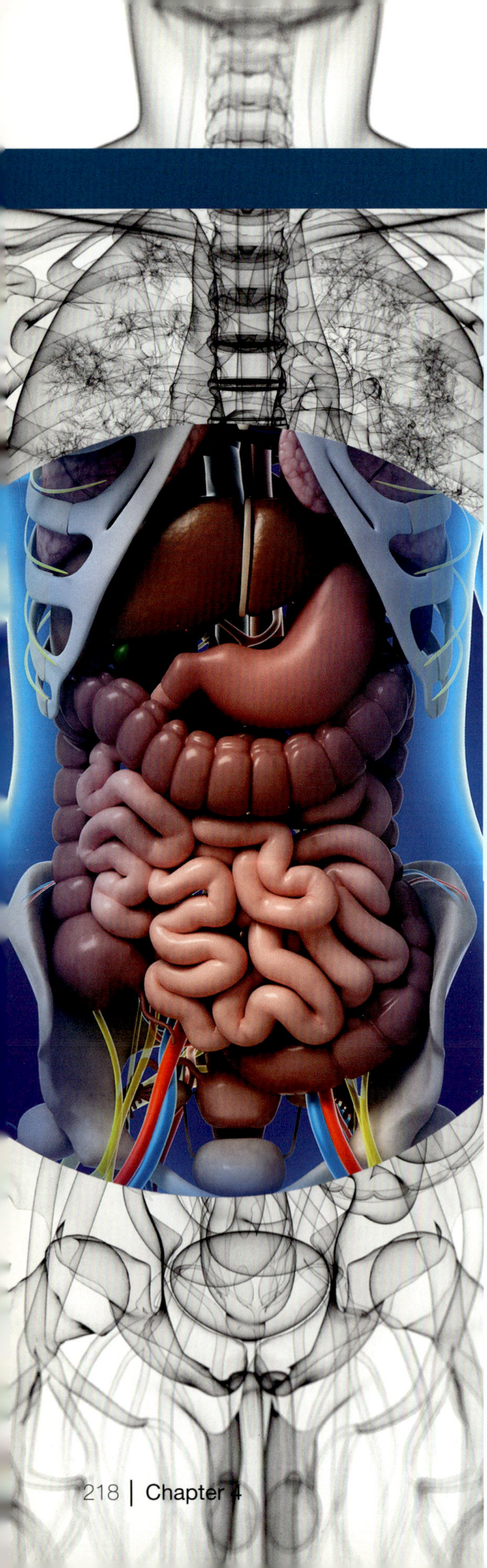

4.3 **REDOX:** The Digestive System

Life demands food (bio-energy) for survival and growth. Food and liquids are introduced on a regular basis into the body. Food undergoes three types of processes in the body, namely, digestion, absorption, and elimination. Digestion and absorption occur in the digestive tract. At all times, the digestive system is busy converting food into basic nutrients (sugars, peptides, fatty acids, etc.). These basic nutrients are absorbed, made available to all cells in the body and are utilized through metabolism. By design – the digestive system is a long, continuous tube called the alimentary canal, or gastrointestinal (GI) tract. The alimentary tract of the digestive system is composed of the mouth, pharynx, esophagus, stomach, small and large intestines, rectum and anus. Muscles in the walls of the alimentary canal move the food along. Most digestive organs are part of the alimentary canal. However, certain associated organs are located outside the GI tract, which includes: salivary glands, liver, gallbladder, and pancreas. These associated organs contribute to chemical digestion by releasing digestive enzymes and juices into the GI tract through tubes called ducts.

The Oral Cavity (Mouth)

In humans, digestion begins in the mouth, which converts food into a soft, moist mass. The mouth opens up the body to an external environment filled with bacteria, fungi, dust, etc. The oral mucosa (or lining of the mouth), along with saliva, act as a primary defense for protection against a variety of microbes. The flow of saliva has a mechanical effect, flushing microorganisms from mucosal and tooth surfaces. The muscular tongue pushes the food against the teeth, which cut, chop, and grind the food. Glands in the cheek lining secrete mucus, which lubricates the food and makes it easier to chew and swallow. The ideal fit of upper and lower teeth with two cogs of a wheel coming together at precisely the right point.

Three pairs of glands empty saliva into the mouth through ducts to moisten the food. The salivary glands produce about 1 to 1½ quarts of saliva. Saliva functions as a solvent in cleansing the teeth and dissolving food chemicals so they can be tasted. Saliva contains enzymes to digest starch, and mucus to lubricate the pharynx that facilitates swallowing. Also, saliva contains amylase enzyme, which digests starch to maltose. By the time thoroughly chewed food and reaches the stomach, the acidity inactivates amylase, and the average chain length of starch reduces from several thousand to fewer than 8 glucose units. Saliva also contains a wide spectrum of antimicrobial agents such as lactoferrin, lysozyme and salivary peroxidase. Saliva has a cleansing action as its constant flow helps to dissolve and remove food particles from the teeth.

A human adult's digestive tract if stretched, would measure ~20 to 30 ft.

More than 400 bacterial species reside in the oral cavity.

The design pattern of teeth - its intricate shape, contour, and angulation - makes it possible to effectively grind and tear food; whereas without proper angulation and contour, teeth would simply shred and poke holes in food. The front teeth, which tear and shred food, can exert a force of up to 80 pounds; while the grinding molars can apply 100-250 pounds of force against food particles!

The presence of food in the pharynx stimulates swallowing, which squeezes the food into the esophagus. The esophagus is a muscular tube (about 10 inch long) that connects the mouth to the stomach. A circular muscle called the esophageal sphincter separates the esophagus and the stomach. As food is swallowed, this muscle relaxes, forming an opening through which the food can pass down into the stomach. Then the muscle contracts, closing the opening to prevent food from moving back up into the esophagus. The esophageal sphincter is the first of several such muscles along the alimentary canal. These muscles act as valves to regulate the passage of food and keep it from moving backward. Food travels the length of the esophagus in 2 to 3 seconds.

The Stomach

Stomach is a J-shaped, expandable sack, located on the left side of the upper abdomen. It has active muscles that expand and contract based on the quantity of food in the stomach. This contraction causes mechanical breakdown of the food. The purpose of this breakdown is to increase the available surface area for

- Normal stomach expands to hold about 2 liters (~ ½ gallon) of food.
- Gastric juices in the stomach are secreted at a rate of 2-3 quarts/day.
- Stomach secretes mucus to protect itself from being digested by its own acid and enzymes.
- Stomach empties at a slow rate of about 3/100 ounce for each peri staltic wave. At 3 waves/min, it can take up to 5-h for 2-lbs of food to leave the stomach.

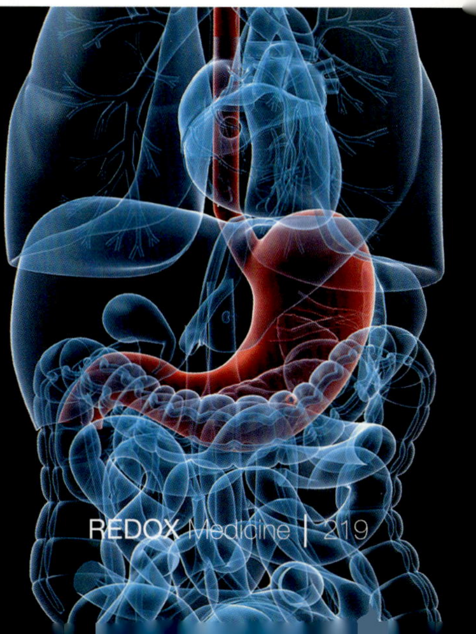

subsequent chemical activity. The gastric glands of the stomach secrete juices that perform chemical breakdown, partly digesting the proteins. These juices contain primarily hydrochloric acid (HCl) and digestive enzymes. HCl makes the stomach a very acid environment with a pH factor between 1.5 and 3.0. This acidic environment serves two functions: (i) it denatures the proteins and ionizes the minerals; and (ii) it kills food-borne microorganisms.

Two primary enzymes are also present in the gastric juices. The first is pepsin, which aids in the hydrolysis of proteins. Pepsin begins breaking down complex food proteins into their simpler forms. The acidic milieu created by HCl is critical for pepsin activity. The second enzyme in gastric juice is lipase, which aids in the hydrolysis of fats. Lipase starts the digestion of fats and breaks them into glycerol and fatty acids. The lipase activity is optimum at neutral pH (in contrast to acidic environment required for pepsin enzyme). These enzymes, along with the gastric juices, are mixed into the food by the mechanical actions of the stomach.

Control of gastric secretion and motility: The movement and flow of chemicals into the stomach are controlled by both the autonomic nervous system and by various hormones. The hormone gastrin causes an increase in HCl secretion from the parietal cells, and pepsinogen from chief cells in the stomach. It also causes increased motility in the stomach. It is inhibited by a pH normally <4 (high acid), as well as, by the hormone somatostatin. Cholecystokinin (CCK) has most effect on the gall bladder contractions, but it also decreases gastric emptying and increases release of pancreatic juice (which is alkaline and neutralizes the chyme). In a different and rare manner, secretin, produced in the small intestine, has most effect on the pancreas, but will also diminish acid secretion in the stomach. Gastric inhibitory peptide (GIP) and enteroglucagon decrease both gastric secretion and motility. Other than gastrin, rest of the above hormones all act to turn off the stomach action. Stomach needs to push food into the small intestine only when the intestine is not busy. When the intestine is full and busy in digesting the food, stomach acts as storage for food.

The emptying time of the stomach also varies with the type of food present. Water and liquids leave the stomach most rapidly. Carbohydrates empty more quickly than proteins; proteins, in turn, leave the stomach more quickly than fats. Not all foods undergo the same digestive processes in the stomach, and not all foods leave the stomach at the same rate. Proteins digest in an acid environment, while fats need a neutral environment. Although absorption is mainly a function of the small intestine, certain small molecules are also absorbed in the stomach through its lining; these includes: water (if the body is dehydrated), simple sugars (e.g. glucose), medication (e.g. aspirin), and amino acids.

Within 5-min after fat enters the stomach, the hormone – enterogastrone, enters the bloodstream and travels to the stomach and inhibits the motion of the stomach and causes it to empty at a much slower rate.

Stomach: The Nutrient Sensor

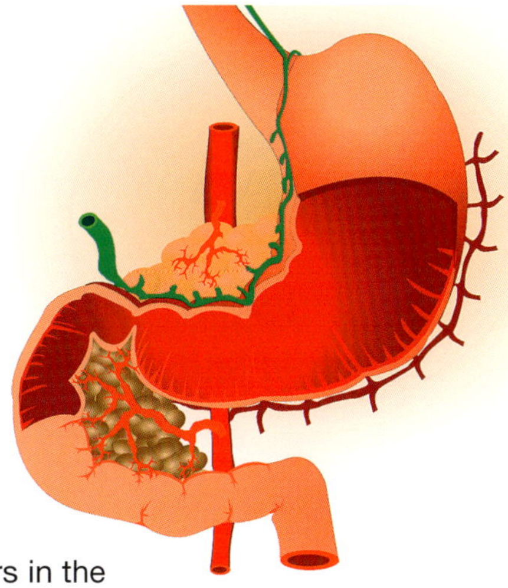

Stomach acts as a nutrition sensor – for example, it can 'taste' sodium glutamate using glutamate receptors and this sensory message is passed to the lateral hypothalamus and limbic system in the brain as a palatability signal through the vagus nerve. The stomach can also sense, independent of to tongue and oral taste receptors - glucose, carbohydrate, proteins and fats. This allows the brain to link nutritional value of foods to their tastes.

The Small Intestine: Most digestion and absorption of digested food, occurs in the small intestine. This narrow, twisting tube of about 2.5 cm (1 in) in diameter, fills most of the lower abdomen, extending about 6 m (20 ft) in length. Over a period of 3 to 6 hours, peristalsis moves chyme through the duodenum into the next portion of the small intestine, the jejunum, and finally into the ileum, last section of the small intestine. The duodenum is the smallest segment of the intestine, being only 8 inches long. Food travels through the small intestine by peristalsis (weak contracting waves of motion) that propels the food toward the large intestine. The other 2 segments of the small intestine are the jejunum, which is 3 ft long and connects the duodenum to the ileum, the final 3 feet of the small intestine. Peristalsis is a manifestation of two major reflexes within the enteric nervous system that are stimulated by a bolus of foodstuff in the lumen.

During this time, the liver secretes bile into the small intestine through the bile duct from liver, the largest glandular organ in the human body. Bile breaks large fat globules into small droplets, which are hydrolyzed by enzymes present in the small intestine. Pancreatic juice, secreted by the pancreas, enters the small intestine through the pancreatic duct. Pancreatic juice contains enzymes that break down sugars and starches into simple sugars, fats into fatty acids and glycerol, as well as, proteins into amino acids. Glands in the intestinal walls secrete additional enzymes that break down starches and complex sugars into nutrients that the intestine absorbs. Structures called Brunner's glands secrete mucus to protect the intestinal walls from the acid effects of digestive juices.

The small intestine's capacity for absorption is increased by millions of finger-like projections called villi, which line the inner walls of the small intestine. Each villus is about 0.5 to 1.5 mm (0.02 to 0.06 in) long and covered with a single layer of cells. Even tinier fingerlike projections called microvilli cover the cell surfaces. The villi of the intestine move back and forth, like thousands of tiny tentacles, passing through the food as it is moved along the intestinal tract. The villi play an important role in the absorption of food from the small intestine. Through the center of each villi is one or more fine white vessels called lacteals. The lacteals are part of the lymphatic system. Their principal function is absorption of fat. As food passes through the small intestine, it is taken up (absorbed) by structures

The combination of villi and microvilli increases the surface area of the small intestine's lining by about 150 times, multiplying its capacity for absorption.

in the wall of the intestines, especially the villi, and is secreted into the lacteals. Beneath the villi's single layer of cells are capillaries (tiny vessels) of the blood stream and the lymphatic system. These capillaries allow nutrients produced by digestion to travel to the cells of the body. Simple sugars and amino acids pass through the capillaries to enter the bloodstream. Fatty acids and glycerol pass through to the lymphatic system. Blood from the intestines (with these products of digestion) is collected in the portal vein, which is connected to the liver.

The liver removes the excess glucose from the blood (glucose being one of the end-products of digestion) and stores it as glycogen. Glycogen is used later in normalizing the blood-sugar level and for supplying energy. The liver also attempts to detoxify harmful elements in the food (such as pesticides), and regulates the level of nutrients available to the body. The liver is one of the master organs in the body. It receives all the end-products of digestion. The bulk that remains behind after the vital elements are extracted by the villi in the intestine and sent to the liver is then pushed down toward the large intestine. Normally, most of the contents of the intestines have been absorbed by the time the food reaches the middle of the jejunum segment of the intestine, or about 3 feet along the 9 feet of tubing that makes up the small intestine.

The tone and motility of the small intestine is increased by foods served at room temperature, fibrous foods, and high-carbohydrate, low-fat foods. Movement is slowed by cold, dry, and high-fat foods.

The Anatomy of Microvilli

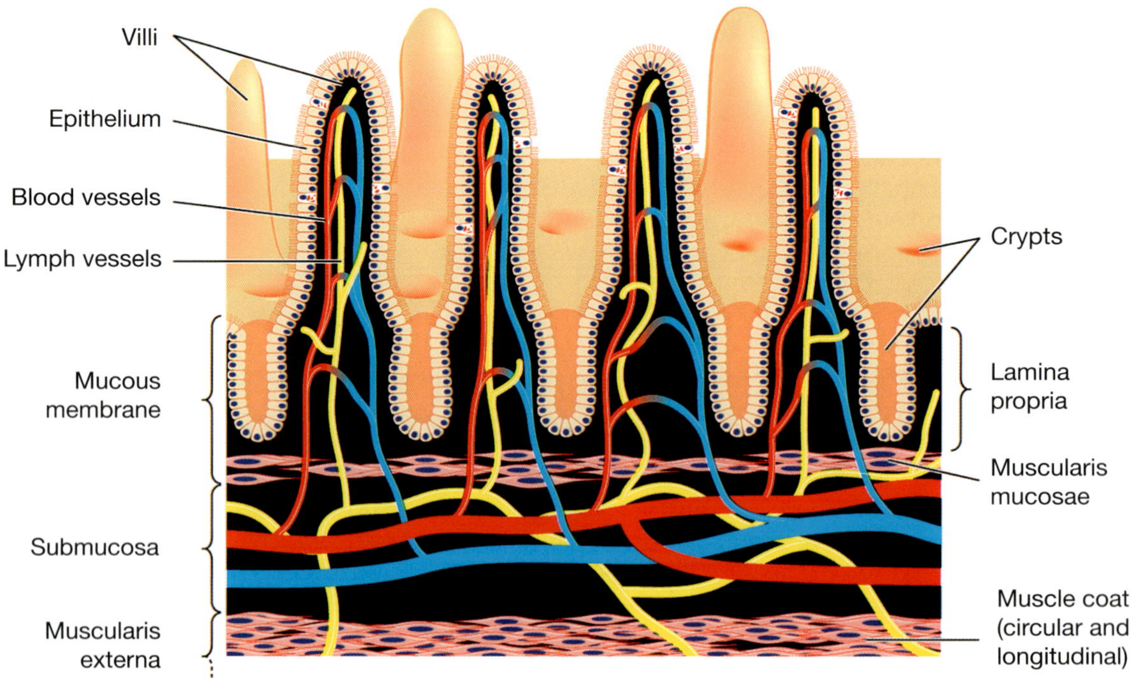

The Large Intestine (Colon): A watery residue of indigestible food and digestive juices remains unabsorbed. This residue leaves the ileum of the small intestine and moves by peristalsis into the large intestine, where it spends 12 to 24 hours.

The large intestine serves several important functions. It absorbs water, about 6 liters (1.6 gallons) daily, as well as, dissolved salts from the residue passed on by the small intestine.

Small intestine joins the colon in the region of the right groin. Large intestine forms an inverted 'U' over the coils of the small intestine. It starts on the lower right-hand side of the body and ends on the lower left-hand side. The large intestine is 1.5 to 1.8 m (5 to 6 ft) long and about 6 cm (2.5 in) in diameter. At this juncture is the ileo-cecal valve that controls the speed of passage of substances from the small intestine. The ileo-cecal valve opens into a pouch in the colon, the cecum, which is the first receptacle for waste residue. At the tip of the cecum is the appendix. Due to the appendix's position near the waste receptacle, toxins from a diet high in meat, heavy starches, etc. can contribute to its inflammation which may result in a condition known as appendicitis. From the cecum, the large intestine ascends on the right side to the middle of the abdomen, then crosses to the left side and descends again. These three sections are called the ascending, transverse and descending colon.

If an excess of water is expelled with the feces, a condition known as diarrhea exists. Any irritation in the stomach and small intestine due to unsuitable food or inflammation develops the onset of diarrhea. In such condition, the colon expels its entire waste residue upon entry without holding it for water re-absorption. On the other hand, if the waste residue moves too slowly through the colon, then excessive water is reabsorbed and the feces become hardened, which leads to constipation. Large intestine moves its remaining contents toward the rectum, which makes up the final 15 to 20 cm (6 to 8 in) of the alimentary canal. The rectum stores the feces – waste material that consists largely of undigested food, digestive juices, bacteria, and mucus – until elimination.

The human body is designed with pelvic muscles that aid in removing waste from the body. During the act of defecation the longitudinal rectal muscles contract to increase rectal pressure, and the internal and external anal sphincter muscles relax. Excretion is aided by contractions of abdominal and pelvic skeletal muscles, which raise the intra-abdominal pressure and help push the feces from the rectum through the anal canal and out the anus. When sphincters between the rectum and anus relax, the feces pass out of the body.

Bacteria in the large intestine promote the breakdown of undigested materials and synthesize several vitamins, notably vitamin K, which the body needs for blood clotting.

Vitamin K: The Nobel Prize in Physiology or Medicine 1943 was divided equally between Henrik Dam for his discovery of vitamin K and Edward Doisy for his discovery of the chemical nature of vitamin K.

The Rectum: Rectum is the terminal portion of the large intestine and functions for storage of the feces, the wastes of the digestive tract, until these are eliminated. The external opening at the end of the rectum is the anus. The anus has two sphincters, one voluntary and one involuntary. The fecal pressure on the involuntary sphincter causes the urge to defecate and the voluntary sphincter controls whether a person defecates or not. Finally, the journey of food through the body is completed.

GI Tract Transit Time: The transit time for food or other ingested substances to pass through the digestive tract depends on several factors. It takes roughly takes 2.5 to 3-h after a meal for 50% of stomach contents to empty into the intestines, while total emptying of the stomach takes about 4 to 5-h. Subsequently, 50% emptying of the small intestine takes 2.5 to 3-h. Finally, transit through the colon takes 30 to 40-h. These estimates of average GI tract transit times vary among individuals, as well as, within the same person at different times and after separate meals. First, there is considerable variability in transit times through different sections of the digestive tract among healthy people. Second, the time required for material to move through the digestive tract is markedly affected by the chemical and nutrient composition of the meal. Finally, transit time is influenced by factors such as age, psychological stress, gender and reproductive status of an individual.

The Digestion Process

Many healthy individuals process the food from the mouth to the anus in about 16 to 24-h. Most adults eating a conventional diet, however, generally take from 48 to 72 hours for their food to complete its journey. Much of this added delay is due to incompatible food combinations and lack of colon vitality.

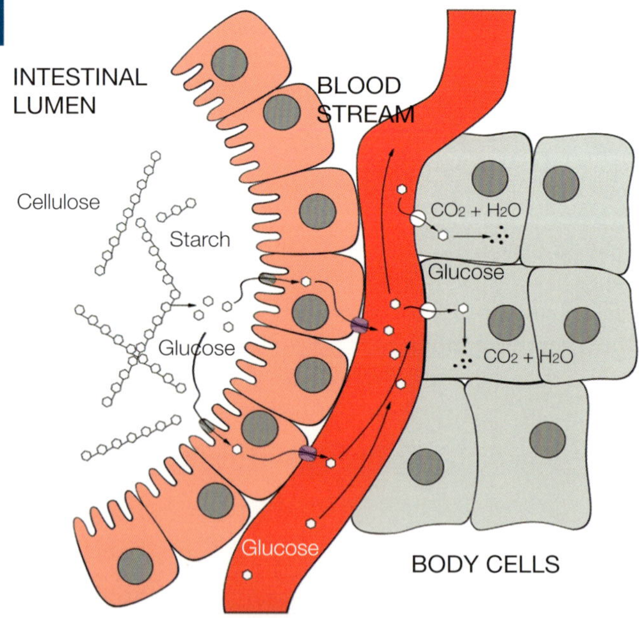

Bile acids: The Nobel Prize in Chemistry 1927 was awarded to Heinrich Wieland for his investigations of the constitution of the bile acids and related substances.

GI Tract and Host Defense

The GI tract is also a prominent part of the host defense. The surface area of the digestive tract is estimated to be the surface area of a football field. With such a large exposure, the immune system must work hard to prevent pathogens from entering into blood and lymph. The low pH (ranging from 1 to 4) of the stomach is fatal for many microorganisms. Similarly, mucus (containing IgA antibodies) neutralizes many pathogens. Enzymes such as *Cyp3A4*, along with the anti-porter activities, also play a key role in detoxification of antigens and xenobiotics, such as drugs, involved in first pass metabolism. Lactoferrin, an metal-binding protein present in the saliva and gastric juices is a broad-sprectrum antimicrobial defense factor in the GI tract.

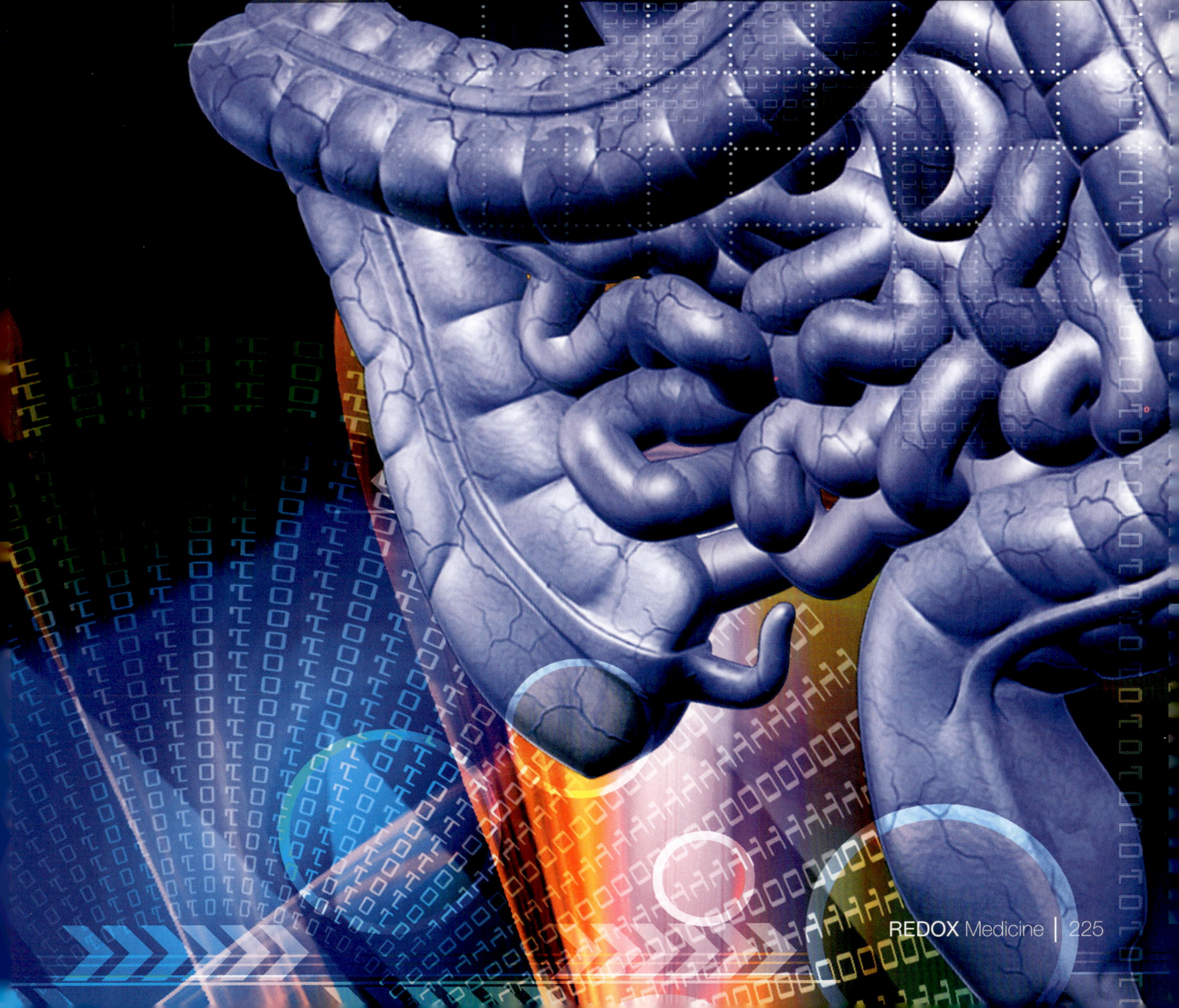

4.3.1. REDOX and Gastrointestinal (GI) Health

Oxidative exposure of the GI tract: The digestive system processes nutrients and evades harmful microbes and toxins. The GI tract needs to tolerate a wide range of chemical exposure, including carcinogens and pro-oxidants. The body may be exposed to various oxidation products such as the aldehydes from cooking oils; peroxidized lipids and cholesterol by-products from cooked meats and vegetables. The systemic toxicity of oxidized dietary lipids is low; however, the intestinal cell lining will be exposed to the full force of these agents, especially in the presence of highly catalytic heme compounds. Heme is well absorbed, but an excess of it might reach the colon and exert prooxidant and proliferative effects.

Other dietary sources that can contribute to oxidative stress in the digestive tract include: (i) Mixtures of iron and ascorbate, especially in the stomach and duodenum, which can catalyze lipid oxidation and promote ·OH generation. Diet-derived copper ions (e.g. Cu^{2+} in drinking water) which can generate ·OH and other ROS from ascorbate. Multi-vitamin pills containing iron and/or copper plus ascorbate can also play a pro-oxidant role. (ii) Nitrite in saliva and foods is converted to reactive nitrogen species (RNS) by gastric acid. (iii) High levels of hydrogen peroxide (H_2O_2) in certain beverages, especially coffee. Oral bacteria also generate H_2O_2. (iv) Food preservatives, such as the widely used sulfites are strong pro-oxidants. These sulfur compounds can be oxidized by transition metal ions and the NADPH oxidase system of intestinal lymphocytes and generate reactive sulfur species (RSS). (v) The GI tract contains part of the mucosal immune system and its resident phagocytes can respond to challenge with antigens in food or bacteria to produce ROS. (vi) The intestinal epithelial cells may release oxidative metabolites. Thus, superoxide ($O_2^{\bullet-}$)-producing NOx enzyme is found in colon. A similar system, dual oxidase-2 (DUOX2), is found in salivary glands and perhaps throughout the GI tract. The ROS produced by NOx/DUOX systems act directly against bacteria and/or by supplying H_2O_2 to peroxidases, such as lactoperoxidase.

Oxidative defense(s) in the GI tract: The intestinal tract has innate capability for self-renewal, which represents the first barrier of defense between the organism and its luminal environment. The thiol/disulfide REDOX systems equipped with the glutathione/glutathione disulfide (GSH/GSSG), cysteine/cystine (Cys/CySS) and reduced and oxidized thioredoxin (Trx/TrxSS) REDOX couples play important roles in preserving intestinal REDOX homeostasis, metabolic functions, and cellular integrity. Control of the thiol-disulfide status at the luminal surface is essential for maintaining mucus fluidity and absorption of nutrients, and protection against chemical-induced oxidant injury. Within intestinal cells, these REDOX couples preserve an environment that supports physiological processes and orchestrate networks of enzymatic reactions against oxidative stress. The GI tract is lined by mucus, a powerful scavenger of the ROS, RNS and RSS. The intestinal epithelial cells contain DNA repair enzymes, proteasomes that remove oxidized proteins, SOD, catalase, peroxiredoxins, glutathione-S-transferase (GST) and glutathione peroxidase (GPx), both GPx1 and the 'intestinal form', GPx2. Certain antioxidative enzyme activities (i.e. SOD and GPx) of the small intestine and colon are located on the cell surface, which can facilitate detoxification of dietary oxidants. The GPx enzymes from the intestinal epithelia can remove the bulk of peroxides from the absorbed dietary lipids. The gut is also rich in GSTs, which may metabolize peroxides and cytotoxic aldehydes.

Saliva can rapidly quench the H_2O_2 exposure, probably by the action of both peroxidases and catalase. The catalse is liberated into saliva as a result of turnover and shedding of buccal cell lining in the oral cavity. The total antioxidant capacity (TAC) of saliva is lower than that of plasma; ascorbate levels are <10 μM and the major antioxidant present is urate. The TAC of saliva varies depending on the source of salivary glands.

Dietary Antioxidants and the GI tract

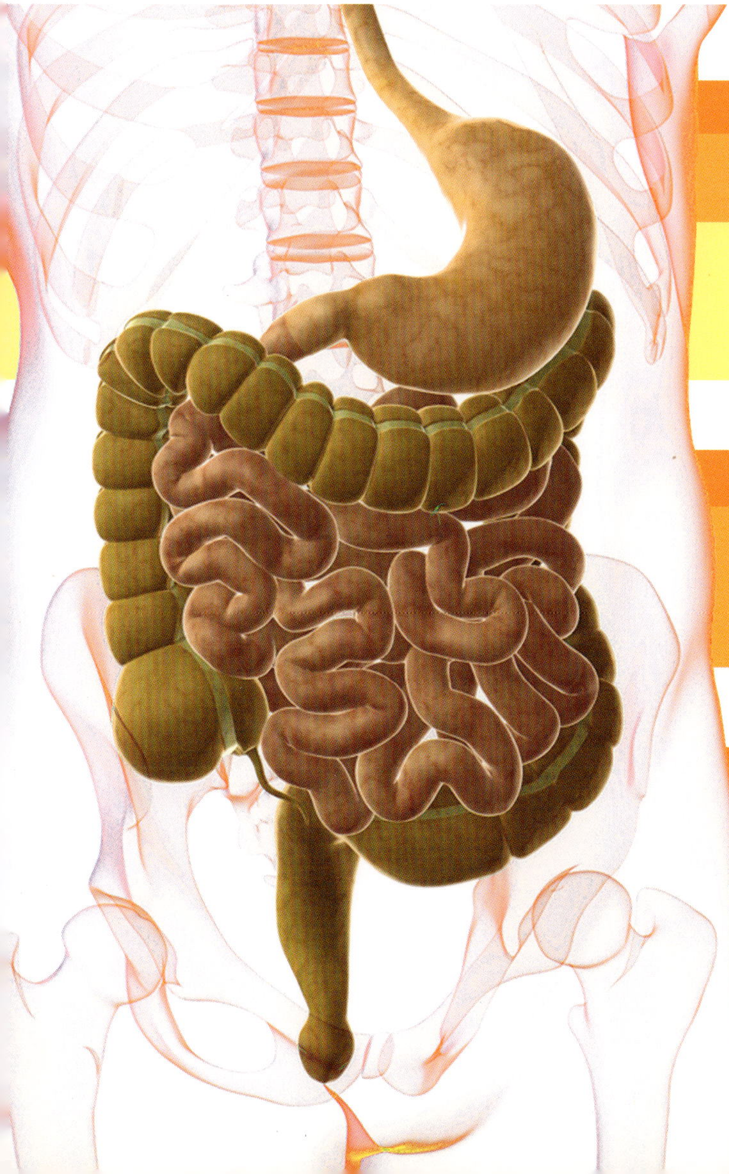

STOMACH

VITAMIN C enters through food into gastric juice

VITAMIN E, Polyphenols, Carotenoids & Flavonoids enters through diets rich in fruit/vegetable/grain

RNS from acidified nitrite
ROS from iron/copper/ascorbate interactions
H_2O_2 from consumed beverages
RO and RO_2 from dietary lipid peroxides

SMALL INTESTINE

VITAMIN C is completely absorbed

VITAMIN E is mostly absorbed - some tocotrienols & tocopherols return to GI tract
Some carotenoids, flavonoids and phenolic compounds are absorbed

LARGE INTESTINE

VITAMIN C - just a fractional amount present

VITAMIN E - only a limited amount left
Extensive metabolism of phenolic compounds by gut flora

Metal-binding proteins protect intestinal epithelia and help delay colon and rectal cancer
Inhibition of LOX, COX-2, angiogenesis, MMPs etc. with polyphenols may help to limit cancer development

4.3.2. Oxidative Stress and Inflammatory Bowel Disease (IBD)

Inflammatory bowel disease (IBD) includes ulcerative colitis and Crohn's disease, with clinical manifestations of active (symptomatic) and inactive (asymptomatic) phases, involving a cascade of REDOX/inflammatory events at the intestinal level. IBD initiates with the infiltration of inflammatory cells into the mucosa and release of pro-inflammatory mediators such as ROS/RNS. Pro-oxidants such as nitric oxide (·NO), superoxide ($O_2^{\bullet-}$), peroxynitrite ($ONOO^-$), hydrogen peroxide (H_2O_2), and hypochlorite (OCl^-) are implicated in the onset of IBD. Increased production of ROS/RNS can stimulate whole colonic mucosa, mucosal macrophages, and peripheral blood monocytes of IBD patients. Furthermore, L-Citrulline, the co-product of ·NO generation, is elevated in the colonic lumen of IBD patients. This leads to intestinal tissue damage and an additional recruitment of inflammatory cells - a vicious cycle that triggers and sustains the REDOX/inflammatory cascade. Consequently, a disrupted intestinal barrier initiates and perpetuates the cascade by exposing luminal factors to the mucosal immune system. Only if this vicious cycle is aborted, tissue damage can be repaired. Left unchecked, inflammation is sustained, leading to intestinal tissue damage and symptoms of the active phase of IBD.

Oxidative stress and its pro-inflammatory factors are critical in maintaining the vicious cycle of IBD. These pro-inflammatory oxidants can overwhelm antioxidant defenses through protein oxidation, DNA strand breaks, and ATP depletion, which cumulatively inflict tissue damage. Oxidative stress in IBD also causes denaturation of cytoskeletal proteins. Oxidative injury correlates with disease severity but is also present in high amounts in normal appearing mucosa of IBD patients. More than 50% oxidation of actin, the cytoskeletal protein, is reported in the inflamed intestinal mucosa. This suggests that oxidant induced cytoskeletal disruption is required for tissue injury, mucosal disruption, and IBD flare up.

Infiltrating macrophages and neutrophils overwhelm the inflamed gut in patients with IBD. There is a growing body of evidence that these infiltrated cells expose the inflamed intestine to extensive oxidative stress through production of ROS. Increased ROS production is also demonstrated by circulating phagocytic cells in patients with IBD and seems to be involved in several models of experimental colitis. The mucosa is endowed with various endogenous antioxidant defense systems to remove ROS resulting from normal metabolism. However, antioxidant enzyme activities of catalase, SOD, and glutathione peroxidase (GPx) were low at only 4%, 8%, and 45%, respectively, in healthy colonic mucosa. Therefore, the free radical formation in excess may overtax the limited intestinal antioxidant defense system and initiate oxidative injury to the gut.

The tripeptide glutathione (GSH) is the most important intra-cellular antioxidant. In the presence of ROS, reduced GSH (rGSH) is oxidized to glutathione disulphide (GSSG). This reaction is catalyzed by GSH peroxidase and reversed by GSH reductase. Deterioration of antioxidative GSH metabolism and increased colonic oxidative damage of proteins and DNA with impaired activity of Cu-Zn-SOD enzyme has been reported in patients with Crohn's disease.

REDOX-based Interventions: The most widely prescribed drug for treatment of IBD is sulfasalazine, which inhibits free radical activity. The active moiety of sulfasalazine, 5-aminosalicylic acid, is a radical scavenger shown to reduce lipid peroxidation. Clinical studies have shown that Cu-Zn-SOD could be beneficial in Crohn's disease and ulcerative colitis. Several antioxidants were reportedly effective in reducing ROS production in ulcerative colitis and experimental colitis. Intestinal inflammation induced by administration of an ROS-generating agent can be reversed by GSH supplementation. It has therefore been proposed that IBD results from an imbalance between pro-oxidant and antioxidant mechanisms – which causes oxidative stress. Decreased activity of key enzymes involved in GSH synthesis, followd by a decreased availability of cyst(e)ine for GSH synthesis, contribute to mucosal GSH deficiency in IBD. Impaired mucosal antioxidative capacity may further promote oxidative damage; therefore, GSH deficiency might be a target for therapeutic intervention in the clinical management of IBD.

Crohn's disease is a chronic IBD that can affect any region of the GI tract. Most commonly, the inflammation involves the distal small bowel, often including segments of the proximal large intestine. Its patho-biology involves uptake of luminal antigens across the mucosal epithelium eliciting an aberrant immune response, culminating in intestinal inflammation, local peroxidation, and systemic circulation of inflammatory cytokines and mediators. The onset of Crohn's disease results in a reduction in antioxidant enzymes in the intestinal mucosa and plasma.

Inflammatory Bowel Diseases

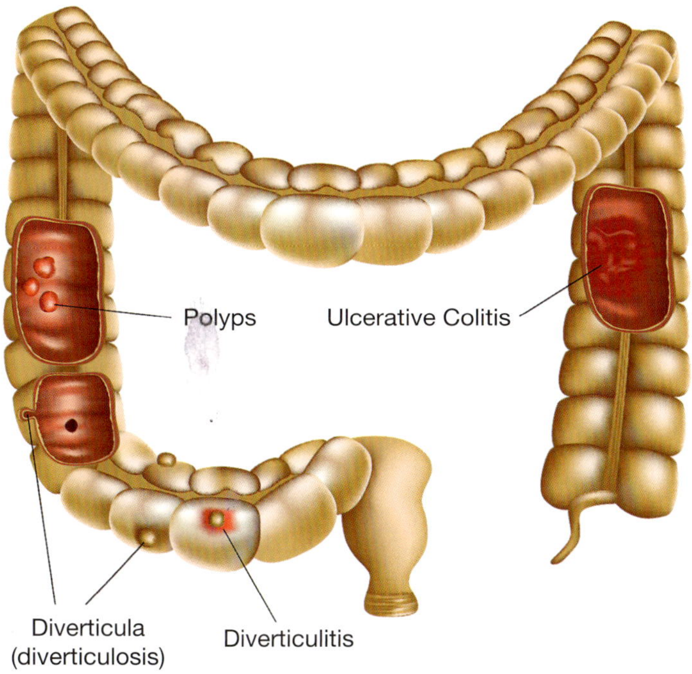

Crohn's

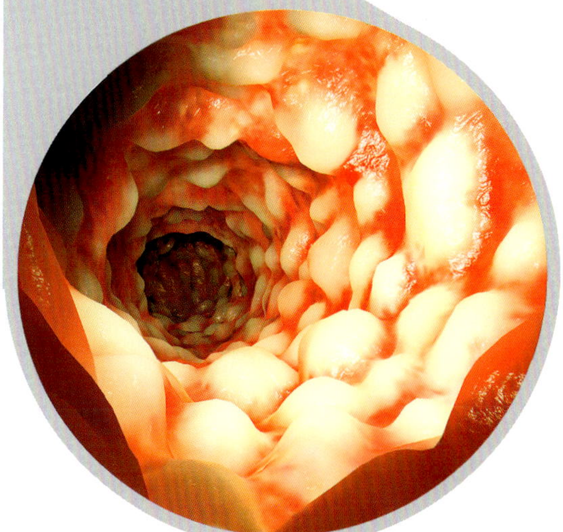

4.3.3. Oxidative Stress and Celiac Disease

Celiac Disease can develop at any point in life, from infancy to late adulthood. People who have a family member with celiac disease are at a greater risk. The disorder is most common in Caucasians and persons of European ancestry. Women are affected more often than men.

Celiac disease is an autoimmune GI disorder caused by permanent intolerance to gluten/gliadin (prolamin). The prolamin fractions in cereal grains (gliadin in wheat and similar alcohol-soluble proteins in other cereals, secalin in rye, hordein in barley) are the external stimuli involved in the development of intestinal damage associated with celiac disease. The classical presentation of celiac disease as a pediatric predominant disorder is characterized by small-intestinal villous atrophy and crypt hyperplasia. However, especially in adult-onset patients, a preserved mucosal architecture characterized by dense lymphocytic infiltration, without any villous atrophy or crypt hyperplasia can also be observed. The only treatment for celiac patients is a permanent gluten-free diet.

Accumulation of certain β-gliadin peptides in lysosomes leads to activation of some signal transduction pathways and to increased levels of ROS and RNS. Thus, it has been suggested that oxidative stress is one of the mechanisms that plays a role in the pathogenesis of celiac disease. Several studies have reported that gliadin exposure could cause an intra cellular oxidative imbalance, which is characterized by an increase in the levels of lipid peroxidation products, an increase in the oxidized (GSSG)/reduced glutathione (GSH) ratio, and a decrease in protein-bound sulfhydryl groups. Significant structural perturbations of the cell plasma membrane were also detected. The increase in oxidative damage can induce alterations to cell morphology, cell proliferation, apoptosis and cell viability.

Alteration in oxidative balance induced by gliadin peptides in enterocytes is involved in the activation of transcription factor NF-κB. This activation induces the transcription of pro-inflammatory cytokines and enzymes such as COX2 and iNOS, with consequent higher production of prostaglandins and ·NO metabolites contributing to the oxidative stress. The relationship between oxidative damage and celiac disease is also supported by several studies on intestinal cells and circulating cells. Increased levels of prostaglandin E2 (PGE2) were demonstrated in homogenized small bowel biopsy specimens from patients with active celiac disease, while the levels of the antioxidant enzyme glutathione peroxidase (GPx) and reductase were decreased in biopsies from celiac disease patients with consequent decreased levels of GSH. Several investigations reported that RNS such as ·NO also play an important role in the pathogenesis of celiac disease with enhanced ·NO. Inducible-nitric oxide synthase (iNOS) is constitutively expressed in human duodenal enterocytes; its activity is increased in patients with untreated celiac disease and is partially corrected when patients are treated with gluten-free diet. Elevated ·NO levels are present in serum and urine of children with celiac disease and correlates with an increased expression of iNOS in the small intestine.

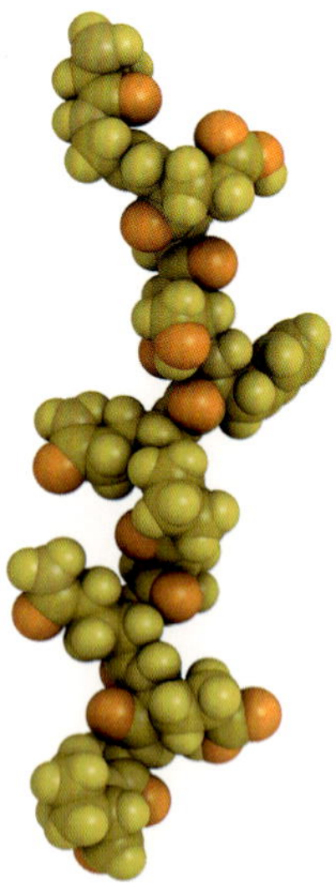

GLIADIN

An increase of markers of oxidative damage of lipids [thiobarbituric acid-reactive substances (TBARS), lipid hydroperoxides], proteins (carbonyl groups) and DNA was reported in intestinal cells and biological fluids of celiac disease patients. Changes in vitamin E and antioxidant enzyme activities were observed in circulating cells and plasma of celiac disease patients. Also, decreased expression of antioxidant and anti-inflammatory enzymes was reported in the intestinal biopsies of celiac disease patients. The decreased antioxidant defenses may compromise the inflamed mucosa (making it more susceptible to oxidative tissue damage) and hinder recovery of the mucosa and the epithelial cell layer integrity. An association between gluten, oxidative stress, gene expression and inflammation is supported by recent studies. The analysis of gene expression in small-bowel mucosal biopsy samples from untreated celiac disease patients showed that about nine of 30 genes involved in cell signaling and differentiation were modified. A significant malfunction of GSH-REDOX cycle, with a concomitant decrease in the capacity to regenerate GSH and detoxify LOOH was observed in celiac patients (even after many years of GFD). Thus, oxidative stress and inflammation due to an increase of ROS and a decrease of antioxidant defenses are involved in the molecular mechanisms of celiac disease.

The average length of time it takes for a symptomatic person to be diagnosed with celiac disease in the US is four years; this type of delay dramatically increases an individual's risk of developing auto-immune disorders, neurological problems, osteoporosis, and even cancer.

REDOX-based Interventions: Various free radical scavenging phyto-nutrients may have a role in the REDOX therapy of celiac disease. These compounds act through a variety of mechanisms including cell signaling, gene expression, and promote gut barrier function by neutralizing pro-oxidant molecules. Antioxidant vitamins (vitamin C and E), polyphenols and carotenoids (from dietary vegetable food and oil) are absorbed by intestinal cells and exert protective effects against oxidative damage. A likely target for these compounds is the signal transduction cascade leading to the activation of transcription factors such as NF-κB. Also, carotenoids are able to inhibit the expression of inflammation-related enzymes/proteins. Phyto-nutrients such as lycopene, quercetin, and tyrosol decrease iNOS and COX-2 gene expression induced by gliadin. Flavonoids ameliorate intestinal barrier dysfunction induced by oxidative stress. The oral administration of GSH and a diet rich in natural antioxidants, as well as, appropriate dietary supplements, could be of great benefit to celiac patients.

Celiac Disease and the United States

- About 1% of healthy, average Americans are affected. That means at least 3 million people are living with celiac disease – 97% of them are undiagnosed.
- Type 1 Diabetes affects 3 million people; 6% of those diagnosed also have celiac disease.
- About 610,000 women experience unexplained infertility; 6% of these women might never learn that celiac disease is the cause.
- About 350,000 people are living with Down Syndrome; 12% of them also have celiac disease.

4.3.4. Oxidative Stress and Gastric Disorders

Excess ROS are produced in H. pylori-colonized human stomachs; this induces oxidative stress to both the gastric mucosa and H. pylori. However, H. pylori has a unique ability to detoxify ROS using a variety of enzymes to establish long-term colonization; thus, the excess ROS leads solely to host cell damage.

The stomach is the foremost food processing organ of the GI tract, susceptible to exogenous pathogens from the diet. During microbial insult, the stomach induces oxidative stress, which could lead to gastric disorders such as gastritis, gastric ulcers, gastric cancer, as well as, functional disorders such as dyspepsia. Psychological stress, in addition to physical stress such as surgical intervention and microbial infection [including *Helicobacter pylori (H. pylori)*], leads to oxidative stress in the stomach. Several gastrointestinal diseases, such as peptic ulcer and gastroparesis, are associated with reduced or dysfunctional antioxidant activity in the stomach.

Gastric (Peptic) Ulcers and *Helicobacter pylori*: *H. pylori* infection leads to gastro-duodenal inflammation, peptic ulceration and gastric carcinoma. *H. pylori* is the major etiological agent that causes chronic gastritis, with other features, including the lymphoid follicles or lymphoid aggregates, surface epithelial degradation with mucous depletion, and intestinal metaplasia. During the pathogenesis of a chronic infection in the gastric mucosa, *H. pylori* induces a strong inflammatory response, leading to generation of a number of ROS and RNS, mediated by neutrophils and macrophages. *H. pylori* stimulates gastric hyper-proliferation, which is a critical step in the development of a gastric carcinoma. The *H. pylori* induced expression of the inflammatory genes, oncogenes and cell-cycle regulators may be mediated by the ROS/RNS-induced activation of oxidant-sensitive transcription factors in the gastric epithelial cells. *H. pylori* combats the oxidative stress generated by the host immune response using an antioxidant protein in order to establish long-term colonization. The mechanisms for ROS detoxification are of particular interest in the clinical management of *H. pylori*-associated pathogenesis. *H. pylori* produces a variety of antioxidant enzymes to combat the toxic effects of ROS including catalase (KatA), iron-cofactored SOD, and alkyl hydroperoxide reductase (AhpC).

ROS/RNS released from activated neutrophils are also potential virulence factors involved in *H. pylori*-infected host cells. In *H. pylori*-infected cells, OCl^- anions are generated from H_2O_2 in the presence of Cl^-. The hypochlorous ($HOCl^-$) anions subsequently react with ammonia (NH_3), which is derived from urea by urease produced by *H. pylori,* ultimately yielding monochloramine (NH_2Cl). NH_2Cl induces mucosal cytotoxicity with its lipophilic properties and freely penetrates biological membranes to oxidize intra-cellular components.

Gastritis and Peptic Ulcer: The Nobel Prize in Physiology or Medicine 2005 was awarded jointly to Barry J. Marshall and J. Robin Warren for their discovery of the bacterium *Helicobacter pylori* and its role in gastritis and peptic ulcer disease.

Alcohol-induced Gastritis: The effect of alcohol (ethanol) on gastric mucosa is associated with disturbances in the balance between gastric mucosal protective factors. Gastric mucosa is exposed to gastric acid, pepsin, and stimulants among others; gastro-protective factors maintain the integrity of the gastric mucous layer, microcirculatory system, bicarbonate anion (HCO_3^-), and epithelial cell restitution. Ethanol injures the vascular endothelial cells of the gastric mucosa and induces microcirculatory disturbance and hypoxia, with overproduction of ROS. Ethanol disturbs the energy metabolism of mitochondria, which could lead to the gastric mucosal injury.

GI function declines with aging, including delayed gastric emptying, decreased peristalsis, and slowed colonic transit. These conditions impair quality of life and increase morbidity and mortality.

Portal Hypertensive Gastropathy (PHG): PHG is a common complication of liver cirrhosis and contributes to impaired gastric mucosal healing. The clinical onset of PHG is related to increased ROS and lipid peroxidation. Increased levels of lipid peroxides and nitrotyrosine (an indicator of nitration of tyrosine residues due to $OONO^-$) are reported in the gastric mucosa of PHG. The gastro-protective, anti-inflammatory agent, rebamipide, prevents free radical production by scavenging ·OH radicals. Rebamipide improves the ulcer index by decreasing the lipid peroxidation and nitrotyrosine levels in the gastric mucosa.

Non-steroidal anti-inflammatory drugs (NSAIDs) and Aspirin: In addition to the inhibition of cyclooxygenase and decreasing prostaglandin production, NSAIDs induce mucosal damage via ROS produced by leukocytes. The ROS mediates mitochondrial damage and oxidation of lipids, proteins, and DNA; which leads to apoptosis and mucosal injury. Proton pump inhibitor therapy has been suggested to primarily protect gastric mucosa by inhibiting gastric acid secretion. These agents, such as lansoprazole, also inhibit NSAID-induced gastropathy by blocking mitochondrial and Fas-mediated apoptosis pathways. On the other hand, aspirin increases the permeability of gastric epithelia and disrupts the barrier integrity.

REDOX-based Interventions: Oxidative stress is a major contributor in the pathogenesis of gastritis. Therapeutic options such as gastro-protective agents with antioxidant properties (e.g. rebamipide) can modulate the level of oxidative stress and enhance anti-inflammatory or antioxidant capacity. The stomach is an organ in direct contact with external pathogens; by presenting a strong acid environment, it has a special biological defense mechanism that eliminates such pathogens. However, *H. pylori* manages to colonize the stomach by compromising this line of host defense. In response to the colonization of this bacterium, gastric mucosa can be exposed to severe oxidative stress with considerable levels of inflammatory cell accumulation; this could be related to the development of gastric mucosal, as well as, neuro-muscular disorders. REDOX-based interventions could offer a promising future in the clinical management of *H. pylori*-induced gastric ulcers.

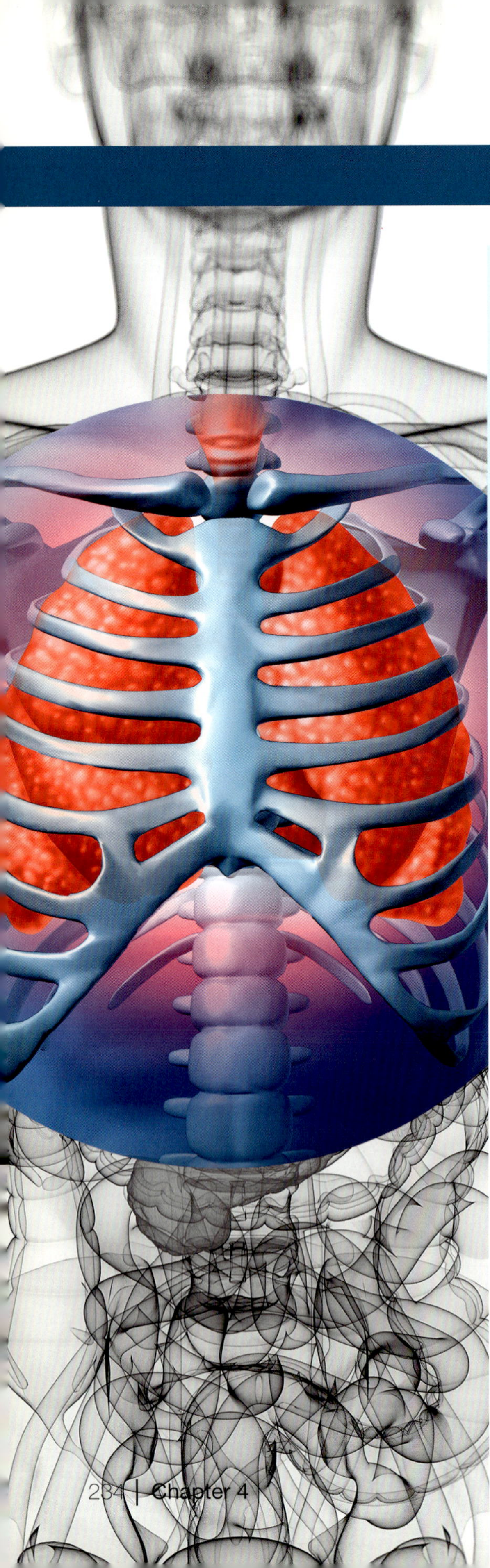

4.4 REDOX: The Respiratory System

The Respiratory (or Ventilatory) System is essential for aerobic metabolism, an oxidative cellular process that generates bio-energy (ATP) from food. Respiration is a gaseous exchange activity that facilitates oxygenation of blood with a synchronized disposal of CO_2 and other gaseous metabolic wastes from the circulation. Respiration is necessary for maintenance of REDOX balance in the body. Any imbalance in homeostasis could lead to respiratory acidosis (oxidative stress) or respiratory alkalosis (reductive stress) - both life threatening conditions.

Respiratory system is divided into two functional zones: (i) *Conducting zone* that consists of nose, pharynx, larynx, trachea, and bronchi. These structures form a continuous path for air to move in and out of the lungs. (ii) *Respiratory zone* is located deep inside the lungs and is made up of respiratory bronchioles, alveolar ducts, and alveoli. These thin-walled structures facilitate diffusion of inhaled O_2 into the lung capillaries and exchange CO_2 from the blood circulation.

In addition to gas exchange, lungs also perform other vital functions. They contain fibrinolytic enzymes that dissolve blood clots in the pulmonary vessels and remove several toxic substances from the venous blood. In addition, the lungs eliminate prostaglandins from circulation. Lungs also activate angiotensin and convert this functionally inactive decapeptide hormone into aldosterone-stimulating octapeptide angiotensin II. The angiotensin-converting enzyme (ACE) is located on the surface of the endothelial cells in pulmonary capillaries. Circulation time through the pulmonary capillaries is <1-second, yet 70% of the angiotensin I reaching the lungs is converted to angiotensin II in a single trip through the pulmonary capillaries.

Human respiration is also known as tidal breathing, since air comes out the same way and it goes in. A normal adult breathes about 9-20 times per minute. Accordingly, each breath takes about half a liter of air, which accounts for an average 6.5 liters of inhaled air every minute. An individual inhales and exhales about 20,000 times daily, while exchanging ~10,000 liters (~350 cubic feet) of air. It is noteworthy that half a liter of water vapor is also exhaled during respiration every day. Breathing is vital for life, an automatic operation that lasts for nearly 600 million breaths in a lifetime.

4.4.1. REDOX and Respiratory Health

Oxygen supply to human tissues is tightly regulated by the allosteric binding of O_2 to hemoglobin in red blood cells. The moist mucosal surface of the lung is directly exposed to inhaled oxygen, as well as, airborne reactive pollutants. Lungs where the O_2 is instantly dissolved into the epithelial lining fluid; this makes the lungs particularly susceptible to environmental oxidant-mediated injury. The respiratory tract is constantly exposed to transient encounters of oxidant stress due to the inhalation of foreign materials including atmospheric pollutants and microorganisms. The ROS and RNS production during episodes of infection and inflammation has been implicated in the pathogenesis of several pulmonary diseases. Also, recruitment of activated inflammatory cells (eosinophils, neutrophils and macrophages) results in the generation of highly cytotoxic ROS, which may lead to tissue damage via direct oxidation of proteins, DNA, or lipids. In order to combat potential encounters of oxidant stress, the airways possess a number of antioxidant defense mechanisms to neutralize ROS. However, REDOX state in the healthy lung is primarily reducing in nature. This is attributed to the diversity and abundance of antioxidant systems in the respiratory system. The vast excess of reduced substances over oxidized ones is maintained by several enzymatic and non-enzymatic antioxidants on the surface of, and within, the epithelial cells in the respiratory tract.

Coughing is a method to remove dust, mucus, saliva, and other debris from the lungs. Inability to cough can lead to serious infections. Deep breathing exercises may help keep finer structures of the lungs clear from particulate matter.

The respiratory system, due to its interface with the external environment, is frequently exposed to airborne oxidant gases and particulates; thus, it is prone to oxidant-mediated cellular damage. Several studies have demonstrated that high concentrations of ozone can be harmful to the lung. Ozone can react directly with unsaturated fatty acids and cell membranes to generate lipid ozonation products, which are small, diffusible, and relatively stable compounds. Particulate matter in the atmosphere is one of the most serious air pollution problems in urban

Human Respiratory System

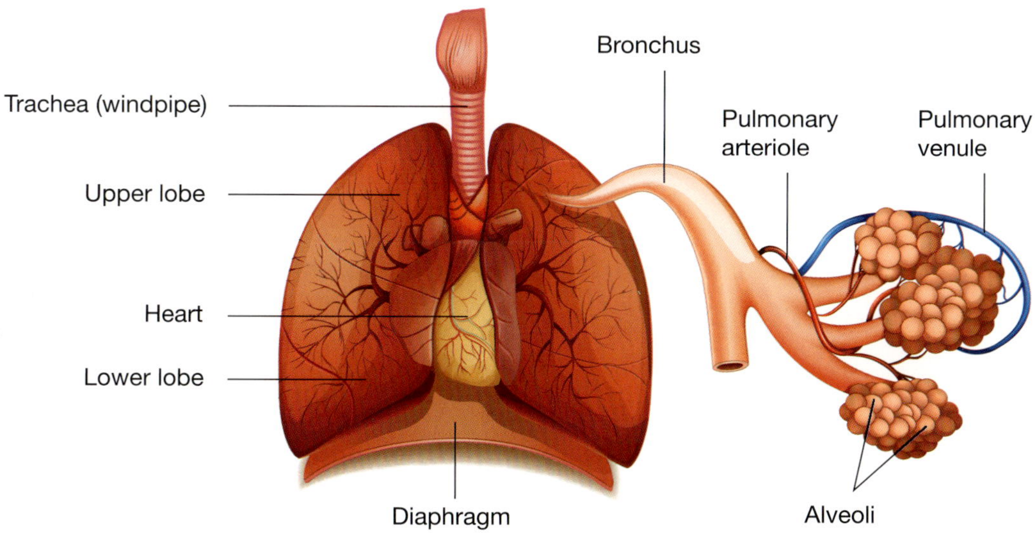

environments. The particle size is the critical factor that determines the target site for the particulate matter in the respiratory tract, when inhaled. One of the most dangerous forms of particulate matter pollution is diesel exhaust. Diesel exhaust particles are hydrophobic polyaromatic hydrocarbon molecules that can diffuse easily through membranes and negatively influence cellular growth and differentiation.

Environmental tobacco smoke or second hand smoke is a complex mixture of gases and particles that include smoke from the burning cigarette (side stream smoke) and exhaled mainstream smoke. Environmental tobacco smoke leads to activation of phagocytes, which augments the release of free radicals. Since free radicals cause oxidative damage to macromolecules such as DNA, lipids, and protein, they are involved in the pathogenesis of many diseases.

The respiratory system consists of several endogenous antioxidant systems that prevent and protect cells from oxidative damage. These systems are divided into enzymatic and non-enzymatic antioxidant groups. The enzymatic antioxidants include superoxide dismutases (SOD), catalase, glutathione peroxidases (GPx), heme oxygenase (HOx), glutaredoxin (GRx), thioredoxin (TRx), and peroxiredoxin (PRx). These antioxidant enzymes usually require trace metal cofactors. SOD, for example, consists of proteins co-factored with copper, zinc, or manganese. Iron is required as a co-factor for catalase. The prominent none-enzymatic antioxidants include lipid-soluble vitamin E (α-tocopherol), vitamin A [carotenoids (β-carotene)], water-soluble vitamin C and glutathione (GSH). Glutathione, which is synthesized intra-cellularly from amino acids cysteine, glycine, and glutamate, is a potent free radical scavenger. In addition, GSH is crucial to the maintenance of enzymes and other cellular components in a reduced state.

Smoking and Emphysema

Environmental tobacco smoke contains a number of components, and many of them are toxic to epithelial cells. Cigarette smoke contains >4,700 toxic chemicals, including 50 compounds that are known to cause cancer. Some of the chemicals in cigarette smoke are CO, cyanide, arsenic, mercury, and NO. Cigarette smoke produce or contains ~10^{15}-10^{17} free radicals per puff such as H_2O_2 and $O_2^{\bullet -}$.

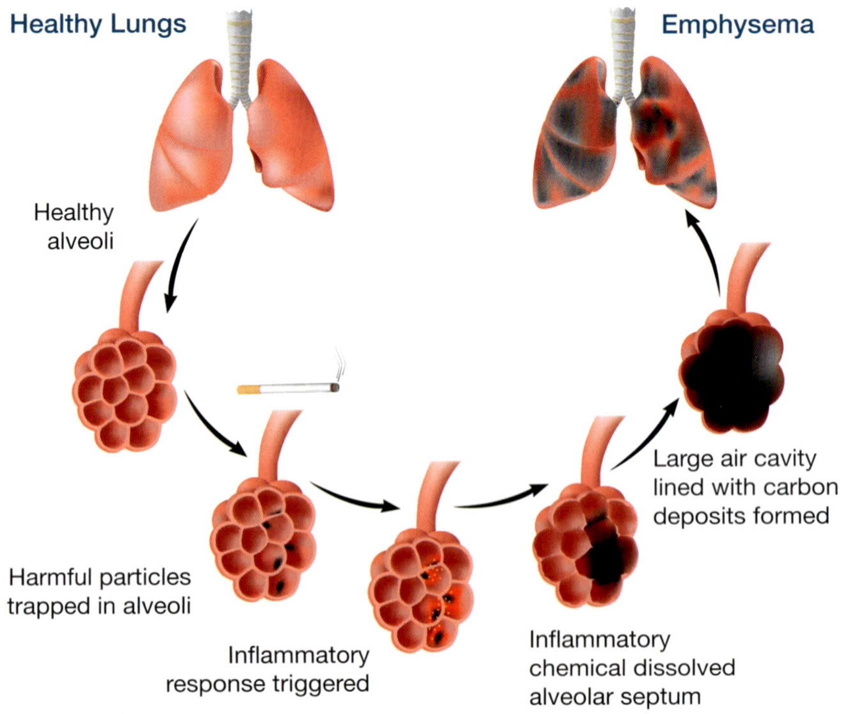

4.4.2. Oxidative Stress and COPD

Chronic Obstructive Pulmonary Disease (COPD) is a major and increasing global health problem. It is a disabling condition associated with progressive breathlessness. COPD will account for over 6 million deaths per year by 2020 - predicted to take a leap from the 6th to the 3rd leading cause of death worldwide. In America, COPD affects 9% of residents aged 60 years and above and is ranked 4th in the recent morbidity survey of the elderly population. It is estimated that about 23.4 million people in the USA have COPD. The typical clinical manifestations of COPD include chronic bronchitis (a condition of large-airway inflammation and remodeling), emphysema of the distal airways, and lung parenchyma (that results in loss of surface area for gas exchange). COPD decreases patients' quality of life due to shortness of breath and chronic cough, which can progress over years. Other manifestations of COPD such as systemic inflammation, alterations of metabolism, cardiovascular events, and cancer contribute to the untimely death of these patients.

Pathogenesis of COPD involves three stages: (i) *Initiation stage:* Cigarette smoke and environmental agents trigger host cell responses, largely dominated by inflammation, oxidative stress, and alveolar cell death. (ii) *Progression:* Cigarette smoke disrupts alveolar maintenance, triggers apoptosis and autophagy; moreover, oxidants in tobacco and activated inflammatory and alveolar cells lead to extra-cellular matrix proteolysis, which further enhances inflammation and promotes a feedback loop with apoptosis. (iii) *Consolidation of emphysema:* Over decades of exposure to cigarette smoke and endogenous amplifiers of destructive processes, the lungs progressively age.

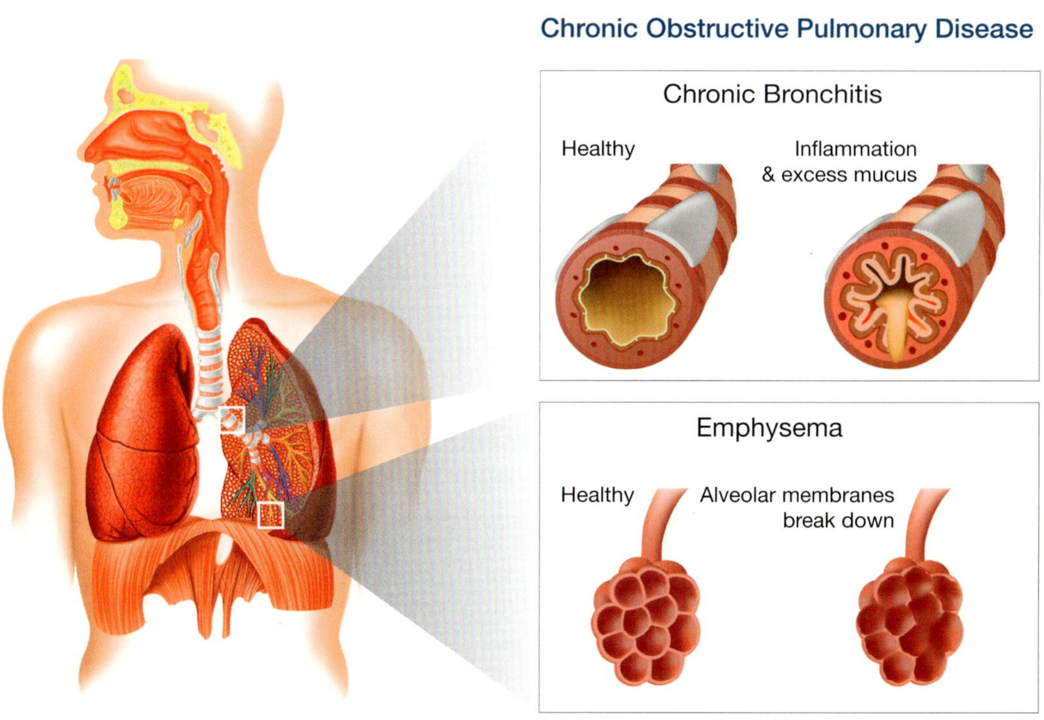

Autophagy is a dynamic process responsible for the turnover of cellular organelles and long living proteins, which serve an essential function to maintain cell homeostasis and confer adaption to adverse environment.

Cigarette smoke is the primary cause of COPD - which is characterized by rapid decline in lung function, inflammation, and premature aging of the lung. However, only 10%–20% of smokers develop COPD, which suggests additional risk factors such as genetic susceptibility. Other noxious environmental gases/particles such as nitric dioxide (NO_2), sulfur dioxide (SO_2), particulate matters, and exposure to second hand tobacco smoke and biomass fuel can also cause oxidative stress and trigger inflammatory responses in the lungs. Formation of ROS such as superoxide anion ($O_2^{\bullet-}$), nitric oxide (·NO), peroxynitrite ($ONOO^-$) and hydroxyl (·OH) radicals leads to a series of chain reactions which yield uncontrolled tissue destruction. The role of oxidative stress is confirmed by the presence of free radical biomarkers in COPD patients, such as elevated level of 8-hydroxy-deoxyguansoine in urine, and increased 3-nitrotyrosine (3-NT) and F2α isoprostanes in the lungs. Furthermore, elevated levels of lipid peroxidation products 4-hydroxynonenal (4-HNE) and malondialdehyde (MDA) were reported in the lungs of patients with COPD. These reactive aldehydes can alter protein function, induce inflammation, autoimmune responses, and inflict lung injury. Thus, resolving the REDOX imbalance and reducing lipid peroxidation could prevent the progression of COPD.

Under normal condition, lungs have well co-ordinated and efficient endogenous antioxidant defense mechanisms, that protect against the injurious effects of oxidants. In patients with COPD, this balance is disturbed due to mutations in genes that encode antioxidant enzymes. Furthermore, accumulation of ROS leads to autophagy, a protective mechanism to overcome oxidative stress. Autophagy facilitates the removal and degradation of damaged mitochrondria and oxidized proteins. However, excessive autophagy could lead to cell death, and COPD patients demonstrate increased autophagy in lungs. Endothelial cells dysfunction also plays a pivotal role in pathogenesis of emphysema. Cigarette smoke-induced emphysematous alveolar septa are almost avascular, associated with reduced expression of endothelial NO-synthase (eNOS) and endothelial dysfunction.

REDOX-based Interventions: The mainstays of pharmacotherapy are broncho dilators (to relieve the symptoms of broncho-constriction), corticosteroids (for reducing airway inflammation), and combination of both. However, current corticosteroid therapy in COPD is poorly effective, which has prompted an intense search for new therapeutic targets. Oxidative stress leads to lung inflammation in COPD. Cigarette smoke/environmental pollutants up-regulate REDOX-sensitive transcription factors, induction of autophagy and unfolded protein response in the clinical manifestation of COPD. Hence, development of antioxidants/thiol agents or other pharmacological agents such as SOD enzyme mimetics to boost the endogenous antioxidant system could be promising to ameliorate chronic inflammatory and injurious responses in COPD.

Regulation of Respiratory System: The Nobel Prize in Physiology or Medicine 1938 was awarded to Corneille Heymans for elucidating the role played by the sinus and aortic mechanisms in the regulation of respiration.

Cigarette smoke is the major risk factor for the development of COPD. It is likely to account for ~80%–90% of COPD cases in USA. Cigarette smoke contains an estimated 10^{15}–10^{17} oxidants/free radicals and ~4,700 different chemical compounds, including reactive aldehydes and quinines – per puff. Given that cigarette smoke contains millions of injurious agents, its pathology cannot be stringently studied with any one compound at a time. Aside from nicotine, heavy metals, and carcinogens; tobacco smoke leads to a significant exposure to oxidants. These include alkyl, alkoyl, and peroxyl organic free radicals that cause lipid peroxidation; unsaturated aldehydes (which cause protein carbonylation and loss of sulfhydryls); $O_2^{\bullet-}$, nitric dioxide (NO_2), and ·NO, which generate peroxynitrite ($ONOO^-$). Tobacco smoke causes airway inflammatory responses within minutes or hours of exposure. One of the earliest manifestations is a breach in the vascular/airway barrier function, with recruitment of circulating inflammatory cells into the lung. Oxidants present in the cigarette smoke trigger NF-κB–dependent inflammatory responses. The acute inflammatory response is transient in nature and mediated by NF-κB, likely counteracted by regulatory networks that dampen NF-κB–dependent responses. Cigarette smoke activates inducible nitric oxide synthase (iNOS), leading to generation of oxidants, such as peroxynitrite ($ONOO^-$), which has been linked to alveolar injury.

Lung function in bartenders improved after legislative ban of smoking in public places; the cohort with preexisting asthma or rhinitis had the greatest increase of lung function after ban of smoking. Environmental tobacco smoke exposure of children related to parental smoking is associated with poor lung functions in asthmatic children; the relative risk of asthma is greater in children exposed to cigarette smoking by both parents compared with smoking of neither parent.

4.4.3. Oxidative Stress and Asthma

Asthma is a chronic inflammatory disease of the lower airways, characterized by reversible airway obstruction and airway hyper-sensitivity. The clinical feature of asthma includes airway inflammation that results due to epithelial cell desquamation, mucus production, and airway remodeling. Inflammatory cells in the airway include mast cells, eosinophils, lymphocytes, activated monocytes, macrophages, and neutrophils. A complex interaction of cells and proinflammatory mediators are responsible for the pathogenesis of asthma. Among these mediators, there is an overwhelming evidence that the airway inflammation is defined by alterations of the airway REDOX and its disequilibrium (reducing state) is a determinant of asthma severity.

The abnormalities in REDOX are magnified in the asthmatic airway in response to exacerbating factors, including microbial infection, exposure to inhaled oxidative pollutants, or allergen triggers in atopic individuals. Increased inflammation and ROS production is characteristic to the pathology of asthma. Asthma attacks and experimental antigen challenges are associated with immediate formation of $O_2^{\bullet-}$ that persists throughout the late asthmatic response. More than two-fold increase in $O_2^{\bullet-}$ generation is observed within 10 min after a local instillation of antigen into

Patho-physiology of Asthma

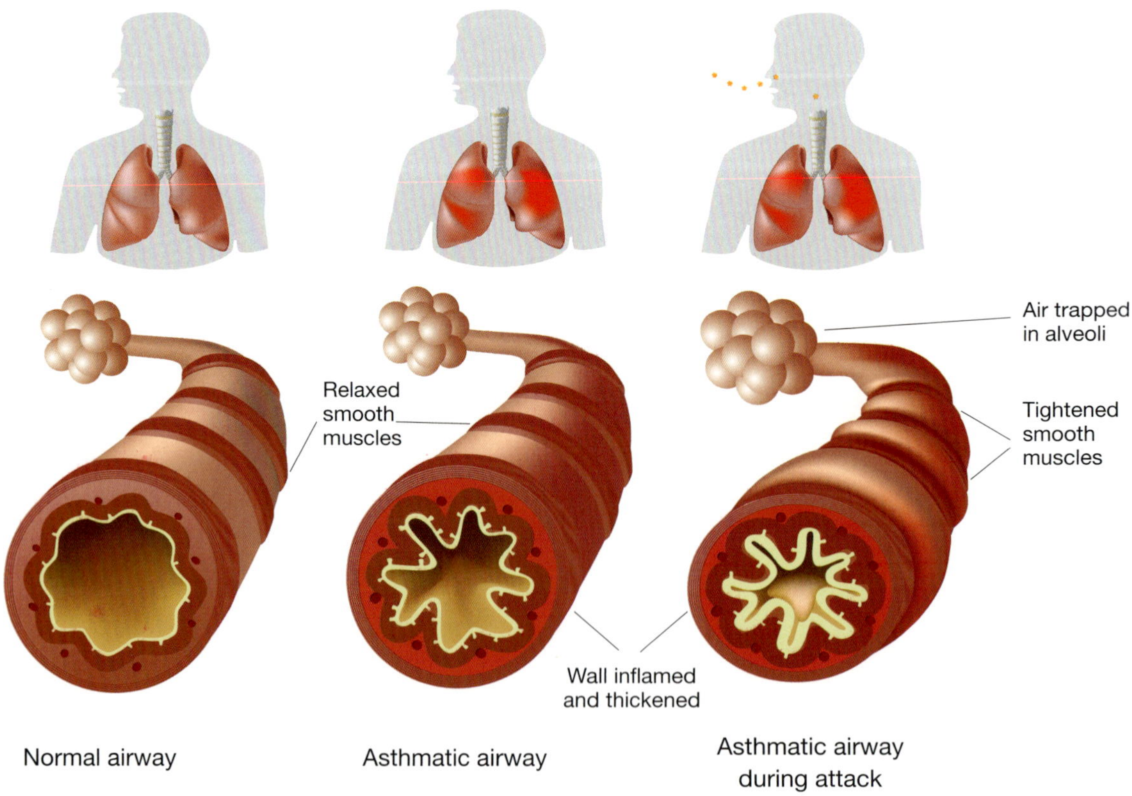

airways of atopic individuals. Spontaneous and experimental allergen-induced asthma attacks lead to leukocyte (eosinophil, neutrophil) activation, during which NADPH oxidase (NOx) is activated and ROS such as $O_2^{\bullet-}$ and its dismutation product, H_2O_2 are rapidly formed. ROS production by neutrophils from asthmatics correlates with severity of reactivity of airways. Myeloperoxidase (MPO)-mediated oxidant modifications also contribute to the patho physiology of severe asthma.

Exogenous oxidants and Asthma: Ozone and diesel exhaust particles have an additive effect on airway hyper-reactivity and inflammation in asthma. Ozone also leads to oxidative modification of surfactant proteins, such as SP-A, which causes the lung to be more susceptible to lipid peroxidation and inflammation resulting in reduced phagocytic activity. Diesel exhaust particles increase the airway hyper-reactivity in asthma. Adult asthmatics, walking for 2 h in a street with only diesel-powered vehicles, show significant reduction in lung function. These changes include an increased MPO activity, 8-isoprostane in sputum, and exhaled breath condensate.

Antioxidant deficiency in Asthma: Both enzymatic and non-enzymatic antioxidants play a major role in lung physiology. The lung epithelial surface lining fluid contains several non-enzymatic antioxidants, such as GSH, ascorbic acid, albumin, and α-tocopherol. Enzymatic antioxidant defenses are present in the epithelial lining fluid as well as in plasma and epithelial cells. Asthma is characterized by loss of one or more of such antioxidant activities. In asthma, SOD enzyme activity is significantly lower in epithelial lining fluid and airway epithelial cells as compared to healthy controls. SOD activity declines within few minutes of an acute asthmatic response as a consequence of the increased oxidative and nitrative stress in the asthmatic airway. Reduction in SOD activity also contributes to apoptosis and shedding of epithelial cells in asthmatic patients. Thus, the REDOX modifications of SOD may contribute to a major component of asthmatic airway remodeling, airway epithelial apoptosis, which leads to removal of the airway surface and predisposes to greater airway hyper-reactivity. Also, catalase activity was found to be 50% lower in broncho-alveolar lavage of asthmatic lungs, as compared to healthy controls.

In contrast to SODs and catalase, extra-celluluar GPx (eGPx) is present at higher than normal levels in lungs of individuals with asthma. Asthma and its exacerbations lead to rapid changes in intra-cellular, as well as, extra-cellular GSH and GSSG. Rapid changes in REDOX potential occur immediately after antigen challenge in epithelial lining fluid of asthmatics. Minutes after challenge, GSH levels drop, and GSSG levels increases in the lung epithelial lining fluid, which confirms the loss of REDOX potential in asthmatic airways. Hence, alterations in the GSSG/GSH ratio and intra-/extra-cellular distribution is likely to contribute to the airway epithelial cell apoptosis in asthma.

REDOX-sensitive molecular targets, such as transcription factors, containing highly conserved cysteine residues, play a crucial role in cell signaling cascade. Nuclear factor-kB (NF-kB) is activated in epithelial cells and inflammatory cells during oxidative stress, which regulates several inflammatory genes responsible for the pathogenesis of asthma.

REDOX-based Interventions: The inhaled glucocorticosteroids are the mainstay of therapies for most asthmatics. Oral corticosteroids are administered when the inhaled therapies are inadequate. However, subsets of patients with severe asthma are refractory to treatment or develop significant side effects to medications. An overwhelming research in the respiratory medicine provides a compelling rationale to develop therapeutic strategies for asthma that aim in correcting the REDOX abnormalities.

Nitric Oxide in the Lungs: Individuals with asthma have 3-fold higher than normal ·NO concentrations. Persistently increased ROS and NO in asthma leads to RNS formation, and subsequent oxidation and nitration of proteins (which may cause alterations in protein function) could lead to airway injury/inflammation. On the other hand, NO synthesis can also decrease airway resistance, an effect mediated in part by formation of the endogenous ·NO oxidation product and bronchodilator, the S-nitroso-glutathione (GSNO).

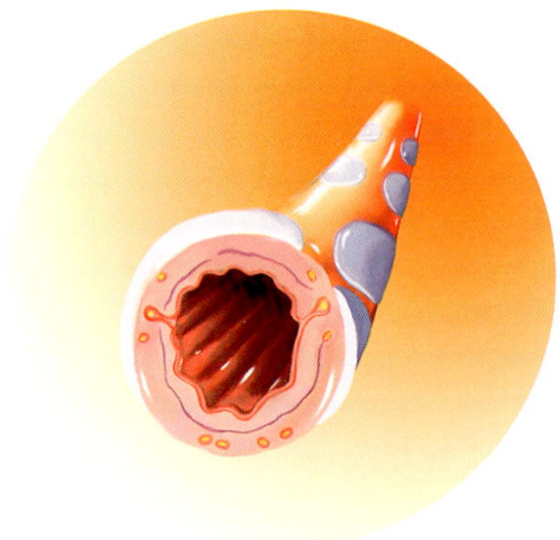

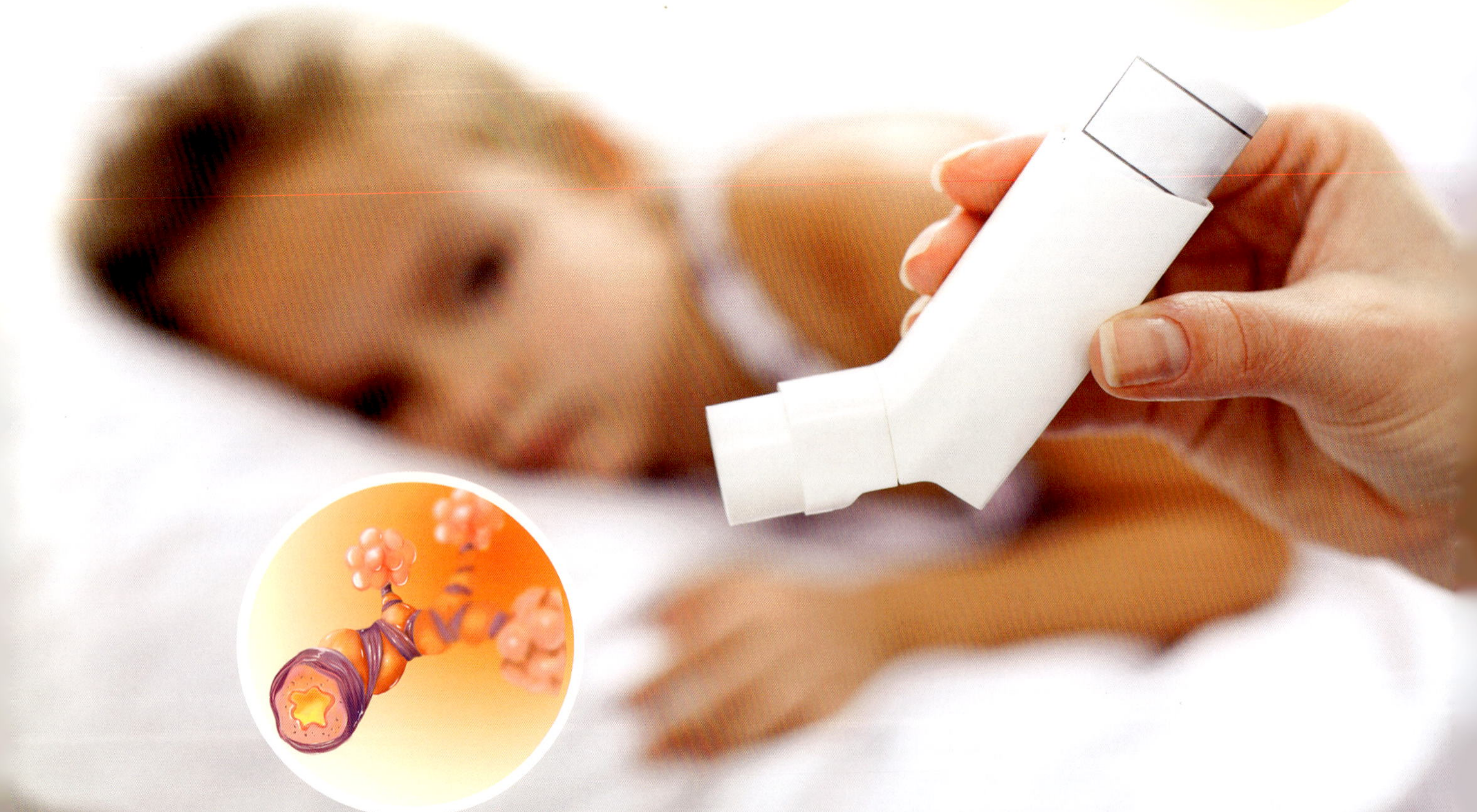

4.4.4. Oxidative Stress and Cystic Fibrosis (CF)

Cystic Fibrosis (CF) is a progressive autosomal recessive disease resulting from mutations in the CF transmembrane conductance regulator gene (CFTR). CFTR is a chloride channel regulator and transporter of GSH; thus, CFTR regulates the pulmonary REDOX status. More than 1,800 mutations have been identified resulting in a defective CFTR protein. The physiological consequences of CF include decreased respiratory tract chloride conductance and increased water resorption, resulting in elevated mucus viscosity and compromised mucociliary clearance. Due to these abnormalities, patients with CF are highly susceptible to respiratory tract colonization and infection, decreased host defense, and overly exuberant inflammatory responses. The magnitude of neutrophilic responses and its accompanying proteolytic and oxidative processes play a major role in the progressive destruction of lung tissues that characterize CF.

Early, massive, and sustained inflammation by neutrophils, is a hallmark of CF airway disease. In particular, human neutrophil elastase (HNE) destroys the elastic fabric of airways. Also, neutrophil-derived ROS oxidize DNA, lipids, proteins and spill actin strings into mucus secretions that increase viscosity; a characteristic of CF airway disease deriving its other common name, the 'mucoviscidosi'. CF is further characterized by endo-bronchial infection, which is often intermittent in early life, but becomes persistent in older patients (particularly after infected with *Pseudomonas aeruginosa*). The endobronchial infection is accompanied by an intense neutrophil dominated inflammatory response and oxidative damage.

During patho-physiological conditions, activated neutrophils and epithelial cells release ROS and RNS, in order to combat and eliminate invasive pathogens. However, these defense systems are ineffective in CF. On the other hand, an increased production of ROS leads to cellular dysfunction and rapid progression of the disease. Nitric oxide (·NO) plays a dual role under such inflammatory conditions. On one hand, ·NO acts as a broncho-dilator that modulates immune responses, antimicrobial activities, and REDOX signaling. One the other hand, reactions of ·NO with ROS can originate intermediate reactive species with deleterious effects on the lung. Despite intense inflammation and up-regulated immune responses, the expired NO (eNO) levels are lower in CF patients compared to those measured in healthy individuals.

Submucosal gland serous cell in the lung maintains the sterile environment within the airways. Serous cell secretions contain potent antimicrobial agents such as lysozyme, lactoferrin, and protease inhibitors that play a multitude of roles in primary host defense. Serous cells also control glandular secretion of salt and water, critical for mucus hydration and required for an efficient mucociliary clearance. The pulmonary serous cell is also the principal site of CFTR expression. Any dysfunction of CFTR leads to the pathogenesis of CF.

Indeed, eNO levels are even lower in patients with more severe CF condition. The mechanisms leading to decreased levels of eNO in CF are due to limited NOS in the respiratory tract. The decline in ·NO bioavailability at lung epithelial cell surface may contribute to compromised antimicrobial host defense and reduced levels of endogenous bronchodilator, S-nitrosothiols, and S-nitroso-GSH (GSNO). Although secretory lactoperoxidase (LPO), H_2O_2 generating systems, and thiocyanate are known to play a cumulative role in the airway; these antimicrobial systems may be defective (dysfunctional) in CF. Early bacterial colonization/infection in the CF respiratory tract (e.g. *Pseudomonas aeruginosa*) results in exuberant phagocytic response releasing large amounts of neutrophilic myeloperoxidase (MPO) in airway secretions. In the presence of H_2O_2 arising from airway epithelium and neutrophil NADPH oxidases (NOx) and the indwelling bacteria, the luminal MPO may further compromise the ·NO generation. GSH is found in extra-cellular fluids, in lung and in cells at high levels in its reduced (GSSG) form. Extra-cellular GSH neutralizes free radicals produced by neutrophils during inflammation. Taking into consideration that CFTR permits the transport of GSH between cells and apical extra-cellular media, possibly the intra-cellular GSH content may be altered in CF. The level of GSH in plasma of CF patients is also reduced, which suggests that GSH deficiency is not limited to the site of inflammation but is rather a systemic condition.

Cystic Fibrosis

Cystic fibrosis is an inherited chronic disease that affects the lungs and digestive system of about 30,000 children and adults in the United States (70,000 worldwide).

- About 1,000 new cases of CF are diagnosed each year.
- More than 70% of patients are diagnosed with CF by age two.
- More than 45% of the CF patient population is age 18 or older.
- The predicted median age of survival for a person with CF is upto the late 30s.

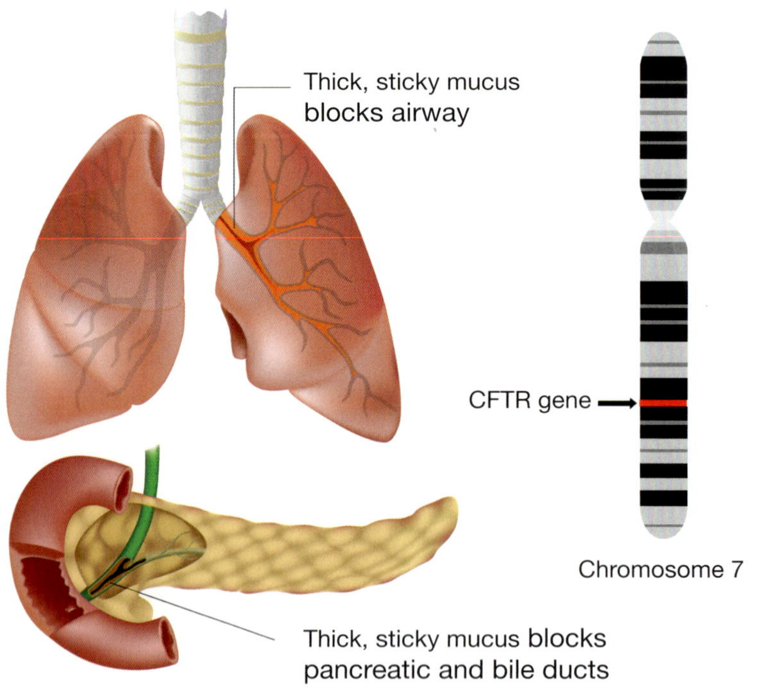

REDOX-based Interventions: Natural antioxidants represent potential therapy in CF management. In particular, flavonoids and other bioactive compounds, with large spectrum antioxidant potential, could affect different stages of CF pathogenesis. For instance, several phytochemical extracts could decrease secretion of inflammatory cytokines from CF cells. Others may correct defective electrolyte transport in CF airways acting in parallel with second messengers and channel activators. Curcumin is a corrector of the CFTR defect by eliciting increased CFTR traffic towards the plasma membrane. Nebulized buffered GSH or the combination of oral GSH and inhaled buffered GSSG attenuates CF disease, which suggests that new approaches against altered cellular REDOX status may represent potential treatments for CF.

Pseudomonas aeruginosa is the major pathogen in the CF lung. Prevalence is high and, once acquired, chronic infection will always ensue. The underlying molecular defects in CF explains the high rate of prevalence of this pathogen. These include abnormalities of airway surface liquid leading to impaired mucociliary clearance or malfunction of antibacterial peptides, increased availability of bacterial receptors, reduced ingestion of pathogens by CF cells and impaired defense related to low levels of molecules such as ·NO or GSH. After colonization in the CF airway, pseudomonads have a wide array of virulence factors that enhance their survival and allow them to evade host defenses and antibiotic agents successfully. Conversion to mucoidy forms and the formation of biofilms are two of the main mechanisms by which this bacterium achieves antimicrobial resistance in the respiratory tract.

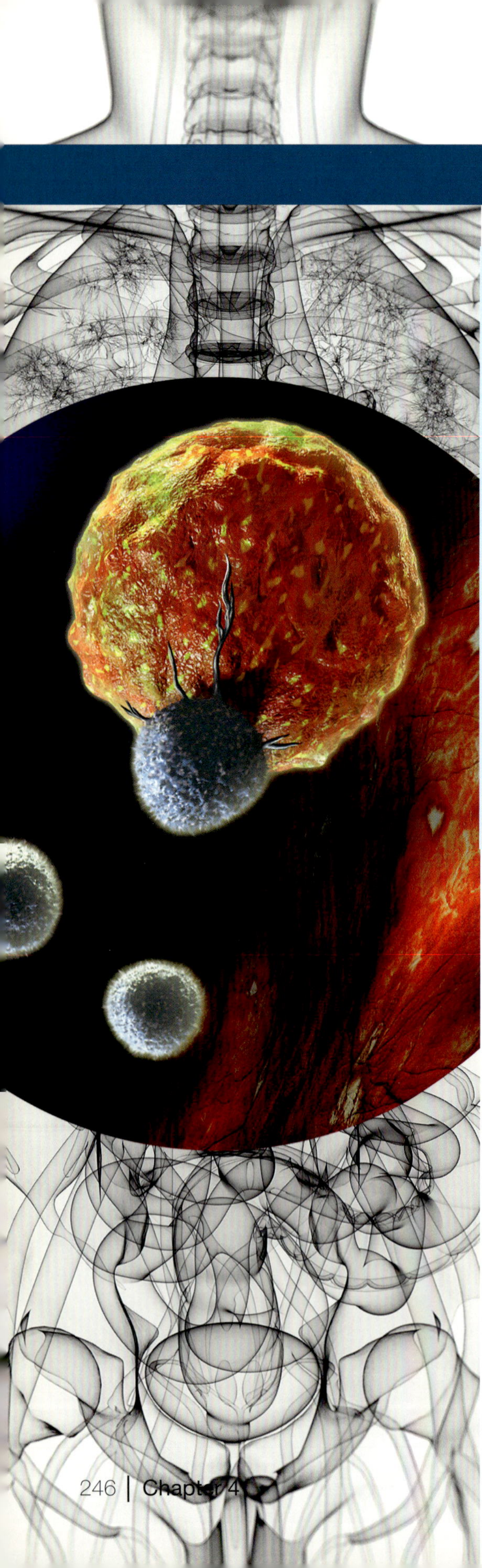

4.5 **REDOX:** The Immune System

Life is a 'self operating program' designed to regulate entry and exit of chemicals into its open frame to sustain a defined organic cellular mass. The ability of this open frame to distinguish its self sustained mass from non-self intrusion is the governing principle for every life form. 'Self' includes chemical structures such as tissue, cells, molecules etc., inherent to the body and 'Non-Self' refers to entities that are encountered by the body through a foreign source. This fundamental distinction for life has necessitated the evolution of a vital 'self defense program' – the immune system. Since life is an open system based on uninterrupted chemical flow, the efficiency of the self operation depends on the ability of the immune system to also distinguish harmful non-self (i.e. bacteria, viruses and toxins) from useful non-self (i.e. food, water and air). The overall success of the self program is contingent upon the ability of the immune system to recognize myriads of foreign bodies. The system also needs to detect and react to harmful organisms that the body has never encountered before.

Immune system is based on a multi-layered arrangement of membranes, cells, and molecules that collectively perform the defensive act – lethal enough to eliminate harmful microbes or cancer cells, yet specific enough not to cause any collateral damage to 'self'. Immunity and host defense are similar to an 'arms race'. An offensive intruder may have an ability to multiply at a rate, several-fold faster than the immune cells. Accordingly, the immune cells should quickly respond to such threats by shifting from a dormant phase to a highly active metabolic state within a short period of time for the combat. Therefore, the immune system is designed with a unique computational program to operate at four strategic levels of host defense:

i) Tolerance to Self: Generally the immune system is tolerant to self and does not attack itself. Tolerance is implemented by a class of lymphocytes, called helper T-cells that mature in thymus, an organ located just behind the breastbone. Self and non-self recognition in human body is also achieved by a cellular marker based on the major histocompatibility complex (MHC). Any cell not displaying this marker is treated as non-self, attacked and eliminated. Seldom the tolerance control may break down and the immune system could attack self-cells, resulting in autoimmune diseases (i.e. rheumatoid arthritis, multiple sclerosis, etc).

ii) Recognition of Non-self: Recognition of non-self by the immune system is based on the detection of tiny fragments from a non-self entity. These specific tiny fragments are known as antigens. Helper lymphocytes are again the immune cells that are responsible for the detection of antigens.

iii) Processing of Non-self: As soon as the non-self is identified, innate defense, comprised of a pre-existing basic level security force, get into immediate action. Simultaneously, a more sophisticated line of defense at another level known as 'acquired' immune response may swing into full action to eliminate foreign threat. The 'innate' defense involves roaming scavenger cells, such as phagocytes, that ingest extra-cellular molecules and substances, clearing the system of both debris and pathogens. The acquired immune response involves a variety of cells, chemicals and molecules. It results in secretion of soluble form of specific receptors, the antibodies, which precisely bind to antigens and remove foreign substances or inactivate pathogens.

Helper T Cell Activation

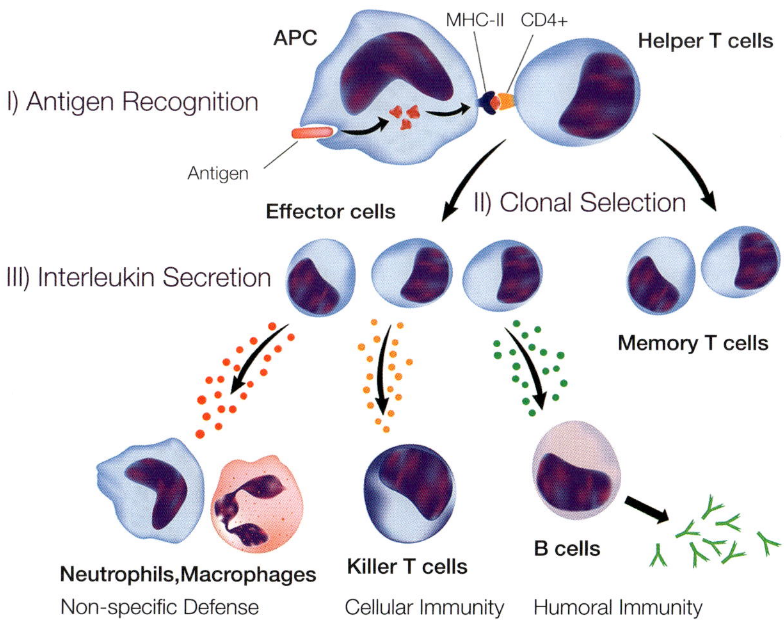

Immune System and Antibodies: The Nobel Prizes in Physiology or Medicine
1919 – Jules Bordet for his discoveries relating to immunity.
1972 – Gerald Edelman and Rodney Porter for discovering the chemical structure of antibodies.

iv) Development of Memory: The immune system can learn and memorize the structures of foreign agents that it comes across. Such memory-induced immune responses are significantly rapid and effectively protect the body from harmful foreign alerts.

Immunity is fundamental for the prevention or elimination of foreign invasion. It is also important for the overall repair and recovery of the host from pathogenic encounters. Regular maintenance of the immune system includes an uninterrupted supply of nutrients for the body to produce immune cells, immunoglobulins and other plasma proteins, and to replenish those that were expended during a routine host defense. This innate process of biosynthesis is collectively known as immune replenishment.

Cytotoxic T Cell Activation

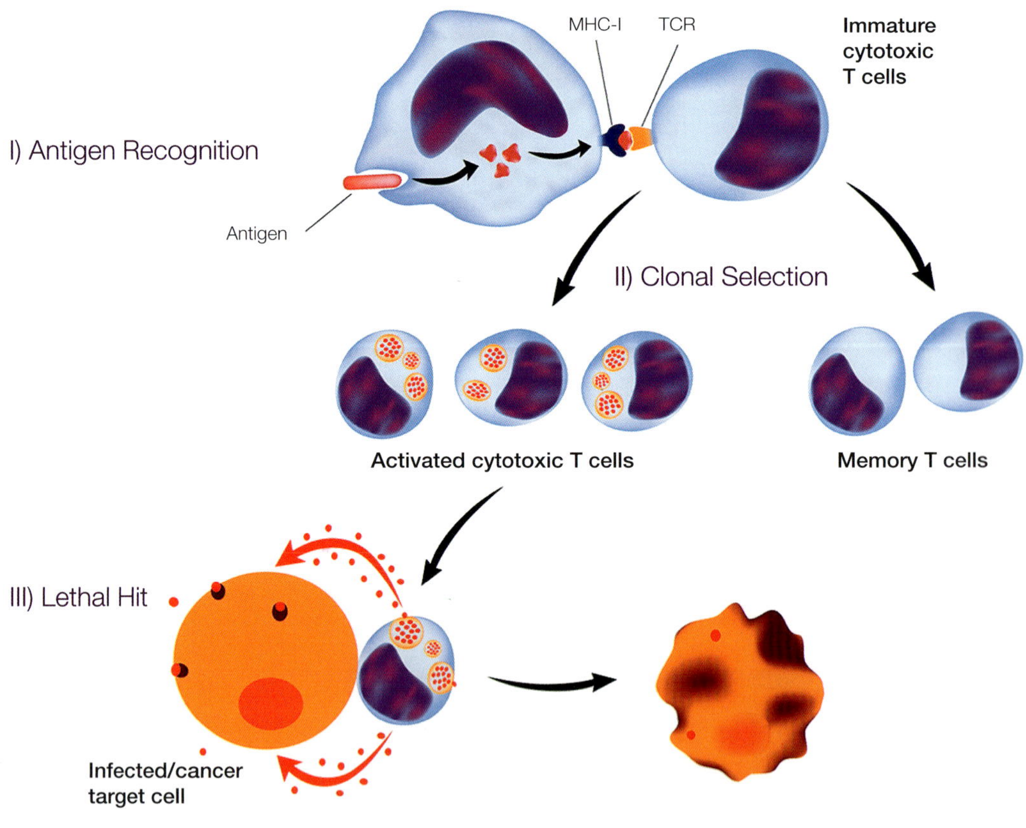

Cell Mediated Immunity: The Nobel Prizes in Physiology or Medicine
1980 – Baruj Benacerraf, Jean Dausset and George Snell for genetic determination of structures on cell surface that regulate immunological reactions.
1996 – Peter Doherty and Rolf Zinkernagel for discovering the specificity of the cell mediated immunity.

Mounting any type of immune response triggers a huge metabolic and nutritional demand on the host physiology. Accordingly, the immune replenishment should meet the energy demand to fuel the host emergency, the sudden process of antigen recognition, as well as, fulfill the nutritional requirements to support the immune response on alert. Life has evolved with a conservative energy budget for its survival, growth and multiplication. During an immune response, however, balancing an energy budget becomes a challenge – after all, it is a question of life or death. Immunity requires substantial energy to trigger an effective immune response. Severity, type, and duration of infection; also gender, age and nutritional status of the host; collectively influence the cost of mounting an immune response. Dietary components can regulate physiological functions of the body. Interaction with the immune response is one of the most vital aspects of nutrient function. The interaction involving nutrition and immunity and ability of the host to deal with infectious agents is a strategic determinant of immune replenishment.

Interestingly, metabolism and immunity share a common purpose; both processes must engulf other particles, however, metabolism does it for energy and immunity does it to stop harmful invaders!

TABLE: *Specific Cells involved in Cellular Immunity*

Designation	Cell Types	Functions
T helper cells	CD4+ T lymphocytes	Secretion of cytokines
Antigen Presenting cells	Mononuclear phagocytes, DC, B cells, Langerhans cells	Processing & presentation of foreign antigens to T_h lymphocytes
Activated Macrophages	Differentiated & stimulated monocytes	Microbicidal & tumoricidal; stimulation of inflammation; production of growth factors & cytokines
T killer cells	CD8+ T lymphocytes	Lysis of virus infected cells or associated with allografts and tumors
Natural Killer (NK) cells	Large granular lymphocytes	Non-specific killing of tumor cells & some infected cells
Killer (K) cells	NK cells with Fc receptors	Antibody-dependent cell killing

Immunoglobulins (Ig) are glycoproteins produced by plasma cells in response to an immunogen (antigen) and which function as antibodies. Igs derive their name from its characteristic migration with globular proteins when antibody-containing serum is placed in an electrical field. Igs bind specifically to one or a few closely related antigens. Igs mediate a variety of functions: (i) *Complement fixation* – which results in cell lysis and release of biologically active molecules. (ii) *Binding to various cell types* - Phagocytic cells, lymphocytes, platelets, mast cells, and basophils have receptors that bind immunoglobulins. This binding can activate the cells to perform specific functions.

Serum Proteins: The Nobel Prize in Chemistry 1948 was awarded to Arne Tiselius for his research on electrophoresis and adsorption analysis, especially for his discoveries concerning the complex nature of the serum proteins.

TABLE: *Immune Cytokines and Their Cellular Activities*

Cytokine	Producing Cell	Target Cell	Function
GM-CSF	Th cells	Progenitor cells	Growth and differentiation of monocytes & DC
IL-1 α IL-1 β	Monocytes Macrophages B cells DC	Th cells	Co-stimulation
		B cells	Maturation and proliferation
		NK cells	Activation
IL-2	Th_1 cells	Activated T & B cells, NK cells	Growth, proliferation, activation
IL-3	Th cells NK cells	Stem cells	Growth and differentiation
		Mast cells	Growth and histamine release
IL-4	Th_2 cells	Activated B cells	Proliferation/differentiation IgG_1/IgE synthesis
		Macrophages	MHC Class II
		T cells	Proliferation
IL-5	Th_2 cells	Activated B cells	Proliferation and differentiation IgA synthesis
IL-6	Monocytes Macrophages Th_2 cells Stromal cells	Activated B cells	Differentiation into plasma cells
		Plasma cells	Antibody secretion
		Stem cells	Differentiation
IL-7	Marrow Stroma Thymus Stroma	Stem cells	Differentiation into progenitor B and T cells
IL-8	Macrophages Endothelial cells	Neutrophils	Chemotaxis
IL-10	Th_2 cells	Macrophages	Cytokine production
		B cells	Activation
IL-12	Macrophages B cells	Activated Tc cells	Differentiation into CTL (with IL-2)
		NK cells	Activation
IFN-α	Leukocytes	Various	Viral replication, MHC I expression
IFN-β	Fibroblasts	Various	Viral replication, MHC I expression
IFN-γ	Th_1 cells Tc cells NK cells	Monocytes, Macrophages	Chemotaxis
		Activated Macrophages	IL-1 synthesis
		Activated B cells	IgA synthesis
MIP-1α	Macrophages	Monocytes, T cells	Chemotaxis
MIP-1β	Lymphocytes	Monocytes, T cells	Chemotaxis
TGF-β	T cells Monocytes	Monocytes, macrophages	Chemotaxis
		Activated macrophages	IL-1 synthesis
		Activated B cells	IgA synthesis
TNFα	Macrophages NK cells	Macrophages	CAM and cytokine expression
		Tumor cells	Apoptosis
TNF-β	Th_1 and Tc cells	Phagocytes	Phagocytosis, NO production
		Tumor cells	Apoptosis

4.5.1. REDOX and Immune Regulation

REDOX signaling plays a key regulatory role in the adaptive immune system, especially with several implications in T cell biology. The T cell activation and proliferation require a reducing milieu. Dendritic cells (DCs – from the antigen presenting cellular cascade), provide the extra-cellular REDOX (a reductive milieu) in the immune synapse for activation and proliferation of T cells. The T cells, in turn, activate the NF-κB pathway in DCs to elicit an antioxidant response. They also increase cystine uptake, which increases glutathione (GSH) synthesis/ export and finally degradation of GSH to cysteine outside the cell. Extra-cellular accumulation of cysteine supports GSH synthesis in T cells and in creating REDOX (more reductive-type) potential to promote T cell proliferation. In addition to extra-cellular cysteine secretion, the interaction between DCs and T cells also results in appearance of extra-cellular thioredoxin (Trx)1. Trx1 interacts with the TNF-α receptor in a REDOX-sensitive manner and triggers proinflammatory effects by stimulating cytokine release and lymphocyte proliferation.

These sequence of events demonstrate a dynamic interplay between the intra- and extra-cellular compartments to maintain REDOX homeostasis by two independent REDOX nodes (GSH and cysteine). The extra-cellular cysteine/ cystine REDOX potential for DCs in culture is about -80 mV, a value consistent for cells that experience growth arrest ('stasis'). Naive T cells that did not receive activation signals would undergo death ('apoptosis') and exhibit an extra-cellular cysteine/cystine REDOX potential of about -45 mV. In contrast, when naive T cells receive activation signals during co-culture with DCs, a more reducing extra-cellular environment is reflected in a REDOX potential of -110 mV, which is conducive for T cell to divide ('proliferation'). Furthermore, the involvement of DCs and T cells during activation leads to dynamic changes in the REDOX status of surface proteins in both cell types. A 30 mV potential shift is expected to induce ~10-fold change in the reduction:oxidation ratio of cysteine moieties of a protein. Therefore, the kinetic of REDOX during the interaction of DCs and T cells are important for stasis, apoptosis, and proliferation of T cells.

At low levels, ROS (H_2O_2 and $O_2^{\bullet-}$) act as mitogenic factors and their downstream effects mediate changes in protein phosphorylation and/or activation/inhibition of transcription factors. Major sources of ROS include membrane-bound NADPH-dependent oxidase (NOx), lipoxygenase (LOx), and the mitochondrial respiratory chain. However, sustained pro-oxidant conditions inhibit T cell proliferation and promote apoptosis. During activation, increased ROS levels launch an early pro-oxidant response, which relay via signaling pathways in antigen presenting cells and in T cells. This results in activation of protein tyrosine kinases and transcription factors (i.e. NF-κB). The NF-κB pathway regulates expression of various inflammatory genes, cytokines, chemokines, and co-stimulatory molecules. Activated NF-κB also stimulates cellular synthesis of GSH, which could trigger an antioxidant wave. The cumulative effects of these REDOX responses provide an optimum condition for T cell activities.

GSH serves as an important proliferative signal for T lymphocytes and is required for cell cycle progression. GSH is concentrated in the nucleus during the early phase of cell proliferation and becomes more evenly distributed in confluent cells. GSH regulates nuclear protein function via glutathionylation and protects DNA from oxidative damage during the key stage of replication. GSH affects ROS levels in cells, which can either activate or inactivate specific REDOX sensitive targets at cell cycle checkpoints, thereby influencing cellular fate. On the other hand, the REDOX status of secondary lymphoid organs (such as lymph nodes and spleen) are more reduced than non-lymphoid organs. The non-protein thiol content in lymphoid tissues is elevated in response to immunization with DCs, B cells, and macrophages, which might contribute to the reductive remodeling. Reducing microenvironment can protect lymphoid organs from oxidative stress during T cell activation and antibody production; where as, low levels of ROS are highly critical for the onset of the immune response. Thus, REDOX signaling during T cell priming involves dynamic and spatially regulated changes in the intra- and extra-cellular compartments and comprises both small molecules (ROS and REDOX-active metabolites) and proteins.

REDOX modulation as a strategy for immune-regulation is integral part of many diseases. HIV infects and kills $CD4^+$ cells that lead to a significant decrease in functional T cell population in AIDS. HIV infected individuals demonstrate lower cellular and plasma GSH levels compared to healthy controls that correlates with low T cell numbers and deficient function. Administration of N-acetyl cysteine, a cysteine precursor, restores intra-cellular GSH levels and has shown to benefit HIV-infected individuals.

HIV / AIDS Facts

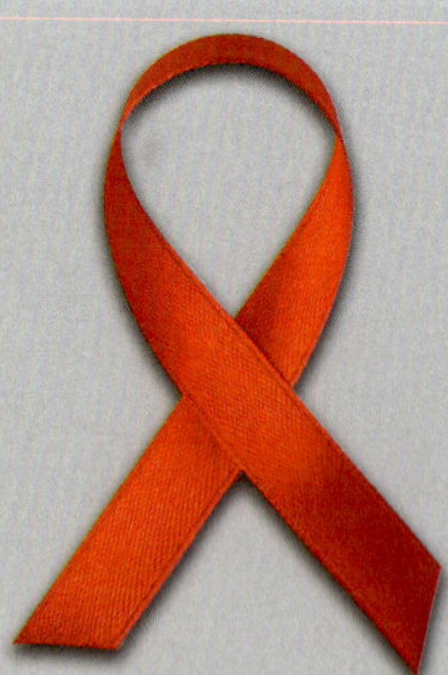

- AIDS refers to the most advanced stages of HIV infection, defined by the occurrence of any of more than 20 opportunistic infections or related cancers.
- 34 million people live with HIV worldwide. The vast majority are in low- and middle-income countries. An estimated 2.5 million people were newly infected with the virus in 2011.
- About 25 million people have died to date. An estimated 1.7 million people died of HIV/AIDS in 2011
- There are some 26 million people who will require access to anti-retroviral drugs under the new 2013 guidelines.
- An estimated 3.34 million children are living with HIV.
- In 2011, approximately 430 000 deaths from tuberculosis occurred among people living with HIV. That is one quarter of the estimated 1.7 million deaths from HIV. The majority of people living with both HIV and TB reside in sub-Saharan Africa (about 79% of cases worldwide).

4.5.2. Oxidative Stress and HIV / AIDS

AIDS is the 6th leading cause of death among people ages 25 to 44 in the United States, down from number one in 1995. Millions of people around the world are living with HIV/AIDS, including many children under age 15.

Human immunodeficiency virus (HIV) causes AIDS. The virus attacks the immune system and leaves the body vulnerable to a variety of life-threatening infections and cancers. During the initial infection, an individual may experience influenza-like illness for a brief period of time - typically followed by a prolonged period without symptoms. As the illness progresses, it interferes more and more with the immune system, making the person more likely to get infections, including opportunistic infections and tumors that do not usually affect people with functional immune systems. Common bacteria, yeast, parasites, and viruses that usually do not cause serious disease in people with healthy immune systems, can cause fatal illnesses in people with AIDS. HIV is a slow replicating retrovirus and its infection occurs by the transfer of blood, semen, vaginal fluid, pre-ejaculate or breast milk. In these body fluids, HIV is present as both free virus particle and as virus within infected immune cells. HIV infects immune cells such as helper T cells (specifically $CD4^+$), macrophages, and dendritic cells. HIV infection triggers a steady decline in $CD4^+$ cell population compromising the cell-mediated immunity, and the body becomes progressively more susceptible to opportunistic infections.

Oxidative Stress and HIV Infection: Oxidative stress plays a pivotal role in HIV-1 pathogenesis, which includes viral replication, inflammatory response, decreased immune cell proliferation, loss of immune function, chronic weight loss and an increased sensitivity to drug toxicity. The disruption of oxidative status contributes to cell damage, which is common during the HIV-1 infection. Monocytes/Macrophages (M/M) serve as viral targets during the whole course of HIV-1 infection. Viral 'Tat' protein induces enhanced ROS production in HIV-infected patients, which in turn may activate nuclear factor κB (NF-κB). This specific HIV-1 regulatory protein also down-regulates the expression and activity of Mn-SOD enzyme (the mitochondrial isoform of SOD), leading to a sustained generation of $O_2^{\bullet-}$ and $ONOO^-$. These ROS are important mediators of HIV-1 replication in M/M. Subsequent activation of $CD4^+$ T-lymphocytes and M/M, during the HIV infection contributes to an increased production of free radicals such as $O_2^{\bullet-}$ anion, preroxynitrite ($ONOO^-$, a by-product of $O_2^{\bullet-}$ and ·NO), and ·OH radical (generated by $ONOO^-$ decomposition). Elevated serum levels of hydroperoxides and malondialdehyde (MDA), which are the markers of ROS-induced lipid peroxidation, have been reported in asymptomatic HIV-infected patients early in the course of the disease. During normal physiology, oxidative stress is effectively controlled by various endogenous antioxidant systems such as SOD, glutathione peroxidase (GPx), xanthine reducto-oxidase (XOR), etc. However, a significant perturbation in the antioxidative status is observed among HIV-infected patients; especially,

HIV / AIDS: The Nobel Prize in Physiology or Medicine 2008 was awarded to Françoise Barré-Sinoussi and Luc Montagnier for their discovery of human immunodeficiency virus (HIV).

decreased synthesis of endogenous glutathione (GSH), and Mn-SOD activities. The REDOX imbalance induced by HIV-1 infection in M/M, with consequent production of toxic factors, is claimed to be the main cause of neuronal damage in AIDS.

Oxidative Stress and Secondary Complications in AIDS: HIV-1 induces a state of chronic oxidative stress, thus, patients demonstrate elevated plasma levels of hydroperoxides and malondialdehyde, indicative of polyunsaturated fatty acid peroxidation. Furthermore, the contribution of other factors such as current anti-retroviral therapy, disease stage, and smoking/drug abuse history could further induce oxidative stress and cardiovascular disease (CVD). The high oxidative stress is a predisposing factor for cardiovascular risk in HIV-1 patients. The HIV-1 protein expression decreases ·NO and causes endothelial dysfunction. Also, diminished antioxidant capacity increases vascular $O_2^{\bullet -}$ levels, which reduce ·NO bioavailability and promote $ONOO^-$ generation. Restoring GSH levels could reverse HIV-1 protein-mediated effects on $O_2^{\bullet -}$, ·NO, and vaso-relaxation.

Oxidative Stress and Highly Active Antiretroviral Therapy (HAART): HAART is currently the therapy of choice for HIV-infected patients. HAART has extended the life expectancy of HIV-1 patients and has significantly reduced viral burden. However, serious HIV-related complications including cardiac and skeletal myopathies, hypertension, hyper-insulinemia, hyper-triglyceridemia, and insulin resistance develop over time. These complications increase the risk of developing diabetes and CVD, which have become an important cause of morbidity in HIV-1 patients. Although much effort focuses on the contribution of HAART, there is increasing clinical evidence that HIV-1 itself increases the risk of developing CVD. Therefore, long-term HAART appears to be of limited use in many patients due to severe side effects and/or regimen adherence issues. Furthermore, poor adherence to HAART can have serious consequences, including loss of serum HIV suppression, development of drug-resistant HIV strains, and increased probability of illness progression.

HAART could increase oxidative stress in circulation through generation of ROS/RNS and more oxidized metabolites that derive from their interaction with infected cells. This process is supported by several oxidative mechanisms, such as mitochondrial interference, following treatment with HAART-NRTI (Nucleoside Reverse Transcriptase Inhibitors), and activation of the P450 hepatic system by HAART, when combined with protease inhibitors. Insufficient intake or malabsorption of nutrients (antioxidants, in particular), which is common in HIV patients, may further worsen the oxidative stress. It has been suggested that immune system reactivation may, itself, be a direct cause of oxidative stress in HAART, when combining with non-NRTI drug (i.e. 'Efavirenz'). Increased serum isoprostane levels, an oxidative metabolic byproduct of arachidonic acid, usually observed in smokers, has also been reported in HIV patients taking efavirenz.

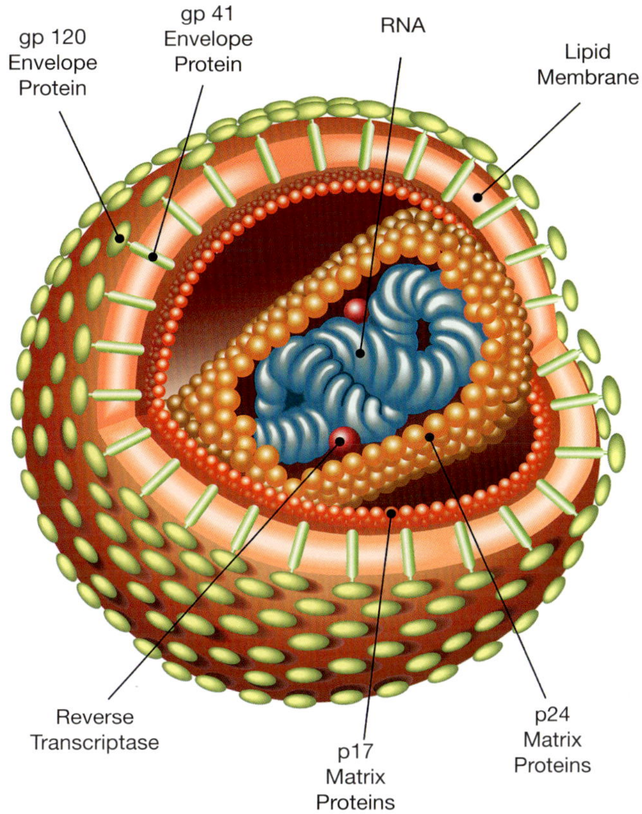

Anatomy of the AIDS Virus

HIV Infection and Propagation

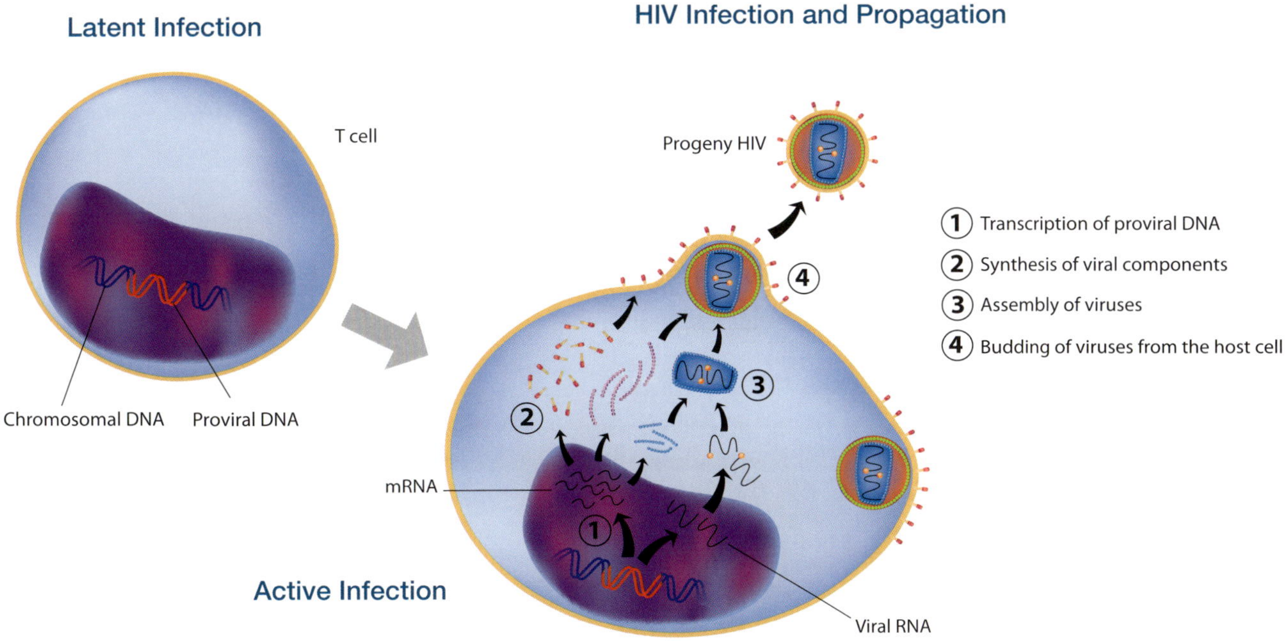

4.5.3. Oxidative Stress and Sepsis

Sepsis is a complex inflammatory host response to microbial infection which frequently occurs after trauma, hemorrhage, burn, or surgery, and manifests as 'Systemic Inflammatory Response Syndrome (SIRS)'. Without an effective clinical management, sepsis often leads to multiple organ dysfunction (or failure) and ultimately causes death. SIRS and sepsis remain as the prime causes of death in intensive care units worldwide, with mortality rates ranging between 30% and 70%. The problem becomes even more serious because the frequency of sepsis is increasing with the emergence of antibiotic-resistant pathogens.

Sepsis is initiated by the release of endotoxins [bacterial lipopolysaccharides (LPS)] from Gram-negative and Gram-positive pathogens. LPS activates inflammatory cells, such as neutrophils, monocytes/ macrophages and lymphocytes. LPS also triggers cellular and humoral aspects of the inflammatory immune response. It is not the bacterial infection, but rather the inflammatory response to the infection that is the critical determinant of outcome in sepsis. Immune system during sepsis has two phases: an initial hyper-inflammatory stage followed by a prolonged hypo-inflammatory phase. The onset of sepsis is characterized by a complex systemic inflammatory response, endothelial dysfunction, and alterations in the coagulation system. This onset leads to perturbations in the delivery of O_2 and metabolic substrates to the tissues, end-organ dysfunction, and ultimately death.

Oxidative stress plays a crucial role as both a promoter and mediator of the systemic inflammatory response. During sepsis, endothelial cells produce ROS at accelerated rates during sepsis. ROS from activated neutrophils act as paracrine agents to alter endothelial cell function. Additionally, endothelial cells increase intra-cellular ROS production.

Oxidative burst and Sepsis: One of the key features of sepsis is tissue infiltration by phagocytic cells. In this scenario, neutrophils, monocytes, and macrophages respond to septic stimulation by producing ROS (i.e. $O_2^{\bullet-}$ and H_2O_2) and RNS (i.e. ·NO). Also, neutrophils release granular enzymes (i.e. elastase, cathepsin) and the myeloperoxidase (MPO)-derived oxidant, hypochlorous acid (HOCl). All these components may contribute to the killing of bacterial pathogens. However, if produced in excess during SIRS and sepsis, the ROS, RNS and proteolytic enzymes cause micro-vascular dysfunction, and subsequently organ failure.

Hypoxia and Sepsis: Tissue hypoxia, the failure to adequately supply O_2 plays a central role in organ dysfunction. It has been postulated that impaired cellular O_2 utilization, rather than inadequate O_2 delivery, contributes to the development of sepsis. An association between mortality in sepsis and mitochondrial dysfunction has been reported. In addition, a decrease in the number of mitochondria has been observed in patients with prolonged septic conditions, implying that sepsis-related dysfunction of mitochondria may result in mitoptosis.

Microvascular Dysfunction and Sepsis: Over-expression of iNOS occurs early in sepsis, almost in every vital organ. Patients with sepsis demonstrate increased ·NO production and elevated nitrite and nitrate levels, which correlates with the severity of illness. Due to increased ·NO levels in plasma, hemoglobin converts to methemoglobin (Met-Hb), resulting in increased rigidity of red blood cells (RBCs) and RBC lysis. In turn, RBC lysis leads to free Met-Hb levels in the plasma with effects on sepsis development: (i) it leads to increased availability of iron to pathogens, which may fuel their proliferation; (ii) it has pro-inflammatory effects by stimulating interleukins (IL-6 and IL-8); and (iii) Met-Hb is an activator of NF-κB, which may lead to further iNOS expression, thus closing a ·NO-generating loop.

Mitochondrial dysfunction represents the fourth and final major REDOX event in the pathogenesis of sepsis. Patients typically die of sepsis within days to weeks after the onset of the initial infection and in some cases even after eradicating the original infection. During later stages of sepsis, the immune system turns into a hypoactive mode. The plasma oxidative burst can no longer be directly responsible for cell/organ dysfunction. However, the infection could activate a self-sustaining and self-promoting REDOX cycle under special conditions. Instead of pathogen removal, the 'sepsis REDOX cycle' continues to drain energy and prevents normalization of the body due to several positive feedback loops, even in the absence of ROS from plasma. During the 'sepsis REDOX cycle' H_2O_2 crosses both cellular and mitochondrial membranes. The patho-physiological REDOX profile of one cell can spill over to surrounding tissue, due to the ability of H_2O_2 and ·NO to cross membranes, thus leading to organ failure.

REDOX-based Interventions: Sepsis is clinically managed by controlling the source of infection with antimicrobial therapy, hemodynamic support with fluid resuscitation, vasopressor drugs, sedation or analgesia (as needed), and by other supportive therapies. Despite these clinical interventions, about one-third of all septic patients die before leaving hospital. Therefore, development of novel treatment approaches, such as REDOX-based interventions is of immense priority. REDOX approach in sepsis treatment should be targeted to specific pathways in the pathogenesis, such as the dysfunctional mitochondria. Targeted delivery of antioxidants into intra-cellular space to maintain normal function of signaling pathways that involve reactive species is a desirable approach.

Potential candidates for REDOX-based interactions are fructose 1,6-(bis) phosphate and ethyl pyruvate, which act as both antioxidants and energy molecules. These compounds have already shown positive effects in sepsis models. An alternative antioxidative approach is to suppress ROS production in mitochondria. Maintenance of glucose in blood within the range 4.4 to 6.1 mM (instead of the normal range of 10.0 to 11.1 mM), could decrease mortality by more than 10%. Therapeutic strategies capable of decreasing the level of ·NO could be beneficial in sepsis. The supplementation of selective iNOS inhibitors, such as aminoguanide, has positive effects in different sepsis models.

The Sepsis REDOX Cycle

Pathogen-activated immune systems, can excessively generate H_2O_2 and HOCl. Inside the cell, H_2O_2 provokes the activation of NF-κB. These events result in the production of ·NO in micro-molar concentrations. ·NO provokes inhibition of the electron transfer chain (ETC), which leads to increased production of $O_2^{\bullet -}$. The interaction between ·NO and $O_2^{\bullet -}$ leads to the generation of peroxynitrite ($ONOO^-$), which is protonated to form peroxynitrous acid (ONOOH) that spontaneously decomposes to two highly reactive species, ·OH and $\cdot NO_2$. These reactive species damage mitochondria, and in cooperation with ETC inhibition, provoke mitochondrial dysfunction resulting in ATP decline. Superoxide is also produced in the cytosol via increased activities of three enzymes: NADPH oxidase (NOx), cyclooxygenase (COX)-2, and xanthine oxido-reductase (XOR). Superoxide is dismutated in mitochondria by manganese superoxide dismutase (Mn-SOD) to H_2O_2, which closes two positive feedback REDOX loops. Intra-cellular ·NO production in excess leads to its leakage into the plasma and provokes RBC lysis. On the other hand, HOCl provokes pore formation in RBC membranes and release its Met-Hb. This results in elevated iron levels in plasma, which could fuel pathogen proliferation. Also, Met-Hb activates NF-κB, which closes the ·NO-generating loop. The uncoupling of ETC during severe sepsis and septic shock is most pronounced feature among non-survivors.

The anatomic site of infection predicts mortality among patients diagnosed with septic shock.

- Disseminated infections, 84.5%
- Spontaneous bacterial peritonitis, 76%
- Toxic megacolon, 68.3%
- Other abdominal infections, 66.7%
- Pulmonary infections, 54%
- Pyelonephritis, 34.5%
- Enterocolitis/diverticulosis, 28%

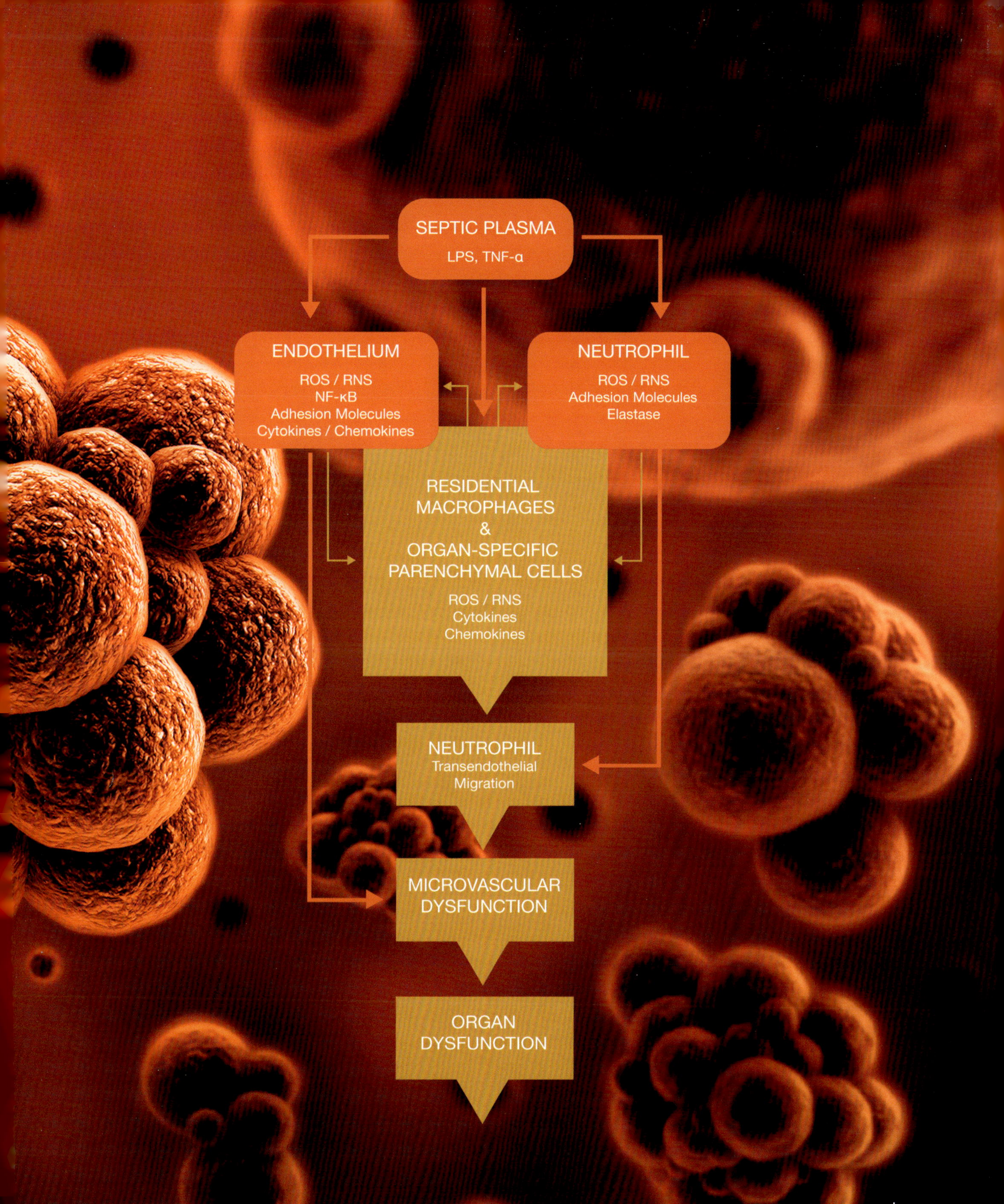
SEPTIC PLASMA
LPS, TNF-α
ENDOTHELIUM
ROS / RNS
NF-κB
Adhesion Molecules
Cytokines / Chemokines
NEUTROPHIL
ROS / RNS
Adhesion Molecules
Elastase
RESIDENTIAL MACROPHAGES & ORGAN-SPECIFIC PARENCHYMAL CELLS
ROS / RNS
Cytokines
Chemokines
NEUTROPHIL
Transendothelial Migration
MICROVASCULAR DYSFUNCTION
ORGAN DYSFUNCTION

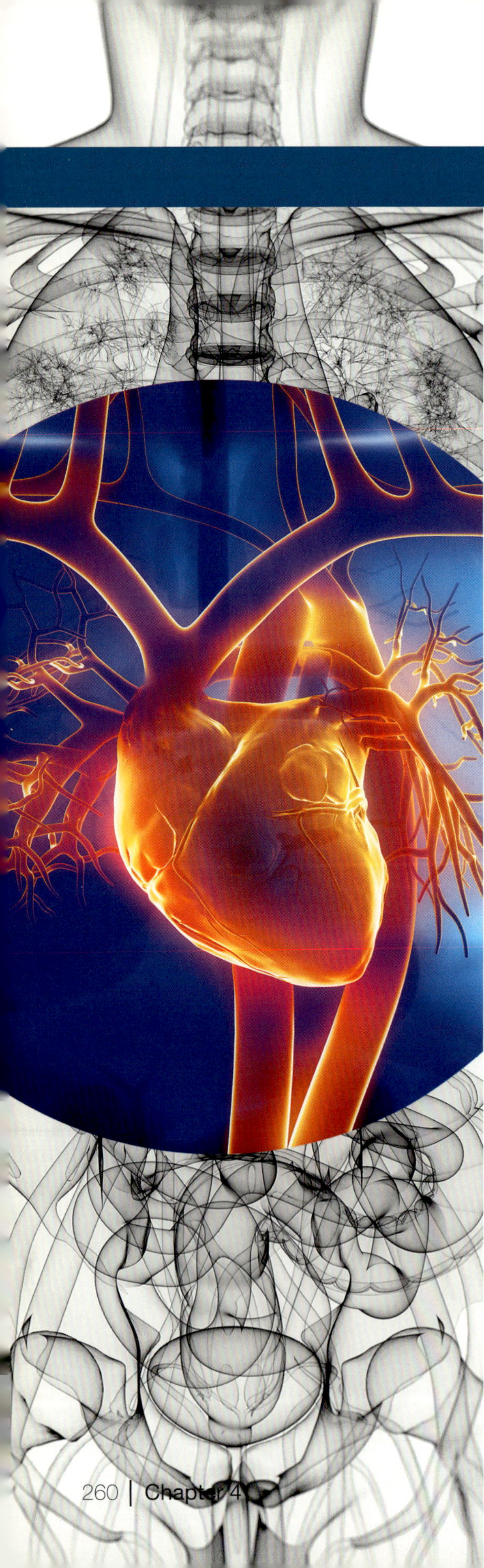

4.6 REDOX: The Circulatory System

The Circulatory (or cardiovascular, CV) system consists of the heart and a closed system of blood vessels, veins, arteries, and capillaries. They transport O_2 and nutrients to organs and tissues throughout the body and dispose of the waste products. Among its vital functions, the CV system increases the flow of blood to meet increased energy demands during exercise and regulates body temperature. Also, when foreign substances or organisms invade the body, this system swiftly conveys disease-fighting elements, such as white blood cells and antibodies, to regions under attack. Also, in the case of injury or bleeding, the system sends clotting cells and proteins to the affected site to stop bleeding and promote healing.

The cardio system works in tandem with the endocrine system, which is a collection of hormone-producing glands. These glands release chemical messengers, called hormones, directly into the bloodstream to be transported to specific organs and tissues. Once they reach their target destination, hormones regulate the body's rate of metabolism, growth, sexual development, and other functions.

The cardio system has some characteristics that make it a unique and complex hydraulic system. It is a closed circle, it is elastic, it is filled with liquid at a positive mean pressure and it is independent of the pumping action of the heart. The heart expands and contracts 100,000 times per day, pumping five or six quarts of blood each minute (about 2000 gallons per day). The human heart undergoes over 3 billion contraction/cardiac cycles during a normal lifetime.

Circulatory Homeostasis: CV system helps balance several internal conditions. Blood circulation helps regulate a stable body temperature with its flow to the skin surface. Blood vessels near the skin's surface open during conditions of excess heat and facilitates thermal transfer into the milieu. During hypothermia, the blood vessels constrict and limit the blood flow only to vital organs. Blood also helps balance the body's pH (with bicarbonate ions). Finally, the blood albumin helps maintain an isotonic cellular environment.

The Circulatory Pump: The heart is a four-chambered 'double pump', where each side (left and right) operates as a separate pump. The left and right sides of the heart are separated by a muscular wall of tissue known as the septum. The right side of the heart receives deoxygenated blood from the systemic veins and pumps it to the lungs for oxygenation. The left side of the heart receives oxygenated blood from the lungs and pumps it through the systemic arteries to the tissues of the body. Each heartbeat results in the simultaneous pumping of both sides of the heart, making the heart a very efficient pump.

The Circulatory Loops: There are two primary circulatory loops in the human body: (i) *Pulmonary Circulation Loop* transports deoxygenated blood from the right side of the heart to the lungs, where the blood picks up O_2 and returns to the left side of the heart. Pumping chambers of the heart that support pulmonary circulation loop are the right atrium and right ventricle. (ii) *Systemic Circulation Loop* carries highly oxygenated blood from the left side of the heart to all of the tissues of the body (with the exception of the heart and lungs). Systemic circulation removes wastes from body tissues and returns deoxygenated blood to the right side of the heart. The left atrium and left ventricle of the heart are the pumping chambers for the systemic circulation loop.

Cardiac Blood Flow

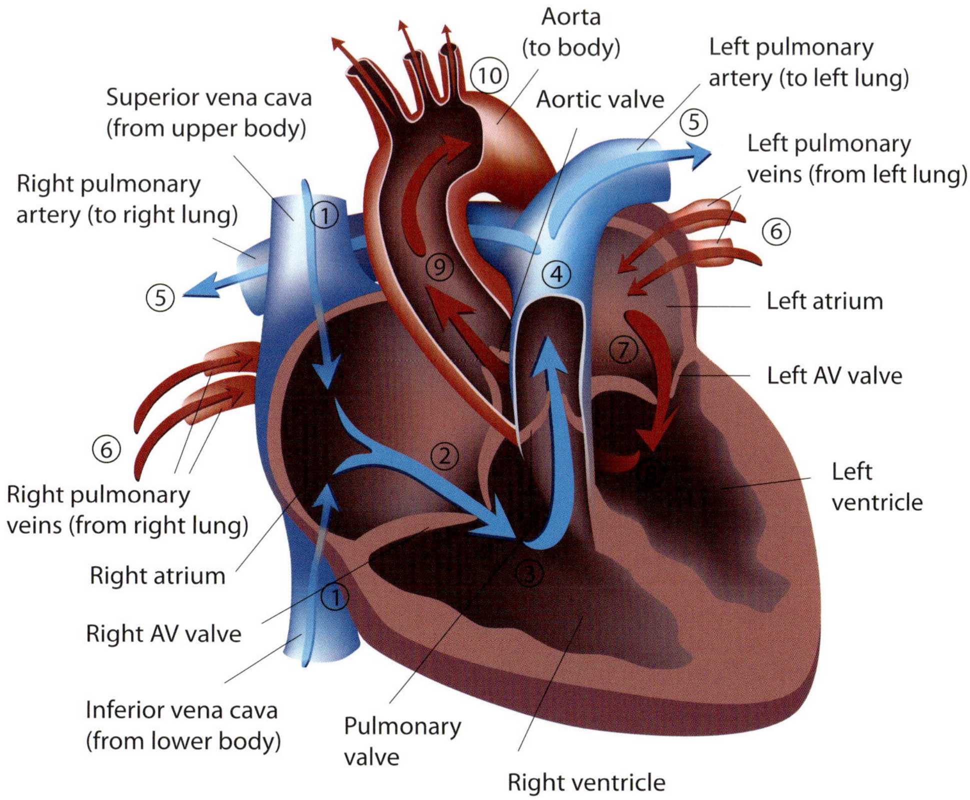

The Cardiac Cycle: The rhythmic contraction of the heart is called heartbeat. A cardiac cycle is defined as a complete cardiac movement, including systole, intervening pause, and diastole. Each cycle requires a certain length of time for its completion. Pressure, volume, electrical, and sound changes occur during each cycle. Heart sounds are created primarily from turbulence in blood flow created by valves closing, not from contraction. In order for the heart to pump efficiently, the individual myocardial fibers must contract and relax in a coordinated, rhythmic fashion. This characteristic of the healthy heartbeat is called synchrony. Synchrony is maintained by the heart's own intrinsic electrical system, which originates and transmits electrical impulses through a specialized conduction pathway.

Cardiac Cycle

REDOX and Heart: Myocardial cell have a negative membrane potential when at rest. Stimulation above a threshold induces the opening of voltage-gated ion channels and a flood of cations into the cell. The positively charged ions entering the cell cause the depolarization, which causes the opening of calcium (Ca^{2+}) channels and and this Ca^{2+} influx causes muscle contraction. After a delay, potassium (K^{+}) channels reopen and this K^{+} flow out of the cell causes repolarization and leads to the resting state of the myocardial cell.

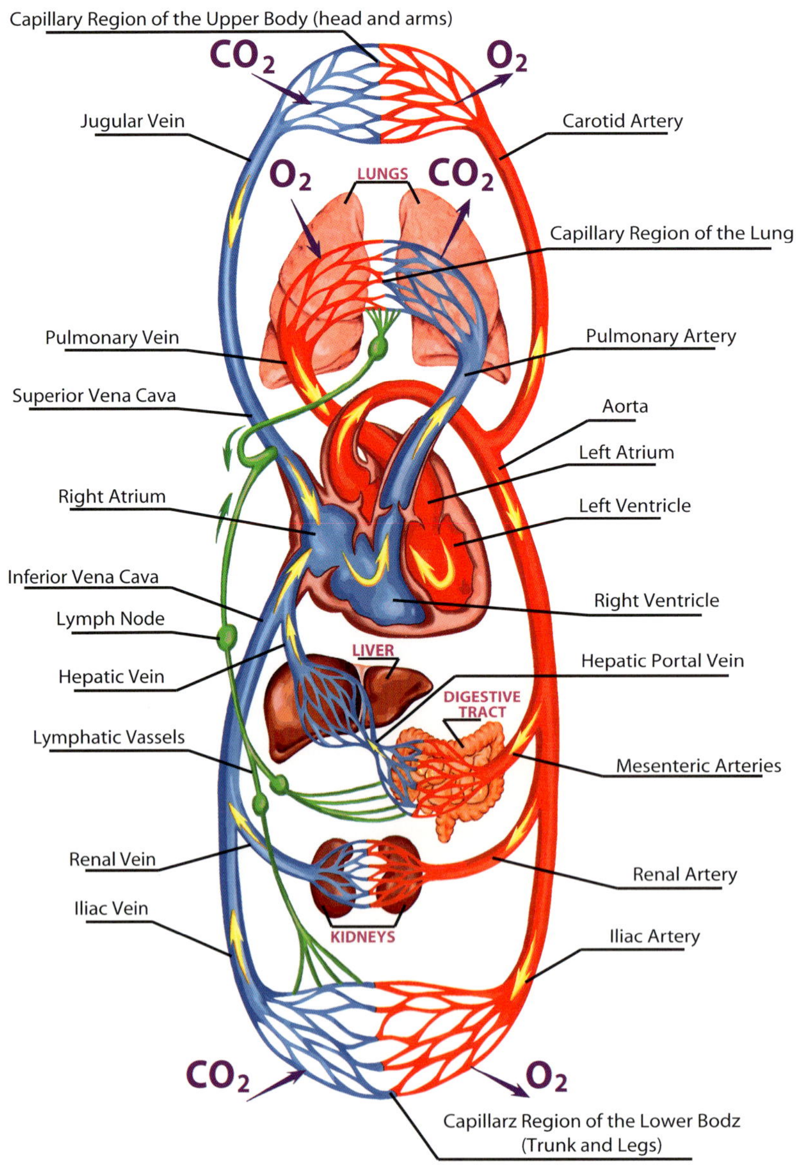

The Cardiac Conduction System: Pulse is evidence for mechanical activity of the heart; whereas, the electrocardiogram or ECG, is evidence of the heart's electrical conductivity. The cardiac conduction or electrical system includes three important components: (i) S-A (sino-atrial) node, known as the heart's natural pacemaker, consists of special cells that create electrical energy to make the heart beat. (ii) A-V (atrio-ventricular) node, a bridge that transmits electrical signals between the atria and ventricles. (iii) His-Purkinje system carries the electrical signals throughout the ventricles to make them contract.

The S-A node produces about 60-100 electrical signals per minute, which is the heart rate, or pulse. With each pulse, signals from the S-A node follow a natural electrical pathway through heart walls. The transmission of electrical signals makes the heart chambers contract and relax. In a healthy heart, the chambers contract and relax in a coordinated way, called the sinus rhythm. The conduction system senses the need for oxygen and responds with the proper heart rate. An abnormal rhythm (or heart beat) due to any disruption in cardiac electrical system is known as arrhythmia.

Blood Pressure: Certain hormones along with autonomic nerve signals from the brain affect the rate and strength of heart contractions. Greater contractile force and heart rate lead to an increase in blood pressure. Blood vessels can also affect blood pressure. Vasoconstriction decreases the diameter of an artery by contracting the smooth muscle in the arterial wall. The sympathetic (autonomic) nervous system causes vasoconstriction, which leads to increase in blood pressure and decrease in blood flow in the constricted region. Vasodilation is the expansion of an artery as the smooth muscle in the arterial wall relaxes. The volume of blood in the body also affects blood pressure. A higher volume of blood in the body raises blood pressure by increasing the amount of blood pumped by each heartbeat.

Cardiac Conduction

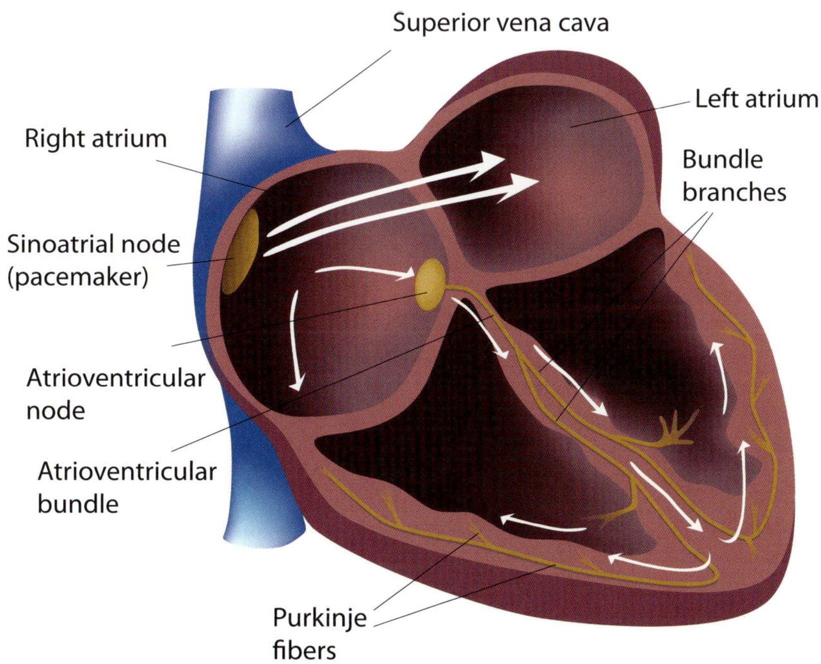

4.6.1. REDOX and Cardiovascular (CV) Health

Oxidants are generated in healthy CV system, and work as signaling molecules. Intra-cellular REDOX status is tightly regulated by oxidant/antioxidant systems. Imbalance between the two causes oxidative or reductive stress that triggers cellular damage or aberrant signaling. ROS can regulate cell signaling through REDOX of specific amino acids within proteins. Structural changes during post-translational modification in proteins result in altered cell function. The molecular basis of REDOX protein modification affects signal transduction in the CV system.

REDOX in Normal Physiology: ROS function is important to modulate many downstream signaling molecules. REDOX potential modulates signal transduction processes and energy metabolism in response to oxidative and nitrosative stresses. Induction of these signaling cascades leads to vascular smooth muscle cell (VSMC) growth and migration, expression of pro-inflammatory mediators, and modification of extra-cellular matrix. In addition, ROS increase intra-cellular free Ca^{2+} concentration, a major determinant of vascular reactivity.

Nitric oxide (·NO) is necessary for the regulation of cardiac physiology, such as coronary vasodilatation, inhibition of platelet/neutrophil adhesion-activation, and modulation of cardiac contractile function. ·NO also has a protective role against the ischemic and/or failing heart. This protective role is mediated by several mechanisms, including the stimulation of soluble guanylyl cyclase, which leads to a decrease of the concentration of intra-cellular Ca^{2+}, and the inhibition of oxidative stress. Therefore, $O_2^{\bullet-}$ can exert cytotoxic effects not only due directly to $O_2^{\bullet-}$, but also are mediated by the inactivation of cytoprotective ·NO and the formation of highly reactive oxidant $ONOO^-$ (which is produced following interaction of ·NO with $O_2^{\bullet-}$). When ROS production exceeds the capacity of antioxidant defense, oxidative stress has a harmful effect on the functional and structural integrity of biological tissue. Specifically, in the heart, excess ROS can cause myocardial remodeling, including contractile dysfunction and structural alterations.

Mechanical forces, comprising of both unidirectional laminar and oscillatory shear, are increasingly being recognized as important inducers of vascular ·NO and ROS generation. In general, laminar flow is associated with upregulation of eNOS and ·NO production and increased expression of antioxidants glutathione peroxidase and superoxide dismutase (SOD), thereby promoting a healthy vascular wall and protecting against oxidative vascular injury.

REDOX in Patho-physiology: The clinical onset and progression of CV diseases (CVD) are strongly linked to the oxidative stress and the dysregulation of REDOX. Accordingly, the oxidation and nitration of cellular proteins, lipids, nucleic acids, and formation of oxidized molecules lead to the loss of cellular function. Evidence shows that sources of ROS, physiological and patho-physiological conditions, and cellular oxidant targets determine the characteristic nature of a

ECG and Electrical Activityof the Myocardium

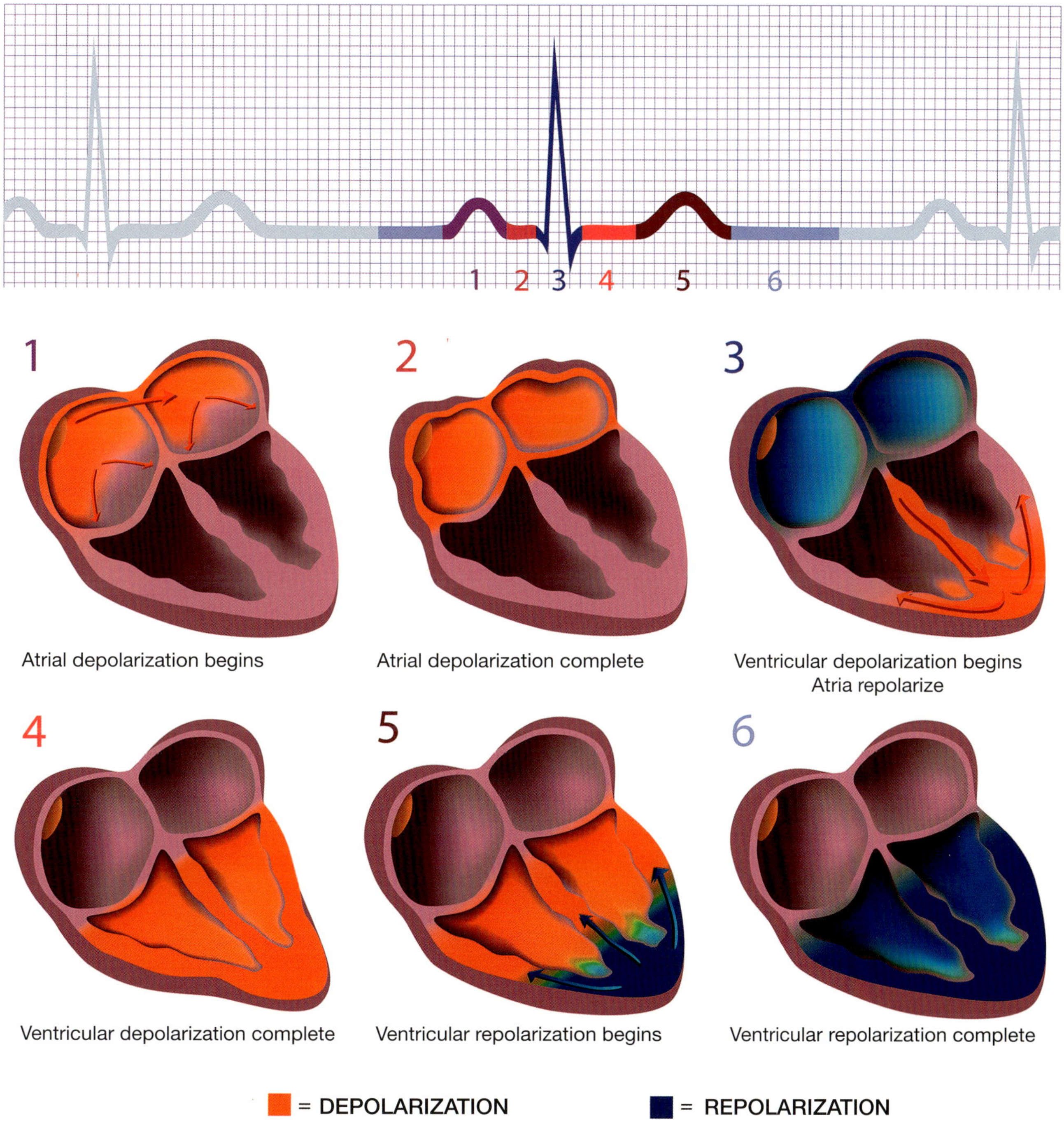

disease process and resultant outcomes.

Fundamental processes underlying the development of vascular diseases such as atherosclerosis, restenosis, and hypertensive vascular remodeling have their origins in an initial insult to the vessel wall. Such an insult may arise from mechanical disruption that occurs during percutaneous coronary angioplasty or at regions of oscillatory shear stress. It can also result from biological conditions, such as hypercholesterolemia, excess free radicals, diabetes, increased concentrations of plasma homocysteine, or infectious agents. Injury promotes disruption of the homeostatic mechanisms of the endothelial protective barrier, changing its natural properties, increasing its adhesiveness to leukocytes and altering its permeability. This endothelial dysfunction also leads to the formation of vasoactive molecules, which in turn induce inflammatory genes, inactivate ·NO (its protective functions), and activate matrix metallo-proteinases (MMPs), leading to extracellular matrix remodeling, and increased smooth muscle cell (VSMC) growth, migration and proliferation. Each of these processes contributes to thickening of the vessel wall or the formation of lesions that can lead to hypertension, acute myocardial infarction and stroke. Under pathological conditions ROS contribute to vascular dysfunction and remodeling through oxidative damage.

REDOX and cellular pathways of CVD

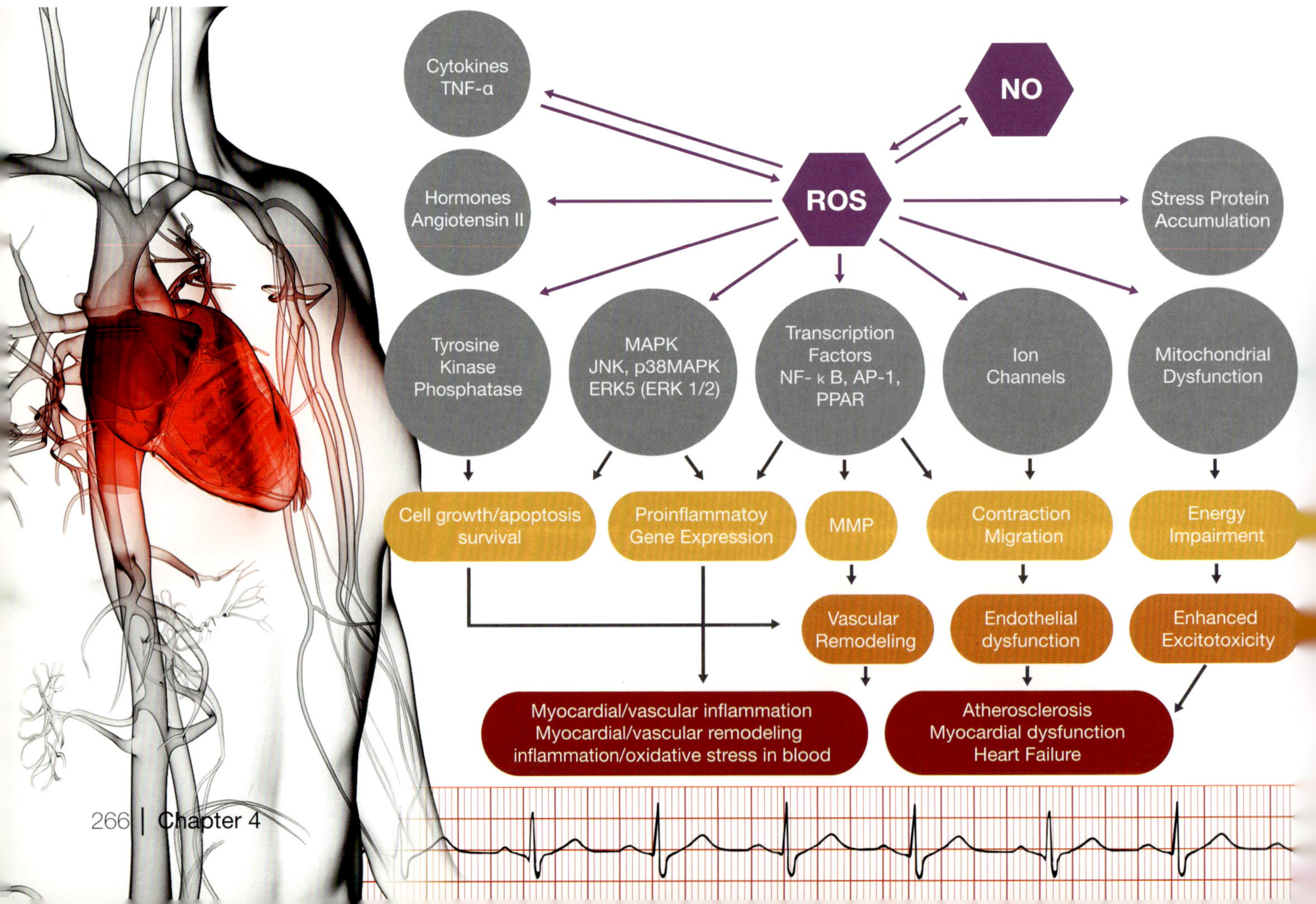

4.6.2. Oxidative Stress and Atherosclerosis

Atherosclerosis is a chronic inflammatory syndrome affecting the walls of arteries, caused largely by the accumulation of macrophage/monocytes. It is promoted by low-density lipoproteins (LDL), which are plasma proteins that carry cholesterol and triglycerides; and inadequate removal of fats and cholesterol from the macrophages by functional high-density lipoproteins (HDL). It is commonly referred to as a hardening of the blood vessels, caused by the formation of multiple plaques within the arteries. Common factors involved in the mechanism of atherosclerosis include endothelial cells, vascular smooth muscle cells, macrophages, platelets, and protein substances (i.e. LDL), growth factors, and cytokines.

In the initial phase of atherogenesis, LDL or other proinflammatory proteins, stimulate endothelial cell expression of adhesion factors, chemokines, and growth factors; this attracts circulating monocytes to the inflamed endothelial cells, and then the monocytes enter the intima where macrophage colony-stimulating factor induces monocyte-macrophage differentiation. The macrophages through the uptake of oxidatively modified LDL and cholesterol lead to foam cell formation, augmenting early lesion development. Meanwhile, T lymphocyte activation, smooth muscle cell proliferation and migration, and extra-cellular matrix deposition occur in the vascular intima. These deposits interact with molecules in the intima and lead to necrosis, causing the formation of fibrous plaques which contain a core of lipid-laden cells and a fibrous cap formed by smooth muscle cells and matrix. Slow progressive enlargement of plaques may eventually cause them to rupture and with blood flow produce stenosis or closure of the lumen of other arteries. In addition, as a result of plaque rupture, platelet activation may lead to thrombosis, in which a thrombus obstructs blood flow, further resulting in clinically acute cardiovascular event or stroke.

Plasma Lipid Profile - Normal Values

Lipid-type	Value
Serum Cholesterol	<200 mg/dL
Serum Triglycerides	10-150 mg/dL
HDL Cholesterol	40-60 mg/dL
LDL Cholesterol	<130 mg/dL
VLDL	2-30 md/dL
Total/HDL ratio	<4

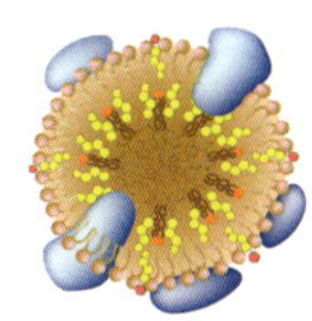

HDL

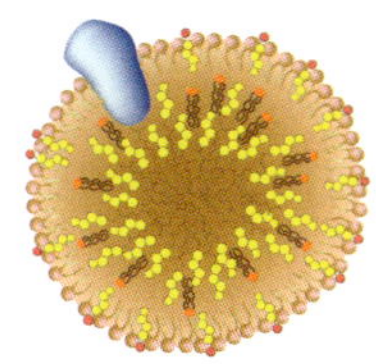

LDL

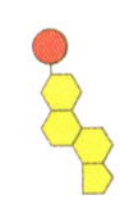

Cholesterol

Triglyceride

Cholesterol ester

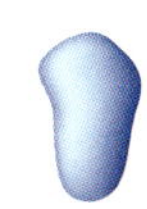

Apoprotein

Cholesterol Metabolism: The Nobel Prize in Physiology or Medicine 1964 was awarded jointly to Konrad Bloch and Feodor Lynen for their the discovery of the mechanism and regulation of the cholesterol and fatty acid metabolism.

Atherosclerosis is a dynamic process, exhibiting constant changes in size and composition of the arteries - known as remodeling of the blood vessel. Two major forms of remodeling may progress within the affected artery. One is thickening and constrictive geometric remodeling, where the lumen of an affected artery becomes narrowed. The other is expansive remodeling in which the artery tends to preserve the lumen by expanding the vessel wall outward. These two remodeling responses by the vessels have been reported to depend on a variety of endogenous and environmental factors. They can vary from vessel to vessel, with age and gender, and are modulated by known cardiovascular risk factors.

Atherogenesis (Plaque Build Up)

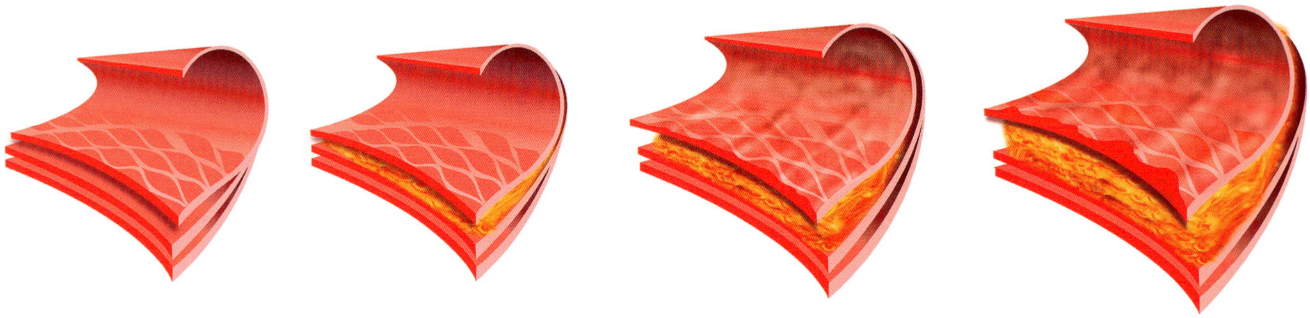

Free Radicals and Atherosclerosis: Reactive oxygen species such as $O_2^{\bullet -}$, H_2O_2, and the highly reactive ·OH radical induce oxidative damage to lipids and proteins in the vascular wall. ROS formation triggers a cascade of events such as oxidative modification of LDL, inflammation, cellular apoptosis and endothelial injury. ROS are produced by phagocytes and non-phagocytic cells through different mechanisms. The enzymes for ROS production in vascular cells include XOR, cytochrome P-450, uncoupled NOS, and NOx. Once produced, ROS play an important role in cell signaling. ROS also attack vital cell components such as polyunsaturated fatty acids (PUFAs), proteins, nucleic acids, and carbohydrates. These reactions can alter cell membrane properties such as fluidity, ion transport, loss of enzyme activity, protein cross-linking, inhibition of protein synthesis, and DNA damage eventually leading to cell apoptosis.

Endothelial Dysfunction and Atherosclerosis: Endothelial dysfunction is an early event in atherosclerosis. ROS are involved in the progression of endothelial cell dysfunction, accompanied by inactivation of eNOS and a decline in NO levels. ROS also induce expression of adhesion factors such as intercellular adhesion molecule-1 (ICAM-1) and vascular adhesion molecule-1 (VCAM-1), which facilitates inflammatory cell recruitment and lipid deposition in the intimal layer. The subsequent ingestion of excess oxidized low density lipoprotein (LDL) particles by macrophages and monocytes leads to release of various inflammatory cytokines and growth factors. Finally, proliferation of vascular smooth muscle cells (VSMCs) is a key step in the development of atherosclerosis. Taken together, ROS are involved in the VSMC proliferation and development of atherosclerosis through various pathways.

Stages of Atherosclerosis Development

Effects of Endothelial Dysfunction

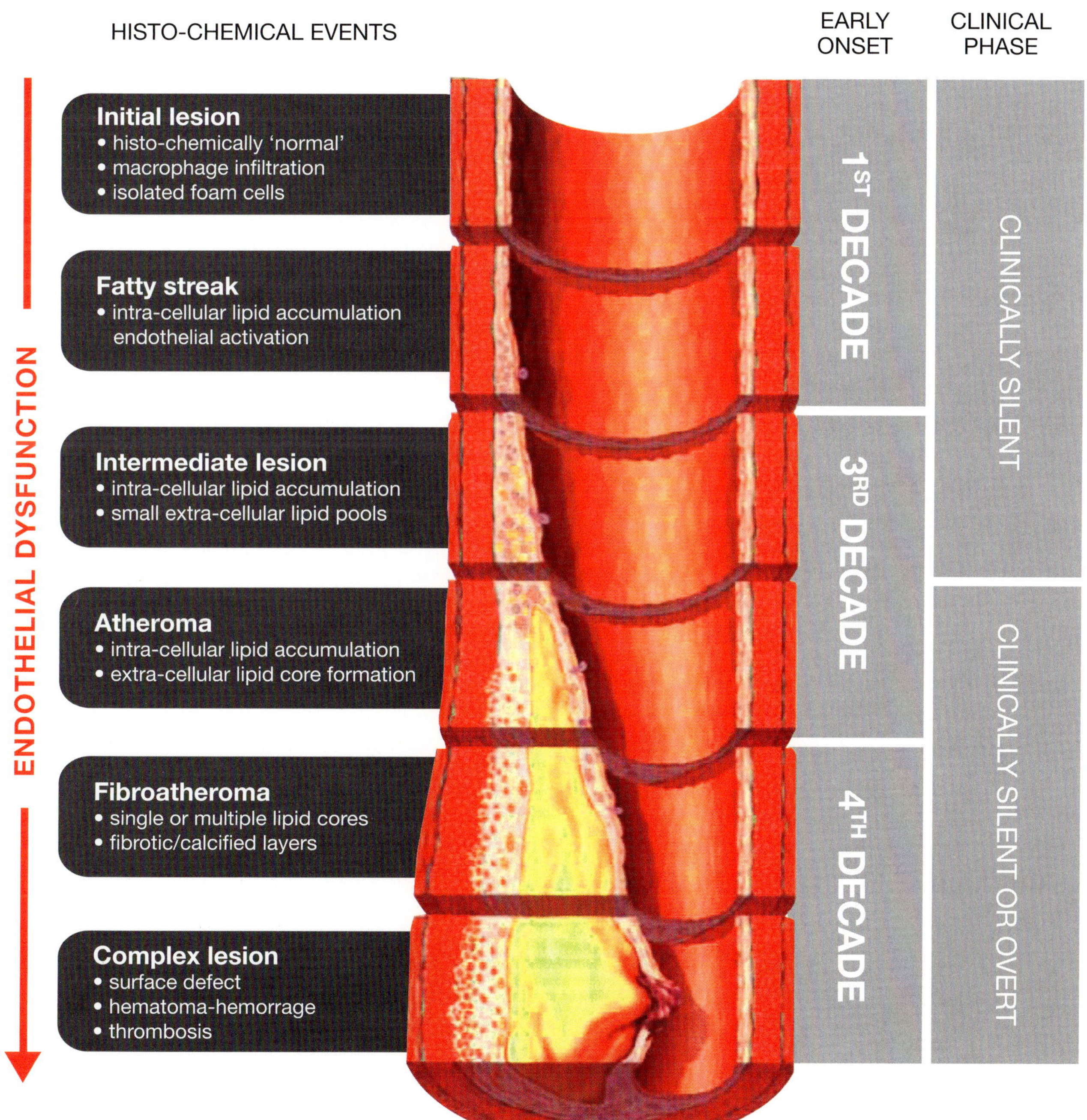

Mitochondrial REDOX and Atherosclerosis: Aging is the driving force in an atherosclerotic cycle, which starts with age-related changes in the blood vessels including arterial thickening and stiffening, which cumulatively increase susceptibility to CVD. Besides aging, the severity of mitochondrial (mt) DNA damage is higher in aortic samples from atherosclerotic patients than in non-atherosclerotic aortic samples from age-matched transplant donors. Net production of mtROS is another crucial mechanism by which mitochondria contribute to a variety of pathologies such as CVD, aging, ischemia/reperfusion, and neurodegenerative diseases. The balance of mitochondrial REDOX reactions is coordinated with the processes and components involved in ROS synthesis, antioxidant defense, and ROS release. Mitochondrial ROS metabolism has a strong influence on the vascular oxidative stress. Under most conditions, the rates of cellular ROS formation and elimination are balanced through mitochondrial and cellular mechanisms that sense relative ROS levels. However, a chronic imbalance in REDOX homeostasis contributes to various chronic diseases, especially atherosclerosis, which is an inflammatory disease associated with the oxidation of lipids and proteins in the vascular wall. Besides the influence of ROS/RNS formation; clinical conditions such as diabetes, dyslipidemia, hypertension, homocysteinemia; and life style practices such as diet, cigarette smoking, and lack of exercise are predisposing factors in the development of atherosclerosis.

REDOX-based Interventions: Clinical management often focuses predominantly on the symptoms. However, over time, clinical trials have shown that treatments focusing on decreasing the underlying atherosclerosis processes (REDOX and oxidative stress) are more effective, as opposed to simply treating symptoms. In summary, the key to more effective approaches has been to better understand the REDOX nature of the disease and to combine multiple treatment strategies - not rely on just one or a few approaches. In addition, for those approaches, such as lipoprotein transport behaviors, which have been shown to produce the most success, adopting more aggressive combination treatment strategies has generally produced better results, both before and especially after individuals become symptomatic. The high incidence of CVD has led to a proposed 'Polypill', a once-daily pill containing aspirin/statin drugs in addition to an ACE inhibitor, diuretic, beta blocker, and folic acid. It was speculated that high uptake by the general population of such a Polypill would reduce cardiovascular mortality. It must be emphasized, however, that this is purely theoretical.

Poly-pill or Combo-pill is commonly a fixed-dose combination drug targeted for treatment or prevention of chronic disease. Poly-pill can be administered to healthy people as a preventive medicine, and/or as a therapeutic to patients. Polypill typically involves lower dosages; therefore, it can reduce the number of pills administered, which may ease oral intake and reduce pill-fatigue.

Cellular Ion Channels: The Nobel Prize in Physiology or Medicine 1991 was awarded jointly to Erwin Neher and Bert Sakmann for their discoveries concerning the function of single ion channels in cells.

4.6.3. Oxidative Stress and Hypertension

Normal blood pressure is less than 120 mmHg systolic and less than 80 mmHg diastolic.

Hypertension is a clinical condition that arises if the blood pressure is persistently at or above 140/90 mmHg.

Hypertension is the clinical term for high blood pressure, which is the force of blood against the walls of arteries. Thus, the higher the pressure, the harder the heart has to pump. Blood pressure is recorded as two numbers - 'systolic pressure' (as the heart beats) over 'diastolic pressure' (as the heart relaxes between beats). The measurement is written one above or before the other, with the systolic number on top and the diastolic number on the bottom. For example, a blood pressure measurement of 120/80 mmHg (millimeters of mercury) is expressed verbally as '120 over 80'. Hypertension is a major risk factor for stroke, myocardial infarction (heart attacks), heart failure, aneurysms of the arteries, and peripheral arterial disease. It is also a cause of chronic kidney disease. Even moderate elevation of arterial blood pressure is associated with a shortened life expectancy. Dietary and lifestyle changes can improve blood pressure control and decrease the risk of associated health complications. Drug treatment is often necessary in people for whom lifestyle changes prove ineffective or insufficient.

Hypertension is a typical outcome of REDOX imbalance. ROS are produced by all types of vascular cells (including endothelial, smooth muscle and adventitial cells); they are also formed by numerous enzymes. The most important sources of ROS, in relation to hypertension, are xanthine oxido-reductase (XOR), uncoupled endothelial NO synthase (eNOS) and NAD(P)H oxidase (NOx). XOR is an active enzyme in the vascular endothelium, which catalyzes the oxidation of hypoxanthine and xanthine to form $O_2^{\bullet-}$. Although XOR-derived $O_2^{\bullet-}$ is established in the ischemia–reperfusion injury and heart failure, there is also evidence of its involvement in hypertension. Nitric oxide synthase (NOS) also contributes to ROS production, as all three NOS isoforms have been shown to be susceptible to the 'uncoupling' that leads to the formation of $O_2^{\bullet-}$ (rather than NO) under certain conditions. Several studies have shown that the major source of ROS in the vascular wall is non-phagocytic NOx, which utilizes NADH/NADPH as the electron donor to reduce molecular oxygen and produce $O_2^{\bullet-}$.

Endothelial dysfunction (resulting from free radical damage) is a major risk factor in the pathobiology of hypertension. Endothelium is an active component of the vasculature; it plays a role in the regulation of vascular tone, platelet activity, thrombosis, inflammation and atherosclerosis. Vasoactive tone of the endothelium is maintained by the release of substances like prostacyclins, endothelins and the endothelium-derived relaxation factor nitric oxide (·NO) or a related compound. Decreased vasorelaxation (due to low bioavailability of ·NO) is a trigger for endothelial dysfunction. A decline in ·NO levels disrupts the non-thrombogenic intimal surface, while promoting adhesion, aggregation and deposition of platelets on to the abnormal endothelial surface. Thrombus formation during acute coronary syndromes results in the release of various vasoactive substances such as thrombin and serotonin. In normal endothelium these substances mediate vasodilatation; whereas, vasoconstriction occurs during an endothelial

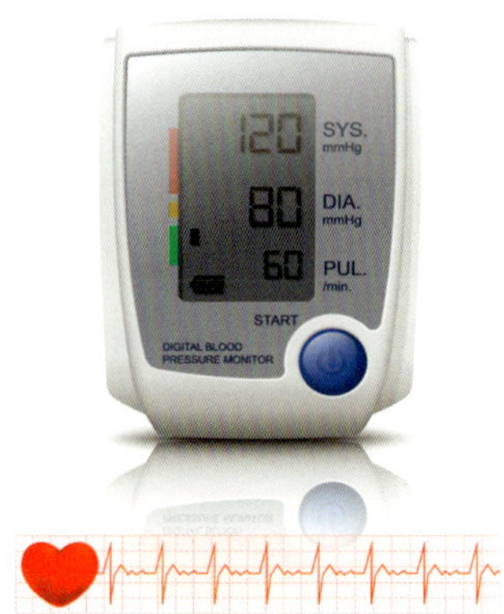

Hypertension:
CDC FastStats (2010)
Non-institutionalized: 31.9%
Ambulatory care: 38.9 million
Number of deaths: 26,634

vasodilator dysfunction. The impairment of vasodilation in response to vasodilator acetylcholine is a measurement of endothelial dysfunction. It correlates with increased local ROS production and reduced SOD activity.

Thus, hypertension is associated with functional and structural changes to the vasculature, including endothelial dysfunction, altered contractility and vascular remodeling. Cellular events underlying these processes involve changes in vascular smooth muscle cell (VSMC) growth, apoptosis/anoikis, cell migration, inflammation, and fibrosis. Many stimuli influence cellular changes, including mechanical forces, such as shear stress and vasoactive agents, of which angiotensin II (Ang II) appears to be amongst the most important. Ang II mediates many of its pleiotropic vascular effects through NOx-derived ROS.

Individuals with hypertension demonstrate increased levels of oxidative stress byproducts together with decreased activity of endogenous antioxidant enzymes in blood and mononuclear cells. These patients also have indications of increased oxidative DNA damage (compared to normotensive individuals). Thus, there is compelling evidence to suggest a role for ROS in the pathogenesis of hypertension. Although all ROS are derived from the reduction of molecular oxygen, the different chemical properties of individual ROS have important implications for their role in cellular signaling. Both $O_2^{\bullet-}$ and ·OH have relatively short biological half-lives – the ·OH radical is particularly reactive, and thus unlikely to mediate effects distant from where it is produced. The charge on the $O_2^{\bullet-}$ anion makes it unable to cross cellular membranes except through ion channels. In contrast, H_2O_2 has a longer biological life span than $O_2^{\bullet-}$ and ·OH and is able to diffuse across lipid bilayers. These distinct properties mean that different species of ROS are capable of activating different signaling pathways, which may then lead to divergent (and potentially opposing) consequences. For example, increased $O_2^{\bullet-}$ levels have long been known to inactivate the vasodilator, leading to endothelial dysfunction and vasoconstriction characteristic of many vascular diseases, including hypertension. H_2O_2; however, has been shown to act as a vasodilator in a number of vascular beds, including cerebral, coronary and mesenteric arteries. Thus, from a etiological standpoint, hypertension can be attributed to the vascular effects of 'oxidative stress'.

REDOX-based Interventions: REDOX technologies have been extensively studied as treatment strategies for the clinical management of hypertension. REDOX processes that control ROS bioavailability (by decreasing production and/or by increasing radical scavenging) may down-regulate cell signaling and prevent further vascular injury and hypertension.

As of 2000, nearly one billion people or ~26% of the adult population of the world had hypertension. The condition was common in both developed (333 million) and undeveloped (639 million) countries. As of 2006, hypertension affected 76 million US adults (34% of the population); African American adults had among the highest rates (at 44%).

4.6.4. Oxidative Stress and Heart Failure (HF)

Heart failure (HF), often called congestive heart failure (CHF) or congestive cardiac failure (CCF), is a complex clinical syndrome that can result from any structural or functional cardiac disorder. HF impairs the ability of the ventricle to fill with or eject blood. Heart failure can cause a number of symptoms including shortness of breath, leg swelling, and exercise intolerance. HF is a leading cause of morbidity and mortality in industrialized countries. It is also a growing public health problem, mainly because of aging of the population and the increased prevalence of HF in the elderly. The major causes of HF are myocardial infarction (MI), hypertension, cardiomyopathy, and valvular heart disease. Following MI, the heart usually adapts through a patho-physiological process known as 'cardiac remodeling', which involves changes in the structure and function of cardiac myocytes, as well as the extra-cellular matrix in the non-infarcted myocardium. These changes lead to substantial alterations in the shape and volume of the heart and progressive ventricular dilatation and impairment of pump function.

Heart Failure:
Risk factors (NHANES-I)
Ischemic heart disease – 62%
Cigarette smoking – 16%
Hypertension – 10%
Obesity – 8%
Diabetes – 3%
Valvular heart disease – 2%.

Oxidative stress is enhanced, which leads to cellular dysfunction and protein and lipid peroxidation; while DNA damage can induce irreversible cell damage and death. The role of oxidative stress is critical in the patho-physiological mechanism of cardiac remodeling, as well as, the development and progression of HF. ROS can directly impair contractile function by modifying proteins that are central to excitation-contraction coupling. ROS also activates hypertrophy signaling kinases and transcription factors that mediate apoptosis. They stimulate cardiac fibroblast proliferation and activate matrix metalloproteinases (MMPs), leading to the extra-cellular matrix remodeling. These cellular events are involved in the progression of maladaptive myocardial remodeling and HF.

Anatomy of a Heart Attack

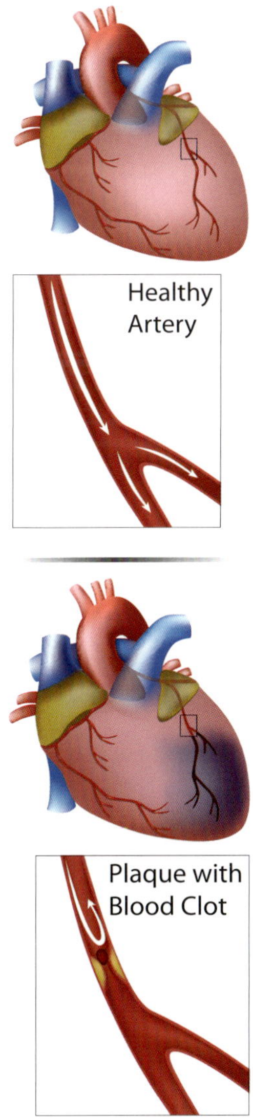

The cellular sources of ROS generation within the heart include cardiac myocytes, endothelial cells, and neutrophils. Within cardiac myocytes, ROS can be produced by several sources, including mitochondria, NAD(P)H oxidase (NOx), xanthine oxidoreductase (XOR), and uncoupled NO synthases (NOS) Mitochondria from the failing heart produce more $O_2^{\bullet-}$ than normal in the presence of NADH, indicating that mitochondrial electron transport could be the predominant source of such $O_2^{\bullet-}$ production. Inside mitochondria, most of the oxygen is reduced to water at the respiratory chain. Therefore, when oxygen availability is reduced in conditions such as ischemia or hypoxia, mitochondrial formation of ROS is increased, which contributes to the induction of myocyte damage or MI. ROS can also be generated via NOx and/or XOR in the vascular endothelial cells, as well as, via NOx in activated leukocytes. Importantly, NOx activity has been shown to be significantly increased by several stimuli that are relevant to the pathophysiology of HF, i.e. mechanical stretch, angiotensin II, endothelin-1, and tumor necrosis factor (TNF)-α. Increased XOR expression and activity were also reported in HF; also, LV contractile function and myocardial efficiency were improved by the treatment of HF animals with the XOR inhibitor, allopurinol.

Endothelial NOS (eNOS) has been shown to be uncoupled and functionally important in pathological cardiovascular remodeling (including HF). Under normal conditions, eNOS consumes NADPH and generates NO and L-citrulline from L-arginine and O_2. When exposed to oxidative stress, or when deprived of L-arginine, eNOS becomes structurally unstable and generates ROS. Since eNOS is expressed in vascular endothelial cells and cardiac myocytes within the heart, these cells are expected to be involved in this process. Alternatively, oxidative stress has direct effects on cellular structure and function; it may activate integral signaling molecules in myocardial remodeling and failure. Oxidative stress also stimulates myocardial growth, matrix remodeling, and cellular dysfunction, which involve the activation of several downstream signaling pathways; (i) ROS activate a broad variety of hypertrophy signaling kinases and transcription factors. (ii) ROS induces apoptosis, another important contributor to remodeling and dysfunction, which is induced by ROS-mediated DNA and mitochondrial damage and activation of proapoptotic signaling kinases. (iii) ROS cause DNA strand breaks, activating the nuclear enzyme poly(ADP-ribose) polymerase-1 (PARP-1). PARP-1 regulates the expression of a variety of inflammatory mediators, which facilitate the progression of cardiac remodeling. (iv) ROS can activate MMPs, which are generally secreted in an inactive form and are activated post-translationally by ROS. MMPs play a pivotal role in normal tissue remodeling processes, such as cell migration, invasion, proliferation, and apoptosis. MMP activity has been shown to be increased in the failing hearts. Furthermore, an MMP inhibitor can limit LV dilatation after an experimental MI. There is also substantial evidence that ROS generation is increased in hypertension.

REDOX-based Interventions: The effects of various antioxidants on HF have been reported. Vitamin antioxidants such as α-tocopherol (vitamin E) and ascorbic acid (vitamin C), that scavenge ROS and prevent free radical chain reactions have been studied extensively in HF. Short-term vitamin E supplementation reduced the levels of oxidative stress biomarkers in patients with HF. However, no significant effects were observed on symptoms or clinical outcomes. Long-term supplementation of vitamin E exerted no effects on primary prevention of cardiovascular events and was even associated with increased risk of developing HF. Uric acid, the metabolic product of XOR, increases in the failing human heart and XOR inhibition with allopurinol has been shown to improve endothelial (as well as cardiac) function in HF. Various drugs, including ACE inhibitors, β-blockers (i.e. carvedilol), and statins, may directly or indirectly modulate oxidative stress in the cardiovascular system. The REDOX approach to regulate oxidative stress in the heart, as well as in the skeletal muscle, may provide an effective therapeutic strategy against HF.

Apoptosis: The Nobel Prize in Physiology or Medicine 2002 was awarded jointly to Sydney Brenner, Robert Horvitz and John E. Sulston for their discoveries concerning genetic regulation of organ development and programmed cell death.

4.6.5. The REDOX Factor: Myocardial Reperfusion Injury

Coronary interventions, including balloon angioplasty and coronary stent implantation, are associated with increased vascular levels of ROS (due to altered endothelial cell and smooth muscle cell function). These alterations potentially lead to restenosis, thrombosis, or endothelial dysfunction in the treated artery. Activation of leukocytes plays an essential role in the mechanism of restenosis. Neutrophils are known to contribute to tissue injury through ROS that nitrate the tyrosine residues. Mac-1-dependent activation of neutrophils causes oxidative burst in the post-stent inflammatory process, possibly leading to restenosis. Following a balloon injury, $O_2^{\bullet-}$ radical-mediated oxidative stress is readily induced by non-endothelial NOx enzymes. The degree of induced oxidative stress is directly correlated with the extent of vessel injury. There is evidence that supports a role for early REDOX processes in apoptotic cell loss and NF-κB activation. Thus, oxidative stress after injury is likely to be an early transient event, which parallels the inflammatory and proliferative phases of the vascular response. Such early REDOX processes act as dose-dependent signal transducers of gene programs that affect the final repair.

Coronary angioplasty is a procedure to open narrow or blocked coronary (heart) arteries. Angioplasty restores blood flow to the heart muscle. During the procedure, a thin, flexible catheter (tube) with a balloon at its tip is threaded through a blood vessel to the affected artery. Once in place, the balloon is inflated to compress the plaque against the artery wall. This restores blood flow through the artery. However, reperfusion can itself further damage the vascular tissue, causing myocardial reperfusion injury. ROS have been implicated in such ischemic-reperfusion cardiac injury.

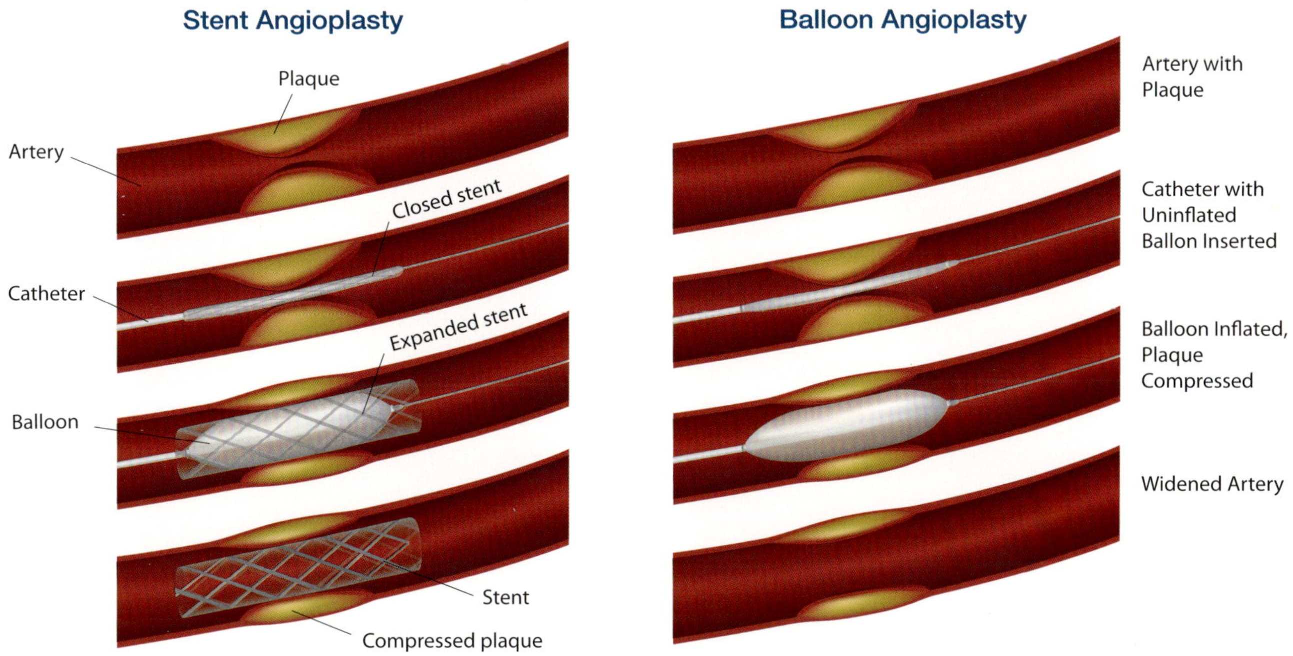

REDOX Indicators: Lipid peroxidation in plasma might predict restenosis after coronary balloon angioplasty. Plasma levels of thiobarbituric acid reactive substances (TBARS, an indicator of lipid peroxidation) have been suggested to be a reliable predictor of restenosis. ROS could also trigger reperfusion injury after rapid revascularization of acute myocardial infarction (AMI) and 8-epi prostaglandin (PG), a free-radical catalyzed product of arachidonic acid, has been proposed as an indicator of oxidative stress in vivo during myocardial reperfusion.

REDOX-based Interventions: Oxidative stress and inflammation in the vessel wall play important roles in the development of restenosis after angioplasty. ROS have been suggested to mediate the proliferative phenotype in smooth muscle cells. For example, circulating thioredoxin (TRx) impairs the chemotactic response to local sites of inflammation. Accordingly, administration of human recombinant TRx has been shown to attenuate ischemic reperfusion injury. Within 4 hours after angioplasty, plasma TRx levels significantly rise and gradually return to baseline within 24 hours. A possible association between elevated levels of TRx after angioplasty, and decreased rate of restenosis at follow-up angiography after 6 months, has also been reported. The role of TRx in the complex processes of vascular injury and restenosis in patients with peripheral arterial disease, suggests that TRx both as a marker of oxidative stress and as a therapeutic agent.

Studies have shown that direct buffering of oxidative stress with superoxide dismutase (SOD) treatment in balloon-injured vessels has also been reported to significantly alter post-injury response and limit neointimal hyperplasia after angioplasty. Also, other REDOX mechanisms regulate vascular cell signaling, vessel repair (or injury); thus they are a key determinant of restenosis post-angioplasty. A protective effect of SOD or antioxidants on vasospasm, neointimal thickening or remodeling after balloon angioplasty (injury) has also been reported. It was also shown that oxidized thiols induce chelatable metal-dependent amplification of the vascular repair reaction. Administration of metoprolol or glucose-insulin-potassium (GIK) solution was also shown to reduce oxidative stress and increase antioxidant activity in patients who have undergone elective angioplasty.

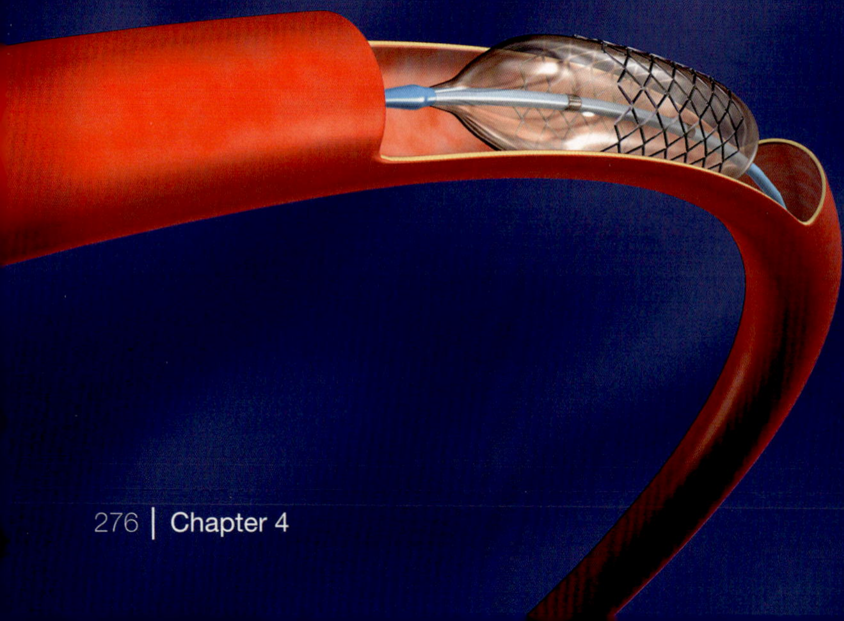

Superoxide ($O_2^{\bullet-}$) anions (whether derived from uncoupled eNOS, NOx, and XOR enzymes, or mitochondria) are the major endogenous source for ROS generation. Superoxide can scavenge NO, modify proteins and nucleotides, and induce pro-inflammatory signaling, which may lead to a chain production of ROS. Current innovations in stent technologies, including bio-degradable stents, NO donor-coated stents, and a new generation of drug-eluting stents, therefore address persistent oxidative stress and reduced NO bioavailability after percutaneous coronary interventions.

Cardiovascular Disease (CVD) Facts

Cardiovascular disease in America is a major health problem and the leading cause of death. According to the most recent statistics released by the American Heart Association:

- Approximately 84 million people in U.S. suffer from some form of CVD, causing about 2,200 deaths a day, averaging 1 death every 40 seconds.
- Almost 1 out of every 3 deaths results from CVD.
- The direct and indirect costs of cardiovascular disease and stroke are about $300 billion. This figure is increasing every year.
- An estimated 16 million U.S. adults have coronary heart disease.
- Approximately 78 million U.S. adults have high blood pressure, and an estimated 20 million have diabetes.
- It is estimated that an additional 8 million adults have undiagnosed diabetes and 87 million have pre-diabetes.
- Heart failure affects well over 5 million U.S. adults.
- CVD is the cause of more deaths than cancer, chronic lower respiratory diseases, and accidents combined.
- It is a myth that heart disease is a man's disease. In fact, CVD is the number one killer of women (and men).
- About one-third of CVD deaths occurred before age 75.
- On average, someone in the U.S. suffers a stroke every 40 seconds.
- Stroke is a leading cause of serious, long-term disability and accounts for more than half of all patients hospitalized for a neurological disease.
- Women have a higher lifetime risk of stroke than men.

When compared with previous trends, the CVD death rates have declined, but there are more people suffering from diabetes and obesity

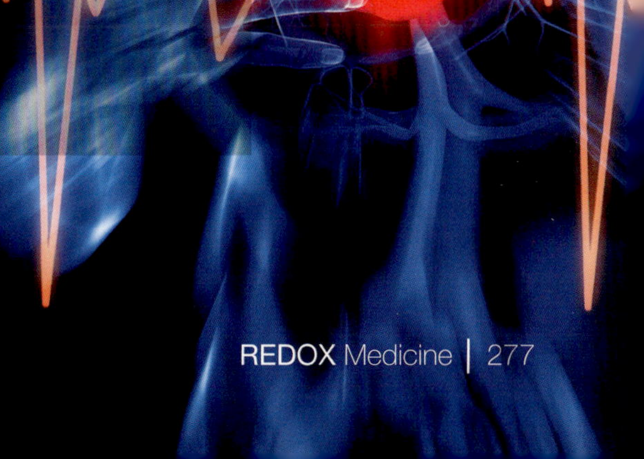

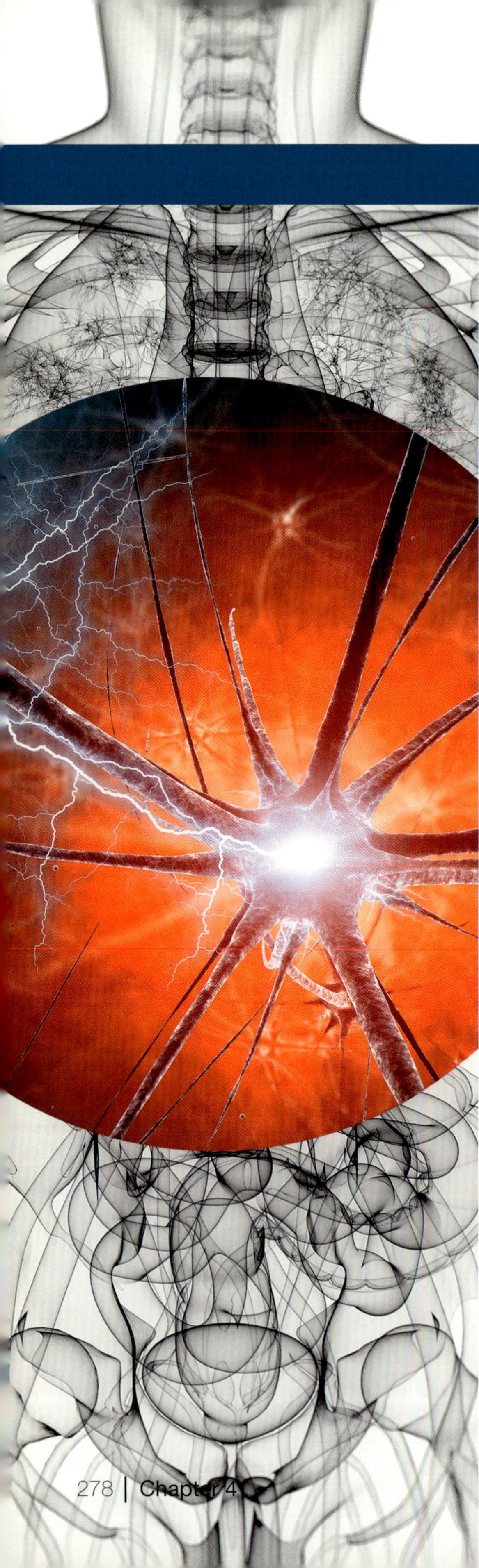

4.7 **REDOX:** The Nervous System

The Nervous (or neuro) system is the major controlling, regulatory, and communicating system in the body. It is the center of all mental activity (including thought, learning, and memory). Together with the endocrine system, the nervous system is responsible for regulating and maintaining homeostasis. Through its receptors, the nervous system keeps the body, both external and internal in touch with the milieu

Like other systems in the body, the nervous system is composed of organs, principally the brain, spinal cord, nerves, and ganglia. These, in turn, consist of various tissues, including nerve, blood, and connective tissue. Together these carry out the complex activities of the nervous system. The various activities of the nervous system can be grouped together as three general, overlapping functions:

i) Sensory function: Millions of sensory receptors detect changes, called stimuli, which occur inside and outside the body. Stimuli monitor such things as temperature, light, and sound from the external environment. Inside the body (the internal environment), receptors detect variations in pressure, pH, carbon dioxide concentration, and the levels of various electrolytes. All of this gathered information is called sensory input.

ii) Integrative function: Sensory input is converted into electrical signals, called nerve impulses that are transmitted to the brain. The signals are brought together to create sensations, to produce thoughts, or to add to memory. Decisions are made each moment based on the sensory input; this is known as integration.

iii) Motor output or function: Based on the sensory input and integration, the nervous system responds by sending signals to muscles (causing them to contract) or to glands (causing them to produce secretions). Muscles and glands are called 'effectors' because they cause an effect in response to directions from the nervous system. This is known as the 'motor output' or 'motor function'.

4.7.1. REDOX and Cognitive Health

The Central Nervous System (CNS) possesses a: (i) high content of unsaturated fatty acids that are susceptible to peroxidation by H_2O_2, (ii) high oxygen consumption per unit weight, (iii) high content of transition metals and other key ingredients that promote free radical formation by Haber-Weiss reaction, (i.e. iron, copper and ascorbate), and (iv) the scarcity of antioxidant defenses systems. Therefore, the CNS is highly susceptible to oxidative stress, more than other tissue in the human body. Since oxidative stress is characterized by an imbalance in radical production of reactive oxygen species (ROS) and antioxidative defense, both are considered to have a major role in the process of age-related neuro-degeneration and cognitive dysfunction. An important aspect of the antioxidant defense in the CNS is glutathione (GSH), which is responsible for the endogenous REDOX potential of the cell. The most important function of GSH is to donate electrons and thereby scavenge the ROS. Intra-cellular GSH concentrations decrease in the aged human brain regions, including the hippocampus. The decrease in GSH leads to an imbalance, where the rate of ROS production exceeds the antioxidant ability, thus generating a situation that favors oxidative stress. A further reason for oxidative stress is caused by an imbalance among the radical detoxifying enzymes in the CNS.

The Brain - Functions

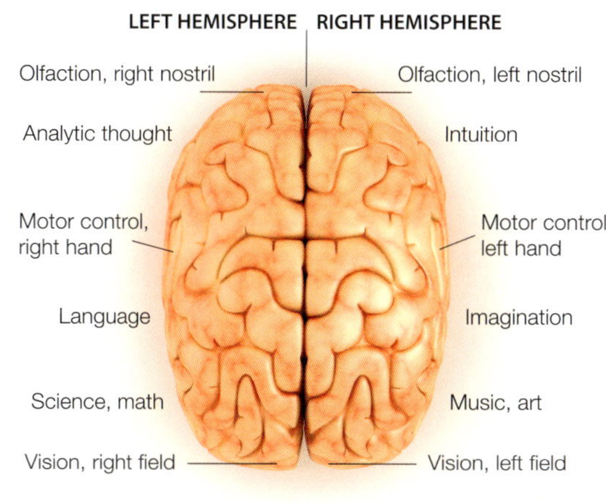

The Brain is one of the body's most metabolically active organs. It consumes oxygen at a rate of 35 ml/min/kg, compared to the heart's oxygen consumption rate of 59 ml/min/kg. Therefore, the brain is highly susceptible to oxidative damage. The brain is also unique in free radical generation with reactive species that are foreign to other body systems. For example, metabolism of excitatory amino acids and neurotransmitters generates ROS, and the constant use of oxygen by neural mitochondria results in high $O_2^{\bullet-}$ levels. The brain is also vulnerable to oxidative damage due to limited access to antioxidant enzymes. For example, SOD is localized primarily in neurons, and GSH and GPx are localized in astrocytes. Also, there is a very little activity of catalase in the brain.

The Brain - Median Section

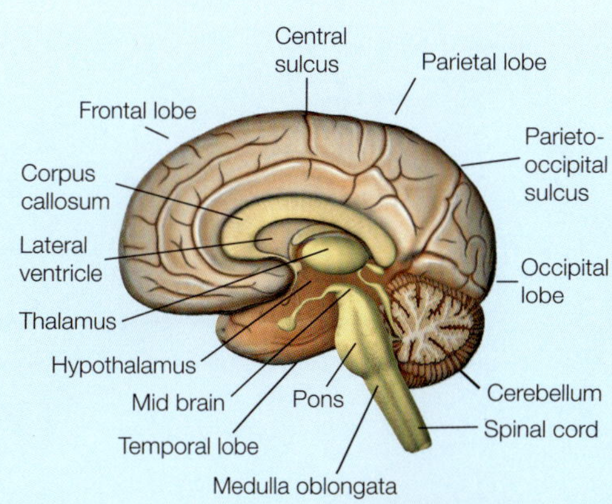

Nerve Functions: The Nobel Prizes in Physiology or Medicine

1932 - Charles Sherrington and Edgar Adrian for discovering nerve functions.

1936 - Henry Dale and Otto Loewi for discovering the chemical transmission of nerve impulses.

1944 - Joseph Erlanger and Herbert Gasser for discovering the differentiated functions of single nerve fibers.

The hypothesis that age-associated and chronic neurodegenerative diseases [like Alzheimer's disease (AD) and Parkinson's disease (PD)] are related to oxidative stress is well documented. First, most patients with neurodegenerative conditions have damaged mitochondria coupled with high levels of oxidative damage. Second, there is a presence of free, non-protein bound Fe^{3+} in the cerebrospinal fluid. Normally, the brain has particularly high levels of iron, which is functionally needed for the optimum binding of neurotransmitters and receptors. In AD, however, elevated neuronal iron levels are observed due to increased neuro-fibrillary tangles. Also, in PD, the total iron content of the substantia nigra was 77% higher than other brain regions. These findings are significant because non-protein bound Fe^{3+}, through the Haber-Weiss reaction, can be reduced by $O_2^{\bullet -}$ to produced the oxidized Fe^{2+} form. In turn, Fe^{2+} reacts with H_2O_2 to form the highly reactive ·OH radical. These effects may contribute to high levels of oxidative stress and the concomitant rapid peroxidation of unsaturated brain lipids – leading to a neurodegenerative condition.

Visual Processing - The Anatomy of a Nerve

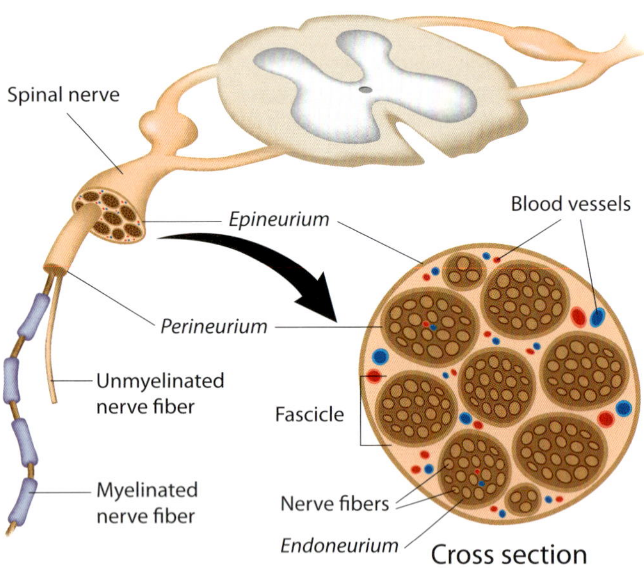

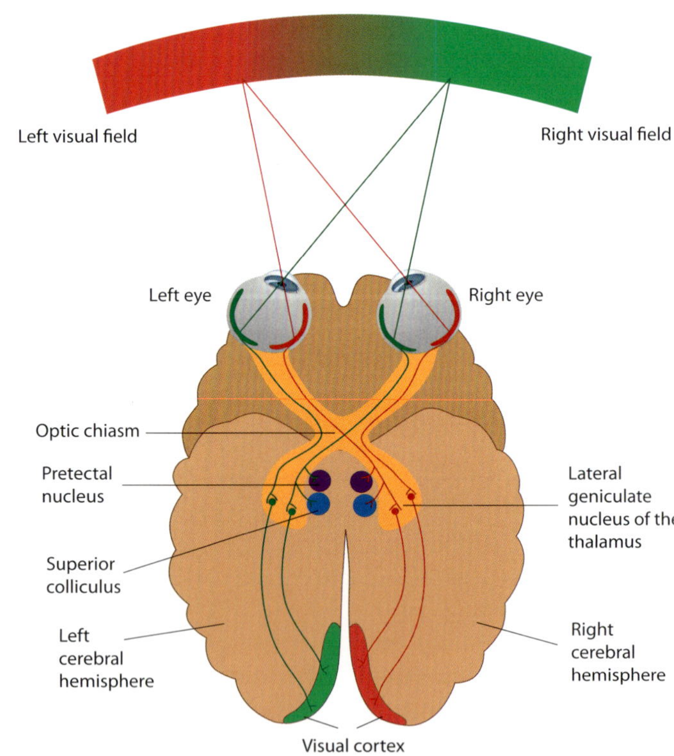

Odor and Visual Processing: The Nobel Prizes in Physiology or Medicine

1981 - Roger Sperry for discovering the function of cerebral hemispheres; jointly to David Hubel and Torsten Wiesel for discovering the information processing of the visual system.

2004 - Richard Axel and Linda Buck for discovering the odorant receptors and the organization of the olfactory system.

4.7.2. Oxidative Stress and Age-related Macular Degeneration

Age-related macular degeneration (AMD), deterioration of the retina around the macular region is the leading cause of permanent blindness in the world. Currently, in the United States alone, 1.75 million people are affected by AMD and 7 million people are at risk of developing AMD. The progression of AMD occurs over an extended time frame, with the incidence of the disease increasing dramatically over the age of 70, mediated by gradual, cumulative damage to the retina from daily oxidative stress. AMD is a multi-factorial disease that affects primarily the photoreceptors and retinal pigment epithelium (RPE), Bruch's membrane, and choriocapillaries. Aging is associated with changes in the retina, including alterations in RPE cellular size and shape, thickening of Bruch's membrane, thickening of the internal limiting membrane, and a decrease in retinal neuronal elements. Alternatively, age-related mitochondrial DNA damage might also play a role in pathogenesis. The molecular pathology of AMD suggests important roles for oxidative stress, inflammation, lipid metabolism, and angiogenesis. The role of oxidative stress is further strengthened by epidemiological studies demonstrating increased risk with smoking, a potent oxidizing stress, and decreased risk with diets rich in antioxidants such as lutein/zeaxanthin.

AMD was described in 1875 by Hutchinson and Tay, as a symmetrical central choroido-retinal disease occurring in senile persons.

Retinal pigment epithelium (RPE) in Visual Function: RPE is a single layer of specialized neural cells that line the posterior segment of the eye. It is located between the light-sensing photoreceptor cells and the choriocapillaris that supplies the retina with O_2 and nutrients. The anatomical positioning of the RPE layer situates these cells for their numerous support functions as guardian and caretaker of the photoreceptors. In conjunction with the endothelium of the retinal vessels, the RPE layer forms the blood–retinal barrier. A primary function of this barrier is to mediate the uptake of ions, water, and nutrients while simultaneously removing metabolic waste products from the sub-retinal space. These exchange processes are vital for maintaining overall metabolic homeostasis and sustenance of the photo-receptor cells. A complementary function of the RPE involves retinoid storage and metabolism. In the classical visual cycle associated with rod photoreceptors, RPE cells convert all-trans-retinol (vitamin A) into 11-cis-retinal. The 11-cis-retinal is then delivered to the photoreceptors for photo-transduction. RPE cells are enriched in numerous pigments, such as melanin, lipofuscin, and flavins, which absorb excess light and thereby function to protect the neuro-retina from phototoxicity. Paradoxically, these same moieties can underlie photochemical damage to the RPE and retina.

RHODOPSIN

An additional key function performed by RPE cells is the maintenance of photoreceptor outer segment length. Each day, the RPE ingests the distal tips of the outer segments and, in doing so, balances the growth of these segments. This growth occurs at the proximal end where new membrane stacks are generated. This trimming function of the RPE ensures that a relatively constant outer segment length is maintained, which is essential for proper photoreceptor function. RPE cells also secrete growth factors, and notably, release the vascular endothelial

Rhodopsin, also known as visual purple, is a biological pigment in the photoreceptor cells of the retina that is responsible for the first events in the perception of light.

growth factor (VEGF) to the choriocapillaris and pigment epithelial-derived growth factor apically to the sub-retinal space. Additional immuno-suppressive factors are also produced and released by RPE cells to impart immune privilege to ocular tissues. RPE cells are also the most active phagocytes that eliminate degraded photoreceptor membranes, which are continuously shed and replaced. The RPE often accumulates lipofuscin (yellow pigment) with age. Degeneration of the RPE leads to the death of its associated rods and cones. The defective RPE can no longer absorb waste products, resulting in abnormal deposits (drusen) in the retina. Drusen are often present (in smaller numbers) in the eyes of elderly people.

Oxidative Stress and the Retina: The retina has a high rate of O_2 consumption due to its intense energy demand for neurotransmission as well as the synthesis and recycling of molecules involved in vision. Even brief ischemia can lead to irreversible vision damage. However, a high rate of O_2 uptake may equate to a high rate of ROS production. The lipids present in the membranes of retinal rods and cones contain a high percentage of PUFA side-chains, especially of docosahexaenoic acid (DHA), and are thus highly susceptible to peroxidation. Rhodopsin, the 'visual pigment' and its product (A2E) readily sensitize formation of singlet O_2. Hence exposure of isolated retinas or animal eyes to high light intensities induces lipid oxidation. Lipid peroxides also increase in the retina after exposure to elevated O_2 or to ionizing radiation, and injection of preformed lipid peroxides into eyes causes severe retinal damage.

Structure of Retina

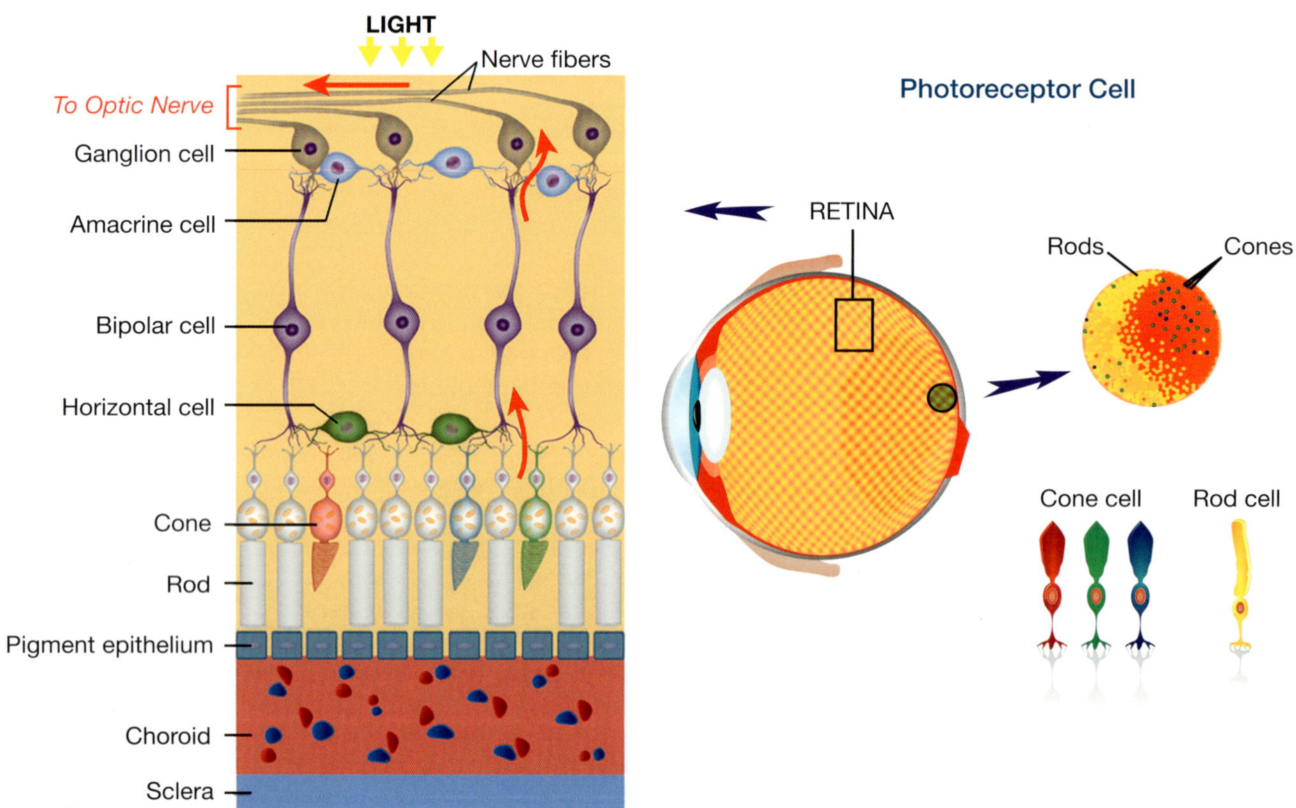

Oxidative Stress and Retinal Pigment Epithelia (RPE): Blood flow rate to choriocapillaris and elevated O_2 saturation levels contribute to high O_2 tension in the RPE layer. Long-term sunlight exposure is also a significant source of free radical stress to the RPE. In particular, blue light (475 nm) causes photo-oxidation of RPE biomolecules (including pigments) and can be especially detrimental. Furthermore, the enriched mitochondrial population of the RPE exhibits robust metabolic activity to cater to the high-energy needs of these cells. This amplified oxidative phosphorylation produces large amounts of ATP, but in the process, also generates high local concentrations of ROS. As the RPE ages, the capacity to utilize and/or neutralize these mitochondrial-derived ROS seems to diminish. Such oxidative stress can lead to mitochondrial dysfunction and impact overall ocular health. In an age-dependent manner, the oxidative stress can elicit compounding deleterious effects by triggering large-scale damage to RPE biomolecules (i.e. proteins, lipids, and DNA) and by promoting destruction of the RPE mitochondrial network. Analysis of postmortem AMD eyes has demonstrated an accumulation of modifications and mutations to RPE mitochondrial DNA that are readily attributable to free radical stress. This 'domino' effect may well represent one, if not the, precipitating event in AMD development.

Development of AMD

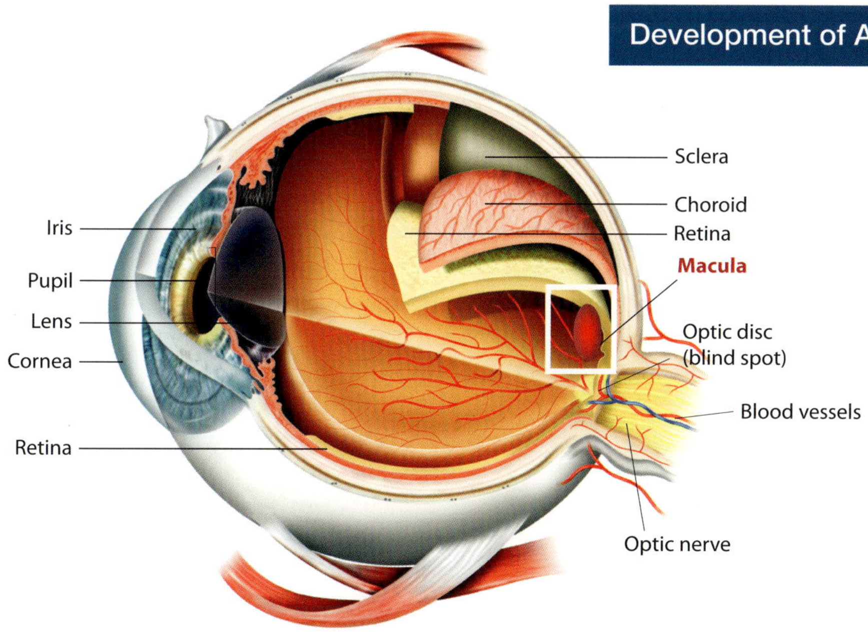

AMD is more prevalent in lightly pigmented individuals where the RPE has fewer melanosomes. The melanosome pigment could block the penetration of the photosensitizing wavelengths to the choriocapillary endothelium. Thus, AMD is less frequently observed in darkly pigmented eyes of Black population versus lightly pigmented eyes of Caucasians.

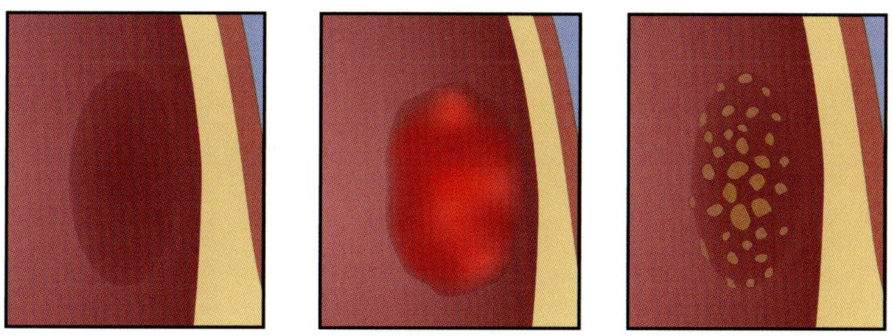

Normal | 'Wet' Macular Degeneration | 'Dry' Macular Degeneration

Oxidative Damage, Lipofuscin and AMD: Another source of intra-cellular oxidative stress in the RPE is the age-dependent accumulation of lipofuscin (aging pigment). Lipofuscin is the generic name given to a heterogeneous group of complex, auto-fluorescent lipid/protein aggregates present in a wide variety of both neuronal and non-neuronal tissues. In the eye, lipofuscin accumulates within the RPE throughout life, eventually occupying up to 19% of cytoplasmic volume by 80 years of age. The spectral dependence of ROS generation by lipofuscin (production being greatest in granules exposed to blue light as compared to other regions of the visible spectrum) may explain the so-called 'blue light hazard' to the retina. An adverse role for lipofuscin in the RPE may explain the association between high levels of lipofuscin and AMD. The RPE is rich in anti-oxidants and these may be sufficient to detoxify any ROS. Conversely, the anti-oxidants may be insufficient to detoxify all the radicals and there may be an insidious buildup of oxidative damage throughout life that only manifests in the aged.

REDOX Chemisty of Lipofuscin: RPE-associated lipofuscin is a photo-inducible free radical generator. White light irradiation of RPE lipofuscin granules results in the production of $O_2^{\bullet-}$ anions, their rate of production increases with higher light intensity. The effect is wavelength dependent; $O_2^{\bullet-}$ anion generation is greatest in granules exposed to blue light (400–520 nm) as compared to red light (660–730 nm) or full white light. Under aerobic conditions, lipofuscin is also capable of photo-generating significant quantities of a variety of ROS including singlet oxygen, hydrogen peroxide and lipid hydroperoxides. The light-induced generation of ROS by lipofuscin supports the hypothetical relationship between light irradiation, aging changes in the retina and retinal degeneration. Rozanowska et al (1995) have made the following extrapolations with respect to $O_2^{\bullet-}$ anion production through photo-induction: (i) that one lipofuscin granule can produce 8 $\times 10^{-19}$ mol $O_2^{\bullet-}$ anion/min; (ii) since 1 mol = 6.02×10^{23} molecules, a single granule is capable of producing 4.8×10^5 $O_2^{\bullet-}$ molecules/min; and (iii) if the average volume of the cell is 2000 μm^3 and up to 19% of that volume is occupied with lipofuscin granules (of 1 µm in diameter) then each RPE cell has the capacity of generating 3.5×10^8 $O_2^{\bullet-}$ anions/cell/min. This high level of free radical production may explain why the RPE contains a high concentration and wide variety of antioxidants.

REDOX-based interventions: The major lipid-soluble antioxidants include vitamin E, retinoids, and carotenoids; the latter group includes lutein and xeaxanthin, which selectively accumulate in the macular and account for the yellow color observed in this region of the retina. A multi-center study found that a decrease in the levels of serum carotenoids is a risk factor for the development of neovascular AMD. Carotenoids are best known for their ability to scavenge singlet oxygen, and photosensitization reactions are a major source of singlet oxygen. Thus, the health of RPE and visual cells is dependent on their REDOX capacity to metabolize free radicals, lipid hydroperoxides and other potentially toxic compounds.

4.7.3. Oxidative Stress and Alzheimer's Disease (AD)

Alzheimer's disease (AD) is a progressive dementia that affects a large group the aging population. The histopathological changes in AD include: widespread neuronal cell death, formation of amyloid plaques, and neurofibrillary tangles (NFTs). The major component of these plaques is amyloid β-peptide (Aβ). Early signs of tangle formation in certain brain regions (such as the entorhinal cortex) precede the clinical diagnosis of AD. There is strong evidence that brain tissue in AD patients is exposed to oxidative stress during the course of the disease.

During dementia, 'sundown' periods increase confusion and agitation as the sun goes down – and sometimes through the night.

Oxidative stress in AD is manifested through high levels of oxidized proteins, advanced glycation end products (AGEs), lipid peroxidation end products, formation of toxic species (such as peroxides, alcohols, aldehydes, free carbonyles, ketones, cholestenone), and oxidative modifications in nuclear and mitochondrial DNA. Lipids are modified by ROS and there is a strong correlation between lipid peroxides, antioxidant enzymes, and amyloid plaques in the brain affected by AD. In the AD pathogenesis, ROS generated in the brain induce calcium influx (and trigger an excitotoxic response leading to cell death. Increased levels of DNA strand breaks have been found in AD. They were first considered to be part of apoptosis, but it is now widely accepted that such oxidative damage is responsible for DNA strand breaks and increased free carbonyls in the nuclei of neurons and glia. The induction of heme oxygenase-1, an antioxidant enzyme involved in the conversion of heme to bilirubin, is increased in AD brains.

Formation of Amyloid Plaques and Neurofibrillary Tangles

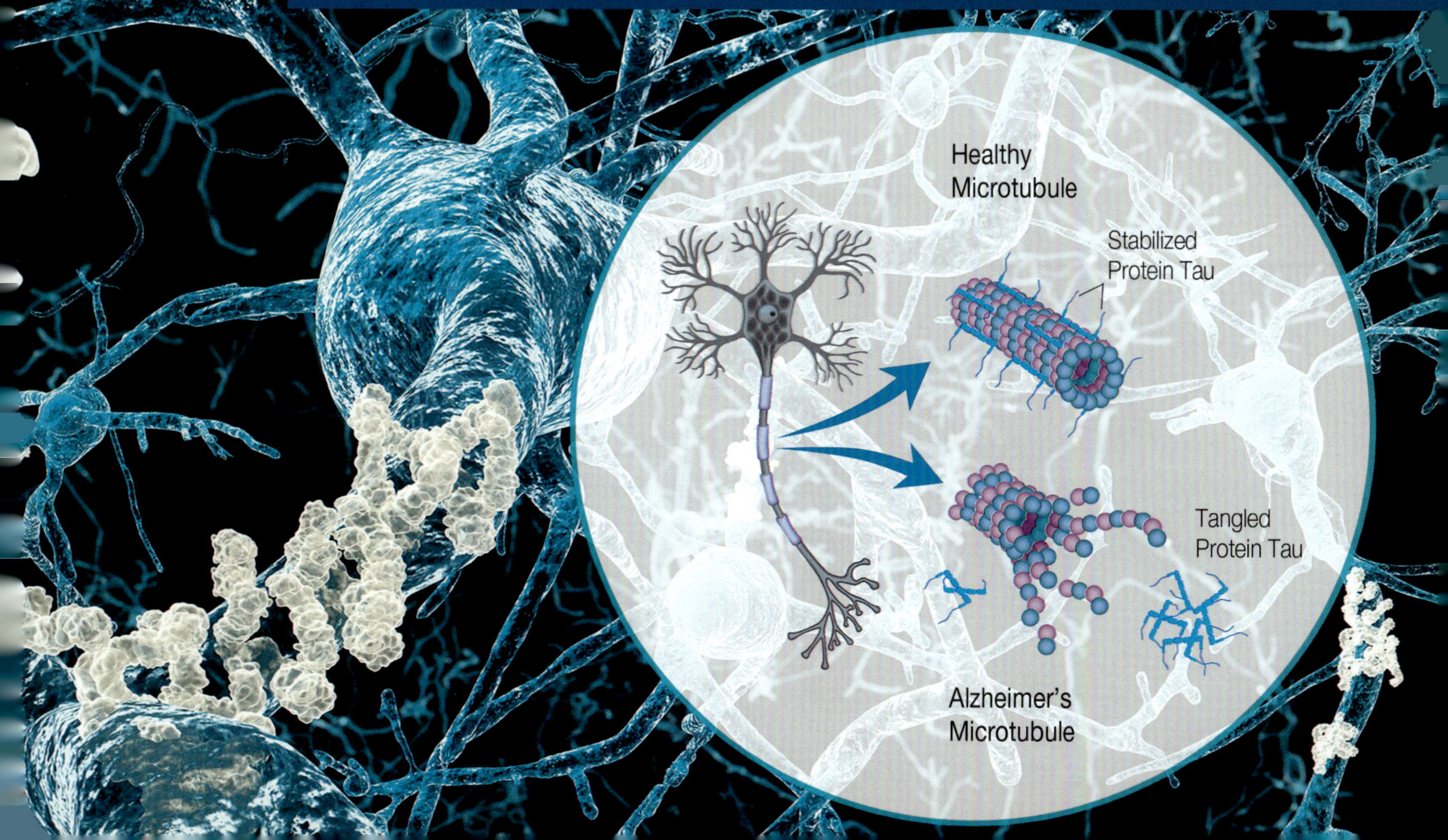

In most people with AD, symptoms first appear after age 60. Estimates vary, but as many as 5.1 million Americans may have AD.

Advanced glycation end products (AGEs) are involved in the pathogenesis of AD. A cascade of reactions lead to the formation of AGEs, which are composed of irreversibly cross-linked heterogeneous protein aggregates. There is increasing evidence that the insolubility of Aβ plaques is caused by extensive covalent protein cross-linking. Accumulation of extra-cellular AGEs has been demonstrated in senile plaques in different cortical areas, in primitive plaques and coronas of classic plaques. Accumulation of extra-cellular AGEs in AD is caused by an accelerated oxidation of glycated proteins ('glycol-oxidation'). Free radicals are involved in glycation processes and can foster the formation of Aβ cross-linking.

REDOX-based Interventions: Treatment approaches which break the vicious cycle of oxidative stress and neuro-degeneration offer new opportunities for the treatment of AD. These approaches include AGE-inhibitors (amino-guanidine, pyridoxamine), antioxidants (thioctic acid, vitamin E, vitamin C, β-carotene) and non-steroidal anti-inflammatory substances (NSAIDs), which not only scavenge radicals passively but also interfere with signal transduction pathways, thereby preventing radical production. AGE inhibitors might be able to stop formation of AGE-modified Aβ deposits or modify their structure with subsequent loss of AGEs binding. Antioxidants are likely to scavenge intra-cellular and extra-cellular $O_2^{\bullet-}$ radicals and H_2O_2 before these radicals damage cell constituents or activate microglia through their action as intra-cellular second messengers. Anti-inflammatory drugs act similarly, attenuating microglial radical and cytokine production. With our growing understanding of the molecular basis of the clinical symptoms of dementia, particularly positive feedback loops involving oxidative stress, it is hoped to further elucidate the etiopathogenesis of AD. This will help to develop novel neuro-protective treatment strategies that are able to interrupt the vicious cycle of oxidative stress and neuro-degeneration. Among such efforts, REDOX cell signaling technologies offer a promising direction

Alzheimer's Disease - Progression Stages

As AD progresses, brain tissue shrinks. Short-term memory declines when the cells in the hippocampus degenerate. As AD spreads into the cerebral cortex, judgment worsens, emotional outbursts may occur and language is impaired. Advancement of AD leads to death of more nerve cells and changes in behavior, such as wandering and agitation. In the final stages, people may lose the ability to speak, recognize people and control bodily functions. Memory worsens and may become almost non-existent. On average, those with AD live for 8 to 10 years after diagnosis, but AD can last for as long as 20 years.

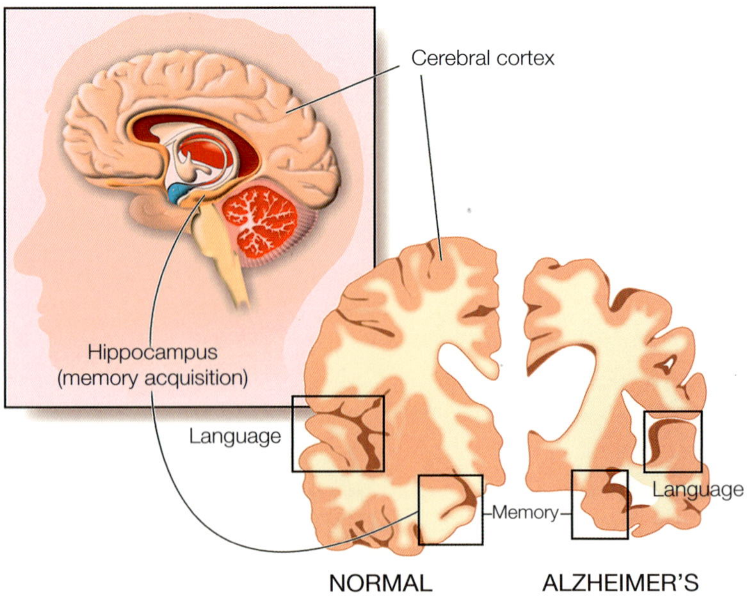

4.7.4. Oxidative Stress and Parkinson's Disease (PD)

In the United States, it is estimated that between 0.1-0.2% of all individuals are at some stage in the progression of PD. This rate is somewhat higher in the elderly and those who have experienced head trauma.

Parkinson's disease (PD) is the 2nd most common disorder of the aging brain (after Alzheimer's). PD is a progressive, age-related movement disorder with neuropathology characterized by degeneration of neurons in the substantia nigra. The primary symptoms of PD are caused by the insufficient formation and action of dopamine within the substantia nigra (SN) region and striatum of the brain. This dopamine deficiency results in tremors, muscle rigidity, cognitive dysfunction, changes in speech patterns and a dynamic progressive dysfunction that can lead to a total loss of all physical movement and ultimately death. Though the main motor manifestations of PD are linked to neuro-degeneration in the dopaminergic systems; non-motor manifestations such as pain, cognitive impairments, and constipation are more likely linked to catecholaminergic and non-catecholaminergic pathways. A profound deficit in brain dopamine is the basic biochemical hallmark of PD; because all of the ascending dopaminergic pathways in the CNS degenerate to variable degrees.

Oxidative stress contributes to the cascade leading to dopamine cell degeneration in PD. However, oxidative stress is intimately linked to other components of the degenerative process, such as mitochondrial dysfunction, excitotoxicity, nitric oxide toxicity and inflammation. Oxidative damage to lipids, proteins, and DNA occurs in PD, and toxic products of oxidative damage, such as 4-hydroxynonenal (HNE), can react with proteins to impair neural cell viability. There is convincing evidence for the involvement of NO that reacts with $O_2^{\bullet -}$ to produce $ONOO^-$ and ultimately ·OH radical production. ROS are elevated in the brains and peripheral blood mononuclear cells of patients with PD. Nitrite in the cerebrospinal fluid as well as increased expression of inducible nitric oxide synthase (iNOS) within the SN has also been found in PD patients, contributing to strong evidence of oxidative stress in the SN of PD patients.

Disrupted iron metabolism in PD was first reported in 1924 by Lehermitte and McAlpine. It is well established that iron levels are increased in the PD substantia nigra, the anatomic region most vulnerable to degenerate in the disease. The neuro-degeneration with iron accumulation in brain is also associated with α-synuclein pathology. Substantial iron deposits are associated with neuronal loss, gliosis, and Lewy body pathology. Iron sequestered in Lewy bodies and other patho-biological pools of iron in PD brain has the potential to promote Fenton chemistry and oxidative damage to macromolecules. The oxidative damage to macromolecules has been reported in postmortem PD tissue, with proteins, nucleic acids, lipids and sugars all showing evidence of oxidative modification. The primary toxic ROS is ·OH radical, which is formed from two substrates,

Nerve functions: The Nobel Prizes in Physiology or Medicine

1932 - Charles Sherrington and Edgar Adrian for discovering nerve functions.

1936 - Henry Dale and Otto Loewi for discovering the chemical transmission of nerve impulses.

1944 - Joseph Erlanger and Herbert Gasser for discovering the differentiated functions of single nerve fibers.

H_2O_2 and a REDOX active metal (e.g., Cu^{+}, Fe^{2+}). There are a number of possible sources of toxic ROS in the PD brain. REDOX active iron in neurons may be the primary source. Since dopaminergic cells are uniquely vulnerable to degenerate, dopamine may provide a source for ·OH radical generation. Dopamine is known to coordinate iron and regenerate Fe^{2+} (the oxidative form), allowing it to donate electrons to H_2O_2. Alternatively, ·OH radical formation may be increased with the availability of $O_2^{\bullet-}$ anion and H_2O_2 from impaired oxidative phosphorylation in mitochondria. There is ample evidence that oxidative phosphorylation becomes increasingly inefficient during the aging process and in PD condition. Abnormal function of complex I of the mitochondrial ETC has been reported from the PD brain, particularly in the substantia nigra. Functionally deficient complex I could likely increase $O_2^{\bullet-}$ anion production and promote oxidative stress through subsequent SOD and Fenton chemistry. Somatic mitochondrial DNA mutations have been reported in the PD brain and also in the aging brain.

Degeneration of Neurons in the Substantia nigra

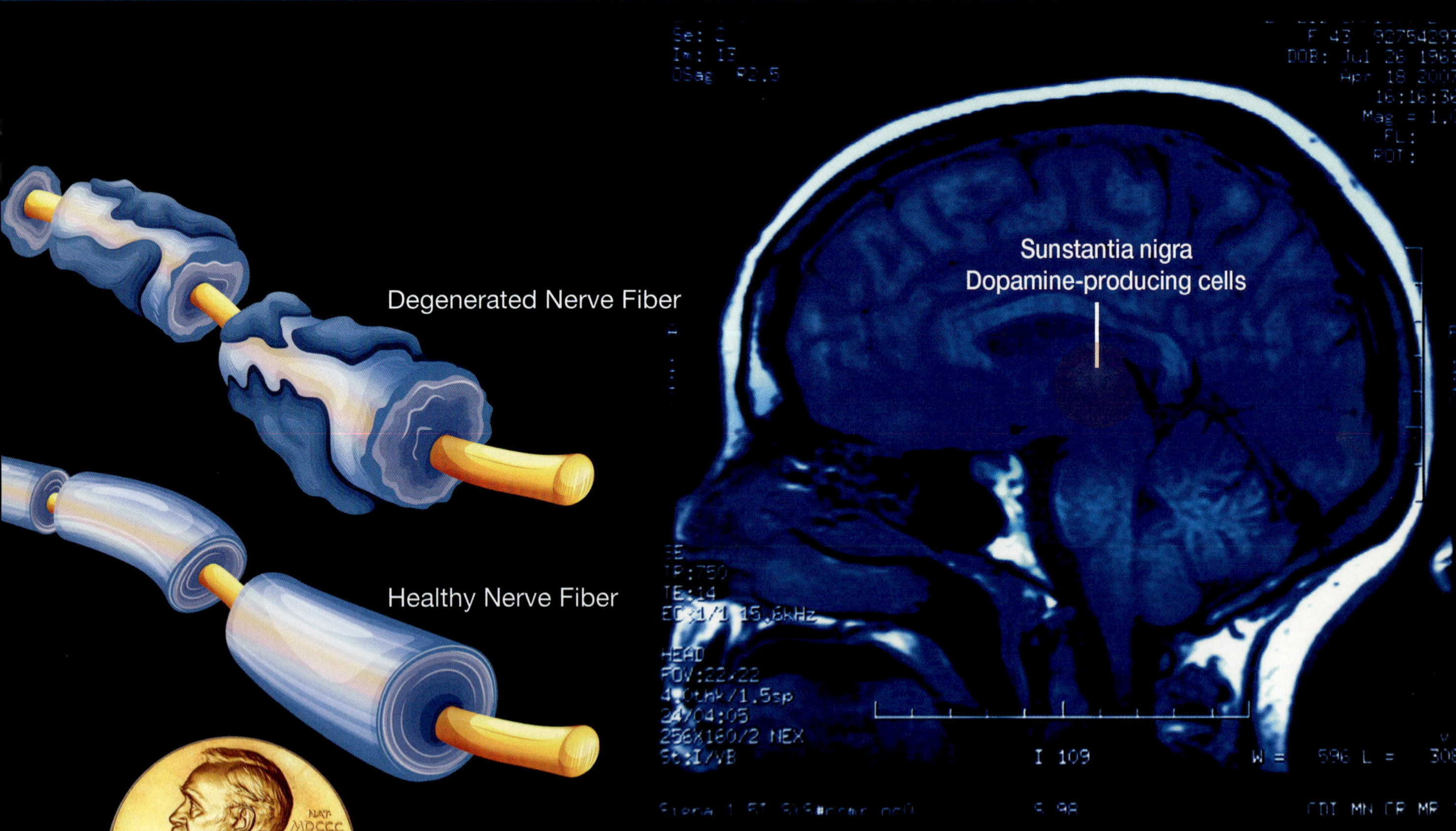

Excitation and Inhibition of Nerves: The Nobel Prize in Physiology or Medicine 1963 was awarded jointly to Sir John Carew Eccles, Alan Lloyd Hodgkin and Andrew Fielding Huxley for their discoveries concerning the ionic mechanisms involved in excitation and inhibition in the peripheral and central portions of the nerve cell membrane.

REDOX-based Interventions: Numerous inflammatory mediators such as ROS, NO, and other products of activated immune cells can play a role in the degeneration of SN dopamine-producing neurons in PD. Therefore, it is likely that treatment with anti-inflammatory reagents (directed at these pro-inflammatory targets) could potentially halt or slow neuro-degeneration. For example, steroidal anti-inflammatory drugs (SAIDs) such as dexamethasone have been reported to show neuro-protection. Nonsteroidal anti-inflammatory drugs (NSAIDS) such as aspirin and ibuprofen reduce inflammation by inhibition of COX activity and have been found to be modestly effective at slowing the progression of neuro-degeneration.

Currently, it is believed that the progression of PD is also due to oxidative stress damage - not only from microglia, but also from within the neurons (due to increased dopamine turnover). The current standard of care for the treatment of PD is the administration of L-3,4-dihydroxyphenylalanine (L-DOPA), which, in the short term, relieves many of the symptoms of PD. It has also been found to induce an oxidative stress response in DA-neurons. It is believed that L-DOPA at high doses actually leads to cell death in neurons due to its activation of apoptosis through the REDOX signaling pathways. However, given the central nature of microglial cells and neuro-inflammation in the pathogenesis of PD, it is important to determine the role of L-DOPA therapy on the REDOX reaction of microglial cells. It has been found that the oxidative stress of peripheral blood mononuclear cells from patients with PD appears to be higher than healthy patients, but L-DOPA therapy appears to lower the REDOX response of these monocytes. Therefore, while L-DOPA-induced oxidative stress may play an important role in the direct toxicity of DA neurons as seen in progressive PD, the effect of L-DOPA therapy on the oxidative stress response of macrophages and microglial cells (and its direct damage on DA neurons) is still not determined. The anti-inflammatory effects of L-DOPA therapy may actually play a role in short-term therapy to help protect DA-neurons from inflammatory damage mediated by activated microglial cells.

PATHO-PHYSIOLOGY

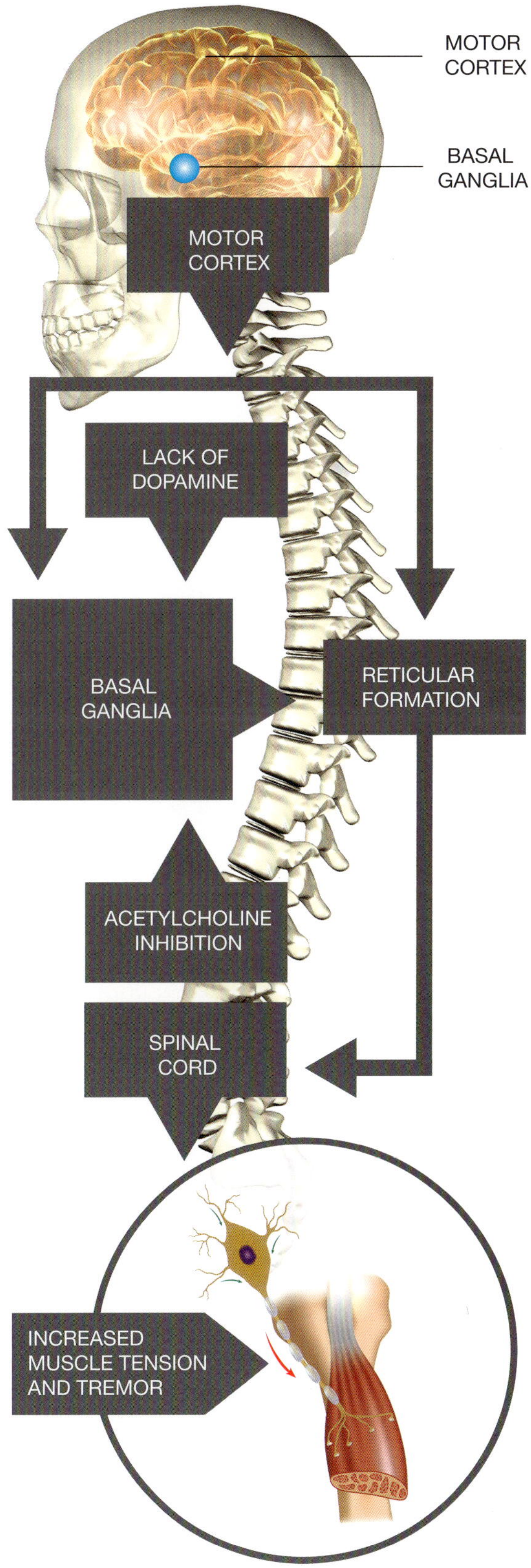

4.7.5. Oxidative Stress and Autism

Autism spectrum disorders (ASD) are complex neuro-developmental behavioral conditions characterized by significant impairments in social interaction, verbal and non-verbal communication, and by restrictive, repetitive and stereotypic behavior patterns. The Centers for Disease Control (2006) estimates that the prevalence of ASD in the United States is 1 in 110 children. Although cognitive and behavioral issues arise from dysfunction of the central nervous system, several studies have documented the involvement of non-neural physiological abnormalities in the development of ASD. These etiological factors include immune dysfunction, inflammation, impaired detoxification, REDOX regulation/oxidative stress, and energy generation/mitochondrial dysfunction. Increased oxidative stress is one of the major features present in autism. Higher rates of mitochondrial hydrogen peroxide (H_2O_2) production (accompanied by lower activities in the pyruvate dehydrogenase complex, Complex I alone, or in combination with other Complexes) from lymphocytic mitochondria and increased markers of oxidative stress (i.e. oxidized glutathione, GSH) have been reported in individuals with autism.

Mitochondrial Dysfunction (MD) and Autism: ASD appears to be associated with systemic metabolic abnormalities. About 50% of children with ASD have biomarkers of MD that are consistently abnormal. About 80% of the children with ASD demonstrated lower than normal electron transport chain (ETC) function in lymphocytes (when compared to controls). However, only 23% of children with ASD/MD have a known mtDNA mutation; some children with ASD/MD have ETC over-activity rather than ETC deficiencies. Furthermore, many ASD/MD cases do not manifest lactate elevation, a key biomarker commonly used to identify individuals with MD. This has raised the idea that children with ASD might have mitochondrial dysfunction that is more prevalent and distinct from classic MD.

In addition to MD, abnormalities in REDOX metabolism and oxidative stress appear to be major metabolic abnormalities that impact a significant portion of children with ASD. Clinical studies have documented glutathione (GSH) deficits in plasma, immune cells and post-mortem brains from ASD children. Furthermore, ROS-mediated damage to mt-DNA and lipids, abnormalities in other ROS biomarkers (such as homocysteine), and epigenetic changes to REDOX homeostasis have been reported with ASD. A deficit in GSH antioxidant capacity is well known to promote oxidative stress and increase vulnerability to prooxidant environmental exposures and oxidative damage. In addition, 3-nitrotyrosine (3-NT), a biomarker of oxidative damage to proteins, and 3-chlorotyrosine (3-CT), a biomarker of chronic inflammation, have been found to be elevated in plasma and brain of children with ASD.

Brain Hormones: The Nobel Prize in Physiology or Medicine 1977 was divided; one half jointly to Roger Guillemin and Andrew V. Schally for their discoveries concerning the peptide hormone production of the brain and the other half to Rosalyn Yalow for the development of radioimmunoassays of peptide hormones.

Oxidative stress may be a key mechanism by which mitochondria are negatively influenced by factors linked to ASD such as pro-oxidant environmental toxicants and immune abnormalities, as ROS can result in mitochondrial dysfunction. Dysfunctional mitochondria can create a self-perpetuating cycle of progressive damage that amplifies functional deficits. Indeed, mitochondria are a major source of ROS as well as the target for ROS-mediated damage. ETC complex I dysfunction can produce high levels of ROS and RNS, which can inhibit ETC function, such as complex I activity, causing MD. Thus, both the exogenous or endogenous sources of ROS can initiate or promote MD and damage, especially if the mitochondria are genetically susceptible to dysfunction.

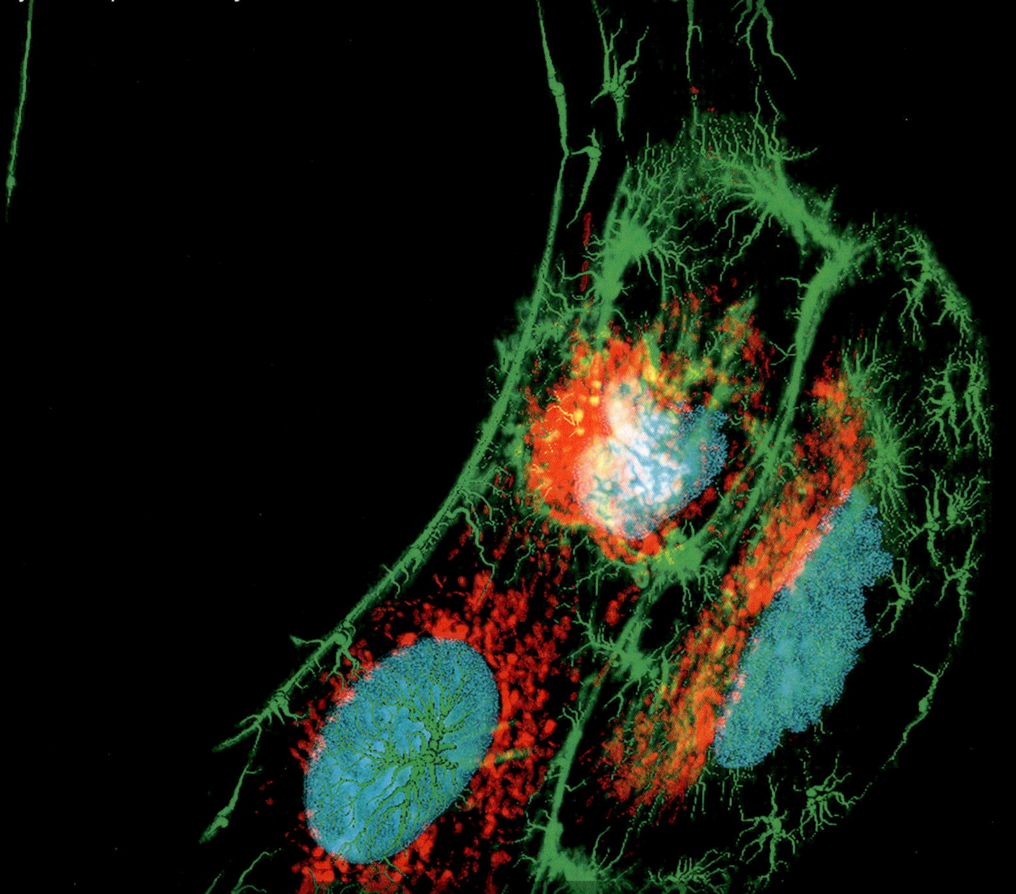

The biological basis for MD is unknown, however, only 23% of children with ASD/MD have mt-DNA abnormality. This suggests that MD may be acquired. It has been suggested that systematic abnormalities in ASD, such as MD, may arise from environmental triggers in genetically sensitive sub-populations. This notion is strongly supported by studies with dizygotic twins, wherein it was estimated that the environment contributed a greater risk of developing autistic disorder (55%) as compared with genetic factors (37%). Mitochondria are central to this concept as polymorphisms in mitochondrial genes can result in susceptibility to disease and MD can result from environmental exposures that have been implicated in the development of ASD such as heavy metals, chemicals, polychlorinated biphenyls or pesticides as well as physiological abnormalities associated with ASD such as elevated TNF-α, GSH deficiency and oxidative stress.

Glutatione (GSH) Depletion and Autism: Oxidative stress occurs when cellular antioxidant defense mechanisms fail to counterbalance endogenous ROS production and/or exogenous prooxidant environmental exposures. GSH (γ-L-glutamyl-L-cysteinylglycine) is a tripeptide that functions as the major intra-cellular antioxidant and REDOX buffer against macromolecular oxidative damage. The GSH thiol/disulfide couple (GSH/GSSG) is the predominant mechanism that maintains the intra-cellular REDOX balance [refer to Chapter Sections 3.2.2(A) & 3.3.1]. GSH depletion and oxidative stress have been implicated in the pathology of numerous neuro-behavioral disorders including schizophrenia, bipolar disorder, and Alzheimer's disease. Also REDOX imbalance and oxidative stress contributes to the patho-physiology of autism. Autistic children show a marked decline in plasma GSH levels and several of its metabolic precursors; while exhibiting an increase in oxidized GSH-disulfide (GSSG) and a decrease in GSH/REDOX ratio (GSH/GSSG). Several interactive polymorphisms in enzymes that regulate GSH synthesis were also found to be more prevalent in children with autism. These observations suggest that the GSH deficit and predisposition to oxidative stress may be genetically regulated in certain autistic children.

REDOX-based Interventions: The interaction between REDOX homeostasis, inflammation and mitochondrial function is vital to maintaining healthy cell physiology, therefore, critical for disease initiation, progression and treatment of ASD. Hyperbaric oxygen therapy (HBOT) has increased in popularity as a treatment for autism. Several studies have demonstrated that HBOT is effective in the clinical management of ASD, especially in reducing oxidative stress and inflammation, along with enhancement of neurological function and cognitive performance. Generally, HBOT involves inhalation of up to 100% oxygen at a pressure greater than one atmosphere (atm) in a pressurized chamber. In a double-blind, placebo-controlled, cross-over study, HBOT at 2.5 atm and 100% O_2, improved cognitive function, including memory in elderly patients. Furthermore, concerns exist that HBOT might increase oxidative stress via the production of ROS. Alternatively, some evidence also suggests that HBOT could actually alleviate oxidative stress in children with autism. Therefore, improving hypoxic areas in the autistic brain might decrease oxidative stress.

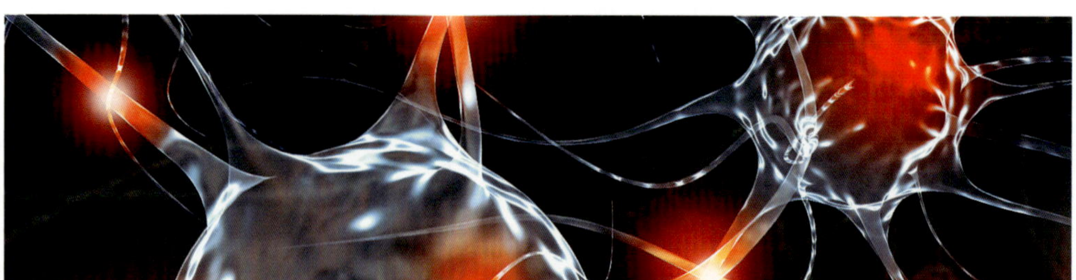

Growth Factors: The Nobel Prize in Physiology or Medicine 1986 was awarded jointly to Stanley Cohen and Rita Levi-Montalcini for their discoveries of growth factors.

Autism Facts

- Genes are a risk factors - children who have a sibling or parent with an ASD are at a higher risk of also having an ASD.
- Autism affects 1 in 110 children. Boys are 4 times more likely to have autism than girls.
- Individuals with autism often suffer from numerous co-morbid medical conditions which may include: allergies, asthma, epilepsy, digestive disorders, persistent viral infections, feeding disorders, sensory integration dysfunction, sleeping disorders, and more.
- Vaccinations can trigger or exacerbate autism in some, if not many, children, especially those who are genetically predisposed to immune, autoimmune or inflammatory conditions.
- *Pesticide Exposure*: Scientists think that chemicals in pesticides may adversely affect those who are genetically predisposed to autism, leading them to develop the full-blown disorder.
- *Parental Age*: Women who are 40 years old have a 50% risk of having a child with autism than women who are between 20 and 29 years old.
- *Pharmaceuticals*: Babies exposed to certain pharmaceuticals in the womb (including SSRI's, valproic acid and thalidomide), may have a higher risk of autism.
- *Freeway Proximity*: Children born to mothers who live within a 1000 feet of freeways have twice the risk of autism.
- *Limited Prenatal Vitamin Intake*: Women not taking prenatal vitamins immediately before and during a pregnancy are twice as likely to have a child with autism.

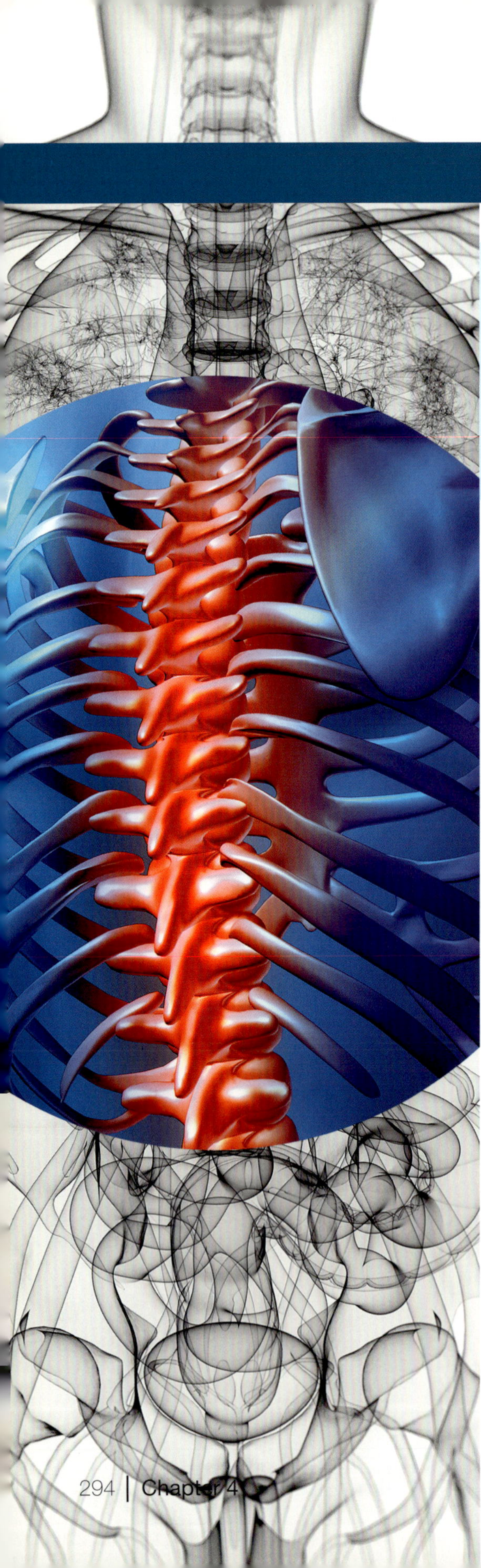

4.8 REDOX: The Skeletal System

The human skeleton consists of bones, cartilage, ligaments and tendons that account ~20% of the total body weight. Bones, muscles and joints work together to constitute the skeleto-muscular system. Muscles pull on the bone joints to facilitate body movement. The human body has more than 650 muscles that make up 50% of the body weight. Joints or articulations provide flexibility and movement.

The ends of bone are covered by cartilage to reduce mechanical friction or grinding of bone joints. The joint is surrounded by a protective capsule called the 'synovium', which produces synovial fluid. Synovial fluid is a clear substance that lubricates and nourishes the cartilage and bone joints. Bone joints are vulnerable to injury and degeneration with aging. Ligaments connect bone to bone and tendons connect muscles to bone; their purpose is to transmit biomechanical forces. The fibers of both ligaments and tendons are made of closely packed collagen fibers. These fibers are held together with other proteins, particularly with proteoglycan, in compressed regions of the tendon. Tendons can passively modulate forces during locomotion, and provide additional stability to bone joints. It also allows tendons to store and recover energy at high efficiency. The mechanical properties of ligaments and tendons increase from early childhood to young adulthood; however, further aging processes affect their stiffness. Immobilization of a joint for long periods of time is detrimental to joint structure and function, including decreased range of motion.

Cartilage is a dense connective tissue that allows bones to slide over one another at all moveable joints with ease. As a result, cartilage reduces friction and prevents damage. Any breakdown of cartilage in the bone joints leads to a severe bone disease – osteoarthritis. Bone mass and its architecture or shape is influenced by the mechanical forces on the skeleton. Genetics and life style factors also play a critical role in the structural outcome of a skeleton. Bone mass and architecture continuously change throughout life in response to mechanical stress and function.

Bone Metabolism: Bones change in size, shape, and position by modeling and remodeling processes. In modeling, a bone is formed at one site and broken down in a different site; this changes its shape and position. In remodeling, the cellular activity of bone removal and replacement occurs at the same site. While remodeling is the predominant process during early adulthood - the bone forming years - modeling continues throughout life (in response to weakening of the bone). Dynamic living tissue is under continuous reorganization; however, the normal bone maintains equilibrium between the old bone being dissolved and the new bone being laid down. This process is called bone turnover. Bone turnover ensures the mechanical integrity of the skeleton throughout life and plays an important role in calcium homeostasis. The processes of bone building and breakdown in response to internal and external signals are carried out by specialized cells that build or break down bone. The cells that form bone are osteoblasts and those that break down the bone are osteoclasts.

Most of the adult skeleton is replaced about every 7 years.

Calcium absorption in the gastrointestinal tract involves two specific cellular pathways: (i) trans-cellular, an active transport that requires magnesium and vitamin-D; and (ii) para-cellular, a passive process that requires acidification of the calcium salts in the stomach. Factors that influence calcium absorption include: age and health status of the individual, vitamin-D availability, calcium levels in food consumption, and type and amount of fiber in the diet. It should be noted that all calcium ingested is not absorbed into the body. Calcium absorption begins in the stomach with hydrochloric acid, which dissolves, ionizes and facilitates mineral assimilation in the gut. Secretion of stomach acid decreases gradually with age.

Several hormones affect calcium mineralization in the bone. The parathyroid hormone (PTH) transports calcium from the bones into the bloodstream. It also signals the kidneys to conserve calcium and other minerals from the urine. Additionally, PTH signals the kidneys to produce calcitrol, which is formed from vitamin D, which regulates the small intestine to absorb more calcium. The thyroid gland secretes calcitonin, which increases bone mineralization, and decreases the rate of bone breakdown.

Calcium homeostasis is a closely regulated process that maintains the levels of Ca^{2+} ions in the body within a tight normal range. During this process, calcium is removed from the bone and released into the blood circulation to maintain Ca^{2+} at 2.5 mmol/L (normal level). The typical calcium content of the adult human body is 1 kg, virtually all of which is found in the skeleton; the amount in body fluids and cells of the soft tissues accounts for the remaining for 1% calcium reserve.

Bone Functions: Bone is a remarkable organ that serves both a structural function (by providing mobility, support, and protection for the body) and a reservoir function (acting as the storehouse for essential minerals). Though delicate in appearance, the bones in our skeletal system are ounce for ounce, stronger than 'mild' steel. The skeletal bone framework is adapted to provide adequate strength and mobility to resist factures upon substantial impact, or during vigorous physical

Up to 40% of post-menopausal women may be severely deficient in natural stomach acid, with a predisposed risk of poor calcium absorption.

activity. Shape and structure of the bone are equally important as its mass in providing strength. The skeletal system performs multi-functional tasks, including:

- Mineral management: The dynamic process of withdrawal and storage of calcium is critical for several vital activities in the body, especially for the heart to beat, the nerve cell to flash an electric impulse, the muscle to twitch, the bowel to move, and the kidneys to filter. Maximum depository of minerals, defined as the 'peak bone mass', largely depends on genetics, environmental factors, and diet.
- Hematopoiesis (blood production): Bone marrow constitutes 4% of total body weight in adults. Bone marrow generates red blood cells (erythrocytes) that oxygenate the tissues, white blood cells (leukocytes) that provide host defense, and platelets that facilitate blood clotting.
- Protective frame: It provides a sturdy 'cage of bones' that houses the delicate internal organs and fragile body tissue. The fused bones of the cranium enclose the brain and make it less vulnerable to injury; the vertebrae surround and protect the spinal cord; and the bones of the rib cage efficiently shield the heart and lungs.
- Acid-base balance: Bone contains large stores of buffer, in the form of mineral salts, to effectively balance the pH changes that occur in the body. Acid-base balance (or titration) in the body has a significant metabolic outcome on the bone turnover, especially on the rates of bone resorption and calcium mobilization.
- Detoxification: Bone scavenges several toxic chemicals, in particular, the heavy metal contaminants (i.e. lead, mercury) in body fluids and effectively disposes of such artifacts via the circulatory route that runs through the skeletal matrix.
- Sound processing: Sound waves are acquired by the external ear and channeled through the ear canal to the eardrum. Three bones in the body, known as the auditory ossicles, help to convert the sound waves (vibrations in air) to mechanical (hydraulic) vibrations in tissues and fluid-filled chambers. Even the smallest vibration of the eardrum results in a significant amplification in the fluid chamber. This allows us to hear even the faintest of whispers.
- Movement: Movement is a basic function, which manifests into postures and motions. Mobility is dependent on several muscles that attach to the bones through tendons, ligaments and cartilage. Muscle contraction, a familiar phenomenon that flexes the 'biceps' or tightens the 'abs', is the quintessential force that enables mobility. Human evolution freed upper limbs from the burden of bearing body weight during locomotion; this enabled us to grasp and manipulate objects with precision.
- Structural frame: Skeletal structure (or size of the bone) grows and changes only up to the point at which a human reaches adulthood; then it remains essentially same for rest of the life. Genetics, gender and lifestyle play a cumulative role in the overall development and appearance of body shape.

Bone replenishment is the homeostatic management of bone turnover, maintenance, and repair of the skeletal system. Multiple factors control bone homeostasis. Humoral factors, such as parathyroid hormone and prostaglandin E, stimulate bone resorption, as well as bone formation, in tandem. Lactoferrin (LF) and ribonuclease (RNase) appear during the early embryonic stage (ossification and vascularization) and regulate bone metabolic homeostasis through each phase of human development. These vital factors are continuously depleted by bone metabolism; and their synthesis (levels) gradually declines with aging. Given the dramatic increase in skeletal size during growth, and the need to preserve skeletal mass during adulthood, and the importance of the bone "bank" having the ability to make mineral deposits and withdrawals – homeostasis takes the center stage. Both intrinsic and extrinsic mechanisms control metabolic homeostasis of the bone. In intrinsic regulation, the metabolic pathway self-regulates in response to changes in the levels of substrates or products; for example, a decrease in the amount of calcium or phosphate can increase the flux through a pathway to compensate the loss.

The inter-reliance between the cardiovascular and skeletal systems manifests as follows: (i) the bones provide the cellular machinery for the circulatory network, whereas, (ii) the circulatory portal supports the bone metabolism with mobilization of chemicals to and from the skeletal tissue. Bone-replenishment plays a significant role in maintaining the fine balance between both systems. Malfunction or breakdown in bone-replenishment can impair several vital pathways and lead to homeostatic imbalance. Abnormal or insufficient angiogenesis (formation of new blood vessels) is the cause of many bone disorders including arthritis, synovitis, osteomyelitis and osteophyte formation. Osteoporosis and impaired healing of bone fractures is linked to insufficient angiogenesis. Age dependent decline of angiogenesis also leads to impaired bone formation. Ribonuclease (RNase), a class of angiogenins derived from milk has recently shown very promising results in promoting bone formation with age-related and post-menopausal bone loss. Bio-replenishment is a potent, physiological process that underlies the natural manner in which bones respond to any decline in nutrient supply through the circulatory transport. The production of new collateral vessels with RNase-replenishment may help overcome the bone metabolic imbalance.

Transport and uninterrupted delivery of nutrients, vitamins, minerals, building blocks (amino acids) and regulatory molecules (enzymes, hormones, etc) to the bone matrix is critical for homeostasis of the skeletal system. Therefore, digestive health is vital for mineral transport and to maintain calcium homeostasis. Lactoferrin (LF), which appears at early embryogenesis and throughout the life, plays a critical role in skeletal development and homeostasis. Both RNase and LF have been proven to be potent anabolic agents for bone formation. A combination of RNase-enriched lactoferrin (R-ELF) could serve as an effective bio-replenishment in restoring and maintaining the balance of bone turnover, as well as, in establishing homeostasis in the skeletal system.

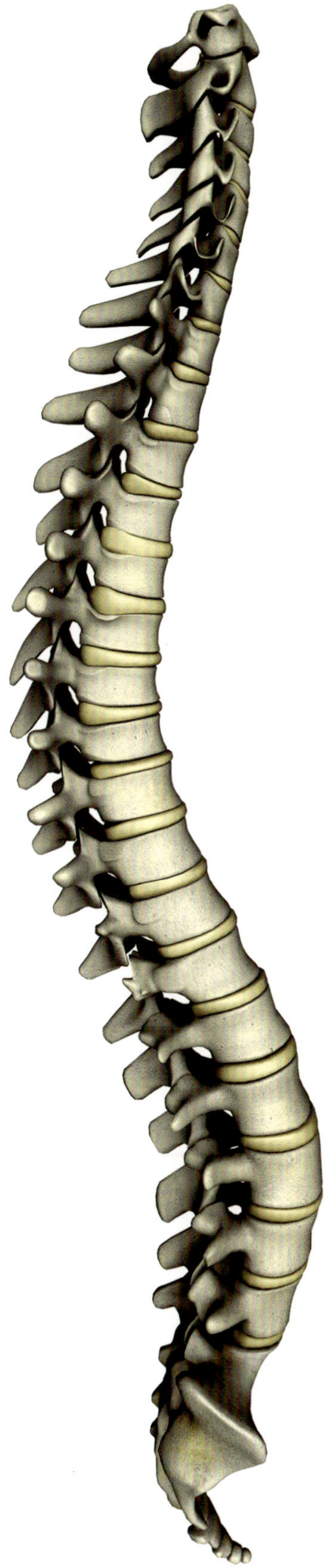

4.8.1. REDOX and the Bone-Joint Health

Bone is a complex tissue (composed of several cell types) that continuously undergoes a process of renewal and repair termed as 'bone remodeling'. The two major cell types responsible for bone remodeling are bone dissolving osteoclasts and new bone forming osteoblasts. Reactive oxygen species (ROS) are generated during normal bone metabolism; however, exessive ROS levels can cause a REDOX imbalance (prooxidants vs. antioxidants), resulting in oxidative stress. For example, ROS play a major role in osteoporosis, in part, by inhibiting osteoblast generation from osteoprogenitors cells. ROS are also involved in cartilage homeostasis and degradation, especially in the development of osteoarthristis (OA). Rheumatoid arthritis (RA) is an autoimmune disease characterized by chronic inflammation leading to joint destruction. Patients with RA demonstrate higher levels of ROS than healthy subjects. Recent studies have also reported that ROS may be a key modulator of bone cell function and play a role in mediating intra-cellular signaling.

Bone remodeling is the continuous process of bone formation and resorption. Any imbalance in favor of the osteoclasts leads to pathological bone resorption, as seen in bone diseases. ROS are involved in apoptosis of osteoblasts, osteocytes, and osteoclastogenesis and therefore, also in bone resorption. ROS also elicit a spectrum of responses ranging from proliferation to growth to differentiation arrest to cell death by activating numerous signaling pathways.

Osteoclasts are acid secreting polykaryons with high energy demands, and therefore, contain abundant mitochondria. ROS, H_2O_2 in particular, generated by the mitochondria, play a crucial role in osteoclast differentiation and function. Also, ROS could increase osteoclast numbers and bone resorption. The molecular triad: osteoprotegerin (OPG), the receptor activator of NF-kB (RANK) and the RANK-Ligand (RANKL), regulate osteoblast and osteoclast activities in the bone remodeling process. Abnormal RANK-RANKL-OPG triad has been implicated in the pathogenesis of various skeletal diseases, including different forms of osteoporosis and inflammatory bone diseases. The elevated osteoclastic activity may increase the $O_2^{\bullet-}$ production and/or inhibit superoxide dismutase (SOD) and glutathione peroxidase (GPx) activities with concomitant bone destruction. TNF-α overproduction is the main contributor of increased ROS release in patients with RA. TNF-α cause cellular damage and inhibits SOD activity in the bone joint. Nitric oxide (·NO) has biphasic effects on osteoclasts. Low concentrations of ·NO potentiate interleukin (IL-1) induced bone resorption. Production of ·NO within osteoclasts is, however, essential for normal osteoclast function. Activation of the iNOS pathway is critical for IL-1-stimulated bone resorption. IL-1 primarily acts on osteoblasts to increase ·NO synthesis by activation of the iNOS pathway, which in turn promotes nuclear translocation of the transcription factor NF-kB in osteoclast progenitors.

Osteoblasts produce antioxidants, such as GPx to protect from ROS; and transforming growth factor (TGF) ß to reduce bone resorption. Nitric oxide has a biphasic effect on osteoblast activity. Small amounts of ·NO, produced in the synovium, acts as an autocrine stimulator of osteoblast growth and cytokine production. However, at high levels, ·NO has potent inhibitory effects on the growth and differentiation of osteoblasts. An increase in intra-cellular ROS could suppress the differentiation of osteoblasts. Hydrogen peroxide or xanthine/xanthine oxidase-induced oxidative stress is also able to inhibit bone marrow stromal cells that undergo osteoblastic differentiation.

Estrogen stimulates eNOS activity, as well as, mRNA levels in osteoblasts. Thus, ·NO derived from the eNOS pathway plays a role in mediating the effects of sex hormones in the bone. Estrogens could decrease oxidative stress and antagonize ROS-induced osteoblast apoptosis and effects of RANKL on osteoclasts. Nevertheless, the REDOX control on osteoblast and osteoclast cell differentiation/function is currently unraveled by several laboratories. Understanding these REDOX mechanisms will make a paradigm shift from the estrogen-based theories, where age-related mechanisms intrinsic to bone are central to the disease process.

Physical activity is recommended as a treatment for the prevention of bone loss. Mechanical loading increases oxidative stress in chondrocytres and osteoblast-like cells. On the other hand, exercise can also increase certain antioxidants in bone (as well as cartilage). Several studies have shown that ·NO synthesis can be induced in osteoblasts and osteocytes by mechanical strain and shear stress. These stimuli cause rapid but transient increases in the production of ·NO and prostaglandins by bone-derived cells and organ cultures. The high concentrations of ·NO induced by these stimuli may augment bone gain by inhibiting bone resorption.

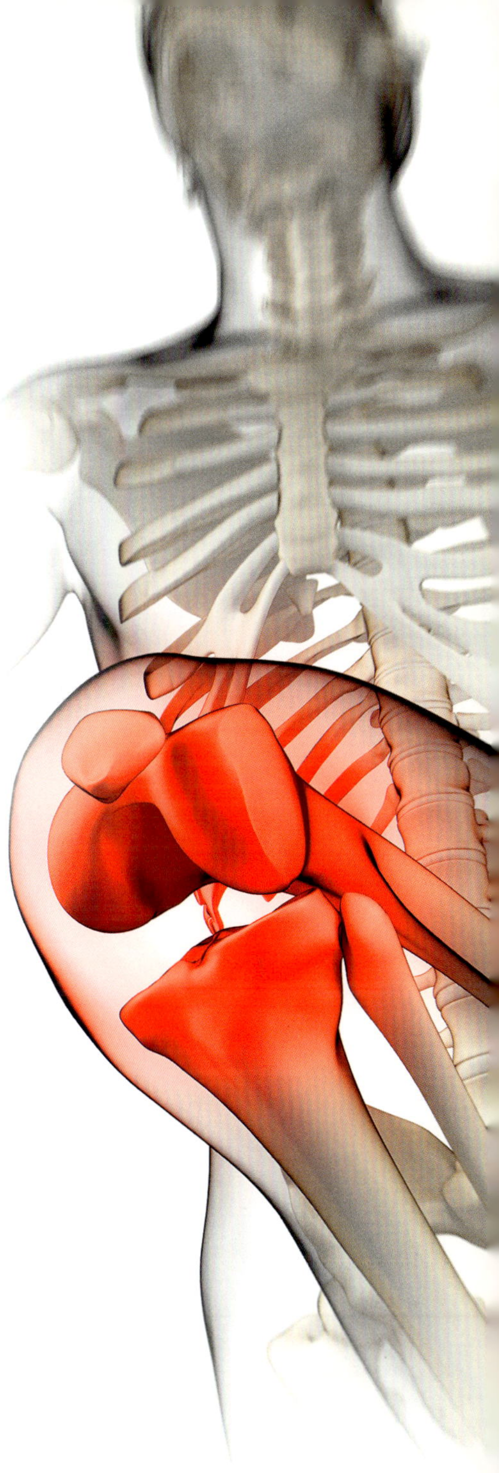

Anatomy of the Knee Joint

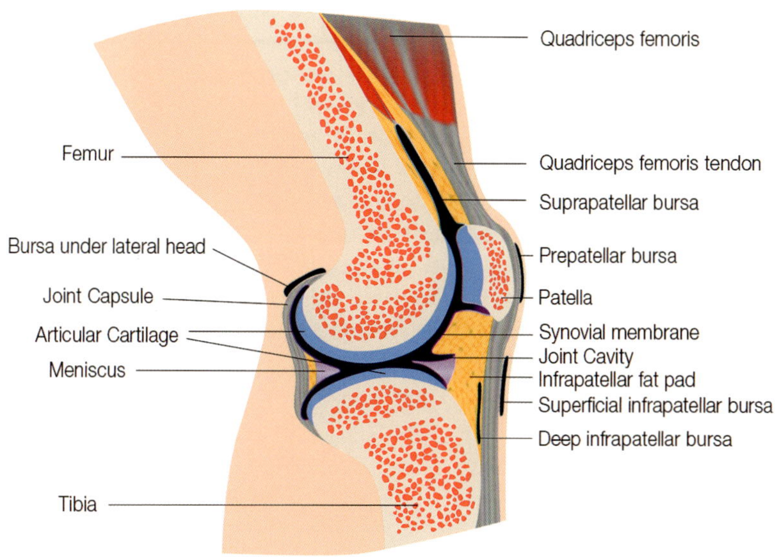

4.8.2. Oxidative Stress and Osteoporosis

Osteoporosis is a skeletal condition characterized by reduced bone mineral density (BMD) and deteriorated bone microarchitecture. The primary type-1 form of osteoporosis is most common in women after menopause. Primary type-2 osteoporosis (or senile osteoporosis) occurs after age 75 and is seen in both females and males at a ratio of 2:1. Secondary osteoporosis may arise at any age and affect men and women equally. This form results from chronic predisposing medical problems, disease, or prolonged use of medications such as glucocorticoids. Osteoporosis - the histological hallmark of this age-related metabolic bone disease is a decline in mean wall thickness—an index of the amount of bone made by each team of osteoblasts during bone remodeling. This is due mainly to a deficit in the number of osteoblasts rather than their biosynthetic capacity, and is thought to result from decreased production of osteoblasts or increased apoptosis. Increase in ROS levels and a decline in GSH-reductase activity in the bone marrow is common in age-dependant loss of bone mass and strength. An association between oxidative stress and a decrease in bone mineral density (BMD), as well as an effect of antioxidants on bone resorption, has been reported in several human clinical studies.

Advancing age and, specifically, increased oxidative stress play a pivotal role in the loss of bone mass and strength. ROS are continuously generated as normal by-products of aerobic metabolism and can cause protein damage and DNA lesions leading to cell death. They also, however, serve as signaling molecules at lower concentrations to control cell proliferation and differentiation in a plethora of cell types. Consistent with this observation, mitochondrial biogenesis and ROS play a very important role in the generation and survival of osteoclasts, osteoblasts, and osteocytes. Loss of estrogens or androgens decreases defense against oxidative stress, which accounts for the increased bone resorption associated with acute sex steroid deficiency, as well as the effects of either class of sex steroids on osteoclastogenesis, apoptosis of osteoclasts, and osteoblasts.

FoxO Transcription Factors: Genetic analysis of the role of the FoxO family of transcription factors in bone indicates that FoxO-dependent oxidative defense provides a mechanism to handle the ROS constantly generated by the aerobic metabolism of osteoblasts. Therefore FoxO is indispensable for bone mass homeostasis. The role of FoxOs in bone physiology provides a strong support on the contribution of ROS in osteoporosis. Sudden increase in ROS activity could adversely affect bone formation. The FoxO proteins defend such an increase by constant up-regulation of the free radical scavenging and DNA-repair enzymes. In addition, FoxOs may control the generation of new osteoblasts from their stem cell progenitors by modulating the proliferation and/or differentiation through their antioxidant properties. However, the antioxidant defense provided by FoxOs is ultimately overwhelmed by high level of oxidative stress or ROS-activated pathways that could lead to osteoporosis.

This FoxO-based phenomenon represents an indispensable homeostatic mechanism for skeletal health.

Sex Hormones (Estrogen): The anti-osteoporotic effect of sex hormones results, at least in part, from their ability to protect against oxidative stress. Estrogens and androgens control osteoblast and osteoclast apoptosis. Loss of estrogens or androgens accelerates the effects of osteoporosis by decreasing defense against oxidative stress. Acute loss of sex steroids (in menopause or after an ovariectomy) leads to an increase in the rate of bone remodeling, due to elevated bone resorption, which exceeds bone formation. This condition is most likely because estrogen deficiency prolongs the lifespan of osteoclasts but decreases the lifespan of osteoblasts. Accordingly, estrogens protect the adult skeleton against bone loss by slowing the rate of bone remodeling (turnover) and by maintaining a focal balance between bone formation and resorption. However, when both aging and loss of sex steroids exert their adverse effects on bone by oxidative damage, the condition leads to osteoporosis.

Diabetes: Similar to aging, both type-1 and type-2 diabetes mellitus adversely affect skeletal homeostasis and increase the risk of osteoporosis; primarily by increasing osteoblast apoptosis and suppressing bone formation. ROS plays a major role in the development of both type-1 and type-2 diabetes, in part by accelerating the death of pancreatic cells. Hyperglycemia and insulin resistance increase ROS production by accelerating mitochondria respiration and activation of NADPH. Increased production of pro-inflammatory cytokines and free fatty acids in type-2 diabetes also contributes to increased oxidative stress, as does the formation of increased advanced glycation endproducts (AGEs) and glucose autoxidation, which cumulatively induce the onset and progression of osteoporosis.

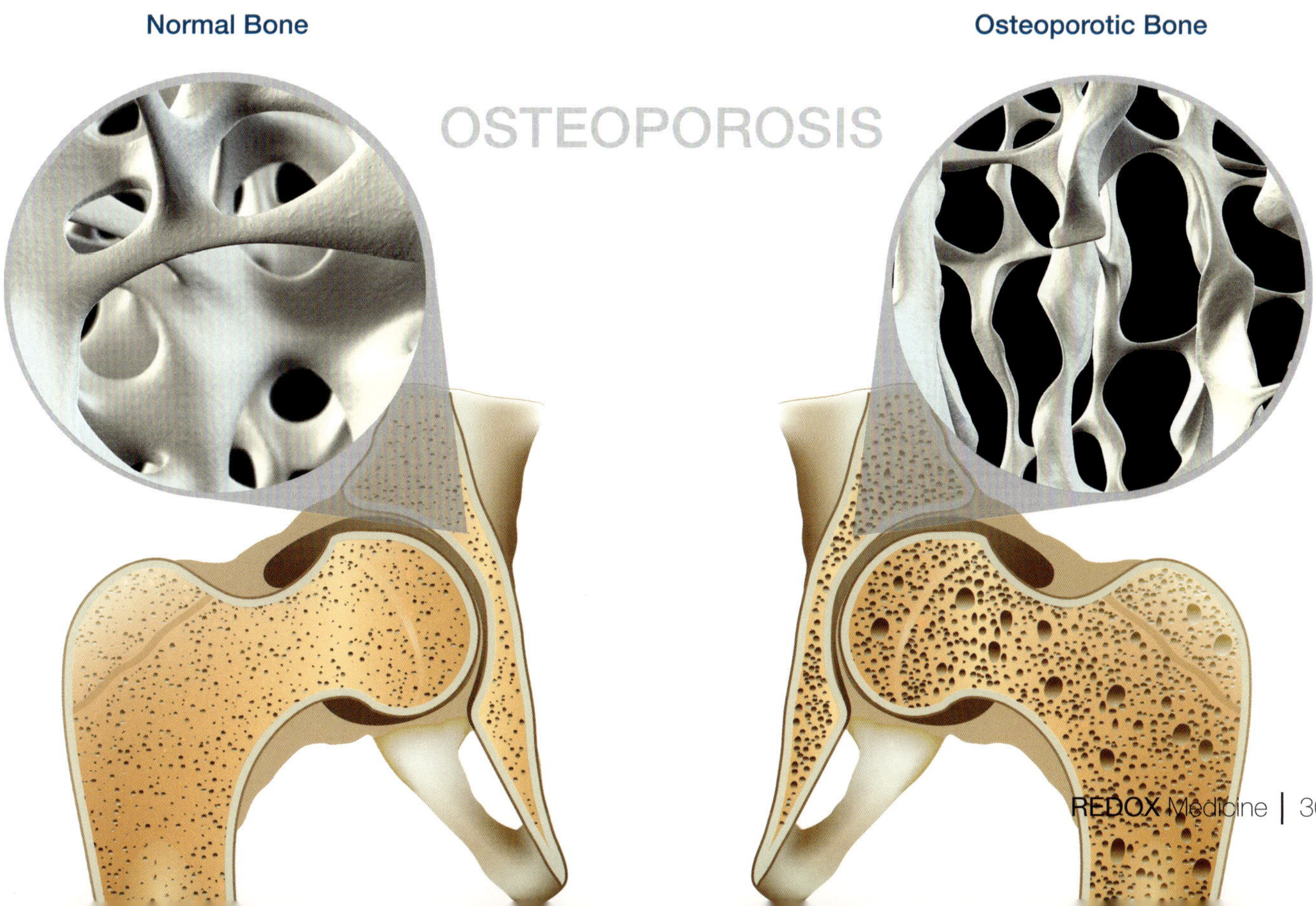

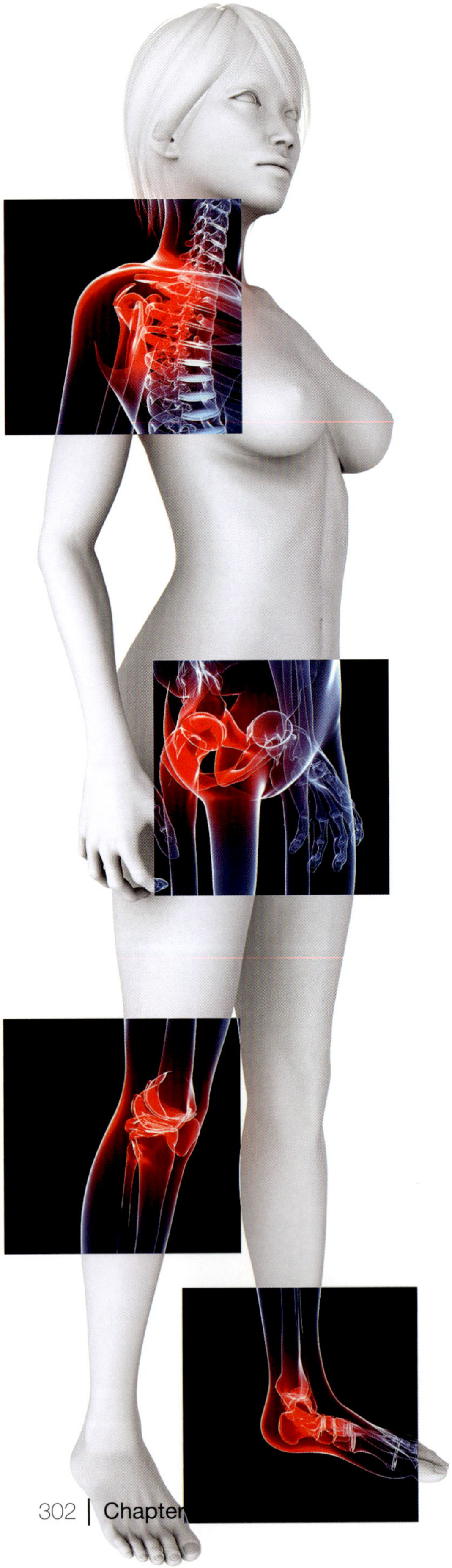

Atherosclerosis: Epidemiological evidence indicates a link between atherosclerosis and osteoporosis development. Bone loss, as well as vascular calcification, progress in parallel with advancing age. Aortic calcification is inversely related to BMD and directly related to osteoporotic fractures in postmenopausal women. Women in the highest quartile for increased atherosclerosis exhibit four times greater bone loss than women in the lowest quartile. Aging increases the differentiation of mesenchymal stem cells (MSCs) into adipocytes in many tissues, including the bone marrow. Increased marrow adiposity, on the other hand, is associated with age-related osteoporosis.

Other Oxidative Factors: Excess of glucocorticoids decrease bone strength disproportionately to its adverse effect on BMD. Hyper-glucocorticoidism induces endothelial oxidative stress and vasoconstriction that both lowers blood flow and the volume of water present in the skeleton. Water represents at least 25%of the wet weight of bone and confers to bone much of its unique strength and resilience by reducing bone stresses during dynamic loading.

REDOX-based Interventions: Elucidation of REDOX mechanisms provides a paradigm shift from the 'estrogen-centric' pathogenesis of osteoporosis (which is considered an age-related process intrinsic to bone). REDOX provides an insight into osteoporosis, which is well aligned with free radical theories of senescence and the pathobiology of several other degenerative disorders. Identifying determinants of bone strength (independent of bone mass) would provide insights into the age-related decline of bone strength. REDOX mechanisms may suggest novel therapeutic opportunities for osteoporosis prevention.

During the last few years, a class of nutrient-sensing nicotinamide adenine dinucleotide (NAD) dependent protein deacetylases, termed 'sirtuins', has been shown to play a crucial role in the resistance against oxidative stress and the promotion of longevity. The prototype of this gene family in vertebrates is the histone deacetylase silent information regulator T-1 (SIRT-1). Sirtuins 1, 2, and 3 prolong lifespan, improve energy expenditure, glucose production, and insulin sensitivity; and protect against high-fat diet-induced metabolic damage via several mechanisms, including the deacetylation of FoxOs and modulation of FoxO-mediated transcription.

4.8.3. Oxidative Stress and Osteo-Arthritis (OA)

Globally ~250 million people have OA of the knees. OA affects nearly 27 million people in the United States, accounting for 25% of visits to primary care physicians, and half of all NSAID prescriptions.

Osteoarthritis (OA) also known as degenerative joint disease is a progressive chronic condition characterized by the breakdown of the joint's cartilage. Cartilage is the part of the joint that cushions the ends of the bones and allows easy movement of joints. The breakdown of cartilage causes the bones to rub against each other, resulting in stiffness, pain and loss of movement in the joint. The prevalence of OA increases significantly with advancing age. Among other factors, the imbalance between free radical burden and cellular scavenging mechanisms (defined as 'oxidative stress') is a relevant part of OA pathogenesis.

Cartilage is a conjunctive tissue, derived from mesenchymal stem cells during embryonic development. Oxidative imbalance in cartilage is an effect of two major mechanisms: (i) Decrease in antioxidative capacity – thus, an increase in oxidative intra-cellular state leads to a pathological change in osteoarthritic cartilage. (ii) Compromised nutrition of articular cartilage. Being an avascular tissue, it contains chondrocytes embedded into a large amount of extra-cellular matrix and nutrition is dependent on diffusion from synovial and subchondral bone compartment. Therefore, oxygen and metabolic end products have to diffuse over relatively long distances, resulting in low O_2 tension in cartilage tissue. A rise of O_2 tension in cartilage tissue can enhance free radical formation, which ultimately leads to cartilage degeneration in chronically irritated joints. Interestingly, ROS produced by the oxidative metabolic turnover of cartilage tissue are a substantial factor for the regulation of intra-cellular signaling.

OA and Mitochondria: There is evidence for mitochondrial DNA damage in OA, which cumulatively could contribute to chondrocyte death. Mechanical injury to cartilage results in increased ROS production from a mitochondrial source. There is also some evidence that Kashin- Back disease, a rare form of endemic OA found primarily in China and Siberia, is associated with mitochondrial dysfunction and cell death. In addition to induced cell death, mitochondrial ROS also act as signaling molecules and promote expression of pro-inflammatory cytokines.

ROS play a major role in the intra-cellular signaling mechanisms of synovia. The activation of nuclear transcription factor (NF)-κB is implicated in the inflammatory response in the synovial lining cells; this is a feature of both RA and OA patients. Synovial inflammation could impair the ability of chondrocytes to balance anabolic and catabolic activities in extra-cellular matrix formation. Increased production of ROS is implicated in chondrocytic death, and in the activation of matrix metalloproteinases (MMPs) in OA.

OA and Mechanical Loading: Excessive repetitive load has been shown to induce chondrocyte apoptosis, which points out the relevance of injurious cartilage compression in OA pathogenesis. Mechanical loading that exceeds the tolerance of the articular surface can cause oxidative stress. Studies on cartilage explants showed an increase of ROS synthesis after induction of mechanical load. Furthermore, RNS act as mediators of cell death after mechanical overuse and injury to cartilage. Thus, excess shear stress could induce oxidative stress and mediate the death of chondrocytes.

Synovial cells are involved in the synthesis of the synovial fluid, an ultra-filtrate of the serum enriched with cellular metabolites. Synovial fluid forms a capillary layer in the healthy joint and minimizes friction. A large amount of cross-linked hyaluronan, an anionic non-sulfated glycosaminoglycan (GAG), contributes substantially to the rheologic properties of joint fluid. Oxidative stress, especially with peroxynitrite ($ONOO^-$) breaks the GAG polymers. Hyaluronan acts as a ROS scavenger in the joint by a competition with other substrates susceptible for ROS attack. A reduced GAG level is indicative of compromised antioxidative capacity in synovial fluid of OA joints. The compromised antioxidative capacity can be emphasized as a disability of the joint to adapt to increased oxidative stress in OA.

In summary, alterations to subchondral bone and synovial cells/fluid are crucial for OA pathogenesis. Therefore, the joint should be considered a dynamic system consisting of three discrete but permanently interacting compartments.

REDOX-based Interventions: REDOX potential considers the joint as a dynamic system of three different tissues interacting by fluid diffusion. Because of their chemical properties, free radicals meet the requirements to mediate and amplify the characteristic sequence of joint degeneration in all tissues affected, making them a crucial factor for the involvement of all three joint compartments in the pathogenesis of OA. Therapeutic modifications of these mechanisms may be promising area of further research to achieve comprehension and therefore approach to modification of disease progression. Mitochondrial dysfunction, with increased production of ROS, may be important and this may be related to changes in energy metabolism. The area of REDOX signaling is beginning to define the aging effects and the response of chondrocytes to growth factor stimulation. Studies of autophagy may reveal new mechanisms relevant to the cell's decision to live or die under conditions of stress.

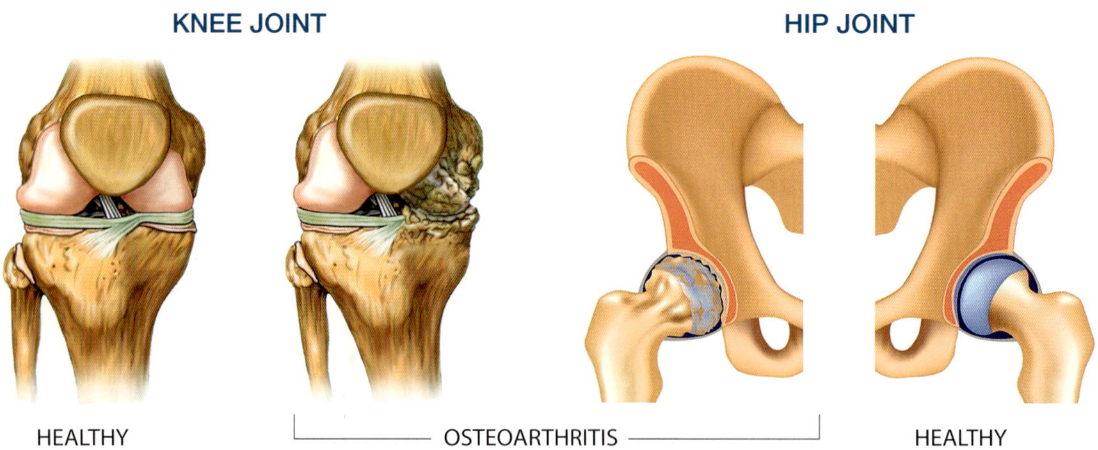

Sports and OA: There is increasing concern that too much physical activity may lead to OA. Continuous physical stress on the joints can result in micro-trauma and degeneration of articular cartilage. The onset of OA depends on intensity, duration and frequency of physical activity. All age groups can tolerate moderate exercise without adverse consequences or accelerated development of OA. However, excessive participation in high impact sports over a long period of time can increase the risk of developing OA. Individuals with abnormal joint anatomy or alignment, joint instability, underlying muscle weakness or imbalance; or if they are overweight and engage in strenuous exercise are at higher risk. Individuals with previous sports injuries to joints, or macro-trauma, may also be at risk of accelerated development of OA. Certain types of surgery for the treatment of sports injuries, particularly to the knee is associated with an increased risk. If knee surgery is required, continuous passive motion is an ideal form of rehabilitative treatment, as it promotes healing of the articular cartilage, ligaments and tendons. Moreover, athletes who have undergone surgery should return slowly to sporting activities to ensure that too much stress is placed on their injured joint(s).

Pathogenesis of Osteoarthritis (OA)

Inter-relation between Oxidative Imbalance, Clinical Symptoms, and the relevant Joint Compartment

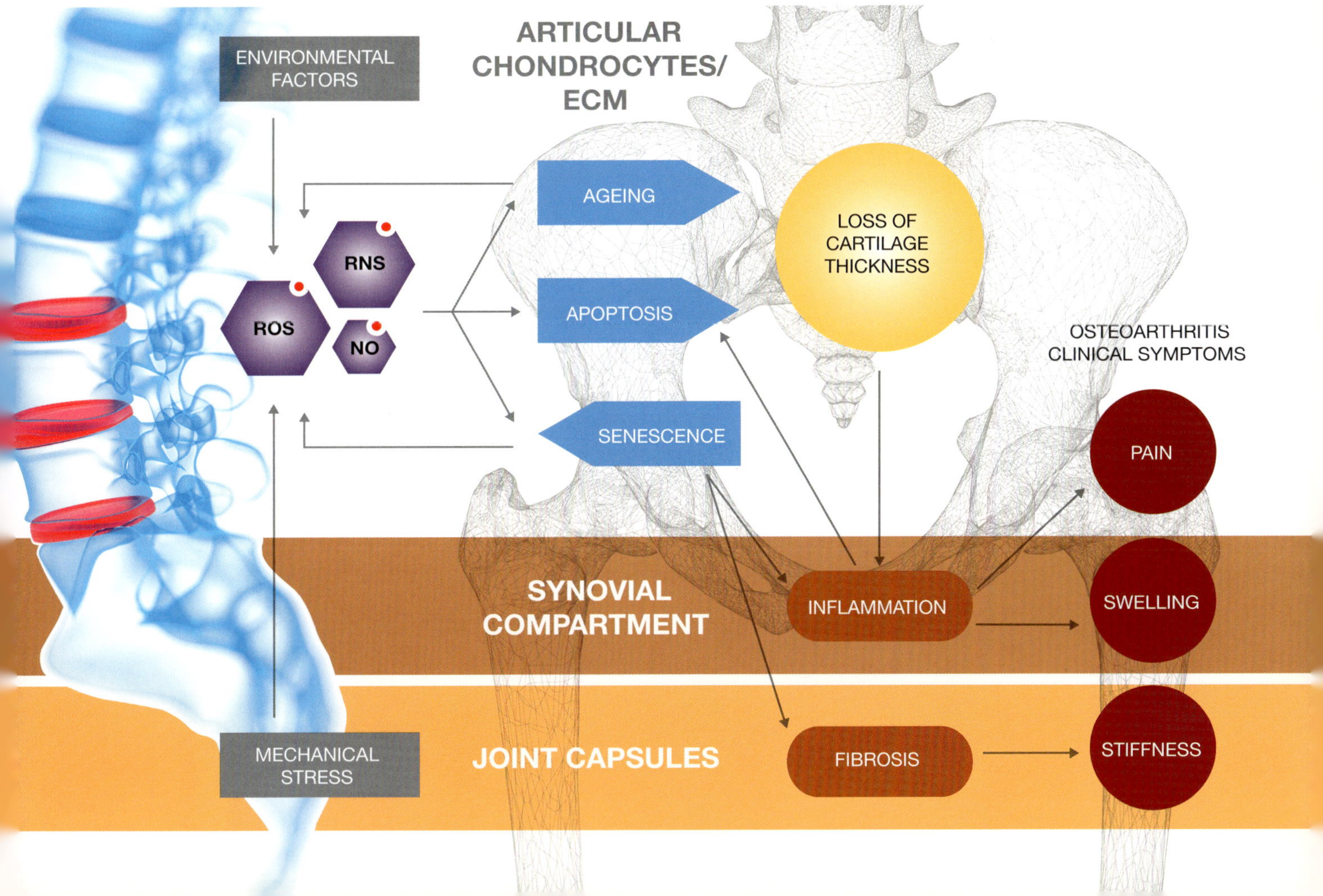

4.8.4. Oxidative Stress and Rheumatoid Arthritis (RA)

RA affects an estimated 1.3 million people in the US, which is about 1% of the nation's adult population. There are nearly 3 times as many women as men with RA. In women, RA most commonly begins between the ages of 30 and 60. It often occurs later in life with men.

Rheumatoid arthritis (RA) is a form of arthritis that causes pain, swelling, stiffness and loss of function of synovial joints. It can affect any joint but is common in the wrist and fingers. RA is different from osteoarthritis (OA), the common arthritis that often comes with older age. RA can affect body parts besides joints, such as the eyes, mouth and lungs. RA is an autoimmune disease, which means the arthritis results from the immune system attacking body's own tissues.

The role of oxidative damage in RA originates from the seminal work of McCord (1985), who noticed a reduced viscosity of synovial fluid in RA patients. He also showed that a similar decrease could be produced by exposing synovial fluid, or solutions of hyaluronic acid, to a system generating $O_2^{\bullet -}$. Later research revealed that hyaluronate degradation is not caused by $O_2^{\bullet -}$, but due to its reactive product ·OH radical generated by Fenton chemistry. Hyaluronate can also degrade by $HOCl^-$ and may additionally be secreted as abnormally short chains by the dysfunctional synovium in RA. It is now established that increased levels of ROS elicit detrimental effects on bone joints. For example, ·OH radicals can degrade cartilage, $HOCl^-$ can attack proteoglycans, and H_2O_2 can inhibit proteoglycan synthesis. Thus a combination of degradation and inhibition of synthesis can

Stages of RA Development

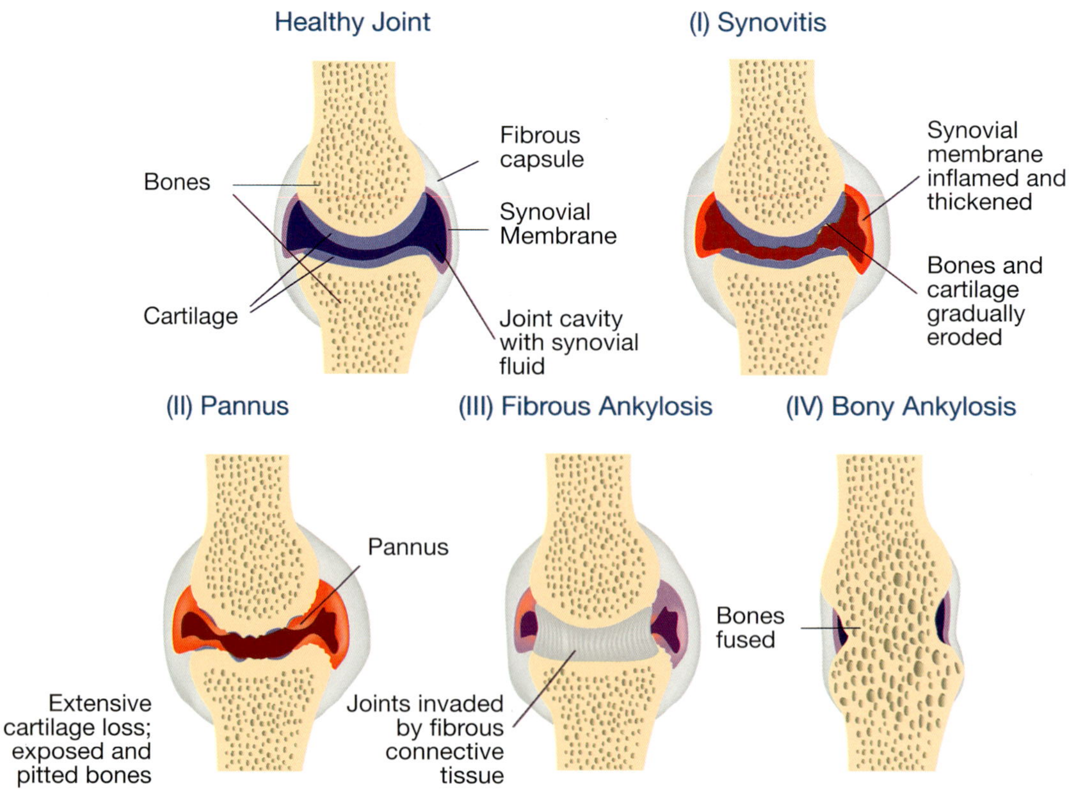

trigger cartilage destruction. Also, structural proteins within synovial cells are modified by ROS. For example, actin is disrupted and vimentin is cleaved by H_2O_2. This cleavage is rapid (<2 h) in RA patients compared to controls (~5 h). This susceptibility is reflected in increased structural damage to rheumatoid synovial cells by ROS and probably explains the changes in vimentin ultra-structure and auto-antibody production in RA. Several observations suggest a role for iron-dependent radical reactions in RA. Iron infusion complexes are known to exacerbate synovitis and bleomycin iron is detectable in synovial fluid; the concentration correlates loosely with indices of RA activity.

Rheumatoid synovial fluid and tissue demonstrate oxidative damage to hyaluronic acid, lipid peroxidation products, oxidized low-density-lipid proteins (LDL), and increased carbonyl groups (indicative of protein damage). Evidence of oxidative damage to cartilage, extra-cellular collagen, and intra-cellular DNA has also been reported in RA. Oxidative stress could induce T cell hypo-responsiveness in RA through effects on proteins and proteosomal degradation.

ROS and RNS damage cellular elements in cartilage. Modification of amino acids by oxidation, nitrosylation, nitration, and chlorination can alter protein structure and impair biological function, leading to cell death. ROS impair chondrocyte responses to growth factors and migration to sites of cartilage injury; RNS, in particular NO, interfere with interactions between chondrocytes and the extra-cellular matrix. NO can also increase chondrocyte apoptosis. Oxygen and nitrogen radicals inhibit the synthesis of matrix components including proteoglycans by chondrocytes. In particular, NO and O_2 seem to inhibit type II collagen and proteoglycan synthesis and destroy newly synthesized glycosaminoglycans (GAGs). Oxygen radicals can cause low levels of collagen fragmentation and enhanced collagen fibril cross-linking. Oxygen radicals have also been shown to fragment hyaluronan and chondroitin sulfate. In addition, ROS and RNS can damage the components of the extra-cellular matrix indirectly through the activation and up-regulation of matrix metalloproteinases (MMPs). Several lines of

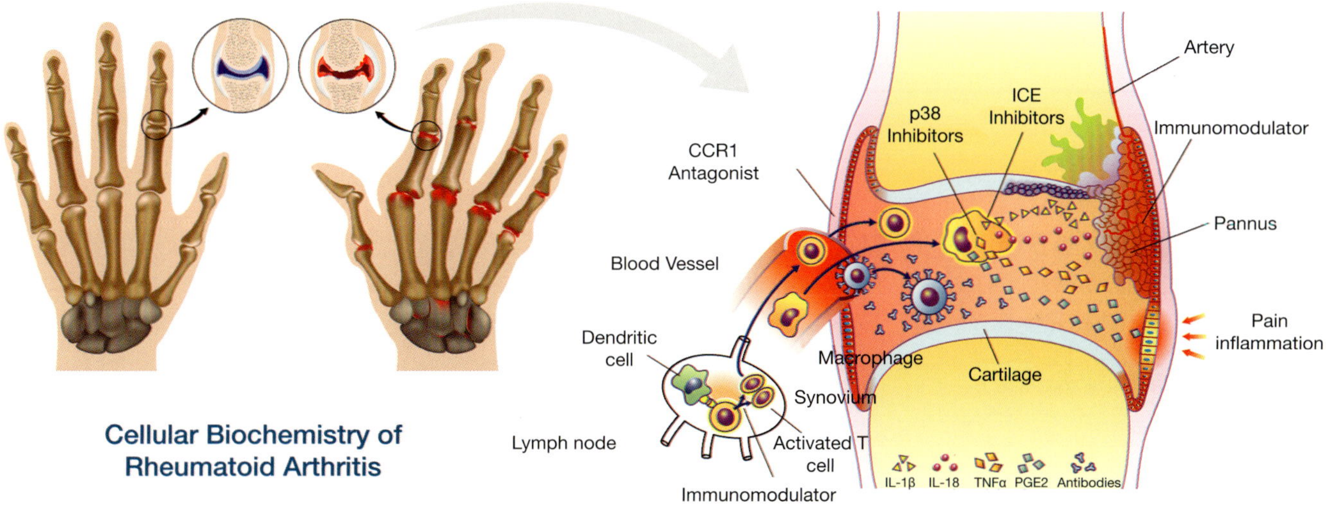

Cellular Biochemistry of Rheumatoid Arthritis

evidence suggest that cycles of hypoxia/reoxygenation are important in the pathobiology of RA synovitis. It has long been known that RA synovial fluids are hypoxic, acidic, and exhibit low glucose and elevated lactate concentrations. This biochemical profile is indicative of anaerobic metabolism in the synovium.

Nitric oxide (·NO) is a potent stimulator of chondrocyte apoptosis and this may contribute to the cartilage damage in RA. Furthermore, cytokine induced ·NO stimulates the production of matrix metallo-proteinases (MMPs) in chondrocytes, which may lead to further destruction of the cartilage. ·NO potentiates cytokine induced bone resorption; high levels can also inhibit bone formation. Therefore, ·NO produced within the inflamed joint may contribute to the peri-articular bone loss observed in RA.

Nitrate levels are elevated in urine and serum of patients suffering from RA, and even higher levels have been detected in the synovial fluid, suggesting that the inflamed joint is the source of the ·NO. Considerable amounts of nitrosylated protein have been detected in the synovial fluid of RA patients, but not in OA patients. Thus, nitrosylation may be an important cause of joint damage in RA. Also, the iNOS enzyme readily detected in synovium and cartilage of RA patients, is expressed at lower levels in OA, but is virtually absent in normal synovium and cartilage. iNOS expression is also associated with apoptosis of synovial cells and chondrocytes.

REDOX-based Interventions: Cyclo-oxygenase (COX) inhibitors, blocking prostaglandin production and hence decreasing pain and swelling, are still commonly used in the day-to-day treatment of RA, although interest is now switching to REDOX interventions. Disease-modifying anti-rheumatic drugs (DMARDs), the newer biologics, have been found to decrease joint damage and improve overall functional abilities. When started very early, DMARDs can result in RA remission in about half of patients and improve overall outcomes. The most commonly used DMARD is methotrexate, known to block oxidative damage and help in REDOX balance. Various forms of antioxidant therapies have demonstrated promising results in experimental arthritis models. The polyphenolic fraction of green tea, containing potent antioxidants, prevents collagen-induced arthritis. The beneficial effects seem to be due to the catechin epigallo-catechin-3-gallate, which inhibits inflammation, blocks ·NO production by chondrocytes, and abolishes MMP activity. In a clinical study of RA patients, supplementation with antioxidant vitamins A, E, and C increased plasma antioxidant levels with a corresponding decrease in malondialdehyde (a oxidative stress marker).

Based on CDC estimates (2005):

- 27 million adults had osteoarthritis.
- 1.5 million adults had rheumatoid arthritis.
- 3.0 million adults had gout.
- 5.0 million adults had fibromyalgia.

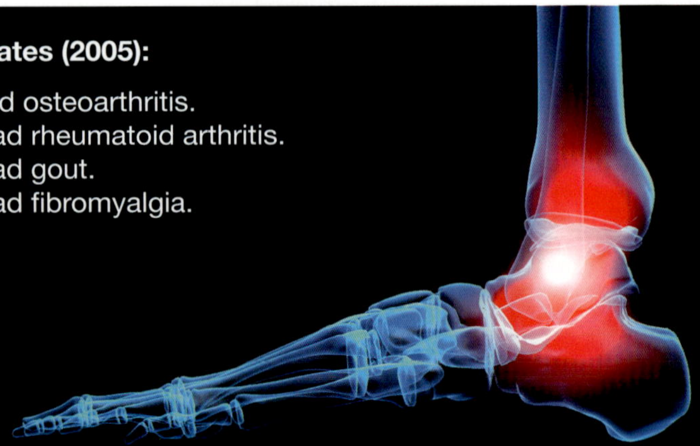

OSTEOPOROSIS

- Globally, osteoporosis causes more than 8.9 million fractures annually, resulting in an osteoporotic fracture every 3 seconds.
- Osteoporosis is estimated to affect 200 million women worldwide - about one-tenth of women aged 60, one-fifth of women aged 70, two-fifths of women aged 80 and two-thirds of women aged 90.
- In the year 2000, there were an estimated 9 million new osteoporotic fractures, of which 1.6 million were at the hip, 1.7 million were at the fore arm and 1.4 million were clinical vertebral fractures. Europe and the Americas accounted for 51% of all these fractures, while most of the remainder occurred in the Western Pacific region and Southeast Asia.
- Worldwide, 1 in 3 women over 50 will experience osteoporotic fractures, as will 1 in 5 men.
- By 2050, the worldwide incidence of hip fracture in men is projected to increase by 310% and 240% in women.
- A prior fracture is associated with an 86% increased risk of any fracture.

ARTHRITIS

- An estimated 50 million adults in the United States reported being told by a doctor that they have some form of arthritis, rheumatoid arthritis, gout, lupus, or fibromyalgia.
- One in five (22%) adults in the United States report having doctor diagnosed arthritis.
- In 2007–2009, about 50% of adults 65 years or older reported an arthritis diagnosis.
- By 2030, an estimated 67 million Americans ages 18 years or older are projected to have doctor-diagnosed arthritis.
- An estimated 294,000 children under age 18 have some form of arthritis or rheumatic condition; this represents approximately 1 in every 250 children in the U.S.

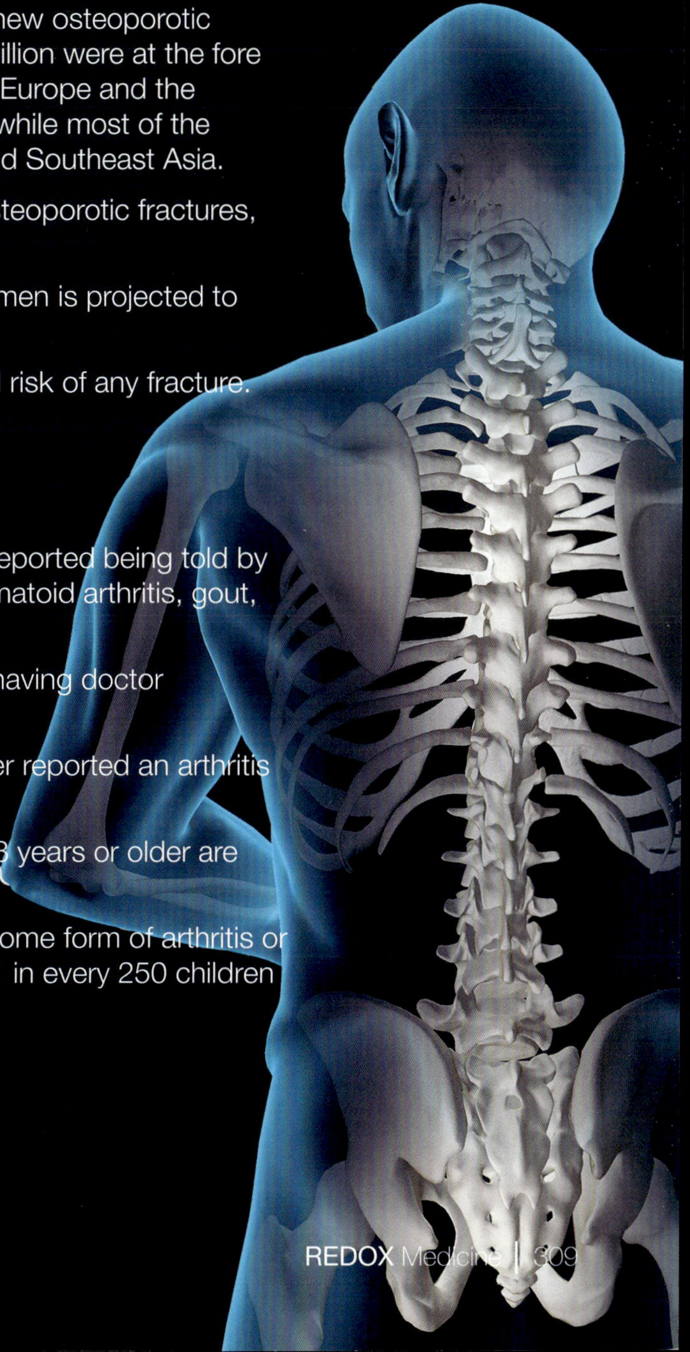

4.9 REDOX: The Muscular System

The Muscular System is composed of over 600 muscles that come in variety of shapes and forms, making up nearly half the weight of the human body. Differences between each muscle are recognized by location, function, structure, and contraction. Muscular system is controlled through the nervous system, although certain muscles (such as the cardiac muscle) can be completely autonomous (involuntary). Muscles provide strength, balance, posture, movement and heat for the body to keep warm.

The muscular system consists of three different tissue types: (i) skeletal, (ii) cardiac, and (iii) smooth. Each of these different muscle types has the ability to contract, which results in movement and function of an organ.

Skeletal muscles are the body's most abundant tissue type, comprising about 23% of a woman's body weight and around 40% of a man's body weight. Skeletal muscles are made of long tissue fibers attached to bones and work in pairs. For example, if a muscle moves the bone in one direction, the other muscle moves it in opposite direction. Skeletal muscles are voluntary muscles, and their contractions could be short or extended depending on the bone-joint function.

Smooth muscles are involuntary in action; this means that they are controlled automatically by the nervous system and hormones. This type of muscle can stretch and maintain tension over extended periods. Smooth muscles are found in the internal organs such as the digestive system, blood vessels, urinary bladder, respiratory organs, and the uterus (in female).

Cardiac muscles are involuntary and found only in the heart. Located at the walls of the heart (myocardium), the main function of this muscle type is to propel blood into circulation. The cardiac muscle is a cross between smooth and skeletal muscle. It is similar to skeletal muscle in its structure, but it is controlled by the nervous system, like smooth muscle.

4.9.1. REDOX and Muscular Health

Skeletal muscle is a highly specialized and adaptive mesodermic tissue capable of rapid remodeling in response to changes in loading and stretch (i.e. mechano-transduction). The subsarcolemmal cytoskeleton ensures mechanical integrity of skeletal muscle and initiates cell signaling in response to alterations in mechanical stress (i.e. mechano-transduction). The subsarcolemmal cytoskeleton is anchored to the sarcolemma where it transfers forces between the muscle cells and extra-cellular matrix (ECM) proteins. It also initiates cell signaling responses to mechanical strain, endocrine influences and the chemical environment.

Another interesting characteristic of cardiac muscle is that it is capable of signal propagation, where when one cell receives a signal to contract, that signal is transmitted to all the cells of the heart, which results in a heartbeat.

Reactive oxygen species (ROS) are a major source of REDOX signals involved in muscle homeostasis, i.e. in maintaining normal skeletal muscle structure and function. ROS play an important role in the regulation of the metabolic pathways that are required to promote skeletal muscle adaptation in response to both exercise and muscle inactivity. Specifically, $O_2^{\bullet-}$ production can occur at multiple locations within the muscle fiber, including the mitochondrion, sarcoplasmic reticulum, transverse tubules, sarcolemma, and the cytosol. The main sites of $O_2^{\bullet-}$ production in the mitochondria are complexes I and III of the electron transport chain (ETC).

Studies indicate that, compared with mitochondria from slow type I muscle fibers, mitochondria from fast type II muscle fibers possess unique properties that promote higher levels of ROS production. The NADPH oxidases (NOx) located within the sarcoplasmic reticulum, transverse tubules, and the sarcolemma also produce $O_2^{\bullet-}$ in skeletal muscle fibers. For example, the transverse tubule NOx produces $O_2^{\bullet-}$ which stimulates calcium release from the triads. Normal skeletal muscle expresses two isoforms of NO-synthase (i.e. NOS1 and NOS3). However, NOS2 can also be expressed in skeletal muscle during inflammatory states.

ROS signaling contributes to muscle fiber adaptation in response to both increased contractile activity (muscular exercise) and prolonged periods of muscle disuse (immobilization). The contradictory signaling function of ROS is probably due to differences in both the magnitude and the temporal pattern of ROS generation. For example, a moderate increase in skeletal muscle ROS production, during a short time period (e.g. minutes), can activate signaling pathways leading to cellular adaptation and protection against future stresses. In contrast, high levels of ROS production over long time periods (e.g. hours) may result in chronic activation of signaling pathways that promote proteolysis and potential cell death.

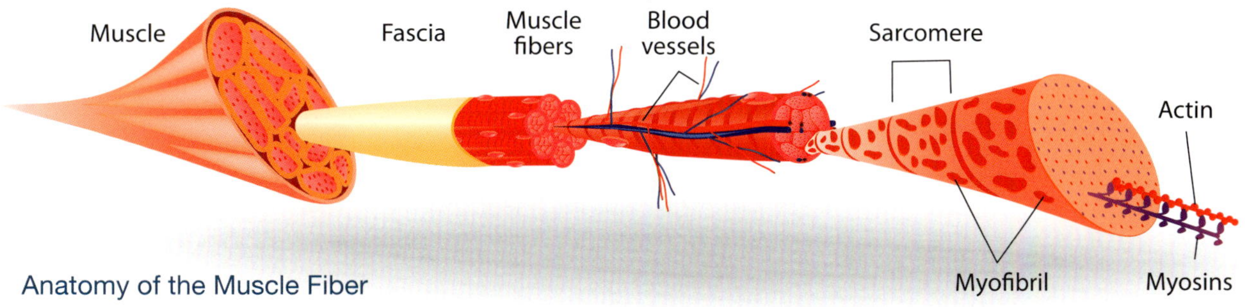

Anatomy of the Muscle Fiber

4.9.2. Oxidative Stress and Physical Exercise

The observation that muscular exercise induces production of free radicals and other ROS in skeletal muscle was first reported by Davies et al. in 1982. Paradoxically, it was later discovered that ROS generation also increases in skeletal muscle fibers following prolonged periods of disuse or immobilization. Physical exercise is associated with numerous health benefits, especially for its role in the prevention of bone loss. Chronic exercise, however, represents a form of oxidative stress to the body and, therefore, can alter the balance between prooxidants and antioxidants. ROS directly affects muscle force, which is necessary for normal muscle homeostasis. ROS play a major role in positive adaptations following moderate exercise, acting as intra-cellular signals that modulate gene expression. Oxidative stress can also induce muscle damage in heavy exercise by modulating the intra-cellular REDOX signals involved in disuse atrophy. ROS can directly oxidize proteins by enhancing their degradation rate. Consequently, it is very likely that differences in the levels, the kinetics and the cellular location of ROS production in different tissues and conditions could trigger different responses.

Several pathways have been suggested by which exercise can increase ROS production. The primary pathway involves electron leakage from mitochondria during aerobic exercise, whereby there is an increase in oxygen flux, resulting in an increase in ROS production. Exercise-associated physiological pathways such as those involved during muscle injury, increase: phagocytic activity, disrupt iron-containing proteins, and production of xanthine oxidase (XOR), all due to increased oxidative stress. Furthermore, auto-oxidation of catecholamines, which are released from the adrenal glands in greater quantities during exercise, induces an increase in radical production. Similarly, heme-proteins such as oxymyoglobin can undergo auto-oxidation as a result of exercise, thus leading to formation of $O_2^{\bullet-}$ radicals in particular.

Exercise-induced ROS production alters muscle gene expression and contributes to exercise-induced adaptations of skeletal muscle *in vivo*. Oxidative stress occurs in skeletal muscle during prolonged disuse and REDOX disturbances regulate signaling pathways that contribute to muscle remodeling. During heavy exercise, disuse, and pathological conditions, the elevated ROS levels cause muscle damage though various mechanisms. The increased ROS production: (i) has a direct, non-specific, large scale oxidative effect on proteins, which makes them more susceptible to proteolysis; (ii) damages plasma membrane, sarcoplasmic reticulum and alters Ca^{2+} homeostasis, which activates proteases (i.e. calpains) that enhance proteolysis, (iii) damages lysosomes and causes a leakage of catabolic enzymes in the cytosol. Oxidized proteins are more susceptible to proteolysis and are easily targeted by the ubiquitin–proteasome system (also up-regulated by ROS).

The Muscular Anatomy

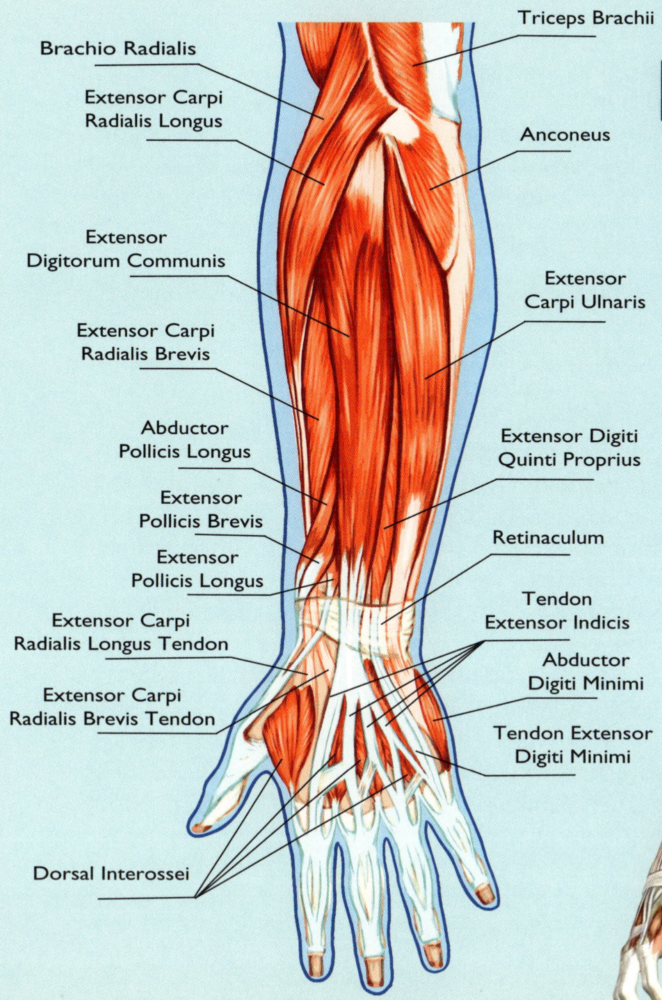

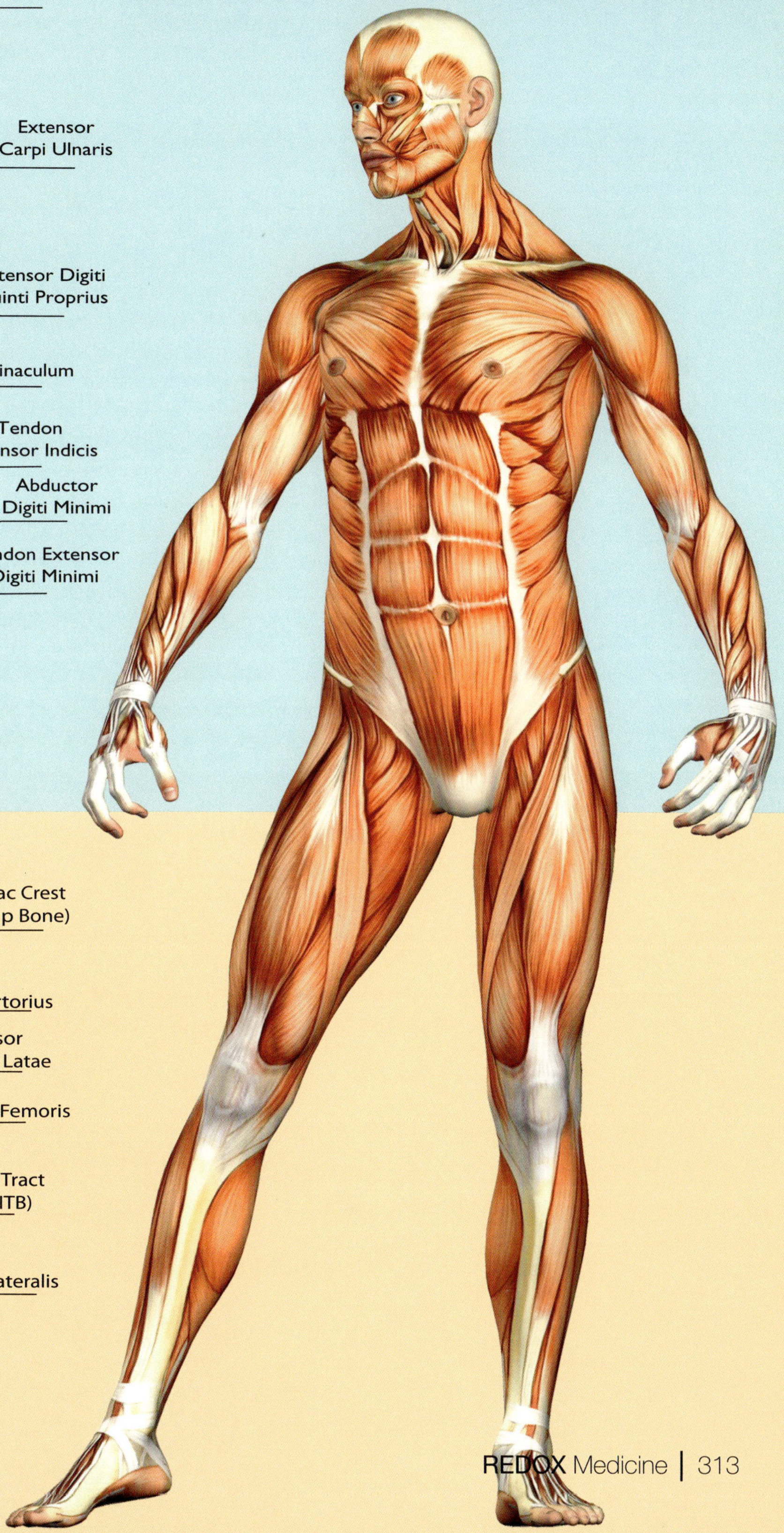

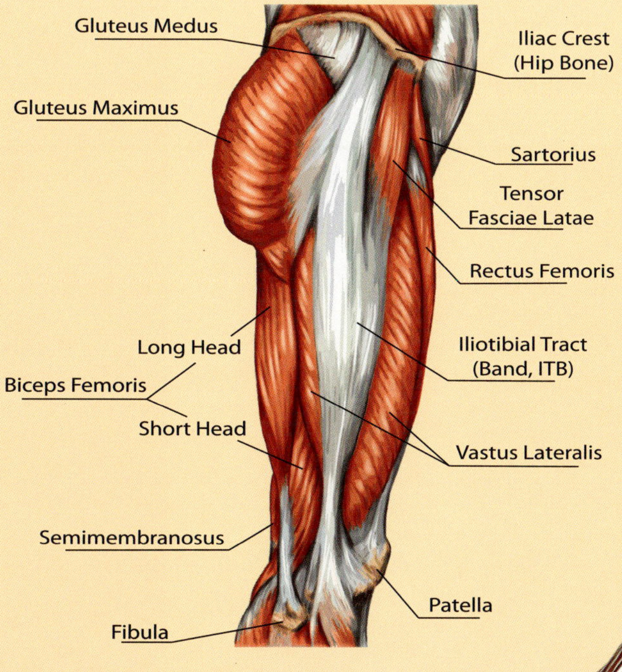

Exercise can have positive or negative effects on oxidative stress depending on training load, specificity, and basal level. Aerobic exercise is accompanied by increased O_2 consumption which may increase ROS activity. However, this phenomenon cannot occur with low exercise intensity (under 60% VO_2 max). In such case, antioxidant capacity is not over-reached and ROS-induced damage does not appear. Regular moderate exercise increases antioxidant defenses in the skeletal muscle and up-regulates SOD and GPx gene expression; thereby inducing adaption to oxidative stress. This adaptation may occur very quickly (about 5 minutes) after ROS production and seems to be specific to oxidative muscular fibers, which are the main site for ROS production during exercise. Regular exercise also increases the activity of proteasome complex in the myocardium, and decreases the level of carbonylated proteins and other damaged proteins. There also are a number of physiological pathways that can increase oxidative stress, such as XOR, NOx, and phagocytic respiratory burst. Another source of ROS production during anaerobic exercise arises from inflammation and cellular damage, which often happen after traumatizing exercise such as impact sports. Iron liberation from hemoglobin or ferritin, may amplify the inflammatory response and the oxidative stress. Antioxidant enzymes may be activated selectively during an acute bout of strenuous exercise depending on the oxidative stress imposed on the specific tissues as well as the intrinsic antioxidant defense capacity.

ROS production due to heavy exercise training can cause muscle damage by increased lipid peroxidation, protein carbonylation, increase in serum creatine kinase and altered glutathione (GSH)-REDOX status. In contrast, ROS production during moderate exercise can cause positive adaptations, such as an increase in insulin sensitivity, mitochondria biogenesis and antioxidant defense. Consequently, antioxidant administration may counter muscle damage following heavy exercise, but also undergo positive adaptations following moderate exercise.

Anatomy of the Foot Muscles & Tendons

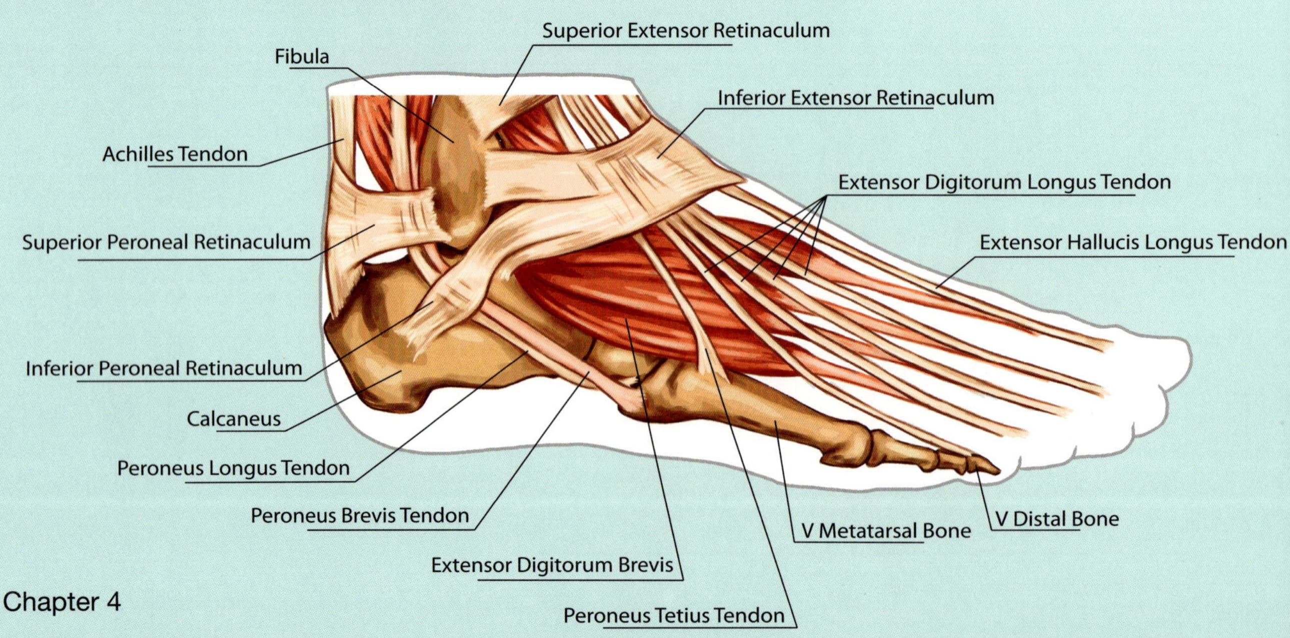

4.9.3. Oxidative Stress and Muscle Atrophy

Oxidative stress is a common phenomenon with physical inactivity (or skeleto-muscular disuse) and many disease conditions. Oxidative stress is considered a major trigger of the imbalance between protein synthesis and degradation leading to muscle atrophy. Oxidative stress can also contribute to disuse muscle atrophy by activation of one or more proteolytic pathways (e.g. calpain, caspase-3 activation and/or the ubiquitin–proteasome pathway) as follows:

Calpains are a family of cellular proteases that are activated by elevated cytosolic levels of free calcium (Ca^{2+}). Activation of calpains is a critical step in the onset of disuse muscle atrophy. Although the mechanism responsible for this muscle condition is unknown, it is possible that intra-cellular generation of ROS could lead to disturbances in the calcium homeostasis. There are at least three mechanisms by which an increase in cellular ROS production can promote increased cytosolic levels of free Ca^{2+}: First, in cell-free systems, oxidation of the ryanodine-sensitive calcium-release channel results in Ca^{2+} release from the sarcoplasmic reticulum. Second, high levels of ROS can inhibit sarcoplasmic Ca^{2+}-ATPase activity and retard Ca^{2+} reuptake into the sarcoplasmic reticulum. A third biochemical mechanism (to link oxidative stress with Ca^{2+} overload) is ROS-mediated formation of reactive aldehydes (i.e. 4-HNE), which can inhibit plasma membrane Ca^{2+}-ATPase activity. These mechanisms may act independently, or interact collectively, to promote oxidative stress induced calcium overload in cells.

Caspase-3 plays an important role in disuse muscle atrophy. Caspase-3 activation promotes degradation of intact actin–myosin complexes. Inhibition of caspase-3 activity has been shown to retard inactivity-induced atrophy and apoptosis. Although the control of caspase-3 activity is complex and involves several inter-connected pathways, the inactivity-associated ROS production in skeletal muscle contributes to caspase-3 activation. Increased cellular levels of ROS have been reported to activate caspase-3 in a wide variety of cell types, including skeletal muscle. Thus, oxidative stress has been postulated to contribute to caspase-3 activation via both mitochondrial and endoplasmic reticulum signaling pathways.

Ubiquitin–Proteasome System: Several studies suggest that ROS signaling is involved in the regulation of the ubiquitin–proteasome system. Oxidative stress stimulates ubiquitin conjugation to muscle proteins through the transcriptional regulation of the enzymes (i.e. E2 and E3 proteins). Up-regulation of E2 and E3 proteins in skeletal muscle leads to accelerated proteolysis via 26S proteasome. Furthermore, the 20S (core) proteasome can degrade oxidized proteins without ubiquitination. Therefore, it is likely that oxidant stress can accelerate muscle protein breakdown via both the 20S and 26S proteasomes.

Oxidative Enzymes: The Nobel Prize in Physiology or Medicine 1955 was awarded to Hugo Theorell for his discoveries concerning the nature and mode of action of oxidation enzymes.

Thus, both contractile activity and prolonged periods of muscle disuse promote increased ROS production in skeletal muscle fibers. Importantly, ROS serve as signaling molecules in each of these conditions to influence biochemical pathways and gene expression. Specifically, many signaling molecules are manipulated by ROS, including REDOX-sensitive kinases, phosphatases and transcription factor NF-κB. These REDOX-sensitive molecules act as downstream effectors of ROS and serve as critical signaling events leading to skeletal muscle remodeling in response to increased contractile activity (i.e. physical exercise) or during disuse (e.g. immobilization).

4.9.4. Oxidative Stress and Muscular Dystroph

Duchenne Muscular Dystrophy (DMD) is the most devastating type of muscular dystrophy, characterized by progressive damage, inflammation, fibrosis, and weakness of respiratory and limb skeletal muscles. Muscle damage and pathology with DMD are proposed to result from high susceptibility to material fatigue injury and chronic inflammation. DMD is caused by X-linked defects in the gene that encodes for dystrophin, a key scaffolding protein of the dystroglycan complex (DGC) within the sarcolemmal cytoskeleton. Due to compromised DGC, mechanical integrity is impaired and important REDOX signaling proteins [i.e. neuronal NO-synthase (nNOS), caveolin-3] and pathways are disrupted. Disruption of DGC leads to increased susceptibility to injury with repeated, eccentric contractions as well as inflammation, which can result in significant damage and necrosis. Elevated oxidative stress may contribute to damage, weakness, and possibly fibrosis during DMD.

Increasing evidence indicates inflammatory processes are highly integrated into muscle wasting with DMD. Circulating inflammatory cytokines may be a 1000 times higher in DMD patients than healthy controls. Inflammatory mediators (i.e. myeloperoxidase, MPO), NAD(P)H oxidase (NOx), mitochondria and decoupling of inducible nitric oxide synthase (iNOS) act as cellular sources of oxidative stress in respiratory and locomotor muscles with DMD. Elevated oxidative stress may promote inflammatory cell invasion, exacerbate damage, and interfere with cell signaling that can promote repair. Markers of oxidative stress and lipid peroxidation are elevated with DMD. Impairment of mitochondrial function with DMD can also lead to oxidative stress, apoptosis and necrosis.

Oxidative stress and DGC imbalance are linked in the patho-physiology of DMD. Uncoupling of nNOS from the DGC increases NOx activity, inflammation, and protein degradation (via activation of ubiquitin ligases), which cumulatively leads to muscle damage. Notably, NOx is an important contributor of skeletal muscle wasting including mechanical ventilation and aging. Loss of nNOS from the sarcolemma appears central to damage and myopathy with DMD. Also, aging results in dissociation of dystrophin from the sarcolemma causing a reduction in nNOS, which is linked with sarcopenia and weakness. Dystroglycan is the mechanical weak link in the transfer of transverse forces linking the contractile apparatus and z-disc with aging. Thus, disruptions of the DGC and subsarcolemmal cytoskeleton possibly share a common signaling mechanism that elicits myopathy across a number of pathologies. The diaphragm muscle, in particular, suffers from significant fibrosis, damage and weakness with DMD, and is also susceptible to oxidative stress.

REDOX-based Interventions: Oxidative stress is a potential upstream contributor to inflammatory signaling and DMD pathology, and thus a therapeutic target. Reduction in oxidative stress correlates with slowed muscle wasting and relief of clinical symptoms associated with DMD, including respiratory muscle distress. Loss of nNOS from the sarcolemma may have a profound effect on oxidative stress, mechanical integrity and capability for satellite cell activation, growth and repair. Thus, interventions that stabilize sarcolemmal localization of nNOS and other DGC proteins may prove effective in ameliorating DMD. Emerging evidence suggests that oxidative stress is an important amplifier of pathology in respiratory and limb muscles with DMD. A cascade of oxidative events involving NOx, caveolin-3 and NF-κB may contribute to damage, inflammation and impaired contractile function in the skeletal muscles of patients suffering from DMD. Pharmaceutical developments that target specific mechanisms elevating oxidative stress with DMD could also attenuate myopathy and provide relief for patients.

REDOX Signaling Model for DMD

ROS-induced Dystroglycan Complex (DGC) disruption, muscle damage, and pathology

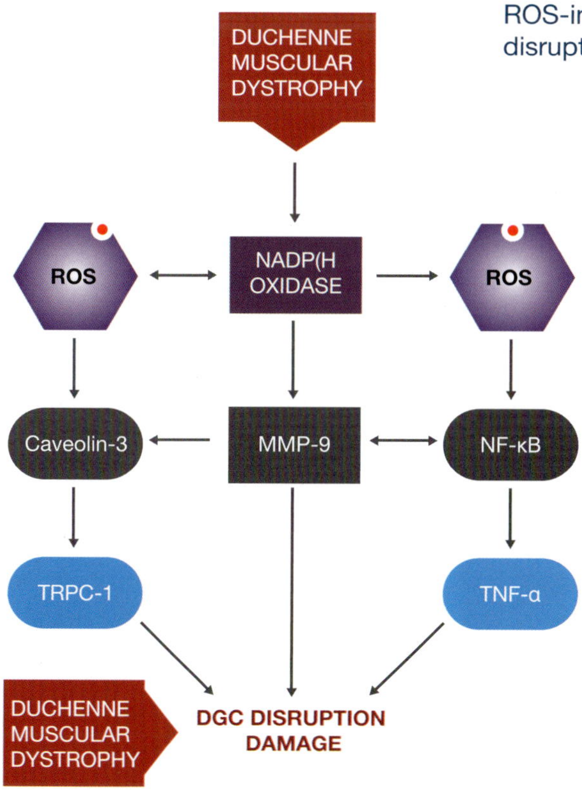

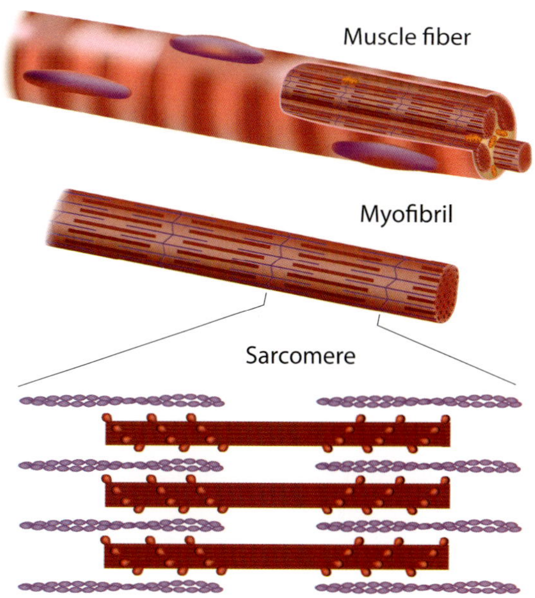

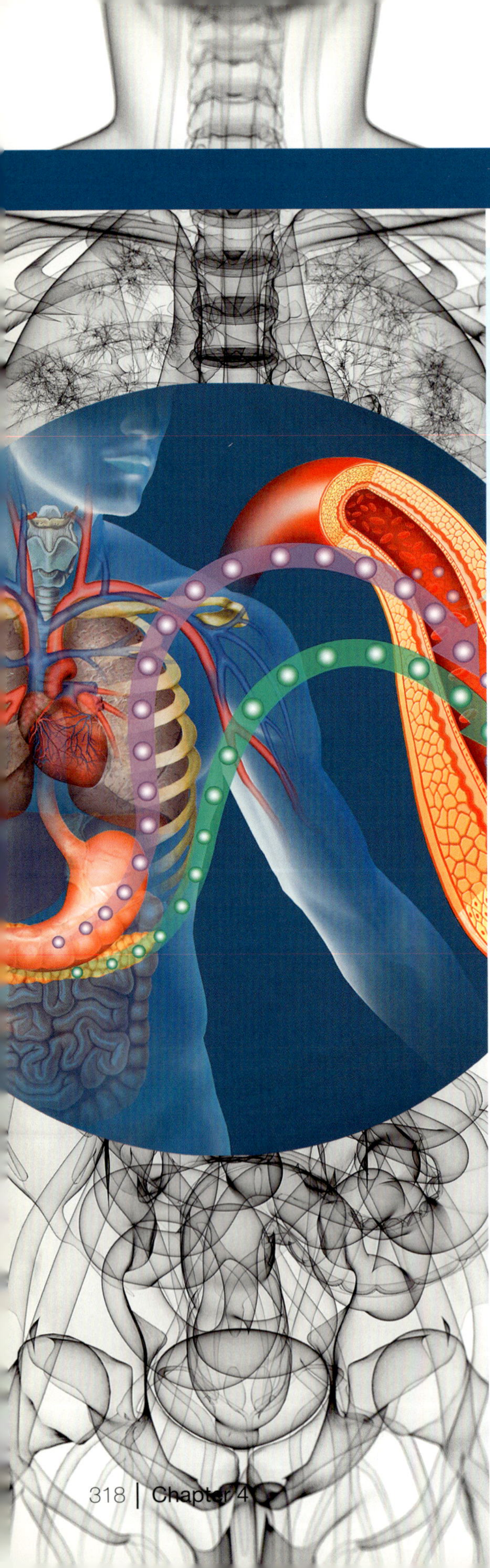

4.10 REDOX: Endocrine System

The Endocrine System is made of a series of glands that produce hormones (chemical mediators). When released into the bloodstream, hormones travel to target tissue and generate a response. Hormones alter the metabolism of target organs by increasing or decreasing their activity. These changes in activity are strictly balanced to maintain homeostasis. The major areas of endocrine control and integration include responses to stress and injury, growth and development, absorption of nutrients, energy metabolism, water and electrolyte balance, reproduction, birth, and lactation.

The endocrine system is a cell signal system, like the nervous system, yet its effects and mechanism(s) are distinct. The endocrine system maintains homeostasis and long-term control using chemical (hormone) signals. These effects are slow to initiate, prolonged in response, and last from a few hours up to weeks. In contrast, the nervous system sends information very quickly, and responses are generally short-lived. The endocrine and nervous systems work closely and the neural centers in the brain control endocrine glands. The hypothalamus, aka the 'master switch board' in the brain, sends messages to the pituitary gland; which in turn, activates the release of hormones.This signal sequence is known as the hypothalamic-pituitary-adrenal axis.

Hormones are grouped into three classes based on their structure, namely, steroids, peptides, and amines. Steroid hormones are lipids derived from cholesterol, and secreted by the gonads, adrenal cortex, and placenta. Peptides are short chains of amino acids (most hormones are peptides) secreted by the pituitary, parathyroid, heart, stomach, liver, and kidneys. Amines are derived from the amino acid tyrosine and are secreted from the thyroid and adrenal medulla. The endocrine system uses cycles and negative feedback to regulate physiological functions. Negative feedback regulates the secretion of almost every hormone. Cycles of secretion maintain physiological and homeostatic control. These cycles can range from hours to months in duration.

There are two mechanisms of hormone action. Non-steroidal (water-soluble) hormones do not enter the cell but bind to plasma membrane receptors and generate an intra-cellular chemical signal (second message). Steroid hormones, on the other hand, pass through the plasma membrane, bind to the DNA and activate specific genes (up-regulate protein synthesis).

TABLE: *Human Endocrine System and its Hormone Functions*

Gland	Hormone	Abbr.	Physiological Function(s)
Pineal	Melatonin		Biological clock.
Hypothalamus	Antidiuretic hormone	ADH	Acts on the kidney to preserve fluid and electrolyte balance.
	Proopiomelanocortin	POMC	Precursor hormone for ACTH and MSH.
Pituitary	Luteinizing hormone	LH	Acts on the ovary to stimulate estrogen production and induce ovulation (in females).
	Follicle stimulating hormone	FSH	In males, acts on the testis to stimulate the production of testosterone.
	Adrenocorticotrophic hormone	ACTH	In females, stimulates ovarian follicle maturation. In males, regulates sperm production.
	Growth hormone	GH	Acts on adrenal to stimulate cortisol production.
	Prolactin	Prl	Acts on various tissues to stimulate growth.
	Melanocyte stimulating hormone	MSH	Milk let down during lactation.
	Thyroid stimulating hormone	TSH	Stimulates skin tone.
Pancreas	Insulin		Acts on thyroid to signal thyroxin production.
Thyroid	Triiodothyronine and thyroxine	T3 and T4	Regulates blood sugar levels.
Adrenal	Cortisol		Development of the brain, reproductive tract, and regulation of metabolism.
	Dehydroepiandrostendione	DHEA	Immune suppression and stress response.
Ovary	Estrogens (estradiol, estrone, estriol)	E2, E1, E3	Menstrual and estrous cycles
	Progesterone	P4	Growth promotion, maintain connective tissues, preserve bone, vascular compliance.
	Testosterone	T	Maintain endometrium for pregnancy.
	Inhibin		Precursor for estrogen and acts on libido.
Testis	Testosterone	T	Feedback regulation on pituitary FSH secretion.
	Dihydrotestosterone	DHT	Growth of male secondary sexual character, sperm production and libido.
	Inhibin		Some male secondary sexual characteristics.
Placenta	Progesterone	P4	Feedback regulation on pituitary FSH secretion.
	Estriol	E3	Maintenance of pregnancy.

4.10.1. REDOX Homeostasis and Hormones

REDOX homeostasis and metabolic/regulatory function of autocrine (i.e. insulin) signaling are the key mechanisms in endocrine function. Excess generation of reactive oxygen species (ROS) causes oxidative stress, which leads to imbalance of REDOX homeostasis, disruption of cell signaling cascade and an ultimate dysfunction of the endocrine system. This failure can manifest into several metabolic disorders such as diabetes, obesity, cardiovascular disease, osteoporosis, and a plethora of pathologies associated with hormonal malfunction. For example, the pathogenesis of type 2 diabetes mellitus (T2DM) might be considered an inevitable result of progressive self-accelerating oxidative stress and concomitantly dysregulated endocrine function (in peripheral tissues and pancreatic β cells). Since REDOX homeostasis is inherent to the insulin receptor signaling mechanism; its impairment ultimately leads to the oxidative and nitrosative stress.

An upsurge in endocrine disrupting chemicals (EDCs), pollutants and other synthetic chemicals in the environment has seriously affected the physiological REDOX homeostasis of both plant and animal kingdoms, including the human species. The biological consequences of this REDOX disturbance could be witnessed as global emergence of endocrine dysfunctions at epidemic proportions. The 'Obesogen' hypothesis, proposed by Bruce Blumberg (2006), explains the weight gain effects of certain chemicals in the environment. A variety of EDCs can influence adipo-genesis and obesity. EDCs are known to impact hormone-dependent metabolic systems and brain function. These chemicals are frequently found in human blood and urine; epidemiological data links the EDC exposure with obesity. EDCs target various endocrine axes and affect adipocyte physiology, in particular, the regulation of energy homeostasis. Several persistent organo-chlorine pesticides and fungicides have also been implicated in obesity. Metabolic alterations, such as metabolic syndrome and type-2 diabetes, are recognized as obesity comorbid conditions, and EDC-exposure may be involved in their pathogenesis.

According to Neel's 'thrifty' gene hypothesis, all humans are double knockouts. Due to genetic mutations, they lack L-gulono-lactone oxidase enzyme (key for vitamin C synthesis) and uricase enzyme (critical for uric acid degradation). Both mutations occurred during late Eocene (34 to 56 million years ago), a period of global cooling associated with a shortage of food supplies. These mutations may have had a survival advantage by augmenting the effects of fructose to increase fat stores. Fructose was the primary nutrient in fruit which was the main staple of early primates, but this food likely became less available during the global cooling that occurred at the time of these mutations. However, the introduction of sugar and high fructose corn syrup in the current human diet has skyrocketed, with a nearly 30- to 50-fold increase in fructose intake since 1700 A.D (while the intake of natural fruits high in vitamin C has plummeted). This dietary shift, coupled with the presence of our double knockout genotype, is largely responsible for the obesity epidemic today.

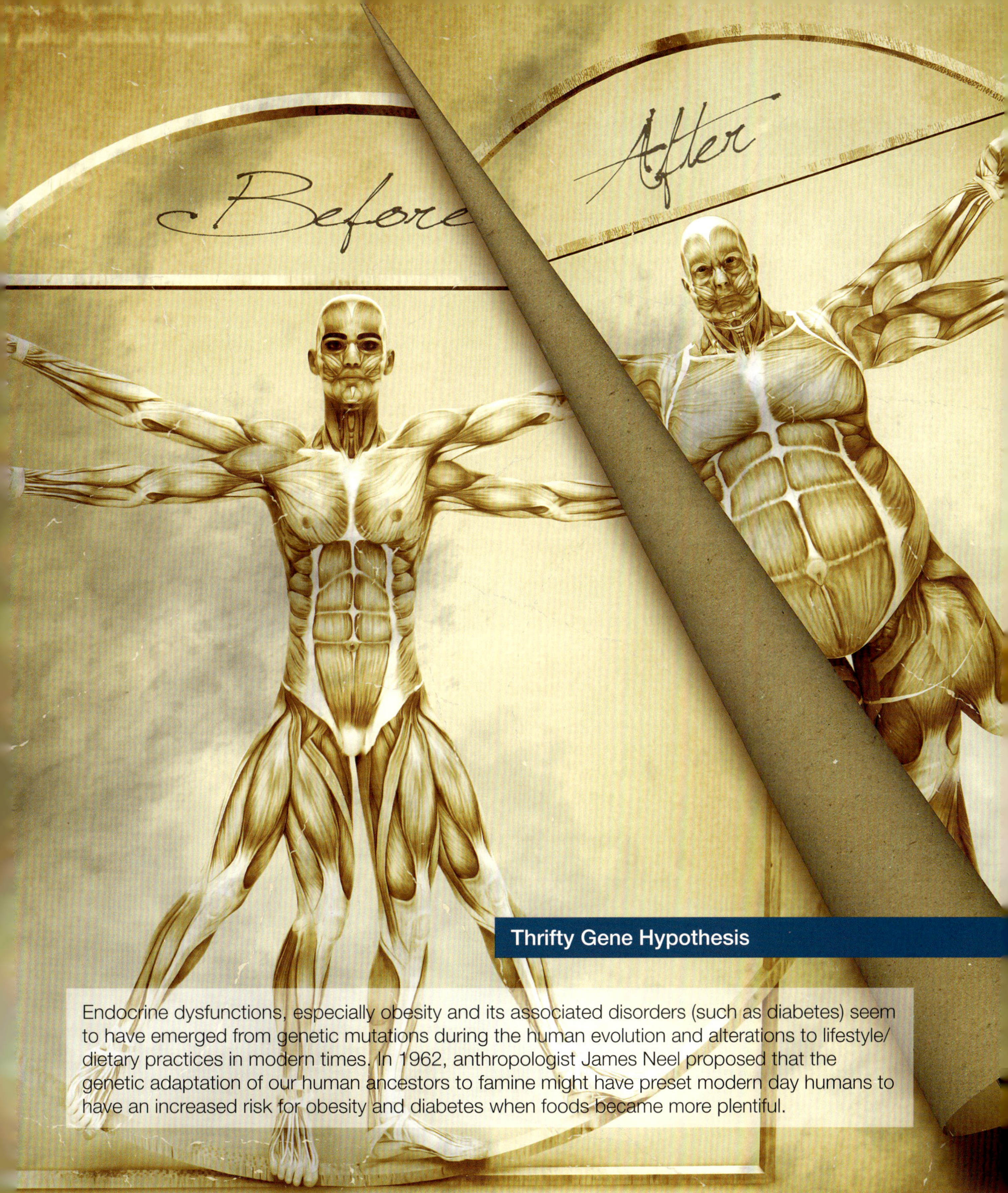

Thrifty Gene Hypothesis

Endocrine dysfunctions, especially obesity and its associated disorders (such as diabetes) seem to have emerged from genetic mutations during the human evolution and alterations to lifestyle/dietary practices in modern times. In 1962, anthropologist James Neel proposed that the genetic adaptation of our human ancestors to famine might have preset modern day humans to have an increased risk for obesity and diabetes when foods became more plentiful.

4.10.2. Oxidative Stress and Diabetes

Diabetes mellitus (DM or Diabetes) is a chronic disease that occurs either when the pancreas does not produce enough insulin, or when the body cannot effectively use the insulin it produces. Insulin is a hormone that regulates blood sugar. Hyperglycemia, or raised blood sugar, is a common effect of uncontrolled diabetes and, over time, leads to serious damage to many of the body's systems, especially the nerves and blood vessels. This high blood sugar produces the classical symptoms of polyuria (frequent urination), polydipsia (increased thirst) and polyphagia (increased hunger). There are three main types of diabetes mellitus.

- Type 1 DM results from the body's failure to produce insulin, and currently requires the person to inject insulin or wear an insulin pump. This form was previously referred to as 'insulin-dependent diabetes mellitus' (IDDM) or 'juvenile diabetes'.
- Type 2 DM (T2DM) results from insulin resistance, a condition in which cells fail to use insulin properly, sometimes combined with an absolute insulin deficiency. This form was previously referred to as non insulin-dependent diabetes mellitus (NIDDM) or 'adult-onset diabetes'.
- Gestational diabetes, the 3rd main form, occurs when pregnant women without a previous diagnosis of diabetes develop a high blood glucose level. It may precede development of T2DM.

Other forms of DM include congenital diabetes due to genetic defects of insulin secretion, cystic fibrosis-related diabetes, steroid diabetes induced by high doses of glucocorticoids, and several forms of monogenic diabetes. Excess nourishment with a sedentary lifestyle, results in over abundance of glucose and fatty acid accumulation within muscle, adipose tissue and pancreatic cells.

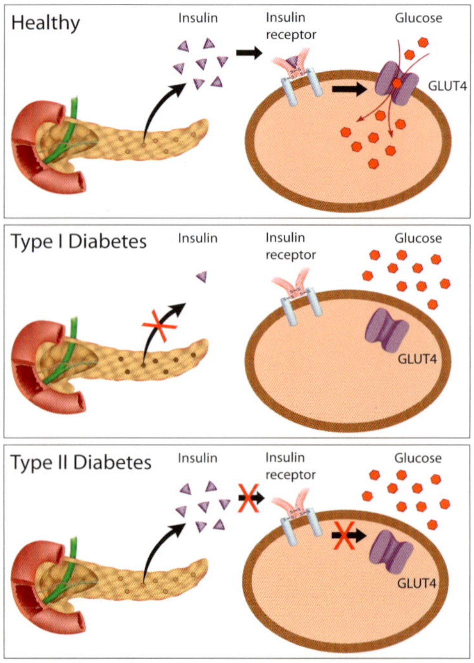

This leads to the generation of excess ROS, particularly $O_2^{\bullet -}$ anion, through the mitochondrial electron transport chain (ETC). Recently, it has been suggested that fluctuating blood glucose concentrations, like those observed during postprandial glycaemic excursions in people with impaired glucose tolerance or T2DM, may contribute significantly to oxidative stress – perhaps even more so than chronically elevated blood glucose.

Reactive Oxygen Species (ROS) and Diabetes: Both types of diabetics display increased levels of ROS; thus, the onset of diabetes is closely associated with oxidative stress. There are multiple sources of oxidative stress in diabetes including non-enzymatic, enzymatic and mitochondrial pathways. Non-enzymatic sources originate from the oxidative biochemistry of glucose metabolism. Hyperglycemia can directly cause increased ROS generation. Glucose can undergo auto-oxidation and generate $O_2^{\bullet -}$ anion and ·OH radicals. This glycation reaction is observed in various tissues and organs, and different kinds of glycated proteins (such as glycosylated hemoglobin, albumin, and lens crystalline) are produced in a non-enzymatic manner. The reaction produces Schiff base, Amadori product, and finally, advanced glycosylation end products (AGEs). ROS is generated at multiple steps during this process. The mitochondrial ETC in mitochondria is also an important pathway to produce ROS. Under diabetic conditions, ETC is activated, which leads to production of larger amounts of ROS. Enzymatic sources of augmented ROS generation in diabetes include NO-synthase (NOS), NAD(P)H oxidase (NOx) and xanthine oxidase (XO). NOx, in particular, is activated by various stimuli such as AGEs, insulin, and angiotensin II; all of which are possibly induced under diabetic conditions. Taken together, it is likely that ROS play an important role in the development of T2DM and atherosclerosis.

Oxidative Stress and Pancreatic β–Cell Dysfunction in T2DM: Oxidative stress plays a role in the progression of pancreatic β-cell dysfunction in type 2 diabetes. ROS are provoked by hyperglycemia and/or hyperlipidemia under diabetic conditions, which leads to activation of the JNK pathway in pancreatic β-cells. ROS and subsequent activation of the JNK pathway induce nucleo-cytoplasmic translocation of PDX-1, which leads to reduction of PDX-1 activity and suppression of insulin. Therefore, it is likely that ROS and activation of the JNK pathway are involved in β-cell dysfunction found in T2DM.

Acute exposure of β-cells to a high glucose concentration stimulates insulin gene expression, but chronic exposure has various adverse effects on various β-cell functions. However, chronic hyperglycemia is a cause of impairment of insulin biosynthesis and secretion. This process is called β-cell glucose toxicity which is often observed under diabetic conditions. In the diabetic state, hyperglycemia per se and subsequent production of ROS decrease insulin gene expression and

Hormonal Activity: The Nobel Prize in Physiology or Medicine
1950 – Edward Kendall, Tadeus Reichstein and Philip Showalter for discovering the hormones of the adrenal cortex, their structure and biological effects.
1971 – Earl Sutherland for discovering the mechanisms of the action of hormones.

secretion and finally bring about apoptosis. The loss of insulin gene expression is accompanied by decreased expression and/or DNA binding activities of transcription factors: pancreatic and duodenal homeobox-1 (PDX-1) and MafA. After chronic exposure to a high glucose concentration, expression and/or DNA binding activities of these two transcription factors are reduced. The PDX-1 plays a crucial role in pancreas development, β-cell differentiation, induction of surrogate β-cells, and maintenance of mature β-cell function. The MafA, a β-cell-specific transcription factor, functions as a potent activator of insulin gene transcription.

Lipotoxicity is also involved in the deterioration of β-cell function found in T2DM. When islets or β-cell-derived cell lines were exposed to free fatty acids (FFAs), ROS were induced, which led to the reduction of insulin secretion and β-cell dysfunction. It was also reported that FFA-mediated induction of inducible nitric oxide synthase (iNOS) and excess ·NO generation were involved in the progression of β-cell dysfunction. Since intra-cellular ·NO is an important mediator of β-cell apoptosis, it is likely that the loss of β-cell mass observed in T2DM is due to ·NO-induced apoptosis.

Oxidative Stress and Insulin Resistance in T2DM: During diabetic conditions, various insulin target tissues such as the liver, muscle, and fat become resistant to insulin. The patho-physiology of insulin resistance involves a complex network of insulin signaling pathways. After insulin binds to insulin receptors on the cell surface, the insulin receptor and its substrates are phosphorylated, which leads to activation of various insulin signaling pathways. ROS are involved in the progression of insulin resistance. These reactive species disrupt insulin-induced cellular redistribution of insulin receptor substrate-1 (IRS-1) and phosphatidylinositol 3-kinase, which could impair insulin-induced translocation in adipocytes.

During hyperglycemia, increased ROS can lead to abnormal activation JNK pathway in the liver, muscle, and adipose tissue in obese T2DM conditions. The Iκ-B kinase β (IKK) pathway is also activated by such factors and is involved in the

Insulin and Glucose Metabolism

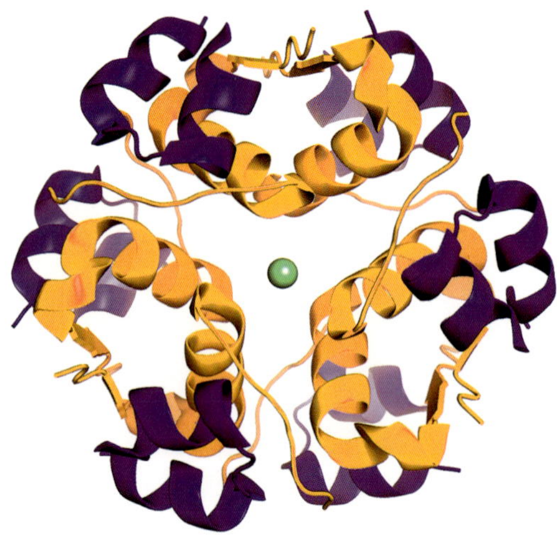

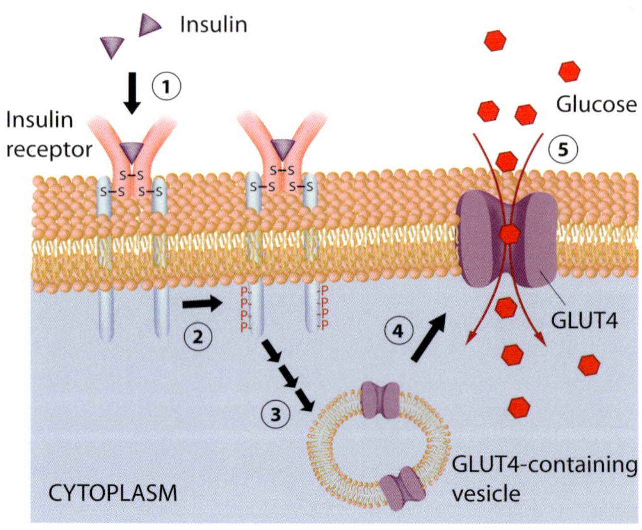

development of insulin resistance. Activation of the IKK pathway increases IRS-1 serine phosphorylation which leads to suppression of insulin signaling. Also, suppression of the IKK pathway decreases insulin resistance and ameliorates glucose intolerance. Therefore, it is likely that activation of stress signaling, such as the JNK and IKK pathways, is involved in the development of insulin resistance and that such pathways could be a therapeutic target for diabetes.

Oxidative Stress and Diabetes in Atherosclerosis: ROS are produced by various metabolic pathways under diabetic conditions and predispose the risk of atherosclerosis. Hyperglycemia induces ROS through activation of the glycation reaction and ETC in mitochondria. Also, AGEs, insulin, and angiotensin II induce ROS through activation of NOx. First, ROS decrease ·NO levels, which leads to endothelial dysfunction. Second, ROS increase expression of various adhesion molecules such as ICAM- 1 and VCAM-1, which leads to inflammatory cell recruitment. Finally, ROS increase expression of various growth factors and activate stress signaling such as JNK and Pim-1, which leads to proliferation of smooth muscle cell. NOx is activated by various factors such as AGEs, insulin, and angiotensin II; all of which are possibly induced under diabetic conditions. In addition, it was shown that high glucose stimulates ROS production through the activation of NOx and that the p22 phox significantly increases in diabetic arteries. Therefore, it is possible that such increased expression of p22 phox contributes to the development of atherosclerosis. The vulnerability to oxidative stress and the progression of atherosclerosis is influenced by genetics. Furthermore, the accumulation of oxidative stress-associated gene polymorphisms is likely associated with the severity and the progression of atherosclerosis in diabetic patients. It was reported that carotid intima-media thickness (IMT) as well as serum 8-OHdG level, a marker of oxidative stress, were closely associated with the accumulation of several oxidative stress-associated gene polymorphisms.

REDOX-based T2DM Interventions: Oxidative stress, through a single unifying mechanism of $O_2^{\bullet-}$ anion production, is the common patho-biological factor leading to insulin resistance, β-cell dysfunction, impaired glucose tolerance (IGT) and ultimately to T2DM. Furthermore, oxidative stress has been implicated as the underlying cause of both the macro-vascular and micro-vascular complications associated with T2DM. It follows that therapies aimed at reducing oxidative stress would benefit patients with T2DM and those at risk for developing diabetes. REDOX factors that attenuate oxidative stress and pro-inflammatory cascades could provide an important mechanism to reduce the burden of chronic diseases linked to these conditions, such as obesity, T2DM, and CVD. Human clinical studies have demonstrated that disrupting the IKKβ/NF-κB pathway improves obesity-related insulin sensitivity. Blocking JNK activity in diabetics improves systemic glucose homeostasis and insulin sensitivity; whereas inhibition of IKK by salicylates could enhance insulin action.

Insulin and Sugar Metabolism: The Nobel Prizes in Physiology or Medicine
1923 – Frederick Banting and John Macleod for the discovery of insulin.
1947 – Carl Cori and Gerty Cori for discovering the catalytic conversion of glycogen; and Bernardo Alberto Houssay for discovering the functional role of hormone of the anterior pituitary lobe in the metabolism of sugar.

The major current therapeutic agents to treat T2DM, sulfonylureas, metformin, and insulin-sensitizing glitazones, all have been associated with the normalization of circulating inflammation and oxidative markers. This effect may be mediated, at least partly, by the action of therapeutic agents on the innate immune system to retain the homeostasis of the oxidative and inflammatory status. Accordingly, glitazones [peroxisome proliferator-activated receptor-γ (PPAR-γ) agonists] improve the activities of superoxide dismutase (SOD), catalase and glutathione peroxidase (GPx) in hyperglycemics. Treatment with repaglinide, an antioxidant often used in T2DM treatment, results in higher glutathione (GSH) levels and normalization of nitrotyrosine. Similarly, several therapeutic agents prescribed in T2DM, such as metformin and sulfonylureas, exhibit antioxidant and anti-inflammatory activity, an effect that may be partially related to their therapeutic efficacy. These observations provide an additional line of evidence to emphasize that targeting the inflammation-related and oxidation-related pathways can provide a feasible approach in the prevention of T2DM.

DIABETES FACTS:

DM occurs worldwide, but is more common (especially type, the T2DM) in the more developed countries. The greatest increase in prevalence is, however, expected to occur in Asia and Africa, where most patients will probably be found by 2030. The increase in incidence in developing countries follows the trend of urbanization and lifestyle changes, perhaps most importantly a 'western-style' diet. Rates of disease incidence increase with age, obesity, and a sedentary lifestyle, and are elevated in certain ethnicities (Hispanics, Africans, and Aboriginals). Rates have been increasingly noted in children.

- Over time, diabetes can damage the heart, blood vessels, eyes, kidneys, and nerves.
- Diabetes increases the risk of heart disease and stroke. 50% of people with diabetes die of CVD (primarily heart disease and stroke).
- Combined with reduced blood flow, neuropathy (nerve damage) in the feet increases the chance of foot ulcers, infection and eventual need for limb amputation.
- Diabetic retinopathy is an important cause of blindness, and occurs as a result of long-term accumulated damage to the small blood vessels in the retina. One percent of global blindness can be attributed to diabetes.
- Diabetes is among the leading causes of kidney failure.
- The overall risk of dying among people with diabetes is at least double the risk of their peers without diabetes.

Globally, an estimated 347 million people have diabetes, with type 2 making up about 90% of the cases. Its incidence is increasing rapidly, and by 2030, this number is estimated to double. It is estimated that 6 people die every minute from T2DM globally, a figure that will soon make the disease one of the most prevalent causes of preventable mortality in the world.

DIABETES

4.10.3. Oxidative Stress and Obesity

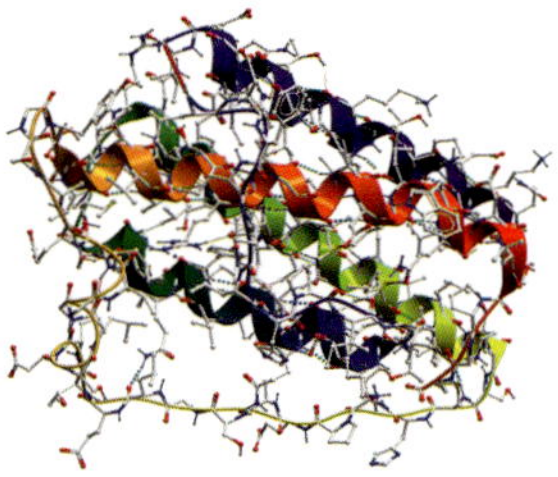

LEPTIN

Obesity or excessive fat accumulation is a consequence of positive energy balance that results from interaction among several factors, including diet (increased intake of energy-dense foods and decreased intake of food rich in micronutrients and bioactive compounds), decreased physical activity (sedentary lifestyle), nutritional and hormonal status in early life; as well as genetic, environmental, cultural, and economic factors. Other etiological factors that are associated with obesity are certain chromosomal aberrations (i.e. Prader-Willi syndrome), hormonal pathologies (i.e. Cushing's disease), hypothalamic lesions or tumors, and drugs (such as steroids and antidepressants). Obesity is a growing global health problem with increasing risk of morbidity from several pathologies such as hypertension, dyslipidemia, type 2 diabetes, coronary heart disease, stroke, non-alcoholic fatty liver disease, osteoarthritis, sleep apnea, and endometrial, breast, prostate, and colon cancers.

Oxidative Stress and Obesity: Oxidative stress can be a consequence, but also a trigger of obesity coupled with altered REDOX state and increased metabolic risk. Chronic hyper-nutrition, high fat high carbohydrate (HFHC) meals, as well as high dietary saturated fatty acids (SFA) and trans-fatty acids, stimulate intra-cellular pathways, leading to oxidative stress through multiple biochemical mechanisms. This includes superoxide ($O_2^{\bullet-}$) generation from NADPH oxidases (NOx), oxidative phosphorylation, glyceraldehyde autoxidation, protein kinase C (PKC) activation, and polyol and hexosamine pathways. Oxidative stress could play a causative role in the development of obesity by stimulating white adipose tissue deposition and altering food intake. In vitro studies have demonstrated that oxidative stress could increase pre-adipocyte proliferation, adipocyte differentiation, and size of mature adipocytes. Also, reactive oxygen species (ROS) seem to be involved in the control of body weight, by exerting different effects on hypothalamic neurons that control satiety and hunger behavior. It has also been demonstrated that obesity can induce systemic oxidative stress; accordingly, fat accumulation increases NOx activity and endoplasmic reticulum (ER) stress in adipocytes that lead to increased ROS production. Other factors that contribute to oxidative stress in obesity are abnormal post-prandial ROS generation, hyper-leptinemia, chronic inflammation, tissue dysfunction, and low antioxidant defenses. Oxidative stress and inflammation appear to be closely interlinked in obesity, although it is difficult to establish the temporal sequence of their relationship. For example, several pro-inflammatory transcription factors, such as nuclear factor-κB (NF-κB) and activator protein-1 (AP-1) are REDOX sensitive; therefore, ROS trigger the release of inflammatory cytokines, which in turn enhance ROS production, thus establishing a vicious circle. Systemic oxidative stress and inflammation are also key factors in the pathogenesis of obesity-related diseases, including atherosclerosis, insulin resistance, type-2 diabetes, and cancer. It has been suggested that increased oxidative stress and inflammation in obesity also enhance the aging processes.

LEPTIN is the human obesity protein the regulates appetite.

Oxidative Stress and Adipocytes: Adipocytes (also known as lipocytes or fat cells) are primarily composed of adipose tissue, specialized in storing energy as fat. There are two types of adipose tissue, white adipose tissue (WAT) and brown adipose tissue (BAT), which are also known as white fat and brown fat, respectively. The presence of beige adipocytes with gene expression pattern distinct from either white or brown adipocytes has also been reported. Adipocytes produce a variety of biologically active molecules, collectively known as adipo-cytokines or adipokines, including plasminogen activator inhibitor-1 (PAI-1), TNF-α, resistin, leptin, and adiponectin. Dysregulated production of these adipo-cytokines participates in the pathogenesis of obesity-associated metabolic syndrome. Increased production of PAI-1 and TNF-α from accumulated fat contribute to the development of thrombosis and insulin resistance in obesity. In contrast, adiponectin exerts insulin-sensitizing and anti-atherogenic effects, and hence a decrease in plasma adiponectin is causative for insulin resistance and atherosclerosis in obesity. Leptin is a hormone produced by adipose tissue, which regulates appetite and exerts protective effects against lipotoxicity in non-adipose tissues. Hyper-leptinemia also induces oxidative stress, mainly by increasing mitochondrial and peroxisomal fatty acid oxidation; it also stimulates proliferation and activation of monocytes/macrophages and production of IL-6 and TNF-α.

Oxidative Stress and Lipotoxicity: Several mechanisms have been suggested to explain the enhanced oxidative stress observed in obese individuals. These include altered lipid and glucose metabolism, chronic inflammation, tissue dys-function, hyperleptinemia, and abnormal post-prandial ROS generation. When adipocytes reach non-physiological limits, they fail to function as an energy storage systems. Furthermore, elevated serum triglycerides and free fatty acids (FFAs) accumulate in non-adipose tissues, including the pancreas, heart, liver, kidney, and blood vessel wall. While FFAs and their metabolites play a key role in membrane structure, intra-cellular signaling, and energy homeostasis, accumu-lation of excess lipid in these organs leads to cellular dysfunction and cell death. This process, termed 'lipotoxicity', has been implicated in β-cell loss during the progression of T2DM, and in the pathogenesis of diabetic complications with loss of cardiomyocytes, hepatocytes, renal parenchymal cells, and endothelial cells.

Since 1980, obesity has doubled worldwide and is now increasing in low- and middle-income countries (principally in urban settings). According the W.H.O. Report (2009-2010), over one-third of U.S. adults (35.7%) were obese; while in European Union countries, obesity affects 23% of women and 20% of men. Furthermore, inadequate dietary patterns and low levels of physical activity have dramatically increased childhood obesity, globally; in 2010, more than 40 million children under five were overweight.

There are several mechanisms through which lipotoxicity may induce oxidative stress. The FFA metabolite ceramide can activate NOx and disrupt respiration, either by inducing release of cytochrome c or through interaction with mitochondrial respiratory chain complex III. Excess FFAs also increase diacylglycerol (DAG), which activates NOx through PKC-dependent pathways. Increased FFA oxidation and oxidative phosphorylation also have the potential to produce singlet electrons, contributing to oxidative stress. It has also been proposed that FFA remodeling of mitochondrial membranes leads to organelle dysfunction that causes or amplifies oxidative stress.

Endocrine Disruption and Obesity: Endocrine disrupting chemicals (EDCs) are exogenous substances that alter function(s) of the endocrine system and consequently cause adverse health effects in an intact organism, or its progeny, or (sub)-populations. EDCs can mimic the action of natural hormones or interfere with their production, release, metabolism, and elimination. These substances may be derived from natural, plant, animal, or synthetic sources. EDCs can bind and activate multiple hormone nuclear receptors, among which, the estrogen receptors regulate many aspects of metabolism, including glucose transport, glycolysis, mitochondrial activity, and lipid metabolism. It is likely that estrogen receptor activation modulates neural networks controlling food intake and adipose tissue.

Neonatal exposure to a low dose of the estrogenic drug diethylstilbestrol (DES) stimulates a subsequent increase in body weight and an increase in body fat. Bisphenol A (BPA), a breakdown product of coatings in food and beverage containers, may act as an estrogen receptor agonist. In the US population, exposure is nearly ubiquitous, and BPA has been detected in fat, blood, and urine. Short exposure to BPA provokes chronic hyper-insulinemia, with perturbations of glucose and insulin tolerance. Furthermore, high- or low-dose exposure to BPA during gestation up to puberty leads to hyper-lipidemia with increased body and adipose tissue weight in both sexes. BPA exposure has been shown to disrupt multiple metabolic mechanisms, suggesting that it may contribute to obesity in humans. All major endocrine organs are vulnerable to endocrine disruption, including the Hypothalamus- Pituitary-Adrenal axis, reproductive organs, the pancreas, and the thyroid gland.

REDOX-based Interventions: Obesity could induce systemic oxidative stress and that increased oxidative stress in accumulated fat is, at least in part, the underlying cause of dysregulation of adipocytokines and development of metabolic syndrome. As an early instigator of obesity-associated metabolic syndrome, increased oxidative stress in accumulated fat should be an important target for the development of new therapies. Strategies to lower oxidative stress in obesity include weight loss, physical activity, and antioxidant-rich diet. It is known that weight reduction decreases oxidation markers, increases antioxidant defenses and improves metabolic and cardiovascular risks associated with human obesity. It is well established that a diet abundant in fruits, vegetables, whole grains, legumes, fish, olive oil, and dairy fermented foods is helpful to maintain weight and reduce the incidence of metabolic diseases. Beneficial components present in these foods are certain REDOX macronutrients [i.e. monounsaturated fatty acids (MUFA) and Ω-3 polyunsaturated fatty acids (Ω-3 PUFA)], vitamin C, vitamin E, and phytoceuticals. In vitro and in vivo studies show that these dietary factors can act through several mechanisms, such as cell signaling and modulation of gene expression, reduction of obesity-induced oxidative stress, production of inflammatory molecules, and lipid accumulation.

Obesity is defined as excessive fat accumulation that may impair health and increase mortality; the current classification of obesity is based on the Body Mass Index (BMI), which is the weight (in kilograms) divided by the square of height (in meters). Definition of BMI 'normal range' is based on Caucasian mortality data; a BMI of 30 kg/m^2 or more is index of obesity, while a BMI over 25 kg/m^2 is index of overweight. For Asian populations, a BMI of 27.5 kg/m^2 or more is index of obesity whereas a BMI >23 kg/m2 is index of overweight. BMI has limitations because it does not distinguish between lean mass and fat; it may overestimate body fat in well-trained body builders and underestimate body fat in older persons; moreover, BMI does not identify fat distribution. It is now well recognized that abdominal fat is a major risk for obesity-related diseases: indeed, visceral fat accumulation contributes to pro-oxidant and pro-inflammatory states, as well as to alterations in glucose and lipid metabolisms. Waist circumference (WC) or 'waist-to-hip ratio' are useful indicators of visceral fat distribution: WC equal to or more than 80 cm (in women) or 94 cm (in men) and waist-to-hip ratio above 0.90 for males and 0.85 for females are associated to high cardiometabolic risk in Europeans. However, these cut offs may not be appropriate for African-American, Hispanic, and Middle Eastern populations.

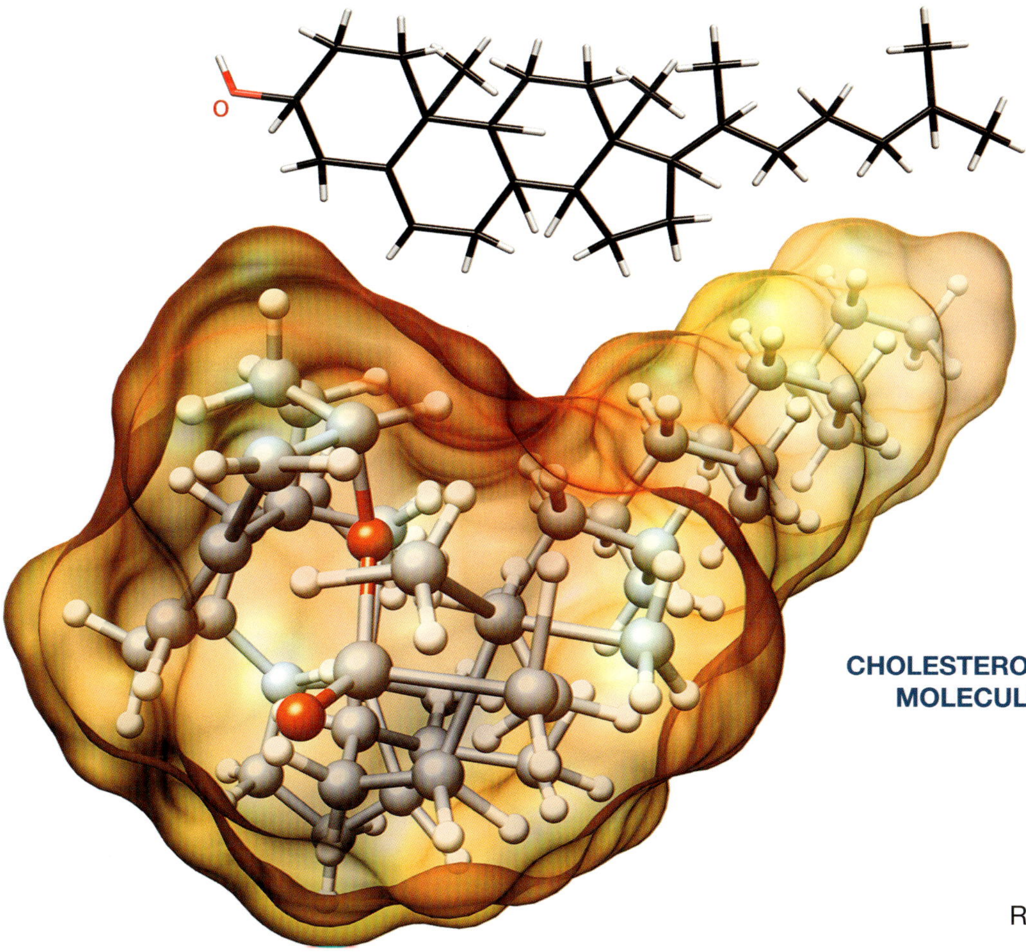

CHOLESTEROL MOLECULE

4.11 REDOX: The Reproductive System

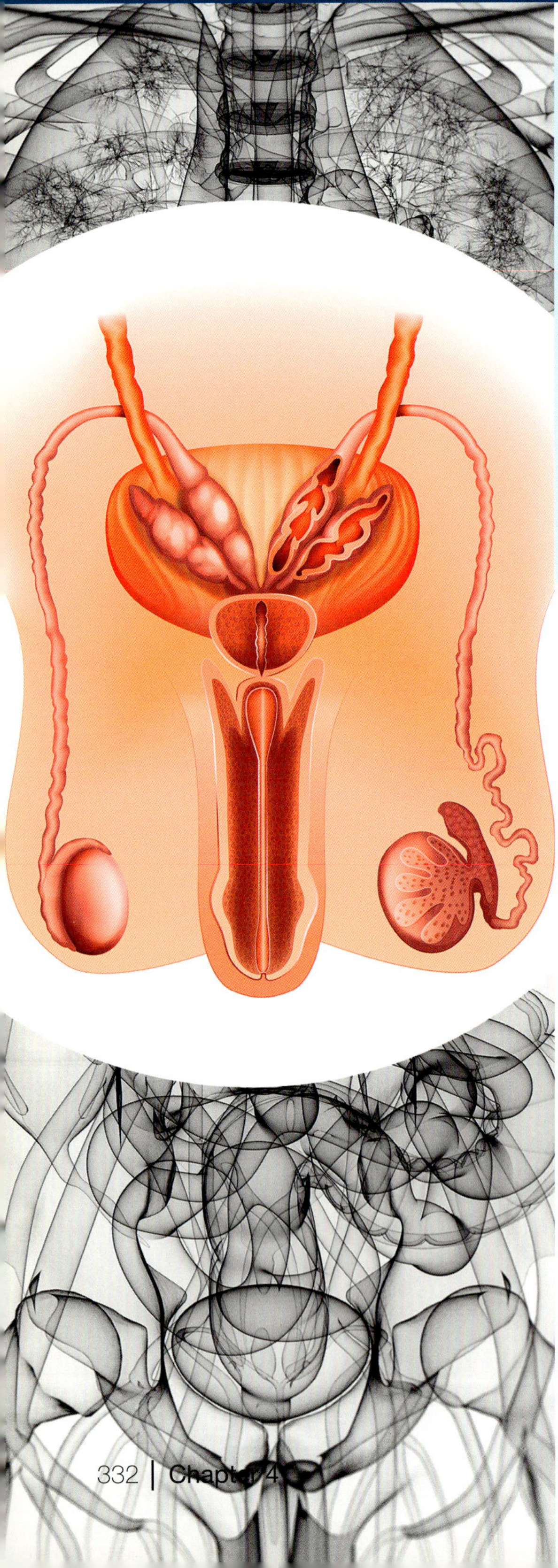

The Reproductive (Genital) System operates through internal fertilization, and depends on the integrated action of hormones and nervous system. Fertilization has three functions: (i) transmission of genes from both parents to offspring, (ii) restoration of the diploid number of chromosomes reduced during meiosis, and (iii) initiation of the offspring development. Sexual reproduction leads to the formation of a new offspring following the union of two distinct gametes that differ in structure (anisogamy) and origin. Male gonads (testes) make sperm and testosterone. Female gonads (ovaries) produce eggs and female sex hormones (i.e. estrogen) that protect and nourish the offspring. This anisogamy allows for merger of genetic material between two individuals with a greater possibility of genetic fitness in the offspring.

Male reproductive system has two major functions: (i) the production of sperm, and (ii) the delivery of sperm to the reproductive tract of the female. Sperm production begins at puberty that continues throughout life, with several hundred million sperm being produced each day. An adult male manufactures over 140 million sperm cells each day. Fluid is added to the sperm by the seminal vesicles, Cowper's glands, and the prostate. These fluids provide a source of energy (fructose) and an alkaline environment to activate the sperm.

In contrast to males, females initial steps in egg production occur prior to birth. By the time the female fetus is 20 weeks old, the stem cells (oogonia) divide by mitosis and produce more than 4 million oocytes. At the time of birth, only about 1 million of these remain (the others eliminated by apoptosis). No further development occurs until years later when the girl becomes sexually mature. A developing egg (secondary oocyte) is released each month from puberty until menopause, a total of 400-500 eggs in a lifetime. After puberty the ovary cycles between a follicular phase (maturing follicles) and a luteal phase (lasts usually 28 days). These cyclic phases are interrupted only by pregnancy and continue until menopause, when reproductive capability ends.

4.11.1. REDOX and Reproductive Health

REDOX mechanisms affect the total reproductive lifespan of a woman and continue through the menopausal phase. It plays a role during pregnancy and normal parturition and in initiation of preterm labor. ROS serve as cell signal molecules in physiological processes. ROS affect multiple physiological processes from oocyte maturation to fertilization, embryo development, and pregnancy. In regular physiology, oxidative stress is essential for normal cycling ovaries, follicular development and cyclical endometrial changes. Nitric oxide (·NO) is involved in the regulation of endometrial and ovarian function, maintenance of uterine quiescence, initiation of labor, and ripening of cervix at parturition. Oxidant status of the cell also modulates angiogenesis, which is critical for follicular growth, corpus luteum formation endometrial differentiation, and embryonic growth.

Each month, a colony of oocytes appears in the ovary; however, with meiosis-I only a single cell matures into a dominant oocyte. This process is targeted by an increase in ROS and inhibited by antioxidants. In contrast, the progression of meiosis-II is promoted by antioxidants that suggest a complex relationship between ROS and antioxidants in the ovary. The increase in steroid production in the growing follicle causes an increase in P450, resulting in ROS formation. ROS produced by the pre-ovulatory follicle are considered important inducers for ovulation. Oxygen deprivation stimulates follicular angiogenesis, which is important for adequate growth and development of the ovarian follicle. Follicular ROS promotes apoptosis, whereas GSH and follicular stimulating hormone (FSH) counter-balance this action in the growing follicle. Estrogen increases in response to FSH, triggering the generation of catalase in the dominant follicle, and thus avoiding apoptosis.

Female Reproductive Cycles

The Ovulation Cycle

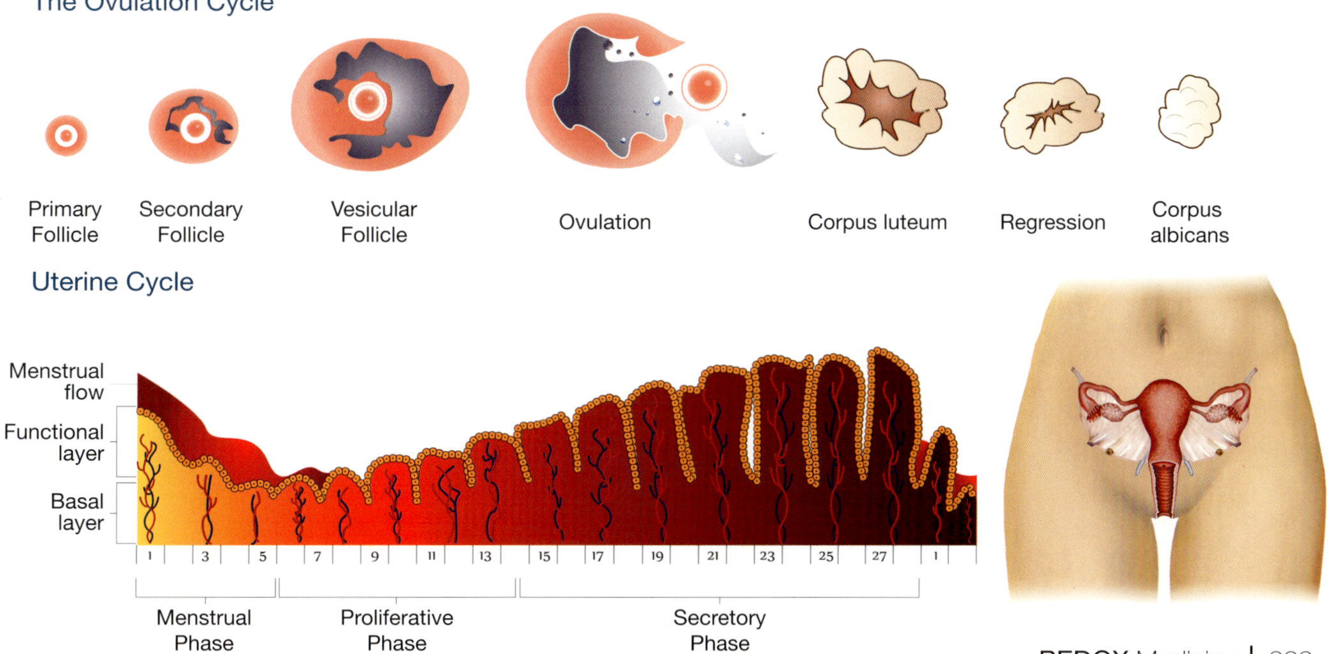

Ovulation is essential for reproduction, which initiates with the luteal hormone (LH) surge. It promotes important physiological changes that result in the release of a mature ovum. An over abundance of post-LH surge inflammatory precursors generates ROS; on the other hand, depletion of these precursors impairs ovulation. In the ovaries, the corpus luteum is produced after ovulation; it produces progesterone, which is critical for a successful pregnancy. ROS are also produced in the corpus luteum that are critical for reproduction. When pregnancy does not occur, the corpus luteum regresses. Conversely, when pregnancy takes place, the corpus luteum persists. A rapid decline in progesterone is needed for adequate development of the follicle in the next cycle. Cu-Zn-SOD increases in the corpus luteum during the early to mid-luteal phase and decreases during the regression phase. The decrease in Cu-Zn-SOD concentration elevates the ROS levels during regression. Other possible explanations for decreased Cu-Zn-SOD are an increase in prostaglandin (PG) F2-α or macrophages, or a decrease in ovarian blood flow. Prostaglandin F2-α stimulates production of the $O_2^{\bullet-}$ anion by luteal cells and phagocytic leukocytes in the corpus luteum. Decreased ovarian blood flow causes tissue damage by ROS production. Mn-SOD in the corpus luteum during regression increase to scavenge the ROS produced in the mitochondria by inflammatory reactions and cytokines. Complete disruption of the corpus luteum causes a substantial decrease of Mn-SOD in the regressed cell. At this point, cell death is imminent. The Cu-Zn-SOD enzyme is intimately related to progesterone production, while Mn-SOD protects luteal cells from oxidative stress-induced inflammation. During normal pregnancy, leukocyte activation produces an inflammatory response associated with increased production of $O_2^{\bullet-}$ anions in the 1st trimester. Importantly, oxidative stress during the 2nd trimester of pregnancy is considered as normal occurrence, and is supported by mitochondrial production of lipid peroxides, free radicals, and vitamin E in the placenta

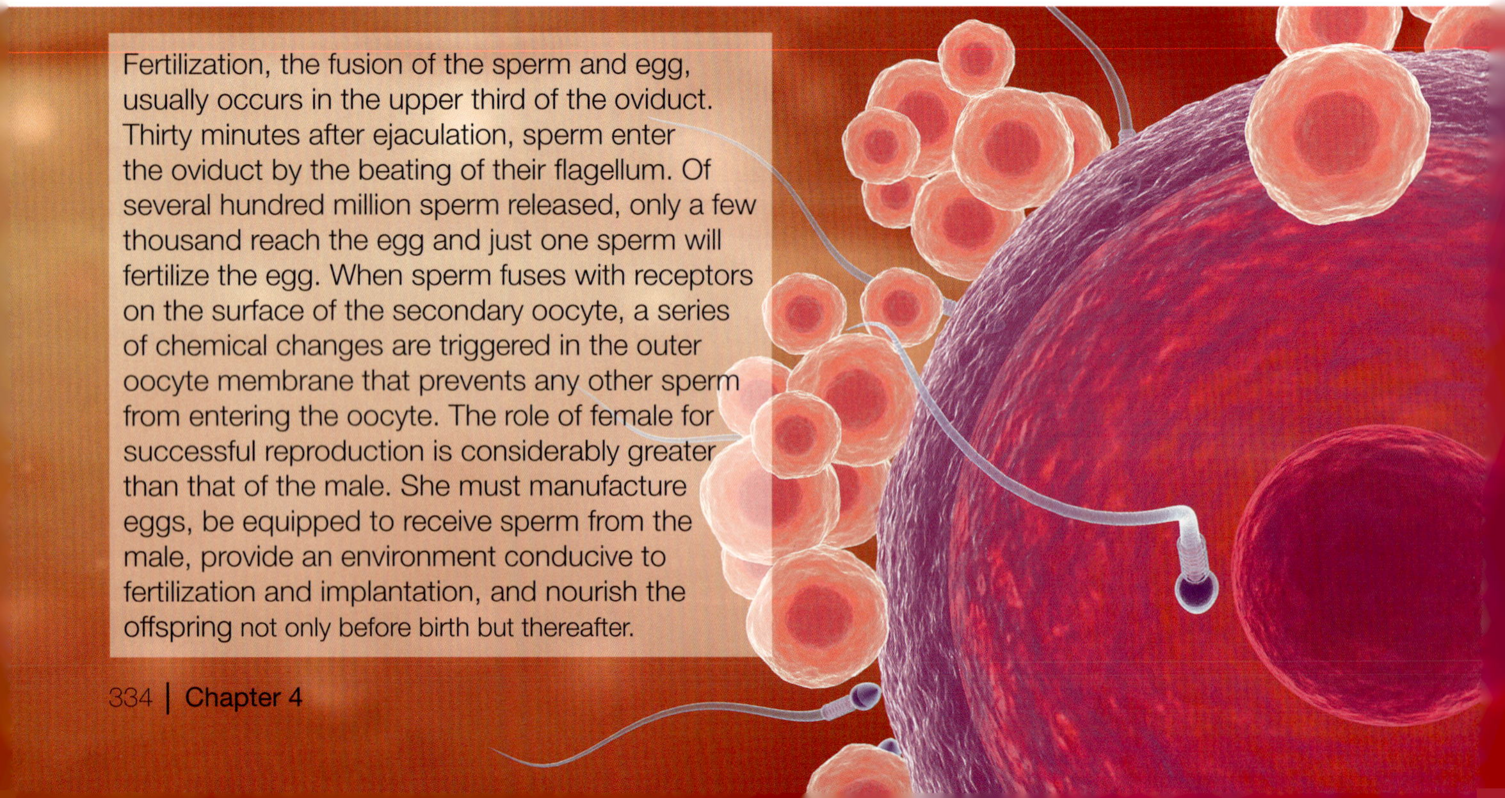

Fertilization, the fusion of the sperm and egg, usually occurs in the upper third of the oviduct. Thirty minutes after ejaculation, sperm enter the oviduct by the beating of their flagellum. Of several hundred million sperm released, only a few thousand reach the egg and just one sperm will fertilize the egg. When sperm fuses with receptors on the surface of the secondary oocyte, a series of chemical changes are triggered in the outer oocyte membrane that prevents any other sperm from entering the oocyte. The role of female for successful reproduction is considerably greater than that of the male. She must manufacture eggs, be equipped to receive sperm from the male, provide an environment conducive to fertilization and implantation, and nourish the offspring not only before birth but thereafter.

4.11.2. REDOX: Fertilization and Embryogenesis

The pro-oxidant (ROS/RNS)-mediated oxidative stress damages the DNA; these reactive species also regulate the oxidation of sulfhydryls (-SH) to disulfide (S=S) in protamines – a critical step for sperm maturation in the epididymis. This regulated sulfoxidation is essential for correct packaging of the nucleus into the small sperm head, so that it can resist oxidative damage during the fertilization process. The glutathione (GSH)-peroxidases (GPx3 and GPx4) present in the epididymal fluid effectively neutralize the accidentally produced peroxides (if any) during fertilization. However, after fertilization, a high REDOX potential is required for male pronuclear formation by reducing S=S bonds in oocytes. The source of this reducing power is GSH, which is present in the nucleus of the oocyte (at 9–10 mM) as the major source of REDOX potential. Oocytes are also rich in GSH-reductase. Since GSH alone is not effective in the reduction of S=S bonds, cross-reactions with thioredoxin (TRx), which has protein disulfide isomerase activity, may occur. Glycolytic activity (which generates NADH) and the hexose monophosphate shunt (a regenerating system for NADPH) are enhanced during the penetration of spermatozoa into the oocyte. An elevated REDOX potential drives the fertilization process. The orchestration of these REDOX and NADPH generating systems enables the early embryonic development. The peri-hatching blastocyst generates a significantly large amount of ROS for an extremely short period of time when compared to the unhatched and hatched blastocysts.

The glutathione (GSH) REDOX system plays a defining role during the embryogenesis. For example, pre-implanted embryos are highly sensitive to conditions that cause oxidative stress. The GSH status changes dramatically during their development. GSH in reproductive tract fluid may help protect pre-implanted embryos from the adverse effects of oxidants. GSH synthesis and turnover increase between the two-cell and blastocyst stages. The increase in the ability of embryos to synthesize GSH on day-3 is dependent on protein synthesis. Hence, the recycling of GSSG must play an important role in maintenance of intra-cellular GSH levels from the oocyte to the two-cell stage.

The placenta is rich in aldose reductase, which in conjunction with sorbitol dehydrogenase catalyzes the conversion of glucose to fructose, the energy source for the spermatozoa. Both enzymes are also abundant in eggs and may participate in the production of fructose. Thus, aldose reductase appears to have additional roles beyond detoxification. The production of certain cytokines, such as IL-1 and TNF-α, are elevated; hence, ROS levels are also elevated. The detoxification of carbonyls present in reproductive tract fluids would be advantageous to embryos at their early developmental stage in oviducts and the uterus. Steroid metabolites (such as isocorticosteoids and progesterone) and lipid peroxidation products (such as 4-hydroxynonenal and acrolein) are all aldose reductase substrates. The GSH conjugate of 4-hydroxy-3-nonenal actually serves as a substrate for aldose reductase. Thus, GSH could facilitate the detoxification of aldose reductase by producing GSH conjugates. The production of carbonyl compounds is caused mainly by ROS, the level of which increases during reproduction processes such as cell proliferation, steroidgenesis, and ovulation.

Generation of reactive species (ROS/RNS) is high in reproductive tissue due to active metabolism and steroidogenesis; thus, the tissue is under continuous oxidative stress. Reactive species either damage or modify susceptible molecules including DNA, lipids, and proteins. Oxidative damage to oocytes increases the risk of hereditable disease; hence, living organisms must eliminate such defective gametes to preserve the species. On the other hand, the reproductive system utilizes reactive species in vital processes that are essential for fertilization and embryonic development. To minimize the risk caused by oxidative stress, living organisms have evolved antioxidative systems, such as SOD and GPx. When the pro-oxidant levels exceed the scavenging capacity of the system, a REDOX system, under such situations, can repair oxidized and damaged molecules using NADPH as an original electron source. Thus, the maintenance of a high REDOX potential is prerequisite for maintaining the reproductive system in a healthy state.

Fertilization / Embryogenesis

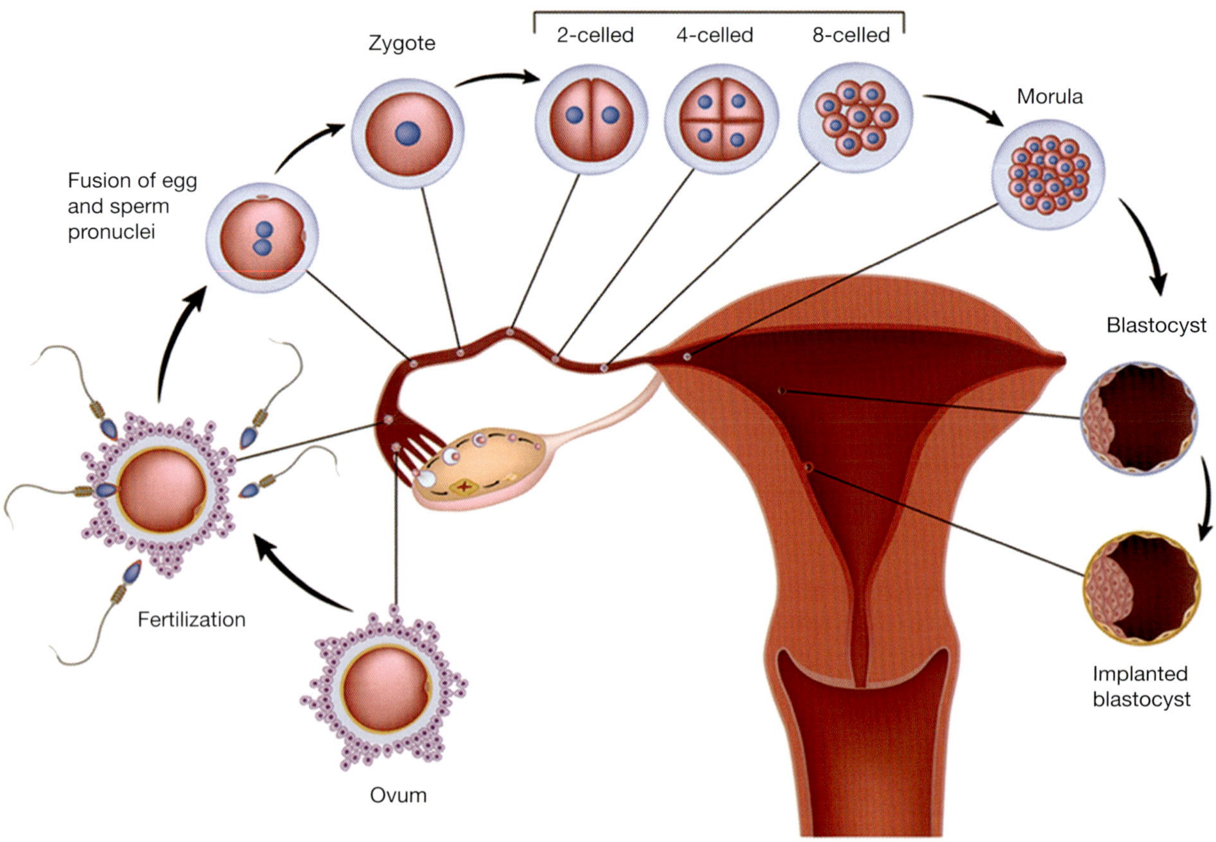

4.11.3. REDOX and Pregnancy

The placenta in mammals is the essential interface between the maternal circulation (carrying O_2-rich blood and nutrients) and the fetal circulation. Human placenta is exposed to major changes in O_2 concentration from conception to delivery. In normal pregnancies, the REDOX status is very well controlled and provides a delicate balance between the metabolic needs of the fetus and its placenta and the potential danger of ROS. Embryonic and placental cells are sensitive to oxidative stress due to extensive cell divisions and the concomitant exposure of their DNA. The placental syncytio-trophoblast is highly sensitive – it is the outermost tissue of the conceptus, with high O_2 exposure coming from the mother. The placenta delivers O_2, nutrients, and antioxidants from the mother to the baby; which includes iron, copper, ascorbate, cysteine and cystine (for GSH synthesis), and α-tocopherol. The human placenta also secretes 'extra-cellular' GPx (GPx3) into the maternal circulation. At term, the utero–placental circulation carries ~600-mL of maternal blood per minute.

Human placenta has low O_2 tension (<20 mmHg) and antioxidant levels for about the first 10 weeks of gestation. Oxygen levels rise to about 50 mmHg at 11-14 weeks as the maternal intra-placental circulation becomes fully established. During this phase, there is a 'burst' of ROS production, which is accompanied by increased synthesis of antioxidant enzymes (Cu-Zn-SOD, GPx and Mn-SOD) to promote proper cell differentiation (although it can cause damage with prolonged exposure). Chlorinated and nitrated proteins and lipid oxidation products (including HNE) are also present, to variable extents, in the healthy placenta, which is rich in macrophages. ·NO plays a role in maintaining normal placental blood flow, but its presence also poses a risk of excessive $ONOO^-$ formation if $O_2^{\bullet-}$ levels are raised. As in other tissues, placental EC-SOD is involved in the regulation of $O_2^{\bullet-}$/·NO interactions.

In normal pregnancies, the earliest stages of development take place in a low O_2 environment. This physiological hypoxia of the early gestational sac protects the developing fetus against the deleterious and teratogenic effects of ROS. In miscarriage, development of the placental interface is severely impaired leading to early and widespread onset of maternal blood flow and major oxidative degeneration. This mechanism is common to all miscarriages; it occurs in the 1st trimester depending on the etiology. Disruptions of placental blood flow later in pregnancy can cause pre-eclampsia, the leading cause of maternal death in advanced countries and a major cause of maternal morbidity and fetal growth restriction. It affects up to 2.8% of pregnancies in developed countries, and more in developing countries. Impaired blood flow through the placenta causes severe hypoxia and release of pre-eclamptic factors into the systemic circulation. These promote vasoconstriction and hypertension, as well as neutrophil and platelet activation in the maternal vascular system. The baby must be delivered as soon as possible to reverse the syndrome, even if it is too premature to survive. Thus, normal pregnancy is based on total REDOX balance; some ROS are needed for optimum placental function; however, in excess, oxidative stress can be detrimental.

Oxidative Stress and the Fetus: The increased risk of fetal malformations observed in mothers with poorly controlled diabetes may involve oxidative stress. Hypercholesterolemia in mothers can lead to atherosclerosis in fetal arteries. Exposure of pregnant females to certain xenobiotics can cause embryonic death, lethal fetal malformations or birth defects. Teratogenicity can sometimes be latent, for example an increased risk of cancer development in the offspring after the mother was exposed to toxins. Fortunately, the innate (maternal) detoxification systems and the barrier action of the placenta perform an excellent job in protecting the fetus against most potential insults.

Miscarriages are the most common human pregnancy disorder with 50% of all conceptions lost before or around implantation and another 20% lost between implantation and completion of the 1st trimester. The difference in miscarriage rates between humans and other mammals including primates is linked to a difference in chromosomal abnormalities that occur at a rate of more than 50% in humans compared with less than 5% in rats, rabbits and hamsters. The high rate of chromosomal abnormalities in humans could be due to longer reproductive spans, which leads to an accumulation of chromosomal errors not observed in short-lived

Embryonic / Fetal Development

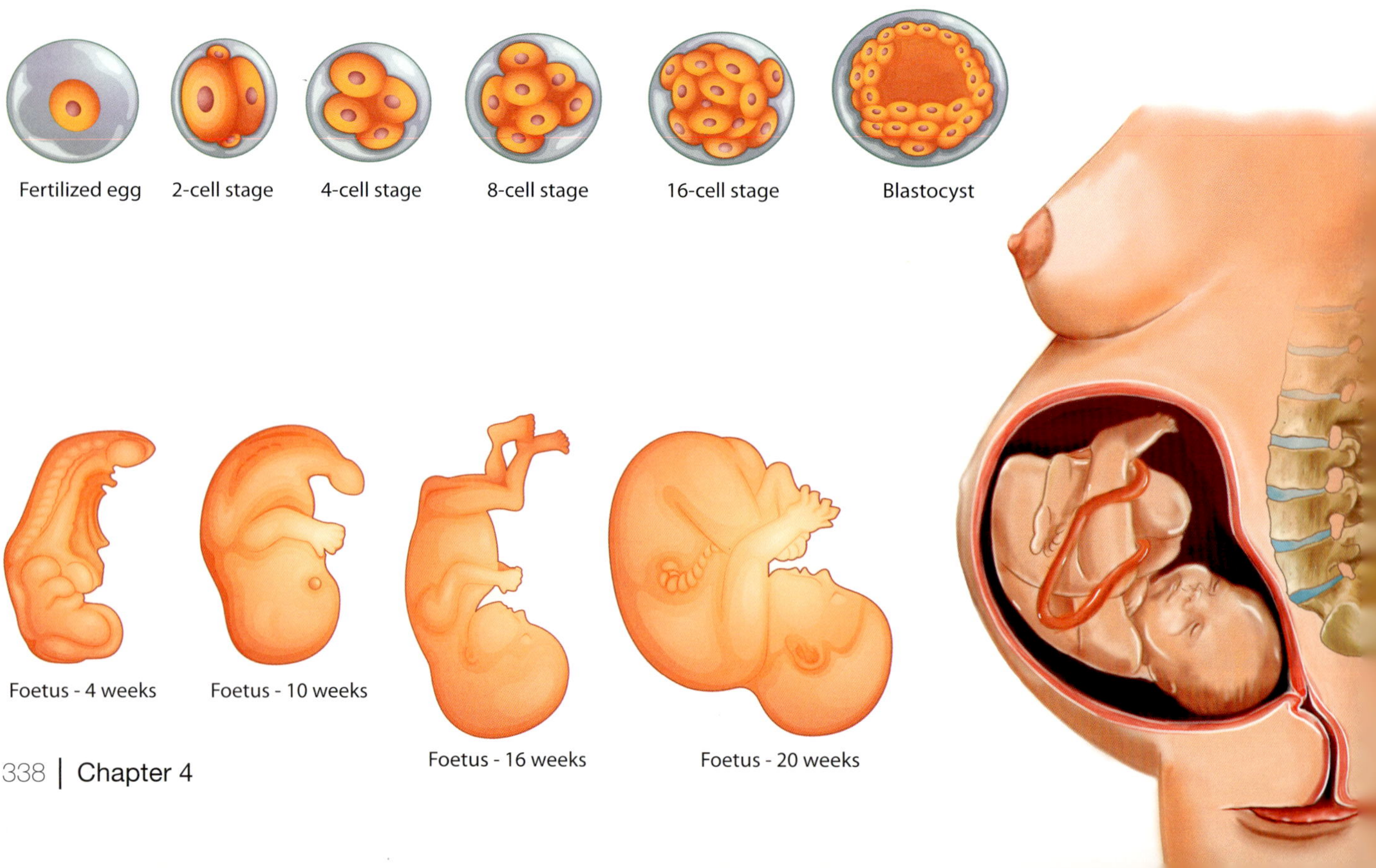

mammalian species. Indeed, the incidence of chromosomal abnormalities in the human conceptus increases with maternal and paternal age. Modern lifestyle and in particular delayed childbirth, maternal smoking, and hyper-caloric diets leading to diabetes, have significantly increased the incidence of miscarriage.

Oxidative Stress and Term Labor: There is increased generation of $O_2^{\bullet -}$ and $\cdot NO$ in pregnancy, which results in oxidative stress. Term labor induces increased lipid peroxidation, as evidenced by increased levels of malondialdehyde. The serum levels of H_2O_2 are elevated in women in labor, compared to the controls, who were not in labor. Term labor was demonstrated to cause an up regulation of the antioxidant reserve in the fetal compartment. However, the role of oxidative stress in initiation of labor is not known.

REDOX and the Birth: Birth involves hypoxia-reoxygenation complicated by an interruption of the nutrient supply until feeding is established. Newborns face a sudden transition from a warm, fluid-filled environment to a cooler, air-breathing environment, with exposure of the lung to O_2 concentrations much greater than those experienced in utero. Often, severe hypoxia occurs during the process of birth itself, making the eventual exposure to 21% more of oxygen as a shock. Preparations for birth occur in the later stages of development in utero and include rises in the levels of antioxidants such as SODs, GSH, catalase, GPx and PRx in the lung. The forkhead transcription factor, FoxA2, plays a key role in controlling these events. In human lung some antioxidants (e.g. Mn-SOD, Cu-Zn-SOD) appear early in gestation, whereas others (e.g. catalase, TRx) do not until the last 4-weeks of pregnancy.

Placenta / Chorionic Villi

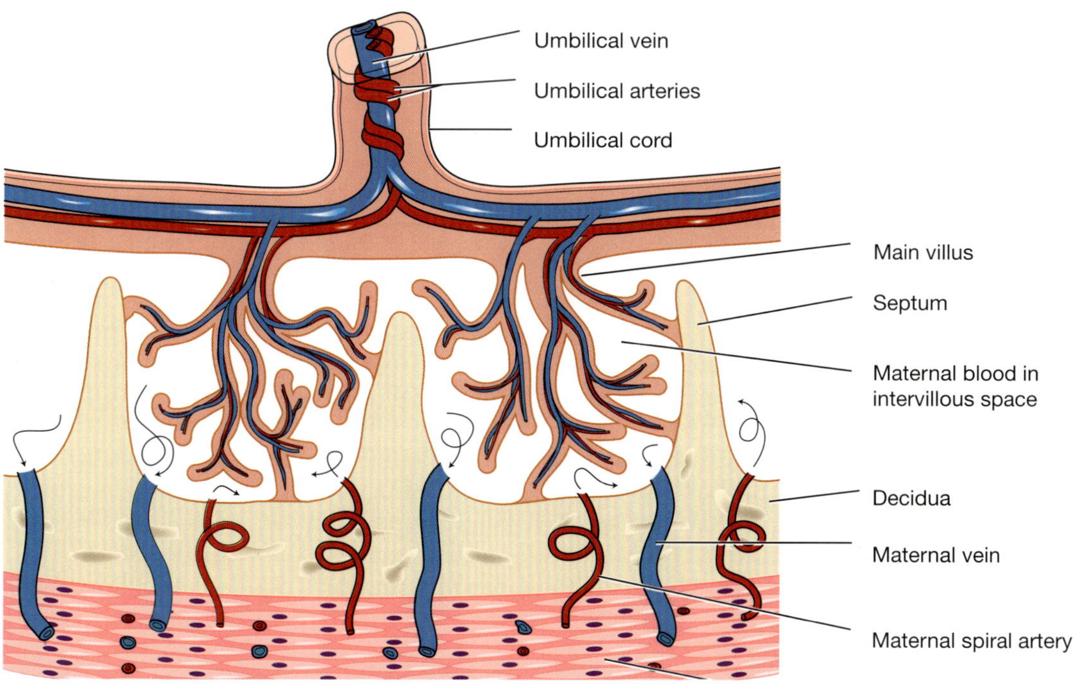

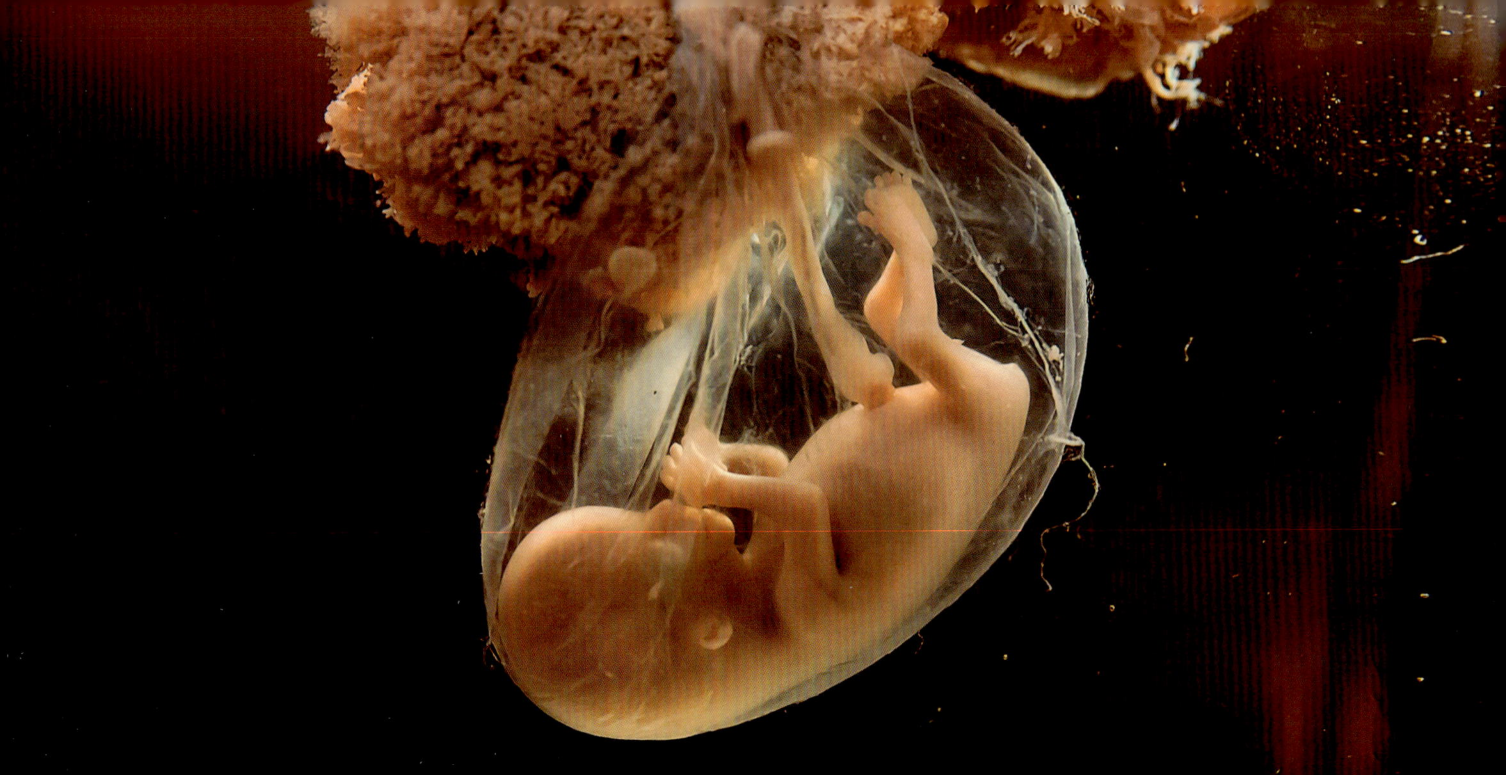

The human birth is a cold hyperoxgenic shock and the survival of a newborn depends on REDOX potential

Increased ROS production at birth may trigger important adaptations. *In utero*, the baby does not breathe and blood bypasses the pulmonary circulation through a fetal artery, the ductus arteriosus. This rapidly closes at birth, first by immediate vasoconstriction. Within a short duration, cell migration into the lumen makes the closure irreversible. How does this artery know when to close? One hypothesis is that a 'REDOX Sensor' (possibly mitochondrial-based) responds to elevated O_2 by generating H_2O_2, which blocks potassium (K^+) channels, depolarizes the membrane, opens calcium (Ca^{2+}) channels and causes vasoconstriction. As the ductus arteriosus closes, pulmonary vascular resistance falls and blood flows through the lungs. Sometimes this fails to occur properly, lung resistance remains high, which results in persistent pulmonary hypertension of the newborn.

Oxidative Stress and Premature Birth: Premature newborns face special challenges; fortunately, recent advances in neonatal care, such as providing respiratory support (exposure to elevated O_2) for extended time, have increased their survival rate. However, exposure of premature neonates to excess O_2 causes retinopathy of prematurity (ROP) and brain damage. Another problem is broncho-pulmonary dysplasia (BPD) with involvement of lung injury and fibrosis that resemble acute respiratory distress syndrome. BPD is a leading cause of morbidity and mortality in premature babies. BPD is accompanied by increased oxidative damage. The premature lung is deficient in both antioxidants and surfactants. Clinically, ROP, BPD, and brain damage, are referred to as different manifestations of oxygen-radical diseases of prematurity.

REDOX and Breast feeding: Colostrum, the earliest breast milk (lacteal secretion) is enriched particularly with lactoferrin, α-tocopherol and coenzyme Q. Accordingly, the tissue and plasma levels of a neonate are elevated with these passively transferred (maternal) antioxidants. Also, the plasma levels of vitamin C in the newborn are generally higher than in adults, but steadily drop after few days. Levels of plasma ceruloplasmin, albumin tends to be lower than in adults, whereas, the plasma xanthine XOR is high. Thus, any neonatal condition of hypoxia will increase plasma hypoxanthine levels and trigger oxidative stress.

Plasma levels of holo-lactoferrin, the iron-saturated glycoprotein, are higher in the newborn. Why should there be more lactoferrin at birth? The fetus receives iron (via lactoferrin) from the mother, especially during the last 3 months *in utero*. About 80% of fetal iron is in hemoglobin (rest is ferritin). Birth is associated with a virtual cessation of red blood cell formation that persists for at least 6 weeks. While the RBC degradation continues, iron is not used up for erythropoiesis. Fetal erythrocytes differ from those of adults in size (fetal RBCs are larger), higher numbers per unit volume of blood, and contain predominantly hemoglobin F. Hemoglobin F has a higher affinity for O_2 under physiological conditions, to optimize O_2 transfer from the maternal to the fetal circulation. Fetal erythrocytes are more susceptible to oxidative stress than adult RBC; therefore, they produce more $O_2^{\bullet-}$ and H_2O_2. Antioxidants such as α-tocopherol, GPx, catalase, and Cu-Zn-SOD are all lower in fetal erythrocytes than in adults. Lactoferrin, the breast milk-replenishment, provides the multi-functional benefit of antioxidant protection, as well as antimicrobial host defense to the fragile neonate. In the later stages of development, a newborn starts producing his or her own lactoferrin protein in gastric/respiratory/reproductive passages, saliva, synovial and cerebrospinal fluids, as well as in various exocrine secretions that bathe the mucosal surfaces.

Human implantation is unique amongst mammals. It is highly invasive - the conceptus is embeded completely within the maternal uterine endometrium and superficial myometrium. The chorionic villi, the basic structures of the early placenta, form during the 4th and 5th week of post-menstruation and surround the entire gestational sac during the 8th to 9th week of gestation. Between the 3rd and 4th month, the villi at the implantation site become elaborately branched and form the definitive placenta; whereas, the villi on the opposite pole degenerate to form the placental membranes. Towards the end of pregnancy, the villi present a surface area of 12-14 m^2, providing an extensive and intimate interface for materno–fetal exchange.

Cellular Channels: The Nobel Prize in Chemistry 2003 was awarded jointly to two scientists for their discoveries concerning channels in cell membranes. Peter Agre won it for the discovery of water channels and Roderick MacKinnon won it for describing the structural and mechanistic properties of ion channels.

4.11.4. Oxidative Stress and Infertility

Infertility is defined as the failure to conceive a recognized pregnancy after 12 months of unprotected intercourse. Human reproduction is inefficient, only one-fourth to one-third of fertilized human embryos may survive to produce a term delivery. The great majority of the failures occur in the early weeks after ovulation, with relatively little loss after clinical detection. Women who experience a delay before achieving a recognized conception also have elevated rates of early unrecognized pregnancy loss and clinical spontaneous abortion.

Oxidative stress plays a key role in the pathogenesis of infertility in both males and females. The prevalence of female infertility ranges from 7% to 28%, depending on the age of the woman. In general, an estimated 84% of couples conceive after 1 year of intercourse, and 92% of the couples conceive after 2 years. Although the frequency and origin of different forms of infertility varies, 40%–50% of the etiology of infertility studied is due to female causes. A primary diagnosis of male factor infertility is made in 30% of infertile couples. Combined female and male factor infertility is responsible for 20%–30% of cases. Finally, unexplained infertility affects 15% of couples. If the results of a standard infertility examination are normal, a diagnosis of unexplained or idiopathic infertility is assigned.

Oxidative Stress and Male Infertility: Oxidative stress is a significant contributor to male infertility. Male reproductive pathologies, congenital or acquired, both impair spermatogenesis and contribute to almost half of infertility cases. Men with high oxidative stress levels or DNA damaged sperm are likely to be infertile. The key

SPERMTOGENESIS

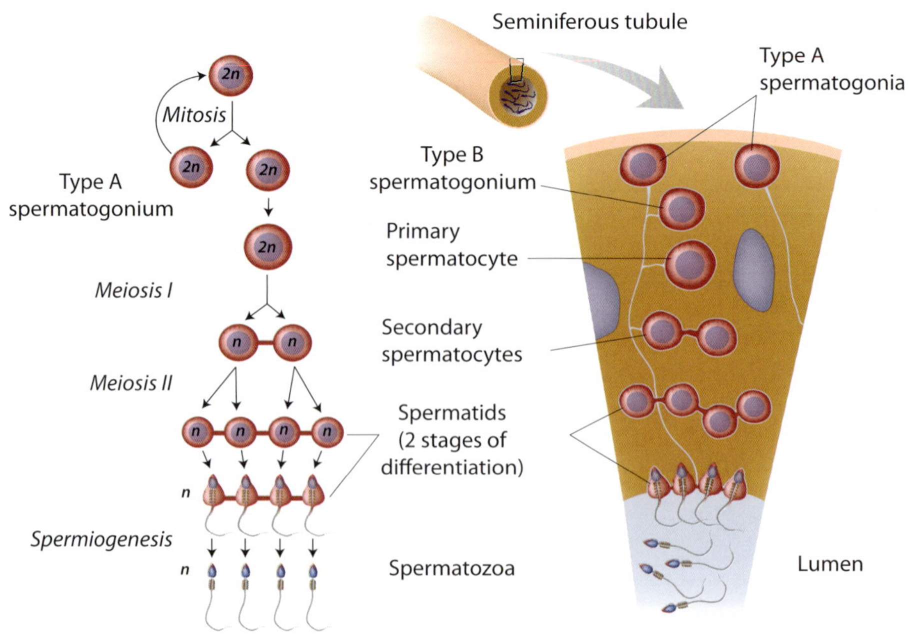

predictors of male fertilization capability are sperm count and motility. These male fertility factors can be disturbed by oxidative stress. Sperm-related dysfunctions associated with ROS are decreased sperm number and motility, and inhibition of sperm-oocyte fusion.

Low levels of ROS are necessary to optimize the maturation and function of spermatozoa. The sources of seminal ROS are immature spermatozoa and leukocytes. In addition, acrosome reactions, motility, sperm capacitation, and fusion of the sperm membrane and the oolemma are especially dependent on the presence of ROS. On the other hand, high levels of ROS produced by spermatozoa trigger lipid peroxidation that could damage the sperm's plasma membrane and cause oxidative stress. Abnormal and non-viable spermatozoa can generate additional ROS and RNS, which can disrupt normal sperm development and maturation and may even result in apoptosis. Specifically, H_2O_2 and $O_2^{\bullet-}$ anions could induce defective sperm function in infertile males. Elevated seminal ROS production may alter sperm motility and morphology, thus impair their ability to fertilize.

Oxidative Stress and Female Infertility: REDOX mechanisms affect the total reproductive lifespan of a woman and continue through her menopausal phase. ROS can affect a variety of physiological functions in the reproductive tract, and excessive levels can result in precipitous pathologies affecting female reproduction. The oxidant status can influence early embryo development by modifying the key transcription factors and hence modifying gene expression. ROS may also play a major role both in the implantation and fertilization of eggs.

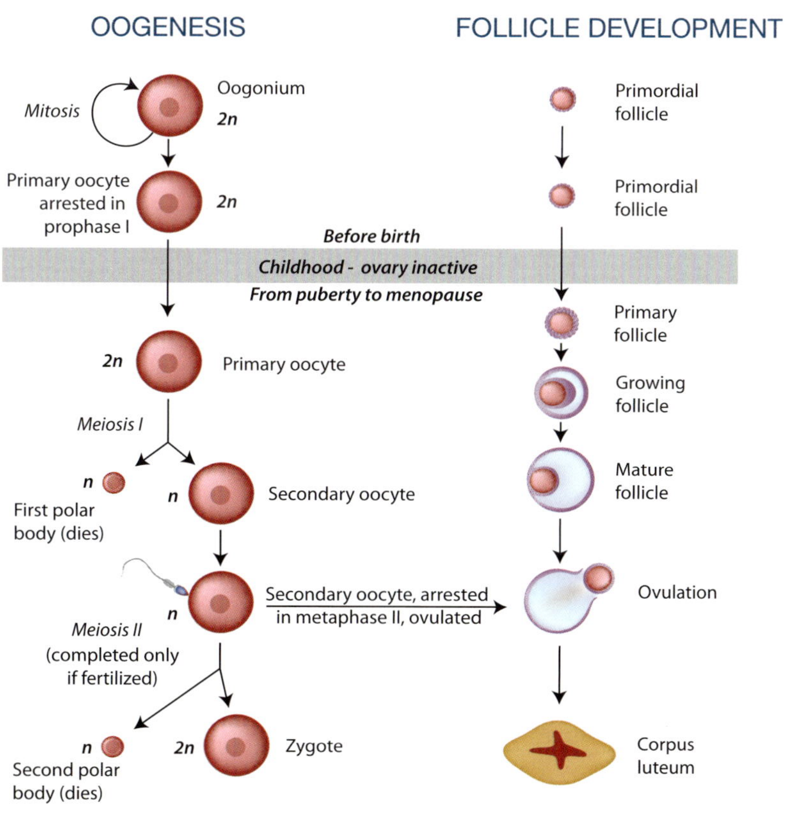

Oxidative stress modulates the age-related decline in female fertility. It plays a role during pregnancy and normal parturition and in initiation of preterm labor. Most ovarian cancers appear in the surface epithelium, and repetitive ovulation has been thought to be a causative factor.

Oxidative stress results from an imbalance between prooxidants (ROS/RNS) and the free radical scavengers (antioxidants). This imbalance could predispose a number of female fertility dysfunctions such as endometriosis, polycystic ovary syndrome (PCOS), and unexplained infertility. Oxidative stress could also trigger various pregnancy complications such as spontaneous abortion, recurrent pregnancy loss, and pre-eclampsia. Studies have shown that extremes in body weight (thinness or obesity) and lifestyle factors (such as cigarette smoking, alcohol consumption, and recreational drug use) can promote excess free radical production - all which could affect fertility. Exposures to environmental pollutants are of increasing concern, as they could trigger oxidative states, possibly contributing to female infertility.

Reactive species serve as cell signal molecules that support normal physiology. ROS affect multiple reproductive pathways from oocyte maturation to fertilization, embryo development, and pregnancy. In regular physiology, oxidative stress is essential for normal cycling ovaries, follicular development, and cyclical endometrial changes. Nitric oxide (·NO) is involved in the regulation of endometrial and ovarian function, maintenance of uterine quiescence, initiation of labor, and ripening of cervix at parturition. Oxidant status of the cell also modulates angiogenesis, which is critical for follicular growth, corpus luteum formation, endometrial differentiation, and embryonic growth. Environmental and occupational exposures to ovo-toxicants can also alter reproductive stability. Infertile couples often turn to assisted reproductive techniques (ART) to improve their chances of conception. The supplementation of ART culture media with antioxidants continues to be of interest to increase the probability for ART success.

Age-related Fertility Decline and Menopause: Aging is defined as the gradual loss of organ and tissue function. Oocyte quality decreases in relation to increasing maternal age. Low quality oocytes contain increased mtDNA damage and chromosomal defects, secondary to age-related dysfunctions. These mitochondrial changes may arise from excessive ROS, which occurs through the opening of ion channels (i.e. loss of Ca^{2+} homeostasis). Levels of 8-oxodeoxyguanosine (8-OHdG), an oxidized derivative of deoxyguanosine, are higher in aging oocytes. Oxidative stress, iron stores, blood lipids, and body fat typically increase with age, especially after menopause. The cessation of menses leads to an increase in iron levels throughout the body. Elevated iron stores could induce oxidative imbalance, which may explain why the incidence of heart disease is higher in post menopausal than pre menopausal women. Menopause also leads to a decrease in estrogen and the loss of its protective effects against oxidative damage to the endometrium. Hormone replacement therapy (HRT) may be beneficial against oxidative stress by combating the effects of lower antioxidant levels that normally occurs with aging.

4.11.5. Oxidative Stress and Erectile Dysfunction (ED)

Erectile Dysfunction (ED) is defined as the inability to attain and/or maintain penile erection. ED becomes increasingly prevalent with age and affects the quality of life of patients and their partners. About 52% of men beyond 40 years of age have some degree of ED and it is projected to affect 322 million men worldwide by 2025. In addition, the presence of chronic illness (e.g. heart disease, hypertension, and diabetes mellitus), alcohol or drug abuse, sedentary lifestyle, and smoking are major risk factors for ED.

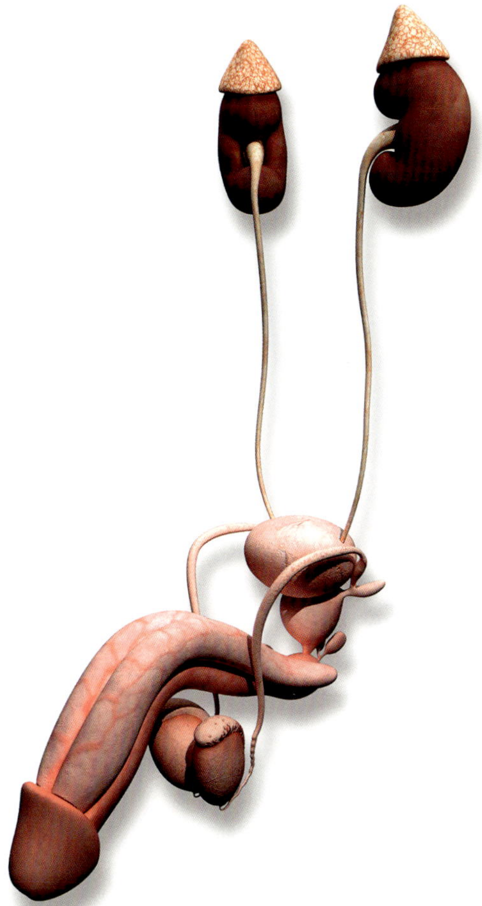

Penile Anatomy

REDOX and Penile Erection: Penile erection is achieved by pressure changes in the cavernosal sinuses. The vasculature of the erectile tissue differs from most vascular beds as it is composed of arterioles and hollow blood-filled sinuses, both of which are lined with smooth muscle and endothelial cells. In the absence of arousal stimuli, cavernosal vasoconstriction maintains the penis in the non-erect state. Contraction of the cavernosal smooth muscle, mainly in response to the excitatory neuro-transmitter (norepinephrine) released from the sympathetic innervation of the penis, narrows the arteriolar lumen and sinusoidal cavities, restricting blood flow to maintain low intra-cavernosal pressure (ICP) and a non-erect (flaccid) penis. During sexual arousal or nocturnal tumescence, the release of ·NO stimulates smooth muscle relaxation. The resulting dilation of the cavernosal arterioles and sinuses results in increased blood flow and a subsequent rise in ICP. This initial rise in ICP activates a veno-occlusive mechanism to limit the outflow of blood and further increase the pressure inside the cavernosum. The erectile response ensues as the force of the elevated pressure expands the outer tunica albuginea of the penis, resulting in the increased penile length and diameter characteristic of erection.

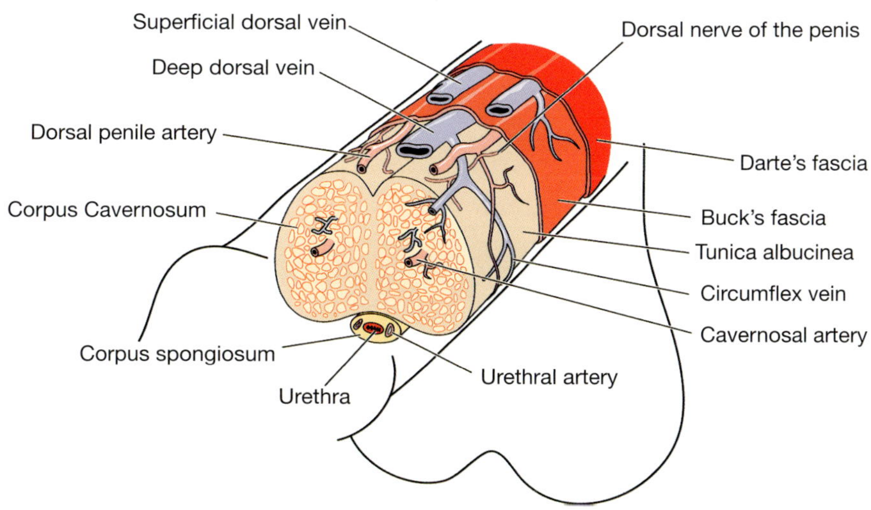

Role of ·NO in Penile Erection: ·NO is formed through conversion of amino acid, L-arginine to L-citrulline, by enzymatic action of NO-synthase (NOS). This enzyme exists in three main isoforms: neuronal NOS (nNOS), inducible (iNOS), endothelial NOS (eNOS); and all these forms have been detected in the penis. Erection is sustained mainly by the endothelium-derived ·NO, in response to strain forces brought on by increased blood flow and pressure in the penis. Upon its release, ·NO diffuses locally into adjacent smooth muscle cells of the corpus cavernosum and binds to soluble guanylyl cyclase, which catalyzes the formation of cGMP, the cyclic nucleotide. The cGMO activates protein kinase G that decreases cytosolic calcium (Ca^{2+}). The decay in cytosolic Ca^{2+} induces relaxation of the vascular and cavernosal smooth muscle cells, leading to dilation of arterial vessels, increased blood flow into the sinuses of the corpora cavernosa, and penile erection.

Oxidative Stress in the Pathogenesis of ED: Oxidative stress is an important factor contributing to endothelial dysfunction associated with aging. Age-related ED is characterized by decreased nerve and endothelium-mediated corpus cavernosum relaxation, decreased androgen levels, pathological remodeling of the pudendal artery, reduction in smooth muscle content, impaired growth factor and cytokine signaling in the penis. Decreased endothelial signaling is attributed to impaired eNOS expression, decreased levels of L-arginine, and increased oxidative stress. Enzymes such as NOx, XOR, COX, mitochondrial electron transport and eNOS uncoupling are potential vascular sources of ROS.

Several epidemiological studies have shown that cigarette smoking not only increases the risk of ED, but also amplifies the risk of ED associated with other risk factors (e.g. hypertension, diabetes, and dyslipidaemia) or with aging. Hypercholesterolemia resulting in atherosclerosis of the penile vasculature is one of the leading causes of vasculogenic ED. Hypercholesterolemia in men impairs endothelium-dependent and independent relaxations of the corpus cavernosum, decreases the cavernosal content of endothelial cells, alters the function of smooth muscle cells, and increases collagen content. Impaired eNOS function and endothelial ·NO availability in the hypercholesterolemic vasculature have been mostly attributed to increased vascular $O_2^{\bullet-}$ production. Oxidative stress seems to contribute significantly to hypercholesterolemia-associated ED. The $O_2^{\bullet-}$ anion may directly inactivate ·NO and decrease its bioavailability. Moreover, the reaction of $O_2^{\bullet-}$ and ·NO results in the formation of highly toxic $ONOO^-$, which may cause oxidative damage to DNA, proteins, lipids, and increase apoptosis.

***In Vitro* Fertilization (IVF):** The Nobel Prize in Physiology or Medicine 2010 was awarded to Robert G. Edwards for the development of in vitro fertilization. Louise Brown, born in England in 1978, was the first 'Test Tube Baby' to be conceived outside her mother's womb.

IVF is a process by which an egg is fertilized by sperm outside the body. IVF is a major treatment for infertility when other methods of assisted reproductive technology have failed. The process involves monitoring the female ovulatory process, removing ovum or ova from the woman's ovaries and performing fertilization with sperm in a laboratory culture medium. The fertilized egg (zygote) is then transferred to the patient's uterus to establish a successful pregnancy.

IVF is a complex procedure; only about 5% of couples with infertility seek it out. However, since its introduction in the U.S. in 1981, IVF and other similar techniques have resulted in more than 200,000 babies. When it comes to infertility, IVF may be an option if one of the sex partners is diagnosed with: endometriosis, low sperm counts, problems with the uterus or fallopian tubes, problems with ovulation, autoimmunity that harm sperm or eggs, inability of sperm to penetrate or survive in the cervical mucus or any unexplained fertility problems.

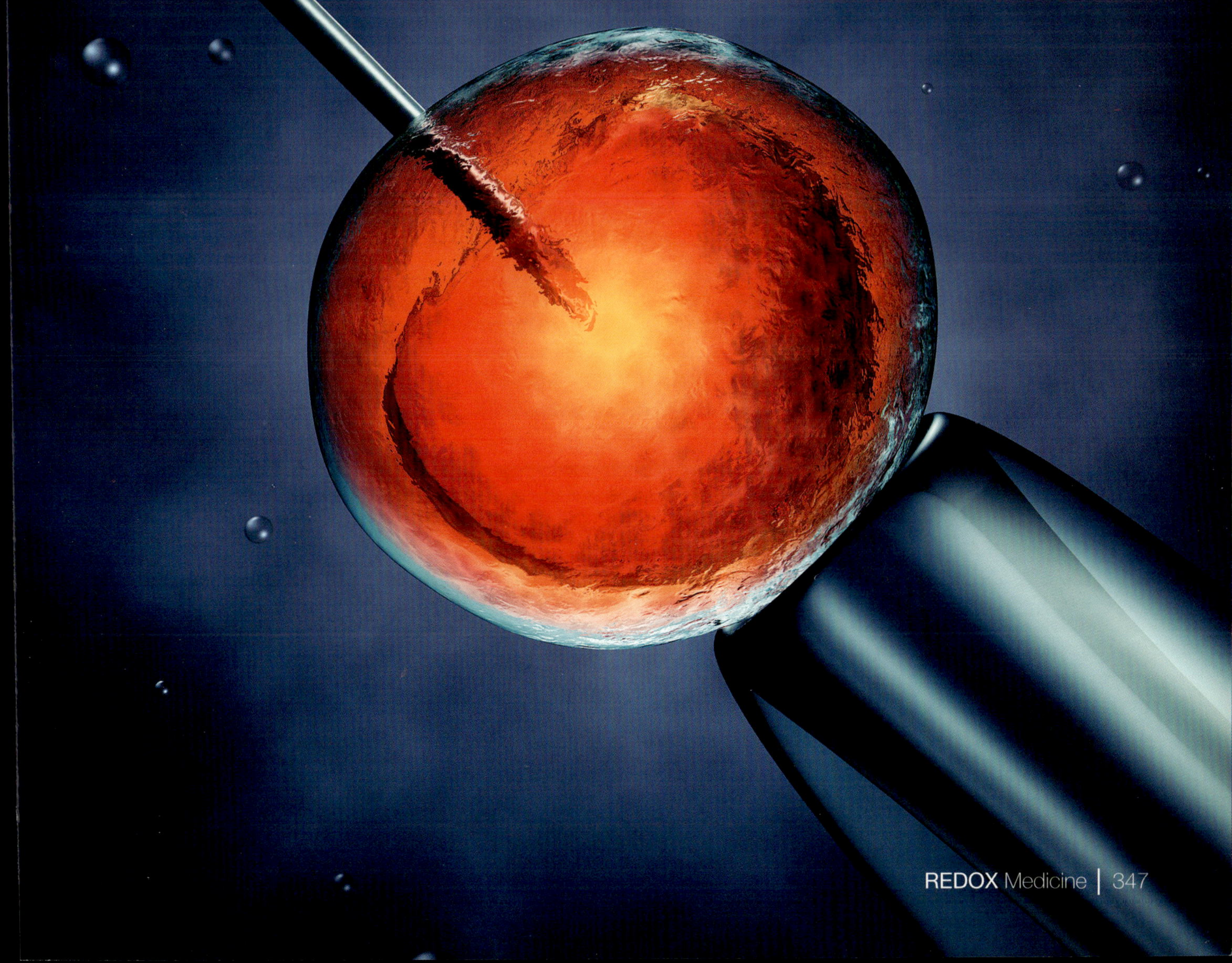

4.12 **REDOX:** The Integumentary System

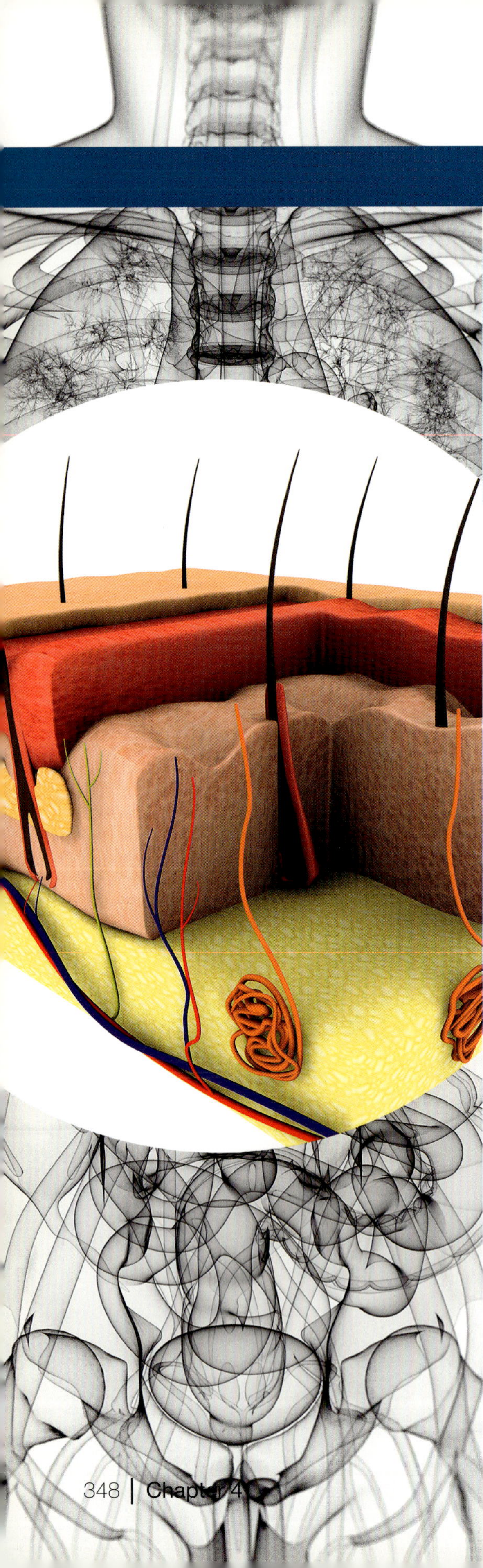

The Integumentary System includes the skin and the skin derivatives hair, nails, and glands. It is the body's largest organ accounting for about 12 to15% of total body weight and covers 1.5-2 m^2 of surface area. It distinguishes, separates, and protects the body from its surroundings. Human skin (integument) is divided into three layers of tissue: epidermis, dermis, and the subcutaneous - hypodermis.

i) Epidermis forms the outermost layer, about 20-30 cells thick, which provides the initial barrier to the external environment. Keratinocyte, the major cell type of this skin layer, produces keratin, a fibrous water-proofing protein. Millions of dead keratinocytes rub off daily.

- Hair is composed of columns of dead, keratinized cells bound together by extra-cellular proteins. Hair functions include protection and sensing light touch.
- Nails are plates of tightly packed, hard, keratinized epidermal cells. Nails participate in the grasping and handling of small objects.

ii) Dermis is the second layer of skin, directly beneath the epidermis. Unlike the epidermis, the dermis has its own blood supply and contains connective tissues, vessels, glands, follicles, hair roots, sensory nerve endings, and muscular tissue. Dermis also contains dense connective tissue, made of collagen fibers, which gives the skin much of its elasticity and strength.

- Glands participate in regulating body temperature. There are three main types of glands associated with the integument: (i) Sebaceous - oil glands located in the dermis that secrete sebum; (ii) Sudoriferous - sweat glands to regulate of body temperature; and (iii) Ceruminous - that secrete cerumen (ear wax).

iii) Hypodermis is the final and deepest layer of skin. It is primarily made up of adipose tissue. The adipose-rich subcutaneous layer provides a cushion for the delicate organs lying beneath the skin. It also functions to insulate the body to maintain body temperature.

4.12.1. Photo-Oxidative Stress and Skin Health

The integumentary system has multi-functional role in homeostasis. Skin acts as the body's first line of defense against infection. Acidic secretions from skin glands retard the growth of fungi. When a microbe penetrates the skin (or when the skin is breached by a cut) the inflammatory response occurs. Melanocytes form a second barrier by shielding from the damaging effects of ultraviolet radiation. Skin also serves to waterproof, cushion, and protects the deeper tissues. It is the attachment site for sensory receptors to detect pain, sensation, pressure, and temperature. Heat and cold receptors are located in the skin. When the body temperature rises, the sweat-producing skin glands release about 1-2 liters of water per hour, cooling the body. When body temperature falls, the sweat glands constrict and triggers thermogenesis (or heat generation), by raising the body's metabolic rate (and by shivering). The integumentary system also supports vitamin D synthesis upon exposure to sunlight.

The skin is directly exposed to atmospheric O_2, although the strateum corneum may hinder O_2 diffusion to the lower layers. The skin plays a role in heat exchange; vasodilation (to lose heat) and vasoconstriction (to conserve heat) can deliver too much or too little O_2 to the skin respectively. Skin must also put up with air pollutants, deodorants (many anti-perspirants are acidic and contain aluminium salts), other applied chemicals (e.g. creams, lotions, cosmetics, greases, soaps, shower gels, shampoos, and oils), industrial chemicals, water-borne toxins (e.g. chlorine, dissolved metal ions), UV light, other radiation, microorganisms from skin, and agents excreted in sweat. Sweat is rich in transition metal ions which is capable of catalytic free-radical reactions and promote microbial growth.

Skin Anatomy

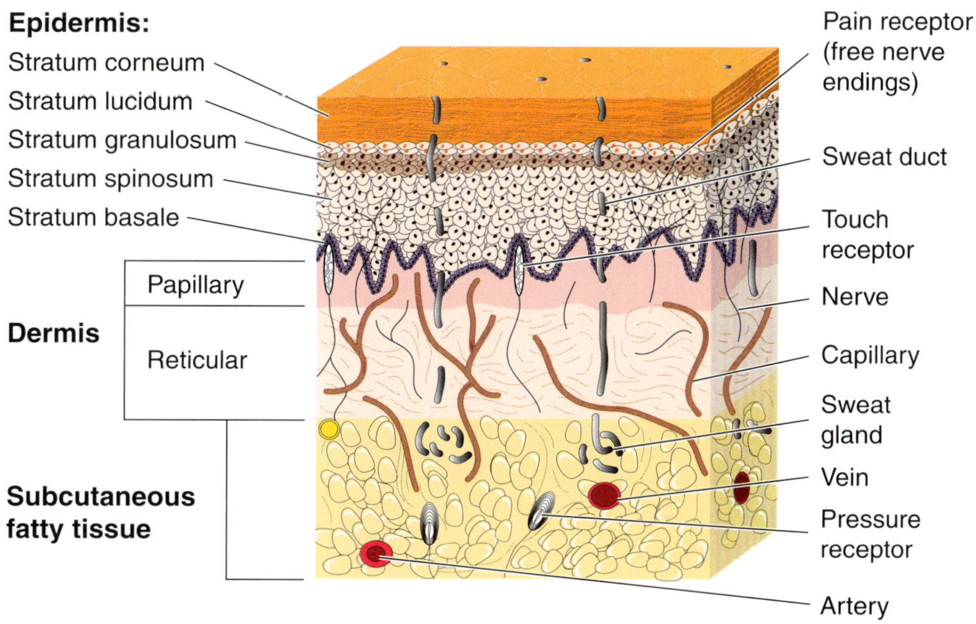

Damaged proteins in the epidermis are degraded by the proteasome system. Proteasome activity declines with age, which may contribute to the photo-ageing phenomenon.

Photo-oxidation and Skin Damage: Skin is susceptible to photo-oxidative damage, both by type I reactions and by ROS generation including singlet oxygen, hydroxyl (·OH) radical, and hydrogen peroxide (H_2O_2). The skin also contains endogenous photo-sensitizers, including catalase and pyridinoline (a compound present in the cross-links of collagen). Exogenous photo-sensitizers are found in certain cosmetics or ingested photo-sensitizing drugs that reach the skin, including some phenothiazines (tranquillizers), as well as fluoroquinolone and tetracycline antibiotics. Individuals affected with porphyria are susceptible to photo-oxidative damage. Light exposure, ultraviolet (UV) light in particular, is the major contributor of photo-oxidative skin damage. Over exposure to sunlight is a risk factor for skin cancers (basal cell carcinoma and malignant melanoma) in humans, and also involved in the ageing of skin. Photo-ageing of the skin leads to wrinkling, loss of elasticity, increased skin fragility, and slower wound healing.

In the UV region of the solar spectrum, it is only UV-A (320-400 nm) and some UV-B (290-320 nm) that reach the surface of the earth. Fortunately, the more damaging UV-C (190-290 nm) is filtered out by the ozone layer in the stratosphere. It has been estimated that a 5% depletion of stratospheric O_3 may result in the raise of UV-B flux at ground level by 10%, which could increase its epidermal penetration by several fold and cause significant oxidative skin damage. In contrast to X-rays and γ-rays, UV light does not deposit sufficient energy to fragment water molecules; however, in the presence of H_2O_2, UV-B forms hydroxyl (·OH) radicals that can directly damage the DNA. The absorption of UV-B by nucleotides (thymine or cytosine) results in excited states that can react with water to form pyrimidine hydrates or cross-links. It has been suggested that sunlight passing through the skin can not only cause DNA damage to white blood cells circulating through skin capillaries, but also induce detrimental effects on mitochondria in the dermal region. UV light also leads to oxidation of polyunsaturated fatty acid (PUFA) side chains and cholesterol in skin lipids and generate reactive aldehydes such as acrolein and HNE from peroxides that can damage proteins and DNA. Thus, DNA damage by UV light through above and other mechanisms contributes to skin ageing and cancer. Mutations in the tumor suppressor gene p53 appear to be early events in the pathobiology of skin cancer and double transition mutations are highly common in mutated p53 genes in skin cancers (versus carcinomas from internal organs) - which is consistent with the deleterious role of sunlight.

Ultraviolet-A or -B induce activation (or via ROS as signaling molecules) of a wide range of transcription factors in skin cells, including AP-1 and NF-κB. This can increase production of matrix metalloproteinases (MMPs) that can degrade collagen and other connective tissue components. Levels of cytokines also elevate the inducible NO synthases (iNOS), which can generate excess ·NO and peroxynitrite ($ONOO^-$) radicals. In photo-damaged skin, Fe(III) found in sweat, can contribute to oxidative damage through Haber-Weiss reaction and via activation of transcription factors.

BLACK RIBBON
Melanoma Awareness

Photo-oxidative Skin damage: Melanoma

Every 8 minutes, someone in the United States will be given a melanoma diagnosis and every hour someone will die from the disease.

- The risk of developing invasive melanoma in the U.S. is 1 in 41 for men and 1 in 61 for women. There is a 1 in 50 chance of developing melanoma throughout your lifetime.
- Melanoma is the most common form of cancer for young adults 25-29 years old and the 2nd most common cancer in adolescents and young adults 15-29 years.
- Although most common in Caucasians, melanoma can strike men and women of all ages and all races. The mean age for diagnosis of melanoma is 50, while for many other cancers it is 65-70 years.

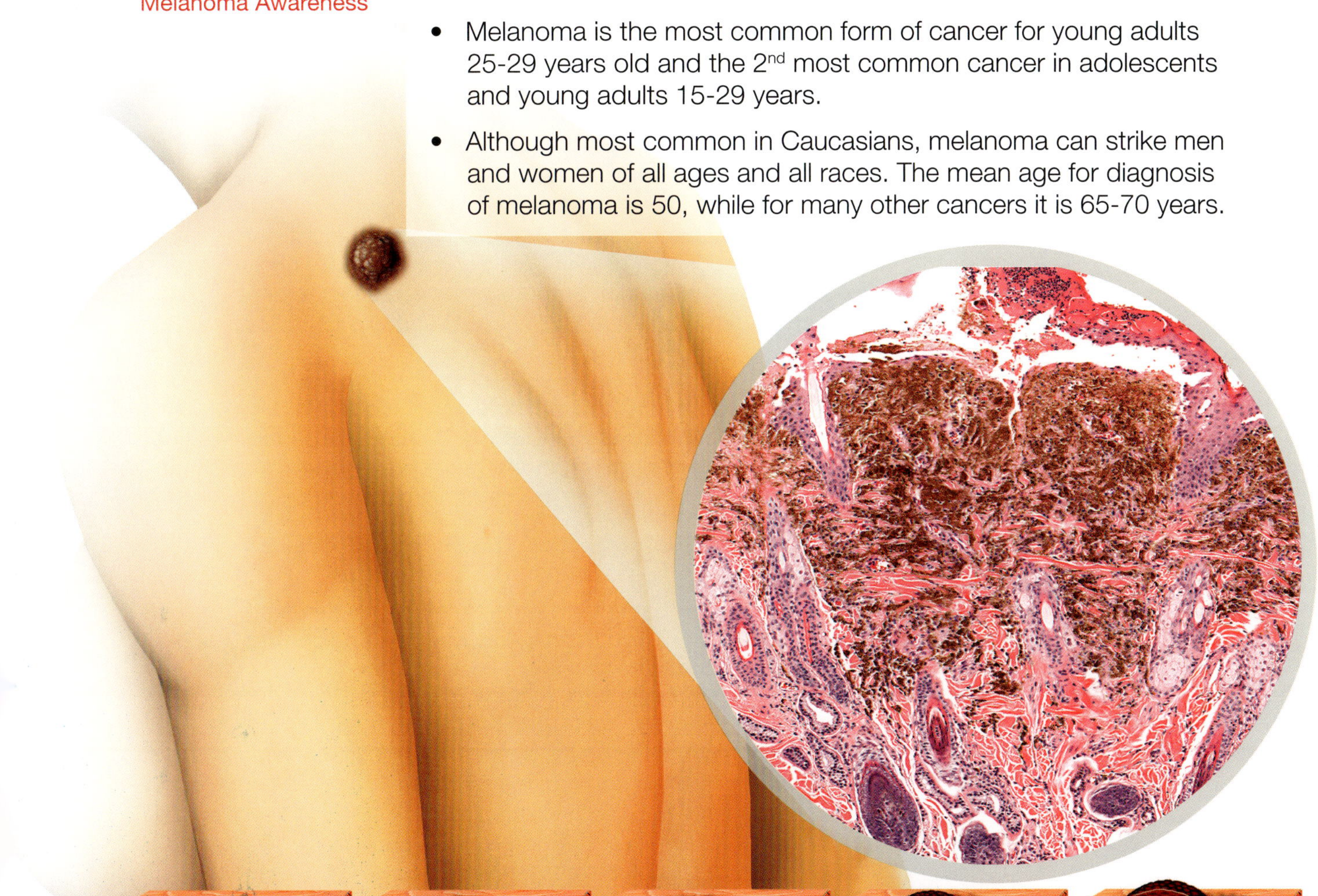

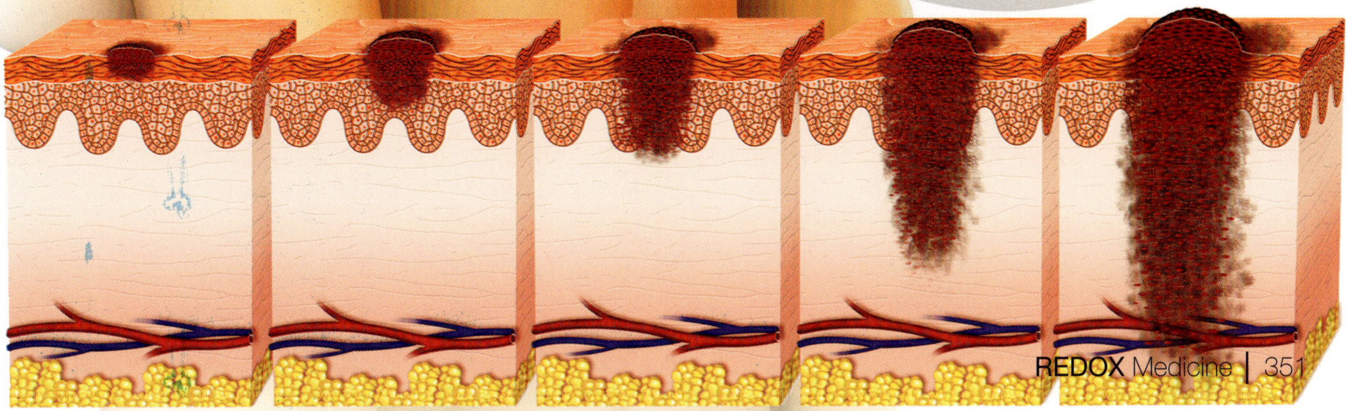

SKIN INFLAMMATION AND ACNE

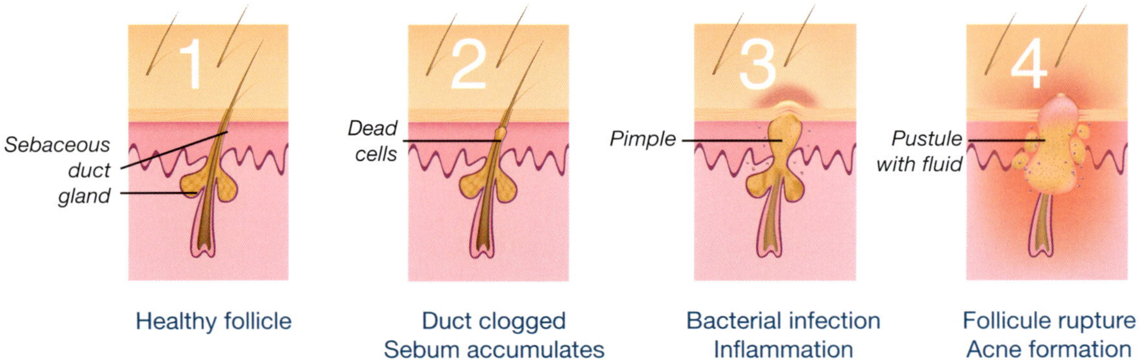

Photo-oxidation and Skin Inflammation: Photo-oxidative damage to skin can also elicit an inflammatory response, including recruitment and activation of lymphocytes and neutrophils (generate ROS through respiratory burst). Several skin diseases involve inflammation, including acne and psoriasis. Psoriasis affects up to 3% of the population of developed countries and is characterized by increased keratinocyte proliferation and abnormal keratinocyte differentiation. TNF-α may be an important pro-inflammatory agent during such disease conditions.

Sunlight induces melanin synthesis, which has a broad absorbance spectrum ranging through the UV-B, UV-A and visible ranges. Melanin is made in melanocytes, which transfers it to keratinocytes and hair. Eumelanin is protective against photodamage, but this is not always true of pheomelanin. In the disease 'vitiligo' there is patchy depigmentation of the skin, decreased catalase activity, and accumulation of peroxides.

REDOX-based Interventions: Most effective measure against detrimental effects of UV light on skin metabolism is avoidance; stay out of the sun or use a sun block. Unfortunately, certain active ingredients in several sun creams in the consumer market contain pro-oxidants (i.e. titanium dioxide, TiO_2), which can generate ROS upon light exposure. Antioxidants such as tocopherols, vitamin C (as hydrophobic esters such as ascorbyl palmitate), and flavonoids (as cosmetic additives) in skin crèmes can provide protection again photo-oxidative damage. However, their ability to penetrate deep into the skin is limited, and a better way of raising antioxidant levels in the dermis is consumption through the diet. Keratinocytes and skin fibroblasts contain glutathione (GSH), antioxidant enzymes (i.e. peroxiredoxins), α-tocopherol, ascorbate (essential for skin collagen synthesis), and DNA repair enzymes. Therefore, keratinocytes are more resistant than most cells to killing by H_2O_2, organic peroxides or $ONOO^-$. Ascorbate, GSH, SOD, catalase and ubiquinol are depleted in both dermis and epidermis in UV-B-exposed skin. Introduction of REDOX nutrients and bioreplenishments (i.e. SOD, GSH, CoQ10, SAMe and lactoferrin) could provide antioxidant protection to the skin surface when secreted as sebum by the sebaceous glands.

4.12.2. Oxidative Stress and Wound Healing

Damage of a tissue triggers a cascade of repair events, which begin with the formation of a fibrin clot as a result of leakage of blood. The clot provides protection to the underlying tissue, serves as a provisional matrix through which cells can move, and also acts as a reservoir for growth factors and cytokines. The growth factors initiate inflammation, epithelial formation, wound contraction and angiogenesis. Platelet-derived growth factor (PDGF) and tissue growth factor (TGF) released from platelets can act as chemo-attractant for neutrophils and monocytes/macrophages. Breakdown of red blood cells in the clot can release heme, which can trigger ROS generation and create a pro-oxidant milieu around the wound. The major role of neutrophils is to kill invading microorganisms by their characteristic respiratory burst activity and also to activate keratinocytes and fibroblasts. The monocytes in the inflamed tissues can mature into macrophages and are responsible for phagocytosis of dying neutrophils, damaged tissue, and microorganisms.

Macrophages play an important role in the long term repair response by releasing a battery of cytokines and growth factors (e.g. FGF, PDGF, TGF and IGF1) to amplify the inflammatory response and also to initiate the proliferative phase of wound healing. The re-epithelization process of wounds is marked by the migration of keratinocytes across the fibrin clot, and at the same time, proliferation of keratinocytes at the wound edge. Angiogenesis leads to rebuilding of the damaged vessels, restoration of the blood flow, and O_2 supply to the tissue. During angiogenesis, proteases released from the activated endothelium degrade the endothelial cells and allow their migration, as well as differentiation into mature capillary blood vessels. About a week after wounding, the clot is totally replaced by fibroblasts to synthesize and remodel a new collagen rich matrix. Also few fibroblasts transform into myo-fibroblasts that resemble smooth muscle and cause wound contraction. Thus, oxidative stress plays an essential role in various pathways of wound healing: (i) the reactive species (i.e. ROS) produced by activated phagocytes elicit antimicrobial activity to prevent infection; and heme accumulation in the blood clot may facilitate phagocyte recruitment; (ii) both ROS and RNS may aid healing process; for example, by promoting fibroblast proliferation or by increasing the production of growth factors.

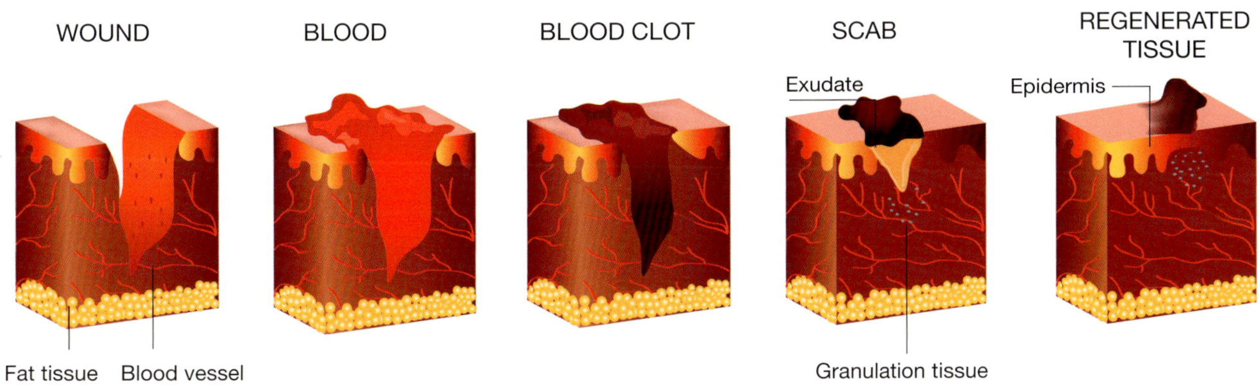

WOUND HEALING PROCESS

Oxidative and Nitrosative Stress in Wound Pathology: In acute wounds the oxidants levels are temporarily elevated; while the antioxidant defense mechanisms depend on steady detoxification of these oxidants and a gradual return of the wound milieu to the state of REDOX homeostasis. However, in chronic wounds the detoxification process is delayed due to persistent and uncontrolled production of ROS and RNS during the inflammatory phase. Though elevated in chronic wounds, antioxidants are not able to quench pro-oxidants that are produced in excess, beyond titration on a molar basis. As a result, the REDOX homeostasis shifts towards oxidative stress. Therefore, chronic wounds become stagnant in the inflammatory phase and show no spontaneous tendency to produce granulation tissue and heal. Clinically, chronic wounds demonstrate no signs of healing even after local basic treatment for several weeks. In the case of chronic wounds, up-regulated expression of inducible NO synthase (iNOS) and oxidants (i.e. ·NO) will result in sustained production of components of oxidative and nitroxidative stress. Accordingly, the indirect effects of ·NO will prevail in chronic wounds due to reaction of ·NO with other oxidants – forming a viscous cycle. Increased iNOS activity and nitrite levels have been shown to be responsible for diabetic foot and chronic venous ulcers.

Oxidative Stress and Wound Healing: Hydrogen peroxide (H_2O_2) plays an important role in wound healing. Non-phagocytes can produce H_2O_2 after wounding, which can attract neutrophils, as well as promote re-innervation of the peripheral sensory axons. Removal of H_2O_2 by catalase over-expression is shown to delay wound closure and retard angiogenesis. There have been suggestions that application of low levels of H_2O_2 may be beneficial for wound healing. Collagen film dressings that contained glucose oxidase were found to promote wound healing in a diabetic animal model by increasing levels of reactive oxygen species (ROS) in the wounds. Glucose oxidase is an enzyme that oxidizes glucose to gluconic acid with the formation of H_2O_2 as a by-product. Interestingly, medicinal grade honey, claimed to promote healing of chronic wounds, has also been shown to contain H_2O_2, possibly mediated via glucose oxidase activity. Thus, low concentrations of H_2O_2 promote healing by acting as a signaling molecule; while elevated levels of it would delay the wound healing by causing oxidative damage.

Nitrosative Stress and Wound Healing: A balance between ·NO level and the degree of oxidative/ nitrosative stress in wound tissue may be crucial for its repair. The reparative effect of ·NO is attributed to scavenging of $O_2^{\bullet-}$ anion, which is the key radical of oxidative stress. Also, the high level of ·NO can influence angiogenesis and endothelial/skeletal muscle cell remodeling and proliferation. However, under conditions of excess and prolonged production of $O_2^{\bullet-}$ in wounds, ·NO can lead to an elevated nitrosative stress due to production of $ONOO^-$ and peroxynitrous acid (ONOOH). Peroxynitrous acid, in particular, can trigger a cascade of events leading to production of highly reactive and damaging radicals and oxidative species. These species (mainly $CO_3^{\bullet-}$, NO_2^+, NO_2, N_2O_3, ·OH) can inflict potential damage in biological milieu and impair the wound healing process. Therefore, an effective strategy for rapid wound healing process may include an intervention(s) leading to the decrease in oxidative stress; for example, treatment with antioxidants and/or prevention of $O_2^{\bullet-}$ generation by uncoupled constitutive NO-synthase (cNOS) and ·NO treatment (i.e. ·NO donors and cNOS gene therapy).

REDOX-based Interventions: REDOX gene therapy is an emerging therapeutic technology (currently in development) for management of chronic wounds. For example, wound healing is slower in diabetic animals; and therapy with genes encoding eNOS or Mn-SOD showed accelerated healing, which suggest that a correct ratio of $O_2^{\bullet-}$ to ·NO is important. In ulcers, however, the ROS balance might tilt in favor oxidative stress and may cause more harm than help. Wound dressings containing desferoxamine (an iron chelator) can decrease iron-dependent oxidative damage. Certain care should be exercised in this treatment approach, since Fe(III) is also essential for cell proliferation; therefore, chelation must not be overdone.

Antioxidants such as glutathione (GSH) and ascorbate are usually depleted in wounds. Thus, antioxidant supplementation could possibly promote or impair wound healing (based on type of the wound). Tocopherol was reported as a beneficial supplement for wound healing. Burns result in phagocyte activation, iNOS induction, protein nitration and oxidative damage to the skin. In severe cases, systemic oxidative stress may increase levels of xanthine oxidase (XO) in plasma. Also, severe burns can lead to extensive fluid loss and shock syndrome; administration of antioxidants such as superoxide dismutase (SOD) or allopurinol encapsulated in liposomes, showed an improvement in the survival rates in experimental animals with severe thermal skin injury.

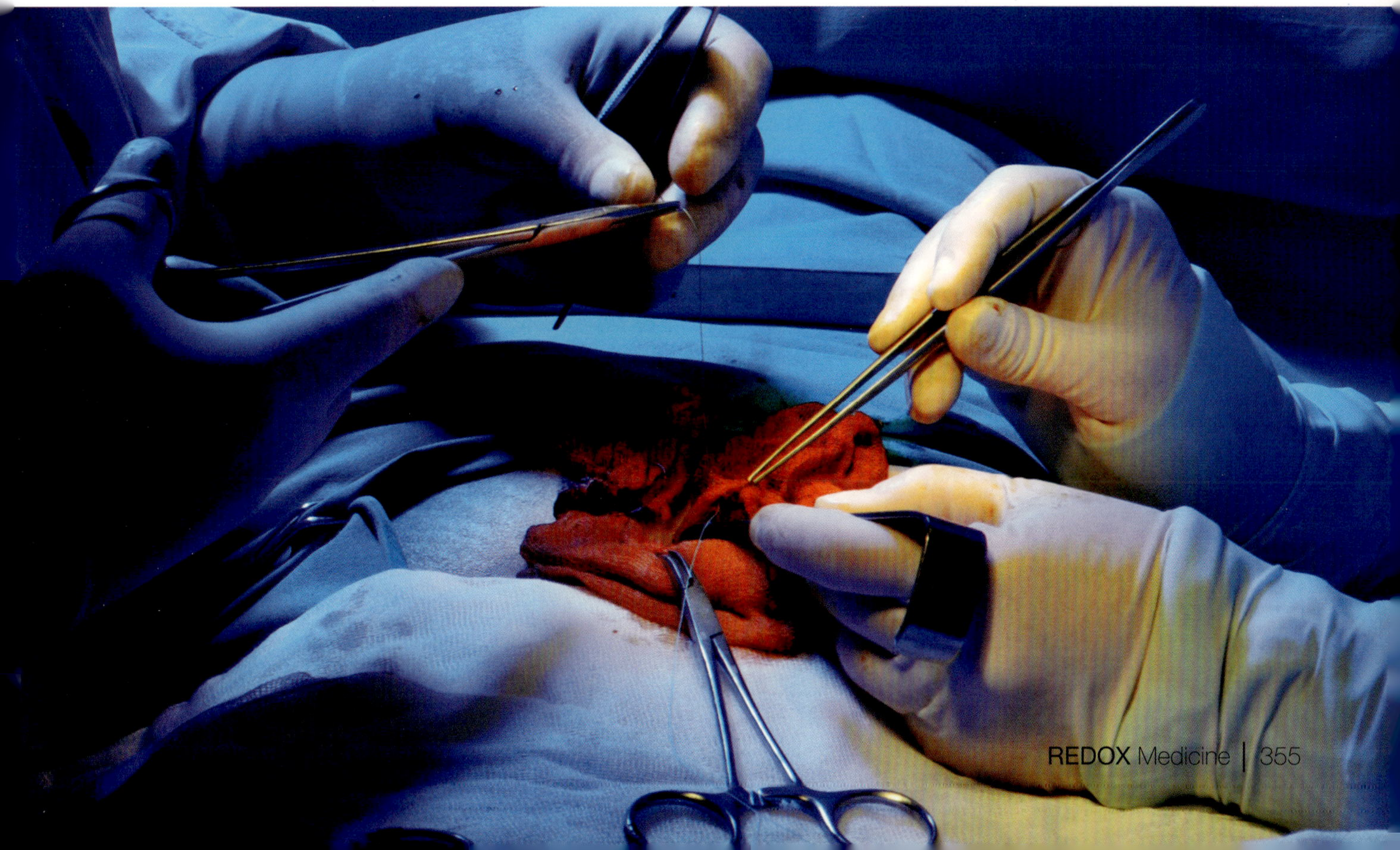

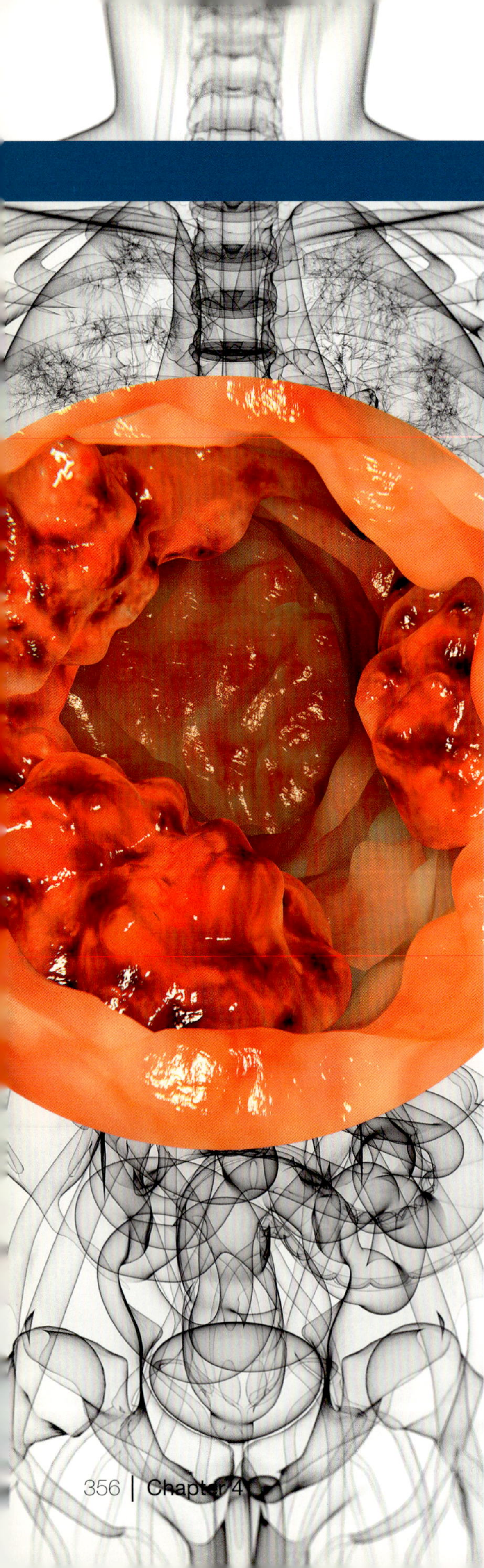

4.13 **REDOX:** Cancer

Cancer cells are known to differ distinctly in their REDOX metabolism from healthy tissues. Thus, enhanced levels of intra-cellular ROS are often observed in tumor cells. The specific milieu of the solid tumor is characterized by high metabolic activity, hypoxia, and, in general, reductive conditions. Consequently, interference with the cellular REDOX homeostasis of cancer cells seems an attractive and promising approach for cancer therapy. Most of the currently used chemotherapeutic drugs have been shown to exert some interaction with the cellular REDOX balance and there are several attempts to specifically target the altered REDOX conditions in cancer cells.

The role of oxidative stress in metastasis and tumor progression involves a number of factors including cell type, cellular milieu, and free radical type and compartmentalization. Tumor survival depends on a number of processes involving proliferation, motility, apoptosis, and senescence - all of which are influenced by changes in REDOX metabolism. Complexity lies in the fact that individual cancers may be characterized by different REDOX-based signaling mechanisms.

REDOX Imbalance and Carcinogenesis: Cancer initiation and progression is linked to the disruption of REDOX balance and oxidative stress. Intra- and extra-cellular REDOX compartments are critical to the etiology and maintenance of oxidative stress associated with malignancy and metastasis. In these compartments, REDOX-sensitive sites (i.e. cysteines) regulate cell signaling events and promote the malignant cells via the activation of survival pathways, disruption of cell-death signaling, and augmenting cell proliferation. Elevated oxidative stress is observed in many solid tumors and carcinoma cell lines. REDOX imbalance in cancer cells may be tied to a number of pathways including mitochondrial dysfunction, activation of oncogenes, aberrant oxidative metabolism, and hypoxia/ reoxygenation cycles. Low levels of oxidative stress may be advantageous for cancer progression, which may contribute to increased rates of genetic mutation and contribute to the acquisition of a malignant phenotype. Higher levels of ROS are more likely to lead to cell senescence and/or death. Certain ROS also act as second messengers in signaling pathways with the potential to regulate cell phenotype directly as effector molecules.

Low levels of O_2 or H_2O_2 can enhance cellular survival and stimulate proliferation. However, when this is concomitant with chronic ROS production, REDOX homeostasis is imbalanced and transform normal cells. Cancer cells are characterized by enhanced ROS generation, increased ROS accumulation, and the deregulation of antioxidant enzymes; thus existing in a state of perpetually elevated stress. Constitutively produced NO-synthases (cNOS) have also been found in several human tumor cell lines, an observation further complicated by the fact that reactive nitrogen species (RNS) can be chemically heterogeneous in different tumor cell types. Since oxidative stress plays a major role in the induced adaptive expression of antioxidants, malignant cells are often characterized by differential expression in the levels of a number of enzymes, including superoxide dismutases (SOD), catalase, and glutathione-S-transferases. Dysregulation in expression of such enzymes can influence cancer therapy and may contribute to metastasis and drug resistance. These characteristics further perpetuate a state of oxidative stress and result in either adaptation or apoptosis. Since oxidative stress may induce apoptosis, cell cycle arrest or cellular senescence, eventual cellular fate may be a delicate balance contingent upon factors such as cell type, tissue microenvironment, and levels of free radical production/accumulation.

According to the American Cancer Society (2013) – about 1.660,290 new cancer cases are expected to be diagnosed in the United States; about 580,350 Americans are projected to die of cancer, and almost 1,600 people a day. Cancer remains the second most common cause of death in the US, accounting for nearly 1 of every 4 deaths.

TABLE: *Different Carcinoma and Their REDOX Markers*

Location	Carcinoma Type	REDOX Markers
Extra-cellular	Prostate	↑H_2O_2; ↑GSH; ↓GSSG; ↑EC-SOD
	Hepatocellular	↑Thioredoxin
	Pancreatic	↑Thioredoxin
	Liver Metastasis	↑·OH Radicals; ↑H_2O_2; ↑$O_2^{\bullet-}$ anions
	Colon	Thiol/disulfide balance
	Lymphoma	↑Glutathione S-transeferase P
	Breast, Gastric. Colorectal	↓Glutathione peroxidase 3 activity
Intra-cellular	Pancreatic	↑Thioredoxin; ↑Glutaredoxin
	Squamous	↑γ-glutamyl-transferase
	Ovarian, Lung, Breast, Colon, Pancreas, Lymphoma	↑Glutathione S-transeferase P
	Breast	↑Thioredoxin; ↑Peroxiredoxins
	Ovarian, Prostate Melanoma	↑H_2O_2
	Colon, Pancreatic, Neuroblastoma, Breast, Ovarian	↑H_2O_2

Oxidative Damage and Carcinogenesis: An elevated ATP requirement is critical for supporting activities such as rapid proliferation of active cancer cells. A combination of energy demand and ROS induced mitochondrial damage may result in amplified electron 'leakage' and ROS generation. Mitochondrial electron-transport chain (ETC) and other oxidizing agents are the prime pathways that generate excess ROS in vivo, leading to several types of DNA damage, including depurination and depyrimidination, single and double-stranded DNA breaks, base and sugar modifications and DNA-protein cross-links. Permanent modification of genetic material resulting from the oxidative damage is one of the vital steps involved in mutagenesis that leads to carcinogenesis. Stimulation of DNA damage can either arrest or induce transcription, signal transduction pathways, replication errors, and genomic instability - all of which are associated with carcinogenesis. In addition to mitochondrial damage, high levels of ROS can produce DNA damage and genetic instability. Paradoxically, this damage further encourages the development of a malignant phenotype when normal repair mechanisms are overwhelmed and result in the accumulation of mutations. Disruption of REDOX homeostasis impacts cell signaling pathways that regulate cell proliferation, apoptosis, metastasis and angiogenesis, and result in cancer progression.

REDOX Signaling and Cancer: REDOX signaling pathways are used by normal and cancer cells to modulate physiological or aberrant cellular functions, respectively. ROS and RNS are mediators of oxidative stress, as well as REDOX signaling; two distinct but related processes which contribute to neoplasia. Despite the differences between both processes, oxidative stress and REDOX signaling are far from being independent events, especially in cancer, where ROS derived from oxidative stress could potentially activate a REDOX signaling pathway, or a deregulated REDOX signaling pathway. REDOX signaling is initiated by physiologically generated ROS to regulate cellular functions or outcomes. Oxidative stress can be initiated by the unregulated production of ROS from either extra-cellular (i.e. ultra-violet irradiation, drugs, xenobiotics) or intra-cellular (i.e. mitochondria, peroxisomes, oncogenes) sources. These ROS are usually present at high levels and can cause damage by irreversibly oxidizing cellular proteins, lipids, and nucleic acids. In normal cells, oxidative stress elicits an antioxidant response that leads to either damage repair or death; however, cancer cells aberrantly tolerate oxidative stress which contributes to cancer progression by driving, for instance, genetic instability. Thus, oxidative stress and REDOX signaling have been implicated in the initiation and/or maintenance of a pro-invasive program in cancer cells.

Cell invasion is considered one of the limiting steps for the progression of primary solid tumors to metastatic lesions. Without local invasion into neighboring tissues, the primary tumor might be resectable; but the initiation and maintenance of a cell invasive program allows the spreading of the cancer cells into the surrounding tissues and may promote metastatic disease. The main cellular mechanisms in the initiation and maintenance of an invasive program includes modulation of the adhesive, migratory, and proteolytic abilities of the cancer cell, as well as its adaptation to the tumor milieu. In cancer cells, cell-cell and cell-substrate adhesion properties change in response to pro-invasive stimulation. Modification of these adhesion complexes not only alters the adhesive and migratory properties of cells by inducing cytoskeletal remodeling, but also modulates survival and cell cycle progression by affecting the cross-talk between adhesion platforms and the cell signaling networks that regulate the communication between cells and their microenvironment. REDOX signaling pathways play an important role in the regulation of all these processes of cancer metastasis.

REDOX and Oncogenes: More than 127 genes and signal transduction proteins have been reported to be directly affected by REDOX state. Transcription factors, tumor suppressors, as well as members of the mitogen activated protein kinase (MAPK) family and anti-apoptotic pathways (such as PI 3-K and NF- κB) are regulated by ROS; they have roles in stimulating cell proliferation, sensitizing cancer cells to electrophilic agents, and contributing to treatment resistance.

BRCA1 and Oxidative Stress: The breast cancer susceptibility gene, BRCA1, encodes a tumor suppressor protein; mutations of it account for about 40–50% of hereditary breast cancers. In sporadic (non-hereditary) breast cancers, BRCA1 expression is absent or decreased in about 30–40% of cases. Environmental and genetic factors are important contributors of the neoplastic transformation leading to breast cancer. BRCA1 has several important functions and one of them is to render protection to the cells during oxidative stress. BRCA1 up regulates the expression of multiple genes involved in the cyto-protective antioxidant response, including GSH-S-transferase (GST), GSH-peroxidase (GPx), oxidoreductases, and other antioxidant genes. BRCA1-expression confers resistance while BRCA1-deficiency predisposes susceptibility to various oxidizing agents (i.e. H_2O_2 and

Breast Cancer and BRCA1 Gene

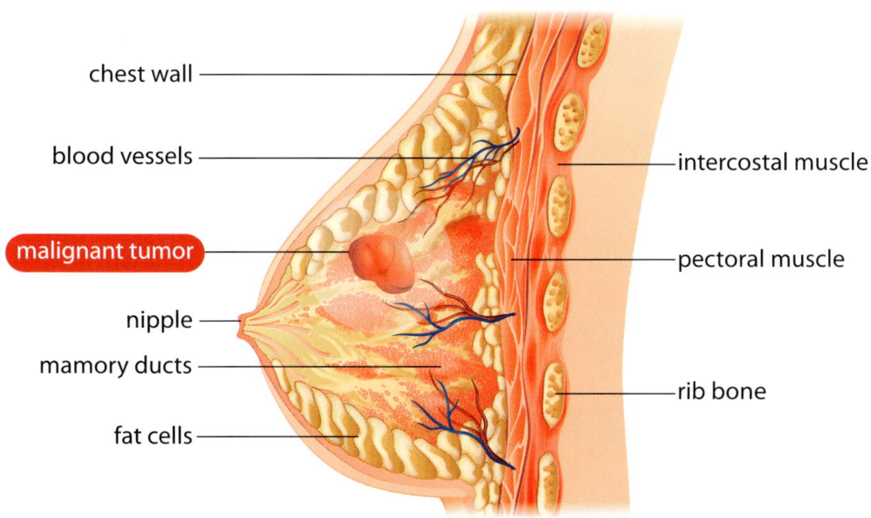

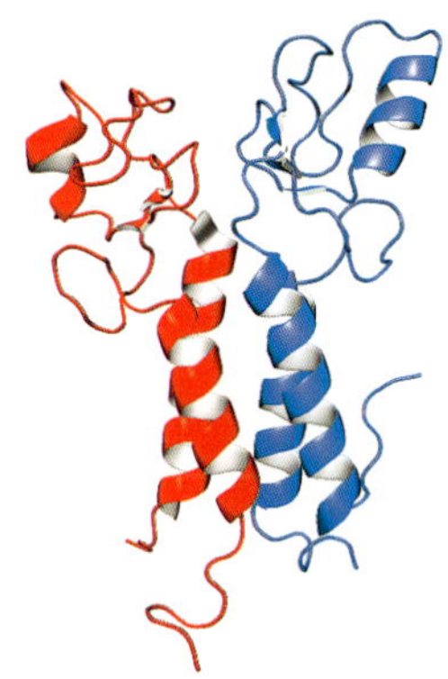

paraquat). During oxidative stress, BRCA1 shifts the cellular REDOX balance to a higher ratio of reduced to oxidized GSH. Higher content of reduced GSH functions in favor of REDOX maintenance. The identification of transcriptional target of several antioxidant genes that are elevated during BRCA1 over expression led to the findings that BRCA1 regulates the activity of the antioxidant response transcription factor Nrf2 and NF-κB and protects cells against oxidative stress. Thus, BRCA1 functions as a caretaker gene in preserving genetic integrity via regulating the REDOX maintenance of the cells.

REDOX-sensitive Transcription Factors/Activators and Cancer: REDOX sensitive transcription factors regulate several pathways that lead to carcinogenesis. Transcription factors/activators are a group of proteins that either activate or inhibit effector gene expression. The change in the expression of the effector genes eventually leads to several biological modifications such as proliferation, growth suppression, differentiation or senescence. Similarly, REDOX-induced biochemical alterations could also affect the biological functions of these proteins. Thus, differential modulation of such transcriptional activators can regulate many target/effector genes, which may provide an additional mechanism for small antioxidant molecules play a protective in anti-cancer processes.

Cervical Cancer

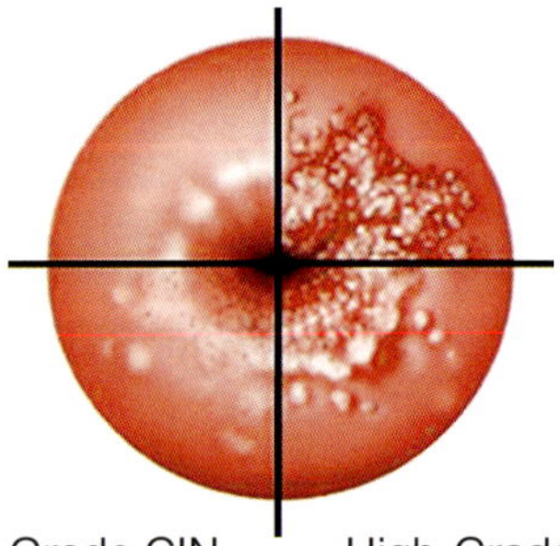

Cervical intraepithelial Neoplasia (CIN)

Cervical cancer is 2nd most common and the 5th deadliest cancer in women. It affects about 16 out of 100,000 women per year and kills about 9 out of 100,000 per year, worldwide. About 80% of cervical cancers occur in developing countries. In 2008, ~473,000 cases of cervical cancer were estimated and about 225,000 deaths were reported in 2010. In the United States, it is the 8th most common cancer of women. The median age at diagnosis is 48. Hispanic women are significantly more likely to be diagnosed with cervical cancer than the general population. In 2008, in the US an estimated 11,000 new cases were expected to be diagnosed, and about 3,870 were expected to die of cervical cancer. Among gynecological cancers, it ranks behind endometrial cancer and ovarian cancer. In the European Union, there were about 34,000 new cases per year and over 16,000 deaths due to cervical cancer in 2004. Cervical cancer deaths decreased by ~74% in the last 50 years, largely due to widespread 'Pap' smear screening. The incidence and mortality in the US is about half those for the rest of the world, which is due in part to the success of screening with the 'Pap' smear.

Cervical Cancer: The Nobel Prize in Physiology or Medicine 2008 was awarded to Harald zur Hausen for his discovery of human papilloma viruses causing cervical cancer.

Nuclear factor-erythroid-2-related factor 2 (Nrf2) and Cancer: The role of Nrf2 transcription factor and its down-regulation by 'Kelch-like ECH-associated protein 1' (Keap1) during oxidative stress has been well established in REDOX biochemistry. Nrf2 is a member of basic leucine zipper transcription factors and is essential for the coordinated induction of genes that encodes stress-responsive or cyto-protective enzymes and related proteins, such as NAD(P)H quinone oxido-reductase 1 (NQO1), SODs, GSTs, GPx, hemeoxygenase (HO)-1, glutamate cysteine ligase (GCL), catalase and thioredoxinin (TRx). Nrf2 plays an essential role in maintaining cellular homeostasis and hence represents a critical target for prevention of oxidative stress- or inflammation-associated carcinogenesis. Under normal physiological conditions, Nrf2 forms an inactive complex with the negative regulator, Keap1, which controls the sub-cellular localization and steady-state levels of Nrf2. Cysteine residues present in Keap1 function as REDOX sensors. Oxidation or chemical modification of some of the highly reactive cysteine residues facilitates the dissociation of Nrf2 from Keap1 and subsequent nuclear translocation. Nrf2 constitutes a unique REDOX regulator that can be modulated in response to REDOX imbalance caused by oxidative stress. In breast cancer cells, Nrf2 plays an important role in suppressing oxidative stress rendered by either paraquat or H_2O_2 by up-regulating several phase II detoxification enzymes. On the other hand expression of BRCA1 also stimulate the expression of Nrf2 in the same cells, thus indicate an important function of Nrf2 in breast cancer.

Nuclear factor kappa-light-chain-enhancer of activated B cells (NF-κB) and Cancer: The NF-κB family of REDOX transcription factors has been activated in many malignancies including breast, prostate, pancreas, colon, leukemia, lymphoma etc. NF-κB activity is regulated by intra-cellular ROS levels, and is important in the development and/or progression of cancer by regulating the expression of genes involved in cell growth and proliferation, anti-apoptosis, angiogenesis and metastasis. Intra-cellular level of reduced-GSH, which may differ from one cell type to another, may be crucial for H_2O_2-induced NF-κB response during oxidative stress. Several lines of evidence suggest a role for ROS as a common and critical intermediate for various NF-κB-activating signals.

Mitogen-activated protein kinases (MAPKs) and Cancer: MAPKs are the key players of kinase cascades that connect extra-cellular signals to specific transcription factors, thereby converting these signals into cellular responses. In mammalian systems, there are three subgroups of MAPKs: (i) ERKs (extra-cellular signal-regulated kinases), (ii) JNKs (c-Jun N-terminal kinases) and (iii) p38 MAPKs. Growth factors and mitogens use the MAPK signaling cascade to transmit signals from their receptors to regulate gene expression and prevent apoptosis in normal condition. The MAPK signaling cascade is activated by a wide variety of receptors involved in growth and differentiation including receptor tyrosine kinases, integrins, and ion channels. This pathway has been reported to be activated in over 50% of acute myelogenous leukemia and acute lymphocytic leukemia. Several natural phytochemicals exploit the signaling cascades of MAP kinases to exert its effects on the modulation of cancer initiation and progression. Kempferol is a flavonoid that reduces the cell viability of breast cancer cells by inducing ROS mediated apoptosis and ERK activation. Proline oxidase, often considered as a 'house-keeping enzyme', induces apoptosis through both intrinsic and extrinsic pathways and is involved in nuclear factor of activated T cell signaling and regulation of the MEK/ERK pathway in colon cancer cells. As a nutrition factor, proline oxidase may modulate apoptosis signals induced by p53 or other anti-cancer agents and enhance apoptosis in stress situations. Activating MAPK pathways in prostate cancer cells is often beneficial to stimulate apoptosis and inhibition of growth. Melatonin, the main secretory product of the pineal gland, inhibits the growth of prostate cancer cells and promotes apoptosis via MAPKs (which are closely associated with apoptosis and survival). Melatonin also activates JNK and p38 kinase whereas extra-cellular signal-regulated kinase (ERK) was not responsive to melatonin. Thus, melatonin may be of promise for anti-prostate cancer strategies that will utilize the MAP kinase pathways. Thus, exploiting the MAP kinase pathway could be an important step towards inhibition of the occurrences of various cancers.

REDOX-based Intervention for Cancer: Cells require tight regulation of the intra-cellular REDOX balance and consequently, of ROS for optimum REDOX signaling and maintenance of metal [i.e. Fe(III) and Cu(II)] homeostasis. In several diseases, including cancer, this balance is disturbed. Therefore, anti-cancer drugs targeting the REDOX systems, for example, GSH and TRx, have entered focus of interest. Anti-cancer metal complexes (e.g. platinum, gold, arsenic, ruthenium, rhodium, copper, vanadium, cobalt, manganese, gadolinium, and molybdenum) strongly interact with or even disturb cellular REDOX homeostasis. In this context, especially the hypothesis of Activation by Reduction, as well as, the Hard and Soft Acids and Bases theory with respect to coordination of metal ions to cellular ligands, represent important concepts to understand the molecular modes of action of anticancer metal drugs.

Arsenic trioxide (ATO) is one of the first metal compounds used in anticancer therapy; during the 18th and 19th century, ATO represented the main treatment for leukemia. The modern era of metal-based anti-cancer drugs began with the discovery of the platinum(II) complex, 'cisplatin', by Barnett Rosenberg in the 1960s. Nowadays, cisplatin and its successors carboplatin and oxaliplatin, are among the most important chemotherapeutics used against a wide variety of different cancers. Stimulated by the success of cisplatin, also other coordination compounds based on ruthenium, gold, titanium, copper, rhodium, vanadium, and cobalt were tested for their anticancer activity and several promising candidates are currently in (pre)clinical evaluation. One of the characteristics of metals is their potential to undergo REDOX processes, as determined by their REDOX potentials. Especially, transition metal ions are usually able to switch between several oxidation states. However, not all oxidation states are observed under physiological conditions in the living organism. Due to the REDOX activity of metals and their ability to disturb the sensitive cellular REDOX homeostasis, a tight regulation of the metal and REDOX balance is crucial for health and survival.

The specific roles of extra-cellular vs. intra-cellular REDOX state; the importance of non-radicals in REDOX metabolism; and the recognition of the impact of tumor microenvironment on metastasis, form the scientific basis for the development and uses of targeted REDOX-modulating therapeutics. REDOX Chemotherapeutics represent a novel class of anticancer agents that selectively target tumors by a combination of direct- or indirect-acting, pro- and anti-oxidants and REDOX modulators. Many of the novel molecularly targeted chemotherapeutic agents that have entered advanced clinical trials or received FDA-approval (including the proteasome inhibitor bortezomib, cyclin-dependent-kinase inhibitors, MAP-kinase antagonists, and mTOR inhibitors) remarkably target molecules essential to normal cells. Therefore, these agents have been referred to as 'neocytotoxics', agents that exert anti-cancer activity based on the preferential sensitivity of rapidly dividing cells. The pharmacodynamics of these REDOX modulators often depends on the presence of chemically reactive pharmacophores that display an obvious potential for uncontrolled reactivity and untargeted cytotoxicity. Covalent adduction and generation of reactive intermediates is associated with numerous reactive pharmacophore contained in REDOX chemotherapeutics. The pleiotropic mode of action associated with numerous REDOX chemotherapeutics (i.e. the simultaneous modulation of multiple REDOX sensitive targets) that seems to be uniquely tailored to overcome cancer cell drug resistance originating from redundancy of oncogenic signaling and rapid mutation.

TABLE: *REDOX-based Anti-Cancer Interventions – Mechanisms / Drug Status (from Tew & Townsend, 2011)*

Anti-Cancer Agent	REDOX Mechanism of Action/Target	Pharmacological Status
Amifostine	Aminothiol protects against radiation toxicity.	FDA approved, for esophageal cancer.
MESNA/Tavocept	Tavocept is disulfide of MESNA, used with cancer drugs as a thiol protective agent.	Tavocept is in early clinical testing.
Disulfiram	Thiol disulfide exchange/ metal [Cu(II)] chelation.	FDA approved as alcohol deterrent. Cu(II) chelation has antitumor effects.
Telcyta	GSTP activated pro-drug for ovarian cancer.	IND in Phase III trial.
Telintra	GSTP inhibitor, myeloproliferative.	IND in Phase II trial.
PABA/NO	GSTP activated NO releasing pro-drug.	Preclinical testing.
NOV-002	Mimetic of GSSG complex with non-therapeutic concentrations of cisplatin.	IND in Phase II/III trial.
Arsenic trioxide	Covalent cross-linking of vicinal thiols.	FDA approved for leukemia.
Menadione	REDOX cycling produces semiquinone radical.	Similar activity to anthracyclines.
Artemisinin	Sesquiterpene with reactive electrophilic.	Anti-malarial, but analogs have longer half-lives and anti-cancer activity.
Organo-sulfur isothiocyanates	Chemo-protective agents that includes sulfora-phane, β-phenyl ethyl isothiocyanate.	Induce Phase II detoxification enzymes via antioxidant response element (ARE).
PX-12	Irreversible inhibitor of thioredoxin (TRx)-1.	Preclinical to clinical transition.
Ethacrynic acid	An example of Michael addition chemistry interface with GSH/GST pathways.	FDA approved as a diuretic; used in Phase I setting in cancer.

Cancer Facts

- Cancer is a leading cause of death worldwide, accounting for 7.6 million deaths (around 13% of all deaths in 2008).
- Lung, stomach, liver, colon and breast cancer cause the most cancer deaths each year.
- About 30% of cancer deaths are due to the five leading behavioral and dietary risks: high body mass index, low fruit and vegetable intake, lack of physical activity, tobacco use, alcohol use.
- Tobacco use is the most important risk factor for cancer causing 22% of global cancer deaths and 71% of global lung cancer deaths.
- Cancer causing viral infections such as HBV/HCV and HPV are responsible for up to 20% of cancer deaths in low- and middle-income countries.
- About 70% of all cancer deaths (in 2008) occurred in low- and middle-income countries.
- Deaths from cancer worldwide is projected to continue rising, with an estimated 13.1 million deaths in 2030.

4.14 REDOX Medicine: **Bibliography**

1. Auten RL, Davis JM (2009) Oxygen toxicity and reactive oxygen species: The devil is in the details. *Pediatric Research* 66:121-127.
2. Banerjee R, Becker D, Dickman M, Gladyshey V, Ragsdale S (2007) *Redox Biochemistry.* New Jersey: John Wiley. ISBN:978-0471786245.
3. Davies KJ (1995) Oxidative stress - the paradox of aerobic life. *Biochemical Society Symposium* 61:1-31.
4. Halliwell B, Gutteridge JMC (2007) *Free Radicals in Biology and Medicine.* Oxford: Oxford University Press.ISBN:978-0198568698.
5. Jacob C, Winyard PG (2009) *Redox Signaling and Regulation in Biology and Medicine.* Weinheim: Wiley-VCN Verlag. ISBN: 978-3527319251
6. Latham F (1951) The oxygen paradox. Experiments on the effects of oxygen in human anoxia. *The Lancet* 1(6646):77-81.

4.1 | REDOX: Homeostasis

1. Allen JF (2010) Redox homeostasis in the emergence of life. On the constant internal environment of nascent living cells. *Journal of Cosmology* 10:3362-3373.
2. Biteau B, Hochmuth CE, Jasper H (2011) Maintaining tissue homeostasis: dynamic control of somatic stem cell activity. *Cell Stem Cell* 9(5):402-411.
3. Foyer CH, Noctor G 2005) Redox homeostasis and antioxidant signaling: a metabolic interface between stress perception and physiological responses. *Plant Cell* 17(7):1866-1875.
4. Greer EL, Brunet A (2008) FOXO transcription factors in ageing and cancer. *Acta Physiologica* (Oxford) 192:19-28.
5. Hedrick SM (2009) The cunning little vixen: Foxo and the cycle of life and death. *Nature Immunology* 10(10):1057-1063.
6. Morrison SF, Nakamura K, Madden CJ (2008) Central control of thermogenesis in mammals. *Experimental Physiology* 93:773-797.
7. Nistala R, Whaley-Connell A, Sowers JR (2008) Redox control of renal function and hypertension. *Antioxidants & Redox Signaling* 10(12):2047-2089.
8. Pellegrino MA, Desaphy JF, Brocca L, Pierno S, et al (2011) Redox homeostasis, oxidative stress and disuse muscle atrophy. *Journal of Physiology* 589(9):2147-2160.
9. Ray PD, Huang BW, Tsuji Y (2012) Reactive oxygen species (ROS) homeostasis and redox regulation in cellular signaling. *Cell Signal* 24(5):981-990.
10. Salih DA, Brunet A (2008) FoxO transcription factors in the maintenance of cellular homeostasis during aging. *Current Opinion in Cell Biology* 20:126-136.
11. Tao GZ, Lehwald N, Jang KY, Baek J, et al (2013)Wnt/β-Catenin signaling protects mouse liver against oxidative stress-induced apoptosis through the inhibition of Forkhead transcription factor FoxO3. *Journal Biology Chemistry* 288(24):17214-17224.
12. Trachootham D, Lu W, Ogasawara MA, Nilsa RD, Huang P (2008) Redox regulation of cell survival. *Antioxidants & Redox Signaling* 10(8):1343-1374.
13. Van der Horst A, Burgering BM (2007) Stressing the role of FoxO proteins in lifespan and disease. *Nature Reviews: Molecular Cellular Biology* 8:440-450.
14. Weigel D, Jäckle H (1990) The fork head domain, a novel DNA-binding motif of eucaryotic transcription factors? *Cell* 63:455-456.
15. Yamamoto M, Clark JD, Pastor JV, Gurnani P, et al (2005) Regulation of oxidative stress by the anti-aging hormone klotho. *Journal of Biological Chemistry* 280(45):38029-38034.

4.2 | REDOX: Oxidative Stress

1. DNA in synovial fluid of patients with rheumatoid arthritis. *Arthritis Research & Therapeutics* 5:R234-R240.
2. Hajizadeh S, DeGroot J, TeKoppele JM, Tarkowski A, Collins LV (2003) Extracellular mitochondrial DNA and oxidatively damaged Halliwell B (2007) Oxidative stress and cancer: Have we moved forward? *Biochemistry Journal* 401(1):1-11.

3. Imlay JA (2003) Pathways of oxidative damage. *Annual Review of Microbiology* 57:395-418.
4. James SJ, Cutler P, Melnyk S, Jernigan S, et al (2004) Metabolic biomarkers of increased oxidative stress and impaired methylation capacity in children with autism. *American Journal of Clinical Nutrition* 80(6):1611-1617.
5. Kennedy G, Spence VA, McLaren M, Hill A, et al (2005) Oxidative stress levels are raised in chronic fatigue syndrome and are associated with clinical symptoms. *Free Radical Biology & Medicine* 39(5):584-589.
6. Nathan C, Shiloh MU (2000) Reactive oxygen and nitrogen intermediates in the relationship between mammalian hosts and microbial pathogens. *Proceedings of the National Academy of Sciences USA* 97(16): 8841-8848.
7. Newkirk MM, Goldbach-Mansky R, Lee J, Hoxworth J, et al (2003) Advanced glycation endproduct (AGE)-damaged IgG and IgM autoantibodies to IgGAGE in patients with early synovitis. *Arthritis Research & Therapeutics* 5:R82-R90.
8. Nunomura A, Castellani RJ, Zhu X, Moreira PI, et al (2005) Involvement of oxidative stress in Alzheimer disease. *Journal of Neuropathology & Experimental Neurology* 65(7):631-641.
9. Singh N, Dhalla AK, Seneviratne C, Singal PK (1995) Oxidative stress and heart failure. *Molecular & Cellular Biochemistry* 147(1):77-81.
10. Sohal RS (2002) Role of oxidative stress and protein oxidation in the aging process. *Free Radical Biology & Medicine* 33(1):37-44.

4.3 | REDOX: The Digestive System

1. Bayliss WM, Starling EH (1899) The movements and innervation of the small intestine. *Journal of Physiology,* 24:99-143.
2. Bik EM, Eckburg PB, et al. (2006) Molecular analysis of the bacterial microbiota in the human stomach. *Proceedings of the National Academy of Science* 103:732-737.
3. Camilleri M, Colemont LJ, Phillips SF, Brown ML, Thomforde GM, Chapman N, Zinsmeister AR (1989) Human gastric emptying and colonic filling of solids characterized by a new method. *Am J Physiol Gastrointest Liver Physiol* 257:G284-290.
4. Charles F, Camilleri M, Phillips SF, Thomforde GM, Forstrom LA (1995) Scintigraphy of the whole gut: clinical evaluation of transit disorders. *Mayo Clinical Proceedings* 70:113-118.
5. Dale BA, Fredericks LP (2005) Antimicrobial peptides in the oral environment: Expression and function in health and disease. *Current Issues in Molecular Biology* 7:119-133.
6. Degen LP, Phillips SF (1996) Variability of gastrointestinal transit in healthy women and men. *Gut* 39:299-305.
7. Kendall KA (2002) Oropharyngeal swallowing variability. *Laryngoscope* 112:547-551.
8. Metcalf AM, Phillips SF, Zinsmeister AR, MacCarty RL, et al (1987) Simplified assessment of segmental colonic transit. *Gastroenterology* 92:40-47.
9. Sugawara SA, Tamai UR, Takada H (2002) Innate immune responses in oral mucosa. *Journal of Endotoxin Research,* 8:465-468.
10. Uematsu A, Tsurugizawa T, Uneyama H, Torii K (2010) Brain-gut communication via vagus nerve modulates conditioned flavor preference. *European Journal of Neuroscience* 31:1136-1143.
11. Walker DM (2004) Oral mucosal immunology: An overview. *Annals of Academy of Medicine Singapore*, 33(Supplement):27S-30S.

4.3.1. REDOX and Gastrointestinal (GI) Health

1. Aw TY (2005) Intestinal GSH: Determinant of mucosal peroxide transport, metabolism and oxidative susceptibility. *Toxicology & Applied Pharmacology* 204:320-328.
2. Biasi F, Leonarduzzi G, Oteiza PI, Poli G (2013) Inflammatory Bowel Disease: Mechanisms, Redox Considerations, and Therapeutic Targets. *Antioxidants & Redox Signal* [Epub ahead of print] PMID:23305298.
3. Circu ML, Aw TY (2011) Redox biology of the intestine. *Free Radical Research* 45(11-12):1245-1266.
4. Circu ML, Aw TY (2012) Intestinal redox biology and oxidative stress. *Seminars in Cell Development Biology* 23(7):729-737.
5. Halliwell B, Gutteridge JMC (2007) *Free Radicals in Biology and Medicine* (4th ed). New York: Oxford University Press. ISBN:978-0-19-856868-1
6. Nagler RM, Klein I, Zarzhevsky N, Drigues N, Reznick AZ (2002) Characterization of the differentiated antioxidant profile of human saliva. *Free Radical Biology & Medicine* 32(3):268-277.

4.3.2. Oxidative Stress and Inflammatory Bowel Disease (IBD)

1. Banan A, Choudhary S, Zhang Y, et al (2000) Oxidant-induced intestinal barrier disruption and it's prevention by epidermal growth factor and transforming growth factor-α in a human colonic cell line:Role of the microtubule cytoskeleton. *Free Radical Biology & Medicine* 28:727-738.
2. Grisham MB, MacDermott RP, Deitch EA (1992) Oxidant defense mechanisms in the human colon. *Inflammation* 6:669-680.
3. Iantomasi T, Marraccini P, Favilli F, et al (1994) Glutathione metabolism in Crohn's disease. *Biochemistry Medicine & Metabolic Biology* 53:87-91.
4. Iborra M, Moret I, Rausell F, Bastida G, et al (2011) Role of oxidative stress and antioxidant enzymes in Crohn's disease. *Biochemical Society Transactions* 39(4):1102-1106.
5. Keshavarzian A, Banan A, Farhadi A, Komanduri S, et al (2003) Increases in free radicals and cytoskeletal protein oxidation and nitration in the colon of patients with inflammatory bowel disease. *Gut* 52:720-728.
6. Keshavarzian A, Sedghi S, Kanofsky J, et al (1992) Excessive production of reactive oxygen metabolites by inflamed colon: Analysis by chemiluminescence probe. *Gastroenterology* 103:177-185.
7. Krzystek-Korpacka M, Neubauer K, Berdowska I, Boehm D, et al (2008) Enhanced formation of ad vanced oxidation protein products in IBD. *Inflammatory Bowel Diseases* 14(6):794-802.
8. Mahida YR,Wu KC, Jewell DP (1989) Respiratory burst activity of intestinal macrophages in normal and inflammatory bowel disease. *Gut* 30:1362-1370.
9. Maor I, Rainis T, Lanir A, Lavy A (2008) Oxidative stress, inflammation and neutrophil superoxide release in patients with Crohn's disease: distinction between active and non-active disease. *Digestive Disease Science* 53(8):2208-2214.
10. Ramchilewitz D, Stamler JS, Bachwich D, et al (1995) Enhanced colonic nitric oxide generation and nitric oxide synthase activity in Ulcerative colitis and Crohn's disease. *Gut* 36:718-123.
11. van der Vliet A, Bast A (1992) Role of reactive oxygen species in intestinal diseases. *Free Radical Biology & Medicine* 12:499-513.
12. Williams JG (1990) Phagocytes, toxic oxygen metabolites and inflammatory bowel disease: implications for treatment. *Annals of Royal College of Surgery England* 72:253-262.
13. Zhu H, Li YR (2012) Oxidative stress and redox signaling mechanisms of inflammatory bowel disease: updated experimental and clinical evidence. *Experimental Biology & Medicine (Maywood)* 237(5):474-480.

4.3.3. Oxidative Stress and Celiac Disease

1. Beckett CG, Dell'Olio D, Ellis HJ, Rosen-Bronson S, Ciclitira PJ (1998)The detection and localization of inducible nitric oxide synthase production in the small intestine of patients with celiac disease. *European Journal of Gastroenterology & Hepatology* 10:641-647.
2. Chiurchiù V, Maccarrone M (2011) Chronic inflammatory disorders and their redox control: from mo lecular mechanisms to therapeutic opportunities. *Antioxidants & Redox Signal* 15(9):2605-2641.
3. Daniels I, Cavill D, Murray IA, Long RG (2005) Elevated expression of iNOS mRNA and protein in celiac disease. *Clinica Chimica Acta* 356:134-142.
4. Dugas B, Dugas N, Conti M, Calenda A et al (2003) Wheat gliadin promotes the interleukin-4-induced IgE production by normal human peripheral mononuclear cells through a redox-dependent mechanism. *Cytokine* 21:270-280.
5. Ferretti G, Bacchetti T, Masciangelo S, Saturni L (2012) Celiac disease, inflammation and oxidative damage: A nutrigenetic approach. *Nutrients* 4:243-257.
6. Holmgren PK, Falth-Magnusson K, Magnusson KE, Stenhammar L, Sundqvist T (1998) Children with celiac disease express inducible nitric oxide synthase in the small intestine during gluten challenge. *Scandinavian Journal of Gastroenterology* 33:939-943.
7. Murray IA, Daniels I, Coupland K, Smith JA, Long RG (2002) Increased activity and expression of iNOS in human duodenal enterocytes from patients with celiac disease. *American Journal of Physiology – Gastrointestinal & Liver Physiology* 283:319-326.
8. Odetti P, Valentini S, Aragno I, Garibaldi S, et al (1998) Oxidative stress in subjects affected by celiac disease. *Free Radical Research* 29(1):17-24.
9. Sido B, Hack V, Hochlehnert A, Lipps H, Herfarth C, Droge W (1998) Impairment of intestinal glutathi one synthesis in patients with inflammatory bowel disease. *Gut* 42:485-492.
10. Stojiljkovic V, Pejic S, Kasapovic J, Gavrilovic L, et al (2012) Glutathione redox cycle in small intestinal mucosa and peripheral blood of pediatric celiac disease patients. *Annals of the Brazilian Academy of Sciences* 84(1):175-84.

11. Szaflarska-Poplawska A, Siomek A, Czerwionka-Szaflarska M, Gackowski D, et al (2010) Oxidatively damaged DNA/oxidative stress in children with celiac disease. *Cancer Epidemiological Biomarkers & Prevention* 19(8):1960-1965.
12. Van Heel DA, West J (2006) Recent advances in celiac disease. *Gut* 36:864-874.

5.3.4. Oxidative Stress and Gastric Disorders

1. Bagchi D, Bhattacharya G, Stohs SJ (1996) Production of reactive oxygen species by gastric cells in association with *Helicobacter pylori*. *Free Radical Research* 24:439-450.
2. Blaser MJ (1990) *Helicobacter pylori* and the pathogenesis of gastro-duodenal inflammation. *Journal of Infectious Diseases* 161:626-633
3. Kinjo N, Kawanaka H, Akahoshi T, et al (2008) Significance of ERK nitration in portal hypertensive gas tropathy and its therapeutic implications. *American Journal of Physiology: Gastrointestinal & Liver Physiology* 295:G1016-G1024.
4. Maity P, Bindu S, Choubey V, et al (2008) Lansoprazole protects and heals gastric mucosa from non-steroidal anti-inflammatory drug (NSAID)-induced gastropathy by inhibiting mitochondrial as well as Fas-mediated death pathways with concurrent induction of mucosal cell renewal. *Journal of Biological Chemistry* 283:14391-14401.
5. Marshall B (2002) *Helicobacter pylori*: 20 years on. *Clinical Medicine* 2(2):147-152.
6. Olczak AA, Olson JW, Maier RJ (2002) Oxidative-stress resistance mutants of *Helicobacter pylori*. *Journal of Bacteriology* 184:3186-3193.
7. Oshima T, Miwa H, Joh T (2008) Aspirin induces gastric epithelial barrier dysfunction by activating p38 MAPK via claudin-7. *American Journal of Physiology: Cell Physiology* 295:C800-C806.
8. Pan JS, He SZ, Xu HZ, et al (2008) Oxidative stress disturbs energy metabolism of mitochondria in ethanol-induced gastric mucosa injury. *World Journal of Gastroenterology* 14:5857-5867.
9. Parsonnet J, Friedman GD, Vandersteen DP, Chang Y, et al (1991) *Helicobacter pylori* infection and gastric carcinoma. *New England Journal of Medicine* 325:1127-1131.
10. Suzuki H, Nishizawa T, Tsugawa H, Mogami S, Hibi T (2011) Role of oxidative stress in stomach disorders. *Journal of Clinical Biochemistry & Nutrition* 50(1):35-39.
11. Suzuki M, Miura S, Mori M, et al (1994) Rebamipide, a novel anti-ulcer agent, attenuates *Helicobacter pylori* induced gastric mucosal cell injury associated with neutrophil derived oxidants. *Gut* 35:1375-1378.
12. Suzuki M, Miura S, Suematsu M, et al (1992) *Helicobacter pylori*-associated ammonia production enhances neutrophil-dependent gastric mucosal cell injury. *American Journal of Physiology* 263:G719-G725.
13. Wang G, Alamuri P, Maier RJ (2006) The diverse antioxidant systems of *Helicobacter pylori*. *Molecular Microbiology* 61:847-860.

4.4 | REDOX: The Respiratory System

4.4.1. REDOX and Respiratory Health

1. Auerbach A, Hernandez ML (2012) The effect of environmental oxidative stress on airway inflammation. *Current Opinion in Allergy & Clinical Immunology* 12(2):133-139.
2. Biswas SK, Rahman I (2009) Environmental toxicity, redox signaling and lung inflammation: the role of glutathione. *Molecular Aspects of Medicine* 30(1-2):60-76.
3. Ciencewicki J, Trivedi S, Kleeberger SR (2008) Oxidants and the pathogenesis of lung diseases. *Journal of Allergy & Clinical Immunology* 122: 456-468.
4. Comhair SA, Erzurum SC (2002) Antioxidant responses to oxidant-mediated lung diseases. *American Journal of Physiology: Lung Cellular & Molecular Physiology* 283: L246–255.
5. Fridovich I, Freeman B (1986) Antioxidant defenses in the lung. *Annual Reviews in Physiology* 48:693-702.
6. Gaston B, Drazen JM, Loscalzo J, Stamler JS (1994) The biology of nitrogen oxides in the airways. *American Journal of Respiratory Critical Care Medicine* 149: 538-551.
7. Kinnula VL (2005) Focus on antioxidant enzymes and antioxidant strategies in smoking related airway diseases. *Thorax* 60:693-700.
8. Koren HS, Bromberg PA (1995) Respiratory responses of asthmatics to ozone. *International Archives in Allergy & Immunology* 107: 236-238.
9. Montuschi P, Barnes PJ (2006) Isoprostanes and asthma. *Drug Discovery Today: Therapeutic Strategies* 3:287-292.

10. Peden DB, Boehlecke B, Horstman D, Devlin R (1997) Prolonged acute exposure to 0.16ppm ozone induces eosinophilic airway inflammation in asthmatic subjects with allergies. *Journal of Allergy & Clinical Immunology* 100: 802-808.
11. Van der Toorn M, Rezayat D, Kauffman HF, Bakker SJ, et al (2009) Lipid-soluble components in cigarette smoke induce mitochondrial production of reactive oxygen species in lung epithelial cells. *American Journal of Physiology: Lung Cellular & Molecular Physiology* 297(1):L109-114.

4.4.2. Oxidative Stress and COPD

1. Barnes PJ, Ito K, Adcock IM (2004) Corticosteroid resistance in chronic obstructive pulmonary disease: inactivation of histone deacetylase. *Lancet* 363:731-733.
2. Church DF, Pryor WA (1985) Free-radical chemistry of cigarette smoke and its toxicological implications. *Environmental Health Perspective* 64:111-126.
3. De Boer WI, Yao H, Rahman I (2007) Future therapeutic treatment of COPD: struggle between oxidants and cytokines. *International Journal of Chronic Obstructive Pulmonary Disease* 2:205-228.
4. Eiserich JP, van der Vliet A, Handelman GJ, Halliwell B, Cross CE (1995) Dietary antioxidants and cigarette smoke-induced biomolecular damage: a complex interaction. *American Journal of Clinical Nutrition* 62(6 suppl):1490S-1500S.
5. Ferrer E, Peinado VI, Diez M, Carrasco JL, et al (2009) Effects of cigarette smoke on endothelial function of pulmonary arteries in the guinea pig. *Respiratory Research* 10:76-87.
6. Kinnula VL (2005) Focus on antioxidant enzymes and antioxidant strategies in smoking related airway diseases. *Thorax* 60:693-700.
7. MacNee W (2001) Oxidants/antioxidants and chronic obstructive pulmonary disease: pathogenesis to therapy. *Novartis Foundation Symposium* 234:169-185.
8. Rahman I, Adcock IM (2006) Oxidative stress and redox regulation of lung inflammation in COPD. *European Respiratory Journal* 28(1):219-242.
9. Rahman I, van Schadewijk AA, Crowther AJ, Hiemstra PS, et al (2002) 4- Hydroxy-2-nonenal, a specific lipid peroxidation product, is elevated in lungs of patients with chronic obstructive pulmonary disease. *American Journal of Respiratory Critical Care Medicine* 166:490-495.
10. Ryter SW, Chen ZH, Kim HP, Choi AM (2009) Autophagy in chronic obstructive pulmonary disease: homeostatic or pathogenic mechanism? *Autophagy* 5:235-237.
11. Scherz-Shouval R, Shvets E, Elazar Z (2007) Oxidation as a post-translational modification that regulates autophagy. *Autophagy* 3:371-373.
12. Scherz-Shouval R, Shvets E, Fass E, Shorer H, et al (2007) Reactive oxygen species are essential for autophagy and specifically regulate the activity of Atg4. *EMBO Journal* 26:1749-1760.
13. Sethi JM, Rochester CL (2000) Smoking and chronic obstructive pulmonary disease. Clinical Chest *Medicine* 21:67–86.
14. Tomaki M, Sugiura H, Koarai A, Komaki Y, et al (2007) Decreased expression of antioxidant enzymes and increased expression of chemokines in COPD lung. *Pulmonary Pharmacology & Therapeutics* 20:596-605.
15. Tuder RM, Petrache I (2012) Pathogenesis of chronic obstructive pulmonary disease. *Journal of Clinical Investigation* 122(8):2749–2755.
16. Yoshida T, Tuder RM (2007) Pathobiology of cigarette smoke-induced chronic obstructive pulmonary disease. *Physiological Reviews* 87(3):1047-1082.

4.4.3. Oxidative Stress and Asthma

1. Bucchieri F, Puddicombe SM, Lordan JL, Richter A, et al (2002) Asthmatic bronchial epithelium is more susceptible to oxidantinduced apoptosis. *American Journal of Respiratory Cell Molecular Biology* 27:179-185.
2. Calhoun WJ, Reed HE, Moest DR, Stevens CA (1992) Enhanced superoxide production by alveolar macrophages and air-space cells, airway inflammation, and alveolar macrophage density changes after segmental antigen bronchoprovocation in allergic subjects. *American Review of Respiratory Diseases* 145:317-325.
3. Caramori G, Papi A (2004) Oxidants and asthma. *Thorax* 59(2):170-173.
4. Comhair SA, Bhathena PR, Dweik RA, Kavuru M, Erzurum SC (2000) Rapid loss of superoxide dis mutase activity during antigen-induced asthmatic response. *Lancet* 355:624.
5. Comhair SA, Bhathena PR, Farver C, Thunnissen FB, Erzurum SC (2001) Extracellular glutathione peroxidase induction in asthmatic lungs: Evidence for redox regulation of expression in human airway epithelial cells. *FASEB Journal* 15:70-78.

6. Comhair SA, Erzurum SC (2010) Redox control of asthma: molecular mechanisms and therapeutic opportunities. *Antioxidants & Redox Signal* 12(1):93-124.
7. Dworski R (2000) Oxidant stress in asthma. *Thorax* 55 (Suppl 2):S51-53.
8. Gaston B, Sears S, Woods J, Hunt J, et al (1998) Bronchodilator S-nitrosothiol deficiency in asthmatic respiratory failure. *Lancet* 351: 1317–1319.
9. Ghosh S, Erzurum SC (2011) Nitric oxide metabolism in asthma pathophysiology. *Biochimica et Bio physica Acta* 1810(11):1008-1016.
10. Ghosh S, Masri F, Comhair S, Andreadis A, et al (2003) Nitration of proteins in murine model of asthma. *American Journal of Respiratory Critical Care Medicine* 167:A889, 2003.
11. Jatakanon A, Uasuf C, Maziak W, Lim S, et al (1999) Neutrophilic inflammation in severe persistent asthma. *American Journal of Respiratory Critical Care Medicine* 160:1532-1539.
12. McCreanor J, Cullinan P, Nieuwenhuijsen MJ, Stewart- Evans J, et al (2007) Respiratory effects of exposure to diesel traffic in persons with asthma. *New England Journal of Medicine* 357:2348–2358.
13. O'Sullivan MP, Tyner JW, Holtzman MJ (2003) Apoptosis in the airways: Another balancing act in the epithelial program. *American Journal of Respiratory Cell Molecular Biology* 29:3-7.
14. Sahiner UM, Birben E, Erzurum S, Sackesen C, Kalayci O (2011) Oxidative stress in asthma. *World Allergy Organization Journal* 4(10):151-158.

4.4.4. Oxidative Stress and Cystic Fibrosis (CF)

1. Basbaum CB, Berthold J, Finkbeiner WE (1990) The serous cell. *Annual Review of Physiology* 52:97–113.
2. Chapman AL, Morrissey BM, Vasu VT, Juarez MM, et al (2010) Myeloperoxidase-dependent oxidative metabolism of nitric oxide in the cystic fibrosis airway. *Journal of Cystic Fibrosis* 9(2):84-92.
3. Cowley EA, Linsdell P (2002) Oxidant stress stimulates anion secretion from the human airway epithelial cell line Calu-3: implications for cystic fibrosis lung disease. *Journal of Physiology* 543(1):201-209.
4. Folkerts G, Kloek J, Muijsers RB, Nijkamp FP (2001) Reactive nitrogen and oxygen species in airway inflammation. *European Journal of Pharmacology* 429:251-262.
5. Forteza R, Salathe M, Miot F, Conner GE (2005) Regulated hydrogen peroxide production by Duox in human airway epithelial cells. *American Journal of Respiratory Cell Molecular Biology* 32(5):462-469.
6. Grasemann H, Michler E, Wallot M, Ratjen F (1997) Decreased concentration of exhaled nitric oxide (NO) in patients with cystic fibrosis. *Pediatric Pulmonology* 24(3):173-177.
7. Kerem B, Rommens JM, Buchanan JA, Markiewicz D, et al (1989) Identification of the cystic fibrosis gene: genetic analysis. *Science* 245(4922):1073-1080.
8. Meng QH, Springall DR, Bishop AE, Morgan K, et al (1998) Lack of inducible nitric oxide synthase in bronchial epithelium: a possible mechanism of susceptibility to infection in cystic fibrosis. *Journal of Pathology* 184(3):323-331.
9. Pilewski JM, Frizzell RA (1999) Role of CFTR in airway disease. *Physiological Reviews* 79:S215–255.
10. Rottner M, Freyssinet JM, Martínez MC (2009) Mechanisms of the noxious inflammatory cycle in cystic fibrosis. *Respiratory Research* 10:23.
11. Rottner M, Tual-Chalot S, Mostefai HA, Andriantsitohaina R, et al (2011) Increased oxidative stress induces apoptosis in human cystic fibrosis cells. *PLoS One* 6(9):e24880.
12. Van Der Vliet A, Cross CE (2000) Oxidants, nitrosants, and the lung. *American Journal of Medicine* 109:398-421.
13. Wine JJ (1999) The genesis of cystic fibrosis lung disease. *Journal of Clinical Investigation* 103:309–312.

4.5 | REDOX: The Immune System

4.5.1. REDOX and Immune Regulation

1. Angelini G, Gardella S, Ardy M, Ciriolo MR, et al (2002) Antigen-presenting dendritic cells provide the reducing extracellular microenvironment required for T lymphocyte activation. *Proceedings of the National Academy of Sciences USA* 99:1491-1496.
2. Castellani P, Angelini G, Delfino L, Matucci A, Rubartelli A (2008) The thiol redox state of lymphoid organs is modified by immunization: role of different immune cell populations. *European Journal of Immunology* 38:2419-2425.

3. Herzenberg LA, De Rosa SC, Dubs JG, Roederer M, et al (1997) Glutathione deficiency is associated with impaired survival in HIV disease. *Proceedings of the National Academy Sciences USA* 94:1967-1972.
4. Markovic J, Borras C, Ortega A, Sastre J, et al (2007) Glutathione is recruited into the nucleus in early phases of cell proliferation. *Journal of Biological Chemistry* 282:20416-20424.
5. Masciarelli S, Sitia R (2008) Building and operating an antibody factory: redox control during B to plasma cell terminal differentiation. *Biochimica et Biophysica Acta* 1783:578-588.
6. Matsue H, Edelbaum D, Shalhevet D, Mizumoto N, et al (2003) Generation and function of reactive oxygen species in dendritic cells during antigen presentation. *Journal of Immunology* 171:3010-3018.
7. Menon SG, Goswami PC (2007) A redox cycle within the cell cycle: ring in the old with the new. *Oncogene* 26:1101-1109.
8. Pani G, Colavitti R, Borrello S, Galeotti T (2000) Redox regulation of lymphocyte signaling. *IUBMB Life* 49:381-389.
9. Schwertassek U, Balmer Y, Gutscher M, Weingarten L, et al (2007) Selective redox regulation of cytokine receptor signaling by extracellular thioredoxin-1. *EMBO Journal* 26:3086-3097.
10. Tse HM, Milton MJ, Schreiner S, Profozich JL, et al (2007) Disruption of innate-mediated proinflam matory cytokine and reactive oxygen species third signal leads to antigen-specific hyporesponsive ness. *Journal of Immunology* 178:908-917.
11. Yan Z, Banerjee R (2010) Redox remodeling as an immunoregulatory strategy. *Biochemistry* 49(6):1059.
12. Yan Z, Garg SK, Kipnis J, Banerjee R. Extracellular redox modulation by regulatory T cells. *Nature Chemical Biology* 5:721-723.

4.5.2. Oxidative Stress and HIV / AIDS

1. Aquaro S, Muscoli C, Ranazzi A, Pollicita M, et al (2007) The contribution of peroxynitrite generation in HIV replication in human primary macrophages. *Retrovirology* 4:76.
2. Aukrust P, Muller F, Svardal AM, Ueland T, et al (2003) Disturbed glutathione metabolism and decreased antioxidant levels in human immunodeficiency virus-infected patients during highly active antiretroviral therapy – potential immune-modulatory effects of antioxidants. *Journal of Infectious Diseases* 188:232-238.
3. Carr A (2003) Cardiovascular risk factors in HIV-infected patients. *Journal of AIDS* 34(1):S73–S78.
4. Droge W, Eck HP, Mihm S (1994) Oxidant-antioxidant status in human immunodeficiency virus infection. *Methods in Enzymology* 233:594-601.
5. Garaci E, Palamara AT, Circolo MR, D'Agostini C, et al (1997) Intra-cellular GSH content and HIV replication in human macrophages. *Journal of Leukocyte Biology* 62:54-59.
6. Hulgan T, Morrow J, D'Aquila RT, et al (2003) Oxidant stress is increased during treatment of human immunodeficiency virus infection. *Clinical Infectious Diseases* 37(12):1711-1717.
7. Kline ER, Kleinhenz DJ, Liang B, Dikalov S, et al (2007) Vascular oxidative stress and nitric oxide depletion in HIV-1 transgenic rats are reversed by glutathione restoration. *American Journal of Physiology: Heart Circulation Physiology* 294(6):H2792-2804.
8. Kruman II, Nath A, Mattson MP (1998) HIV-1 protein 'tat' induces apoptosis of hippocampal neurons by a mechanism involving caspase activation, calcium overload, and oxidative stress. *Experimental Neurology* 154(2): 276-288.
9. Montagnier L (2010). 25 years after HIV discovery: prospects for cure and vaccine. *Virology* 397(2):248-254.
10. Olinski R, Gackowski D, Foksinski M, Rozalski R, et al (2002) Oxidative DNA damage: assessment of the role in carcinogenesis, atherosclerosis, and acquired immunodeficiency syndrome. *Free Radical Biology & Medicine* 33:192-200.
11. Palamara AT, Perno CF, Aquaro S, Buè MC, (1996) Glutathione inhibits HIV replication by acting at late stages of the virus life cycle. *AIDS Research in Human Retroviruses* 12:1537-1541.
12. Porter K (2000) Survival after introduction of HAART in people with known duration of HIV-1 infection. *The Lancet* 355(9210):1158-1159.
13. Schreck R, Rieber P, Baeuerle PA (1991) Reactive oxygen intermediates as apparently widely used messengers in the activation of the NF-κB transcription factor and HIV-1. *The EMBO Journal* 10(8):2247–2258.
14. Tang AM, Lanzillotti J, Hendricks K, et al (2005) Micronutrients: current issues for HIV care providers. *AIDS* 19(9):847-861.

4.5.3. Oxidative Stress and Sepsis

1. Brealey D, Brand M, Hargreaves I, Heales S, et al (2002) Association between mitochondrial dysfunction and severity and outcome of septic shock. *The Lancet* 360:219-223.
2. Bullen J, Griffiths E, Rogers H, Ward G (2000) Sepsis: the critical role of iron. *Microbes & Infection* 2:409-415.
3. Cavaillon JM, Annane D (2006) Compartmentalization of the inflammatory response in sepsis and SIRS. *Journal of Endotoxin Research* 12:151-170.
4. Crimi E, Sica V, Slutsky AS, Zhang H, et al (2006) Role of oxidative stress in experimental sepsis and multisystem organ dysfunction. *Free Radical Research* 40:665-672.
5. Fink MP (2001) Cytopathic hypoxia: mitochondrial dysfunction as mechanism contributing to organ dysfunction in sepsis. *Critical Care Clinic* 17:219-237.
6. Fink MP (2007) Ethyl pyruvate: a novel treatment for sepsis. *Current Drug Targets* 8:515-518.
7. Galley HF (2010) Bench-to-bedside review: targeting antioxidants to mitochondria in sepsis. *Critical Care* 14:230.
8. Hotchkiss RS, Karl IE (2003) The pathophysiology and treatment of sepsis. *New England Journal of Medicine* 348:138-150.
9. Jean-Baptiste, E (2007) Cellular mechanisms in sepsis. *Journal of Intensive Care Medicine* 22:63-72.
10. King C, Tytgat S, Delude RL, Fink MP (1999) Ileal mucosal oxygen consumption is decreased in endotoxaemic rats but is restored toward normal by treatment with aminoguanidine. *Critical Care Medicine* 27:2518-2524.
11. Lange M, Connelly R, Traber DL, Hamahata A, et al (2010) Time course of nitric oxide synthases, nitrosative stress, and poly(ADP ribosylation) in an ovine sepsis model. *Critical Care* 14:R129.
12. O'Brien JM, Ali NA, Aberegg SK, Abraham E (2007) Sepsis. *American Journal of Medicine* 120:1012-1022.
13. Riedemann NC, Guo RF, Ward PA (2003) The enigma of sepsis. *Journal of Clinical Investigation* 112:460-467.
14. Russell JA (2006) Management of sepsis. *New England Journal of Medicine* 355:1699-1713.
15. Taylor DE, Piantadosi CA (1995) Oxidative metabolism in sepsis and sepsis syndrome. *Journal of Critical Care* 10:122-135.
16. Victor VM, Rocha M, De la Fuente M (2004) Immune cells: Free radicals and antioxidants in sepsis. *International Immunopharmacology* 4:327-347.

4.6 | REDOX: The Circulatory System

4.6.1. REDOX and Cardiovascular (CV) Health

1. Ago T, Liu T, Zhai P, Chen W, et al (2008) A redox-dependent pathway for regulating class II HDACs and cardiac hypertrophy. *Cell* 133:978-993.
2. Ago T, Sadoshima J (2007) Thioredoxin1 as a negative regulator of cardiac hypertrophy. *Antioxidant & Redox Signaling* 9:679-687.
3. Baines CP, Goto M, Downey JM (1997) Oxygen radicals released during ischemic preconditioning contribute to cardioprotection in the rabbit myocardium. *Journal of Molecular & Cellular Cardiology* 29:207-216.
4. Fu XY, Kassim SY, Parks WC, Heinecke JW (2001) Hypochlorous acid oxygenates the cysteine switch domain of pro-matrilysin (MMP-7). A mechanism for matrix metalloproteinase activation and atherosclerotic plaque rupture by myeloperoxidase. *Journal of Biological Chemistry* 276:41279-41287.
5. Fuster V, Narula J (2012) *Coronary Risk Factors Update.* Vol. 90(1) Medical Clinics of North America. ISSN:0025-7125.
6. Kuroda J, Ago T, Matsushima S, Zhai PY, et al (2010) NADPH oxidase 4 (Nox4) is a major source of oxidative stress in the failing heart. *Proceedings of the Natiional Academy of Sciences USA* 107:15565-15570.
7. Linke K, Jakob U (2003) Not every disulfide lasts forever: Disulfide bond formation as a redox switch. *Antioxidant & Redox Signaling* 5:425-434.
8. Shao D, Oka S, Brady CD, Haendeler J et al (2012) Redox modification of cell signaling in the cardio vascular system. *Journal of Molecular & Cellular Cardiology* 52(3):550–558.
9. Shen X, Zheng SR, Metreveli NS, Epstein PN (2006) Protection of cardiac mitochondria by over-ex pression of MnSOD reduces diabetic cardiomyopathy. *Diabetes* 55:798-805.

10. Sun JH, Murphy E (2010) Protein S-Nitrosylation and Cardioprotection. *Circulation Research* 106:285-296.

4.6.2. Oxidative Stress and Atherosclerosis

1. Ballinger SW, Patterson C, Knight-Lozano CA, Burow DL, et al (2002) Mitochondrial integrity and function in atherogenesis. *Circulation* 106:544-549.
2. Brown MS, Goldstein JL (1983) Lipoprotein metabolism in the macrophage: implications for cholesterol deposition in atherosclerosis. *Annal Reviews in Biochemistry* 52:223-261.
3. Carreira RS, Facundo HT, Kowaltowski AJ (2005) Mitochondrial K+ transport and cardiac protection during ischemia/reperfusion. *Brazilian Journal of Medicine & Biological Research* 38:345-352.
4. Chang JC, Kou SJ, Lin WT, Liu CS (2010) Regulatory role of mitochondria in oxidative stress and atherosclerosis. *World Journal of Cardiology* 2(6):150-159.
5. Delafontaine P, Ku L (1997) Reactive oxygen species stimulate insulin-like growth factor I synthesis in vascular smooth muscle cells. *Cardiovascular Research* 33(1):216-222.
6. Galis ZS, Khatri JJ (2002) Matrix metalloproteinases in vascular remodeling and atherogenesis: the good, the bad, and the ugly. *Circulation Research* 90: 51-262.
7. Heinecke JW (1997) Mechanisms of oxidative damage of low density lipoprotein in human atherosclerosis. *Current Opinion in Lipidology* 8(5):268-274.
8. Irani K (2000) Oxidant signaling in vascular cell growth, death, and survival: a review of the roles of reactive oxygen species in smooth muscle and endothelial cell mitogenic and apoptotic signaling. *Circulation Research* 87:179-183.
9. Madamanchi NR, Vendrov A, Runge MS (2005) Oxidative stress and vascular disease. *Arteriosclerosis, Thrombosis & Vascular Biology* 25(1):29-38.
10. Minamiyama Y, Bito Y, Takemura S, Takahashi Y, et al (2007) Calorie restriction improves cardiovas cular risk factors via reduction of mitochondrial reactive oxygen species in type II diabetic rats. *Journal of Pharmacology & Experimental Therapeutics* 320:535-543.
11. Ross R (1999) Atherosclerosis - An Inflammatory Disease. *New England Journal of Medicine* 340:115-126.
12. Victor VM, Rocha M (2007) Targeting antioxidants to mitochondria: a potential new therapeutic strategy for cardiovascular diseases. *Current Pharmaceutical Design* 13:845-863.
13. Victor VM, Rocha M, Sola E, Banuls C, et al (2009) Oxidative stress, endothelial dysfunction and atherosclerosis. *Current Pharmaceutical Design* 15:2988-3002.
14. Virmani R, Narula J, Leon MB, Willerson JT (2007) *The Vulnerable Atherosclerotic Plaque: Strategies for Diagonsis and Management.* Massachusetts:BlackWell Publishing. ISBN: 13-978-1405-158596.
15. Wald NJ, Law MR (2003). A strategy to reduce cardiovascular disease by more than 80%. *British Medical Journal* 326(7404):1419.

4.6.3. Oxidative Stress and Hypertension

1. Cai H, Harrison DG (2000) Endothelial dysfunction in cardiovascular diseases: the role of oxidant stress. *Circulation Research* 87:840-844.
2. Celermajer DS, Sorensen KE, Georgakopoulos D, Bull C, et al (1993) Cigarette smoking is associated with dose-related and potentially reversible impairment of endothelium-dependent dilation in healthy young adults. *Circulation* 88:2149-2155.
3. Heitzer T, Schlinzig T, Krohn K, Meinertz T, Munzel T (2001) Endothelial dysfunction, oxidative stress and risk of cardiovascular events in patients with coronary artery disease. *Circulation* 104:2673-2678.
4. Liu Y, Zhao H, Li H, Kalyanaraman B, et al (2003) Mitochondrial sources of H_2O_2 generation play a key role in flow-mediated dilation in human coronary resistance arteries. *Circulation Research* 93:573-580.
5. Paravicini TM, Chrissobolis S, Drummond GR, Sobey CG (2004) Increased NAD(P)H-oxidase activity and Nox4 expression during chronic hypertension is associated with enhanced cerebral vasodilatation to NAD(P)H in vivo. *Stroke* 35:584-589.
6. Paravicini TM, Touyz RM (2006) Redox signaling in hypertension. *Cardiovascular Research* 71:247-258
7. Redon J, Oliva MR, Tormos C, Giner V, et al (2003) Antioxidant activities and oxidative stress by products in human hypertension. *Hypertension* 41:1096-1101.
8. Sakai T, Inoue S, Matsuyama TA, Takei M, et al (2009) Eosinophils may be involved in thrombus growth in acute coronary syndrome. *International Heart Journal* 50:267-277.
9. Vasquez-Vivar J, Kalyanaraman B, Martasek P, Hogg N, et al (1998) Superoxide generation by endothelial nitric oxide synthase: the influence of cofactors. *Proceedings of the Nationale Academy of Sciences USA* 95:9220-9225.

10. Zeiher AM, Drexler H, Wollschlager H, Just H (1991) Endothelial dysfunction of the coronary microvas culature is associated with coronary blood flow regulation in patients with early atherosclerosis. *Circulation* 84:1984-1992.

4.6.4. Oxidative Stress and Heart Failure

1. Braunwald E (2008) Biomarkers in heart failure. *New England Journal of Medicine* 358:2148-2159.
2. Doehner W, Schoene N, Rauchhaus M, Leyva-Leon F, et al (2002) Effects of xanthine oxidase inhibition with allopurinol on endothelial function and peripheral blood flow in hyperuricemic patients with chronic heart failure: results from 2 placebo-controlled studies. *Circulation* 105:2619-2624.
3. Harrison D, Griendling KK, Landmesser U, Hornig B, Drexler H (2003) Role of oxidative stress in atherosclerosis. *American Journal of Cardiology* 91:7A-11A.
4. Hill MF, Singal PK (1997) Right and left myocardial antioxidant responses during heart failure subsequent to myocardial infarction. *Circulation* 96:2414-2420, 1997.
5. Li JM, Shah AM (2003) Mechanism of endothelial cell NADPH oxidase activation by angiotensin II. Role of the p47phox subunit. *Journal of Biological Chemistry* 278:12094-12100.
6. Lonn E, Bosch J, Yusuf S, Sheridan P, et al (2005) Effects of long-term vitamin E supplementation on cardiovascular events and cancer: a randomized controlled trial. *JAMA* 293:1338-1347.
7. Mallat Z, Philip I, Lebret M, Chatel D, Maclouf J, Tedgui A (1998) Elevated levels of 8-iso-prostaglandin F2alpha in pericardial fluid of patients with heart failure: a potential role for in vivo oxidant stress in ventricular dilatation and progression to heart failure. *Circulation* 97: 1536-1539.
8. Sawyer DB, Colucci WS (2000) Mitochondrial oxidative stress in heart failure: "oxygen wastage" revisited. *Circulation Research* 86:119-120.
9. Takimoto E, Kass DA (2007) Role of oxidative stress in cardiac hypertrophy and remodeling. *Hypertension* 49: 241-248.
10. Tsutsui H, Kinugawa S, Matsushima S (2011) Oxidative stress and heart failure. *American Journal of Physiology - Heart & Circulatory Physiology* 301:H2181-H2190.
11. Ukai T, Cheng CP, Tachibana H, Igawa A, et al (2001) Allopurinol enhances the contractile response to dobutamine and exercise in dogs with pacing-induced heart failure. *Circulation* 103:750-755.
12. Young JB, Narula J (2008) Heart Failure Vol. 26(1) New York:Saunders. ISSN: 0733-8651

4.6.5. The REDOX Factor: Myocardial Reperfusion Injury

1. Azevedo LC, Pedro MA, Souza LC, de Souza HP, et al (2000) Oxidative stress as a signaling mechanism of the vascular response to injury: the redox hypothesis of restenosis. *Cardiovascular Research* 47(3):436-445.
2. Demircan S, Yazici M, Diraman E, Demircan G, et al (2008) The effect of glucose-insulin-potassium treatment on myocardial oxidative stress in patients with acute coronary syndromes undergoing percutaneous coronary intervention. *Coronary Artery Disease* 19(2):99-104.
3. Guan W, Osanai T, Kamada T, Ishizaka H, et al (1999) Time course of free radical production after primary coronary angioplasty for acute myocardial infarction and the effect of vitamin C. *Japanese Circulation Journal* 63(12):924-928.
4. Imai K, Matsubara T, Kanashiro M, Ichimiya S, Hotta N (2001) Lipid peroxidation may predict restenosis after coronary balloon angioplasty. *Japanese Circulation Journal* 65(6):495-499.
5. Inoue T, Kato T, Hikichi Y, Hashimoto S, et al (2006) Stent-induced neutrophil activation is associated with an oxidative burst in the inflammatory process, leading to neointimal thickening. *Thrombosis Haemostasis* 95(1):43-48.
6. Juni RP, Duckers HJ, Vanhoutte PM, Virmani R, Moens AL (2013) Oxidative stress and pathological changes after coronary artery interventions. *Journal of American College of Cardiology* 61(14):1471-1481.
7. Kuo MD, Bright IJ, Wang DS, Ghafouri P, et al (2004) Local resistance to oxidative stress by over expression of copper-zinc superoxide dismutase limits neointimal formation after angioplasty. *Journal of Endovascular Therapy*11(6):585-594.
8. Wahlgren CM, Pekkari K (2005) Elevated thioredoxin after angioplasty in peripheral arterial disease. *European Journal of Vascular & Endovascular Surgery* 29(3):281-286.

4.7 | REDOX: The Nervous System

4.7.1. REDOX and Cognitive Health

1. Barnham KJ, Masters CL, Bush AI (2004) Neuro-degenerative diseases and oxidative stress. *Nature Reviews Drug Discovery* 3:205-214.
2. Beal MF (2002) Oxidatively modified proteins in aging and disease. *Free Radical Biology & Medicine* 32(9):797-803.
3. Benzi G, Moretti A (1995) Age- and peroxidative stress-related modifications of the cerebral enzymatic activities linked to mitochondria and the glutathione system. *Free Radical Biology & Medicine* 19(1):77-101.
4. Delacourte A, Defossez A, Ceballos I, Nicole A, Sinet PM (1988) Preferential localization of copper zinc superoxide dismutase in the vulnerable cortical neurons in Alzheimer's disease. *Neuroscience Letters* 92(3):247-253.
5. Friedlich AL, Beal MF (2000) Prospects for redox-based therapy in neurodegenerative diseases. *Neurotoxicity Research* 2:229-237.
6. Gerlach M, Ben-Shachar D, Riederer P, Youdim MB (1994) Altered brain metabolism of iron as a cause of neurodegenerative diseases? *Journal of Neurochemistry* 63(3):793-807.
7. Halliwell B, Jenner P (1998) Impaired clearance of oxidized proteins in neurodegenerative diseases. *Lancet* 351(9114):1510.
8. Mohanakumar KP, de Bartolomeis A, Wu RM, Yeh KJ, et al (1994) Ferrous-citrate complex and nigral degeneration: evidence for free-radical formation and lipid peroxidation. *Annals of New York Academy of Sciences* 738:392-399.
9. Perry G, Avila J, Espey MG, Wink DA, et al (2001)Biochemistry of neurodegeneration. *Science* 291:595-597.
10. Volicer L, Crino PB (1990) Involvement of free radicals in dementia of the Alzheimer type: a hypothesis. *Neurobiology of Aging* 11(5):567-571.

4.7.2. Oxidative Stress and Age-related Macular Degeneration (AMD)

1. Boulton M, Rozanowska M, Rozanowski B (2001) Retinal photodamage. *Journal of Photochemistry & Photobiology B*. 64:144-161.
2. Boulton ME (1991) Ageing of the retinal pigment epithelium. *Progress in Retina & Eye Research* 11:125-151.
3. Ding X, Patel M, Chan CC (2009) Molecular pathology of age-related macular degeneration. *Progress in Retinal & Eye Research* 28(1):1-18.
4. Feher J, Kovacs I, Artico M, Cavallotti C, et al (2006) Mitochondrial alterations of retinal pigment epithe lium in age-related macular degeneration. *Neurobiological Aging* 27:983-993.
5. Friedman DS, O'Colmain BJ, Munoz B, Tomany SC, et al (2004) Prevalence of age-related macular degeneration in the United States. *Archives in Ophthalmology* 122:564-572.
6. Gehrs KM, Anderson DH, Johnson LV, Hageman GS (2006) Age-related macular degeneration-emerg ing pathogenetic and therapeutic concepts. *Annals of Medicine* 38:450-471.
7. Hutchison J, Tay W (1875) Symmetrical central choroidoretinal disease occurring in senile persons. *Royal London Ophthalmology Hospital Reports* 8:231–244.
8. Kaarniranta K, Hyttinen J, Ryhanen T, Viiri J, et al (2010) Mechanisms of protein aggregation in the retinal pigment epithelial cells. *Frontiers of Bioscience* 2:1374–1384.
9. Karunadharma PP, Nordgaard CL, Olsen TW, Ferrington DA (2010) Mitochondrial DNA damage as a potential mechanism for age-related macular degeneration. *Investigative Ophthalmology & Visual Sciences*. 51:5470-5479.
10. Nandrot EF, Kim Y, Brodie SE, Huang X, et al (2004) Loss of synchronized retinal phagocytosis and age-related blindness in mice lacking alpha/beta-5 integrin. *Journal of Experimental Medicine* 200:1539-1545.
11. Plafker SM, O'Mealey GB, Szweda LI (2012) Mechanisms for countering oxidative stress and damage in retinal pigment epithelium. *International Review of Cell & Molecular Biology* 298:135-177.
12. Rozanowska M, Jarvis-Evans J, Korytowski W, Boulton ME, et al (1995) Blue light-induced reactivity of retinal age pigment. In vitro generation of oxygen-reactive species. *Journal of Biological Chemistry* 270:18825-18830.
13. Rozanowska M, Wessels J, Boulton M, Burke JM, et al (1998) Blue lightinduced singlet oxygen generation by retinal lipofuscin in non-polar media. *Free Radical Biology & Medicine* 24:1107-1112.

14. Sparrow JR, Gregory-Roberts E, Yamamoto K, Blonska A, et al (2012) The bisretinoids of retinal pigment epithelium. *Progress in Retina & Eye Research* 31:121-135.
15. Strauss O (2005) The retinal pigment epithelium in visual function. *Physiological Reviews* 85:845-881.
16. Winkler BS, Boulton ME, Gottsch JD, Stenberg P (2007) Oxidative damage and age-related macular degeneration. *Molecular Vision* 5:32.
17. Young RW (1988) Solar radiation and age-related macular degeneration. *Survey in Ophthalmology* 32:252-269.

4.7.3. Oxidative Stress and Alzheimer's Disease (AD)

1. Christen Y (2000) Oxidative stress and Alzheimer disease. *American Journal of Clinical Nutrition* 71:621-629.
2. Gabbita SP, Lovell MA, Markesbery WR (1998) Increased nuclear DNA oxidation in the brain in Alzheimer's disease. *Journal of Neurochemistry* 71:2034-2040.
3. Gella A, Durany N (2009) Oxidative stress in Alzheimer disease. *Cell Adhesion & Migration* 3:88-93.
4. Hensley K, Carney JM, Mattson MP, Aksenova M, et al (1994) A model for beta-amyloid aggregation and neurotoxicity based on free radical generation by the peptide: relevance to Alzheimer disease. *Proceeding of the National Academy of Sciences USA* 91(8):3270-3274.
5. Mattson MP, Chan SL (2003) Neuronal and glial calcium signaling in Alzheimer's disease. *Cell Calcium* 34:385-397.
6. Nunomura A, Perry G, Aliev G, Hirai K, et al (2001) Oxidative damage is the earliest event in Alzheimer's disease. *Journal of Neuropathology & Experimental Neurology* 60:759-767.
7. Retz W, Gsell W, Münch G, Rosler M, Riederer P (1998) Free radicals in Alzheimer's disease. *Journal of Neural Transmission* 54:221-236.
8. Rösler M, Retz W, Thome J, Riederer P (1998) Free radicals in Alzheimer's dementia: currently available therapeutic strategies. *Journal of Neural Transmission* 54:211-219.
9. Smith MA, Perry G, Richey PL, Sayre LM, et al (1996) Oxidative damage in Alzheimer's. *Nature* 382:120-121.
10. Smith MA, Sayre LM, Monnier VM, Perry G (1995) Radical AGEing in Alzheimer's disease. *Trends in Neuroscience* 18(4):172-176.
11. Vitek MP, Bhattacharya K, Glendening JM, Stopa E, et al (1994) Advanced glycation end products contribute to amyloidosis in Alzheimer's disease. *Proceedings of the National Academy of Sciences USA* 91:4766-4770.

4.7.4. Oxidative Stress and Parkinson's Disease (PD)

1. Bender A, Krishnan KJ, Morris CM, et al (2006) High levels of mitochondrial DNA deletions in substantia nigra neurons in aging and Parkinson disease. *Nature Genetics* 38:515-517.
2. Gandhi S, Wood NW (2005) Molecular pathogenesis of Parkinson's disease. *Human Molecular Genetics* 14(18):2749-2755.
3. Henchcliffe C, Beal MF (2008) Mitochondrial biology and oxidative stress in Parkinson disease pathogenesis. *Nature Clinical Practice in Neurology* 4(11):600-609.
4. Jellinger KA (2001) The pathology of Parkinson's disease. *Advances in Neurology* 86:55-72.
5. Jenner P (2003) Oxidative stress in Parkinson's disease. *Annals of Neurology* 53 (Suppl 3):S26-36.
6. Jin CM, Yang YJ, Huang HS, Kai M, Lee MK (2010) Mechanisms of L-DOPA-induced cytotoxicity in rat adrenal pheochromocytoma cells: implication of oxidative stress-related kinases and cyclic AMP. *Neuroscience* 170(2):390-398.
7. Lehermitte JKW, McAlpine MA (1924) On the occurrence of abnormal deposits of iron in the brain in Parkinson's disease with special reference to its location. *Journal of Neurology & Psychopathology* 5:119-208.
8. Sabens Liedhegner EA, Steller KM, Mieyal Jj (2011) Levodopa activates apoptosis signaling kinase 1 (ASK1) and promotes apoptosis in a neuronal model: implications for the treatment of Parkinson's disease. *Chemical Research in Toxicology* 24(10)1644-1652.
9. Schapira AH (2008) Mitochondria in the aetiology and pathogenesis of Parkinson's disease. *Lancet Neurology* 7:97-109.
10. Singleton AB, Farrer M, Johnson J, et al (2003) Alpha-Synuclein locus triplication causes Parkinson's disease. *Science* 302:841.
11. Smith MA, Harris PL, Sayre LM, Perry G (1997) Iron accumulation in Alzheimer disease is a source of redox-generated free radicals. *Proceedings of the National Academy of Sciences USA* 94:9866-9868.
12. Youdim MB (2001) Deficiency and excess of iron in brain function and dysfunction. *Nutrition Review* 59:S83-85.

4.7.5. Oxidative Stress and Autism

1. Frye RE, Rossignol DA (2011) Mitochondrial dysfunction can connect the diverse medical symptoms associated with autism spectrum disorders. *Pediatric Research* 69:41R-47RR.
2. Giulivi C, Zhang YF, Omanska-Klusek A, Ross-Inta C, et al (2010) Mitochondrial dysfunction in autism. *JAMA* 304:2389-2396.
3. Herbert MR (2005) Autism: a brain disorder or a disorder that affects the brain. *Clinical Neuro-psychology* 2: 354-379.
4. Herbert MR, Russo JP, Yang S, Roohi J, et al (2006) Autism and environmental genomics. *Neurotoxicology* 27:671-684.
5. Jacobs EA, Winter PM, Alvis HJ, Small SM (1969) Hyperoxygenation effect on cognitive functioning in the aged. *New England Journal of Medicine* 281(14):753-757.
6. James SJ, Rose S, Melnyk S, Jernigan S, et al (2009) Cellular and mitochondrial glutathione redox imbalance in lymphoblastoid cells derived from children with autism. *FASEB Journal* 23:2374-2383.
7. Ming X, Brimacombe M, Chaaban J, Zimmerman-Bier B, Wagner GC (2008) Autism spectrum disorders: concurrent clinical disorders. *Journal of Child Neurology* 23:6-13.
8. Napoli E, Wong S, Giulivi C (2013) Evidence of reactive oxygen species-mediated damage to mito chondrial DNA in children with typical autism. *Molecular Autism* 4(1):2.
9. Pardo CA, Eberhart CG (2007) The neurobiology of autism. *Brain Pathology* 17:434-447.
10. Rice C (2009) Prevalence of autism spectrum disorders—Autism and Developmental Disabilities Monitoring Network, United States, 2006. *MMWR Surveillence Summary* 58:1-20.
11. Rose S, Melnyk S, Pavliv O, Bai S, et al (2012) Evidence of oxidative damage and inflammation associated with low glutathione redox status in the autism brain. *Translational Psychiatry* 2:e134
12. Rossignol DA, Frye RE (2012) A review of research trends in physiological abnormalities in autism spectrum disorders: immune dysregulation, inflammation, oxidative stress, mitochondrial dysfunction and environmental toxicant exposures. *Molecular Psychiatry* 17:389-401.
13. Rossignol DA, Frye RE (2012) Mitochondrial dysfunction in autism spectrum disorders: a systematic review and meta-analysis. *Molecular Psychiatry* 17:290-314.
14. Rossignol DA, Rossignol LW, James SJ, Melnyk S, Mumper E (2007) The effects of hyperbaric oxygen therapy on oxidative stress, inflammation, and symptoms in children with autism: an open-label pilot study. *BMC Pediatrics* 7:36.

4.8 | REDOX: The Skeletal System

4.8.1. REDOX and Bone-Joint Health

1. Afonso V, Champy R, Mitrovic D, et al (2007) Reactive oxygen species and superoxide dismutases: Role in joint diseases. *Joint, Bone & Spine* 74:324-329.
2. Altindag O, Erel O, Soran N, et al (2008) Total oxidative/anti-oxidative status and relation to bone mineral density in osteoporosis. *Rheumatology International* 28:317-321.
3. Bai XC, Lu D, Liu AL, et al (2005) Reactive oxygen species stimulates receptor activator of NF-κB ligand expression in osteoblast. *Journal of Biological Chemistry* 29:17497–17506.
4. Banfi G, Iorio EL, Corsi MM (2008) Oxidative stress, free radicals and bone remodeling. *Clinical Chemistry & Laboratory Medicine* 46:1550-1555.
5. Damoulis PD, Hauschka PV (1997) Nitric oxide acts in conjunction with pro-inflammatory cytokines to promote cell death in osteoblasts. *Journal of Bone Mineral Research* 12:412-423.
6. Mody N, Parhami F, Saraflan TA, et al (2001) Oxidative stress modulates osteoblastic differentiation of vascular and bone cells. *Free Radical Biology & Medicine* 31:509-519.
7. Ralston SH, Ho LP, Helfrich MH, Grabowski PS, et al (1995) Nitric oxide: a cytokine-induced regulator of bone resorption. *Journal of Bone Mineral Research*10:1040-1049.
8. Riancho JA, Salas E, Zarrabeitia MT, Olmos JM, et al (1995) Expression and functional role of nitric oxide synthase in osteoblast-like cells. *Journal of Bone Mineral Research* 10:439-446.
9. van't Hof RJ, Armour KJ, Smith LM, Armour KE, et al (2000) Requirement of the inducible nitric oxide synthase pathway for IL-1- induced osteoclastic bone resorption. *Proceedings of the National Academy of Sciences USA* 97:7993-7998
10. Wauquier F, Leotoing L, Coxam V, et al (2009) Oxidative stress in bone remodelling and disease. *Trends in Molecular Medicine* 15:468-476.

11. Zaman G, Pitsillides AA, Rawlinson SC, Suswillo RF, et al (1999) Mechanical strain stimulates nitric oxide production by rapid activation of endothelial nitric oxide synthase in osteocytes. *Journal of Bone Mineral Research* 14:1123=1131.
12. Ziskoven C, Jäger M, Zilkens C, et al (2010) Oxidative stress in secondary osteoarthristis: from cartilage destruction to clinical presentation. *Orthopedic Reviews* 2:95-101.

4.8.2. REDOX and Osteoporosis

1. Almeida M (2011) Unraveling the role of FoxOs in bone-insights from mouse models. *Bone* 49:319-327.
2. Brandi ML, Collin-Osdoby P (2006) Vascular biology and the skeleton. *Journal of Bone Mineral Research* 21:183-192.
3. Finkel T, Deng CX, MostoslavskyR (2009) Recent progress in the biology and physiology of sirtuins. *Nature* 460:587-591.
4. Guarente L (2006) Sirtuins as potential targets for metabolic syndrome. *Nature* 444:868-874.
5. Hofbauer LC, Brueck CC, Singh SK, Dobnig H (2007) Osteoporosis in patients with diabetes mellitus. *Journal of Bone Mineral Research* 22:1317-1328.
6. Jia D, O'Brien CA, Stewart SA, Manolagas SC, Weinstein RS (2006) Glucocorticoids act directly on osteoclasts to increase their life span and reduce bone density. *Endocrinology* 147:5592–5599.
7. Jilka RL, Hangoc G, Girasole G, Passeri G, et al (1992) Increased osteoclast development after estrogen loss: mediation by interleukin-6. *Science* 257:88-91.
8. Manolagas SC (2000) Birth and death of bone cells: basic regulatory mechanisms and implications for the pathogenesis and treatment of osteoporosis. *Endocrine Reviews* 21:115–137.
9. Manolagas SC (2010) From estrogen-centric to aging and oxidative stress: a revised perspective of the pathogenesis of osteoporosis. *Endocrine Reviews* 31:266-300.
10. Rached MT, Kode A, Xu L, et al (2010) FoxO1 is a positive regulator of bone formation by favoring protein synthesis and resistance to oxidative stress in osteoblasts. *Cellular Metabolism* 11:147-160.
11. Sanchez-Rodriguez MA, Ruiz-Ramos M, Correa-Munoz E, et al (2007) Oxidative stress as a risk factor for osteoporosis in elderlyMexicans as characterized by antioxidant enzymes. *BMC Musculo-skeletal Disorders* 19:124.
12. Schulz E, Arfai K, Liu X, Sayre J, Gilsanz V 2004 Aortic calcification and the risk of osteoporosis and fractures. *Journal of Clinical Endocrinology & Metabolism* 89:4246-4253.

4.8.3. Oxidative Stress and Osteo-arthritis (OA)

1. Abramson SB (2008) Nitric oxide in inflammation and pain associated with osteoarthritis. *Arthritis Research & Therapy* 10:S2.
2. Alcaraz MJ, Megias J, Garcia-Arnandis I, et al (2010) New molecular targets for the treatment of osteoarthritis. *Biochemical Pharmacology* 2010;80:13-21.
3. Benito MJ, Veale DJ, FitzGerald O, et al (2005) Synovial tissue inflammation in early and late osteoar thritis. *Annals of Rheumatic Diseases* 64:1263-1267.
4. Blanco FJ, Guitian R, Vazquez-Martul E, et al (1998) Osteoarthritis chondrocytes die by apoptosis. A possible pathway for osteoarthritis pathology. *Arthritis & Rheumatology* 41:284-289.
5. D'Ambrosia RD (2005) Epidemiology of osteoarthritis. *Orthopedics* 28:201-205.
6. Davies CM, Guilak F, Weinberg JB, Fermor B (2008) Reactive nitrogen and oxygen species in interleukin-1-mediated DNA damage associated with osteoarthritis. *Osteoarthritis Cartilage* 16:624-630.
7. Goldring SR (2009) Role of bone in osteoarthritis pathogenesis. *Medical Clinics of North America* 93:25-35.
8. Goodwin W, McCabe D, Sauter E, et al (2010) Rotenone prevents impact-induced chondrocyte death. *Journal of Orthopedic Research* 28:1057-1063.
9. Henrotin YE, Bruckner P, Pujol JP (2003) The role of reactive oxygen species in homeostasis and degradation of cartilage. *Osteoarthritis Cartilage* 11:747-755.
10. Kim J, Xu M, Xo R, et al (2010) Mitochondrial DNA damage is involved in apoptosis caused by proin flammatory cytokines in human OA chondrocytes. *Osteoarthritis Cartilage* 18:424-432.
11. Lane NE, Michel B, Bjorkengren A, et al (1993) The risk of osteoarthritis with running and aging: a 5-year longitudinal study. *Journal of Rheumatology* 20:461-468.
12. Liu JT, Guo X, Ma WJ, et al (2010) Mitochondrial function is altered in articular chondrocytes of an endemic osteoarthritis, Kashin-Beck disease. *Osteoarthritis Cartilage* 18:1218-1226.
13. Loeser RF (2006) Molecular mechanisms of cartilage destruction: mechanics, inflammatory mediators, and aging collide. *Arthritis & Rheumatology* 54:1357-1360.

14. Martin JA, Buckwalter JA (2006) Post-traumatic osteoarthritis: the role of stress induced chondrocyte damage. *Biorheology* 43:517-521.
15. Mendoza G, Alvarez AI, Pulido MM, et al (2007) Inhibitory effects of different antioxidants on hyaluronan depolymerization. *Carbohydrate Research* 342:96-102.
16. Sharma L, Song J, Felson DT, et al (2001) The role of knee alignment in disease progression and func tional decline in knee osteoarthritis. *JAMA* 286:188-195.
17. Terkeltaub R, Johnson K, Murphy A, Ghosh S (2002) Invited review: the mitochondrion in osteoarthritis. *Mitochondrion* 1:301-319.
18. Tiku ML, Shah R, Allison GT (2000) Evidence linking chondrocyte lipid peroxidation to cartilage matrix protein degradation - Possible role in cartilage aging and the pathogenesis of osteoarthritis. *Journal of Biological Chemistry* 275:20069-76.

4.8.4. Oxidative Stress and Rheumatoid Arthritis (RA)

1. Dai L, Lamb DJ, Leake DS, Kus ML, et al (2000) Evidence for oxidised low density lipoprotein in synovial fluid from rheumatoid arthritis patients. *Free Radical Research* 32:479-486.
2. Deighton C, O'Mahony R, Tosh J, Turner C, Rudolf M, and Guideline Development Group (2009) Management of rheumatoid arthritis: Summary of NICE guidance. *British Medical Journal* 338:710-712.
3. Grootveld M, Henderson EB, Farrell A, Blake DR, et al (1991) Oxidative damage to hyaluronate and glucose in synovial fluid during exercise of the inflamed rheumatoid joint. Detection of abnormal low-molecular-mass metabolites by proton-NMR spectroscopy. *Biochemistry Journal*, 273:459-467.
4. Haqqi TM, Anthony DD, Gupta S, Ahmad N, et al (1999) Prevention of collagen-induced arthritis in mice by a polyphenolic fraction from green tea. *Proceedings of the National Academy of Sciences USA* 96:4524-4529.
5. Hitchon CA, El-Gabalawy HS (2004) Oxidation in rheumatoid arthritis. *Arthritis Research & Therapy* 6:265-278.
6. Jaswal S, Mehta HC, Sood AK, Kaur J (2003) Antioxidant status in rheumatoid arthritis and role of antioxidant therapy. *Clinica Chimica Acta* 338:123-129.
7. McCord JM (1985) Oxygen-derived free radicals in postischaemic tissue injury. *New England Journal of Medicine* 312:159-163.
8. Panasyuk A, Frati E, Ribault D, Mitrovic D (1994) Effect of reactive oxygen species on the biosynthesis and structure of newly synthesized proteoglycans. *Free Radical Biology & Medicine* 16:157-167.
9. Rowley D, Gutteridge JM, Blake D, Farr M, Halliwell B (1984) Lipid peroxidation in rheumatoid arthritis: thiobarbituric acid-reactive material and catalytic iron salts in synovial fluid from rheumatoid patients. *Clinical Sciences (London)* 66:691-695.
10. Stichtenoth DO, Fauler J, Zeidler H, Frolich JC (1995) Urinary nitrate excretion is increased in patients with rheumatoid arthritis and reduced by prednisolone. *Annals of Rheumatic Diseases* 54:820-824,
11. Tak PP, Zvaifler NJ, Green DR, Firestein GS (2000) Rheumatoid arthritis and p53: how oxidative stress might alter the course of inflammatory diseases. *Immunology Today* 21:78-82.
12. Taysi S, Polat F, Gul M, Sari RA, Bakan E (2002) Lipid peroxidation, some extracellular antioxidants, and antioxidant enzymes in serum of patients with rheumatoid arthritis. *Rheumatology International* 21:200-204.
13. Ueki Y, Miyake S, Tominaga Y, Eguchi K (1996) Increased nitric oxide levels in patients with rheumatoid arthritis. *Journal of Rheumatology* 23:230-236.

4.9 | REDOX: The Muscular System

4.9.1. REDOX and Muscular Health

1. Anderson EJ, Neufer PD (2006) Type II skeletal myofibers possess unique properties that potentiate mitochondrial H2O2 generation. *American Journal of Physiology: Cell Physiology* 290: C844-C851.
2. Droge W (2002) Free radicals in the physiological control of cell function. *Physiological Reviews* 82:47–95.
3. Hidalgo C, Sanchez G, Barrientos G, Aracena-Parks P (2006) A transverse tubule NADPH oxidase activity stimulates calcium release from isolated triads via ryanodine receptor type 1 S-glutathionylation. *Journal of Biological Chemistry* 281:26473–26482.
4. Ji LL, Gomez-Cabrera MC, Vina J (2006) Exercise and hormesis: activation of cellular antioxidant signaling pathway. *Annals of New York Academy of Sciences* 1067:425–435.

5. Moylan JS. Reid MB (2007) Oxidative stress, chronic disease and muscle wasting. *Muscle Nerve* 35:411-429.
6. Musaro A, Fulle S & Fano G (2010) Oxidative stress and muscle homeostasis. *Current Opinion in Clinical Nutrition & Metabolic Care* 13:236–242.
7. Pellegrino MA, Desaphy JF, Brocca L, Pierno S, et al (2011) Redox homeostasis, oxidative stress and disuse muscle atrophy. *Journal of Physiology* 589(9):2147-2160.
8. Smith MA, Reid MB (2006) Redox modulation of contractile function in respiratory and limb skeletal muscle. *Respiratory Physiology & Neurobiology* 151:229-241.

4.9.2. Oxidative Stress and Physical Exercise

1. Davies KJ, Quintanilha AT, Brooks GA, Packer L (1982) Free radicals and tissue damage produced by exercise. *Biochemistry Biophysics Reseach Communications* 107:1198-1205.
2. Deaton CM, Marlin DJ (2003) Exercise-associated oxidative stress. *Clinical Technology & Equine Practice* 2:278-291.
3. Filairea E, Toumib H (2012) Reactive oxygen species and exercise on bone metabolism: Friend or enemy? *Joint Bone Spine* 79:341-346.
4. Finaud J, Lac G, Filaire E (2006) Oxidative stress: relationship with exercise and training. *Sports Medicine* 36:327-358.
5. Jackson MJ (2009) Redox regulation of adaptive responses in skeletal muscle to contractile activity. *Free Radical Biology & Medicine* 47:1267-1275.
6. Kondo H, Miura M, Itokawa Y (1991) Oxidative stress in skeletal muscle atrophied by immobilization. *Acta Physiologica Scandinavica* 142:527-528.
7. Pellegrino MA, Desaphy JF, Brocca L, Pierno S, et al (2011) Redox homeostasis, oxidative stress and disuse muscle atrophy. *Journal of Physiology* 589(9):2147-2160.
8. Powers SK, Duarte J, Kavazis AN, Talbert EE (2009) Reactive oxygen species are signaling molecules for skeletal muscle adaptation. *Experimental Physiology* 95(1):1-9.
9. Powers SK, Jackson MJ (2008) Exercise-induced oxidative stress: cellular mechanisms and impact on muscle force production. *Physiological Reviews* 88:1243-1276.
10. Ristow M, Zarse K, Oberbach A, Kloting N, et al (2009) Antioxidants prevent health-promoting effects of physical exercise in humans. *Proceedings of the National Academy of Sciences USA* 106:8665-8670.
11. Smuder AJ, Kavazis AN, Hudson MB, Nelson WB, Powers SK (2010) Oxidation enhances myofibrillar protein degradation via calpain and caspase-3. *Free Radical Biology & Medicine* 49:1152-1160.
12. Vincent HK, Powers SK, Stewart DJ, Demirel HA, et al (2000) Short-term exercise training improves diaphragm antioxidant capacity and endurance. *European Journal of Applied Physiology* 81:67-74.

4.9.3. Oxidative Stress and Muscle Atrophy

1. Du J, Wang X, Miereles C, Bailey JL, et al (2004) Activation of caspase-3 is an initial step triggering ac celerated muscle proteolysis in catabolic conditions. *Journal of Clinical Investigation* 113:115–123.
2. Goll DE, Thompson VF, Li H,WeiW, Cong J (2003) The calpain system. *Physiological Reviews* 83:731–801.
3. Grune T, Merker K, Sandig G & Davies KJ (2003) Selective degradation of oxidatively modified protein substrates by the proteasome. *Biochemical Biophysical Research Communications* 305:709-718.
4. Kandarian SC, Stevenson EJ (2002) Molecular events in skeletal muscle during disuse atrophy. *Exercise Sport Science Reviews* 30:111–116.
5. Li YP, Chen Y, Li AS, Reid MB (2003) Hydrogen peroxide stimulates ubiquitin-conjugating activity and expression of genes for specific E2 and E3 proteins in skeletal muscle myotubes. American Journal of Physiology: *Cell Physiology* 285:C806-C812.
6. McClung JM, Kavazis AN, DeRuisseau KC, Falk DJ, et al (2007) Caspase-3 regulation of diaphragm myonuclear domain during mechanical ventilation-induced atrophy. *American Journal of Respiratory Critical Care Medicine* 175:150-159.
7. Powers SK, Kavazis AN, DeRuisseau KC (2005) Mechanisms of disuse muscle atrophy: role of oxidative stress. *American Journal of Physiology: Regulatory Integrative & Comparative Physiology* 288:R337-R344.
8. Powers SK, Kavazis AN, McClung JM (2007) Oxidative stress and disuse muscle atrophy. *Journal of Applied Physiology* 102:2389-2397.
9. Primeau AJ, Adhihetty PJ, Hood DA (2002) Apoptosis in heart and skeletal muscle. *Canadian Journal of Appled Physiology* 27:349-395.

10. SiemsW, Capuozzo E, Lucano A, Salerno C, Crifo C (2003) High sensitivity of plasma membrane ion transport ATPases from human neutrophils towards 4-hydroxy-2,3-transnonenal. *Life Sciences* 73:2583–2590.

4.9.4. Oxidative Stress and Muscular Dystrophy

1. Bernardi P (1999) Mitochondria in muscle cell death. *Italian Journal of Neurological Sciences* 20:395-400.
2. Bonafati MD, Ruzza G, Bonometto P, Bernardinelli A, et al (2000) A multicenter, double-blind, randomized trial of deflazacort versus prednisone in Duchenne muscular dystrophy. *Muscle Nerve* 23:1344-1347.
3. Lawler JM (2011) Exacerbation of pathology by oxidative stress in respiratory and locomotor muscles with Duchenne muscular dystrophy. *Journal of Physiology* 589 (9): 2161-2170.
4. McClung JM, Kavazis AN, DeRuisseau KC, Falk DJ, et al (2007) Caspase-3 regulation of diaphragm myonuclear domain during mechanical ventilation-induced atrophy. *American Journal of Respiration* 175:150-159.
5. Ramaswamy KS, Palmer ML, van der Meulen JH, Renoux A, et (2011) Lateral transmission of force is impaired in skeletal muscles of dystrophic mice and very old rats. *Journal of Physiology* 589 (5):1195-1208
6. Rando TA (2001) The dystrophin-glycoprotein complex, cellular signaling, and the regulation of cell survival in muscular dystrophies. *Muscle Nerve* 24:1575-1594.
7. Song W, Kwak HB, Lawler JM (2009) Exercise training modulates the NOS profile in skeletal muscle from old rats. *Journal of Gerontology Series A Biological Sciences & Medical Sciences* 64:540–549.
8. Suzuki N, Motohashi N, Uezumi A, Fukada S, et al. (2007) NO production results in suspension induced muscle atrophy through dislocation of neuronal NOS. *Journal of Clinical Investigation* 117:2468-2476.
9. Tidball JG, Wehling-Henricks M (2004) Expression of a NOS transgene in dystrophin-deficient muscle reduces muscle membrane damage without increasing the expression of membrane-associated cytoskeletal proteins. *Molecular Genetics & Metabolism* 82:312-320.
10. Tidball JG, Wehling-Henricks M (2007) The role of free radicals in the pathophysiology of muscular dystrophy. *Journal of Applied Physiology* 102:1677-1686.

4.10 | REDOX: The Endocrine System

4.10.1. REDOX and Hormones

1. Berthoud HR, Lenard NR, Shin AC (2011) Food reward, hyperphagia, and obesity. *American Journal of Physiology: Regulatory, Integrative & Comparative Physiology* 300(6):R1266-1277.
2. Heindel RJJ, vom Saal FS (2009) Role of nutrition and environmental endocrine disrupting chemicals during the perinatal period on the aetiology of obesity. *Molecular and Cellular Endocrinology* 304(1-2):90–96.
3. Johnson RJ, Andrews P, Benner SA, Oliver W (2010) Theodore E. Woodward award. The evolution of obesity: insights from the mid-Miocene. *Transactions of the American Clinical & Climatological Association* 121:295-305.
4. Lubrano C, Genovesi G, Specchia P, Costantini D, et al (2013) Obesity and metabolic comorbidities: environmental diseases? *Oxidative Medicine & Cell Longevity PMID*: 23577225.
5. Neel JV (1962) Diabetes mellitus: a 'thrifty' genotype rendered detrimental by progress? *American Journal of Human Genetics* 14:353-362.

4.10.2. Oxidative Stress and Diabetes

1. Brownlee M (2001) Biochemistry and molecular cell biology of diabetic complications. *Nature* 414:813-820.
2. Danaei G, Finucane MM, Lu Y, Singh GM, et al (2011) National, regional, and global trends in fasting plasma glucose and diabetes prevalence since 1980: systematic analysis of health examination surveys and epidemiological studies with 370 country-years and 2.7 million participants. *Lancet* 378(9785):31-40.
3. Dandona P, Aljada A, Chaudhuri A, Bandyopadhyay A (2003) The potential influence of inflammation and insulin resistance on the pathogenesis and treatment of atherosclerosis-related complications in type 2 diabetes. *Journal of Clinical Endocrinology & Metabolism* 88(6):2422-2429.

4. Dandona P, Thusu K, Cook S, et al (1996) Oxidative damage to DNA in diabetes mellitus. *The Lancet* 347(8999):444-445.
5. Delerive P, Fruchart JC, Staels B (2001) Peroxisome proliferator-activated receptors in inflammation control. *Journal of Endocrinology* 169(3):453-459.
6. Evans JL, Goldfine ID, Maddux BA, Grodsky GM (2002) Oxidative stress and stress-activated signaling pathways: a unifying hypothesis of type 2 diabetes. *Endocrine Reviews* 23(5):599-622.
7. Guzik TJ, Mussa S, Gastaldi D, Sadowski J, et al (2002) Mechanisms of increased vascular superoxide production in human diabetes mellitus: role of NAD(P)H oxidase and endothelial nitric oxide synthase. *Circulation* 105(14):1656-1662.
8. Haber CA, Lam TK, Yu Z, Gupta N, et al (2003) N-acetylcysteine and taurine prevent hyperglycemia induced insulin resistance in vivo: possible role of oxidative stress. *American Journal of Physiology* 285(4):E744-E753.
9. Harmon JS, Stein R, Robertson RP (2005) Oxidative stress-mediated, post-translational loss of MafA protein as a contributing mechanism to loss of insulin gene expression in glucotoxic beta cells. *The Journal of Biological Chemistry* 280(12):11107-11113.
10. Hirosumi J, Tuncman G, Chang L, Görgün CZ, et al (2002) A central role for JNK in obesity and insulin resistance. *Nature* 420(6913):333-336.
11. Hundal RS, Petersen KF, Mayerson AB, et al (2002) Mechanism by which high-dose aspirin improves glucose metabolism in type 2 diabetes. *Journal of Clinical Investigation* 109(10):1321-1326.
12. Kaneto H, Matsuoka TA, Nakatani Y, Kawamori D, et al (2005) Oxidative stress, ER stress, and the JNK pathway in type 2 diabetes. *Journal of Molecular Medicine* 83(6):429-439.
13. Katakami N, Sakamoto K, Kaneto H, Matsuhisa M, et al (2009) Cumulative effect of oxidative stress-related gene polymorphisms on myocardial infarction in type 2 diabetes. *Diabetes Care*. 32(5):e55.
14. Maechler P, Jornot L, Wollheim CB (1999) Hydrogen peroxide alters mitochondrial activation and insulin secretion in pancreatic beta cells. *The Journal of Biological Chemistry* 274(39):27905-27913.
15. Nakae J, Biggs III WH Kitamura T, Cavenee WK, et al (2002) Regulation of insulin action and pancreatic β -cell function by mutated alleles of the gene encoding forkhead transcription factor Foxo1. *Nature Genetics* 32(2):245-253.
16. Oprescu AI, Bikopoulos G, Naassan A, Allister EM, et al (2007) Free fatty acid-induced reduction in glucose-stimulated insulin secretion: evidence for a role of oxidative stress in vitro and in vivo. *Diabetes* 56(12):2927-2937.
17. Robertson RP (2004) Chronic oxidative stress as a central mechanism for glucose toxicity in pancreatic islet beta cells in diabetes. *The Journal of Biological Chemistry* 279(41):42351-42354.
18. Sakai K, Matsumoto K, Nishikawa T, Suefuji M, et al (2003) Mitochondrial reactive oxygen species reduce insulin secretion by pancreatic β-cells. *Biochemical & Biophysical Research Communications* 300:216-222.
19. Saltiel AR, Kahn CR (2001) Insulin signalling and the regulation of glucose and lipid metabolism. *Nature* 414(6865):799-806.
20. Tanaka Y, Tran PO, Harmon J, Robertson RP (2002) A role for glutathione peroxidase in protecting pancreatic β cells against oxidative stress in a model of glucose toxicity. Proceedings of the National Academy of Sciences of the United States of America 99(19):12363-12368.
21. Turko IV, Marcondes S, Murad F (2001) Diabetes-associated nitration of tyrosine and inactivation of succinyl-CoA:3-oxoacid CoA transferase. *American Journal of Physiology: Heart & Circulation Physiology* 281(6):H2289-2294.

4.10.4. Oxidative Stress and Obesity

1. Borradaile NM, Han X, Harp JD, Gale SE, et al (2006) Disruption of endoplasmic reticulum structure and integrity in lipotoxic cell death. *Journal of Lipid Research* 47(12):2726-2737.
2. Brown LM, Gent L, Davis K, Clegg DJ (2010) Metabolic impact of sex hormones on obesity. *Brain Research* 1350:77-85.
3. Bryan S, Baregzay B, Spicer D, Singal PK, et al (2013) Redox-inflammatory synergy in the metabolic syndrome. *Canadian Journal of Physiology and Pharmacology* 91(1):22-30.
4. Casals-Casas C, Desvergne B (2011) Endocrine disruptors: from endocrine to metabolic disruption. *Annual Review of Physiology* 73:135-162.
5. Colborn T, vom Saal FS, Soto AM (1993) Developmental effects of endocrine-disrupting chemicals in wildlife and humans. *Environmental Health Perspectives* 101(5):378-384.
6. Friedman JM (2000) Obesity in the new millennium. *Nature* 404(6778):632-634.
7. Friedman JM, Halaas JL (1998) Leptin and the regulation of body weight in mammals. *Nature* 395(6704):763-770.

8. Grün F, Blumberg B (2006) Environmental obesogens: organotins and endocrine disruption via nuclear receptor signaling. *Endocrinology* 147(6):S50-S55.
9. Hatch EE, Nelson JW, Stahlhut RW, Webster TF (2010) Association of endocrine disruptors and obesity: perspectives from epidemiological studies. *International Journal of Andrology* 33(2):324-332.
10. Heindel JJ, vom Saal FS (2009) Role of nutrition and environmental endocrine disrupting chemicals during the perinatal period on the aetiology of obesity. *Molecular and Cellular Endocrinology* 304:90-96.
11. Horvath TL, Andrews ZB, Diano S (2009) Fuel utilization by hypothalamic neurons: roles for ROS. *Trends in Endocrinology and Metabolism* 20(2):78-87.
12. Lee H, Lee YJ, Choi H, Ko EH, et al (2009) Reactive oxygen species facilitate adipocyte differentiation by accelerating mitotic clonal expansion. *The Journal of Biological Chemistry* 284(16):10601-10609.
13. Listenberger LL, Han X, Lewis SE, Cases S, et al (2003) Triglyceride accumulation protects against fatty acid-induced lipotoxicity. *Proceedings of the National Academy of Sciences of the United States of America* 100(6):3077-3082.
14. Newbold RR (2008) Prenatal exposure to diethylstilbestrol (DES). *Fertility and Sterility* 89(2):e55-e56.
15. Nishida C, Ko GT, Kumanyika S (2010) Body fat distribution and noncommunicable diseases in populations: Overview of the 2008 WHO expert consultation on waist circumference and waist-hip ratio. *European Journal of Clinical Nutrition* 64(1):2-5.
16. Savini I, Catani MV, Evangelista D, Gasperi V, et al (2013) Obesity-associated oxidative stress:strategies finalized to improve redox state. *International Journal of Molecular Sciences* 14(5):10497-10538.
17. Serra D, Mera P, Malandrino MI, Mir JF, et al (2013) Mitochondrial fatty acid oxidation in obesity. *Antioxididants and Redox Signaling* 19(3):269-84.
18. Tzanetakou IP, Katsilambros NL, Benetos A, Mikhailidis DP, et al (2012) "Is obesity linked to aging?": adipose tissue and the role of telomeres. *Ageing Research Reviews* 11(2):220-229.
19. Vom Saal FS, Nagel SC, Coe BL, Angle BM, et al (2012) The estrogenic endocrine disrupting chemical bisphenol A (BPA) and obesity. *Molecular and Cellular Endocrinology* 354(1-2):74-84.
20. Winter Y, Sankowski R, Back T (2013) Genetic determinants of obesity and related vascular diseases. *Vitamins and Hormones* 91:29-48.

4.11 | REDOX: The Genital System

4.11.1. REDOX and Reproductive Health

1. Behrman HR, Kodaman PH, Preston SL, Gao S (2001) Oxidative stress and the ovary. *Journal of Society for Gynecological Investigation* 8:S40–S42.
2. Diamant S, Kissilevitz R, Diamant Y (1980) Lipid peroxidation system in human placental tissue: general properties and the influence of gestational age. *Biological Reproduction* 23:776-781.
3. Fait V, Sela S, Ophir E, Khoury S, et al (2002) Hyperemesis gravidarum is associated with oxidative stress. *American Journal of Perinatology* 19:93-98.
4. Holthe MR, Staff AC, Berge LN, Lyberg T (2004) Leukocyte adhesion molecules and reactive oxygen species in preeclampsia. *Obstetrics & Gynecology* 103:913-922.
5. Little RE, Gladen BC (1999) Levels of lipid peroxides in uncomplicated pregnancy: a review of the literature. *Reproductive Toxicology* 13:347-352.
6. Myatt L, Cui X (2004) Oxidative stress in the placenta. *Histochemistry & Cell Biology* 122:369-382.
7. Ruder EH, Hartman TJ, Goldman MB (2009) Impact of oxidative stress on female fertility. *Current Opinion in Obstetrics & Gynecology* 21:219-222.
8. Shkolnik K, Tadmor A, Ben-Dor S, Nevo N, et al (2011) Reactive oxygen species are indispensable in ovulation. *Proceedings of the National Academy of Sciences USA* 108:1462-1467.
9. Sugino N (2006) Roles of reactive oxygen species in the corpus luteum. *Animal Science Journal* 77:556-565.
10. Wang YP, Walsh SW, Guo JD, Zhang JY (1991) Maternal levels of prostacyclin, thromboxane, vitamin E, and lipid peroxides throughout normal pregnancy. *American Journal of Obstetrics & Gynecology* 165:1690–1694.

4.11.2. REDOX: Fertilization and Embryogenesis

1. Agarwal A, Saleh RA, Bedaiwy MA (2003) Role of reactive oxygen species in the pathophysiology of human reproduction. *Fertility & Sterility* 79:829-843.
2. Bedaiwy MA, Falcone T, Mohamed MS, Aleem AA, et al (2004) Differential growth of human embryos in vitro: role of reactive oxygen species. *Fertility & Sterility* 82:593-600.

3. Chapman JC, Michael SD (2003) Proposed mechanism for sperm chromatin condensation/deconden sation in the male rat. *Reproductive Biology & Endocrinology* 1:20-26.
4. Fujii J, Iuchi Y, Okada F (2005) Fundamental roles of reactive oxygen species and protective mechanisms in the female reproductive system. *Reproductive Biology & Endocrinology* 3:43
5. Knapen MF, Zusterzeel PL, Peters WH, Steegers EA (1999) Glutathione and glutathione-related enzymes in reproduction. A review. *European Journal of Obstetrics, Gynecology & Reproductive Biology* 82:171-184.
6. Perreault SD, Barbee RR, Slott VL (1988) Importance of glutathione in the acquisition and maintenance of sperm nuclear decondensing activity in maturing hamster oocytes. *Developmental Biology* 125:181-186.
7. Perreault SD, Wolff RA, Zirkin BR (1984)The role of disulfide bond reduction during mammalian sperm nuclear decondensation in vivo. *Developmental Biology* 101:160-167.
8. Sanocka D, Kurpisz M: Reactive oxygen species and sperm cells. *Reproductive Biology & Endocrinology* 2:12-18.
9. Thomas M, Jain S, Kumar GP, Laloraya M (1997) A programmed oxyradical burst causes hatching of mouse blastocysts. *Journal of Cell Science* 110:1597-1602.
10. Urner F, Sakkas D: Involvement of the pentose phosphate pathway and redox regulation in fertilization in the mouse. *Molecular Reproduction & Development* 70:494-503.

4.11.3. REDOX and Pregnancy

1. Agarwal A, Gupta S, Sharma RK (2005) Role of oxidative stress in female reproduction. *Reproductive Biology & Endocrinology* 3:28.
2. Archer SL, Wu XC, Thébaud B, Moudgil R, et al (2004) O2 sensing in the human ductus arteriosus: redox-sensitive K+ channels are regulated by mitochondria-derived hydrogen peroxide. *Biological Chemistry* 385(3-4):205-216.
3. Burton GJ, Hempstock J, Jauniaux E (2003) Oxygen, early embryonic metabolism and free radical-mediated embryopathies. *Reproductive Biomedicine Online* 6:84-96.
4. Hassold T (1986) Chromosome abnormalities in human reproductive wastage. *Trends in Genetics* 2:105-110.
5. Hempstock J, Jauniaux E, Greenwold N, Burton GJ (2003) The contribution of placental oxidative stress to early pregnancy failure. *Human Pathology* 34(12):1265-1275.
6. Jain SK (1989) The neonatal erythrocyte and its oxidative susceptibility. *Seminars in Hematology* 26:286.
7. Jauniaux E, Poston L, Burton GJ (2006) Placental-related diseases of pregnancy: Involvement of oxidative stress and implications in human evolution. *Human Reproduction Update* 12(6):747-755.
8. Norwitz ER, Schust DJ, Fisher SJ (2001) Implantation and the survival of early pregnancy. *New England Journal of Medicine* 345:1400-1408.
9. Speakman JR (2008) The physiological costs of reproduction in small mammals. *Philosophical Trans actions of the Royal Society of London (B) Biological Sciences* 363(1490):375-398.
10. Wan H, Xu Y, Ikegami M, Stahlman MT, et al (2004) Foxa2 is required for transition to air breathing at birth. *Proceedings of the National Academy Sciences USA* 101(40):14449-14454.

4.11.4. Oxidative Stress and Infertility

1. Cummins JM, Jequier AM, Kan R (1994) Molecular biology of human male infertility: links with aging, mitochondrial genetics, and oxidative stress? *Molecular Reproductive Development* 37:345-362.
2. Desai N, Sharma R, Makker K, Sabanegh E, Agarwal A (2009) Physiologic and pathologic levels of reactive oxygen species in neat semen of infertile men. *Fertility & Sterility* 92:1626-1631.
3. Lekamge DN, Lane M, Gilchrist RB, Tremellen KP (2008) Increased gonadotrophin stimulation does not improve IVF outcomes in patients with predicted poor ovarian reserve. *Journal of Assisted Reproduction & Genetics* 25:515–521.
4. Makker K, Agarwal A, Sharma R (2009) Oxidative stress & male infertility. *Indian Journal of Medical Research* 129:357-367.
5. Rowe T (2006) Fertility and a woman's age. *Journal of Reproductive Medicine* 51:157-163.
6. Ruder EH, Hartman TJ, Blumberg J, Goldman MB (2008) Oxidative stress and antioxidants: exposure and impact on female fertility. *Human Reproduction Update* 14(4):345–357.

4.11.5. Oxidative Stress and Erectile Dysfunction (ED)

1. Andersson KE (2001) Pharmacology of penile erection. *Pharmacology Reviews* 53:417-450.
2. Andersson KE, Steif CG (1997) Neurotransmission and the contraction and relaxation of penile erectile tissues. *World Journal of Urology* 15:14-20.
3. Burnett AL (2006) Nitric oxide in the penis-science and therapeutic implications from erectile dysfunction to priapism. *Journal of Sex Medicine* 3:578-582.
4. Burnett AL, Tillman SL, Chang TS, Epstein JI, et al (1993) Immunohistochemical localization of nitric oxide synthase in the autonomic innervation of the human penis. *Journal of Urology* 150:73-76.
5. Hannan JL, Blaser MC, Oldfield L, Pang JJ, et al (2010) Morphological and functional evidence for the contribution of the pudendal artery in aging-induced erectile dysfunction. *Journal of Sex Medicine* 7:3373–3384.
6. Hurt KJ, Musicki B, Palese MA, Crone JK, et al (2002) Akt-dependent phosphorylation of endothelial nitricoxide synthase mediates penile erection. *Proceedings of the National Academy of Sciences USA* 99:4061-4066.
7. Johnson JM, Bivalacqua TJ, Lagoda GA, Burnett AL, Musicki B (2011) eNOS-uncoupling in age-related erectile dysfunction. *International Journal of Impotence Research* 23(2):43-48.
8. Leite R, Giachini FRC, Carneiro FS, Nunes KP, et al (2007) Targets for the treatment of erectile dysfunction: Is NO still the answer? *Recent Patents in Cardiovascular Drug Discovery* 2:119-132.
9. Magee TR, Kovanecz I, Davila HH, Ferrini MG, et al (2007) Anti-sense and short hairpin RNA (shRNA) constructs targeting PIN (Protein Inhibitor of NOS) ameliorate aging-related erectile dysfunction in the rat. *Journal of Sex Medicine* 4:633-643.
10. Musicki B, Champion HC, Becker RE, Liu T, et al (2005) Erection capability is potentiated by chronic sildenafil treatment: role of blood flow-induced endothelial nitric oxide synthase phosphorylation. *Molecular Pharmacology* 8:26–232.
11. Musicki B, Liu T, Lagoda GA, Strong TD, et al (2010) Hypercholesterolemiainduced erectile dysfunc tion: endothelial nitric oxide synthase (eNOS) uncoupling in the mouse penis by NAD(P)H oxidase. *Journal of Sex Medicine* 7:3023–3032.
12. Tostes RC, Carneiro FS, Lee AJ, Giachini FRC, et al (2008) Cigarette smoking and Erectile Dysfunction: Focus on NO bioavailability and ROS generation. *Journal of Sex Medicine* 5(6):1284-1295.

4.12 | REDOX: The Integumentary System

4.12.1. Photo-Oxidative Stress and Skin Health

1. Brezová V, Gabcová S, Dvoranová D, Stasko A (2005) Reactive oxygen species produced upon photoexcitation of sunscreens containing titanium dioxide (an EPR study). *Journal of Photochemistry and Photobiology B: Biology* 79(2):121-134.
2. Bulteau AL, Moreau M, Nizard C, Friguet B (2002) Impairment of proteasome function upon UVA- and UVB-irradiation of human keratinocytes. *Free Radical Biology & Medicine* 32(11):1157-1170.
3. Nishigori C, Hattori Y, Toyokuni S (2004) Role of reactive oxygen species in skin carcinogenesis. *Antioxididants & Redox Signaling* 6(3):561-570.
4. Reelfs O, Tyrrell RM, Pourzand C (2004) Ultraviolet a radiation-induced immediate iron release is a key modulator of the activation of NF-κB in human skin fibroblasts. *Journal of Investigative Dermatology* 122(6):1440-1447.
5. Tanaka N, Tajima S, Ishibashi A, Uchida K, et al (2001) Immunohistochemical detection of lipid peroxidation products, protein-bound acrolein and 4-hydroxynonenal protein adducts, in actinic elastosis of photodamaged skin. *Archives of Dermatological Research* 293(7):363-367.
6. Trouba KJ, Hamadeh HK, Amin RP, Germolec DR (2002) Oxidative stress and its role in skin disease. *Antioxididants & Redox Signaling* 4(4):665-673.
7. Tyrrell RM (2004) Solar ultraviolet A radiation: an oxidizing skin carcinogen that activates heme oxygenase-1. *Antioxididants & Redox Signaling* 6(5):835-840.

4.12.2. Oxidative Stress and Wound Healing

1. Clark RAF (1995) *The Molecular and Cellular Biology of Wound Repair* (2nd Ed.) New York:Plenum Press, New York. ISBN-13: 978-0306427169

2. Gorlach A, Brandes RP, Bassus S, Kronemann N, et al (2000) Oxidative stress and expression of p22phox are involved in the up-regulation of tissue factor in vascular smooth muscle cells in response to activated platelets. *FASEB Journal* 14:1518-1528.
3. Hubner G, Brauchle M, Smola H, Madlener M, et al (1996) Differential regulation of pro-inflammatory cytokines during wound healing in normal and glucocorticoid-treated mice. *Cytokine* 8:548-556.
4. Loo AEK, Wong YT, Ho R, Wasser M, Du T, et al. (2012) Effects of hydrogen peroxide on wound healing in mice in relation to oxidative damage. *PLoS One* 7(11): e49215.
5. Niethammer P, Grabher C, Look AT, Mitchison TJ (2009) A tissue-scale gradient of hydrogen peroxide mediates rapid wound detection in zebra fish. *Nature* 459: 996–999.
6. Novo E, Parola M (2012) The role of redox mechanisms in hepatic chronic wound healing and fibrogenesis. *Fibrogenesis & Tissue Repair* 5(Suppl 1):S4.
7. Roy S, Khanna S, Nallu K, Hunt TK, Sen CK (2006) Dermal wound healing is subject to redox control. *Molecular Therapy* 13:211–220.
8. Schafer M, Werner S (2008) Oxidative stress in normal and impaired wound repair. *Pharmacological Research* 58:165-171.
9. Sen CK (2003) The general case for redox control in wound repair. *Wound Repair & Regeneration* 11:431-438.
10. Sen CK (2009) Wound healing essentials: Let there be oxygen. *Wound Repair & Regeneration* 17(1): 1–18.
11. Sen CK, Roy S (2008) Redox signals in wound healing. *Biochimica et Biophysica Acta* 1780(11): 1348-1361.
12. Soneja A, Drews M, Malinski T (2005) Role of nitric oxide, nitroxidative and oxidative stress in wound healing. *Pharmacological Reports* 57:108-119.

4.13 | REDOX: Cancer

1. Acharya A, Das I, Chandhok D, Saha T (2010) REDOX regulation in cancer: a double-edged sword with therapeutic potential. *Oxidative Medicine & Cellular Longevity* 3(1):23-34.
2. Allen RG, Tresini M (2000) Oxidative stress and gene regulation. *Free Radical Biology & Medicine* 28(3):463-499.
3. Bae I, Fan S, Meng Q, Rih JK, et al (2004) BRCA1 induces antioxidant gene expression and resistance to oxidative stress. *Cancer Research* 64:7893-7909.
4. Díaz B, Courtneidge SA (2012) Redox signaling at invasive microdomains in cancer cells. *Free Radical Biology & Medicine* 52(2):247-256.
5. Giles GI (2006) The redox regulation of thiol dependent signaling pathways in cancer. *Current Pharma ceutical Design* 12(34):4427-4443.
6. Grek CL, Tew KD (2010) Redox metabolism and malignancy. *Current Opinion in Pharmacology* 10(4):362-368.
7. Gupta GP, Massague J (2006) Cancer metastasis: building a framework. *Cell* 127:679-695.
8. Gupte A, Mumper RJ (2009) Elevated copper and oxidative stress in cancer cells as a target for cancer treatment. *Cancer Treatment Reviews* 35:32-46.
9. Hagemann T, Biswas SK, Lawrence T, Sica A, Lewis CE (2009) Regulation of macrophage function in tumors: the multifaceted role of NF-kappa-B. *Blood* 113:3139-3146.
10. Hall MD, Mellor HR, Callaghan R, Hambley TW (2007) Basis for design and development of platinum(IV) anticancer complexes. *Journal of Medicinal Chemistry* 50:3403-3411.
11. Harvey CJ, Thimmulappa RK, Singh A, Blake DJ, et al. Nrf2-regulated glutathione recycling ind pendent of biosynthesis is critical for cell survival during oxidative stress. *Free Radical Biology & Medicine* 46:443-453.
12. Jackson AL, Loeb LA (2001) The contribution of endogenous sources of DNA damage to the multiple mutations in cancer. *Mutation Research* 477(1-2):7-21.
13. Jones DP (2008) Radical-free biology of oxidative stress. *American Journal of Physiology: Cell Physiology* 295(4):C849-868.
14. Joo SS, Yoo YM (2009) Melatonin induces apoptotic death in LNCaP cells via p38 and JNK pathways: therapeutic implications for prostate cancer. *Journal of Pineal Research* 47:8-14.
15. Jungwirth U, Kowol CR, Keppler BK, Hartinger CG, et al (2011) Anti-cancer activity of metal complexes: involvement of redox processes. *Antioxidants & Redox Signaling* 15(4):1085-1127

16. Kamb A, Wee S, Lengauer C (2008) Why is cancer drug discovery so difficult? Nat Rev Drug Discov 6: 115–120, 2007. 111. Fojo T. Commentary: Novel therapies for cancer: Why dirty might be better. *Oncologist* 13:277-283.
17. Kim BW, Lee ER, Min HM, Jeong HS, et al (2008) Sustained ERK activation is involved in the kaemp ferol-induced apoptosis of breast cancer cells and is more evident under 3-D culture condition. *Cancer Biology & Therapy* 7:1080-1089.
18. Kim SK, Yang JW, Kim MR, Roh SH, et al (2008)Increased expression of Nrf2/AREdependent anti-oxidant proteins in tamoxifen-resistant breast cancer cells. *Free Radical Biology & Medicine* 45:537-546.
19. Kobayashi M, Yamamoto M (2006) Nrf2-Keap1 regulation of cellular defense mechanisms against electrophiles and reactive oxygen species. *Advances in Enzyme Regulation* 46:113-40.
20. Marnett LJ (2000) Oxyradicals and DNA damage. *Carcinogenesis* 21:361-370.
21. Matsuzawa A, Ichijo H (2008) Redox control of cell fate by MAP kinase: physiological roles of ASK1-
22. MAP kinase pathway in stress signaling. *Biochimica et Biophysica Acta* 1780:1325-1336.
23. Mauro C, Zazzeroni F, Papa S, Bubici C, Franzoso G (2009) The NFkappaB transcription factor pathway as a therapeutic target in cancer: methods for detection of NFkappaB activity. *Methods in Molecular Biology* 512:169-207.
24. Ott I, Gust R (2007) Non platinum metal complexes as anti-cancer drugs. *Archives in Pharmacology (Weinheim)* 340:117-126.
25. Rosenberg B, VanCamp L, Trosko JE, Mansour VH (1969) Platinum compounds: a new class of potent anti-tumor agents. *Nature* 222:385-386.
26. Saha T, Rih JK, Rosen EM (2009) BRCA1 downregulates cellular levels of reactive oxygen species. *FEBS Letters* 583:1535-1543.
27. Sarkar FH, Li Y, Wang Z, Kong D (2008) NFkappaB signaling pathway and its therapeutic implications in human diseases. *International Reviews in Immunology* 27:293-319.
28. Schafer ZT, Grassian AR, Song L, Jiang Z, et al (2009) Antioxidant and oncogene rescue of metabolic defects caused by loss of matrix attachment. *Nature* 461:109-113.
29. Tew KD, Townsend DM (2011) Redox platforms in cancer drug discovery and development. *Current Opinion in Chemical Biology* 15(1):156-161.
30. Trachootham D, Lu W, Ogasawara MA, Nilsa RD, Huang P (2008) Redox regulation of cell survival. *Antioxidants & Redox Signaling* 10(8):1343–1374.
31. Valko M, Rhodes CJ, Moncol J, Izakovic M, Mazur M (2006) Free radicals, metals and antioxidants in oxidative stress-induced cancer. *Chemical & Biological Interactions* 160:1-40.
32. Wondrak GT (2009) Redox-directed cancer therapeutics: molecular mechanisms and opportunities. *Antioxidants & Redox Signaling* 11(12):3013-3069.

CHAPTER

5

REDOX Nutrition

Nutrition (also called nourishment or aliment) is the provision, to cells and organisms, of the chemicals necessary to support metabolism. Since metabolism/energy production requires O_2, potentially cell damaging free radical species are generated. A poor diet is detrimental to health. Several diseases and dysfunctions can be prevented or alleviated with proper diet enriched with antioxidant minerals and vitamins; free radical scavengers and bio-replenishments.

5

REDOX: Nutrition

The amounts of nutrients contained in the adult human body represent the integration of the day to day accretions from the time of conception to the termination of growth. These accretions are usually determined by balance experiments carried out at different periods of growth and are assumed to measure the net requirements of the respective nutrients with due consideration of the synthesis and transformation of organic nutrients in metabolism.

Exposure to free radicals has led living organisms to develop a series of defense strategies involving prevention mechanisms, repair mechanisms, physical combats and antioxidant reactions. Enzymatic antioxidant defenses include superoxide dismutase (SOD), catalase (CAT) and glutathione peroxidase (GPx); these defenses can be modified by exercise, training, nutrition and aging. Non-enzymatic antioxidants include a variety of free radical quenchers such as ascorbic acid, α-tocopherol, carotenoids, flavonoids, thiols which include glutathione (GSH), ubiquinones (i.e. CoQ10), uric acid, bilirubin, ferritin and micronutrients which act as enzymatic cofactors.

Physiological fate of Dietary Antioxidants: Foods and beverages deliver both pro-oxidants as well as a complex mixture of antioxidants to the gastrointestinal tract (GI). Some of these antioxidants are well absorbed, while others not so well; however, even the unabsorbed ones can exert extra-cellular antioxidant protective effects in the GI tract. For example, flavonoids and certain polyphenols are powerful ROS scavengers that inhibit lipid oxidation by heme proteins. Initially, the stomach encounters the highest levels, followed by a substantial exposure of the small intestine; thus, many flavonoids and polyphenols pass into the colon. Hence ROS scavenging within the GI tract through diet is possible. For instance, nitrite (from saliva and food) forms nitrosylation and DNA-deamination species, but damage by these species is inhibited by flavonoids, *in vitro*, at levels similar to those that can occur in the gastric juice. Also, scavenging of various ROS by dietary antioxidants further generates oxidized, chlorinated, nitrosylated or nitrated products that exert a plethora of additional biological effects. In addition, several phenolic compounds interact with iron, often in complex ways, although the resulting effects are usually antioxidant. Finally, the unabsorbed dietary iron and heme enter the feces, where phenolics can chelate iron, scavenge ROS, and help in alleviating prooxidant activity. Flavonoid levels in human fecal fluid can be up to 1 μM or higher.

The potential contribution of vitamin C and E in the GI tract should not be neglected. Ascorbate can scavenge ROS and decrease nitrosamine formation. When tocopherols and tocotrienols are absorbed in the GI tract, α-tocopherol is preferentially selected by the liver for incorporation into plasma lipoproteins, whereas other tocopherols are excreted in bile and returned to the GI tract. High concentrations of α- and γ-tocopherols and tocotrienols are found in fecal matter (compared with plasma levels); they might be taken up by colonic cells or exert extra-cellular protective effects, for example tocopherol-mediated scavenging of reactive nitrogen species (RNS) and COX-2 inhibition.

Numerous polyphenolic natural substances, commonly considered antioxidants, are true oxidants due to the presence of electron acceptor groups in their molecules. Their beneficial effects support the importance of oxidative reactions in maintaining optimal health and/or preventing degenerative diseases. Oxidation of sulfhydryl groups on the cell surface of cancer cells by sodium selenite is another example of the importance of oxidative processes in health homeostasis. Phenolic compounds and/or polyphenols are the important groups of compounds occurring in plants, comprising at least 8000 different known structures including simple phenols, phenolic acids, coumarins and isocoumarins, naphthoquinones, xanthones, stilbenes, flavonoids, and lignins. Flavonoids, constituting the most important polyphenolic class, are the natural substances exhibiting a wide range of biological effects including: antibacterial, anti-inflammatory, antiallergic, antithrombotic, and vasodilatory actions. There are also a plethora of other plants and herbs that contain polyphenolic substances that can potentially scavenge hydroxyl radicals and/or chelate free iron.

According to Halliwell and Gutteridge (2007), the following points need to be noted preceding the use of an antioxidant in humans: (i) Is it non-toxic? (ii) Which biomolecules will the antioxidant protect, and are such molecules implicated in the pathogenesis of the disease? (iii) How does the antioxidant work - by scavenging reactive species, increasing endogenous defense, or repairing the damage? (iv) Can the antioxidant-derived products cause damage? (v) Does the antioxidant interfere with REDOX signaling?

Table: *Selected Phytochemicals, Their Sources, and Beneficial Health Effects.*

Name	Source	Chemical class	Beneficial effect
Ajoene	Garlic	Polysulfide	Cancer, CVD
Apigenin	Fruits, vegetables	Flavonoid	Cancer
Aronia	Chokeberries	Anthocyanin	Cancer, inflammation
Caffeic acid	Coffee, cabbage	Flavonoid	Inflammation
Capsicum	Solanaceae plant	Capsaicin	Pain relief
Carvone	Caraway, dill	Terpenoid	Insecticide
Catechin	Chocolate, apples	Flavan-3-ol	Ischemic stroke
Chrysin	Passion flower	Flavones	Inflammation
Crocin	Honey	Polyphenols	Cold and flu
Devil's claw	African plant	Iridoid glycoside	Inflammation
Ellagic acid	Green tea	Cyclic polyphenol	Cancer
Farnesol	Acacia flowers	Aliphatic alcohol	Chemoprevention
Ferulic acid	Brown rice, apples	OH-cinnamic acid	Inflammation
Feverfew	Asteracea plant	Sesquiterpene lactone	Headaches
Galangin	Propolis	Trihydroxy-flavone	Antifungal
Ganoderic acid	Reishi	Triterpenoid	Breast cancer
Genistein	Lupin, soybeans	Isoflavone	Anticancer
GLA	Borage	Gamma-linolenic acid	Inflammation
Kaempferol	Tea, broccoli	Flavonoid	Cancer
Naringenin	Grapefruit	Flavonone	Hepatitis
Nasunin	Egg plant	Anthocyanin	Antiproliferative
Quercetin	Red onions	Polyphenol	Antihistamine
Resveratrol	Red grapes	Trihydroxystilbene	Anti-aging
Sabinene	Caraway	Bicyclin monoterpene	Hypertension
Safranal	Saffron	Aromatic ketone	Antidepressant
Salicin	Willow bark	Salicylic acid	Antipyretic
Silybin	Silymarin	Polyphenol	Liver diseases
Turmeric yellow	Indian plant	Curcumin	Diabetes, AD

Medicinal Phytochemicals : The 1947 Nobel Prize in Chemistry was awarded to Robert Robinson for his investigations on plant products of biological importance, especially alkaloids.

5.1. REDOX-Active Metals

Transition metals are essential to many biological processes in almost all living organisms. Their versatility, which arises from an ability to undergo reduction–oxidation (REDOX) chemistry, enables them to act as critical cofactors of enzymes throughout the cell. Accumulation of metals, however, can also lead to oxidative stress and cellular damage. The importance of metals to both enzymatic reactions and oxidative stress makes them key players in cellular metabolism. The systemic and cellular levels of these metals must be actively monitored and regulated though, as their deficiency leads to loss of crucial enzyme activities, while their accumulation leads to inappropriate enzymatic reactions that generate ROS, oxidative damage and, eventually, cell death.

REDOX-Active Metals and Site-specific Formation of Radicals: Transition metals tend to undergo hydrolytic polymerization and precipitation in aqueous media at a neutral pH. These metals can remain in solution in physiological media only by association with certain high- or low-molecular-weight cellular components, such as nucleotides, peptides, polypeptides, proteins or DNA. Intra-cellular free iron, for example, is more likely to associate with low molecular weight ligands, such as nucleotides, citrate, glycine and glucose; whereas copper is more likely to associate with macromolecular structures, such as DNA, carbohydrates, enzymes, peptides and proteins. After forming complexes, movement of the metal is hindered, and the association site can serve as a locus for repetitive free radical formation through REDOX cycling of the metal. This is known as 'site-specific' formation of radicals. Since the metal is fixed in one place, any ·OH formed, due to its high reactivity, will attack adjacent structures and cause severe localized damage.

REDOX-Active Metals and Oxidative Damage: Metal-catalyzed ·OH formation can initiate destructive processes. Both copper and iron, for instance, can play critical roles in the initiation and propagation of lipid peroxidation. Metal-catalyzed formation of ·OH can also result in the abstraction of a hydrogen from an unsaturated fatty acid and cause lipid radical formation. This can initiate a propagative cascade of cyclical reactions that lead to repetitive formation of short-chain alkanes and lipid acid aldehydes, which can damage lipid bilayers. Another possible process of protein oxidation is initiated by the binding of a reduced REDOX-active transition metal to an enzyme to form a coordination complex, which can react with H_2O_2 to form ·OH. The ·OH radical can then abstract a hydrogen from the amino-bearing carbon, leading to the formation of a carbon-centered protein radical that undergoes a series of reactions, resulting in hydrolysis of the amino group and the formation of an aldehyde or a protein carbonyl. Once the amino group is hydrolyzed, the metal-binding site is disrupted, leading to dissociation of the metal. Because the oxidative modifications occur around the metal-binding site, the process of protein oxidation is a clear example of a site-specific reaction.

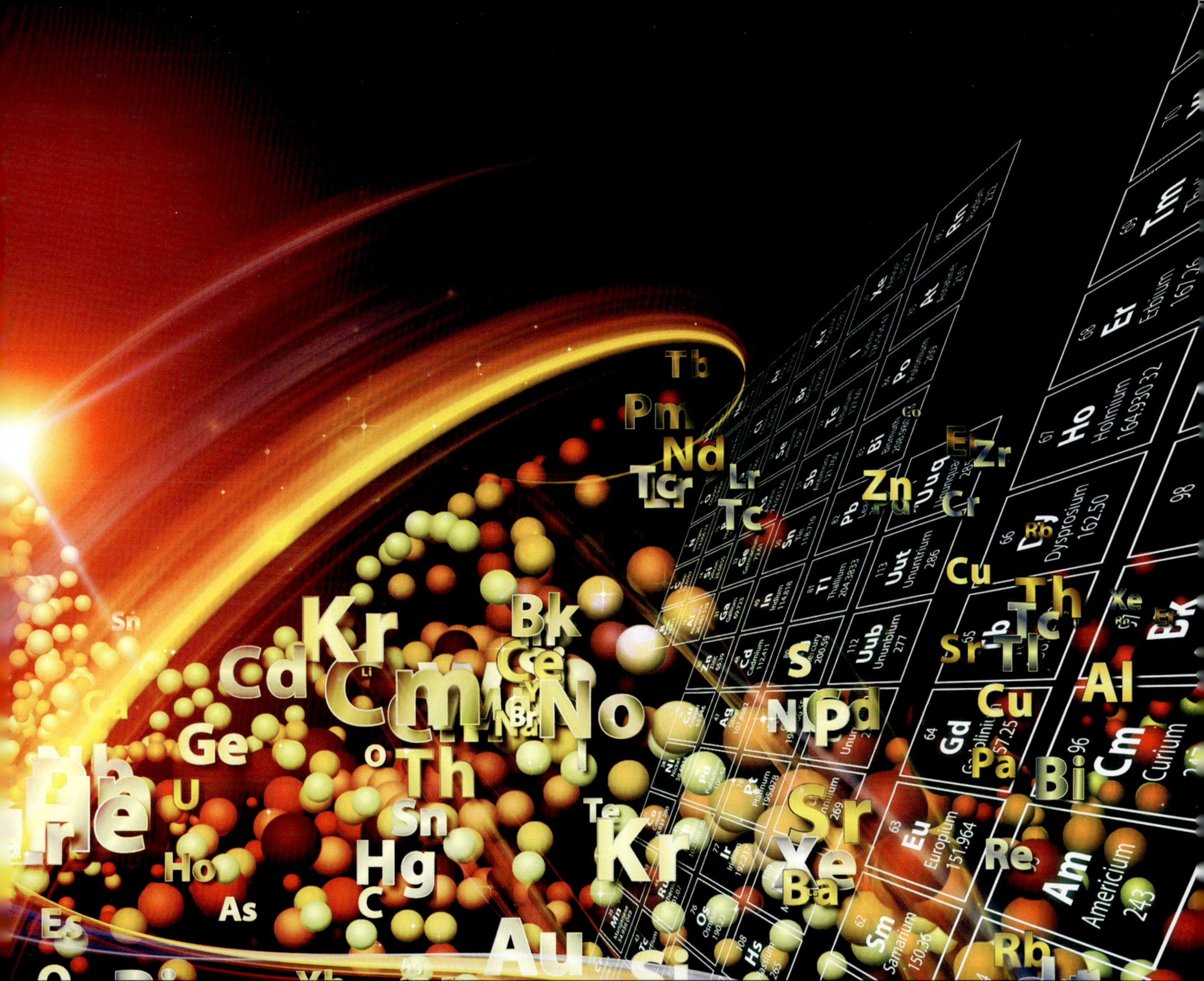

Table: *First-row Transition Metal ions in Human Physiology*

Element	Mass/mg	Biological Roles
Vanadium (V)	0.11	Enzymes (nitrogenases, haloperoxidases), essential for humans
Chromium (Cr)	14	Involved in carbohydrate utilization, essential for higher animals
Manganese (Mn)	12	Essential for mitochondrial function. Part of metalloenzymes including: arginase, glutamine synthetase, pyruvate carboxylase, and superoxide dismutase (SOD)
Iron (Fe)	4200	Vital element (e.g. heme), hemoglobin, electron transfer systems, ferredoxins
Cobalt (Co)	3	Vitamin B12 coenzyme contains Co(III) in corrin
Nickel (Ni)	15	Nickel containing enzymes include urease and some hydrogenases
Copper (Cu)	72	Some vertebrates use Cu-O_2 carrier, hemocyanin. Cu(II) containing enzymes include: tyrosinase, amine oxidase and cytochrome oxidase

Mass of the first row transition metal ions present in a 70 kg human.

5.1.1. REDOX-Active Metal: Iron (Fe)

Iron (Fe) is an essential nutritional element for all life forms. Iron plays a critical role in electron transport and cellular respiration, cell proliferation and differentiation, and regulation of gene expression. Iron is integral to the catalysis of the REDOX reactions of oxidative phosphorylation in the mitochondrial respiratory chain. It also provides a specific binding site for oxygen (O_2) in the heme binding moiety of hemoglobin, which allows O_2 transport in the blood. These same properties also render iron toxic when its REDOX-active chelatable 'labile' form exceeds the normal binding capacity of the cell. Indeed, in contrast to iron bound to proteins, intra-cellular labile iron can be potentially toxic especially in the presence of ROS, as it can lead to catalytic formation of oxygen-derived free radicals such as the ·OH radical that can ultimately overwhelm the cellular antioxidant defense mechanisms and cause damage to the macromolecular components of cells, including DNA and proteins. The handling of iron in the body must therefore be very carefully regulated.

Iron transport, storage and homeostasis: Most environmental iron is in the Fe^{3+} state, which is almost insoluble at neutral pH. To overcome the virtual insolubility and potential toxicity of iron, however, a myriad of specialized transport systems and associated proteins have evolved to mediate its regulated acquisition, transport, and storage of iron in a soluble, biologically useful, and non-toxic form. These proteins individually and, in concert, serve to maintain cellular and whole body homeostasis of this essential (yet potentially harmful) metal ion. Iron metabolism (both deficiency or overload) is associated with the specific pathologies of many organs.

Iron is an essential micronutrient. However, because humans have no means to control iron excretion, excess iron (regardless of the route of entry) accumulates in the parenchymal organs and threatens cell viability. Indeed, when iron-buffering capability is overwhelmed, oxidative stress-induced cell damage and fibrogenesis may arise, mainly in the liver. The liver is the main body site for iron storage and central in the regulation of iron homeostasis. Important iron-proteins, such as hepcidin, the iron regulatory hormone, are specifically produced by the liver. Mutations in hepatic iron transporters and regulators lead to hereditary iron overload diseases. Increased REDOX-active iron (present in heme-proteins and the cytosolic iron pool) can catalyze oxidative damage to lipids, proteins, and nucleic acids, either by ROS-dependent or -independent mechanisms. Iron-catalyzed injury results in damage to cell constituents, including mitochondria, lysosomes, and the sarcolemmal membrane. These mechanisms of iron-mediated damage are involved in the pathogenesis of organ dysfunction in primary hemochromatosis, transfusion-related iron overload and ischemia-reperfusion injury.

Iron and Oxidative Stress: Intra-cellular iron homeostasis and body iron balance are tightly regulated to minimize the presence of potentially toxic labile iron.

An altered iron homeostasis, however, can lead to oxidative stress and certain pathologies. During such pathological conditions, iron may serve as a potential catalyst, particularly in the form of REDOX-active iron or labile iron, for free radical reactions in oxidative stress and cellular damage. Ferritin is one of the major proteins of iron metabolism. It is almost ubiquitous and tightly regulated by iron. However, the reaction between iron and ferritin can result in the production of H_2O_2. The most frequent toxic role of iron is linked with the catalytic decomposition of H_2O_2 (Fenton-type reaction) leading to the formation of reactive oxygen species (ROS) which causes damage to biomolecules, including lipids, proteins and DNA. Thus, ferritin-dependent oxidative stress may be involved in the pathogenesis of many diseases.

Hemoglobin is an O_2 transporter, dependent on the reversible binding of oxygen to Fe(II). The partially oxygenated hemoglobin formed with the release of O_2 is susceptible to REDOX reactions. The reduction of O_2 to $O_2^{\bullet-}$ starts a cascade of oxidative reactions, which are a source for red blood cell-induced oxidative stress. The reduction of nitrite to nitric oxide (·NO) can be a source for oxidative stress, but may also have important physiological functions. Depending on the concentrations formed, such species can cause molecular damage or influence REDOX signaling mechanisms.

Cellular DNA damage under prooxidant conditions can be mediated by iron. In fact, iron is an important element in the establishment of a prooxidant status in the cell. Aberrant regulation of iron metabolism, and/or deficient antioxidant protection, is also associated with diseases, such as acute respiratory distress syndrome (ARDS). Iron, with the consequent production of ROS, microbial growth promotion, and adverse signaling are strongly implicated as major contributors to the pathogenesis of numerous diseases.

The binding of iron-designed chelators via nitrogen, oxygen or sulfur donors blocks the ability of iron to catalyze the formation of free radicals. Thus, the design of various metal chelators to prevent free radical reactions is an important approach in the treatment of many iron-related diseases. An effective dual functioning bio-replenishment such as lactoferrin, which possesses both metal-chelating and free radical-scavenging properties, is ideal for this purpose.

Iron and Cell Signaling: Iron-generated ROS may also function in normal intra-cellular signaling. Therefore, roles of iron are both essential and extraordinarily diverse. The transport of electrons between different forms of iron makes it essential for many fundamental cell functions. Emerging evidence indicates that cellular iron may participate in various cellular signaling pathways. These include iron homeostatic control signaling, iron-induced oxidative-responsive transcription factors, iron-induced activation of inflammatory cytokines, and iron-dependent signaling in cell growth and apoptosis. Protein iron-sulfur (FeS) centers have been implicated in the regulation of gene expression. In REDOX-sensing proteins, the oxidation state of the FeS centers controls its activity as a transcription activator

IRON (Fe)-RICH FOODS

Food Source	Serving	% DV
Chicken Liver (3-oz)	11 mg	61
Soy Beans (1-cup)	8.8 mg	48
Lentils (1-cup)	6.6 mg	37
Roast Beef (3-oz)	3.1 mg	17

independent of DNA-binding ability. Thus, FeS centers allosterically link cellular oxidative stress to the expression of defense genes. Iron is a REDOX-active metal critical for the transcription of certain key genes required for cell growth and function (i.e. nitric oxide synthase – NOS, and protein kinase C-β); as well as for hematopoietic cell differentiation.

Iron in Health and Disease: Most iron in the human body is contained in red blood cells. Iron deficiency results in impaired production of iron-containing proteins and inhibition of cell growth. In contrast, abnormal iron uptake has been related to the most common hereditary disease, hemochromatosis, and leads to tissue damage derived from free radical toxicity. In addition, disruption of iron regulation plays a key role in the etiology of Alzheimer's disease, Parkinson's disease, Huntington's disease, Friedreich's ataxia and other neurological disorders, carcinoma (lung, breast and colon cancers), Fanconi anemia, stroke and aging. Thus the control of this essential but potentially toxic element is an important aspect of human health and disease.

Iron-Deficiency: Severe iron deficiency results in anemia. Anemia in pregnancy increases the risk of premature and low birth weight babies. In young children, iron deficiency is associated with behavioral abnormalities (such as reduced attention span) and reduced cognitive performance that may not be fully reversible by iron replacement. In adults, severe iron deficiency anemia impairs physical work capacity. Moderate iron deficiency without anemia is most common in children 1-2 years old (9%) and females 12-49 years old (9-11%), reflecting rapid growth or menstrual iron loss, but is less common in other groups.

Table: *Iron - Recommended Dietary Allowances*

AGE GROUP	MALE	FEMALE	PREGNANCY
0 to 6 months	0.27 mg	0.27 mg	
7 to 12 months	11 mg	11 mg	
1 to 3 years	7 mg	7 mg	
4 to 8 years	10 mg	10 mg	
9 to 13 years	8 mg	8 mg	
14 to 18 years	11 mg	15 mg	27 mg
19 to 50 years	8 mg	18 mg	27 mg
51+ years	8 mg	8 mg	

5.1.2. REDOX-Active Metal: Copper (Cu)

Copper plays an essential role in human physiology. It is required for respiration, radical defense, neuronal myelination, angiogenesis, and many other processes. Copper ions can adopt distinct REDOX states and form a range of ligands that include carboxylate oxygen, imidazole nitrogen, cysteine thiolate, and methionine thioether groups and engage in cationic interactions. Copper also serves as a catalytic and structural cofactor for enzymes that function in energy generation, iron acquisition, oxygen transport, cellular metabolism, peptide hormone maturation, blood clotting, signal transduction and a host of other processes.

Copper as a Metabolic Co-factor: All cells in the human body require copper for their growth and development. Copper also serves as an essential cofactor for the activity of cytochrome C oxidase (CCOx) in mitochondria, the enzyme that is central for respiration; and for Cu, Zn-dependent superoxide dismutase (SOD), which plays an important role in the detoxification of oxygen radicals in the cytosol. In addition, many cells require copper in the secretory pathway, where enzymes including dopamine-β-hydroxylase, peptidyl-α-mono-oxygenase, ceruloplasmin, and tyrosinase incorporate copper as a co-factor in their catalytic sites. These enzymes then critically contribute to a number of key physiological processes, such as iron export from the cells, the production of neuro-endocrine peptides and neurotransmitters, pigmentation, blood clotting and others. Copper deficiency decreases the activity of these enzymes and thus adversely affects the corresponding physiological processes. Heart function is particularly sensitive to inadequate copper supply. Accumulating evidence points to the involvement of copper in lipid metabolism, antimicrobial defense, neuronal activity, resistance of tumor cells to platinum-based chemotherapeutic drugs, kinase-mediated signal transduction, and other essential cellular processes.

Copper and Mitochondrial Function: Aerobic life depends on cellular copper homeostasis and distribution since this element is a critical component of the enzymes involved in primary metabolism. Copper ions can undergo unique chemistry due to their ability to adopt distinct REDOX states, either oxidized cupric [Cu(II)] or reduced cuprous [Cu(I)], and they serve as important catalytic co-factors in REDOX chemistry for proteins that carry out fundamental biological functions. A copper-containing metalloenzyme, mitochondrial cyt-c oxidase (CCOx), is the final electron acceptor in the mitochondrial electron transport chain and is required for aerobic ATP production. CCOx is a heme/copper terminal oxidase which uses cytochrome c as an electron donor. The enzyme contains four REDOX-active metal centers which sequentially aid in reducing molecular oxygen to water: two iron centers, 'Heme a' and 'Heme a3' (also referred to as cytochromes 'a' and 'a3'), and two copper centers, 'CuA' and 'CuB'. Copper complexation of mitochondrial subunits is essential for the maturation and stability of CCOx. In the absence of the copper centers (as well as in the absence of the heme prosthetic groups), proteins are rapidly degraded and the CCOx holoenzyme fails to assemble. Thus, copper is not only required for CCOx catalytic function but also for its biogenesis, assembly and stability. CCOx biogenesis is vital for mitochondrial function; defective CCOx biogenesis results in mitochondrial diseases frequently involving brain, skeletal muscle, and heart. In particular, mutations in the genes required for CCOx copper metallation result in severe hepatopathies, encephalo-myopathies, hypertrophic cardiomyopathy, and muscular atrophy.

Copper and REDOX: The REDOX properties that render copper useful for several important metalloproteins can also lead to oxidative damage in cells. Reaction of Cu(I) with H_2O_2 and re-reduction of Cu (II) by $O_2^{\bullet-}$ via Fenton and Haber-Weiss chemistry yields ·OH radicals that can damage proteins, lipids, and nucleic acids. Thus, intra-cellular copper concentrations must be controlled such that copper ions are provided to essential enzymes, but do not accumulate to deleterious levels. If present in excess, free copper ions can cause damage to cellular components. A delicate balance between the uptake and efflux of copper ions determines the amount of cellular copper. In biological systems, copper homeostasis has been characterized at the molecular level. It is coordinated by several proteins such as glutathione (GSH), metallothionein, Cu-transporting P-type ATPases, Menkes and Wilson proteins, and by cytoplasmic transport proteins (called 'copper chaperones') to ensure that it is delivered to specific sub-cellular compartments and thereby to copper-requiring proteins.

Table: *Copper in Human Physiology*

Enzyme/Protein	Function	Consequence of Deficiency
Cu-Zn-Superoxide Dismutase	Superoxide ($O_2^{\bullet-}$) disproportionation, Signaling	Oxidative stress, Hepato-cellular carcinoma, Neuro-degeneration
Cytochrome Oxidase (CCOx)	Mitochondrial oxidative phosphorylation; ATP production	Respiratory deficiency, Cardiomyopathy, Lethality
Tyrosinase/Laccase	Melanin synthesis, Innate immunity	Pigmentation Defects/Albinism
Peptidylglycine α-amidating monooxygenase	Peptide amidation/maturation	Heart rate defects, Endocrine dysfunction, Lethality
Dopamine β-Hydroxylase	Norepinephrine synthesis	Hypoglycemia, Hypotension
Ceruloplasmin	Ferroxidase: Iron loading onto transferrin	Aceruloplasminemia, Progressive anemia, Neuro-degeneration, Diabetes
Hephaestin	Ferroxidase for ferroportin-mediated iron efflux	Anemia, Impaired iron absorption by peripheral tissues
Lysyl Oxidase	Covalent cross-linking of collagens and elastin	Aortic aneurisms, Cardiovascular dysfunction
Lysyl Oxidase-like Proteins	Oxidation of snail transcription factor resulting in E-cadherin silencing and promoting EMT	Altered cell-cell contacts
Coagulation Factor V & VIII	Blood Clotting	Hemophilia
XIAP	Inhibitor of apoptosis	Copper deficiency
Copper Amine Oxidase	Deamination of primary amines to aldehyde	Impaired immune responses, Reduced fat deposition

Copper Physiology in Health and Disease: Adequate copper intake helps normal utilization of dietary iron, including its intestinal absorption, release from stores (e.g. liver and spleen), and incorporation into hemoglobin. In addition to preventing anemia, copper assists in blood coagulation and blood pressure control; cross-linking of connective tissues in arteries, bones, and heart; defense against oxidative damage; energy transformation; myelination of brain and spinal cord; reproduction; and synthesis of hormones.

Copper is involved in normalized function of many enzymes, such as Cytochrome C oxidase (CCOx), which is complex IV in the mitochondrial electron transport chain (ETC), ceruloplasmin, Cu/Zn-SOD, and in amine oxidases. These enzymes catalyze reactions for oxidative phosphorylation, iron transport, antioxidant and free radical scavenging and neutrotransmitter synthesis.

Copper Deflciency: Inadequate copper causes adverse effects on the metabolism of cholesterol and glucose, blood pressure and heart function, mineralization of bones, and immunity. Moreover, Cu-deficient individuals can also suffer from osteoporosis (which is reversible with extra copper). Severe dietary deficiency in humans is rare; however, marginal Cu-deficiency is not uncommon, particularly in pregnancy. Pathological states with Cu-deficiency are becoming more recognized; Menkes disease is one such best known disorder. Copper deficiency can also be an underlying complication of celiac disease or be acquired as a result of excessive zinc ingestion or general malabsorption.

Dietary Cu-deficiency causes several cardiovascular conditions. Systemic effects include high blood pressure, enhancement of inflammation, anemia, reduced blood clotting, and possibly arteriosclerosis. Effects on specific organs or tissues include weakened structural integrity of the heart and blood vessels, impairment of energy use by the heart, reduced ability of the heart to contract, and altered structure and function of circulating blood cells. In some instances, the cause of a defect can be directly attributed to reduced activity of a specific copper-dependent enzyme. However, three nonspecific mechanisms of damage have been implicated in cardiovascular defects of copper deficiency. These include: ROS induced damage of proteins (possibly DNA); lipid peroxidation, which is the non-enzymatic glycosylation of proteins; and nitration, the interaction of nitric oxide and its metabolites with peptides and proteins. Though independently these mechanisms present great potential for damage, the possibility that they may interact presents an added reason for concern. Furthermore, the fact that at least two of these mechanisms are associated with diabetes and aging suggests that Cu-deficiency may exacerbate deficits associated with these mechanisms.

Menkes disease is a hereditary condition caused by a defective gene responsible for the metabolism of copper in the body. Menkes disease involves a wide variety of symptoms including floppy muscle tone, seizures, abnormally low temperatures, and a peculiar steel color hair that feels very rough. Menkes disease is usually a fatal disease with most children dying within the first ten years of life.

COPPER (Cu)-RICH FOODS

Food Source	Serving	% DV
Cashews (1-oz)	629 mcg	31
Sunflower Seeds (1-oz)	519 mcg	26
Hazel Nuts (1-oz)	496 mcg	25
Almonds (1-oz)	332 mcg	17

Since 2001, Cu-deficiency has been associated with neurological manifestations such as sensory ataxia (irregular coordination due to proprioceptive loss), spasticity, muscle weakness, and (rarely) visual loss due to peripheral neuropathy, myelopathy (disease of the spinal cord), and rarely optic neuropathy. The deficiency in copper can also lead to several hematological manifestations, such as myelodysplasia (when a blood profile has indicators of possible future leukemia development), anemia, leucopenia (low white blood cell count) and neutropenia (low count of neutrophils associated with compromised immunity).

In conclusion, copper is vital for normal physiological functions. The human body has complex homeostatic mechanisms to ensure a constant supply of bio-available copper, and eliminate excess copper when neccesary. However, like all essential elements and nutrients, too much or too little nutritional ingestion of copper can result in a imbalanced condition of excess or deficiency in the body, each of which has its own unique set of adverse health effects.

A regular diet contains a variable amount of copper, usually about 5 mg/day, of which only 20-50% is absorbed. The diet of the elderly may contain a lower copper content than the recommended daily intake. Dietary copper can be found in whole grain cereals, legumes, oysters, organ meats (particularly liver), cherries, dark chocolate, fruits, leafy green vegetables, nuts, poultry, prunes, and soybean products like tofu.

Table: *Copper - Recommended Dietary Allowances*

AGE GROUP	MALE	FEMALE	PREGNANCY
0 to 6 months	200 mcg	200 mcg	
7 to 12 months	220 mcg	220 mcg	
1 to 3 years	340 mcg	340 mcg	
4 to 8 years	440 mcg	440 mcg	
9 to 13 years	700 mcg	700 mcg	
14 to 18 years	890 mcg	890 mcg	1000 mcg
19 to 50 years	900 mcg	900 mcg	1000 mcg
51+ years	900 mcg	900 mcg	

5.1.3. REDOX-Active Metal: Zinc (Zn)

Zinc [Zn(II)] is one of the most abundant and nutritionally essential elements in the human body. It is found in all body tissues, with 85% of Zn(II) in muscle and bone, 11% in the skin and the liver, and the remaining in all the other tissues. Zinc is critical to the structure and function of a large number of macromolecules and for over 300 enzymatic reactions. Zn(II) can interact strongly with electronegative sulfur, nitrogen, and oxygen moieties in multiple coordination geometries, serving catalytic and structural roles in maintaining active peptide conformations. Zn(II) is most known for its ability to bind and stabilize the proteins involved in gene regulation in domains called zinc fingers, zinc clusters, and zinc twists. Thus, Zn(II) ions exist primarily in the form of complexes (with proteins and nucleic acids) and participate in all aspects of intermediary metabolism, transmission and regulation of the expression of genetic information, storage, synthesis and action of peptide hormones, and structural maintenance of chromatin and bio-membranes. Zn(II) is also critical for the function of a number of metallo-proteins, inducing members of the oxido-reductase, hydrolase ligase, lyase families and has co-activating functions with copper in SOD or phospholipase C.

The paramount importance of zinc for a wide range of biological functions is based on its occurrence in thousands of known zinc proteins. About 3% of the ~32,000 identified genes in the human genome encode proteins with zinc-finger domains. This suggests that prokaryotes, eukaryotes and multi-cellular organisms increasingly acquired the potential to use zinc in their protein structures. To regulate the physiological availability of Zn(II), eukaryotes have compartmentalized zinc and the metallothionein/ thionein pair (which controls the pico- to nanomolar concentrations of metabolically-active cellular zinc).

Zn(II) as an Antioxidant: The antioxidant effects of Zn(II) can be divided into acute and chronic effects. Chronic effects involve exposure of an organism to Zn(II) on a long-term basis, resulting in induction of some other substance that is the ultimate antioxidant, such as metallothionein. Chronic zinc deprivation generally results in increased sensitivity to oxidative stress. The acute effects involve two mechanisms: protection of protein sulfhydryls (-SH) or reduction of ·OH radical formation from H_2O_2 through the antagonism of REDOX-active transition metals, such as iron and copper. Zinc is capable of reducing post-ischemic injury to a variety of tissues and organs through a mechanism that might involve the antagonism of copper reactivity.

Zn(II) and REDOX Balance: Zinc does not participate in REDOX reactions, which makes it a stable Zn(II) ion in a biological medium whose potential is in constant flux. However, REDOX-regulated signaling is exerted through several mechanisms based on the unique chemical and functional properties of zinc. Overall, zinc contributes to maintaining cellular REDOX balance through various mechanisms including: (i) the regulation of oxidant production and metal-induced oxidative damage; (ii) the dynamic association of zinc with sulfur in protein cysteine clusters, from which the metal can be released by nitric oxide (NO·), peroxides, oxidized glutathione (GSH), and other thiol oxidant species; (iii) zinc-mediated induction of the zinc-binding protein metallothionein, which releases the metal under oxidative conditions and acts per se as a scavenging oxidant; (iv) the involvement of zinc in the regulation of GSH metabolism and of the overall protein thiol REDOX status; and (v) a direct or indirect regulation of REDOX signaling.

A critical aspect of cellular zinc metabolism is the tight control of the picomolar concentrations of free Zn(II) ions (and their fluctuations) to balance folding and misfolding of proteins, supply of thousands of zinc-requiring proteins with Zn(II), and dual functions of zinc as either a prooxidant or a antioxidant. Zinc/sulfur (cysteine) bonds in proteins have a key role in this control because they generate REDOX-active coordination environments. Metallothionein (MT) is a REDOX-active zinc protein which couples biochemically to the cellular REDOX state. The coordination dynamics and REDOX state of its zinc/thiolate clusters determine cellular Zn(II) availability. REDOX stress affects the Zn(II) and REDOX buffering capacity of metallothioneins and elicits fluctuations of Zn(II) ions, which are potent effectors of multiple metabolic and signaling pathways.

Zinc-fingers and Gene Regulation: Zinc fingers act as REDOX-sensitive molecular switches controlling several crucial cellular processes. Oxidative or nitrosative stress, via modification of zinc finger cysteine thiols, leads to a release of Zn(II) from these structures. This results in the loss of zinc-finger function and an increase in cytoplasmic or nuclear free Zn(II) that may, in turn, stimulate and interfere with cellular signaling cascades. A growing body of evidence suggests that post-translational modifications of transcription factors serve a regulating role on gene transcription, particularly after changes to the cellular REDOX state. Zinc fingers are the most prevalent transcription factor DNA-binding motif. ·NO is able to S-nitrosate thiols of Zn-S clusters leading to reversible disruption of zinc finger structures, and provide a molecular mechanism to regulate gene transcription.

Zn(II) as a Protein Co-factor: Zinc is more widely used as a cofactor in proteins than any other transition metal ion. In addition to catalytic and structural functions, Zn(II) ions have a role in information transfer and cellular control. Transient zinc-binding sites employ the same donors of amino acid side chains as catalytic and structural sites. However, they differ in their coordination chemistry in that they can modulate zinc affinities over at least ten orders of magnitude. REDOX activity of the cysteine ligands, multiple binding modes of oxygen, sulfur and nitrogen donors, and protein conformational changes induce coordination dynamics in zinc sites and zinc ion mobility. A considerable number of proteins participate in cellular zinc homeostasis (e.g. membrane transporters, cellular storage, sensor, and trafficking proteins).

Zn(II) and Cell Signaling: Free Zn(II) ions are potent effectors of proteins. Their tightly controlled fluctuations ('zinc signals') in the picomolar range of concentrations modulate cellular signaling pathways. Sulfur (cysteine) donors in proteins generate REDOX-active coordination environments for REDOX-inert Zn(II) ions and make it possible for REDOX signals to induce zinc signals. Amplitudes of zinc signals are determined by the cellular zinc buffering capacity, which itself is REDOX-sensitive. By interfering with zinc and REDOX buffering, reactive species, drugs, toxins, and metal ions can elicit zinc signals that initiate physiological

Zinc-sensitive Targets in the Central Nervous System

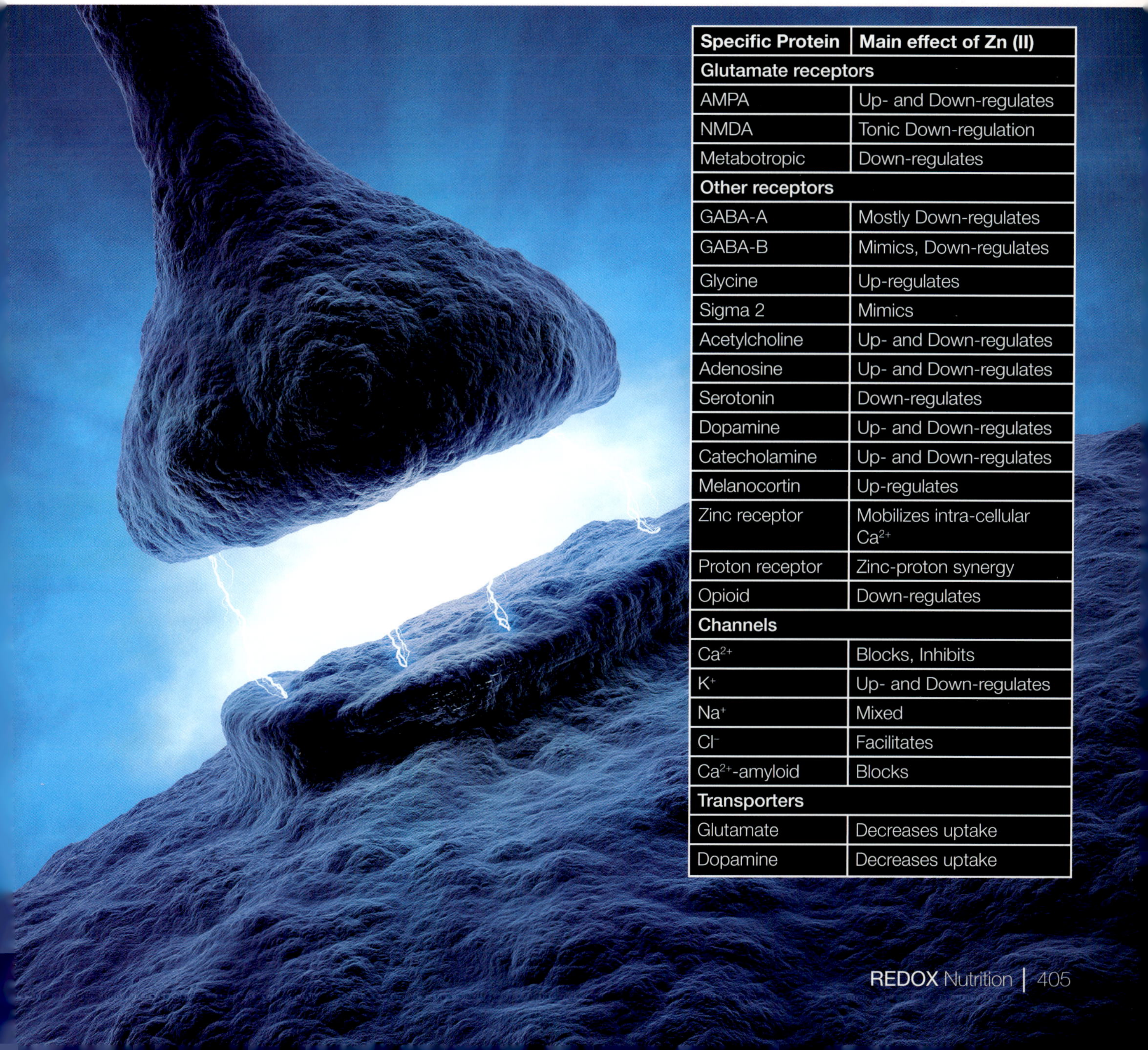

Specific Protein	Main effect of Zn (II)
Glutamate receptors	
AMPA	Up- and Down-regulates
NMDA	Tonic Down-regulation
Metabotropic	Down-regulates
Other receptors	
GABA-A	Mostly Down-regulates
GABA-B	Mimics, Down-regulates
Glycine	Up-regulates
Sigma 2	Mimics
Acetylcholine	Up- and Down-regulates
Adenosine	Up- and Down-regulates
Serotonin	Down-regulates
Dopamine	Up- and Down-regulates
Catecholamine	Up- and Down-regulates
Melanocortin	Up-regulates
Zinc receptor	Mobilizes intra-cellular Ca^{2+}
Proton receptor	Zinc-proton synergy
Opioid	Down-regulates
Channels	
Ca^{2+}	Blocks, Inhibits
K^{+}	Up- and Down-regulates
Na^{+}	Mixed
Cl^{-}	Facilitates
Ca^{2+}-amyloid	Blocks
Transporters	
Glutamate	Decreases uptake
Dopamine	Decreases uptake

ZINC (Zn)-RICH FOODS

Food Source	Serving	% DV
Oysters (3-oz)	74 mg	493
King Crab (3-oz)	6.5 mg	43
Beef Patty (3-oz)	5.3 mg	35
Breakfast Cereal (¼-cup)	3.8 mg	25

and patho-biochemical changes. They can also lead to cellular injury when free Zn(II) ions are sustained at higher concentrations. These interactions establish REDOX-inert zinc as an important factor in REDOX signaling. At the center of Zn(II)/REDOX signaling are the Zn(II)/thiolate clusters of metallothionein. They can transduce zinc and REDOX signals and thereby attenuate or amplify these signals. For example, Zn(II) ions have been shown to influence insulin and leptin signaling, potentially linking an imbalance of zinc (at the cellular level) to insulin resistance and dyslipidemia.

Zn (II) Supplementation: The catalytic, structural and regulatory functions of zinc in proteins account for its importance in maintaining a wide variety of biological processes. Free Zn(II) ions are not readily available at micro-molar or higher concentrations within a cell; their distribution is tightly controlled and regulated by multiple protein sensors, transporters and other types of proteins with novel mechanisms of action. With ~3 g in the human body, zinc is similar to iron in amount. In multi-cellular organisms, virtually all Zn(II) is intra-cellular; 30-40% is located in the nucleus, while 50% is in the cytoplasm, organelles and specialized vesicles (for digestive enzymes or hormone storage). The remainder is in the cell membrane. Zn(II) intake ranges from 107 to 231 μM/d depending on the source, and human Zn(II) requirement is estimated at 15 mg/d. Because of the critical role Zn(II) plays in protein structure and function, chronic dietary Zn(II) deficiency manifests in multi-systemic dysfunction, including growth failure, skin changes, delayed wound healing, immune system impairment, neurosensory disorders, and lack of sexual development.

Table: *Zinc - Recommended Dietary Allowances*

AGE GROUP	MALE	FEMALE	PREGNANCY
0 to 6 months	2 mg	2 mg	
7 to 12 months	3 mg	3 mg	
1 to 3 years	3 mg	3 mg	
4 to 8 years	5 mg	5 mg	
9 to 13 years	8 mg	8 mg	
14 to 18 years	11 mg	9 mg	12 mg
19+ years	11 mg	8 mg	11 mg

5.1.4. REDOX-Active Metal: Manganese (Mn)

Manganese (Mn), the 12th most abundant element in the Earth's crust and the 3rd most abundant transition element, occurs in seawater as the Mn(II) aqua ion. Manganese has a rich REDOX chemistry with several oxidizing states. The divalent Mn(II) is easily taken up by cells, remains REDOX-inactive in simple aqueous media and oxidizes only in the presence of suitable ligands (to provide an effective system for the oxygen-evolving centre). Manganese is an essential trace element, necessary for development and growth of organisms. An adequate content of this element in the body determines proper metabolism of amino acids, cholesterol and carbohydrates. As a constituent of numerous enzymes and a cofactor, manganese plays an important role in a number of physiological processes. As a REDOX-active metal, manganese plays a key role in cellular adaptation to oxidative stress. As a co-factor for manganese superoxide dismutase (Mn-SOD), or through formation of non-proteinaceous manganese antioxidants, this metal can combat oxidative damage without the deleterious side effects of Fenton chemistry. Other manganese-containing enzymes include oxidoreductases, transferases, hydrolases, lyases, isomerases, ligases, and glutamine synthetase. Manganese may affect thyroid hormone homeostasis and neurodevelopmental processes as a result of both direct dysregulation at the level of the thyroid gland and thyroid hormones, or indirectly via alterations in dopaminergic control of the thyroid gland and its hormones. Manganese also plays a vital role in various stages of skeletal development and bone turnover.

Mn-SOD and Superoxide ($O_2^{\bullet-}$) Scavenging: Oxygen-based respiration has evolved as an extraordinarily efficient mode of energy metabolism, but the efficiency comes at a price. The aggressive O_2 reactivity results in the formation of a variety of ROS within the cell, $O_2^{\bullet-}$ in particular. The cellular reactions of $O_2^{\bullet-}$ include destruction of Fe-S clusters in metallo-proteins and mutagenesis. In more general terms, ROS are capable of breaking the bonds of proteins, lipids and nucleic acids, altering their biological activity, and creating genetic mutations. In cells the dismutation reaction of $O_2^{\bullet-}$ is catalyzed by a family of enzymes known collectively as superoxide dismutases (SODs). SODs serve as the front-line defense against ROS in living cells, catalyzing the REDOX disproportionation of $O_2^{\bullet-}$ into hydrogen peroxide (H_2O_2) and molecular oxygen (O_2). All known SODs are metallo-enzymes containing a REDOX-active metal cofactor (Cu, Ni, Fe, or Mn) in their active site. During turnover, the active site metal ion undergoes cyclic oxidation and reduction, storing an electron and avoiding the electrostatic repulsion barrier that prevents direct reaction of the two superoxide anions. Mn-SOD is an important antioxidant defense that protects cells from damage by the toxic oxygen metabolite, $O_2^{\bullet-}$ free radical, formed as an unavoidable by-product of aerobic metabolism. Acquisition of the manganese metal cofactor is an essential step in the maturation of Mn-SOD and expression of its catalytic function.

MANGANESE (Mn)-RICH FOODS

Food Source	Serving	% DV
Pecan (1-oz)	1.28 mg	64
Brown Rice (½-cup)	1.07 mg	54
Pineapple Raw (½-cup)	0.77 mg	39
Peanuts (1-oz)	0.55 mg	28

Manganese and Bone Metabolism: Manganese is an essential co-factor in several metabolic and enzymatic reactions. Divalent Mn(II) cations can influence the integrin affinity to ligands and, accordingly, the adhesion to cells and extra-cellular matrix proteins. Thus, Mn supplementation could be a promising approach to improve in-growth and integration of bone implants. Mn(II) ions affect cellular functions such as spreading, proliferation and gene expression in human osteoblasts. Mn is a precursor for chondroitin biosynthesis, an essential component of articular cartilage. Mn(II) deficiency leads to cartilage metabolic disorder, a condition that resembles 'Mseleni joint disease', an osteoarthritis-like disease endemic to a remote part of Zululand. Furthermore, low Mn(II) levels in plasma is linked to the development of osteoporosis, and bone mineral density (BMD) seems to improve when Mn(II) is added to Ca(II) supplementation. Also, serum osteocalcin, one of the bone anabolic biomarkers, significantly increases with Mn supplementation.

Manganese Physiology: Disturbances in Mn absorption and retention may participate in the etiopathogenesis of certain diseases and disorders. Environmental or occupational exposure to high levels of manganese can cause a neuropathy resembling idiopathic Parkinson's disease, commonly referred to as manganism. Manganism was initially characterized as a psychiatric disorder, 'locura manganica', with symptoms including compulsive or violent behavior, emotional instability, and hallucinations. Other early manifestations of manganese neurotoxicity include fatigue, headache, muscle cramps, loss of appetite, apathy, insomnia, and diminished libido. As exposure to Mn continues and the disease progresses, patients may develop prolonged muscle contractions (dystonia),

Table: *Manganese - Recommended Dietary Allowances*

AGE GROUP	MALE	FEMALE	PREGNANCY
0 to 6 months	0.003 mg	0.003 mg	
7 to 12 months	0.6 mg	0.6 mg	
1 to 3 years	1.2 mg	1.2 mg	
4 to 8 years	1.5 mg	1.5 mg	
9 to 13 years	1.9 mg	1.6 mg	
14 to 18 years	2.2 mg	1.6 mg	2 mg
19 to 50 years	2.3 mg	1.8 mg	2 mg
51+ years	2.3 mg	1.8 mg	

decreased muscle movement (hypokinesia), rigidity, and muscle tremors. These signs are associated with damage to the dopaminergic neurons (within brain structures) that control muscle movement. In adult humans, manganese toxicity is a well-recognized occupational hazard. Inhalation of particulate Mn compounds [e.g., manganese dioxide (MnO_2) or manganese tetroxide (Mn_3O_4)] can result in an inflammatory response in the lungs. Respiratory symp-toms include cough, bronchitis, pneumonitis, and impaired pulmonary function.

Manganese deficiency can result in a host of ailments, including impaired growth, poor bone formation and skeletal defects, reduced fertility and birth defects, abnormal glucose tolerance, and altered lipid and carbohydrate metabolism. Sub-optimal manganese status may occur in humans with epilepsy, osteoporosis, or exocrine pancreatic insufficiency. It has also been observed in individuals under-going chronic hemodialysis, and in children with Perthes' disease or phenyl-ketonuria. In one study, young men placed on a manganese-depleted diet developed an erythematous rash on their torsos.

Manganism - Mn(II) Disorder

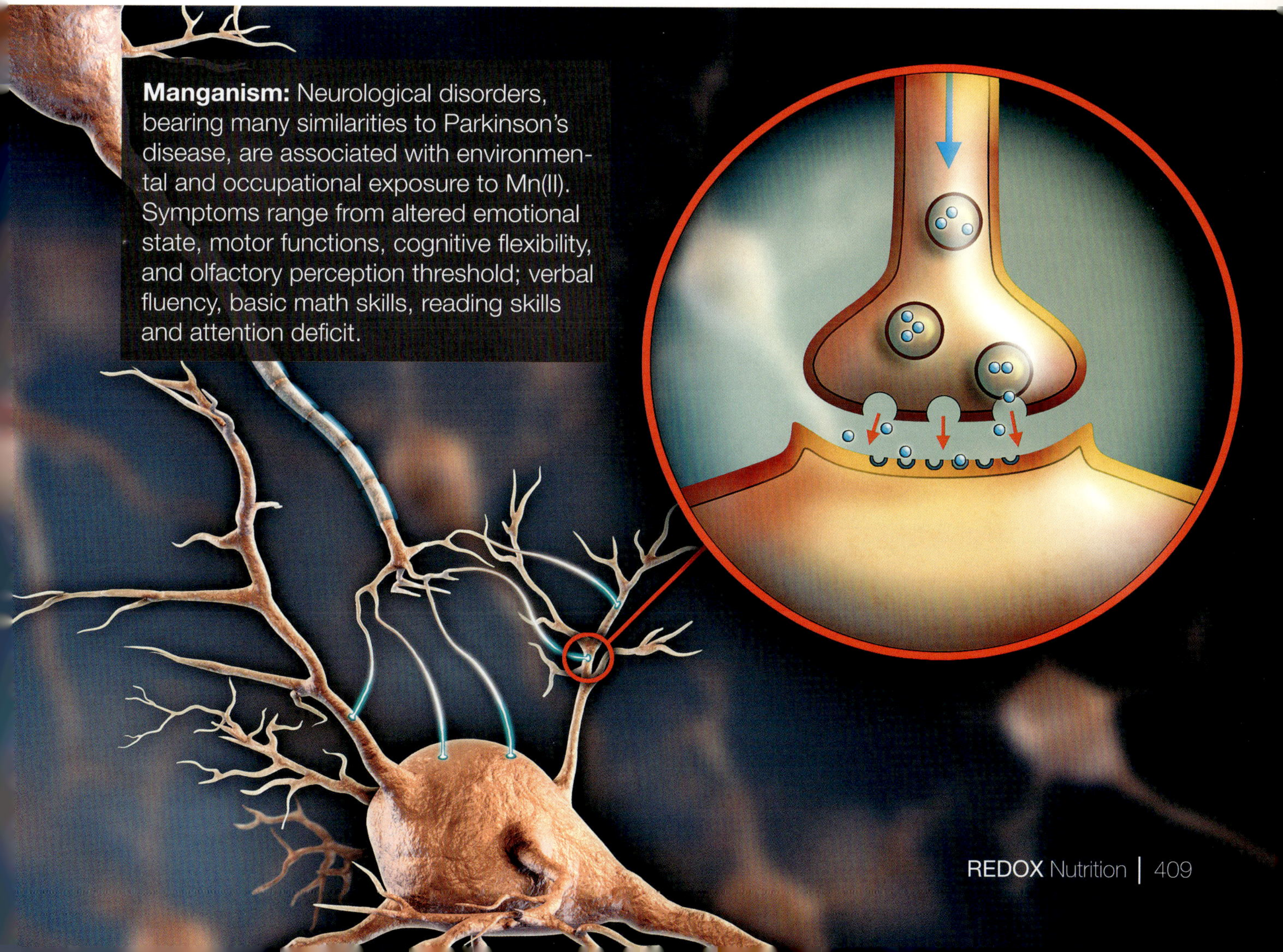

5.1.5. REDOX-Active Metal: Selenium (Se)

Selenium (Se) is an essential trace element with antioxidant, immunological, and anti-inflammatory properties. These properties are attributed to its presence in seleno-proteins, as the 21st amino acid selenocysteine (*Sec*). *Sec* and other Se-containing amino acids, such as selenocystine and selenomethionine (*SeMet*, the major dietary form), can provide benefits by acting both as direct antioxidants, as well as a source of Se for synthesis of selenium-dependent antioxidant and repair proteins (e.g. glutathione peroxidases, thioredoxin reductases, methionine sulfoxide reductases).The biological roles ascribed to Se include the prevention of cancer, cardiovascular disease, and viral mutation. Seleno-proteins are involved in bone metabolism as well as functions of the endocrine pancreas and adrenal glands. Furthermore, spermatogenesis depends on adequate Se supply, whereas excess Se may impair ovarian function. In addition, the trace element is essential for optimal endocrine and immune function and moderating the inflammatory response. These biological actions are mediated in most cases through the expression of at least 30 seleno-proteins coded by 25 seleno-protein genes in humans.

The chemical similarity of selenium with sulfur results in a number of interactions mainly connected to thiols and REDOX processes. Since the tRNAs cannot discriminate the selenium from the sulfur form, *Sec* and *SeMet* are indiscriminately incorporated into proteins instead of methionine (*Met*) and cysteine (*Cys*). The presence of *Sec* (in place of cysteine) may alter the tertiary structures and the activity of REDOX proteins if *Sec* is incorporated at critical positions (i.e. active sites or regulatory units). *SeMet* is incorporated mainly in albumin, and stored as an inactive form, with the consequent delay of metabolism. Supplementation with *SeMet* will thus lead to a prolonged increase in total plasma selenium but a small or no fraction of active selenium. Eventually, *SeMet* may be converted to *Sec*.

Mammalian seleno-proteins can be classified mainly into two groups (according to the location of *Sec*). One group of seleno-proteins possesses *Sec* in a site very close to the C-terminus of protein; these include TrxRs and seleno-proteins S, R, O, I, and K. The other group (including GPxs; DIOs; seleno-proteins H, M, N, T, V, and W) has *Sec* in the N-terminal part and, in most cases, possesses TRx fold structure. Overall, TrxRs, GPxs and DIOs are the three best characterized seleno-protein families. These seleno-proteins have different enzymatic activities, but all require reductants to provide the electrons to run their catalytic REDOX cycle. These seleno-proteins, also called seleno-enzymes, are involved in cell growth, apoptosis, and modifying the action of cell signaling systems and transcription factors. Thus, Se can influence at least three broad areas of cell biochemistry, namely antioxidant function, REDOX status and thyroid hormone metabolism.

Selenocysteine in Thioredoxin Reductases (TRxRs): TRxRs in mammalian cells are members of the oxido-reductase family, and three TrxRs have been identified in mammals: TRxR1 in the cytosol/nucleus, TrxR2 in mitochondria, and TRx-glutathione reductase in the testis (with the latter also possessing glutathione and glutaredoxin reductase activity).TRxR is the only enzyme able to reduce oxidized TRx. TRx provides electrons to ribonucleotide reductase, which is essential for DNA synthesis by converting ribonucleotide to deoxyribonucleotides. The TRx system participates in many cellular signaling pathways by controlling the activity of transcription factors containing critical cysteines in their DNA-binding domains, such as NF-κB and the glucocorticoid receptor.

Selenocysteine in Glutathione Peroxidases (GPxs): GPxs are major components of human antioxidant defense. In humans, there are five Sec-containing GPxs: the ubiquitous cytosolic GPx (GPx1), the gastrointestinal-specific GPx (GPx2), the plasma GPx (GPx3), the ubiquitous phospholipid hydroperoxide GPx (GPx4), and the olfactory epithelium- and embryonic tissue-specific GPx (GPx6). GPx1–3 catalyzes the reduction of H_2O_2 and organic hydroperoxides, whereas GPx4 directly reduces phospholipid and cholesterol hydroperoxides. GPx4 is also involved in sperm maturation and male fertility because it has been found to be a main structural component of the sperm mitochondrial capsule in mature spermatozoa as an enzymatically inactive, oxidatively cross-linked, insoluble protein.

Selenocysteine in Iodothyronine Deiodinases (DIOs): Three DIOs have been identified with tissue and subcellular localization. DIO1 is found primarily in the liver, kidney, and thyroid; DIO2 is in the brain, pituitary gland, thyroid, skeletal muscle, and brown adipose tissue; and DIO3 is found in the cerebral cortex and skin and is expressed at a very high level in the placenta and pregnant uterus. DIO1 and DIO2 catalyze the deiodination of T4, the major thyroid hormone secreted by the thyroid gland, into the active hormone T3; DIO3 converts T4 into reverse T3 and also T3 into di-idothyronine. DIO1 and DIO2 can also convert reverse T3 into di-iodothyronine.

Selenium and Thyroid function: The selenium content in endocrine tissues (thyroid, adrenals, pituitary, testes, ovary) is higher than in many other organs. Selenium, like iodine, is essential for normal thyroid function and thyroid hormone homeostasis. Several seleno-proteins participate in protecting thyrocytes from damage by the H_2O_2 produced for thyroid hormone biosynthesis. Iodothyronine deiodinases are seleno-proteins that contribute to systemic or local thyroid hormone homeostasis. In selenium deficiency, the apoptotic response to H_2O_2 is increased. When selenium intake is adequate, the intra-cellular GPx and TRxR systems protect the thyrocyte from the peroxide damage. In iodine deficiency or Grave's disease, hyper-stimulation of the TSH receptor signals increases H_2O_2 production. The activation of the calcium-phosphoinositol cascade stimulates GPx1 and TRxR1 to provide enhanced antioxidant protection.

Selenium and Male Fertility: The testes contain high concentrations of selenium, an mineral that essential for testicular function. Selenium-deficiency is associated with low sperm production and poor sperm quality including impaired motility with flagella defects localized primarily to the mid-piece. GPx4 provides the pivotal link between selenium, sperm quality and male fertility. GPx4 is critical for the synthesis of the mid-piece structure in spermatozoa. In the testes, GPx4 is present in all three isoforms and is found in the cytosol, mitochondria and nucleus. The nuclear form differs from the other forms because it has an arginine-rich N-terminus. In the spermatogenesis, the seleno-enzyme GPx4 provides protection from harmful ROS; during sperm maturation it takes on a structural role. GPx4 is a major component, present as a polymeric form with no enzymatic activity, in the mid-piece of mature sperm. GPx4 forms covalent cross-links with itself and other proteins that ultimately build up as a keratin-like material. This material is largely incorporated into the helix of mitochondria in the mid-piece of spermatozoa and ultimately forms up to 50% of the capsule material. Many individuals with infertility due to low sperm count and poor sperm quality have marked decreases in polymerized GPx4 in their sperm. The loss in GPx4 is particularly evident in oligo-astheno-zoospermic specimens. In a clinical study, the sperm quality and fertility of patients improved after selenium supplementation (30-40 µg/day). However, selenium deficiency has no significant effect on female reproduction.

Selenium and Diabetes: Selenium has insulin-mimetic properties in vitro and in vivo. Insulin-stimulated glucose metabolism is impaired in adipocytes isolated from selenium-deficient animals. Selenium also elicits insulin-like effects on cultured adipocytes including glucose transport stimulation, phosphor-diesterase activity and ribosomal S6 protein phosphorylation. Selenium restores glycemic control and modifies the activity of a range of enzymes involved in hepatic glycolysis and gluconeogenesis. These changes are not linked to changes in insulin levels. Selenium also prevents or alleviates the adverse effects that diabetes has on cardiac, renal and platelet function. Selenium may exert these insulin-like effects on glucose metabolism by stimulating the tyrosine kinases involved in the distal signaling of the insulin regulation cascade.

Selenium and Anti-Cancer Effects: Selenium can affect carcinogenesis at different stages; effects vary depending on the form of selenium compound administered. Anti-tumor effects, which occur at the systemic, cellular and molecular levels, include gene expression and direct protein alterations. The inhibitory effects are tumor specific and selenium induces apoptosis in malignant cells at concentrations that do not affect the viability of normal cells.

Pre-initiation Effects: The antioxidant role of selenium is the most important preventive effect at very early pre-nepolastic onset. Selenium exerts its antioxidant function as a constituent of cysteine containing seleno-proteins. These proteins play a central role in upholding REDOX homeostasis within the cell by acting as ROS detoxifying agents, thus reducing its presence in the intracellular environment. In addition to the reduction of ROS, they play a crucial role in reducing intra- and intermolecular disulfides or mixed disulfides/selenides. Another important property of selenium is its chemical ability to interact with the metals that increase the risk of cancers. Several of these metals can react and inhibit essential proteins like thioredoxin reductase (TRxR), and may thus exert their toxic effect by modulating the REDOX balance in the cell. Cadmium, for instance, is one key element for developing prostate and breast cancer; selenium has been reported to protect against cadmium-induced peroxidative damage.

Effects During Initiation and Promotion: The major effects of selenium compounds in the latter stages of carcinogenesis can be ascribed to its prooxidative effects (such as REDOX cycling and ROS production). The REDOX-active selenium metabolites, like selenide, also cause thiol oxidation. The oxidation of structural cysteine residues leads to a thiol modification, which directly inhibits several proteins, including signaling molecules, enzymes, tumor suppressors and transcription factors. REDOX-active selenium compounds can also induce apoptosis with different mechanisms. Selenide induces a caspase-independent apoptosis that generates DNA strand breaks.

Table: *Seleno-proteins in Human Physiology*

Seleno-protein	Proposed Function
Glutathione Peroxidase (GPx)	
GPx1	Antioxidant in cell cytosol; Possible selenium store
GPx2	Antioxidant in gastrointestinal tract
GPx3	Antioxidant in extra-cellular space and plasma
GPx4	Antioxidant; Structural protein in sperm; Apoptosis
GPx5	Unknown
GPx6	Possible GPx1 homolog
Thioredoxin Reductase (TRxR)	Multiple roles including dithiol-disulphide oxoreductase Detoxifies peroxides, reduces thioredoxin (TRx), Maintains REDOX state of transcription factors
TRxR1	Mainly cystosolic, Ubiquitous
TRxR2	Expressed in testes
TRxR3	Mitochondrial, Ubiquitous
Iodothyronine Deiodinase (DIO)	
Type DIO1 and DIO2	Converts thyroxine (T4) to bioactive T3
Type DIO1 and DIO3	Converts thyroxine (T4) to bio-inactive T3
Seleno-protein P	Se-transport protein. Antioxidant on endothelium
Seleno-protein W	Possibly an antioxidant in cardiac and skeletal muscle
Seleno-phosphate Synthetase	Synthesis of seleno-phosphate for seleno-protein
15 kDa Seleno-protein (Sep 15)	Possibly protects against cancer
H, I, K, M, N, O, R, S, T, V	Role largely unknown

SELENIUM (Se)-RICH FOODS

Food Source	Serving	% DV
Brazil Nuts (1-oz)	544 mcg	777
Tuna (3-oz)	92 mcg	131
Sardines (3-oz)	45 mcg	64
Ham Roasted (3-oz)	42 mcg	60

Effects during Progression and Metastasis: Selenium inhibiits tumors during the progression phase of carcinogenesis. Selenium treatment can also impair microvascular development and affect the vascular endothelial growth factor (VEGF) and contribute to anti-angiogenesis. Through down regulation of genes such as osteopontin and collagen, selenium also counteracts metastasis. The tumor specific toxicity together with the anti-angiogenic and anti-metastatic effects of Se makes it a potent agent in the arena of cancer chemotherapy.

Selenium in Health and Disease: Selenium has a two-sided effect depending on its concentration. A selenium-deficient diet is associated with various endemic diseases, including cardio-muscular malfunctions, osteoarthritis, cancer and viral infections that lead to premature death. These defects are prevented when dietary intake of selenium is adequate. Conversely, an overdose of selenium generates ROS and leads to apoptotic cell death by inducing oxidation and cross-linking of protein thiol groups. A lower REDOX state caused by selenium may be implicated in toxic conditions, such as alkali disease and blind staggers. Inadequate supply of selenium has been associated with predisposition for, or manifestation of, various human diseases such as Keshan and Kashin-Beck disease, cancer, impaired immune function, neurodegenerative and age-related disorders, and disturbances of the thyroid hormone axis. The current recommended dietary intake of selenium in humans is between 55 and 75 µg per day. These amounts are based on selenium intake that induces the GPx activity in plasma or erythrocytes. However, the anti-cancer property of selenium has been observed at intakes of 200 µg/day, which suggest that a re-appraisal of dietary selenium intake may be useful.

Table: *Selenium - Recommended Dietary Allowances*

AGE GROUP	MALE	FEMALE	PREGNANCY
0 to 6 months	15 mcg	15 mcg	
7 to 12 months	20 mcg	20 mcg	
1 to 3 years	20 mcg	20 mcg	
4 to 8 years	30 mcg	30 mcg	
9 to 13 years	40 mcg	40 mcg	
14 to 18 years	55 mcg	55 mcg	60 mcg
19 to 50 years	55 mcg	55 mcg	60 mcg
51+ years	55 mcg	55 mcg	

5.2 **REDOX** Vitamins

Vitamins are organic compounds required by an organism as vital nutrients in limited amounts. An organic chemical compound (or related set of compounds) is called a vitamin only when it cannot be synthesized in sufficient quantities by an organism and, as a result, must be obtained from the diet. Thus, the term is conditional both on the circumstances and on the particular organism. By convention, the term 'vitamin' does not include dietary minerals, essential fatty acids, essential amino acids (which are needed in larger amounts than vitamins), or the large number of other nutrients that promote health but are otherwise required less often. Currently, thirteen vitamins are universally recognized.

In 1897, Christiaan Eijkman, discovered that the consumption of unpolished rice (instead of the polished variety) helped to prevent beriberi. Eijkman believed that there were special nutrients contained in the husk of the rice. The following year, Frederick Hopkins postulated that some foods contained 'accessory factors' (in addition to proteins, carbohydrates, and fats) that are necessary for the functions of the human body. In 1929, both Hopkins and Eijkman were awarded the Nobel Prize for Physiology or Medicine for their discovery of several vitamins.

In 1910, the first vitamin complex was isolated by Japanese scientist Umetaro Suzuki, who succeeded in extracting a water-soluble complex of micronutrients from rice bran and named it aberic acid (later Orizanin). In 1912, Casimir Funk isolated the same complex of micronutrients and proposed the complex be named 'vitamine' – 'vita' meaning life and 'amine' from com-pounds found in the thiamine he isolated from rice husks. The name soon became synonymous with Hopkins' 'accessory factors'. Together, Hopkins and Funk formulated the hypothesis of vitamin deficiency disease - that a lack of vitamins could make people sick. By the time it was shown that not all vitamins are amines, the word was already ubiquitous. In 1920, Jack Cecil Drummond proposed that the final 'e' be dropped to de-emphasize the 'amine' reference, after researchers began to suspect that not all 'vitamines' (in particular, vitamin A) have an amine component.

In 1931, Albert Szent-Györgyi and a fellow researcher, Joseph Svirbely, suspected that 'hexuronic acid' was actually vitamin C; they gave a sample to Charles Glen King, who proved its anti-scorbutic activity. In 1937, Szent-Györgyi was awarded the Nobel Prize for his discovery. In 1943, Edward Adelbert Doisy and Henrik Dam were awarded the Nobel Prize in Physiology or Medicine for their discovery of vitamin K and its chemical structure. In 1967, George Wald was awarded the Nobel Prize (along with Ragnar Granit and Haldan Keffer Hartline) for his discovery that vitamin A could participate directly in physiological processes.

The Food Pyramid

A 'food pyramid' consists of healthy foods divided into sections with recommended intake for each group. The first food pyramid was published in Sweden in 1974. The most widely known food pyramid was introduced by the United States Department of Agriculture (USDA) in 1992, was updated in 2005, and then replaced in 2011. Over 25 other countries and organizations have also published their own food pyramids.

Table: *Vitamins in Human Physiology - Deficiencies and Overdosing*

Vitamin	Chemical Name(s)	Deficiency Condition	Overdose Condition
Vitamin A	Retinol, retinal, 4 carotenoids including beta Carotene	Night-blindness, hyperkeratosis, and keratomalacia	Hypervitaminosis A
Vitamin B1	Thiamine	Wernicke-Korsakoff syndrome, Beriberi	Drowsiness or muscle relaxation
Vitamin B12	Cyano-cobalamin, hydroxy-cobalamin, methyl-cobalamin	Megaloblastic anemia	Acne-like rash
Vitamin B2	Riboflavin	Ariboflavinosis	Not known
Vitamin B3	Niacin, niacinamide	Pellagra	Liver damage (doses > 2g/day)
Vitamin B5	Pantothenic acid	Paresthesia	Diarrhea, Nausea and Heartburn
Vitamin B6	Pyridoxine, pyridoxamine, pyridoxal	Anemia, peripheral neuropathy.	Impairment of proprioception, Nerve damage (doses > 100 mg/day)
Vitamin B7	Biotin	Dermatitis, enteritis	Not known
Vitamin B9	Folic acid, folinic acid	Megaloblastic anemia: deficiency during pregnancy (birth defects and neural tube defects)	May mask symptoms of vitamin B12 deficiency
Vitamin C	Ascorbic acid	Scurvy	Vitamin C megadosage
Vitamin D	Cholecalciferol	Rickets and Osteomalacia	Hypervitaminosis D
Vitamin E	Tocopherols, Tocotrienols	Deficiency is very rare; mild hemolytic anemia in newborn infants	Not known
Vitamin K	Phylloquinone, Menaquinones	Bleeding diathesis	Increases coagulation with warfarin

Vitamins are classified by their biological and chemical activity, not their structure. Thus, each vitamin refers to a number of vitamer compounds that show the biological activity associated with a particular vitamin. Such a set of chemicals is grouped under an alphabetized vitamin 'generic descriptor' title, such as 'vitamin A' (which includes the compounds retinal, retinol, and four known carotenoids). Vitamers by definition are convertible to the active form of the vitamin in the body, and are sometimes inter-convertible to one another.

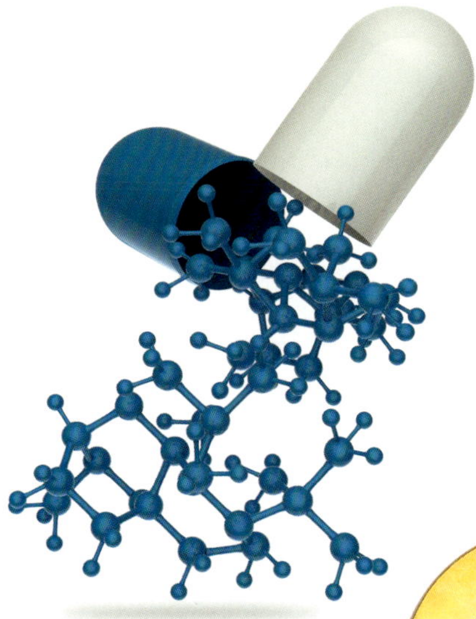

Vitamins: The 1929 Nobel Prize in Physiology or Medicine was divided equally between Christiaan Eijkman for his discovery of the anti-neuritic vitamin and Sir Frederick Gowland Hopkins for his discovery of the growth-stimulating vitamins.

5.2.1. REDOX Vitamin A

Vitamin A is a generic term for a large number of related compounds. Retinol (an alcohol) and retinal (an aldehyde) are often referred to as preformed vitamin A. All vitamin A forms have a β-ionone ring to which an isoprenoid chain (retinyl group) is attached. Both structural features are essential for vitamin activity. Vitamin A can be found in two principal forms in foods: (i) Retionol, the form of vitamin A absorbed when eating animal food sources, is a yellow, fat-soluble substance. Since the pure alcohol form is unstable, the vitamin is found in tissues in a form of retinyl ester. It is also commercially produced and administered as esters such as retinyl acetate or plamitate; (ii) Carotenes: α-carotene, β-carotene, γ-carotene; and the xanthophyll β-cryptoxanthin (all of which contain β-ionone rings). Retinal can be converted by the body to retinoic acid, the form of vitamin A known to affect gene transcription. Retinol, retinal, retinoic acid, and related compounds are known as retinoids. β-carotene and other carotenoids that can be converted by the body into retinol are referred to as provitamin A carotenoids. More than 600 carotenoids have been isolated in nature (β-carotene being the best known). Approximately 50 carotenoids have been found in the human diet and 20 have been identified in the plasma and tissues. Vitamin A is stored as long chain fatty esters and as provitamin carotenoids in the liver, kidney, and adipose tissue.

Vitamin A and REDOX: The antioxidant activity of vitamin A and carotenoids is conferred by the hydrophobic chain of polyene units that can quench singlet oxygen, and neutralize thiol and peroxyl radicals. The longer the polyene chain length, the higher the peroxyl radical neutralization ability. Because of their structures, vitamin A and carotenoids can auto-oxidize when O_2 tension increases, and thus are most effective antioxidants at low O_2 tensions (a typical physiological condition found in tissues). Also, both sets of compounds can participate in a biological antioxidant network. Under physiological conditions, vitamin A esters are transported and stored in a lipid matrix that contains other antioxidants, and retinol and its active metabolites are largely bound in clefts of specific retinoid-binding proteins. *In vivo*, vitamin A is protected by other anti-oxidants (such as vitamin E). Carotenoids are largely distributed in lipoproteins, membranes, and the lipid phases of intra-cellular structures, usually together with vitamin E.

Carotenoids are red, yellow, and orange in color and substantial in number (over 400 types). It is estimated that only 10% of pigments have vitamin A activity, with β-carotene having the greatest activity, followed by the α and γ forms.

Carotenoids are particularly vulnerable to free radical attack due to their long chains of conjugated double bonds. Carotenoids interact with cigarette smoke-containing oxidants. Depletion of carotenoids and tocopherol, particularly trans-β-carotene, α-tocopherol, and lycopene, has been demonstrated in cigarette smoke-exposed human plasma. β-carotene acts as a scavenger of nitrogen oxides in cigarette smoke. The β-apo-carotenals and other carotene oxidation products that are created, if not effectively neutralized by other antioxidants (specifically tocopherol and ascorbate), may initiate cell damage that can lead to neoplasm. Carotenoids act as oxidizing agents selectively in tumor cells by increasing the production of heat shock proteins, which ultimately enhance tumor cytotoxicity.

Vitamin A and Human Physiology: Vitamin A is vital for a variety of physiological functions including vision, reproduction, growth and maintenance of epithelial tissue and bone structures. Carotenoids exhibit biological activities as anti-oxidants; they can affect cell growth regulation and modulate gene expression. Adequate vitamin A is required for normal organogenesis, immune competence, tissue differentiation, and the visual cycle. Deficiency, which is widespread throughout the developing world, is responsible for a million or more instances of unnecessary death and blindness each year. β-carotene is an important, but insufficient, source of vitamin A among poor populations, which accounts for the widespread nature of vitamin A deficiency. The other major carotenoids, particularly lycopene, lutein, and zeaxanthin, have been found to have important biological properties, including antioxidant and photoprotective activity; high intake has been linked in observational studies with reduced risk of a number of chronic diseases.

Retinol, the active form of vitamin A, is rarely found in foods. Instead, precursors to retinol, fatty acid retinyl esters, are found in the human diet. The esters are commonly found in foods of animal origin, such as egg yolks, liver, fish oil, whole milk and butter.

Vitamin A and Vision: The role of vitamin A in the visual cycle is specifically related to the retinal form. The retina is located at the back of the eye. When light passes through the lens, it is sensed by the retina and converted into a nerve impulse for interpretation by the brain. Retinol is transported to the retina via circulation and accumulates in the retinal pigment epithelial cells. Here, retinol is esterified to form a retinyl ester, which can be stored. When needed, retinyl esters are broken apart (hydrolyzed) and isomerized to form 11-*cis*-retinol, which can be oxidized to form 11-*cis*-retinal. 11-*cis*-retinal can be shuttled across the inter-photoreceptor matrix to the rod cell, where it binds to a protein called opsin to form the visual pigment, rhodopsin (also known as visual purple). Rod cells with rhodopsin can detect very small amounts of light, making them important for night vision. Absorption of a photon of light catalyzes the isomerization of 11-*cis*-retinal to all-trans-retinal and results in its release. This isomerization triggers a cascade of events, leading to the generation of an electrical signal to the optic nerve. The nerve impulse generated by the optic nerve is conveyed to the brain where it can be interpreted as vision. Once released, all-trans retinal is converted to all-trans-retinol, which is transported across the inter-photoreceptor matrix to the retinal epithelial cell, thereby completing the visual cycle. Inadequate retinol available to the retina results in impaired dark adaptation, known as 'night blindness'.

Vitamin A and Dermatology: Vitamin A and retinoic acid, appears to maintain normal skin health by switching on genes and differentiation of keratinocytes (immature skin cells) into mature epidermal cells. For the treatment of acne, the most prescribed retinoid drug is 13-cis retinoic acid (iso-tretinoin). It reduces the size and secretion of the sebaceous glands. Isotretinoin also reduces bacterial numbers in both the ducts and skin surface by reducing sebum levels, a nutrient source for the bacteria. Isotretinoin controls inflammation via inhibition of chemotactic responses of monocytes and neutrophils. Isotretinoin also initiates remodeling of the sebaceous glands; triggering changes in gene expression that selectively induce apoptosis. Despite its benefits, however, isotretinoin is a 'teratogen' (an agent that can cause birth defects) with a number of potential side-effects; therefore, its use requires medical supervision.

Plants can synthesize carotenoids, but cannot convert them to retinoids; this process occurs in the human body.

Fruits and vegetables that appear bright orange or yellow are usually high in carotenoids. All green vegetables also contain substantial amounts of carotenoids, but the orange or yellow color is masked by chlorophyll.

Vitamin A and Immune Defense: Vitamin A is commonly known as the anti-infective vitamin, because it is required for normal functioning of the immune system. The skin and mucosal cells (cells that line the airways, digestive tract, and urinary tract) function as a barrier and form the body's first line of defense against infection. Retinol and its metabolites are required to maintain the integrity and function of these cells. Vitamin A and retinoic acid play a central role in the development and differentiation of white blood cells, such as lymphocytes, which are involved in the immune response. Furthermore, β-carotene can protect phagocytic cells from auto-oxidative damage, enhance T and B lymphocyte proliferative responses, stimulate effector T cell functions, and enhance macrophage, cytotoxic T cell and natural killer cell tumoricidal capacities. β-carotene has also been shown to increase the production of certain interleukins. Many of these effects have also been seen with carotenoids lacking provitamin A activity (but having the antioxidant and singlet oxygen-quenching capacities) of β-carotene.

Vitamin A Deficiency: Vitamin A deficiency among children in developing nations is the leading preventable cause of blindness. The earliest evidence of vitamin A deficiency is impaired dark adaptation or 'night blindness'. Mild vitamin A deficiency may result in changes in the conjunctiva (corner of the eye) called Bitot's spots. Severe or prolonged vitamin A deficiency causes a condition called xerophthalmia (dry eye), characterized by changes in the cells of the cornea (clear covering of the eye) that ultimately results in corneal ulcers, scarring, and blindness. Vitamin A deficiency can be considered a nutritionally acquired immunodeficiency disease. Even children who are only mildly deficient in vitamin A have a higher incidence of respiratory disease and diarrhea, as well as a higher rate of mortality from infectious disease (compared to children who consume sufficient vitamin A). Vitamin A supplementation has been found to decrease both the severity and incidence of deaths related to diarrhea and measles in developing countries, where vitamin A deficiency is common.

Vitamin A in Pharmacology: Retinoids are used at pharmacologic doses to treat several conditions, including retinitis pigmentosa, acute promyelocytic leukemia, and various skin diseases. It is important to note that treatment with high doses of natural or synthetic retinoids overrides the body's own control mechanisms; therefore, retinoid therapies are associated with potential side effects and toxicities. Both natural and synthetic retinoids have been used as pharmacologic agents to treat disorders of the skin. Etretinate and acitretin are retinoids that have been useful in the treatment of psoriasis, while tretinoin (Retin-A) and isotretinoin (Accutane) have been used successfully to treat severe acne. Retinoids most likely affect the transcription of skin growth factors and their receptors. Use of pharmacological doses of retinoids by pregnant women may cause birth defects.

Vitamin A and Carotenoids: The 1937 Nobel Prize in Chemistry was awarded to Paul Karrer for his investigations on carotenoids, flavins and vitamins A and B2.

Two forms of vitamin A are available in the human diet: preformed vitamin A (retinol and its esterified form, retinyl ester) and provitamin A carotenoids. By far the most important provitamin A carotenoid is β-carotene. Both provitamin A and preformed vitamin A must be metabolized intra-cellularly to retinal and retinoic acid, the active forms of vitamin A, to support the vitamin's important biological functions. Other carotenoids found in food, such as lycopene, lutein, and zeaxanthin, are not converted into vitamin A. Concentrations of preformed vitamin A are highest in liver and fish oils. Other sources of preformed vitamin A are milk and eggs, which also include some provitamin A. Most dietary provitamin A, however, comes from leafy green vegetables, orange and yellow vegetables, tomato products, fruits, and some vegetable oils. The top food sources of vitamin A in the diet include dairy products, liver, fish, and fortified cereals; the top sources of provitamin A include carrots, broccoli, cantaloupe, and squash.

Table: *Vitamin A - Recommended Dietary Allowances*

AGE GROUP	MALE	FEMALE	PREGNANCY
0–6 months	400 mcg RAE	400 mcg RAE	
7–12 months	500 mcg RAE	500 mcg RAE	
1–3 years	300 mcg RAE	300 mcg RAE	
4–8 years	400 mcg RAE	400 mcg RAE	
9–13 years	600 mcg RAE	600 mcg RAE	
14–18 years	900 mcg RAE	700 mcg RAE	750 mcg RAE
19–50 years	900 mcg RAE	700 mcg RAE	770 mcg RAE
51+ years	900 mcg RAE	700 mcg RAE	

1 IU retinol = 0.3 mcg RAE
1 IU beta-carotene from dietary supplements = 0.15 mcg RAE
1 IU beta-carotene from food = 0.05 mcg RAE
1 IU alpha-carotene or beta-cryptoxanthin = 0.025 mcg RAE

A RAE cannot be directly converted into an IU without knowing the source(s) of vitamin A.

VITAMIN A-RICH FOODS

Food Source	Serving	% DV
Sweet Potato (1-whole)	28,058 IU	561
Spinach (3-oz)	22,175 IU	229
Carrots (½-cup)	9,189 IU	184
Pumpkin Pie (1-piece)	3,743 IU	249

5.2.2. REDOX Vitamin C (Ascorbic Acid)

Vitamin C (or ascorbic acid) is a 6-carbon lactone, synthesized from glucose in the liver of most mammalian species (except primates). Vitamin C is an essential nutrient for the biosynthesis of collagen, L-carnitine, and the conversion of dopamine to norepinephrine. Under physiological conditions, it functions as a potent reducing agent that efficiently quenches potentially damaging free radicals produced by normal metabolic respiration of the body. Though most animals are able to synthesize large quantities of vitamin C endogenously, humans have lost this capability. This loss has been attributed to a series of inactivating mutations of the gene encoding gulonolactone oxidase (GULO), a key enzyme in the vitamin C biosynthetic pathway. These mutational events are estimated to have occurred about 40 million years ago, rendering all descending species, including humans, ascorbic acid deficient. Acute lack of vitamin C leads to scurvy, which manifests into blood vessel fragility, connective tissue damage, fatigue, and, ultimately, death.

Vitamin C and REDOX: Vitamin C is an antioxidant (due to its electron donor or reducing property) that prevents other compounds from oxidation. Ascorbic acid donates two electrons from a double bond between the 2nd and 3rd carbons of the 6-carbon molecule. However, the very nature of this reaction also oxidizes vitamin C. It is noteworthy that when vitamin C donates electrons, it forms semidehydroascorbic acid or ascorbyl, a free radical species. Compared to other free radicals, ascorbyl radical is relatively stable (non-reactive) with a half-life of 10^{-5} seconds. This unique property makes ascorbate the most preferred antioxidant in biological systems. In simple terms, when a harmful free radical interacts with ascorbate, the reactive free radical is reduced, and a less reactive ascorbyl radical is formed in its place, a phenomenon known as free radical scavenging or quenching. Ascorbyl radical has a short life due to its unpaired electron. Upon the loss of a 2nd electron, dehydro-ascorbic acid is formed. The stability of dehydroascorbic acid depends on factors such as temperature and pH. Formation of both ascorbyl radical and dehydro-ascorbic acid is mediated by wide variety of oxidants in biological systems, including molecular oxygen, superoxide, hydroxyl radicals, hypochlorous acid, reactive nitrogen species and specific trace metals (iron and copper). Both ascorbyl radical and dehydroascorbic acid can be reduced back to ascorbic acid by at least three separate enzyme pathways or by reducing compounds in biological systems (such as glutathione). In humans, there is only partial reduction and recovery of ascorbate. In vivo factors that regulate the efficiency of the ascorbate recovery process are not known. If dehydroascorbic acid is not reduced back to ascorbic acid, it is hydrolyzed irreversibly to 2, 3-diketogulonic acid. This compound is formed by irreversible rupture of the lactone ring structure that is a part of ascorbic acid, the ascorbyl radical, and dehydroascorbic acid. 2,3-diketogulonic acid is further metabolized into xylose, xylonite and oxalate. The formation of oxalate has clinical significance because hyperoxaluria (over excretion of oxalate) can result in oxalate kidney stones in some people.

Vitamin C - Antioxidant and Prooxidant Activities: At physiological concentrations, vitamin C is a potent free radical scavenger in the plasma, protecting cells against oxidative damage caused by ROS. The antioxidant property of ascorbic acid is attributed to its ability to reduce potentially damaging ROS, forming relatively stable ascorbate free radicals. This mechanism is attributed to a number of cyto-protective functions of ascorbate under physiological conditions, including prevention of DNA mutations induced by oxidation, protection of lipids against peroxidative damage, and repair of oxidized amino acid residues to maintain protein integrity. As DNA mutations are major contributors to the age-related development of cancer, attenuation of oxidation-induced mutations by vitamin C constitutes a potential anti-cancer mechanism. Plasma vitamin C, at normal to high physiological levels (60–100 mmol/L), decreases oxidative stress-induced DNA damage by neutralizing potentially mutagenic ROS. Consumption of vitamin C-rich foods is inversely related to the level of oxidative DNA damage *in vivo*.

Paradoxically, ascorbic acid may also function as a prooxidant, promoting oxidative damage to DNA. This occurs in the presence of free transition metals, such as copper and iron, which are reduced by ascorbate and, in turn, react with H_2O_2, leading to the formation of highly reactive and damaging ·OH radicals. However, the relevance of this, under normal physiological conditions *in vivo*, has been questioned (as most transition metals exist in inactive, protein-bound form *in vivo*). When administered at pharmacological concentrations (0.3–20 mmol/L), ascorbic acid displays transition metal-independent prooxidant activity, which is more profound in cancer cells and causes cell death. This tumor cell-killing response is dependent upon ascorbate incubation time and extra-cellular ascorbate concentration.

Vitamin C and Endothelial Function: Vitamin C protects normal ·NO synthesis by modulating the REDOX states of its components. It stabilizes endothelial tetrahydrobiopterin (BH4) levels by regenerating BH4 from its oxidized form through a series of reduction reactions. BH4 is an essential cofactor for eNOS; its oxidative inactivation is the major contributing factor to NO pathway aberration and consequent clinical manifestations. The stabilization and reactivation of the endothelial BH4 by ascorbic acid, and the consequent restoration of the normal biological activities of endothelial NO synthase (eNOS) activity and endothelial ·NO accumulation, may represent a key mechanism by which vitamin C impacts overall endothelial health.

Vitamin C attenuates the activity of NAD(P)H oxidase (NOx) and the production of $O_2^{\bullet -}$ *in vitro* and *in vivo*. NOx is the major source of endothelial $O_2^{\bullet -}$ responsible for the initial uncoupling of eNOS. Its inactivation by vitamins C and E suggests that these two vitamins may prevent the onset of eNOS uncoupling-induced

Vitamin C: The Nobel Prize in Chemistry 1937 was awarded to Walter Norman Haworth for his investigations on carbohydrates and vitamin C.

endothelial dysfunction by inhibiting the early events of this process. As a free radical scavenger, ascorbic acid, at high concentrations, reacts directly with $O_2^{\bullet-}$, decreasing its availability and inhibiting $O_2^{\bullet-}$ -mediated ·NO inactivation. Ascorbic acid may also preserve the normal enzymatic activity of eNOS by preventing the S-nitrosylation of the regulatory cysteine residues. In addition, ascorbate directly mediates the production of ·NO by reducing nitrite compounds, thereby maintaining high levels of local tissue ·NO concentrations (independent of the ·NO synthesis). It potentiates the responsiveness to ·NO stimulation by increasing the sensitivity of guanylyl cyclase to ·NO signaling.

The protective role for vitamin C on ·NO and its biosynthetic pathway is supported by clinical evidence. Vitamin C improves endothelium-dependent vasodilatation in patients with endothelial dysfunction. Because endothelial dysfunction generally marks the onset of atherosclerosis and cardiovascular disease (CVD), vitamin C supplementation may be beneficial during early stages of these conditions.

Vitamin C and Bone Metabolism: Vitamin C can induce embryonic stem cells to differentiate into osteoblasts (the bone forming cells). Its pre-osteoblast proliferation is mediated through the synthesis of collagen type I, interaction with α2- and β1-integrin, activation of the mitogen-activated protein kinase pathway, and phosphorylation of osteoblast-specific transcription factors. Mouse calvaria-derived cell lines (MC3T3-E1) are a well-defined in vitro model for pre-osteoblast differentiation. By using DNA micro-arrays containing 15,000 genes, it was found that ascorbate can regulate the expression of several genes in MC3T3-E1. The differentially expressed genes covered a broad range of functional activities including cell growth, metabolism, morphogenesis, apoptosis and cell signaling. Elucidation of these molecular mechanism have important clinical implications; they may facilitate the correct use of vitamin C to accelerate bone regeneration. Vitamin C is also an essential cofactor for lysyl- and prolyl-hydroxylases, enzymes involved in collagen biosynthesis. Ascorbate is required for hydroxylation of proline residues in procollagen and hydroxyproline that stabilize the triple helical structure by cross-linking of collagen fibrils in the bone joint. Vitamin C deficiency is associated with defective connective tissue, especially during bone and wound healing.

Vitamin C can help both bone formation and bone resorption, the dual aspects of bone turnover. Several studies have reported a beneficial effect of vitamin C intake on the bone mineral density (BMD). The Postmenopausal Estrogen/Progestin Interventions (PEPI) study showed a positive association of vitamin C with BMD in postmenopausal women with dietary Ca(II) intakes of at least 500 mg. On the other hand, ascorbate may offer the clinical benefit of preventing alveolar bone resorption in cases of hyperlipidemia. High-cholesterol diets stimulate alveolar bone resorption, which may be induced via tissue oxidative damage. Dietary vitamin C can suppress alveolar bone resorption, induced by high dietary cholesterol, by decreasing osteoclast differentiation and 8-hydroxydeoxyguanosine expression in periodontal tissue.

Vitamin C and Cell Signaling: The intra-cellular REDOX changes caused by oxidants and antioxidants can modulate the expression of genes involved in signal transduction pathways (leading to cell cycle progression, cell differentiation, and apoptosis). For example, cells treated with ascorbic acid at low pharmacologic concentration (1 mM/ L) modulate the expression of apoptotic genes and induce DNA damage. Ascorbate also mediates the mitochondrial release of cytochrome C and activates caspases leading to apoptotic processes. Vitamin C, at low intra-cellular concentrations, inhibits the activation of nuclear factor kappa B (NF-κB), a rapid response transcription factor, by preventing TNFα-mediated degradation of its inhibitor through independent mechanisms. NF-κB induces the transcription of genes involved in both inhibition of apoptosis and promotion of cell proliferation; its over expression directly contributes to malignancy. Repression of constitutive activation of NF-kB by vitamin C can induce cell cycle arrest and apoptosis in these cells and attenuate tumor progression in different types of cancer. Ascorbate and its lipophilic derivatives attenuate cell proliferation, arrest cell cycle, and induce apoptosis in human glioblastoma tumor and pancreatic cancer cells by reducing the expression of insulin-like growth factor-I receptor. Cell cycle arrest induced by vitamin C is also attributable to its ability to prevent the activation and nuclear accumulation of the mitosis-inducing phosphatase enzymes. The above described cell signaling mechanisms provide a basis for an adjuvant role for vitamin C in the treatment of cancers. However, the concurrent use of antioxidants (such as ascorbic acid) as chemotherapeutic agents is still controversial.

Table: *Vitamin C - Recommended Dietary Allowances*

AGE GROUP	MALE	FEMALE	PREGNANCY
0–6 months	40 mg	40 mg	
7–12 months	50 mg	50 mg	
1–3 years	15 mg	15 mg	
4–8 years	25 mg	25 mg	
9–13 years	45 mg	45 mg	
14–18 years	75 mg	65 mg	80 mg
19+ years	90 mg	75 mg	85 mg

VITAMIN C-RICH FOODS

Food Source	Serving	% DV
Red Pepper (½-cup)	95 mg	158
Orange (1-medium)	70 mg	117
Grapefruit Juice (¼-cup)	70 mg	117
Kiwi Fruit (1-medium)	64 mg	107

5.2.3. REDOX Vitamin E

Vitamin E encompasses a group of potent, lipid-soluble, chain-breaking antioxidants. Structural analyses have revealed that molecules having vitamin E antioxidant activity include: 4 tocopherols (α, β, γ, δ) and 4 tocotrienols (α, β, γ, δ). One form, α-tocopherol, is the most abundant form in nature and has the highest biological activity; it has been shown to reverse vitamin E deficiency symptoms in humans. Tocotrienols penetrate rapidly through skin and efficiently combat oxidative stress induced by UV or ozone. Tocotrienols also have beneficial effects on cardiovascular diseases, both by inhibiting LDL oxidation and by down-regulating 3-hydroxyl-3-methylglutaryl-coenzyme A (HMG CoA) reductase, a key enzyme of the mevalonate pathway. Tocotrienols also elicit antiproliferative and neuroprotective effects, which may be independent of their antioxidant activity.

Vitamin E, as a dietary antioxidant, helps reduce oxidation of lipid membranes and unsaturated fatty acids; it also prevents the breakdown of other nutrients by oxygen. Vitamin E stabilizes cell membranes and protects the tissues of the skin, eyes, liver, breast, and testes, which are more sensitive to oxidation. Vitamin E plays a role in maintaining the biological activity of vitamin A, and prevents the oxidation of some hormones released from the pituitary and adrenal glands. With regard to its wound healing functions, vitamin E is used most commonly (both internally and externally) to assist in the repair of skin lesions, ulcers, burns, abrasions, and dry skin and to heal and/or diminish scars caused from injury or surgery. Vitamin E, with the help of vitamins C and P (bioflavonoids), may be useful in preventing the progression of varicose veins (more so than treating them once they have occurred).

Vitamin E and Antioxidant Activity: As the most important lipid-soluble antioxidant; vitamin E quenches the free radicals and acts as a terminator of lipid peroxidation. The vitamin E radical may be reduced back to vitamin E by ascorbate and GSH. Vitamin E protects cell membranes from oxidative damage by removing ROS intermediates and preventing the progression of oxidative reactions. α-TP elicits antioxidant activity via the GSH-peroxidase (GPx) pathway. It scavenges lipid peroxyl radicals and breaks peroxyl chain propagation reactions. The radical form is converted back to α-TP in a REDOX cycle involving coenzyme Q10 or vitamin C. The regeneration of α-TP from tocopheroxyl radical greatly enhances the turnover efficiency of α-TP in its role as a lipid antioxidant. Vitamin E forms complexes with the lysophospholipids and free fatty acids released during membrane lipid hydrolysis. Both these products form 1:1 stoichiometric complexes with vitamin E and restore the overall balance of hydrophobic: hydrophillic affinity within the membrane. Thus, vitamin E negates the detergent-like properties of the hydrolytic products that can otherwise disrupt membrane stability.

A role distinct from ROS scavenging has been proposed for γ-tocopherol. In contrast to α-tocopherol, γ-tocopherol is a potent nucleophile that traps electrophilic mutagens in lipophilic compartments. It thus complements GSH, which similarly scavenges electrophilic mutagens in the aqueous phase of the cell. An electrophilic mutagen prone to react with γ-tocopherol is peroxynitrite ($ONOO^-$). Thus, γ-tocopherol may protect lipids, DNA, and proteins from peroxynitrite-dependent damage.

Vitamin E and Cellular Signaling: The role of vitamin E in cellular signaling, especially in relation to protein kinase C (PKC), has been studied intensively by Azzi's group. α-Tocopherol inhibits smooth muscle cell proliferation, decreases protein kinase C (PKC) activity, increases phosphoprotein phosphatase 2A activity, and controls expression of the α-tropomyosin gene. These functions are not related to vitamin E antioxidant action because β-tocopherol, which has a similar antioxidant activity, does not perform any of these functions; it actually abrogates the α-tocopherol effect. α-tocopherol effects on PKC inhibition have also been reported in human platelets and monocytes. The mechanism of PKC inhibition by α-tocopherol may be attributable (in part) to its attenuation of the generation of membrane-derived diacylglycerol, a lipid that activates PKC translocation and activity. The PKC inhibition activity is not due directly to the antioxidant capacity of α-tocopherol; rather, it requires the integration of α-tocopherol into a membrane structure, and is likely due to the direct interaction between α-tocopherol and PKC in the cell membrane. α-tocopherol modulates the *in vitro* expression of some significant proteins/enzymes in various cell types involved in atherogenesis. Vitamin E enrichment of endothelial cells down-regulates the expression of intercellular cell adhesion protein and vascular cell adhesion molecule-1, thereby reducing the oxidized LDL-induced adhesion of white cells to the endothelium. α-tocopherol can regulate the pathways of arachidonic acid cascade, and its effect is not always shared by other vitamin E forms. Vitamin E up-regulates the activities of cytosolic phospholipase A2 and cyclooxygenase. The enhanced activity of these two rate-limiting enzymes in the arachidonic acid cascade provides a mechanism for the observation that vitamin E dose-dependency enhances release of prostacyclin, a potent vasodilator and inhibitor of platelet aggregation.

Vitamin E and Cardiovascular Health: One of the key functions of vitamin E is to modify and stabilize plasma lipids so that the blood vessels, heart, and entire body are more protected from free radical-induced injury. Vitamin E also has some anti-clotting (anti-thrombotic) properties which protect the red blood cells from oxidative damage. Because it helps heart and muscle cell respiration, by improving their functioning with less oxygen, vitamin E may help improve stamina and reduce cardiovascular disease (CVD) risk, especially in those with

In 1922, Evans and Bishop discovered vitamin E—a micronutrient essential for reproduction in rats. It was rediscovered in the 1950s as factor 2 and placed in the context of cellular antioxidant systems, together with sulfur amino acids (factor 1) and selenium (factor 3). Vitamin E subsequently proved to be effective in preventing lipid peroxidation and other radical-driven oxidative events.

existing CVD. Vitamin E has recently been shown to reduce platelet aggregation and platelet adhesiveness to collagen, even more so than aspirin. These platelet functions are linked to an increased risk of atherosclerosis and cardiovascular disease, especially in high-risk groups.

Vitamin E has also been shown to neutralize free radicals generated during surgery, particularly cardio-pulmonary bypass surgery. It can also protect against the toxicity of some of the gases used in anesthesia. Decreased blood clotting and increased tissue oxygenation may also help reduce symptoms of heart and vascular limitations, such as angina pectoris, intermittent claudication (leg pain with walking due to insufficiency of blood and oxygen), and problems of arterial spasm. In both congenital and rheumatic heart diseases, vitamin E may help reduce symptoms caused by impaired tissue oxygenation. Vitamin E may also be of help in the prevention of atherosclerosis. Its antioxidant effect reduces thrombin formation and thus helps decrease blood clotting. Vitamin A and E together can help to decrease cholesterol and general fat accumulation.

In 1996, the Cambridge Heart Antioxidant Study (CHAOS) reported that, in over 2000 patients with coronary atherosclerosis, vitamin E supplementation (400–800 IU/day) for slightly under 2 years significantly reduced the incidence of cardiovascular death and non-fatal myocardial infarction by 77%. Decreases in lipid peroxidation of low density lipoproteins (LDL) have been assumed to be the mechanisms for this result. However, tissue responses to oxidative stress may be important. In a Japanese trial (n=60) involving coronary spastic angina, treatment for 30 days with 300 mg/day α-tocopherol resulted in a significant improvement of impaired endothelium-dependent vasodilation concomitant with a reduction of plasma TBARS (thiobarbituric acid-reactive substrate), a marker of lipid peroxidation.

Vitamin E and Genital Health: Vitamin E prevents loss of spermatogenesis in males and the failure to retain zygotes. Vitamin E is known to readily reduce alkyl peroxy radicals of unsaturated lipids, thereby generating H_2O_2 that is reduced by the selenoperoxidases [in particular by phospholipid hydroperoxide glutathione peroxidase (GPx)]. Correspondingly, vitamin E and selenium can substitute for each other, or at least act synergistically, in pathogenic phenomena arising from oxidative stress. Vitamin E cannot, however, replace selenium in the prevention of functional and structural alterations of spermatozoa. Although vitamin E was first discovered as the fertility (or at least the anti-sterility) nutrient, there is no clear evidence that it enhances fertility if there is not a specific deficiency prior to its use. Many people, especially men, take vitamin E with some claimed success in regard to sexual function and vitality. Much of this effect, however, may be due to antioxidant function, improved circulation, and oxygenation. Among women, Vitamin E may help relieve menstrual pains. Other problems of menopause, such as headaches, hot flashes, or vaginal itching due to dryness, may also be reduced with the use of supplemental vitamin E. Supplemental estrogen or estrogen imbalance in women increases the need for vitamin E.

Vitamin E Deficiency: In humans, deficiency of vitamin E may produce diminished erythrocyte life span, neurological dysfunction, and myopathies in patients with various malabsorption syndromes, including abetalipoproteinemia. Vitamin E deficiency has also been linked to the development of broncho-pulmonary dysplasia, intra-ventricular hemorrhage in the brain, retrolental fibroplasia, and anemia in children born prematurely. In addition, alterations of vitamin E status has been associated with the development of certain forms of cancer, cardiovascular diseases, and impaired immune response.

Overt vitamin E deficiency occurs only rarely in humans and virtually never as a result of dietary deficiencies. Vitamin E deficiency occurs as a result of genetic abnormalities in the α-tocopherol transfer protein (α-TTP) and various fat malabsorption syndromes. Patients with familial-isolated vitamin E deficiency, an inborn genetic defect in the gene for the α-TTP, have dramatically reduced plasma vitamin E levels and neurological disorders characteristic of vitamin E deficiency (such as cerebellar ataxia, dysarthria, absence of deep tendon reflexes, vibratory and proprioceptive sensory loss, and positive Babinski sign). The deficiency symptoms associated with this syndrome can be ameliorated when these patients are given doses of vitamin E of up to 2000 mg per day. Also, symptoms of vitamin E deficiency caused by chronic liver disease, fat malabsorption, or abetalipoproteinemia are ameliorated by high doses of vitamin E. None of the curative dosages can be achieved by an optimized dietary regimen; the patients must consume vitamin E supplements.

Vitamin E Toxicity: Vitamin E is not stored as readily as are the other fat-soluble vitamins. Excess intake is usually eliminated in the urine and feces, and most doses clear the body within a few days. For these reasons, toxicity from vitamin E use is unlikely. High intakes of the vitamin E oil can cause nausea, diarrhea, or flatulence in some people. Higher doses of vitamin E should also be avoided in people with rheumatic heart disease and/or anticoagulant therapy, since it may increase the anticoagulant effects of these medicines. Deficiencies are more of a concern in premature babies, since there is no maternal-fetal vitamin E transfer; vitamin E depletion may also appear in newborns fed on cow's milk (instead of breast milk), which contains no vitamin E. Deficiency is also more likely in adults with gastrointestinal disease, poor fat digestion, and metabolism issues.The first sign of vitamin E deficiency may be loss of red blood cells due to fragility caused by the loss of cell membrane protection. Oxidized polyunsaturated fatty acids (PUFAs) may also weaken the red blood cell membranes and cause rupture. The generalized decrease in cell and tissue protection from free radicals may lead to abnormal fat deposits in muscles, muscle wasting, and problems in the kidneys and liver because of the circulating dead cells and released toxins.

Several claims have been made about vitamin E's potential to promote health and prevent and treat disease. The mechanisms by which vitamin E might provide this protection include its function as an antioxidant and its roles in anti-inflammatory processes, inhibition of platelet aggregation, and immune enhancement.

VITAMIN E-RICH FOODS

Food Source	Serving	% DV
Wheatgerm Oil (1-tbs)	20.3 mg	100
Roasted Almonds (1-oz)	6.8 mg	34
Sunflower oil (1-tbs)	5.6 mg	28
Safflower oil (1-tbs)	4.6 mg	25

Dietary sources of Vitamin E: Vitamin E, in its various tocopherol forms, is found in both plant and animal foods. The best sources of vitamin E are vegetable, seed, or nut oils. Vitamin E was first isolated from wheat germ oil, which is still a commonly used and rich source. The content of active α-tocopherol varies among the different foods and oils. Safflower oil is one of the best sources, with about 90 percent of its Vitamin E being the alpha variety. Some other foods that contain significant amounts of vitamin E are soybeans and certain vegetables, such as uncooked green peas, spinach, asparagus, kale, and cucumber (tomato and celery also have a little). The amount of vitamin E required depends upon body size and the amount of polyunsaturated fats in the diet, since vitamin E is needed to protect these fats from oxidation. More Vitamin E is needed when any refined oils, fried foods, or rancid oils are consumed.

Serum concentrations of vitamin E depend on the liver, which takes up the compound after the various forms are absorbed from the small intestine. The liver preferentially releases only α-tocopherol via the hepatic α-TP transfer protein; the liver metabolizes and secretes the other vitamin E forms. Specific α-TP-binding proteins are responsible for retention of this isomer in the cells and tissues of the body. Because of the lipophilic properties of the vitamin, it partitions into lipid storage organelles and cell membranes. It is, therefore, widely distributed throughout the body.

Table: *Vitamin E - Recommended Dietary Allowances*

AGE GROUP	MALE	FEMALE	PREGNANCY
0–6 months	4 mg (6 IU)	4 mg (6 IU)	
7–12 months	5 mg (7.5 IU)	5 mg (7.5 IU)	
1–3 years	6 mg (9 IU)	6 mg (9 IU)	
4–8 years	7 mg (10.4 IU)	7 mg (10.4 IU)	
9–13 years	11 mg (16.4 IU)	11 mg (16.4 IU)	
14+ years	15 mg (22.4 IU)	15 mg (22.4 IU)	15 mg (22.4 IU)

5.3 | **REDOX** Amino Acids

Amino acid residues in proteins represent one of the major targets of ROS and cellular REDOX. All amino acids can be oxidized; these oxidative modifications include the polypeptide backbone as well as the amino acid side chain. Oxidative abstraction of a hydrogen atom from the α-carbon atom can lead to peptide bond cleavage. Side chains of amino acids in proteins are readily oxidized to markedly alter the overall properties of the amino acids, thus potentially modifying protein function.

The sulfur-containing amino acids, cysteine and methionine, are especially sensitive to ROS mediated oxidation. The side chains of arginine, lysine, proline, histidine, tryptophan, and tyrosine are also readily susceptible to oxidative modification. The overall extent of protein oxidation is often estimated by measuring the carbonyl content. A significant fraction of the total protein may be oxidized in humans with consequences in health and aging.

Cysteine (*Cys*) and methionine (*Met*) possess reactive sulfur-containing side chains that represent prime targets of ROS. Oxidation of cysteine and methionine residues in proteins is noteworthy in that the reactions are physiologically reversible. Oxidation of thiol-containing cysteine, often promoted by the presence of trace amounts of metal ions [such as Cu(II), Fe(II), Co(II) and Mn(II)], leads to a variety of products including sulfenic ions, disulfide and sulfonic ions. Physiologically, the disulfide formation may be the most likely consequence of cysteine oxidation. Disulfides can be easily reduced back to thiols using glutathione (GSH) *in vivo.* Reversible oxidation of cysteine has been postulated to work as an important cellular REDOX sensor in some proteins. Stronger experimental oxidants may also produce sulfonic and eventually sulfenic acid. Functional modulation of proteins in cellular excitability by cysteine oxidation has been suggested in many different REDOX signaling pathways.

Disulfide bonds in proteins are formed between the thiol groups of cysteine residues. The other sulfur-containing amino acid, methionine cannot form disulfide bonds. The disulfide bond stabilizes the folded form of a protein in several ways: (i) It holds the two portions of the protein together, biasing the protein towards the folded topology. As a result, the disulfide bond destabilizes the unfolded form of the protein by lowering its entropy. (ii) It may form the nucleus of a hydrophobic core of the folded protein, i.e., local hydrophobic residues may condense around the disulfide bond and onto each other through hydrophobic interactions. (iii) It increases the effective local concentration of protein residues and lowers the effective local concentration of water molecules. Since water molecules attack amide-amide hydrogen bonds and break up secondary structures, a disulfide bond stabilizes the secondary structure in its vicinity.

Standard Amino Acids: Biological Functions

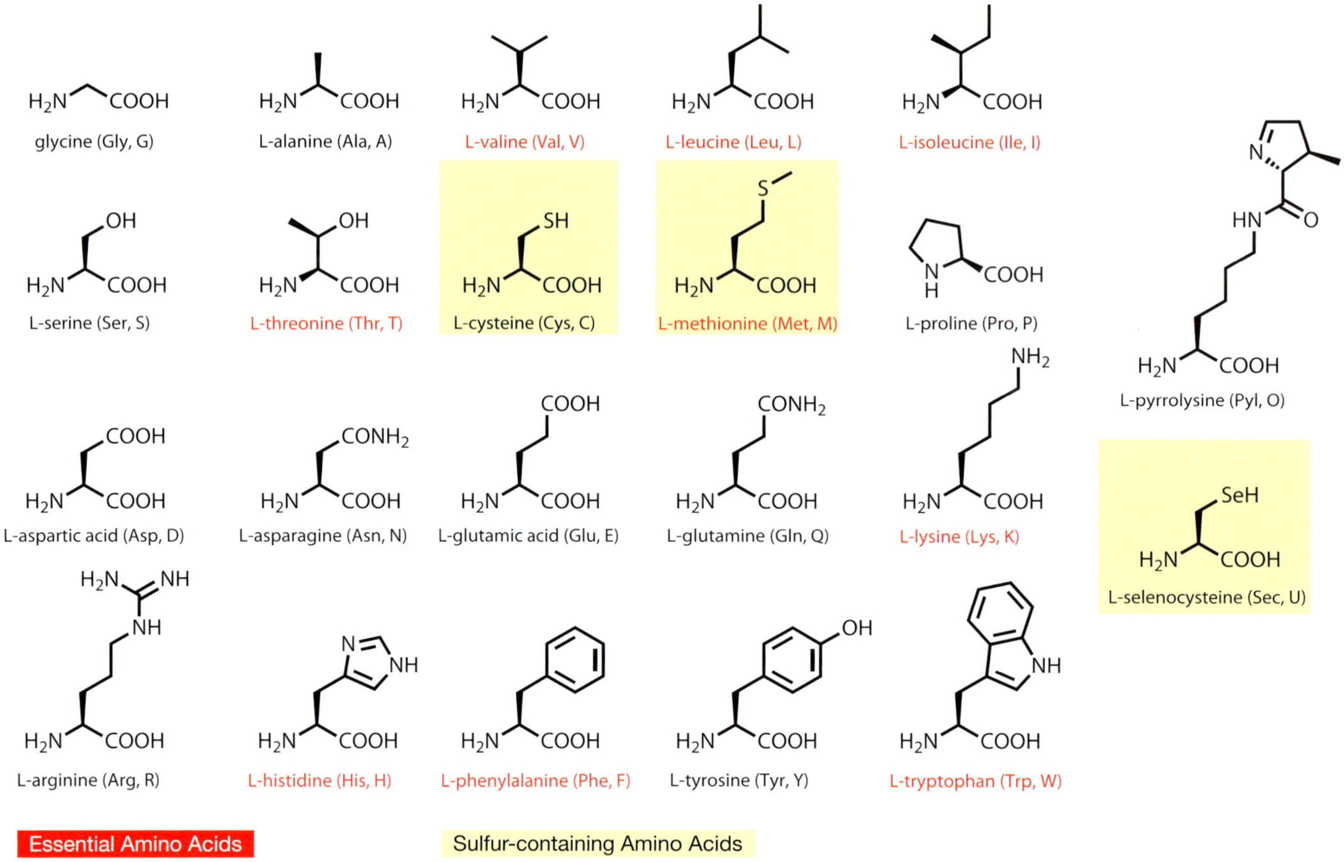

Amino Acid	Biological Function
ALANINE	Major component of connective tissue, key intermediate in the glucose alanine cycle that allows muscles and other tissues to derive energy from amino acids.
ARGININE	Immune modulator; promotes wound healing and liver regeneration; helps release growth hormone; vital for optimal muscle growth and tissue repair; increases secretion of insulin, glucagon; collagen synthesis; precursor of creatine and GABA; increases sperm count and boosts the immune system.
ASPARTATE	Aids in expulsion of harmful ammonia from the body; increases resistance to fatigue and provides endurance; converts carbohydrates into muscle energy, and builds immunity.
CYSTEINE	Detoxifies harmful chemicals in combination with L-aspartic acid and L-citruline; helps prevent damage from alcohol and tobacco use, stimulates white blood cell activity.
CYSTINE	Antioxidant, protects the body from radiation and pollution; slows the aging process, deactivates free radicals, neutralizes toxins; aids in protein synthesis, necessary for the formation of skin and connective tissue. Aids in the healing processes and recovery from burns and wounds. Hair and skin are made up of 10-14% cystine. Stimulates white blood cell activity and helps diminish pain from inflammation.
GLUTAMATE	Considered to be nature's 'brain food', improves mental capacities, helps speed the healing of ulcers, gives a 'lift' from chronic fatigue, helps control alcoholism, schizophrenia and the craving for sugar. A major precursor of glutamine, proline, ornothine, arginine, glutathione, and GABA.

Amino Acid	Biological Function
GLUTAMINE	Critical for immune function, important source of energy (especially for kidneys and intestines during caloric restrictions). A brain fuel that aids in memory and a stimulant for intelligence and concentration.
GLYCINE	Releases oxygen for cellular multiplication, makes hormones for immune system, helps make other amino acids, part of hemoglobin and cytochromes (enzymes involved in energy production), has a calming effect, used in the treatment of manic depressive and aggressive individuals, produces glucagon that mobilizes glycogen, and inhibits sugar cravings.
HISTIDINE	Part of hemoglobin; used for arthritis, allergies, ulcers and anemia; major UV absorbing compound in skin, makes red and white blood cells, used for anemia, allergic diseases, and digestive ulcers.
(OH)-PROLINE	From proline makes collagen; strengthens ligaments, tendons, blood vessels; gives elasticity to skin.
ISOLEUCINE	A branched chain amino acid readily taken up and used for energy by muscle tissue; prevents muscle wasting in debilitated individuals and forms hemoglobin.
LEUCINE	Stimulates the upper brain to be more alert; used as a source of energy; reduces muscle protein breakdown. Modulates uptake of neurotransmitter precursors by the brain as well as the release of encephalins that inhibit the passage of pain signals into the nervous system; promotes healing of skin and broken bones.
LYSINE	Helps Ca(II) absorption; helps collagen synthesis; aids in the production of antibodies, hormones and enzymes, effective against herpes; reduces viral growth. Forms L-carnitine, with lysine and ascorabate, which enables muscle tissue for efficient use of oxygen; delays chronic fatigue.
METHIONINE	Supplies sulfur; prevents disorders of hair, skin and nails; lowers cholesterol levels by increasing lecithin production in the liver; reduces liver fat and protects the kidneys; a natural chelator for heavy metals; regulates ammonia formation and creates ammonia-free urine to reduce bladder irritation; influences hair follicles and promotes hair growth; precursor of cystine and creatine; increases GSH levels.
ORNITHINE	Increases growth hormone secretion, aids in immune and liver function; promotes healing.
PHENYL-ALANINE	Produces norepinephrine that helps nerve transmission to the brain; keeps awake and alert; reduces hunger and pain; functions as an antidepressant and helps improve memory; precursor of tyrosine; enhances learning, memory, mood and alertness; vital for collagen synthesis; suppresses appetite.
PROLINE	Constituent of joints, tendons, and connective tissue; helps maintain and strengthen heart muscle; mobilizer of muscular energy; major constituent of collagen.
SERINE	Storage source of glucose by the liver and muscles; strengthens the immune system by providing antibodies and immunoglobulins; synthesizes fatty acid sheath around nerve fibers; vital for the production of cellular energy production; aids in memory and nervous system functions.
TAURINE	Helps stabilize membrane excitability, which is important in the control of epileptic seizures; taurine and sulfur are vital for the control of aging process; aids in the clearance of free radical species; aids in the absorption and elimination of fats; acts as a neurotransmitter in some areas of the brain and retina.
THREONINE	Constituent of collagen, elastin, and enamel protein; helps prevent fat build-up in the liver; helps in the smooth function of the GI tract; assists in metabolism and assimilation; one of the amino detoxifiers.
TRYPTOPHAN	Natural relaxant, induces normal sleep, reduces anxiety and depression; reduces migraine headaches; immune modulator; reduces the risk of heart spasms; works with lysine in reducing cholesterol level; precursor of neurotransmitter serotonin that exerts a calming effect; releases growth hormones.
TYROSINE	Transmits nerve impulses to the brain; anti-depressant, elevates mood, memory, and mental alertness; helps thyroid, adrenal and pituitary gland functions; precursor of the neurotransmitters dopamine, norepinephrine, and epinephrine; helps in the synthesis of thyroid growth hormones and melanin.
VALINE	Promotes mental vigor, muscle coordination and calm emotions; influences brain uptake of other neurotransmitter precursors (i.e. tryptophan, phenylalanine and tyrosine).

5.3.1. REDOX Amino Acid: N-Acetyl-Cysteine (NAC)

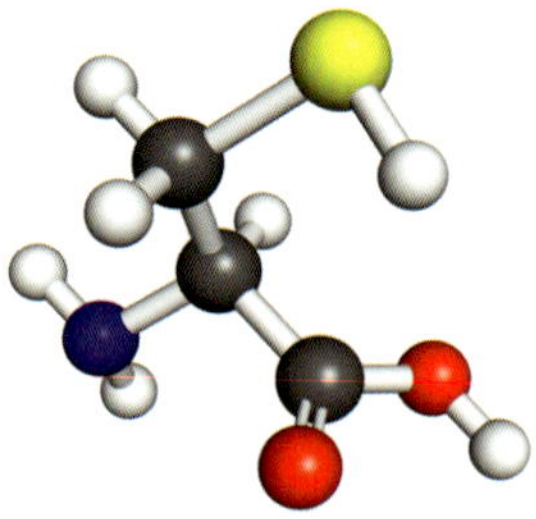

CYSTEINE

N-acetyl-cysteine (NAC), the acetylated variant of the amino acid L-cysteine, is an excellent source of sulfhydryl (SH) groups that converts in the body into metabolites capable of stimulating glutathione (GSH) synthesis. GSH, the most abundant low-molecular-weight thiol in animal cells, plays a central role in the antioxidant defense against ROS. For tissue GSH synthesis, the availability of cysteine is generally the limiting factor, and one of the effective precursors of cysteine is its synthetic derivative, the NAC. The sulfhydryl (SH) group is responsible for a great deal of the metabolic activity of NAC, while the acetyl-substituted amino group makes the molecule more stable against oxidation.

NAC is produced both naturally in the body and also obtained from the diet. NAC is rapidly absorbed following an oral dose; however, extensive first pass metabolism by the cells of the small intestine and the liver results in the incorporation of NAC into protein peptide chains and the formation of a variety of metabolites of NAC. Only a small percentage of the intact NAC molecule arrives in the plasma, and subsequently in tissue. In addition to antioxidant function, NAC has other mechanisms of action, such as inhibition of neutrophil activation, decreased microbial attachment and vasodilation. NAC as a designated (but not approved) orphan drug, plays an important therapeutic role in many clinical conditions in which GSH deficiency and/or oxidative stress are involved. NAC is an immune enhancer; prevents and treats cancer; detoxifies heavy metals; treats smoker's cough and bronchitis; prevents heart disease; slows progression of HIV; relieves hangover symptoms; and reduces exercise fatigue. It is used as a supplement for bronchitis, HIV infection, cystic fibrosis, cancer chemotherapy support, acetaminophen poisoning, heavy metal poisoning, and septic shock.

NAC and Antioxidant Activity: In addition to providing cysteine for production of GSH, NAC has been shown to scavenge oxidants directly, particularly the reduction of the ·OH radical and hypochlorous acid. Oral administration of GSH alone does not adequately restore GSH levels. It is rapidly hydrolyzed by the liver and intestines, and penetration through the blood–brain barrier is poor. Similarly, oral administration of L-cysteine has also been shown to have little effect on brain GSH levels owing to first-pass metabolism. Oral NAC administration results in increased plasma cysteine levels, ultimately leading to increases in plasma GSH. NAC has been shown to successfully penetrate the blood–brain barrier and raise brain GSH levels

NAC and Anti-inflammatory Activity: NAC modulates the cellular REDOX status that affects several pathways leading to decrease in NF-κB activity. Inflammatory responses are strongly influenced by REDOX-sensitive nuclear transcription factor-κB (NF-κB), which regulates a variety of pro-inflammatory genes. Thus, NAC-mediated inhibition of NF-κB activity can decrease several inflammatory mediators. NAC also affects thioredoxin (TRx) and glutaredoxin (GRx) expression and the ubiquitously expressed thiol-reducing system next to GSH. By decreasing the release of TRx and GRx, NAC may attenuate NF-κB binding to DNA resulting in an impaired inflammatory response.

NAC was introduced as a mucolytic agent for COPD some 50 years ago. Its effect is based on breaking the high-molecular-weight glycoproteins of the mucus, resulting in reduced viscosity.

NAC and Respiratory Diseases: NAC is useful as a mucolytic agent for treatment of chronic bronchitis and other pulmonary diseases. Administration of NAC decreases cough severity and diaphragm fatigue. NAC's sulfhydryl group reacts and splits disulfide bonds in the mucous bronchial secretions. Because the mucus is broken down into smaller and less viscous units, NAC is referred to as 'slime loosener. NAC is known, to provide its protective effects, in part, by the replenishment of the intra-cellular stores of GSH in activated granulocytes. NAC was reported to improve lung function when administered for a longer period of time. Six months of treatment with NAC (600 mg/day) showed an improvement in forced expiratory volume in one second (FEV1). Maximal expiratory flow at 50% of forced vital capacity (MEF50) in stable COPD patients NAC could influence sputum cytokine level, as well as sputum chemo-attractant properties. NAC acts at the baseline level of cytokines (IL-6 and IL-8) that limits the inflammatory status and modulates the exacerbation frequency. Moreover, by decreasing the viscosity of the sputum, NAC may facilitate the airway clearance and influence the course of the exacerbation. NAC supplementation also increases GSH levels in the liver, plasma, and bronchioles of the lungs.

NAC and HIV Treatment: HIV-positive individuals usually exhibit low GSH and cysteine levels. The net loss of sulfur in asymptomatic HIV+ patients is equivalent to a mean loss of about 10 g *Cys*/day. Observational studies have linked GSH deficiency directly to patient survival; NAC was shown to restore normal GSH levels among HIV patients and seems to improve their survival rates. Also low intra-cellular GSH levels increase NF-κB activation and consequently, HIV replication. In vitro, NAC causes splenocyte proliferation and protects lymphocytes against mitogen-induced cytotoxicity. NAC could enhance the immune response of peripheral blood T cells. In addition, NAC blocks the suppression of T cell mitogenesis and cytokine production by protease inhibitors. NAC supplementation can protect hematopoietic progenitor cells from zidovudine (AZT)-induced toxicity.

NAC and Acetaminophen Overdose: Acetaminophen (Tylenol®, paracetamol) overdose is a common form of poisoning, capable of injuring the liver, kidneys, heart, and central nervous system. Overdose can also produce fatal results, typically in individuals who intentionally consume greater than 10 g. Liver damage usually develops within several hours of ingestion as a result of oxidation of acetaminophen to toxic metabolites such as N-acetyl-benzoquinonimine. These metabolites deplete hepatic intra-cellular GSH stores and subsequently damage the liver. In cases of acetaminophen overdose, NAC is the antidote of choice. The mainstay of treatment for acetaminophen-induced hepatotoxicity is currently an enteral 18-dose course of NAC. NAC administered intravenously or orally within 24 hours of overdose is effective at preventing liver toxicity; however, improvement is most notable if treatment is initiated within 8-10 hours of acetaminophen

Many drugs and poisons are detoxified through conjugation with GSH (e.g. paracetamol) and additional GSH, provided by administering NAC, has been shown to be effective in intoxication by such drugs and poisons.

NAC is a powerful antioxidant and potential therapeutic agent in the treatment of cancer, heart disease, HIV infection, heavy metal toxicity, and other diseases characterized by oxidative damage. NAC is of value in Sjogren's syndrome, influenza, hepatitis C, and myoclonus epilepsy.

overdose. NAC's effectiveness declines when treatment is delayed beyond 10 hours, at which time the risk of mortality significantly increases.

NAC and Heavy Metal Poisoning: NAC has also been effective for heavy metal poisoning by gold, silver, copper, mercury, lead, and arsenic, as well as in cases of poisoning by carbon tetrachloride, acrylonitriles, halothane, paraquat, acetaldehyde, coumarin, and interferon. NAC appears to have clinical usefulness as a chelating agent in the therapy of acute heavy metal poisoning. NAC has been shown to be more effective than calcium EDTA or dimercapto-succinic acid as an agent to increase the urinary excretion of chromium and boron. NAC has also been shown to be protective against arsenite and copper poisonings.

NAC and Cardiovascular Disease (CVD): NAC appears to have several possible therapeutic roles associated with CVD. NAC administration seems to positively impact homocysteine (and possibly lipoprotein levels), protect against ischemic and reperfusion damage, and enhance aspects of the effectiveness of nitroglycerine. High levels of plasma lipoprotein or homocysteine are associated with an increased risk for CVD. NAC might be useful in the treatment of ischemic and reperfusion injury in acute myocardial infarction. Myocardial ischemia is generally characterized by a decline of cellular SH groups; reperfusion can result in oxidative damage. Under experimental conditions, infusion with NAC for 60 minutes before ischemia increased the tissue content of GSH by 38 percent. Administration of NAC, in combination with nitroglycerin and streptokinase, was associated with significantly less oxidative stress, a trend toward more rapid reperfusion, and better preservation of left ventricular function in patients with evolving acute myocardial infarction.

Evidence suggests pretreatment with NAC might attenuate impaired tissue oxygenation and preserve myocardial performance in cardiac risk patients undergoing hyperoxic ventilation. Intravenous administration of NAC (150 mg/kg) was reported to slightly increase the cardiac index and left ventricular stroke work index, preserve whole-body oxygen consumption, and decrease systemic vascular resistance during brief hyperoxia in cardiac risk patients. NAC potentiates the coronary dilating and anti-platelet effects of nitroglycerin and limits the development of hemodynamic tolerance to nitroglycerin. NAC and nitroglycerin appear to be an effective combination for the treatment of unstable angina; however, the high incidence of side effects (about 35%; mostly severe headaches) limits the clinical usefulness of this therapeutic strategy.

NAC and Cancer: The administration of NAC might have a role in the prevention of cancer and in an integrated approach to the treatment of some forms of cancer. NAC could block experimentally-induced DNA damage and show antimutagenic activity towards various genotoxic agents. In vivo, NAC can inhibit the mutagenicity of a number of compounds and can inhibit the induction of tumors by some carcinogens. Pre-treatment with NAC increases the non-protein SH content of leukemia cells nearly three-fold, without negatively affecting the chemotherapeutic activity of doxorubicin against this tumor. Evidence also indicates that, although NAC can help protect against toxicity resulting from X-rays or chemotherapeutic treatment, it does not interfere with the efficient killing of tumor cells by X-rays or by bleomycin. Moderate fatigue, anorexia, and weight loss were the main toxic effects observed. NAC has been used successfully as a radioprotective agent and apparently exerts part of this effect by modulating cytokine concentrations. Interleukin-1, tumor necrosis factor-α, and interferon-gamma endogenously provide protection to the hematopoietic system against radiation.

NAC is anti-mutagenic and anti-carcinogenic. NAC also inhibits the mutagenicity of certain compounds in vivo.

NAC- Pharmacology and Pharmacokinetics: NAC is rapidly absorbed following an oral dose of 600 mg with a peak of 4.6 μM after 60 min. The plasma half-life has been reported to be 2.5 hours,with no NAC detectable 10–12 hours after administration. It has been estimated that the oral bioavailability of the intact NAC molecule is about 10%. Following an oral dose, the majority of NAC is metabolized into another compound (since accompanying increase in non-protein and protein SH groups were found in plasma). Normalized maximal

Table: *NAC – Protective/Therapeutic Effects on Various Physiological Systems*

Physiological System	Protective/Therapeutic Effect
Immune system	Attenuating, alleviating, adding resistance, healing and strengthening the immune response, including Leishmania infection and influenza
Cardiovascular system	Favorable effects in hypertension, hypertensive cardiac injury, cardioprotection in ischaemia-reperfusion injury/ myocardial infarct and cardiopulmonary bypass
Respiratory system	Protective effect in injuries by different toxins (including tobacco smoke), lung transplantation, lung reperfusion injury, attenuation of pulmonary fibrosis
Nervous system	Neuroprotection in transient hypoxic injury (Alzheimer-type) to cortical neurons, prevention of primary sensory neuronal death, beneficial effects in deficits of learning, memory, and hearing loss
Gastrointestinal system	Favourable effect in acute pancreatitis and liver injuries Inhibition of the growth of *Helicobacter pylori*
Endocrine system	Protection against complications of diabetes, insulin resistance, and estrogen-deficiency bone loss
Reproductive system	Restores NO-mediated effects in the feto-placental circulation in pre-eclampsia
Endothelial cells	Protection against several hazardous factors, including urban air fine particles and tobacco smoke
Articular structures	Protection against various injuries
Miscellaneous	Protection against acute high-dose irradiation, photoageing of human skin, decrease of plasma total homocysteine levels in healthy volunteers

NAC FOOD SOURCES

include meat, fish, seafood, chicken, turkey or eggs. A number of plant foods such as garlic, onions, broccoli, red peppers, legumes and brussels sprouts are also rich-sources of NAC.

plasma concentration was reported to be significantly higher after a 600 mg dose than after a 200 mg dose. Moreover, increasing the dose also increased NAC bioavailability and time for maximal plasma concentration. During absorption, NAC is rapidly metabolized to cysteine, whose thiol group has reducing and antioxidant properties, which is a direct precursor of the GSH.

Side Effects and Toxicity: NAC is generally safe and well tolerated even at high doses. The most common side effects associated with high oral doses are nausea, vomiting, and other gastrointestinal disturbances; therefore, oral administration is contraindicated in persons with active peptic ulcers. In addition, as with any single antioxidant nutrient, NAC at therapeutic doses (even as low as 1.2 grams daily) has the potential to have prooxidant activity and is not recommended at these doses in the absence of significant oxidative stress.

NAC Dosage: The typical oral dose for NAC (as a mucolytic agent and for most other clinical indications) is 600-1,500 mg daily in three divided doses. In patients with cancer or heart disease the therapeutic dosage is higher, usually in the range of 2-4 grams daily. For acetaminophen poisoning, NAC is administered orally with a loading dose of 140 mg/kg and 17 subsequent doses of 70 mg/kg every four hours. In acetaminophen poisoning it is important to begin administering NAC within 8-10 hours of overdose to ensure effectiveness.

The biological activity of NAC is attributed to its sulfhydryl (-SH) group. NAC reduces extra-cellular cystine to cysteine and stimulates GSH synthesis. NAC function as a mucolytic agent results from the SH-groups interacting with disulfide bonds in mucoproteins and subsequently breaking mucus into smaller and less viscous units. NAC also acts as an expectorant by stimulating ciliary action and the gastro-pulmonary vagal reflex; thereby clearing mucus from the airways. NAC alleviates the symptoms of CVD by lowering homocysteine and lipoprotein levels (via dissociation of disulfide bonds), protecting against ischemia and reperfusion damage (via replenishment of the GSH-REDOX system), as well as potentiating the activity of nitroglycerin.

5.3.2. REDOX Amino Acid: Methionine (Met)

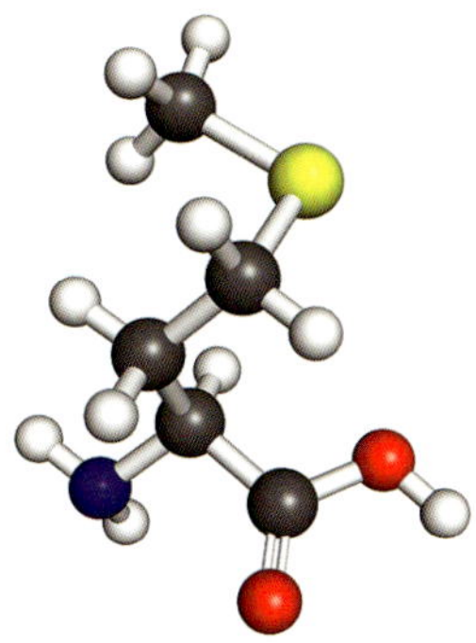

METHIONINE

Methionine (*Met*) is a sulfur-containing, non-polar, α-amino acid genetically coded by the initiation codon AUG. Methionine is one of only two amino acids encoded by a single codon (AUG) in the standard genetic code (tryptophan, encoded by UGG, is the other). The codon AUG is also the most common eukaryote 'Start' message for a ribosome that signals the initiation of protein translation from mRNA. As a consequence, methionine is often incorporated into the N-terminal position of proteins in eukaryotes during translation, although it can be removed by post-translational modification. Therefore, methionine can often be a limiting factor in protein synthesis, especially under conditions of *Met* deficiency.

Methionine is an essential amino acid involved in several critical metabolic processes. In mammals, it can be obtained only from diet and is susceptible to oxidation at its sulfur atom under physiological conditions. Methionine is also a crucial metabolite that influences REDOX homeostasis through sulfur metabolism and the trans-sulfuration pathway. It is also the source of several antioxidants and other sulfur compounds, which function in the defense against oxidative stress, such as glutathione (GSH), taurine, and cysteine (*Cys*). Methionine, together with choline, and inositol, belongs to a class of compounds known as 'lipotropics', which help the liver process fat in the body. Once in the liver, methionine is converted into SAMe (s-adenosyl methionine); about 8 grams of SAMe is produced in the liver each day. However, the amount of SAMe produced in the body can be reduced significantly when liver function is compromised.

Methionine is a core amino acid that supplies antioxidants to balance the cellular REDOX status against the attack by ROS. Methionine is positioned at the entry point of several metabolic processes and serves as an important regulator of those pathways. Methionine is required for protein synthesis, sulfur metabolism, methylation, REDOX regulation and signal transduction. Since, mammals do not possess an enzymatic system to synthesize this amino acid (their only sources are diet and recycling), a delicate control of methionine metabolism is important in order to supply this amino acid for diverse cellular needs. In this regard, susceptibility to oxidation and availability of repair enzymes may be important factors that regulate methionine flux, especially under conditions of oxidative stress.

Methionine and Oxidation: Methionine is highly susceptible to oxidation by ROS and reactive nitrogen species (RNS). Generally, all amino acids are subject to free radical-mediated oxidation by radiation, metal-catalyzed reactions, mitochondrial respiration, and many other processes generating ROS. Methionine, however, is among the most sensitive to oxidation. In the case of metal ion-catalyzed oxidation, α-carbon of amino acids, including that of methionine, can undergo oxidative deamination. The major product of *Met* oxidation is methionine sulfoxide. Sulfur (of methionine) forms a prochiral center and its oxidation results in two diastereomer products, methionine-S-sulfoxide (Met-S-SO) and methionine-R-sulfoxide (Met-R-SO).

Methionine sulfoxide reductases (Msrs) are thiol (or selenol)-dependent oxidoreductases that reduce protein-based and/or free forms of Met-S-SO or Met-R-SO to methionine. This function makes these enzymes important antioxidant proteins that can protect against oxidative stress through reversible oxidation and reduction of methionine. Also, Msrs may regulate protein function by controlling the REDOX state of critical methionine residues. Three Msr classes are known (they differ with regard to substrate utilization): MsrA, MsrB and free methionine-R-sulfoxide reductase (fRMsr). Several studies have provided evidence that MsrA and MsrB activities decrease during aging and in disease conditions, such as Alzheimer's disease. This activity decline might affect protein-based and free methionine sulfoxide levels. Therefore, methionine sulfoxides may represent useful biomarkers of REDOX changes during aging and disease.

Methionine Sulforeductase A (MsrA): The mitochondria is believed to be a primary site of MsrA action because it is a major source of endogenous ROS and is a central mediator of several critical cellular processes including respiration and apoptotic induction. MsrA has been localized to the mitochondria in mammalian cells where it is found in the inner mitochondrial membrane. MsrA has also been shown to associate with all five mitochondrial respiratory complexes. Importantly, MsrA has been shown to directly protect mitochondria against exogenously added oxidative stress. Deletion of MsrA from mammalian cells results in loss of mitochondrial membrane potential in combination with increased mitochondrial ROS production and ultimately death of mammalian cells.

Methionine and REDOX Homeostasis: Methionine (*Met*) is located at the key junction in sulfur metabolism. It has a central role in maintaining cellular REDOX homeostasis, serves as a donor of methyl groups, and is a precursor for other sulfur-containing amino acids and compounds. Being an indispensible dietary amino acid, *Met* cannot be synthesized or replaced by other amino acids in mammals, while cysteine (*Cys*), which is another sulfur amino acid, can be generated from *Met* through the transsulfuration pathway. To defend against oxidative stress, this pathway provides, among other compounds, glutathione (GSH). GSH participates in REDOX regulation and scavenging ROS by being a substrate for several enzymes, such as GSH-reductase that reduces oxidized glutathione (GSSG) and glutathione peroxidase (GPx) (which reduces organic and lipid peroxides in a GSH-dependent manner). GSH also can be used for the reduction of glutaredoxin, where deglutathionylates target proteins and transfers reducing equivalents to enzymes and compounds such as dehydroascorbic acid, peroxiredoxin (PRx), and Msrs. The GSH/GSSG ratio varies among tissues and cellular compartments and serves as an indicator of REDOX status. To avoid collapse of REDOX homeostasis upon oxidative stress, organisms can support de novo GSH synthesis from methionine. For example, inhibition of methionine synthase activity leads to accumulation of homocysteine, enhancing *Cys* synthesis through the trans-sulfuration pathway. In addition, the activity of cystathionine-β-synthase may be stimulated to generate *Cys* from homocysteine. Ultimately, this pathway produces GSH from *Met* to regulate REDOX homeostasis.

Methionine Supplementation: Methionine is an essential amino acid with multi-functional health benefits. It helps in ameliorating the symptoms of depression, inflammation, liver diseases, and certain muscle pains. Methionine is also beneficial for individuals suffering from estrogen dominance, where the amount of estrogen in the body is excessively high when compared to its opposing hormone, progesterone. Methionine is also effective for individuals on oral contraceptives or estrogen replacement therapy; since this amino acid helps the clearance of estrogen by the liver. Specifically, methionine converts the stronger and carcinogenic estradiol (E2) into estriol (E3), which is the desirable estrogen.

Strict vegetarians and individuals who follow a low protein diet should consider methionine supplementation. Also, people consuming diet high in soy should also consider methionine supplementation as soy is low in amino acids. During Met-supplementation, intake of taurine, cysteine, and other sulfur containing amino acids (as well as B6 and folic acid) should also be included. Recommended dosage ranges from 500 mg to 4,000 mg, in divided dosages, throughout the day.

Dietary methionine is a powerful antioxidant needed for tissue repair and rebuilding processes, especially for liver function. During pregnancy, methionine, along with folate, plays a vital role in neural-tube (part of the fetus that forms the nervous system) development in the fetus. Methionine is essential for the normal closure of the neural tube. *Met*-deficiency is associated with a higher occurrence of neural-tube defects (i.e. spina bifida) in newborns. Extra methionine supplementation should be avoided for individuals with conditions of elevated cholesterol, heart disease, or rheumatoid arthritis. Methionine converts to homocysteine, a known risk factor for heart disease. Homocysteine is a by-product of methionine breakdown. Elevated homocysteine levels in the blood results in hardening of the arteries posing a risk of heart attack or stroke. Elevated homocysteine may also predispose the risk of other degenerative processes in the eyes, nervous system, and kidneys.

Met occurs in most protein-based foods but only in small amounts in legumes, peanuts, soybeans, and vegetables in general. Methionine levels are five to ten times lower in vegetables than in animal foods. Vegetarians are much more likely to be deficient in methionine than meat-eaters.

METHIONINE FOOD SOURCES

Food Source	g/100 g
Eggs, white, dried	3.204
Sesame seeds, flour	1.656
Parmesan cheese, shredded	1.114
Beef, cured, dried	0.801

5.3.3. REDOX Amino Acid: Glutamate / Glutamine

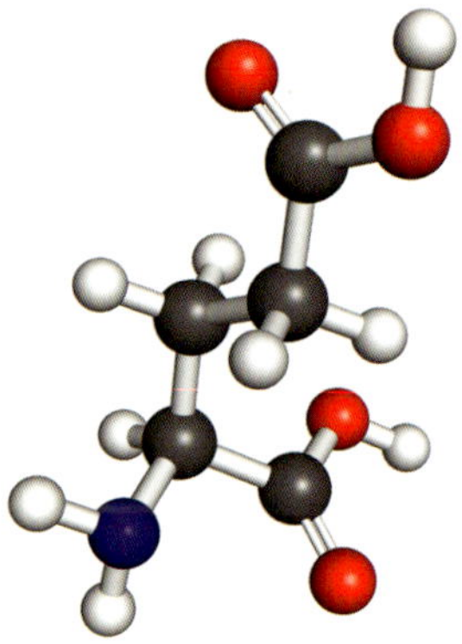
GLUTAMATE

L-Glutamic acid (*Glu*) is a dicarboxylic acid grouped under branched chain amino acids with acidic chains. *Glu* is an aliphatic amino acid undergoing degradation in the body to glutamine (*Gln*), which is an amino acid with uncharged polar side chains. At physiological pH, *Glu* exists in anionic form and is referred to as glutamate. It is found on surfaces of proteins and it plays a central role in transamination reactions; it also equilibrates with corresponding keto acid. The human brain is the only tissue which has the highest content of glutamate. It is obtained from glucose via Krebs cycle intermediate and is stored in synaptic vesicles (to be released by calcium-dependent exocytosis). Glutamate is also synthesized in the body through transamination of α-ketoglutarate, by accepting amino groups from amino acids, the first step in metabolism of amino acids.

L-Glutamine (*Gln*) is the most abundant amino acid present in the body. It is also known as levoglutamide and is synthesized in the body (from glutamate and ammonia) in an energy-requiring reaction, catalyzed by glutamine synthase. Although non-essential in health, glutamine is conditionally essential in stress and illness. Glutamine accounts for 30-35 percent of the amino acid nitrogen in the plasma. It contains two ammonia groups, one from its precursor, glutamate, and the other from free ammonia in the bloodstream. One of glutamine's roles is to protect the body from high levels of ammonia by acting as a 'nitrogen shuttle'. Thus, glutamine can act as a buffer, accepting and releasing excess ammonia, when needed, to form other amino acids, amino sugars, nucleotides, and urea. This capacity to accept and donate nitrogen makes glutamine the major vehicle for nitrogen transfer among tissues. *Gln* is heavily concentrated in skeletal muscle making up to 60% of free amino acid pool in skeletal muscle. Glutamine either acts as a nitrogen donor for synthesis of purines and pyrimidines or undergoes deamidation in kidney to produce glutamate and ammonia. This second reaction plays a central role in maintaining acid–base balance in the body and in controlling urinary pH. *Gln* participates in a variety of physiological functions. It is a major fuel source for enterocytes; as a substrate for neogluco-genesis in kidney, lymphocytes, and monocytes; and a nutrient/substrate in muscle protein metabolism in response to infection, inflammation, and muscle trauma.

Glutamine is one of the three amino acids involved in glutathione (GSH) synthesis. GSH, an important intra-cellular antioxidant and hepatic detoxifier, is comprised of glutamic acid, cysteine, and glycine. GSH is depleted in children with low birth weight and kwashiorkor. Therefore, GSH supplementation with *Glu/Gln* could be useful in maintaining the integrity of the immune system and normal health.

Glutamate and the Nervous System: Glutamate is well established as the principal and classical excitatory neurotransmitter. Multiple subtypes of glutamate receptors have been characterized. Glutamate receptors mediate fast excitatory synaptic transmission in neurons. On the other hand, through NMDA receptors, *Glu* mediates a slow excitatory response, which plays a role in long-term adaptive and pathological changes in the brain, involving synaptic plasticity. However, elevated levels of extra-cellular glutamate can induce seizures and excito-toxic neuronal cell death. Imbalance of glutamatergic excitatory control of motor neurons is reported to be responsible for neuronal damage in motor neuron disease.

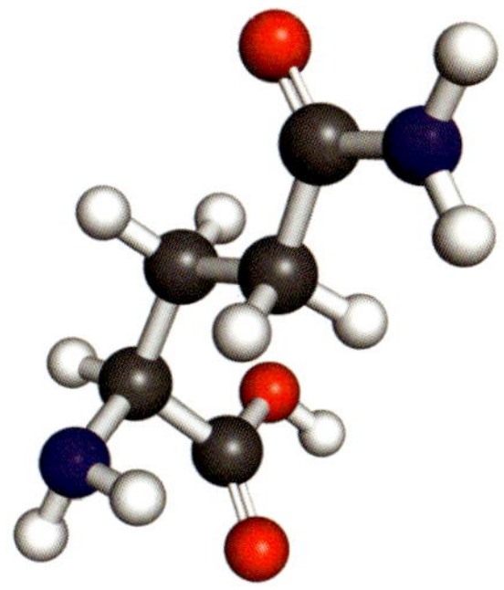

GLUTAMINE

Glu/Gln in Ammonia Metabolism and Nitrogen Balance: Both glutamate and glutamine play an important role in ammonia metabolism. They are involved in the disposal of ammonia and thus regulate nitrogen balance in the body. Glutamate, *Glu*, plays a central role in the transamination reaction in which amino acids funnel their amino group to α-ketoglutarate resulting in glutamate formation. *Glu*, therefore, acts as an acceptor of amino group from other amino acids and further undergoes either oxidative deamination in liver or is used as an amino group donor in synthesis of nonessential amino acids.

Glutamine (*Gln*) provides a non-toxic storage and transport form of ammonia. The formation of *Gln* occurs primarily in the muscle and liver. At all times, there is net output of *Gln* from muscle representing the disposal of amino groups from branched chain amino acids. *Gln* is a major mechanism for the removal of ammonia in the brain. *Gln* supplementation has been shown to improve nitrogen balance in patients who underwent major surgeries or bone marrow transplants. *Gln* also reduces the incidence of infection and duration of hospitalization. *Gln* plays an important role in protein regeneration, repair, and growth. It is identified as a precursor for purine/ pyrimidine and a substrate for gluconeogenesis in kidney, lymphocytes, and monocytes. *Glu/Gln* transporters (in the skeletal muscle and heart) seem to play a role in the control of the steady-state concentration of amino acids in the intra-cellular space, probably through osmotic signaling mechanisms to regulate whole body protein metabolism. *Glu/Gln* play a key role in fetal and placental metabolism. High levels of *Glu* (in free form) are present in breast milk and serve as a vital factor in postnatal development. Further, supplementation with *Gln* could provide significant benefits in low birth weight infants and critically ill patients, especially in preventing nosocomial infections.

Glutamine and the Gastrointestinal (GI) System: Glutamine is important for the maintenance of gut barrier function, modulated through the synthesis of nucleotides, glucosamine, and mucus glycoprotein. The GI system is the principal organ of *Gln* utilization with 12–13% of *Gln* circulating and 50–60% of Gln enterally supplied. Glutamine is thought to act as a major fuel and primary source of energy for enterocytes and to enhance mucosal immunity. Its presence is shown to stimulate blood flow to the gut, maintain mucosal integrity, prevent villous atrophy, bacterial translocation, and subsequent pro-inflammatory responses. The gut, the main organ for nitrogen processing in the body, significantly contributes to the supply of *Gln* to peripheral tissues. Intestinal dysfunction is associated with a significant fall in *Gln* levels, resulting in impaired ability to incorporate amino acids into protein. *Gln*-deficiency is associated with mucosal atrophy and loss of function. The *Gln*-uptake in the GI tract markedly increases during surgical stress.

Glutamine and the Immune System: *Gln* is one of the major sources of nutrition for cells of the immune system that is released from muscles, adipose tissue and other sources. *Gln* helps in the maintenance of general immune-competence by supporting synthesis of cellular proteins of the immune system and hepatic acute phase protein response. During sepsis, the uptake and utilization of *Gln* declines with a concurrent increase in bacterial translocation. L-Glutamine, as precursor of GABA, contributes to anti-stress activity and improves cell-mediated immunity. Decline in *Gln* levels may result in an increased rate of infection in certain stressed patient populations. Critically ill newborn infants also frequently display protein-calorie malnutrition due to the demands of sepsis and respiratory failure.

Glutamine and HIV/AIDS

HIV infection could induce *Gln*-deficiency, resulting in muscle protein wasting, particularly in the AIDS stage of infection. Approximately 20 percent of AIDS patients also have abnormal intestinal permeability. Clinical studies have shown that *Gln* supplementation could provide significant benefits to these patients.

Glu/Gln Supplementation: Glutamate is abundant in both 'free' and 'bound' forms in all natural foods – meat, poultry, fish, cheese, milk (including human breast milk), tomatoes, mushrooms, and many other vegetables, including peas and broccoli. The bound form of *Glu* is linked to other amino acids in proteins present in muscles, hair, and skin. Glutamate in human body is considered essential for protein repair, regeneration, and growth. In an average adult, the body content of *Glu* is around 1.5 to 2 kg, mainly in bound form. The free form of *Glu* (i.e. not linked to protein) is known to enhance food flavor and its content is said to increase during natural ripening of fruits. It is not a tenderizer or preservative and cannot make inferior food taste good. It has no distinctive smell.

Glutamine administration, even at high doses, has no side effects and is well tolerated. *Gln* dosages vary greatly depending on the clinical situation, but are in the range of 2-4 gm/daily in divided doses from general wound healing and intestinal support. For critically ill adults, cancer, and HIV patients, the *Gln* dosage is much higher, ranging from 10-40 gm/day in divided doses. For these patients, the bulk powder form of glutamine eases administration of large doses.

Glutamate is a prominent excitatory neurotransmitter in the central nervous system. Glutamate is a seaweed ingredient, (identified in 1908 by Japanese scientists) responsible for enhancing the flavor for food. It is now best known as mono-sodium glutamate (MSG). MSG is an odorless compound and is best classified as food additive by the Joint Expert Commi-ttee on Food Additives, an advisory board to the WHO and FAO. MSG is known to produce a unique taste sensation termed 'UMAMI' the 5th taste (which is a savory or brothy taste present in tomatoes and cheese). Free glutamate levels increases during the process of natural ripening and bring about a fuller taste in many foods. It is also called as 'palate pleaser.' Studies on taste have shown that infants to adults enjoy the taste of foods containing glutamate. There is said to be 10 times more in breast milk compared to cow's milk.

GLUTAMINE FOOD SOURCES

include meat, sausages, dairy and eggs along with vegetables like cabbage, beets, celery, kale, brussel sprouts, spinach, parsley and carrots. Even fruits like papaya are high in glutamine.

The Bio-Replenishment Theory

LIFE is a physico-chemical phenomena, where an open or continuous system is able to decrease its internal entropy at the expense of substances or free energy obtained from the environment. Subsequently, this biochemical system is rejected in a degraded form. Specific chemical reactions result in biological transformation and adaption of life with its environment. These biological processes are collectively referred to as 'metabolism'. The ability of a living organism to continuously refill the expended (depleted) chemicals that are vital for the maintenance of steady-state metabolism ('negative entropy') is 'Bio-Replenishment'. Any breakdown in bio-replenishment can lead to cellular dysfunction, system failure and, in severe cases, even death.

5.4 REDOX Bio-Replenishment (Bio-Rep)

THE BIO-REPLENISHMENT THEORY, proposed by Naidu (in 2008), elucidates the nature of regulatory molecules in reversible and irreversible functional modes. The reversible pathways run by feed-back mechanism(s) to maintain a steady-state metabolism (with optimal 'survival' body mass) is known as 'cellular homeostasis'. This biological phenomenon is in line with Schrodinger's postulate that life is a result of negative entropy. Whereas, the irreversible pathways force a living system to move forward as a function of optimal 'reproductive' phase across time (aging or senescence).This aspect of life conforms to the 2nd Law of thermodynamics, which states that the entropy of an isolated system tends to increase and approach its peak value over time (in metabolic terms – 'cellular death'). Bio-replenishment plays a conservative role in the renewable use of ΔG by pushing back negative entropy to keep an organism alive by maintaining homeostasis.

Living organisms are self-replicating and self-sustaining thermodynamic chemical systems. The ability to self-replicate resides within their complementary sequences of nucleotide bases, which hold 'bio-informatics' in DNA polymers. However, these DNA polymers cannot sustain life without a continuous influx of 'bio-energetics'- which is a support system for synthesis and replication of DNA itself, as well as, an energy source for metabolism and molecular transport. Both these biological operating domains of life. bio-informatics and bio-energetics, are governed by REDOX principles.

Bio-replenishment is an intricately designed 'biological maintenance program' that keeps the basal metabolic rate (BMR) at optimum, while aging.

Living organisms are self-sustaining; the bio-informatics thus encoded, selectively operate and produce specific biochemicals that regulate bio-energetic pathways and their metabolic outcomes. These regulatory molecules control physiological pathways that lead to spontaneous dissipation of thermodynamic disequilibria. These regulatory pathways also harness free energy and assimilate chemical substrates to build more cells, as well as cellular components. Thus, life gains energy from internal and external chemical resources and uses this energy to perform 'work' (metabolism), in assimilation (catabolism) and in synthesis (anabolism) of the same nucleotides that encoded the machinery for exploring and exploiting resources of energy and raw materials (nutrition). The regulatory molecules that control the bio-informatics and bio-energetics of a living organism by feed-back and feed-forward mechanisms are collectively known as 'Bio-Replenishments'.

Life systems can function only within a narrow range of milieu conditions of temperature, pH, ion concentrations, and nutrients. Homeostasis is the maintenance of equilibrium or steady-state conditions in a living organism with automatic (self-regulating) mechanisms that counteract influences that drive towards imbalance.

An organism stays alive by taking energy from outside, and processing it to generate energy within (at lower entropy) to reach a more organized state.

THE DESIGN OF LIFE

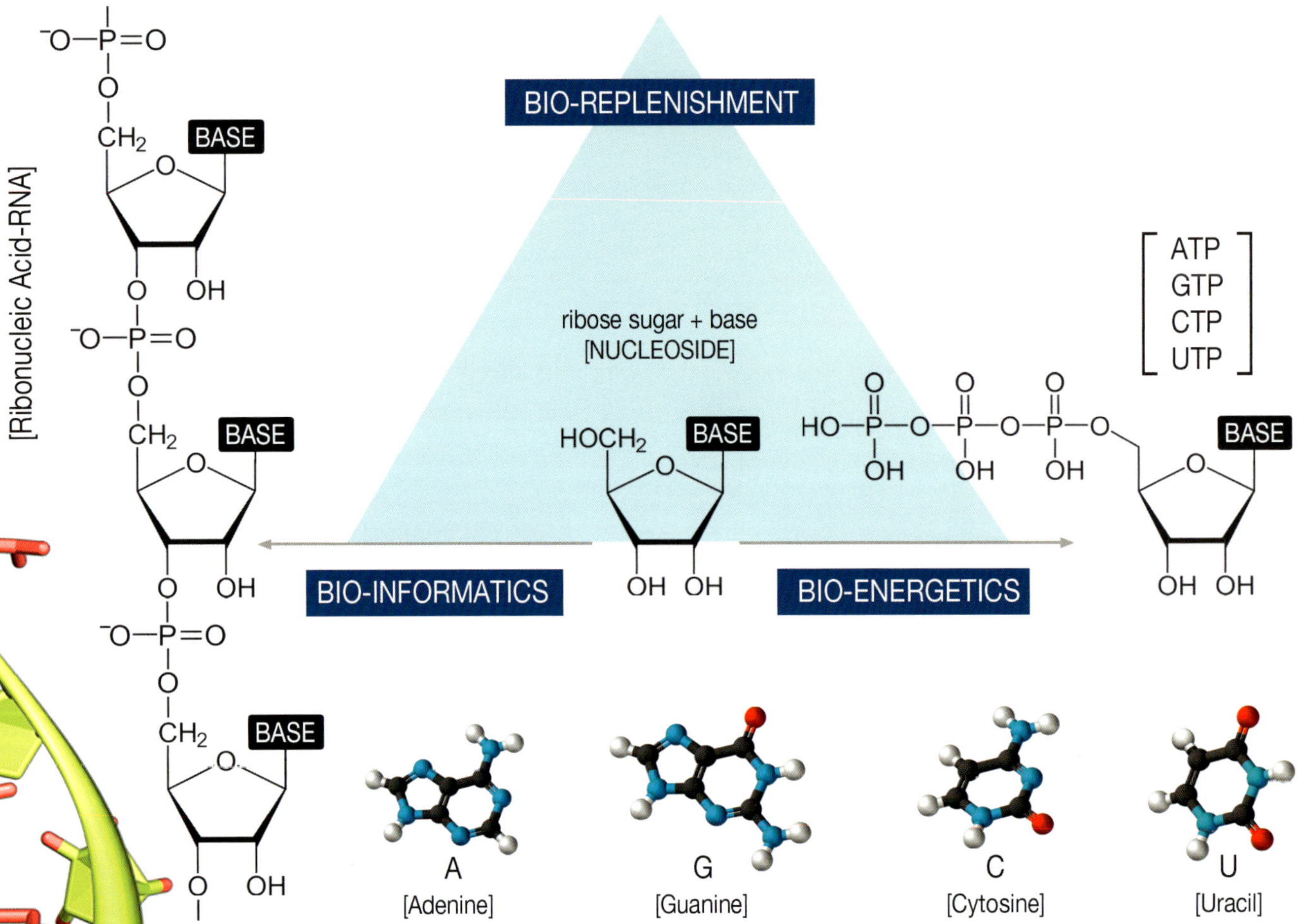

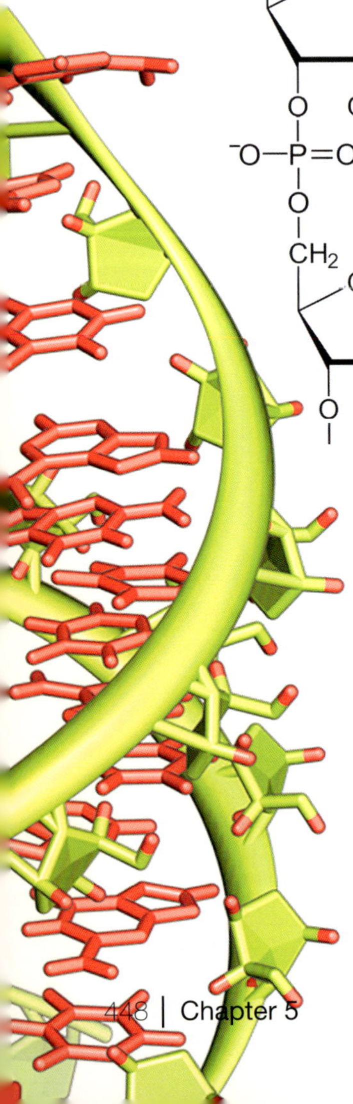

Life is a REDOX phenomenon. The design of life requires a steady supply of energy and electrons in the form of free radicals – the *Bio-Energetics*. A basic feature of this biological design is an extraordinary ability of environmentally adapted and coded replication. Living systems grow and multiply so that the replicates are similar to their progenitor(s), based on a genetic blue print – the *Bio-Informatics*. However, just a self-sustained chemical process cannot sustain life; there is an internal program that directs how the chemistry operates, which is the *Bio-Replenishment*.

Bio-Replenishment and REDOX: Life is about REDOX chemistry; therefore, both cellular bio-informatics and bio-energetics need to be replenished with REDOX reactions in compartmentalized units. Cellular compartmentation from the environment is essential for the evolutionary transition of inorganic matter to the first living systems to enable the origin of replicating systems. Replication means the doubling of mass, hence the precursors for any doubling need to be in steady supply, ideally at their site of synthesis. Also, life cannot have emerged against the laws of thermodynamics. Bio-replenishment regulates the specific transport of REDOX-active transitional elements such as iron, copper, manganese, cobalt from the milieu (i.e. lactoferrin, ceruloplasmin, etc). Thus, in the absence of bio-replenishment, no cell can survive.

Bio-Replenishment and Bio-Informatics: The nature of a living system is determined by the coded pattern (i.e. DNA and RNA) contained within that organism and and its stability in a variety of environments. The living system converts energy and a variety of materials into more living systems – the cells. Growth and reproduction are chemically dependent processes. There is a code, program, or blueprint that describes the structure-function of each species. The genetic code is not a law of nature, but a bio-informatics program that is interpreted by the organism. The implementation of the plan is adaptable, so that propagation can continue under changed circumstances. Because no permanent information copying process can ever be 100% accurate, replication inevitably involves mutation, and hence candidature for evolution. Bio-replenishments (i.e. RNases) help maintain the stability of nucleic acids and regulate the function of several transcription factors

Bio-Replenishment and Bio-Energetics: Mitochondrial bio-energetics is pivotal for cellular functions (including survival, proliferation and death). The mitochondrial inner membrane hosts the most important REDOX reactions that convert the energy of nutrients into the pyrophosphate bond of ATP. These reactions are catalyzed by the mitochondrial electron transport chain (ETC), which transports electrons from NADH (or several other substrates such as ubiquinones and cytochromes) to oxygen. This complex multi-step process is termed 'mitochondrial respiration'. Cellular life needs energy not only to support its vital functions but also to die gracefully. Execution of an apoptotic program contains energy-dependent steps, including kinase signaling, formation of the apoptosome, and effector caspase activation. Under conditions of bio-energetic collapse, cells are diverted toward necrotic demise. Mitochondrial outer membrane permeability is a decisive event in the execution of apoptosis. Bio-replenishment of membrane protective antioxidant enzymes (eg. SOD, catalases) and ETC components (i.e. CoQ10, NADH) is essential for mitochondrial function and bio-energetic pathways.

Life evolved on Earth about 3.8 billion years ago using 6 elements: H, C, N, O, S and P, which are fundamental to all the major biological molecules. Biochemical synthesis requires input of energy and rearrangement of chemical bonds. Bio-energetics, the process of energy transduction, is governed by principles of REDOX chemistry and forms the basis for evolution of life on our Blue Planet.

5.4.1. REDOX Bio-Replenishment: Lactoferrin (LF)

Lactoferrin (LF) is a metal-binding glycoprotein present in milk and many exocrine secretions that bathe the mucosal surface. Although the term 'LF' implies an iron-binding component in milk, this molecule coordinately binds to various metal ions. LF also occurs in divergent biological milieu including saliva, tears, seminal fluids, mucins, and the secondary granules of neutrophils. The ability of LF to bind two Fe(III) ions with high affinity in cooperation with two HCO_3^- ions is an essential characteristic that contributes to its major structure-functional properties including antimicrobial activities.

LF and the Antimicrobial Spectrum: Nature has evolved the LF molecule as one of the potent microbial blocking hurdles in intestinal food processing. Accordingly, LF has a multifunctional role in a variety of physiological pathways and is considered an important milk-borne host defense factor in the gut. When orally consumed, LF enters the gastric lumen and binds with high affinity to heparan sulfate layers on the intestinal mucosa and also to specific enterocytes receptors. This physiological immobilization process triggers a variety of antimicrobial pathways: (i) microbial stasis via its iron deprivation mechanism, which selectively inhibits the growth-multiplication of several pathogens; (ii) inhibition of expression of specific colonization factors on the surface of pathogens that promote tissue attachment; and (iii) competition for tissue-binding sites on the epithelial surface to block or detach microbial colonization. Because of this multi-fold antimicrobial effect, pathogens are selectively inhibited and excreted from the gut. Therefore, LF belongs to a selective group of physiological microbial blocking agents (MBA) that naturally occur as innate defense factors. Thus, LF as a physiological MBA, interferes with adhesion/colonization, detach live or dead microorganisms from biological surfaces, inhibit microbial growth/multiplication, and neutralize activity of endotoxins.

LF is found in the exocrine glands (located mainly at the gateways of the digestive, respiratory and reproductive systems) where it provides mucosal protection against invading pathogens and toxic insults. LF occurs in three different physiological pools: (i) the secretory (exocrine) pool, (ii) the circulatory pool and (iii) the stationary (tissue-borne) pool. In the secretory pool, normal levels of LF are reported at 1-2 mg/mL in breast milk, tears and gastric mucins; 0.1-1 mg/mL in vaginal, cervical and bronchial mucus; 0.01-0.1 mg/mL in seminal plasma, pancreatic juice, saliva and crevicular fluids; <0.01 mg/mL in plasma, cerebrospinal and synovial fluids. Neutrophils contain LF at about 0.01 mg/10^6 cells, which contributes to the plasma levels of LF in the circulatory pool. In the stationary pool, LF is localized in several tissues. LF also co-exists with an array of molecules in different mucosal secretions with varying milieu conditions. Molecular dysfunction or deficiency of LF levels in the body can cause several physiological disorders and predispose various infections. Several studies have established that LF supplementation could provide exceptional health benefits and powerful protection against several illnesses.

LF - Metal-Binding/Transport and Iron Regulation: The ability of LF to bind ferric iron with high affinity (kD approximately 10^{-20} M) and to retain it to low pH gives the protein its antioxidant and antimicrobial properties. LF contains varying numbers of sites for potential glycosylation, mostly on the surface of the molecule. The degree of glycosylation determines the rate of resistance to proteases and acids. The ability to keep iron bound (even at low pH) is important, especially at sites of inflammation where the pH drops below 4.5 due to neutrophil degranulation. In such conditions, LF also binds to iron released from transferrin and prevents iron-acquisition for proliferation of pathogens. LF demonstrates remarkable resistance to proteolytic degradation by trypsin and trypsin-like enzymes. The level of resistance is proportional to the degree of iron saturation. LF binds iron twice higher than that of serum transferrin; it covalently binds two Fe^{3+} ions concurrently with two CO_3^{2-} ions. Although this bond is very strong and can resist pH as low as 4, LF saturation does not exceed 10% in total. There are three forms of LF according to its iron saturation: apo-LF (iron free), mono-ferric form (one ferric ion), and holo-LF (binds two Fe^{3+} ions). Besides iron, LF is also capable of binding a large amount of other compounds and substances such as lipopolysacharides, heparin, glycosaminoglycans, DNA, or other metal ions including aluminum (Al^{3+}), gallium (Ga^{3+}), manganese (Mn^{3+}), cobalt (Co^{3+}), copper (Cu^{2+}), zinc (Zn^{2+}), etc. However, its affinity for these other ions is much lower. Apart from CO_3^{2-}, LF can bind a variety of other anions such as oxalates, carboxylates, and others. In this way, LF could regulate REDOX homeostasis with transport and distribution of various minerals, anionic and cationic substances into the extra-cellular matrix.

LF plays an important role in the intestinal absorption of iron, zinc, copper, manganese and other essential trace elements. LF has also been suggested to protect the gut mucosa from excess uptake of heavy metal ions. Involvement of specific LF binding receptors in the human duodenal brush border has been suggested in iron absorption. An intestinal receptor for LF (M_r 110-kDa) with a cellular density of 4.3 x 10^{14} sites per milligram of solubilized human fetal intestinal brush-border membranes (IBBM) has been isolated. Furthermore, increased iron absorption via this LF receptor from IBBM during the neonatal period has been reported. LF is transported into the circulation from the intestinal lumen and excreted into the bile, suggesting the possibility of entero-hepatic circulation of LF. Following oral administration in neonatal pigs, bovine LF appeared in the blood circulation and reached a peak level after 2 hours. After oral administration, LF was endocytosed via the epithelial cells, absorbed into the blood, detected in the bile and reached a peak value within 12 hours.

LF and REDOX Chemistry: In 1981, Ambruso and Johnston reported the ability of neutrophil LF to provide iron efficiently to the oxygen radical-generating systems and suggested a role for LF as regulator of ·OH production. As such, lactoferrin is an important component in the microbicidal activity of neutrophils. At the same time, Gutteridge and co-workers demonstrated that LF, at physiological amounts, inhibited lipid peroxidation by Fe(II) salts and ascorbate. Thus, LF could help protect neutrophils, inflammatory foci and secretions from metal-ion-dependent oxidative damage. Also apo-LF was shown to alleviate H_2O_2-induced oxidative damage and provide antioxidant protection to the gastrointestinal tract.
During the past three decades, a prominent role for LF in REDOX chemistry has been established by various laboratories worldwide.

Lactoferrin Structure

Lactoferrin (LF) is an 80-kDa (703 amino acid residues) glycoprotein comprised of a single polypeptide chain folded into two globular N- and C-lobes connected by a short peptide helix. Each lobe is further divided into two sub-domains (N1 and N2 in the N-lobe, C1 and C2 in the C-lobe) that form a cleft (inside of which the iron is bound). The iron-coordinating residues in the N- and C-lobes of LF are highly conserved, consisting of an aspartic acid, two tyrosines, a histidine, as well as an arginine that coordinates a requisite anion. Based on iron-bound and iron-free status, each lobe and its sub-domains move as rigid bodies around a few hinge residues that lie between the sub-domains, and open-twist as iron is released and close as iron is bound.

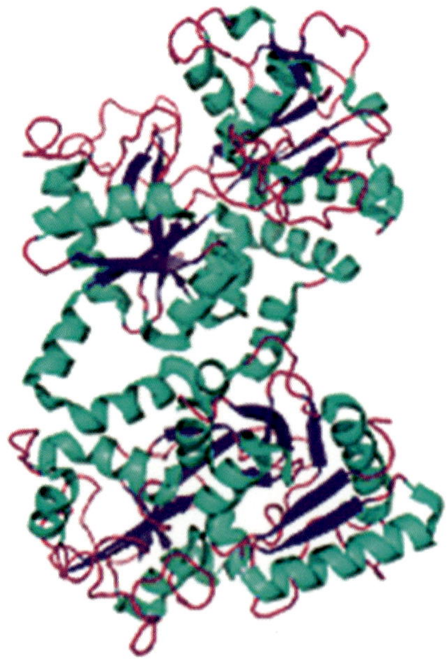

LF is shown to inhibit ROS generation by normal human monocytes, an effect associated with its metal-binding property. The substitution of Fe(II) by another transition metal [such as Cu(II), Zn(II), Pt(II) or Au(II)], could affect the antioxidant outcome. Among the different substituted transition metals, LF saturated with Cu(II) showed the highest inhibition of ROS generation by monocytes. Copper-bound LF (Cu-LF) inhibits the stimulation-dependent generation of ROS; at low concentrations, however, Cu-LF exhibits superoxide dismutase (SOD)-like activity. Cu-LF inhibits cytochrome c (*Cyt c*) in peritoneal neutrophils without affecting NADPH oxidase-mediated respiratory burst. Similar but weaker inhibition of *Cyt c* has also been observed with manganese-LF (Mn-LF) but not with Fe-LF or apo-LF.

Nitric oxide (·NO along with TNF-α and IL-8) is released when macrophages are activated by LF. Bovine LF was shown to reverse the endotoxic effects on enterocytes through its REDOX mechanism. Anti-diarrheal activity was achieved by LF through modulatiion the cyclooxygenase [·NO and PGE2] pathway in the gut. A tetrapeptide (PEP1261) corresponding to 39-42 residues of human LF showed a potent induction of ·NO production. LF also provides antioxidant protection to erythrocytes through various mechanisms, such as: (i) influencing the levels and activity of intra-cellular antioxidants, reduced GSH, in particular; and, (ii) by lowering the levels of lipid peroxidation products, as well as preventing hemolysis under normal conditions and oxidative stress.

LF and Oxidative Stress: LF is a primary constituent of REDOX homeostasis, functioning to reduce oxidative stress at the molecular level, and thus controlling excess inflammatory response. Oxidative stress occurs when the production of potentially destructive ROS exceeds the body's own natural antioxidant defenses and results in cellular damage. A cell is able to overcome small perturbations and regain its original state; but, severe oxidative stress can cause cell death. While moderate oxidation can trigger apoptosis, more intense stresses can cause necrosis within tissue.

Transitional metals may be considered key factors in oxidative stress. In particular, traces of Fe(II) can be detrimental to physiological processes under reactive oxygen conditions. Iron is crucial in modulating production of ROS, by virtue of its ability to catalyze a two step process known as the 'Haber-Weiss reaction'. In normal physiological conditions, the production and neutralization of these ROS largely depends on the efficiency of key enzymes, including superoxide dismutase (SOD), catalase and GSH-peroxidase. Inefficiency of these enzymes leads to over-expression of hydroxyl radical via iron-dependent Haber-Weiss reaction, and subsequent increase in lipid peroxidation. It is hypothesized that endogenous LF can prevent lipid peroxidation by virtue of Fe(II) sequestration.

TABLE: *Occurence of LF in Biological Fluids*

Biological Fluid	LF level (mg/mL)
Breast milk, tears and gastric mucins	1.0-2.0
Vaginal, cervical and bronchial mucus	0.1-1.0
Seminal plasma, pancreatic juice, saliva and crevicular fluids	0.01-0.1
Plasma, cerebrospinal and synovial fluids	<0.01

LF Regulates Fe(II)-mediated Inflammation: The concentration of LF in blood is normally low (0.2–0.6 μg/mL), but at sites of inflammation, this concentration is significantly raised (~200 μg/mL) due to its release from neutrophils. LF regulates synovial Fe(II) loading to reduce inflammation. Iron accumulates during an inflammatory response and macrophages release hydrogen peroxide and superoxide, which can then be converted to damaging hydroxyl radicals by free iron via the Fenton/Haber-Weiss reaction. In inflammatory conditions, such as rheumatoid arthritis, the disease activity correlates with the degree of Fe(II) accumulation in the synovium, while Fe(II) administration to anemic RA patients causes a flare in peripheral synovitis. The failure of normal iron regulatory proteins (IRPs i.e. transferrin and ferritin) to remove Fe(II) from the RA synovium aggravates the disease condition. Iron chelators were shown to reduce inflammation in patients and animal models, but in the former gave rise to severe side effects. Nevertheless, these findings suggest that elimination of toxic Fe(II) from the joint may be beneficial, and of potential clinical use if the side effects of iron chelators can be avoided.

An increased risk of iron loading in osteoporosis has also been indicated. Iron may suppress bone remodeling apparently by decreasing osteoblast formation and new bone synthesis. Thus, LF supplementation could be beneficial in prevention of osteoporosis. LF released by neutrophil degranulation into the inflammatory site, is a potent anti-inflammatory agent due its ability to bind Fe(II). LF can prevent iron-mediated tissue damage by reducing 'free' iron concentration when inflammatory stimuli have deregulated IRP (i.e. transferrin) mediated Fe(II) homeostasis. Iron, after binding to LF, loses its ability to catalyze toxic free radical reactions that trigger tissue damage. Thus, LF, if present in adequate quantities in the inflammatory site, could effectively reduce tissue damage by preventing Fe(II)-catalyzed free radical formation.

LF Regulates Zn(II)-mediated MMP Activity: Matrix metallo-proteinases (MMPs) constitute a family of Zn(II)-dependent proteolytic enzymes that degrade several components of extra-cellular matrix. LF, though an iron-binding protein, is also capable of chelating other divalent metal cations, including Zn(II). LF levels are elevated in various disease conditions in which MMPs, particularly MMP-2, are also elevated, suggesting a possible interplay between the two proteins. Pre-incubation with apo-LF, but not holo-LF, could effectively reduce the hydrolysis of a peptide substrate by MMP-2. This inhibition was specific for the 42-kDa catalytic domain fragment of MMP-2 lacking the hemopexin domain. The inhibition of the MMP-2 catalytic domain was strongly temperature sensitive, indicating that the conformation of one or both proteins is critical for the interaction. To ascertain the mechanism of inhibition, increasing concentrations of $ZnCl_2$ and $FeCl_2$ were added to the reaction. While addition of Fe(II) did not reverse inhibition, the addition of Zn(II) resulted in a recovery of MMP-2 activity. Furthermore, Zn-LF did not inhibit MMP-2. Together, this data suggests that LF could remove the catalytic Zn(II) from the active site of MMP-2 and prevent proteolytic tissue damage. This inhibitory activity is a novel function for LF that represents a potent mechanism that regulates the degradation of extra-cellular matrix by MMP-2 *in vivo*.

LF and the Immune System: LF is considered a first-line defense that protects the host against several microbial infections and prevents systemic inflammation. The ability of LF to bind large quantities of Fe(II) provides protection against pathogens and their metabolites by enhancing phagocytosis, cell adherence, and controlling the release of pro-inflammatory cytokines. Other direct effects of LF include anti-viral, anti-parasitic, and anti-fungal activities. Additionally, LF possesses indirect activity, often through prevention of pathogen invasion by blocking interaction with receptors used for entry on host cells. Sequestration of Fe(II) by LF reduces insult-induced oxidative stress, thus altering the magnitude and specific production of cytokines. LF has a profound modulatory action on the adaptive immune system by promoting the maturation of T-cell precursors into competent helper cells and by the differentiation of immature B-cells into efficient antigen presenting cells. In addition, LF augments the delayed type hypersensitivity response to antigens, leading to a strong induction of cell-mediated immunity in mice.

LF is one of the mediators that naturally bridges the innate and adaptive immune functions by regulating target cell response, including those involved in oxidative stress and systemic inflammatory responses. It is also recognized as a significant contributor in regulation of antigen presentation and development of productive T helper cell response. LF affects immunological functions by modulating the production of cytokines TNF-α, IL-1β, and IL-6, by regulating NK cell activity, inhibiting antibody synthesis, and maturation of T cells. LF-mediated immune modulation of cytokine production, such as down-regulation of TNF-α and up-regulation of IL-10, could be important locally at the site of inflammation. A single administration of bovine LF could suppress TNF-α production and increase IL-10 production in the LPS-stimulated adjuvant arthritis rats. Orally administered LF once daily, starting 3 hr before (preventive mode) or 19 days after (therapeutic mode) adjuvant injection was shown to suppress the development of arthritis and hyperalgesia in the adjuvant-injected paw. Thus, LF has both preventive and therapeutic effects on the development of adjuvant-induced inflammation and pain. LF (1 mg/ml) could also reduce articular inflammation, particularly in septic arthritis, in which anti-inflammatory effects may be achieved without promoting bacterial survival.

LF and Antimicrobial Effects: Different mechanisms are involved in the anti-microbial functions of LF. The first mechanism is a bacteriostatic effect related to the high iron-binding affinity of the protein, which deprives iron-requiring microbes of this essential growth nutrient. Since the microbial stasis properties of LF are due to its iron binding ability, the protein is capable of retarding the growth of a broad range of microorganisms including a variety of Gram negative and Gram positive bacteria and certain yeasts. LF expressed in mothers' milk is about 5% to 8% saturated with iron and therefore, expected to function as an efficient bacteriostatic agent.

Intra-cellular (or phagosome) killing of pathogens by LF is associated its ability to donate iron for catalyzing ·OH radical production. The iron bound to LF is remarkably stable to pH; the prooxidant activites of this protein are likely restricted to the intra-cellular environment of the activated neutrophil where the acidic phago-lysosome fuses with LF-containing specific granules. In the less acidic environment of mucosal secretions, the iron sequestration activity of LF likely prevails, preventing against host cell damage by protecting against free radical production. These antimicrobial effects highlight the many possible mechanisms of LF in host protection against microbial infections at the mucosal surfaces.

Microbicidal effect, distinct from the iron-chelation mechanism, is associated with direct interaction of LF with microbial cell surfaces. LF could damage the outer membrane of Gram negative bacteria by causing the release of lipopolysaccharides (LPS), which leads to lysis of the target cell. A cationic domain at the N-terminus of LF is responsible for its bactericidal properties. LF has also been shown to interact with 'porins' in the outer membrane of Gram-negative bacteria and elicit a variety of antimicrobial effects, including; bacterial detachment from tissue surface, selective inhibition of pathogens (with exclusion of healthy probiotic bacteria) and endotoxin neutralization.

Table: *Lactoferrin (LF) – Multifunctional Bio-Replenishment*

Physiological System	Mechanism of Action
Metal-binding and Transport	Regulates the bio-availability of Fe(III), Zn(II), Cu(II), Co(II), Mn(II), and V(II). LF has specific receptors on enterocytes.
Antimicrobial Activity (Bacteria, Virus, Fungi & Parasites)	As a broad spectrum agent with stasis, cidal and adhesion-blockade mechanisms acts against bacteria, viruses, fungi/yeast and parasites.
Prebiotic Activity	Selectively promotes growth of probiotic bacteria in the intestinal tract.
Inflammation Modulator	Promotes neutrophils aggregation, degranulation.
Immune Modulator	Specific growth stimulation of lymphocytes, up-regulates thymocyte maturation, cytotoxicity of monocytes, cytokine production, etc.
Adjuvant Effects	Regulates antigen presentation and stimulates antibody production. Activates complement via C3 convertase inhibition.
Antioxidant Activity	Play a role in Fenton-type reactions (Haber-Weiss pathways) and prevents ·OH mediated tissue injury.
Anti-tumor Activity	Down-regulates myelopoiesis, activates/regulates NF-κB and TNF-α.
Cellular Growth Promotion	Regulates enterocyte proliferation and gut maturation.
Bone Metabolic Regulation	Up-regulates bone forming osteoblasts and down-regulates bone resorbing osteoclasts, and thus, plays a role in bone turnover.
Protects Extra-cellular Matrix (ECM)	Regulates matrix metallo-proteinase (MMP) mediated tissue hydrolysis, especially bone-joint cartilage degradation.
Endocrine Regulation	Regulates estrogen cycle and various stages of pregnancy.
Promotes Male Fertility	As an integral part of sperm-coating antigen, LF promotes the survival of sperm in the female reproductive tract.
Macro-molecular Interaction	Influences biological activities of molecules such as DNA, RNAses, immunoglobulins, lysozyme, lactoperoxidase, osteopontin, etc.

5.4.2. REDOX Bio-Replenishment: Ribonuclease (RNase)

Ribonuclease (RNase or angiogenin) is a 14-kDa, basic heparin-binding protein and a member of the pancreatic RNase superfamily. It is a key mediator in nearly all phases of angiogenesis, which makes it an important precursor in skeleto-muscular development. Human plasma contains about 0.3 µg/mL RNase with a fast turnover rate (half-life <5 min). RNase can induce most pathways necessary for the formation of new blood vessels. It binds avidly to endothelial cells and stimulates cell migration and invasion. RNase promotes cell proliferation and differentiation; mediates cell adhesion and activates cell associated proteases; and also induces plasminogen activator, thereby, promoting migration and tubular morphogenesis of endothelial cells. Transportation of RNase, from the cell surface into the nucleus and subsequently to the nucleolus, is critical for its angiogenic activity. The import of RNase from cytosol to the nucleus is a signal-dependent, carrier-mediated and energy-dependent, active transport process.

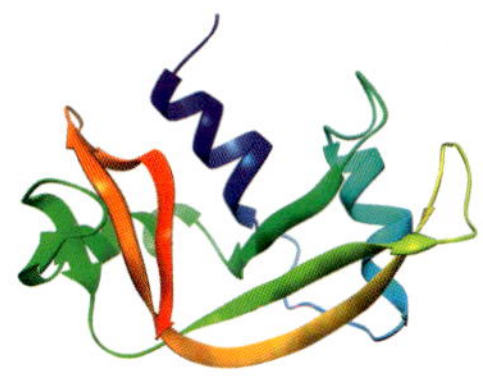

RNase

RNase types 4 and 5 are active secretory proteins found in milk. Bovine milk contains about 2 mg/L RNase 4 and 1 to 8 mg/L RNase 5. Bovine milk RNase is a single-chain protein of 125 amino acids; it contains six cysteines and has an estimated molecular weight of 14.6 kDa. Bovine milk RNase has 65% sequence homology with human plasma RNase and 34% homology with bovine pancreatic RNase A. Three major active site residues involved in the catalytic process, His-14, Lys-41 and His-115, are conserved in bovine milk RNase with ribonucleolytic activity (comparable to that of the human RNase).

RNase and Angiogenesis: Angiogenesis and vasculogenesis are two primary pathways in the development and maintenance of mammalian physiology. During embryonic development, vasculature is among the first systems that forms and maintains metabolic homeostasis by supplying oxygen, nutrients, soluble factors and numerous cell types, as well as by removing waste products.The growth and development of a mature vascular structure is one of the earliest events in organogenesis. In mammalian embryonic development, the nascent vascular networks develop by aggregation of de *novo* forming angioblasts into a primitive vascular plexus (vasculogenesis). This undergoes a complex remodeling process in which sprouting, bridging and growth from existing vessels (angiogenesis) leads to the onset of a functional circulatory system. Angiogenesis is a strictly regulated, multi-step process that occurs during normal bone remodeling and regeneration, pregnancy, and tissue repair conditions such as fractures and wound healing.

RNase and the Skeletal System: Bone is a highly vascularized tissue reliant on the close spatial and temporal connection between blood vessels and bone cells to maintain skeletal integrity. Skeletal development includes the coordination of multiple events such as migration, differentiation, and activation of multiple cell types and tissues. The development of a micro-vasculature and micro-circulation is critical for the homeostasis and regeneration of living bone, without which, the tissue would simply degenerate and die. *In vitro* and in *vivo* models of osteogenesis and fracture repair have elucidated the vascular biology in skeletal development and repair. Exercise could enhance angiogenesis and bone strength. An imbalance in angiogenesis or disruption to normal bone vasculature results in deterioration of bone health, leading to conditions such as osteopetrosis, metastatic bone disease, osteolysis, osteoporosis, inflammatory bone loss (in rheumatoid arthritis), and periodontal disease. Angiogenesis thus plays a pivotal role in skeletal development and bone physiology.

Bovine milk lactoperoxidase (LPO), a co-isolate from milk basic protein fraction, was also reported as the predominant inhibitor of osteoclastogenesis. The LPO treatment seems to down-regulate levels of ROS in osteoclasts. Signaling by RANKL/RANK was down-regulated in LPO-treated cells, and, in particular, the tumor necrosis factor receptor associate factor 6 (TRAF6) and activation of downstream signaling cascades (i.e. NF-κB). Recent studies have demonstrated potent synergism between RNase and lactoferrin (LF), the predominant protein co-isolate from milk.

RNase interactions with Lactoferrin (LF): The role of LF in human physiology, including bone metabolism is well characterized. LF effects include differentiation of precursor cells, osteoblast proliferation, stimulation of matrix synthesis, and angiogenesis. These regulatory effects have a prominent role in augmenting bone repair. RNase-enriched lactoferrin (R-ELF) technology has provided clinical evidence that a combination of RNase with LF could have a promising application in bone regeneration.

RNase and Bone Engineering: Blood vessels have a fundamental role both in skeletal homeostasis and in bone repair; therefore, RNase is important for successful bone engineering. Vascularization of the growth plate region represents a key mechanism for the coupling of two fundamental processes determining the rate of bone growth: chondrogenesis (cartilage production) and osteogenesis (bone formation). Precise coupling is crucial during periods of rapid bone growth or fracture repair in adults, and changes in the balance might induce pathologic conditions such as osteoarthritis and ectopic bone formation. During the formation of the growth plates of long bones, there is close and dynamic interaction between developing vascular structures and the cartilage, which is one of the least vascular tissues in the body. Recent experimental findings provide an explanation of why the close proximity of cartilage and vasculature is mutually exclusive: vascular invasion of cartilage is associated with chondrocyte apoptosis and consequently, inhibition of angiogenesis in the growth plate delays chondrocyte cell death, resulting in a massive expansion in the number of hypertrophic cartilage cells in the growth plate. The fundamental importance of chondrocytes in the growth, development and repair of the skeleton has led to intense investigation of the mechanisms that regulate chondrocyte maturation and apoptosis. RNase, LF and their complex, R-ELF, may have a potential role in bone engineering.

Bone tissue engineering in particular has benefited from the application of pro-angiogenic strategies, considering the need for an adequate vascular supply during healing and the challenges associated with the vascularization of scaffolds implanted *in vivo*. Conventional methods of assessing the *in vivo* angiogenic response to tissue-engineered constructs tend to rely on a two-dimensional assessment of microvessel density within representative histological sections without elaboration of the true vascular tree.The lack of a functional vascular supply has hampered the whole range of clinical applications of successful bone tissue engineering strategies, especially in animal bone grafts. These grafts depend on post-implant vascularization, therefore, the development of RNase and LF based, R-ELF technology could effectively induce the establishment of a microcirculation in the engineered constructs.

RNase-enriched LF and Bone Health: Both RNase and LF co-exist in milk and other biological fluids as key regulatory molecules with specific functions in various physiological pathways. From a bone metabolism standpoint, both the proteins elicit potential effects individually as anabolic and anti-resorptive agents during the aging process, disease and repair conditions. For the first time, a synergistic effect of these two bio-active molecules mixed together (as an R-ELF complex) to advance bone health has been reported.

Osteoporosis is one of the most critical bone metabolic disorders affecting postmenopausal women; it is characterized by a reduced BMD and an increased risk of fracture. Postmenopausal bone loss is a consequence of an estrogen deficiency that leads to an imbalance between bone formation and bone resorption. Reduced angiogenesis could also contribute to this bone metabolic imbalance. Dietary R-ELF supplementation significantly enhances bone formation (as measured by increase in serum bone-specific alkaline phosphatase levels) and reduces osteoclastic bone resorption (as measured by plasma NTx and urine dPD levels); thereby, restores balance of bone turnover within a short duration. Also, postmenopausal osteoporosis is a clinical outcome of an inflammatory process. Since bone and immune systems are genetically and functionally linked, R-ELF could also influence bone remodeling. R-ELF supplementation also decreases serum levels of pro-inflammatory cytokines (IL-6 and TNF-α) and increase anti-inflammatory IL-10. Thus, R-ELF could improve the inflammatory status in postmenopausal women and provide considerable bone health benefits.

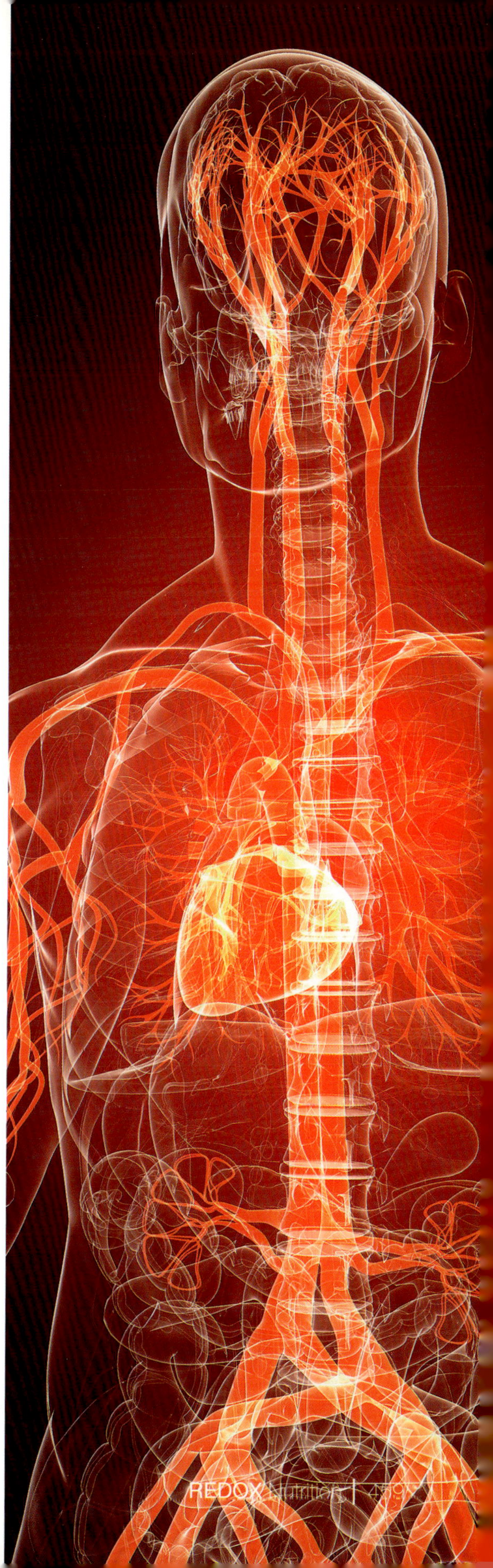

5.4.3. REDOX Bio-Replenishment: Coenzyme Q

CoQ-10

Coenzyme Q or ubiquinone (2,3-dimethoxy,5-methyl, 6-polyisoprene para-benzoquinone) has a structure similar to that of vitamin K and is a fat-soluble quinone. The predominant form of ubiquinone in the human body is coenzyme-Q10 (CoQ10), a polyisoprene chain containing 10 isoprene units (5 carbons each) or a total of 50 carbons. The polyisosoprene moiety of ubiquinone ensures an affinity for the interior of cell membranes, whereas, the two methoxy groups contribute to its enzymatic specificity. The fully substituted quinone ring does not allow additive cellular reactions with thiol groups such as glutathione (GSH), thioredoxin or thioctic acid. The functional quinone ring reduces to quinol, which creates a carrier for protons and electrons.

CoQ10 is found throughout the body, especially in the mitochondrial membranes of the heart, lungs, liver, kidneys, spleen, pancreas, and adrenal glands. *In vivo*, CoQ10 is synthesized from tyrosine (or phenylalanine) and mevalonate; this initial step requires pyridoxal 5'-phosphate (vitamin B6) as a cofactor. The total body content of CoQ10 is about 500-1500 mg. The normal CoQ10 level in blood is ~1 µg/mL, but an increase to 2 µg/mL in blood can be therapeutic for various conditions. Diet and lifestyle markedly influence CoQ10 synthesis *in vivo*. CoQ10 is absorbed from the small intestine, passes into the lymphatics, and finally reaches the blood and tissues.

Several factors affect levels or function of CoQ10, in addition to its decreased biosynthesis. These factors include increase in degradation or changes in membrane lipids to impede quinone movement. CoQ10 levels decline with age, and such decline might contribute to certain manifestations of age-related metabolic dysfunctions. For example, aging rats decrease CoQ10 selectively in skeletal muscle mitochondria but increase CoQ10 levels in mitochondria withi the brain. Physiological systems with active metabolism (such as cardiovascular, skeleto-muscular and immune systems) have the highest CoQ10 demand and such systems are also most susceptible to CoQ10 deficiency.

CoQ and Cellular Bioenergy: CoQ is an essential component of the ETC for ATP synthesis to provide energy for vital cellular functions. Most of the ATP generation occurs in the mitochondrial inner membrane, where CoQ is located. CoQ transfers electrons from primary substrates to the oxidase system and also mobilizes protons outside of mitochondria, resulting in a proton gradient across the membrane. When protons return intra-cellular via the enzymatic machinery (oxidative phosphorylation), ATP is formed. CoQ10 transports protons across the inner mitochondrial membrane by transverse diffusion, resulting in the conservation of energy ('proton-motive Q cycle').

Coenzyme Q10 is required for mitochondrial ATP synthesis and functions as an antioxidant in cell membranes and lipoproteins.

There are two protein complexes in the mitochondrial membrane where electrons and protons are transferred through CoQ. The first is the primary reductase where

CoQ is reduced by NADH (complex-I). During the reduction, 4 protons transfer across the membrane for every CoQ reduced. In the process, CoQ is reduced and re-oxidized in the complex twice before electrons are recycled to a second loosely-bound CoQ to form quinol. Quinol is mobilized via membrane lipids to a second complex where it is re-oxidized (complex-III) with proton transfer across the membrane. As in complex-I, there is a cyclic oxidation-reduction-reoxidation with the oxidation and proton release step always on the outside so that protons are released in the right direction. Again, the oxidation-reduction cycle allows 4 protons to cross the membrane for each quinol oxidation cycle. The quinone cycle thus doubles the CoQ efficiency in building up the proton charge across the membrane, which allows twice as much ATP production as a simple one step oxidation of quinol. A simple energy conversion based on CoQ reduction-oxidation is found in lysosomes. In this case, the quinol transfers a proton across the lysosomal membrane to acidify the interior, which involves energy to work against a proton gradient. Since the lysosomal membrane does not have a proton driven ATP synthase, ATP is not formed. Lysosomal acidification activates hydrolytic enzymes and digests the cellular debris. Thus, CoQ energizes cell house cleaning.

CoQ and the REDOX Chemistry: Another possible activity of CoQ is acidification of pinocytotic endosomes to release the engulfed iron-bound transferrin from receptors and transport iron into the cell. There is a REDOX process in the membrane vesicles for intra-cellular acidification. Reduction of transferrin-bound iron in an acid environment releases the iron for uptake into the cytoplasm. However, the role of CoQ in iron reduction and proton transfer in these membranes has not been established.

Another role for CoQ in proton transfer at the plasma membrane level has been suggested. In this process, CoQ activates Na^+/H^+ exchange across plasma membrane via the Na^+/H^+ anti-port. The energy for this pathway is derived from the high concentration of extra-cellular Na^+ levels that exchange for intra-cellular protons. Subsequently, Na^+ is pumped out of the cell by Na^+/K^+ ATPase, the ATP-dependant enzyme. When excess Na^+ is released during this ATP activity, the cell develops an inside negative membrane potential that supports various cellular functions. CoQ is involved in plasma membrane electron transport, where cytoplasmic NADH transfers electrons (via CoQ) to extra-cellular electron

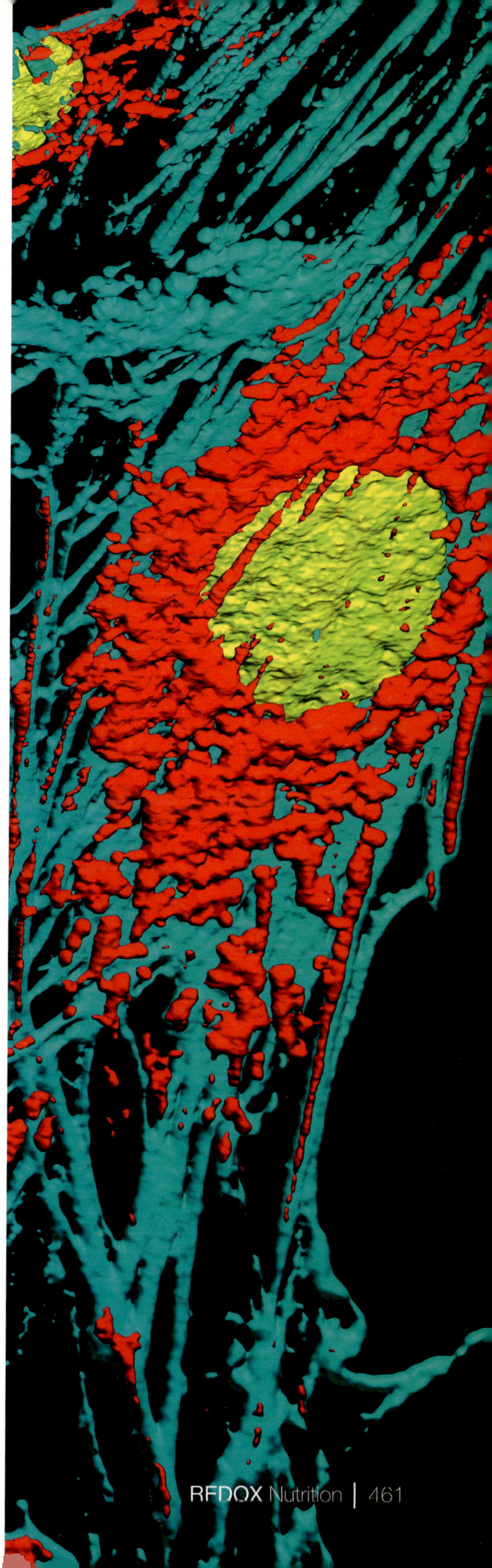

Endogenous synthesis and dietary intake appear to provide sufficient CoQ10 to prevent deficiency in healthy people, although tissue levels of CoQ10 decline with age.

acceptors such as Fe(II) or O_2. During activation of this pathway, the proton release through the H^+/Na^+ anti-port is markedly enhanced. When the system is blocked by CoQ analogs, however, the anti-port is inhibited and the cell interior turns more alkaline. The mechanism by which the CoQ-dependent electron transport activates the anti-port is not known.

CoQ and Antioxidant Activity: CoQ10 is a membrane-bound chemical, located in close proximity to the unsaturated lipid chains, designed to function as a primary scavenger of free radicals. CoQ levels in most membranes range from 3 to 30 times than the tocopherol (vitamin-E). Since most of the CoQ10 in cell membranes is in quinol form, it is a very effective antioxidant. CoQ10 is one of the most significant lipid antioxidants that prevents free radical generation and modifications of proteins, lipids, and DNA. CoQ10 is an effective antioxidant both on its own and in fusion with vitamin E and is fundamental in powering the body's energy production ATP cycle.

Specific enzymes (membrane-bound) readily reduce CoQ-quinone radicals generated by reaction with lipid or oxygen radicals. At least three enzymes are known to maintain CoQ in a reduced-form in plasma and endomembranes, which are: (i) NADH cytochrome b5 reductase, (ii) NADH/NADPH oxidoreductase (DT diaphorase), and (iii) NADPH coenzyme Q reductase. In mitochondria, the NADH and succinate dehydrogenases maintain CoQ in partly reduced form. Reductases 1 and 3 in the endo-membranes are especially important for single electron transfer to re-reduce any semiquinone generated by reaction of quinol. The DT diaphorase is unique for the ability to directly reduce, by 2-electron transfer, any quinine formed without intermediate formation of the semiquinone.

During oxidative stress induced by dietary deficiency of Se and α-tocopherol, the CoQ level in membranes increases significantly. Consequently, the membrane-bound DT diaphorase levels also increase to facilitate the reduction of CoQ. Similarly, any decrease in α-tocopherol induced by peroxisomal proliferator also induce a marked increase in CoQ. In addition to direct antioxidant radical scavenging, the quinol could rescue tocopheryl radicals (produced by reaction with lipid or ROS) by direct reduction back to tocopherol. Without CoQ in a membrane, regeneration of tocopherol is very slow. Tocopherol is also regenerated in low density lipoprotein where a small amount of CoQ protects a large amount of vitamin E. This function is favored by the high quinol content in the blood. CoQ-dependent electron transport across the plasma membrane could also regenerate extra-cellular ascorbate (vitamin-C) from ascorbate radical. Intra-cellular vitamin-C is regenerated by a glutathione (GSH) based system. Extra-cellular regeneration requires electron transfer through the plasma membrane, some of which depends on the presence of membrane-bound CoQ.

Use of cholesterol-lowering medications known as HMG-CoA reductase inhibitors (statins) decrease circulating levels of CoQ10.

CoQ and anti-inflammatory activity may specially benefit during wound healing. Regeneration of damaged skeletal muscle has been observed in animals supplemented with CoQ10 and vitamin E. CoQ10 could prevent oxidative stress, possibly due to its ability to inhibit the arachidonic acid pathway and subsequent formation of inflammatory prostaglandins. Anecdotal reports indicate that pain decreases rapidly when CoQ10 is administered without concurrent non-steroidal anti-inflammatory drugs. Exhausting exercise is also known to induce muscle damage associated with high production of free radicals and pro-inflammatory mediators. Clinical studies showed that CoQ10 supplementation before strenuous exercise could decrease oxidative stress, modulate inflammatory signaling, and reduce any subsequent muscle damage. Exercise was associated with an increase in TNF-α, IL-6, 8-OH-2'-deoxyguanosine (8-OHdG), and isoprostane levels, revealing the degree of inflammation and the induced oxidative stress. Oral supplementation of CoQ10 during exercise efficiently reduced oxidative stress (decreased membrane hydroperoxides, 8-OHdG and isoprostanes generation, increased catalase, and total antioxidant status) and maintained cellular integrity. Data also indicated that CoQ10 averted over-expression of TNF-α after exercise. Furthermore, dietary CoQ10 reduced creatinine production.

CoQ10 Supplementation: Nutritional replenishment of CoQ10 requires a higher level than is available in most food. The normal level in blood is around 1 μg/ml. To increase the concentration significantly requires at least 100 mg/day which can increase the level in blood up to 2 μg/ml or more. An increase to 2 μg/mL in blood can be therapeutic for various conditions; this may indicate that a high blood level is needed to get coenzyme Q into deficient tissues. Even with large amounts of heart or herring in the diet, it would be difficult to supply 100 mg/day. Meat and fish are the richest source of dietary CoQ10 and levels over 50 mg/kg can be found in beef, pork, and chicken heart, and in chicken liver. Within vegetables, parsley and perilla are the richest CoQ10 sources. Broccoli, grape, and cauliflower are also modest sources of CoQ10. Most fruit and berries represent a poor source of CoQ10, with the exception of avocado, which has a relatively high CoQ10 content.

METHIONINE FOOD SOURCES

Food Source	mg/kg
Chicken Heart	115-130
Herring	27
Pistachio Nuts	20
Avocado	10

5.4.4. REDOX Bio-Replenishment: S-Adenosyl-Methionine

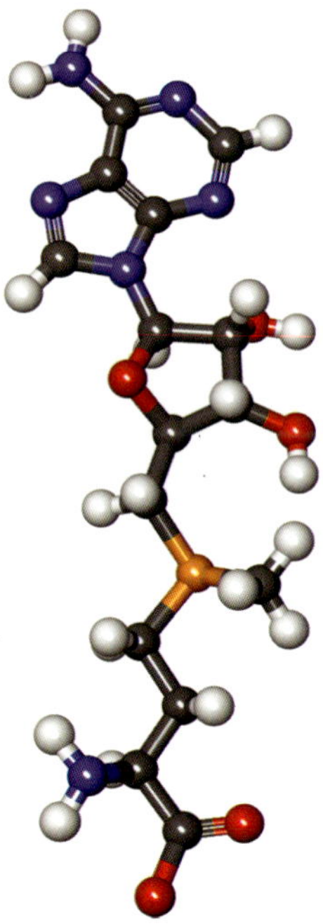

SAMe

S-Adenosylmethionine (SAMe, also abbreviated as SAM and AdoMet) is the precursor of glutathione (GSH) in the liver. Although SAMe biosynthesis occurs in all mammalian cells, in a reaction catalyzed by the essential enzyme methionine-adenosyl-transferase (MAT), the liver is the major site of its synthesis and utilization. Patients with chronic liver disease have impaired SAMe biosynthesis, which may contribute to the pathogenesis of liver disease and predisposition to malignant degeneration. SAMe plays an important role in hepatocyte growth, death and malignant degeneration; it also regulates the activities of various enzymes. SAMe is the principal biological methyl donor made in the cytosol of every cell, but the liver plays a central role in the homeostasis of SAMe as the major site of its synthesis and degradation. In the liver, up to half of the daily intake of methionine is converted to SAMe and up to 85% of all methylation reactions takes place. MAT is the enzyme responsible for the formation of SAMe from methionine and ATP.

SAMe and REDOX Chemistry: SAMe is the link to three key metabolic pathways, polyamine synthesis, trans-methylation and trans-sulfuration. During polyamine synthesis, SAMe is decarboxylated and the remaining propylamino moiety is donated to putrescine to form spermidine and methylthioadenosine (MTA) and to spermidine to form spermine and a second molecule of MTA.

SAMe and Trans-methylation Reactions: SAMe donates its methyl group to a large variety of acceptor molecules in reactions catalyzed by methyltransferase. S-Adenosylhomocysteine (SAH) is generated as a product of trans-methylation and is hydrolyzed to form homocysteine and adenosine through a reversible reaction catalyzed by SAH hydrolase. SAH is a potent competitive inhibitor of methylation reactions; prompt removal of adenosine and homocysteine is required to prevent accumulation of SAH. Homocysteine can be remethylated to form methionine by two enzymes: methionine synthase (MS), which requires normal levels of folate and vitamin B12; and betaine homocysteine methyltransferase (BHMT), which requires betaine (a metabolite of choline). Remethylation of homocysteine, via MS, requires 5-methyl-tetrahydrofolate (THF), which is derived from 5,10-methylene-THF in a reaction catalyzed by methylene-THF-reductase. 5-methyl-THF is then converted to THF as it donates its methyl group and THF is converted to 5,10-methylne-THF to complete the folate cycle.

SAMe donates a methyl group to a wide variety of substrates including DNA, proteins, neurotransmitters, and phospholipids. SAMe plays an important role as a methyl donor in more than a hundred methyltransferase reactions. SAMe functions as a methyl donor in the liver and possesses lipotropic (promoting the utilization of fat) activity. Studies have demonstrated that SAMe is effective in promoting bile flow (necessary for the digestion of fat). SAMe also functions in the liver to inactivate estrogens, thereby protecting the liver from hormonal damage. The process of SAMe-methylation assists the body to grow and repair cells and helps maintain phospholipids in the cell membrane.

About 50% daily intake of methionine is converted to SAMe in the liver and an average adult produces 7-8 g of SAMe each day.

SAMe and Trans-sulfuration Pathway: Homocysteine can also undergo the trans-sulfuration pathway to form cysteine (the rate-limiting precursor for GSH) via a two-step enzymatic process catalyzed by cystathionine β-synthase (CBS) and cystathionase, both requiring vitamin B6. The trans-sulfuration pathway is particularly active in the liver, making SAMe an important precursor of GSH. All mammalian tissues express methionine-adenosyl-transferase and methionine synthase, whereas BHMT is limited to the liver and kidney. In the liver, SAMe inhibits methylene-THF-reductase and methionine synthase and activates CBS. Thus, when SAMe is depleted, homocysteine is channeled to remethylation to regenerate SAMe, whereas when the SAMe level is high, homocysteine is channeled to the trans-sulfuration pathway.

In trans-sulfuration reactions, SAMe is converted to cysteine in a series of enzymatic steps. Cysteine is a precursor for glutathione (GSH), a major cellular antioxidant. SAMe can upregulate the production of the most important detoxification system in cells by increasing the GSH production. SAMe also stimulates the synthesis of various sulfur-containing proteoglycans critical for cartilage regeneration. SAMe is involved in synthesis of the polyamines – spermidine and spermine via the aminopropylation pathway. Polyamines are involved in regulation of cell growth. Spermidine and spermine have anti-inflammatory and analgesic properties.

SAMe and Neural Physiology: SAMe influences brain function is several ways: (i) SAMe facilitates the conversion of norepinephrine to epinephrine and serotonin to melatonin; (ii) SAMe helps in the preservation of GSH antioxidant function; (iii) GSH helps creatine synthesis, an important energy reservoir in muscle tissue. Creatine synthesis could account for consumption of >70% SAMe-derived methyl groups in humans. Furthermore, SAMe is involved in the formation of myelin, the white sheath that surrounds nerve cells, which improves brain cell membrane fluidity. SAMe supports the nervous system in the synthesis and recycling of various neurotransmitters and enhances the sensitivity of nerve receptors. SAMe could positively affect a number of neurotransmitters, including serotonin, dopamine, noradrenaline and norepinephrine. SAMe could improve binding of neurotransmitters to receptor sites; thereby increasing functional activity of serotonin.

SAMe Supplementation: SAMe is a bioreplenishment found in every living cell, with its greatest concentrations located in brain and liver. Since adequate amounts of SAMe are not readily available through diet, human physiology has evolved specific pathways of SAMe biosynthesis via methionine metabolism. SAMe levels decline with certain lifestyles and aging. SAMe is an intermediate metabolite of the emethionine and ATP; in this form, it is sometimes referred to as 'active methionine'. SAMe biosynthesis is closely associated with folate and vitamin B12 metabolism, and deficiencies of both these vitamins have been found to reduce SAMe concentrations in the central nervous system. Both folate and vitamin B12 deficiency could trigger neurological and psychiatric disturbances including depression, dementia, and peripheral neuropathy. SAMe and folate work together to beneficially affect monoamine systems, which directly influence mood and cognitive function.

SAMe has a significant first-pass absorption rate, with approximately 50% metabolized in the liver. Oral administration increases SAMe levels in cerebrospinal fluid, which indicates that SAMe is able to cross the blood brain barrier. SAMe expresses bioactivity in the dopaminergic system. These effects suggest that SAMe acts as a natural reuptake inhibitor to support healthy neurotransmitter balance in the brain. Taking SAMe at bedtime may interfere with restful sleep. Clinical studies indicate that benefits may be evident within a week of SAMe supplementation.

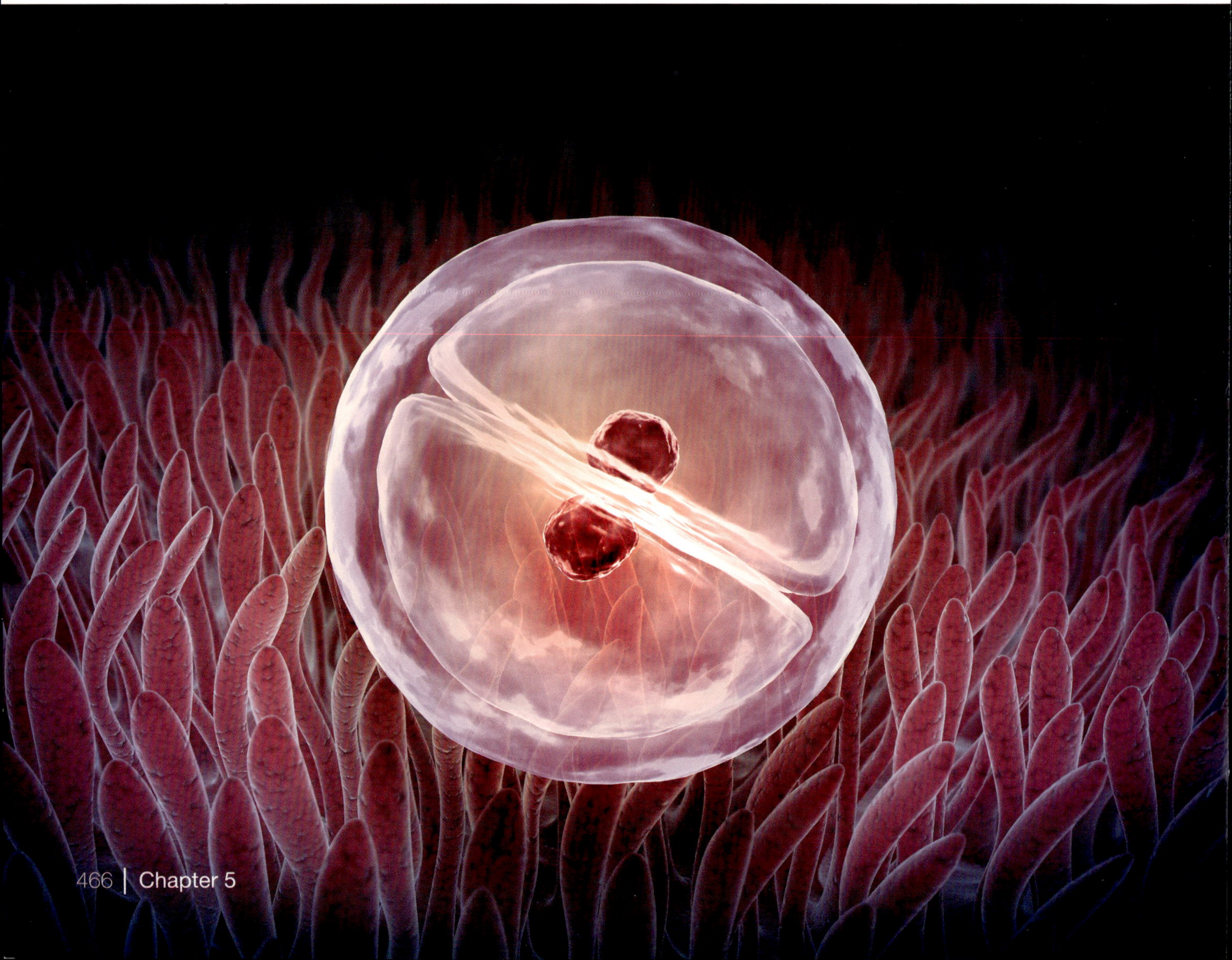

5.5 | REDOX Nutrition: **Bibliography**

5.1 | REDOX-Active Metals

1. Chevion M (1988) A site-specific mechanism for free radical induced biological damage: the essential role of redox-active transition metals. *Free Radical Biology & Medicine* 5(1):27-37.
2. Nelson N (1999) Metal ion transporters and homeostasis. *EMBO Journal* 18:4361-4371.
3. Rines AK, Ardehaili H (2013) Transition metals and mitochondrial metabolism in the heart. *Journal of Molecular & Cellular Cardiology* 55:50-57.
4. Stadtman ER (1990) Metal ion-catalyzed oxidation of proteins: biochemical mechanism and biological consequences. *Free Radical Biology & Medicine* 9(4):315-325.

5.1.1. REDOX-Active Metal: Iron (Fe)

1. Arosio P, Levi S (2002) Ferritin, iron homeostasis, and oxidative damage. *Free Radical Biology & Medicine* 33(4):457-463.
2. Comporti M, Signorini C, Buonocore G, Ciccoli L (2002) Iron release, oxidative stress and erythrocyte ageing. *Free Radical Biology & Medicine* 32(7):568-576.
3. Goswami T, Rolfs A, Hediger MA (2002) Iron transport: emerging roles in health and disease. *Biochemistry & Cell Biology* 80(5):679-689.
4. Hidalgo E, Ding H, Demple B (1997) Redox signal transduction via iron-sulfur clusters in the SoxR transcription activator. *Trends in Biochemical Sciences* 22(6):207-210.
5. Jomova K, Valko M (2011) Importance of iron chelation in free radical-induced oxidative stress and human disease. *Current Pharmaceutical Design* 17(31):3460-3473.
6. Lawen A, Lane DJ (2013) Mammalian iron homeostasis in health and disease: uptake, storage, transport, and molecular mechanisms of action. *Antioxidants & Redox Signaling* 18(18):2473-2507.
7. Meneghini R (1997) Iron homeostasis, oxidative stress, and DNA damage. *Free Radical Biology & Medicine* 23(5):783-792.
8. Mole DR (2010) Iron homeostasis and its interaction with prolyl hydroxylases. *Antioxidants & Redox Signaling* 12(4):445-458.
9. Rifkind JM, Nagababu E, Ramasamy S, Ravi LB (2003) Hemoglobin redox reactions and oxidative stress. *Redox Reports* 8(5):234-237.
10. Yuan XM, Li W (2008) Iron involvement in multiple signaling pathways of atherosclerosis: a revisited hypothesis. *Current Medical Chemistry* 15(21):2157-2172.

5.1.2. REDOX-Active Metal: Copper (Cu)

1. Boal AK, Rosenzweig AC (2009) Structural biology of copper trafficking. *Chemical Reviews* 109(10):4760-4779.
2. Collins JF, Klevay LM (2011) Copper. *Advances in Nutrition* 2(6):520-522.
3. Gupta A, Lutsenko S (2009) Human copper transporters: mechanism, role in human diseases and therapeutic potential. *Future Medicinal Chemistry* 1(6):1125-1142.
4. Halfdanarson TR, Kumar N, Hogan WJ, Murray JA (2009) Copper deficiency in celiac disease. *Journal of Clinical Gastroenterology* 43(2):162-164.
5. Horn D, Barrientos A (2008) Mitochondrial copper metabolism and delivery to cytochrome c oxidase. *IUBMB Life* 60(7):421-429.
6. Kaler SG, Holmes CS, Goldstein DS, et al (2008) Neonatal diagnosis and treatment of Menkes disease. *New England Journal of Medicine* 358(6):605-614.
7. Kaplan JH, Lutsenko S (2009) Copper transport in mammalian cells: special care for a metal with special needs. *Journal of Biological Chemistry* 284(38):25461-25465.
8. Kim BE, Nevitt T, Thiele DJ (2008) Mechanisms for copper acquisition, distribution and regulation. *Nature Chemical Biology* 4(3):176-185.
9. Saari JT (2000) Copper deficiency and cardiovascular disease: role of peroxidation, glycation, and nitration. *Canadian Journal of Physiology & Pharmacology* 78(10):848-855.
10. Tapiero H, Townsend DM, Tew KD (2003) Trace elements in human physiology and pathology. Copper. *Biomedical Pharmacotherapy* 57(9):386-398.

11. Turski ML, Thiele DJ (2009) New roles for copper metabolism in cell proliferation, signaling, and disease. *Journal of Biological Chemistry* 284(2):717-721.
12. Whittaker JW (1999) Oxygen reactions of the copper oxidases. *Essays in Biochemistry* 34:155-172.

5.1.3. REDOX-Active Metal: Zinc (Zn)

1. Aras MA, Aizenman E (2011) Redox Regulation of Intracellular Zinc: Molecular Signaling in the Life and Death of Neurons. *Antioxidants & Redox Signaling* 15:2249–2263.
2. Krezel A, Hao Q, Maret W (2007) The zinc/thiolate redox biochemistry of metallothionein and the control of zinc ion fluctuations in cell signaling. *Archives in Biochemistry & Biophysics* 463(2):188-200.
3. Kröncke KD (2001) Zinc finger proteins as molecular targets for nitric oxide-mediated gene regulation. *Antioxidants & Redox Signaling* 3(4):565-575.
4. Kröncke KD, Klotz LO (2009) Zinc fingers as biologic redox switches? *Antioxidants & Redox Signaling* 11(5):1015-1027.
5. Maret W (2005) Zinc coordination environments in proteins determine zinc functions. *Journal of Trace Elements in Medicine & Biology* 19(1):7-12.
6. Maret W (2008) Metallothionein redox biology in the cytoprotective and cytotoxic functions of zinc. *Experimental Gerontology* 43(5):363-369.
7. Maret W (2012) New perspectives of zinc coordination environments in proteins. *Journal of Inorganic Biochemistry* 111:110-116.
8. Oteiza PI (2012) Zinc and the modulation of redox homeostasis. *Free Radical Biology & Medicine* 53(9):1748-1759.
9. Powell SR (2000) The antioxidant properties of zinc. *Journal of Nutrition* 130:1447S-1454S.
10. Tapiero H, Tew KD (2003) Trace elements in human physiology and pathology: zinc and metallothioneins. *Biomedical Pharmacotherapy* 57(9):399-411.

5.1.4. REDOX-Active Metal: Manganese (Mn)

1. Aguirre JD, Culotta VC (2012) Battles with iron: manganese in oxidative stress protection. *Journal of Biological Chemistry* 287(17):13541-13548.
2. Armstrong FA (2008) Why did Nature choose manganese to make oxygen? *Philosophical Transactions of the Royal Society B* 363:1263-1270.
3. Bae YJ, Kim MH (2007) Manganese supplementation improves mineral density of the spine and femur and serum osteocalcin in rats. *Biology of Trace Element Reseach* 124:28-34.
4. Holley AK, Dhar SK, Xu Y, St Clair DK (2012) Manganese superoxide dismutase: beyond life and death. *Amino Acids* 42(1):139-158.
5. Keen CL, Ensunsa JL, Watson MH, Baly DL, et al (1999) Nutritional aspects of manganese from experimental studies. *Neurotoxicology* 20:213-223.
6. Lüthen F, Bulnheim U, Müller PD, Rychly J, et al (2007) Influence of manganese ions on cellular behavior of human osteoblasts in vitro. *Biomolecular Engineering* 24:531-536.
7. Oikawa S, Hirosawa I, Tada-Oikawa S, Furukawa A, et al (2006) Mechanism for manganese enhance ment of dopamine-induced oxidative DNA damage and neuronal cell death. *Free Radical Biology & Medicine* 41:748-756.
8. Pal PK, Samii A, Calne DB (1999) Manganese neurotoxicity: a review of clinical features, imaging and pathology. *Neurotoxicology* 20:227-238.
9. Soldin OP, Aschner M (2007) Effects of manganese on thyroid hormone homeostasis:potential links. *Neurotoxicology* 28(5):951-956.
10. Wasserman GA, Liu X, Parvez F, Ahsan H, et al (2006) Water manganese exposure and children's intellectual function in Araihazar, Bangladesh. *Environmental Health Perspective* 114:124-129.
11. Whittaker JW (2010) Metal uptake by manganese superoxide dismutase. *Biochimica et Biophysica Acta* 1804(2):298-307.

5.1.5. REDOX-Active Metal: Selenium (Se)

1. Ayaz M, Ozdemir S, Ugur M, Vassort G, Turan B (2004) Effects of selenium on altered mechanical and electrical cardiac activities of diabetic rat. *Archives of Biochemistry & Biophysics* 426:83–90.
2. Beckett GJ, Arthur JR (2005) Selenium and endocrine systems. *Journal of Endocrinology* 184(3):455-465.
3. Björnstedt M, Fernandes AP (2010) Selenium in the prevention of human cancers. *EPMA Journal* 1(3):389-395.

4. Drake EN (2006) Cancer chemoprevention: selenium as a prooxidant, not an antioxidant. *Medical Hypotheses* 67:318-322.
5. Foresta C, Flohe L, Garolla A, Roveri A, et al (2002) Male fertility is linked to the selenoprotein phospholipid hydroperoxide glutathione peroxidase. *Biology of Reproduction* 67:967–971.
6. Jacob C, Giles GI, Giles NM, Sies H (2003) Sulfur and selenium: the role of oxidation state in protein structure and function. *Angewandte Chemistry International Edition England* 42(39):4742-4758.
7. Jiang C, Ganther H, Lu J (2000) Monomethyl selenium-specific inhibition of MMP-2 and VEGF expression: implications for angiogenic switch regulation. *Molecular Carcinogenesis* 29:236-250.
8. Kryukov GV, Castellano S, Novoselov SV, Lobanov AV, et al (2003) Characterization of mammalian selenoproteomes. *Science* 300:1439-1443.
9. Lu J, Holmgren A (2009) Selenoproteins. *Journal of Biological Chemistry* 284(2):723-727.
10. Mukherjee B, Anbazhagan S, Roy A, Ghosh R, Chatterjee M (1998) Novel implications of the potential role of selenium on antioxidant status in streptozotocin-induced diabetic mice. *Biomedicine and Pharmacotherapy* 52:89–95.
11. Nilsonne G, Sun X, Nyström C, Rundlöf AK, et al (2006) Selenite induces apoptosis in sarcomatoid malignant mesothelioma cells through oxidative stress. *Free Radical Biology & Medicine* 41:874-885.
12. Olm E, Fernandes AP, Hebert C, Rundlof AK, et al (2009) Extra-cellular thiol-assisted selenium uptake dependent on the x(c)-cystine transporter explains the cancerspecific cytotoxicity of selenite. *Proceedings of the National Academy of Sciences USA* 106:11400-11405.
13. Rahmanto AS, Davies MJ (2012) Selenium-containing amino acids as direct and indirect antioxidants. *IUBMB Life* 64(11):863-871.
14. Schomburg L (2011) Selenium, selenoproteins and the thyroid gland: interactions in health and disease. *Nature Reviews Endocrinology* 8(3):160-171.
15. Schrauzer GN (2009) Selenium and selenium-antagonistic elements in nutritional cancer prevention. *Critical Reviews in Biotechnology* 29:10-17.
16. Steinbrenner H, Sies H (2009) Protection against reactive oxygen species by selenoproteins. *Biochimica et Biophysica Acta* 1790:1478-1485.
17. Tramer F, Caponecchia L, Sgro P, Martinelli M, et al (2004) Native specific activity of glutathione peroxidase (GPx-1), phospholipid hydroperoxide glutathione peroxidase (PHGPx) and glutathione reductase (GR) does not differ between normo- and hypomotile human sperm samples. *International Journal of Andrology* 27:88–93.

5.2 | REDOX: Vitamins

5.2.1 REDOX Vitamin A

1. Baker DL, Krol ES, Jacobsen N, Liebler DC (1999) Reactions of β-carotene with cigarette smoke oxidants. Identification of carotenoid oxidation products and evaluation of the prooxidant/ antioxidant effect. *Chemical Research in Toxicology* 12:535-543.
2. Combs F (2008) *The Vitamins: Fundamental Aspects in Nutrition and Health* (3rd ed.). Burlington: Elsevier Academic Press ISBN: 978-0-12-183493-7.
3. Duester G (2008) Retinoic acid synthesis and signaling during early organogenesis. *Cell* 134: 921-931.
4. Fuchs E, Green H (1981)Regulation of terminal differentiation of cultured human keratinocytes by vitamin A. Cell 25(3):617-625.
5. Kennedy TA, Liebler DC (1992) Peroxyl radical scavenging of beta-carotene in lipid bilayers: effect of oxygen partial pressure. *Journal of Biological Chemistry* 267:4658-4663.
6. Nelson AM, Zhao W, Thibouto DM (2008) Neutrophil gelatinase–associated lipocalin mediates 13-cis retinoic acid–induced apoptosis of human sebaceous gland cells. *Journal of Clinical Investigation* 118 (4):1468-1478.
7. Omenn GS, Goodman GE, Thornquist MD, et al (1996) Effects of a combination of beta carotene and vitamin A on lung cancer and cardiovascular disease. *New England Journal of Medicine* 334:1150-55.
8. Palozza P (1998) Prooxidant actions of carotenoids in biologic systems. *Nutrition Reviews* 56:257-265.
9. Russell RM (2000) The vitamin A spectrum: from deficiency to toxicity. *American Journal of Clinical Nutrition* 71(4):878-884.
10. Semba RD (1998) The role of vitamin A and related retinoids in immune function. *Nutrition Reviews* 56:S38-48.
11. Stipanuk MH (2006) *Biochemical, Physiological and Molecular Aspects of Human Nutrition* (2nd ed.). Philadelphia: Saunders. ISBN: 9781416002093.

12. Thurnham DI, Northrop-Clewes CA (1999) Optimal nutrition: vitamin A and the carotenoids. *Nutrition Society* 58(2):449-457
13. Underwood BA, Arthur P (1996) The contribution of vitamin A to public health. *FASEB Journal* 10(9):1040-1048.
14. Wang XD, Russell RM (1999) Procarcinogenic and anticarcinogenic effects of beta-carotene. *Nutrition Reviews* 57:263-272.
15. West CE (2000) Vitamin A and measles. *Nutrition Reviews* 58:S46-54.
16. Wolf G (2001) The discovery of the visual function of vitamin A. *Journal of Nutrition* 131(6):1647-1650

5.2.2. REDOX Vitamin C (Ascorbic acid)

1. Carcamo JM, Pedraza A, Borquez-Ojeda O, Golde DW (2002) Vitamin C suppresses TNF alpha-induced NF-κB activation by inhibiting IκB-α phosphorylation. *Biochemistry* 41:12995–13002.
2. Carinci F, Pezzetti F, Spina AM, Palmieri A, Laino G, De Rosa A, Farina E, Illiano F, Stabellini G, Perrotti V, Piattelli A (2005) Effect of Vitamin C on pre-osteoblast gene expression. *Archives in Oral Biology* 50:481-496.
3. Carr A, Frei B (1999) Does vitamin C act as a pro-oxidant under physiological conditions? *FASEB Journal* 13:1007-1024.
4. Chen Q, Espey MG, Krishna MC, Mitchell JB, et al (2005) Pharmacologic ascorbic acid concentrations selectively kill cancer cells: action as a pro-drug to deliver hydrogen peroxide to tissues. *Proceedings of the National Academy of Sciences USA* 102:13604-13609.
5. Hall SL, Greendale GA (1998) The relation of dietary vitamin C intake to bone mineral density: results from the PEPI study. *Calcified Tissue International* 63:183-189.
6. Halliwell B, Wasil M, Grootveld M (1987) Biologically significant scavenging of the myeloperoxidase-derived oxidant hypochlorous acid by ascorbic acid. Implications for antioxidant protection in the inflamed rheumatoid joint. *FEBS Letters* 213:15-17.
7. Heller R, Unbehaun A, Schellenberg B, Mayer B, et al (2001) L-ascorbic acid potentiates endothelial nitric oxide synthesis via a chemical stabilization of tetrahydrobiopterin. *Journal of Biological Chemistry* 276:40-47.
8. Li Y, Schelhorn HE (2007) New developments and novel therapeutic perspectives for vitamin C. *Journal of Nutrition* 137:2171-2184.
9. Lutsenko EA, Carcamo JM, Golde DW (2002) Vitamin C prevents DNA mutation induced by oxidative stress. *Journal of Biological Chemistry* 277:16895-16899.
10. May JM (2000) How does ascorbic acid prevent endothelial dysfunction? *Free Radical Biology & Medicine* 2000;28:1421-1429.
11. Padayatty SJ, Katz A, Wang Y, Eck P, Kwon O, Lee JH, Chen S, Corpe C, Dutta A, Dutta SK, Levine M (2003) Vitamin C as an antioxidant: evaluation of its role in disease prevention. *Journal of American College of Nutrition* 22:18-35.
12. Peregoy J, Wilder FV (2010) The effects of vitamin C supplementation on incident and progressive knee osteoarthritis: a longitudinal study. *Public Health & Nutrition* 14:709-715.
13. Wu F, Schuster DP, Tyml K, Wilson JX (2007) Ascorbate inhibits NADPH oxidase subunit p47phox expression in microvascular endothelial cells. *Free Radical Biology & Medicine* 42:124-131.

5.2.3. REDOX Vitamin E

1. Azzi A, Aratri E, Boscoboinik D, Clement S, et al (1998) Molecular basis of α-tocopherol control of smooth muscle cell proliferation. *Biofactors* 7:3-14.
2. Boscoboinik D, Szewczyk A, Azzi A (1991) α-Tocopherol (vitamin E) regulates vascular smooth muscle cell proliferation and protein kinase C activity. *Archives in Biochemistry & Biophysics* 286:264-269.
3. Boscoboinik D, Szewczyk A, Hensey C, Azzi A (1991) Inhibition of cell proliferation by alpha-toco pherol. Role of protein kinase C. *Journal of Biological Chemistry* 266:6188-6194.
4. Cachia O, Benna JE, Pedruzzi E, Descomps B, et al (1998) α-tocopherol inhibits the respiratory burst in human monocytes. Attenuation of p47(phox) membrane translocation and phosphorylation. *Journal of Biological Chemistry* 273:32801-32805.
5. Christen S, Woodall AA, Shigenaga MK, Southwell-Keely PT, et al (1997) γ-Tocopherol traps muta genic electrophiles such as NOx and complements α-tocopherol: physiological implications. *Proceedings of the National Academy of Sciences USA* 94:3217-3222.
6. Evans HM, Bishop KS (1922) On the existence of a hitherto unrecognized dietary factor essential for reproduction. *Science* 56:650-651.

7. Gotoda T, Arita M, Arai H, Inoue K, et al (1995) Adult-onset spino-cerebellar dysfunction caused by a mutation in the gene for the α-tocopherol-transfer protein. *New England Journal of Medicine* 333:1313-1318.
8. Maiorino M, Wissing JB, Brigelius-Flohe R, Calabrese F, et al (1998) Testosterone mediates expression of the selenoprotein PHGPx by induction of spermatogenesis and not by direct transcriptional gene activation. *FASEB Journal* 12:1359-1370.
9. Motoyama T, Kawano H, Kugiyama K, Hirashima O, et al (1998) Vitamin E administration improves impairment of endothelium-dependent vasodilation in patients with coronary spastic angina. *Journal of American College of Cardiology* 32:1672-1679.
10. Ouahchi K, Arita M, Kayden H, Hentati F, et al (1995) Ataxia with isolated vitamin E deficiency is caused by mutations in the alpha-tocopherol transfer protein. *Nature Genetics* 9:141-145.
11. Packer L, Weber SU, Rimbach G (2001) Molecular aspects of alpha-tocotrienol antioxidant action and cell signalling. *Journal of Nutrition* 131(2):369S-373S.
12. Ricciarelli R, Tasinato A, Clement S, Ozer NK, et al (1998) α-Tocopherol specifically inactivates cellular protein kinase C alpha by changing its phosphorylation state. *Biochemistry Journal* 334:243-249.
13. Stephens NG, Parsons A, Schofield PM, Kelly F, et al (1996) Randomized controlled trial of vitamin E in patients with coronary disease – Cambridge Heart Antioxidant Study (CHAOS). *Lancet* 347:781-786.
14. Traber MG (2007) Vitamin E regulatory mechanisms. *Annual Review of Nutrition* 27:347-362.
15. Traber MG, Sokol RJ, Ringel SP, Neville HE, et al (1987) Lack of tocopherol in peripheral nerves of vitamin E-deficient patients with peripheral neuropathy. *New England Journal of Medicine* 317:262-265.
16. Wang X, Quinn PJ (1999) Vitamin E and its function in membranes. *Progress in Lipid Research* 38:309-336.

5.3 | REDOX: Amino Acids

1. Berlett BS, Stadtman ER (1997) Protein oxidation in aging, disease, and oxidative stress. *Journal of Biological Chemistry* 272:20313-20316.
2. Brosnan JT, Brosnan ME (2006) The sulfur-containing amino acids: An overview. *Journal of Nutrition* 136:1636S-1640S.
3. Finkel T (2000) Redox-dependent signal transduction. *FEBS Letters* 476:52-54.
4. Sevier CS, Kaiser CA (2002) Formation and transfer of disulphide bonds in living cells. *Nature Reviews: Molecular & Cellular Biology* 3 (11): 836–847
5. Stadtman ER (1992) Protein oxidation and aging. *Science* 257:1220-1224.

5.3.1. REDOX Amino Acid: N-Acetyl-Cysteine (NAC)

1. Aitio M-L (2005) N-acetylcysteine – passé-partout or much ado about nothing? *British Journal of Clinical Pharmacology* 61(1):5-15.
2. Akerlund B, Jarstrand C, Lindeke B, Sonnerborg A, et al (1996) Effect of N-acetylcysteine(NAC) treatment on HIV-1 infection, a double-blind placebo-controlled trial. *European Journal of Clinical Pharmacology* 50:457-461.
3. Aruoma OI, Halliwell B, Hoey BM, et al (1989) The antioxidant action of N-acetylcysteine: its reaction with hydrogen peroxide, hydroxyl radical, superoxide, and hypochlorous acid. *Free Radical Biology & Medicine* 6:593-597.
4. Bagshaw SM, Ghali WA (2004) Acetylcysteine for prevention of contrast-induced nephropathy after intravascular angiography: a systematic review and meta-analysis. *BMC Medicine* 22;2:38.
5. Borgstrom L, Kagedal B, Paulsen O (1986) Pharmacokinetics of N-acetylcysteine in man. *European Journal of Clinical Pharmacology* 31:217-222.
6. Ceconi C, Curello S, Cargnoni A, et al (1988) The role of glutathione status in the protection against ischemic and reperfusion damage: effects of N-acetyl cysteine. *Journal of Molecular & Cellular Cardiology* 20:5-13.
7. De Flora S, Bennicelli C, Camoirano A, et al (1985) In vivo effects of N-acetylcysteine on glutathione metabolism and on the biotransformation of carcinogenic and/or mutagenic compounds. *Carcinogenesis* 6:1735-1745.
8. De Vries N, De Flora S (1993) N-Acetyl-l-cysteine. *Journal of Cell Biochemistry* 17F:S270-S277.
9. Dean O, Giorlando F, Berk M (2011) N-acetylcysteine in psychiatry: current therapeutic evidence and potential mechanisms of action. *Journal of Psychiatry & Neuroscience* 36(2):78-86.

10. Gavish D, Breslow JL (1991) Lipoprotein(a) reduction by N-acetylcysteine. *Lancet* 337:203-204.
11. Heard KJ (2008) Acetylcysteine for acetaminophen poisoning. *New England Journal of Medicine* 359(3):285-292.
12. Horowitz JD, Henry CA, Syrjanen ML, et al (1988) Nitroglycerine/N-acetylcysteine in the management of unstable angina pectoris. *European Heart Journal* 9:95-100.
13. Kelly GS (1998) Clinical applications of N-acetylcysteine. *Alternative Medicine Reviews* 3:114–127.
14. Lappas M, Permezel M, Rice GE (2003) N-acetylcysteine inhibits phospholipid metabolism, proinfl ammatory cytokine release, protease activity and nuclear factor-κB deoxyribonucleic acid-binding activity in human fetal membranes in vitro. *Journal of Clinical Enocrinology & Metabolism* 88:1723-1729.
15. Pela R, Calcagni AM, Subiaco S, et al (1999) N-acetylcysteine reduces the exacerbation rate in patients with moderate to severe COPD. *Respiration* 66:495-500.
16. Sadowska AM, Verbraecken J, Darquennes K, De Backer WA (2006) Role of N-acetylcysteine in the management of COPD. *International Journal of Chronic Obstructructive Pulmonary Disease* 1(4):425-434.
17. Dodd S, Dean O, Copolov DL, et al (2008) N-acetylcysteine for antioxidant therapy: pharmacology and clinical utility. *Expert Opinion in Biological Therapy* 8:1955-1962.
18. Sjodin K, Nilsson E, Hallberg A, Tunek A (1989) Metabolism of N-acetyl-l-cysteine. *Biochemical Pharmacy* 38:3981-3985.
19. van Overveld FJ, Demkow U, Gorecka D, et al (2005) New developments in the treatment of COPD: comparing the effects of inhaled corticosteroids and N-acetylcysteine. *Journal of Physiological Pharmacology* 56(Suppl 4):135-142.
20. Wiklund O, Fager G, Andersson A, et al (1996) N-acetylcysteine treatment lowers plasma homo cysteine but not serum lipoprotein(a) levels. *Atherosclerosis* 119:99-106.
21. Zimet I (1988) Acetylcysteine: A drug that is much more than a mucokinetic. *Biomedical Pharmacotherapy* 42:513-520.

5.3.2. REDOX Amino Acid: Methionine (Met)

1. Ball RO, Courtney-Martin G, Pencharz PB (2006) The in vivo sparing of methionine by cysteine in sulfur amino acid requirements in animal models and adult humans. *Journal of Nutrition* 136:1682S-1693S.
2. Boschi-Muller S, Gand A, Branlant G (2008) The methionine sulfoxide reductases: Catalysis and substrate specificities. *Archives in Biochemistry & Biophysics* 474:266-273.
3. Chao CC, Ma YS, Stadtman ER (1997) Modification of protein surface hydrophobicity & methionine oxidation by oxidative systems. *Proceedings of the National Academy of Sciences USA* 94:2969-2974.
4. Hoshi T, Heinemann SH (2001) Regulation of cell function by methionine oxidation and reduction. *Journal of Physiology* 531:1-11.
5. Lee BC, Dikiy A, Kim HY, Gladyshev VN (2009) Functions and evolution of selenoprotein methionine sulfoxide reductases. *Biochimca et Biophysica Acta* 1790:1471-1477.
6. Lee BC, Gladyshev VN (2011) The biological significance of methionine sulfoxide stereochemistry. *Free Radical Biology & Medicine* 50(2):221-227.
7. Mato JM, Martinez-Chantar ML, Lu SC (2008) Methionine metabolism and liver disease. *Annual Review of Nutrition* 28:273-293.
8. Metayer S, Seiliez I, Collin A, Duchene S, et al (2008) Mechanisms through which sulfur amino acids control protein metabolism and oxidative status. *Journal of Nutritional Biochemistry* 19:207-215.
9. Sharov VS, Schoneich C (2000) Diastereoselective protein methionine oxidation by reactive oxygen species and diastereoselective repair by methionine sulfoxide reductase. *Free Radical Biology & Medicine* 29:986-994.
10. Vogt W (1995) Oxidation of methionyl residues in proteins: Tools, targets, and reversal. *Free Radical Biology & Medicine* 18:93-105.
11. Weissbach H, Resnick L, Brot N (2005) Methionine sulfoxide reductases: History and cellular role in protecting against oxidative damage. *Biochimca et Biophysica Acta* 1703:203-212.

5.3.3. REDOX Amino Acid: L-Glutamic acid (Glu) and L-Glutamine (Gln)

1. Andrews FJ, Griffiths RD (2002) Glutamine: essential for immune nutrition in the critically ill. *British Journal of Nutrition* 87(1):S3-8.

2. Bellisle F (1999) Glutamate and the UMAMI taste: Sensory, metabolic, nutritional and behavioural considerations. A review of the literature published in the last 10 years. *Neuroscience Biobehavior Reviews* 23:423-38.
3. dos Santos RG, Viana ML, Generoso SV, Arantes RE, et al (2010) Gutamine supplementation decreases intestinal permeability and preserves gut mucosa integrity in an experimental mouse model. *Journal of Parenteral & Enteral Nutrition* 34:408-413.
4. Haynes TE, Li P, Li X, Shimotori K, et al (2009) L-Glutamine or L-alanyl-L-glutamine prevents oxidant- or endotoxin-induced death of neonatal enterocytes. *Amino Acids* 37:131-142.
5. Kim H (2011) Glutamine as an immunonutrient. *Yonsei Medical Journal* 52:892-897.
6. Kulkarni C, Kulkarni KS, Hamsa BR (2005) L-Glutamic acid and glutamine: Exciting molecules of clinical interest. *Indian Journal of Pharmacology* 37(3):148-154.
7. Rennie MJ, Ahmed A, Khogali SE, Low SY, et al (1996) Glutamine metabolism and transport in skeletal muscle and heart and their clinical relevance. *Journal of Nutrition* 126(4):1142-1149.
8. Rhoads JM, Argenzio RA, Chen W, Rippe RA, et al (1997) L-glutamine stimulates intestinal cell proliferation and activates mitogen-activated protein kinases. *American Journal of Physiology* 272:G943-53.
9. Shabert JK, Wilmore DW (1996) Glutamine deficiency as a cause of human immunodeficiency virus wasting. *Medical Hypotheses* 46:252-256.
10. Stehle P, Zander J, Mertes N, Albers S, et al (1989) Effect of parenteral glutamine peptide supplements on muscle glutamine loss and nitrogen balance after major surgery. *Lancet* 1:231-233.
11. Van der Hulst RR, van Kreel BK, von Meyenfeldt MF, Brummer RJ, et al (1993) Glutamine and the preservation of gut integrity. *Lancet* 341:1363-1365.
12. Wang Y, Tao YX, Cai W, Tang QY, et al (2010) Protective effect of parenteral glutamine supplementa tion on hepatic function in very low birth weight infants. *Clinical Nutrition* 29:307-311.
13. Wischmeyer PE (2007) Glutamine: mode of action in critical illness. *Critical Care Medicine* 35:S541-544.

5.4 | REDOX Bio-Replenishment (Bio-Rep)

1. Bains W (2001) The parts list of life. *Nature Biotechnology* 19:401-402.
2. Gaidos EJ, Nealson KH, Kirschvink JL (1999) Life in ice-covered oceans. *Science* 284:1631-1633.
3. Koonin EV (2007) The biological Big Bang model for major transitions in evolution. *Biology Direct* 2:21.
4. Kushnareva Y, Newmeyer DD (2010) Bioenergetics and cell death. *Annals of New York Academy of Sciences* 1201:50-57.
5. Lander ES, Linton LM, Birren B, Nusbaum C, et al (2001) Initial sequencing and analysis of the human genome. *Nature* 409:860-921.
6. Lane N (2009) *Life Ascending: The Ten Great Inventions of Evolution.* Norton:Profile Books. ISBN: 978-0393338669
7. Martin WF (2011) Evolution without a tree of life. *Biology Direct* 6:36
8. Mitchell P (1961) Coupling of phosphorylation to electron and hydrogen transfer by a chemi-osmotic type of mechanism. *Nature* 191:144-148.
9. Naidu AS (2009) *Bio-replenishment for Bone Health.* Pomona:Bio-Rep Media. ISBN: 9780982445105
10. Skulachev VP (2006) Bioenergetic aspects of apoptosis, necrosis and mitoptosis. *Apoptosis* 11:473-485.

5.4.1. REDOX Bio-Replenishment: Lactoferrin

1. Actor JK, Hwang S-A, Kruzel ML (2009) Lactoferrin as a natural immune modulator. *Current Pharma cology Design* 15:1956-1973.
2. Anderson BF, Baker HM, Norris GE, Rice DW, Baker EN (1989) Structure of human lactoferrin: crystal lographic structure analysis and refinement at 2.8 Å resolution. *Journal of Molecular Biology* 209:711-734.
3. Ambruso DR, Johnston RB Jr (1981) Lactoferrin enhances hydroxyl radical production by human neutrophils, neutrophil particulate fractions, and an enzymatic generating system. *Journal of Clinical Investigation* 67(2):352-360.
4. Baker HM, Baker EN (2004) Lactoferrin and iron: structural and dynamic aspects of binding and release. *Biometals* 17:209-216.

5. Bharadwaj S, Naidu TA, Betageri GV, Prasadarao NV, Naidu AS (2010) Inflammatory responses improve with milk ribonuclease-enriched lactoferrin supplementation in postmenopausal women. *Inflammation Research* 59:971-978.
6. Blakeborough P, Salter DN (1987) The intestinal transport of zinc studied using brush-border-membrane vesicles from the piglet. *British Journal of Nutrition* 57:45-55.
7. Choe YH, Lee SW (1999) Effect of lactoferrin on the production of tumor necrosis factor-alpha and nitric oxide. *Journal of Cellular Biochemistry* 76:30–36.
8. Conneely OM (2001) Anti-inflammatory activities of lactoferrin. *Journal of American College of Nutrition* 20:389S-395S.
9. Cox TM, Mazurier J, Spik G, Montreuil J, Peters TJ (1979) Iron-binding proteins and influx of iron across the duodenal brush border. Evidence for specific lactotransferrin receptors in the human small intestine. *Biochemica et Biophysica Acta* 588:120-128
10. Damiens E, Mazurier J, Yazidi I, et al (1998) Effect of human LF on NK cell cytotoxicity against haematopoietic and epithelial tumour cells. *Biochemica et Biophysica Acta* 1402:277–287.
11. Davidson LA, Lönnerdal B (1989) Fe-saturation and proteolysis of human lactoferrin: Effect on brush-border receptor-mediated uptake of Fe and Mn. *American Journal of Physiology* 257:G930-934.
12. Guillen C, McInnes IB, Kruger H, Brock JH (1998) Iron, lactoferrin and iron regulatory protein activity in the synovium; relative importance of iron loading and the inflammatory response. *Annals of Rheumatic Disease* 57:309-314.
13. Gutteridge JM, Paterson SK, Segal AW, Halliwell B (1981) Inhibition of lipid peroxidation by the iron-binding protein lactoferrin. *Biochemistry Journal* 199(1):259-261.
14. Haversen L, Ohlsson BG, Hahn-Zoric M, et al (2002) Lactoferrin down-regulates the LPS-induced cytokine production in monocytic cells via NF-kB. *Cellular Immunology* 220:83–95.
15. Ishisaka R, Kanno T, Kanematsu H, Utsumi T, et al (1998) Superoxide dismutase-like activity of metal substituted lactoferrin derivatives. *Physiological Chemistry Physics Medicine NMR* 30(1):1-13.
16. Kumar DA, Raju KV, Settu K, Kumanan K, Puvanakrishnan R (2006) Effect of a derivatized tetrapeptide from lactoferrin on nitric oxide mediated matrix metalloproteinase-2 production by synovial fibroblasts in collagen-induced arthritis in rats. *Peptides* 27(6):1434-1442.
17. Maneva A, Taleva B, Maneva L (2003) Lactoferrin-protector against oxidative stress and regulator of glycolysis in human erythrocytes. *Zentralblatt Naturforschung C* 58(3-4):256-262.
18. Mazurier J, Spik G (1980) Comparative study of the iron-binding properties of human transferrins. I. Complete and sequential iron saturation and desaturation of the lactotransferrin. *Biochemica et Biophysica Acta* 629:399–408.
19. Naidu AS, Chen J, Martinez C, Tulpinski J, et al (2004) Activated lactoferrin's ability to inhibit Candida growth and block yeast adhesion to the vaginal epithelial monolayer. *Journal of Reproductive Medicine* 49(11):859-866.
20. Erdei J, Forsgren A, Naidu AS (1994) Lactoferrin binds to porins OmpF and OmpC in Escherichia coli. *Infection & Immunity* 62(4):1236-1240.
21. Naidu SS, Svensson U, Kishore AR, Naidu AS (1993) Relationship between antibacterial activity and porin binding of lactoferrin in Escherichia coli and Salmonella typhimurium. *Antimicrobial Agents & Chemotherapy* 37(2):240-245
22. Naidu AS (2000) Lactoferrin. In: *Natural Food Antimicrobial Systems* (ed) AS Naidu. Pp.17-102. Boca Raton:CRC Press. ISBN: 978-084-09320477
23. Naidu AS (2005) Ultra-cleansing of lactoferrin: Nutraceutical implications. *Agro Food Industry hi-tech* 16(2):7-13.
24. Newsome AL, Johnson JP, Seipelt RL, Thompson MW (2007) Apolactoferrin inhibits the catalytic domain of matrix metalloproteinase-2 by zinc chelation. *Biochemistry & Cellular Biology* 85:563-572.

5.4.2. REDOX Bio-Replenishment: Ribonuclease (RNase)

1. Acharya KR, Shapiro R, Allen SC, Riordan JF, Vallee BL (1994) Crystal structure of human angiogenin reveals the structural basis for its functional divergence from ribonuclease. *Proceedings of the National Academy of Sciences USA* 91:2915-9.
2. Acharya KR, Shapiro R, Riordan JF, Vallee BL (1995) Crystal structure of bovine angiogenin at 1.5-A resolution. *Proceedings of the National Academy of Sciences USA* 92:2949-2953.
3. Amini AA, Nair LS (2011) Lactoferrin: a biologically active molecule for bone regeneration. *Current Medicinal Chemistry* 18:1220-1229.

4. Bharadwaj S, Naidu AG, Betageri GV, Prasadarao NV, Naidu AS (2009) Milk ribonuclease-enriched lactoferrin induces positive effects on bone turnover markers in postmenopausal women. *Osteoporosis International* 20:1603-1611.
5. Bharadwaj S, Naidu TA, Betageri GV, Prasadarao NV, Naidu AS (2010) Inflammatory responses improve with milk ribonuclease-enriched lactoferrin supplementation in postmenopausal women. *Inflammmation Research* 59:971-978.
6. Carano RAD and Filvaroff E (2003) Angiogenesis and bone repair. *Drug Discovery Today* 8:980–989.
7. Gerber HP, Ferrara N (2000) Angiogenesis and bone growth. *Trends in Cardiovascular Medicine* 10:223-228.
8. Glowacki J (1998) Angiogenesis in fracture repair. *Clinical Orthopedics & Related Research* 355S:S82–S89.
9. Hu GF, Riordan JF, Vallee BL (1997) A putative angiogenin receptor in angiogenin-responsive human endothelial cells. *Proceedings of the National Academy of Sciences USA* 94:2204-2209.
10. Moroianu J, Riordan JF (1994) Nuclear translocation of angiogenin in proliferating endothelial cells is essential to its angiogenic activity. *Proceedings of the National Academy of Sciences USA* 91:1677-1681.
11. Naot D, Grey A, Reid IR, Cornish J (2005) Lactoferrin-a novel bone growth factor. *Clinical Medicine* Research 3:93-101.
12. Santos MI, Reis RL (2010) Vascularization in bone tissue engineering: physiology, current strategies, major hurdles and future challenges. *Macromolecules in Bioscience* 10:12-27.
13. Shestenko OP, Nikonov SD, Mertvetsov NP (2001) Angiogenin and its functions in angiogenesis. *Molecular Biology* 35:294–314
14. Strydom DJ (1998) The angiogenins. *Cellular & Molecular Life Sciences* 54:811-824.
15. Velazquez OC (2007) Angiogenesis and vasculogenesis: inducing the growth of new blood vessels and wound healing by stimulation of bone marrow-derived progenitor cell mobilization and homing. *Journal of Vascular Surgergy* 45:39-47.

5.4.3. REDOX Bio-Replenishment: Coenzyme Q

1. Arroyo A, Kagan VE, Tyurin VA, Burgess JR, et al (2000) NADH and NADPH dependent reduction of coenzyme Q at the plasma membrane. *Antioxidant Redox Signal* 2:251-262.
2. Bentinger M, Brismar K, Dallner G (2007) The antioxidant role of coenzyme Q. *Mitochondrion* 7:S41–S50
3. Brandt U (1999) Proton translocation in the respiratory chain involving ubiquinone—a hypothetical semiquinone switch mechanism for complex I. *Biofactors* 9:95–102.
4. Crane FL (2001) Biochemical functions of coenzyme Q10. *Journal of American College of Nutrition* 20:591-598.
5. Díaz-Castro J, Guisado R, Kajarabille N, García C, et al (2012) Coenzyme Q(10) supplementation ameliorates inflammatory signaling and oxidative stress associated with strenuous exercise. *European Journal of Nutrition* 51(7):791-799.
6. Gille L, Nohl H (2000) The existence of a lysosomal redox chain and the role of ubiquinone. *Archives in Biochemistry & Biophysics* 375:347-354.
7. Lass A, Kwong L., Sohal RS (1999) Mitochondrial coenzyme Q content and ageing. *Biofactors* 9:199-205.
8. Navarro F, Arroyo A, Martin SF, Bello RI, et al (1999) Protective role of ubiquinone in vitamin E and selenium deficient plasma membranes. *Biofactors* 9:163-170.
9. Quinn PJ, Fabisiak JP, Kagan VE (1999) Expansion of the antioxidant function of vitamin E by coenzyme Q. *Biofactors* 9:149-154.
10. Quinzii CM, Lopez LC, Naini A, DiMauro S, Hirano M (2008) Human CoQ10 deficiencies. *Biofactors* 32:113-118.
11. Schneider D, Elstner EF (2000) Coenzyme Q10, vitamin E and dihydrothioctic acid cooperatively prevent diene conjugation in isolated low density lipoprotein. *Antioxidant Redox Signal* 2:327-333.
12. Turunen M, Olsson J, Dallner G (2004) Metabolism and function of coenzyme Q. *Biochimica et Biophysica Acta* 1660:171-199.
13. Willis R, Anthony M, Sun L, House Y, Qiao G (1999) Clinical implications of the correlation between coenzyme Q10 and vitamin B6 status. *Biofactors* 9:359-363.

5.4.4. REDOX Bio-Replenishment: S-Adenosyl-Methionine (SAMe)

1. Bottiglieri T (2002) S-adenosyl-L-methionine (SAMe): from the bench to the bedside—molecular basis of a pleiotrophic molecule. *American Journal of Clinical Nutrition* 76(suppl):1151S–1157S.
2. Kotb M, Mudd SH, Mato JM, et al (1997) Consensus nomenclature for the mammalian methionine adenosyltransferase genes and gene products. *Trends in Genetics* 13:51-52.
3. Lieber CS (2002) S-adenosyl-L-methionine: its role in the treatment of liver disorders. *American Journal of Clinical Nutrition* 76(suppl):1183S–1187S.
4. Lu SC (2000) S-adenosylmethionine. *International Journal of Biochemistry & Cell Biology* 32:391-395.
5. Lu SC, Mato JM (2008) S-Adenosylmethionine in cell growth, apoptosis and liver cancer. *Journal of Gastroenterology & Hepatology* 23(Suppl 1):S73-S77.
6. Mato JM, Alvarez L, Ortiz P, Pajares MA (1997) S-adenosylmethionine synthesis: molecular mecha nisms and clinical implications. *Pharmacology & Therapeutics* 73:265-280.
7. Mato JM, Lu SC (2007) Role of S-adenosyl-L-methionine in liver health and injury. *Hepatology* 45:1306-1312.
8. Mischoulon D, Fava M. Role of S-adenosyl-L-methionine in the treatment of depression: a review of the evidence. *American Journal of Clinical Nutrition* 76(suppl):1158S–1161S.
9. Prudova A, Bauman Z, Braun A, Vitvitsky V, et al (2006) AdoMet regulation of cystathionine β-synthase: relevance to oxidative stress in liver disease. *Proceedings of the National Academy of Sciences USA* 103:6489-6494.

Nutrigenomics is the study of the molecular-level interaction between nutrients and other dietary bioactives with the genome. Based on their chemical composition, nutrients can induce REDOX signals. The cellular sensor system interprets such signals and evaluates the dietary milieu with the help of several transcription factors. The nutrient interaction with the cellular sensory system can alter genes, protein expression and metabolite production. Nutrigenomics is an emerging field of nutritional science. It explores the REDOX patterns (known as dietary signatures) of various food chemicals that can influence homeostasis of specific cells, tissues and organisms. This type of data can be effectively used in the prevention and treatment of chronic diet-related diseases, such as obesity and type 2 diabetes.

REDOX Life
Synopsis

REDOX is an integral part of creation of the Universe, and a governing principle in the genesis of life. Life on earth is based on energy transfer reactions via chemical bonds regulated by REDOX. Without a constant supply (capture and release) of energy, life on earth would cease to exist. One of the most common ways to define energy in living systems is the ability to do 'work (metabolism), which is the movement of molecules in and out of the cells. REDOX principles play a vital role in facilitating these metabolic pathways. The following points emphasize the basic relationship between REDOX, energy and life: (i) All biological processes require energy, from cell division, to secretion, to cell migration in an early embryo. Thus, living systems have strategies to harness and utilize energy. (ii) Energy does not arise from nothing. Since, it can neither be created nor destroyed; it transforms from one source to another. Biological processes depend on REDOX principles to control the transformation and use of energy. (iii) The biology of a cell is a complex and continuous web of REDOX couples that build new molecules and dismantle existing ones. Synthesis (anabolism) of new compounds requires energy; breakdown (catabolism) of the old structures releases energy – The REDOX coupling mechanism connects both these metabolic processes. (iv) Adenosine triphosphate (ATP), through high-energy phosphate bonds, temporarily stores energy that a cell uses for a variety of activities. (v) Without enzymes supported by REDOX, biochemical reactions cannot move fast enough for living organisms to survive; enzymes lower the energy demand (requirement) to start these reactions.

Quantum Theory: REDOX (oxidation-reduction) reactions involve the escape of electrons (charged particles) from molecules or ions in one chemical substance and their capture by other substances. Free energy is the key to understanding the electron escape phenomenon. REDOX reactions are important because they are the principal sources of energy on this planet, both natural (biological) and artificial (engineered). Oxidation of molecules, by the removal of hydrogen or combination with oxygen, normally liberates large quantities of energy. Current advances in free radical chemistry and REDOX principles are rooted in the 'Quantum Theory', which explains energy movement via small irreducible charged particles or parcels (the quanta).

The Laws of Thermodynamics deal with energy transfers; therefore, they are fundamental for understanding free radical biology. The evolution of higher life forms became possible with the conjecture that energy flows through biological systems are effective in organizing simple chemicals into complex arrangements associated with life. In existing living systems, the coupling of energy flow to the organizing 'work' (metabolism) is regulated by REDOX principles. REDOX operates the metabolic motor of life that includes DNA, enzymes, and a number of biochemical pathways. Interestingly, the same REDOX principles also apply to automobiles that convert chemical (gasoline) energy into mechanical torque on wheels. Thus, living systems use simple bio-monomers to construct complex polymers (such as DNA and proteins) through thermal, electrical, chemical, or solar energy flows regulated by REDOX. Accordingly, REDOX mechanisms provide a 'radical' explanation of the physio-chemical operation of life's metabolic motor and its underlying laws of thermodynamics.

REDOX and Bio-Energetics: Life is a highly ordered and intricately designed biochemical system. For instance, photons from the Sun do not directly incorporate into the chemical bonds of the cell; there has to be an intermediate 'high energy' carrier to store solar energy until the cell is ready to use it. The accrual of high-energy intermediates,[1] such as adenosine triphosphate (ATP), depends on the generation of proton gradients across membranes. The Chemiosmotic Theory explains the movement of ions across a selectively permeable membrane, down their electrochemical gradient. More specifically, it relates to the generation of ATP by the movement of protons (H^+ ions) across a membrane during cellular respiration. According to this theory, the transfer of electrons down an electron transport chain (ETC), through a series of REDOX reactions, releases energy. This energy allows certain carriers in the ETC to move protons across a membrane. As the protons accumulate on one side of a membrane, this H^+ concentration creates an electrochemical gradient, or REDOX potential (voltage), across the membrane. The energized REDOX state of the membrane with this charge separation is called 'Proton Motive Force' (PMF). PMF provides the energy necessary for synthesis of ATP(from ADP and phosphate). Thus, the ETC consists of a spatially separated series of REDOX reactions. When protons flow back through the membrane, they enable mechanical work using the chemical energy from ATP; for example, movement of a cardiac muscle to support a heartbeat.

Adenosine triphosphate (ATP) is used for four major biological functions: (i) transport compounds across cell membranes; (ii) provide mechanical energy for muscle contraction; (iii) serve as chemical energy for biosynthesis of various types of macromolecules; and (iv) also be used as an 'on-off' switch to control chemical reactions and send cell signals. Thus, ATP is mainly involved in reactions concerned with electrical and mechanical work (ion pumps and contractile proteins), whereas UTP is involved in carbohydrate synthesis, GTP in protein synthesis, and CTP in lipid synthesis.

REDOX and Bio-Informatics: Bio-Informatics, the intricate genetic design and Bio-Energetics, the mechanics of the electron transfer system, interplay to accomplish three vital activities: (i) capture chemical energy to produce complex bio-molecules, (ii) expend energy to transport chemicals from the milieu into the cell, and (iii) generate mechanical energy to enable cells to change their physical location and mobilize and reorganize their interior structure. These activities require energy, as well as a genetic blue print, to direct the spatial arrangement of molecules.

DNA Replication and REDOX Couples: Before cell division, DNA needs to be duplicated. DNA replication is regulated by various coupled REDOX reactions. It begins with the unzipping of the parent molecule as the hydrogen bonds between the base pairs are broken. Several enzymatic processes act as REDOX couples to complete DNA duplication.

All living species require energy for survival, growth and multiplication. The ability to harness energy from a variety of metabolic pathways is a fundamental property of all biological forms. Life is dependent on energy transformations; organisms survive because of their ability to exchange energy within and without. However, just having a source of energy alone is not sufficient to sustain life. REDOX is the driving force for life on Earth. The universe is always moving towards disorder (an increase in 'entropy' or 'randomness'), which is governed by REDOX principles. Life is a minute protest against this 2nd law of thermodynamics! Life is a constant battle with entropy; however, it ultimately loses and dies.

REDOX
Principles

The term REDOX originates from two opposite chemical reactions, REDuction and OXidation. REDOX is based on an electron exchange where a reductant transfers electrons to an oxidant. Oxidants or oxidative agents can remove electrons from another substance. Since they receive electrons, they are also known as electron acceptors. Reductants or reductive agents are able to transfer electrons to another substance. Since, they give away electrons; they are also called as electron donors. In more general definitions, REDOX reactions refer to a change in the oxidation state, since the actual transfer of electrons may never occur. Thus, oxidation is an increase in oxidation state, and reduction is a decrease in oxidation state. Many changes that take place within living organisms are also REDOX reactions. For example, the digestion of food is an oxidation process. Food molecules react with O_2 in the body to form CO_2 and H_2O. Energy is also released in the process. The CO_2 and H_2O are eliminated from the body as waste products, but the energy is used in metabolism to help maintain the survival, growth and multiplication states of an organism.

Free Radicals: Free radicals are molecules with unpaired electrons. They are everywhere – in the air, our bodies, and the materials around us. In search of finding another electron, free radicals are very reactive and can cause damage or affect surrounding molecules. Most free radicals are short-lived species and quickly react with other molecules.

The average adult human body contains detectable traces of at least 60 chemical elements; 20 of these are known to play an active positive role in REDOX life. The adult human body averages ~53% water. Due to water content, the human body has more oxygen (O) by mass than any other element, and more hydrogen (H) by atom-fraction than any element. The 10 other major elemental constituents include: carbon (C), nitrogen (N), calcium, phosphorus, sulfur, potassium, chlorine, sodium, magnesium and iron. The amounts of HOCN contained in the adult human body represent the integration of the day to day accretions from the time of conception to the termination of growth.

REDOX and Transition Metals: Most of the transition metals can participate in the chemistry of radicals and convert relatively stable oxidants into powerful radicals. The products of radical reactions with transition metal complexes, especially those resulting from the decomposition of the transient complexes, are probably the cause of many deleterious effects in biological systems. Thus, metal ions play essential roles in about one third of enzymes. These ions can modify electron flow in a substrate or enzyme, and effectively control enzyme-catalyzed reactions. They can also bind and orient substrate, with respect to functional groups in the active site, and provide a site for REDOX activity (if the metal has several valence states). Without the appropriate metal ion, an enzymatic reaction would proceed at a very slow pace, if at all.Their inter-changeable prevalent ionic forms and unique transitional properties make these metals useful in biochemical REDOX reactions (i.e. hemoglobin for oxygen transfer).

Reactive Oxygen Species (ROS): Molecular oxygen is present in all compartments of the human body. Many major classes of organic molecules in living organisms, such as proteins, nucleic acids, carbohydrates and fats contain oxygen. Elemental oxygen is used in cellular respiration for all complex life. The reactivity of O_2 is driven by REDOX chemistry. Activation of molecular oxygen (to generate ROS) may occur either by electron transfer (which is reduction of oxygen) or by energy transfer reactions (such as those that occur under the influence of light in the presence of appropriate photosensitizers). Singlet oxygen ($^1O_2$), a charge-excited and non-radical form of molecular oxygen, is generated in the latter type of reaction; whereas the superoxide anion ($O_2^{\bullet -}$) results from the monovalent reduction of oxygen. An estimated 1% of oxygen is reduced to $O_2^{\bullet -}$ and contributes to endogenous free radical formation. Some ROS are highly reactive with a short half-life. The milieu factors such as pH and the presence of other ionic species, influence the half-life and activity of ROS. Generally, a longer half-life of a species might imply a higher toxicity of the compound. It allows adequate time to diffuse and reach a sensitive location to interact and cause damage. For example, the relatively long half-life of $O_2^{\bullet-}$ radicals allows movement to locations and interactions with other molecules. On the other hand, a highly reactive species with an extremely short life span, like the hydroxyl (·OH) radical, is produced in locations where it can cause damage by interacting with its immediate surroundings. For example, ·OH survives for 10^{-10} (ten billionth) of a second in biological systems.

Reactive Nitrogen Species (RNS): are various ·NO-derived compounds, that include nitroxyl (·HNO) anions, nitrosonium (NO^+) cations, higher oxides of N_2, S-nitrosothiols (RSNO), and dinitrosyl iron complexes. RNS are known to play a crucial role in the physiological regulation of several cellular functions, such as smooth muscle cells, cardiomyocytes, platelets, and nerve cells.

Reactive Sulfur Species (RSS): Sulfur is an essential constituent of many organic molecules, amino acids (such as cysteine and methionine), and the small tripeptide glutathione (GSH). Sulfur is also important in the formation of Fe-S clusters for many enzymes, particularly, those that are involved in REDOX reactions. Sulfur (in the form of thiol groups) is central to the REDOX control of cellular metabolism. High cellular concentrations of thiols and disulfides make sulfur a prime target for oxidation under conditions of oxidative stress. Inorganic sulfur species, such as hydrogen sulfide (H_2S) and the ·HS anion, have been implicated in the control of cellular processes, such as vasodilation and inflammation.

Free Radical Formation: In the human body, free radicals are generated by both endogenous and exogenous sources. (i) *Endogenous sources:* Cellular respiration (mitochondrial electron transport chain, ETC) is the primary endogenous source of free radicals. A majority of O_2 (~98%) is used during mitochondrial ATP (energy) generation; however, ~2% of O_2 ends up as electron leakage and creates free radicals. The presence of toxic heavy metals in the body, certain processed foods, oxidative stress, and low blood supply (which generally occurs during heart attacks and strokes), are other sources of endogenous free radicals. (ii) *Exogenous sources:* Free radical production is stimulated by various external factors including chemical pollutants (i.e. pesticides, asbestos), cancer chemotherapeutics, environmental exposure to carcinogens (i.e. cigarette smoke, petroleum-based products), over exposure to sunlight or X-rays, irradiation, over-exercising, and prescription drugs (i.e. anti-inflammatory agents and diuretics).

During their highly charged motion path, ROS, RNS and RSS randomly hit various biological molecules, which sometimes triggers chain reactions that generate other reactive species. For example, when ROS and RNS react with the unsaturated fatty acids of phospholipids, a cascade of lipid peroxidation reactions is initiated, which may act as REDOX signals.

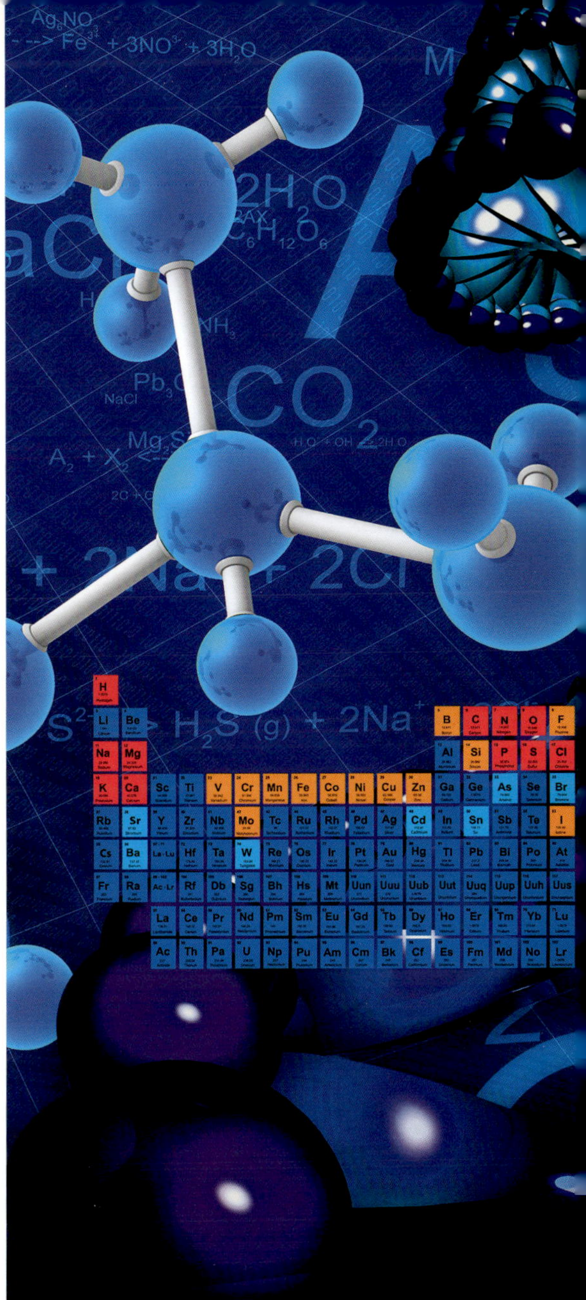

REDOX **Chemistry**

REDOX biology takes on a pivotal role in the various metabolic pathways of living organisms. These biochemical processes include mechanisms of energy production (ATP generation), host defense (phagocytic killing of pathogens) and cell signaling. Several vital processes in biological systems are dependent on the smooth interplay of REDOX-active organic molecules and inorganic ions and complexes. These cellular REDOX mechanisms are tightly controlled by various regulatory proteins and enzymes.

REDOX Prooxidants: The Mitochondrion is the 'cellular power house', the active site for most ATP generation though oxidative phosphorylation, which accounts for 90% to 95% of cellular oxygen consumption. The major bulk of mitochondrial ROS generation occurs at the electron transport chain, as a byproduct of respiration. Quantity and quality of this cellular organelle determines our physiological performance and endurance.

During phagocytosis, the phagocytic cells (neutrophils, eosinophils, and mononuclear phagocytes) undergo an increase in glucose and oxygen consumption, termed the 'respiratory burst'. The respiratory burst generates several oxygen-containing compounds that kill the microbes undergoing phagocytosis – oxygen-dependent intracellular killing. Reactive oxygen species (ROS) are critical weapons in the phagocyte arsenal.

Specific oxidative enzymes are widely distributed in biological systems for ROS generation. Such enzyme catalyzed, kinetic regulation of free radical formation is important for several vital functions including the digestive process in the gastrointestinal lumen, microbial killing of phagocytes, bone turnover of the skeletal system, vascular tone of the circulatory system and synaptic transmission of the central nervous system. This highly controlled REDOX generation provides molecular movement, cell signaling capacity and the metabolic grid for the body. Three major ROS generating enzyme systems, xanthine oxido-reductase (XOR), NADPH oxidase (NOx) and nitric oxide synthase (NOS), are fundamental for the REDOX-based function of the human body. Different isotypes of these enzymes exist in the human body based on tissue location and system function. Any regulatory disruption (qualitative or quantitative) of these enzymes results in severe pathological conditions ranging from inflammatory diseases (Crohn's disease, atherosclerosis, arthritis, asthma) to chronic neuro-degenerative disorders such as Alzheimer's and Parkinson's. These ROS generators, therefore, are vital for REDOX homeostasis. A shift in balance between the oxidant and antioxidant ratio in favor of oxidants is termed 'oxidative stress'. Normal cells maintain REDOX homeostasis through proper balance between ROS generation and elimination.

REDOX Antioxidants: Living organisms are equipped with antioxidant systems to eliminate free radicals, as part of cellular defense, and to maintain REDOX homeostasis. Broadly, antioxidant systems can be divided into different categories based on their functions (enzymatic or non-enzymatic), their physical properties (water-soluble or lipid-soluble), and their source (endogenous or exogenous). Some antioxidants are enzymes or serve as co-factors for enzymes. The major non-enzymatic cellular REDOX buffer systems rely on water-soluble antioxidants such as glutathione (GSH) and ascorbic acid. However, several other less abundant antioxidants also serve important functions in REDOX homeostasis and are important for cellular function and disease prevention. These antioxidant molecules function not only in quenching chain reactions, scavenging ROS, and preventing lipid peroxidation, but also influence cellular functions. These functions include binding to specific receptors to initiate signal transduction cascades, inhibiting or activating enzymes, and regulating gene expression.

REDOX Coupling Systems: The human body consists of three major REDOX coupling systems and each couple functions as unique rheostat 'on/off switches' in the regulation of cellular pathways. Any disruption of these distinct 'REDOX control switches or circuitry' can dramatically affect several vital body functions. The regulatory role of REDOX in a wide range of vital functions, including gene regulation, cell signaling, protein activation/ deactivation, cellular differentiation and apoptosis has been established. Furthermore, among the various classes of ROS-induced macromolecular structural modifications, protein oxidation has a functional consequence, since proteins act as receptors, transporters, enzymes, transcription factors and cytoskeletal components, among others. Moreover, proteins are the largest constituents of the cell and probably the most frequent targets of free radical attacks. Also the cellular REDOX state becomes increasingly more pro-oxidizing in the latter part of life. Aging has a strong genetic component, but it can be viewed partly as a degenerative process dictated by REDOX-based free radical reaction.

REDOX Signaling: REDOX signaling is cellular communication either using short-lived ROS, RNS, RSS or through changes in the electrochemical potential of REDOX couples within biological systems. REDOX signaling is a vital feature of normal physiology, as well as of disease and aging. The cellular operation of REDOX signaling requires ROS, RNS or RSS that either act directly on a target, or convert to another signal, or change the REDOX potential of the biological milieu with orchestration of one or several REDOX couples – thereby altering the functions of a cellular target. Once a REDOX signal is completed, biological processes operate to reverse the effect of the signal and return the system to its basal (ground) state.

REDOX signaling is involved in many biological processes from the sensing of ROS, RNS, RSS and lipid peroxidation products to protection against damage caused by oxidative stress. REDOX signals are important for gene regulation. In the human body, cellular DNA is constantly attacked by free radicals. Such hits may induce mutations and play a role in the evolution process. Moderate levels of DNA damage triggers cell cycle arrest and initiates DNA-repair processes. In contrast, excessive damage or failure in DNA repair can induce apoptosis.In the nucleus, REDOX affects histones that regulate inflammatory gene expression by activation of transcription factors. REDOX signaling plays a key role in signal transduction related to contraction and relaxation of blood vessels, growth, death and migration of vascular cells, extracellular matrix (ECM) modifications and angiogenesis.Thiol-based REDOX regulation provides a checkpoint for antibodies for eliciting rapid and sustained responses to pathogens and blocking any unwanted immune response. REDOX is vital for intercellular signaling of the adaptive immune system. Apoptosis, or programmed cell death, represents a physiologically conserved mechanism of cell regulation by REDOX signaling. This cellular process is critical for embryonic development and tissue homeostasis.

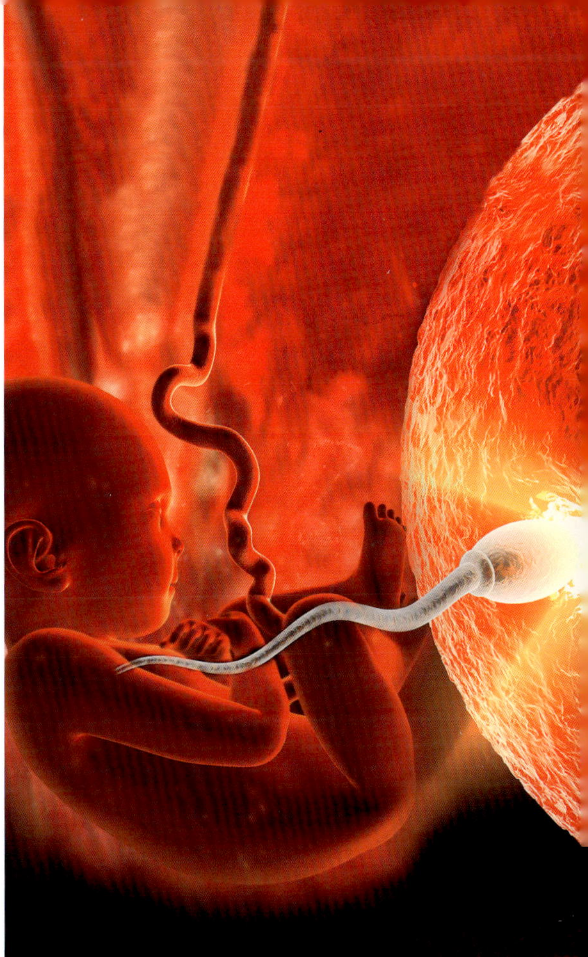

REDOX
Biology

REDOX Medicine deals with the metabolic states of health and disease based on the movement of oxygen (O_2) and its reactive oxygen and nitrogen species (ROS/RNS) through cellular pathways. The movement of ROS/RNS can be controlled and equilibrated by innate systems to establish physiological homeostasis – a sign of healthy status. REDOX homeostasis can be defined as a state of equilibrium between different but inter-dependent elements in an organism. This REDOX optimization is vital for various signaling pathways that control cellular proliferation, maturation and apoptosis.

REDOX homeostasis ensures that cells respond properly to endogenous and exogenous stimuli. However, when REDOX homeostasis is disturbed, the electro-chemical imbalance may led to aberrant cellular dysfunction and contribute to disease development or death. Therefore, REDOX chemistry and homeostasis can be considered as the two most fundamental properties of life. REDOX homeostasis is controlled by two opposite chemical pathways in a living organism. From one end, a regulated production of prooxidative reactants (i.e. ROS, RNS and RSS) that ensure a continuous flow of charged particles (electrons and protons) for breaking or forming chemical bonds. The influx and efflux of this free energy flow through chemical bonds imparts reactivity; in other words, an ionic instability due to loss or gain of charged particles to molecules. This highly reactive nature sets these molecules in a kinetic motion to find a suitable chemical partner to relieve their acquired charge either by covalent interaction or through an ionic discharge. This basic principle of energy transfer in small packets ('quanta') is the driving force for all chemical reactions and every metabolic pathway of life. On the other end, free flow of charged particles obeys the laws of energy (or thermodynamics) and transfers the 'quanta' to any random encounter in its motion path. An organized capture of such 'quanta' is vital for life force; therefore, organisms have evolved specialized REDOX couples to avoid randomness in the quantum energy transfer – even for a millionth fraction of a second! Thus, living organisms have evolved antioxidants and radical quenchers, alongside the REDOX couples, with an array of genetic controls, based on the cellular mass and complexity of an organism. No system can ever hold an organization against the universal nature of randomness (the entropy) for too long – accordingly, free radical damage, aging and death are absolute consequences of the 2nd Law of Thermodynamics. In an amazing twist, however, life has also evolved pathways to utilize certain low thresholds of oxidative stress to its own advantage; for example, cell signaling, programmed cell death (apoptosis) and gene mutations to enhance the chance of survival. In essence, REDOX homeostasis is the monumental effort of a living system to maintain the narrow margin of its survival through its genetic design between two variable effectors – the milieu (environment) and time.

The Oxygen (O_2) Paradox: The human body and its metabolic pathways are critically dependent on O_2. The physiological impact of O_2 in the human body is reflected by the involvement of three major organ systems that have exclusively evolved to transport O_2 from the surrounding environment to the cell systems – namely, the cardiovascular, respiratory, and hematopoietic systems. Cells do not have the ability to store O_2 and are therefore reliant upon its continuous supply to meet the body's metabolic demands. If the supply of O_2 is not aligned with the normal metabolic demand, tissue hypoxia will ensue, eventually resulting in cellular injury and/or death. Ironically, O_2 is inherently toxic to the same cells that are critically dependent upon a steady supply of O_2 for normal cellular function – the so-called oxygen paradox.

Oxidative Stress: An uncontrolled or random movement of prooxidant species can lead to a toxic condition, known as oxidative stress. Oxidative stress results from an imbalance between ROS/RNS (reactive nitrogen species) generation and the body's antioxidant defenses. This imbalance is due to (i) increased generation of ROS/RNS; (ii) decreased synthesis or use of the body's natural antioxidants; or (iii) a combination of both processes. Both ROS and RNS contribute to the host inflammatory response either via direct cellular toxicity or through the activation of pro-inflammatory gene expression pathways. ROS and RNS have direct effects (via oxidation or nitration reactions, respectively) on the structure and function of many enzymes, proteins, lipids, and DNA. These effects disrupt the function of the cellular machinery.

Therefore, the ability to adapt to ROS stress determines the overall fate of the cells. A successful adaptation to increased survival signals (in combination with ROS-mediated mutations and loss of critical regulatory mechanisms) leads to defective cell death and aberrant cell proliferation. These cellular conditions may contribute to the development of cancer. A failure in adaptation to ROS stress (while the cells accumulate oxidative DNA, protein, and lipid damage) may also result in excessive cell death, leading to degenerative disorders and aging. Oxidative stress is implicated in cellular processes of senescence, angiogenesis, apoptosis, and a myriad of diseases including atherosclerosis, hypertension, hypercholesterolemia, obesity, cancer, diabetes mellitus, and neurodegenerative disorders. Most pathology is the result of excessive ROS formation that promotes inflammation in the surrounding tissues and accelerates cell death or senescence. It could also cause necrosis, ATP depletion, and may interfere with controlled apoptotic death. Oxidative stress is strongly linked to cardiovascular disease, since oxidation of LDL in the vascular endothelium is a precursor to plaque formation. Oxidative stress also plays a role in the ischemic cascade due to O_2 reperfusion injury following hypoxia. This cascade includes both stroke and myocardial infarction. Oxidative stress has also been associated with chronic fatigue syndrome. Oxidative stress is likely involved in the age-related development of neurodegenerative diseases (i.e. Alzheimer's disease, Parkinson's disease), macular degeneration and cancer. The reactive species produced in oxidative stress can cause direct damage to the DNA (therefore is mutagenic). It may also suppress apoptosis and promote proliferation, invasiveness and metastasis. Interestingly, excessively low levels of ROS can also invoke pathological changes by interrupting the physiological role of oxidants in proliferation, vasodilation, host defense and may promote carcinogenesis.

Oxidative stress could be beneficial, when used by the immune system as defense mechanism to combat pathogens. At optimum physiological doses, ROS can function as REDOX signaling molecules and support cell growth, protein phosphorylation, and cell migration. This REDOX dose dependency has a significant relevance in the treatment and prevention of several diseases. Accordingly, the REDOX system controls key transcription factors, signal-transduction pathways, and cell-death regulators that affect cell survival. Basic knowledge of REDOX regulation of cell survival can be applied in the development of strategies for the clinical management of various pathological conditions.

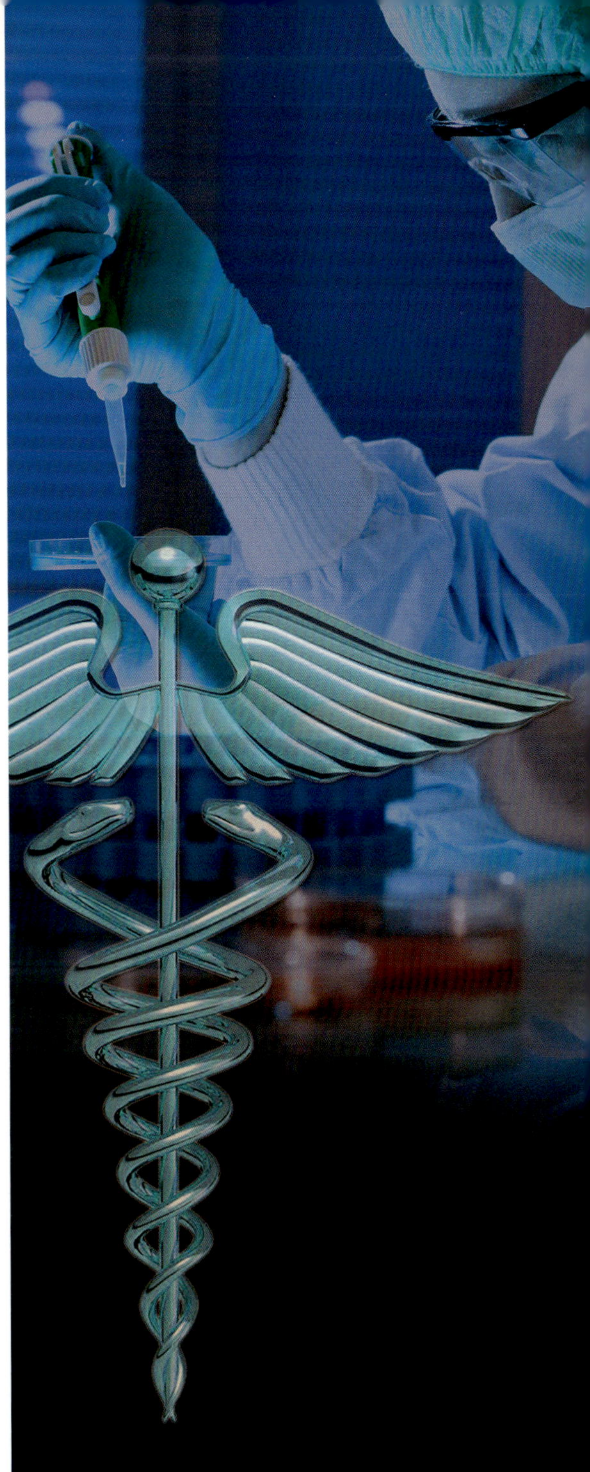

REDOX **Medicine**

5

Nutrition (also called nourishment or aliment) is the provision of chemicals necessary to support the metabolism of cells and organisms. Since metabolism/energy production requires oxygen (O_2), potentially cell damaging free radical species are generated. An optimum dietary intake of antioxidants and other REDOX supportive nutrients is critical to sustain good health. Accordingly, several diseases and dysfunctions can be prevented or alleviated with a proper diet enriched with antioxidant minerals and vitamins, free radical scavengers, and bio-replenishments.

REDOX and Diet: Foods and beverages deliver both prooxidants, as well as a complex mixture of antioxidants, to the gastrointestinal tract. Some of these antioxidants are well absorbed, while others not so well; however, certain unabsorbed ones can also exert extra-cellular antioxidant protective effects in the gut. Also, scavenging of various ROS by dietary antioxidants can generate oxidized, chlorinated, nitrosylated or nitrated products that exert a plethora of additional biological effects. Numerous polyphenolic natural substances, commonly considered antioxidants, are conversely oxidants due to the presence of electron acceptor groups in their molecules. Their beneficial effects support the importance of oxidative reactions in maintaining optimal health and/or preventing degenerative diseases. Flavonoids, constituting the most important polyphenolic class, are natural substances that exhibit a wide range of biological effects including: antibacterial, anti-inflammatory, antiallergic, antithrombotic, and vasodilatory actions. There are also a plethora of other plants and herbs that contain polyphenolic substances that can potentially scavenge hydroxyl radicals and/or chelate free iron.

REDOX Metals: Transition metals are essential to many biological processes in almost all living organisms. Their versatility, which arises from an ability to undergo reduction-oxidation chemistry, enables them to act as critical cofactors of enzymes throughout the cell. Accumulation of metals, however, can also lead to oxidative stress and cellular damage. The importance of metals (to both enzymatic reactions and oxidative stress) makes them key players in cellular metabolism. The systemic and cellular levels of these metals must be actively monitored and regulated though diet. Their deficiency leads to loss of crucial enzyme activities, while their accumulation leads to inappropriate enzymatic reactions that generate ROS, oxidative damage and, eventually, cell death.

Iron (Fe) is an essential nutritional element that plays a critical role in electron transport, cellular respiration, cell proliferation and differentiation, and the regulation of gene expression. Iron is integral to the catalysis of the REDOX reactions of oxidative phosphorylation in the mitochondrial respiratory chain. It also provides a specific binding site for O_2 in the heme binding moiety of hemoglobin, which allows O_2 transport in the blood. Copper (Cu) plays an essential role in human physiology including respiration, radical defense, neuronal myelination, angiogenesis, and many other processes. Copper also serves as a catalytic and structural cofactor for enzymes that function in energy generation, iron acquisition, O_2 transport, cellular metabolism, peptide hormone maturation, blood clotting, signal transduction and a host of other processes. Zinc (Zn) is critical for the function of a number of metallo-proteins; stimulates oxido-reductase, hydrolase ligase, lyase families and co-activates (with copper) superoxide dismutase (SOD) or phospholipase C. Selenium (Se) is a powerful free radical scavenger. The biological roles ascribed to selenium include the prevention of cancer, cardiovascular disease (CVD), and viral mutation. Seleno-proteins are involved in bone metabolism, as well as functions of the endocrine pancreas and adrenal glands.

REDOX Vitamins and Amino acids: Vitamins are required by an organism as vital nutrients in limited amounts. Carotenoids (vitamin A), ascorbate (vitamin C) and tocopherol/tocotrienols (vitamins E) are not only considered essential cofactors for human physiology; but also potent free radical scavengers in aqueous (water-soluble) and lipid (fat-soluble) matrices. Amino acid residues in proteins represent one of the major targets of ROS and cellular REDOX. The sulfur-containing amino acids, cysteine and methionine, are especially sensitive to ROS-mediated oxidation. A significant fraction of the total protein may be oxidized in humans with consequences in health and aging. N-acetyl cysteine (NAC) is a powerful antioxidant and potential therapeutic agent in the treatment of cancer, heart disease, HIV infection, heavy metal toxicity, and other diseases characterized by oxidative damage.

REDOX and Bio-Replenishment: The ability of a living organism to continuously refill its expended (depleted) chemicals that are vital for the maintenance of steady-state metabolism and REDOX homeostasis is known as 'Bio-Replenishment'. Any breakdown in bio-replenishment can lead to cellular dysfunction, system failure and, in severe cases, even death. Bio-replenishment is an intricately designed 'biological maintenance program' that keeps the REDOX state and basal metabolic rate (BMR) at optimum while aging. Lactoferrin (LF), ribonuclease (RNase), coenzyme Q (CoQ) and S-adenosyl-methionine (SAMe) are some noteworthy bio-replenishments.

During the past three decades, a prominent role for lactoferrin (LF) in REDOX chemistry has been established by various laboratories worldwide. LF plays an important role in the intestinal absorption of iron, zinc, copper, manganese and other essential trace elements. LF has also been suggested to protect the gut mucosa from excess uptake of heavy metal ions. LF provides antioxidant protection to erythrocytes through various mechanisms, such as: (i) influencing the levels and activity of intra-cellular antioxidants, reduced GSH, in particular; and, (ii) by lowering the levels of lipid peroxidation products, as well as preventing hemolysis under normal conditions and oxidative stress. LF is a primary constituent of REDOX homeostasis, functioning to reduce oxidative stress at the molecular level, and thus controlling excess inflammatory response. Intra-cellular (or phagosome) killing of pathogens by LF is associated with its ability to donate iron for catalyzing $\cdot OH$ radical production. RNase-enriched lactoferrin (R-ELF) technology has provided clinical evidence that a combination of RNase and LF has a promising applications in bone regeneration. Coenzyme Q (CoQ) is an essential component of the ETC for ATP synthesis to provide energy for vital cellular functions. CoQ10 transports protons across the inner mitochondrial membrane by transverse diffusion, resulting in the conservation of energy ('proton-motive Q cycle'). CoQ10 is a membrane-bound chemical, located in close proximity to the unsaturated lipid chains, designed to function as a primary scavenger of free radicals. SAMe is the link to three key metabolic pathways, polyamine synthesis, trans-methylation and trans-sulfuration. SAMe donates a methyl group to a wide variety of substrates including DNA, proteins, neurotransmitters, and phospholipids. SAMe plays an important role as a methyl donor in more than a hundred methyltransferase reactions.

Abbreviations

Selected Reactive Species / Radicals

CH_4	Methane
CO	Carbon monoxide
CO_2	Carbon dioxide
$C_6H_{12}O_6$	Glucose
Fe_2O_3	Iron oxide
HCl	Hydrochloric acid
HNO	Nitroxyl
HNO_3	Nitric acid
$HO_2^{\bullet -}$	Hydroperoxyl
$HOCl$	Hypochlorous acid
$H_2N_2O_2$	Hyponitrous acid
H_2O	Water
H_2O_2	Hydrogen peroxide
H_2S	Hydrogen sulfide
H_2SO_4	Sulfuric acid
H_3PO_4	Phosphoric acid
NH_3	Ammonia
NH_4^+	Ammonium
$\cdot NO$	Nitric oxide
NO_2	Nitrogen dioxide
NO_2^-	Nitrite
NO_3^-	Nitrate
N_2	Nitrogen
N_2O	Nitrous oxide
$\cdot OH$	Hydroxyl radical
$ONOO^-$	Peroxynitrite
O_2	Oxygen
$O_2^{\bullet -}$	Superoxide
$^1O_2$	Singlet oxygen
O_3	Ozone
SO_2	Sulfur dioxide
SO_3	Sulfur trioxide

Abbreviations

Complete List

Aβ	Amyloid β-peptide
ACE	Angiotensin converting enzyme
AD	Alzheimer's disease
ADP	Adenosine diphosphate
AGE	Advanced glycation end product
AIDS	Acquired immunodeficiency syndrome
AMD	Age-related macular degeneration
AMI	Acute myocardial infarction
Ang II	Angiotensin II
AP-1	Activator protein-1
Apaf-1	Apoptotic protease-activating factor-1
ARDS	Acute respiratory distress syndrome
ARE	Antioxidant response element
ART	Assisted reproductive techniques
ASD	Autism spectrum disorders
ATO	Arsenic trioxide
ATP	Adenosine triphosphate
BH4	Tetrahydrobiopterin (THB)
BHMT	Betaine homocysteine methyltransferase
BMD	Bone mineral density
BMI	Body mass index
BMR	Basal metabolic rate
BPA	Bisphenol A
BPD	Broncho-pulmonary dysplasia
BRCA1	Breast cancer susceptibility gene
cal / kcal	Calorie / kilo-calorie
CAT	Catalase
CBS	Cystathionine β-synthase
CCK	Cholecystokinin
CCOx	Cytochrome c oxidase
CDP	Cytidine diphosphate
CF	Cystic fibrosis

CFC	Chloroflurocarbon
CFTR	CF transmembrane conductance regulator gene
CGD	Chronic granulomatous disease
CHD	Chronic heart disease
CNS	Central nervous system
COPD	Chronic obstructive pulmonary disease
CoQ	Coenzyme Q (ubiquinone)
COX	Cyclooxygenase
CR	Caloric restriction
CRP	C-reactive protein
CSF	Cerebrospinal fluid or Colony-stimulating factor
3-CT	3-chlorotyrosine
CTGF	Connective tissue growth factor
CTP	Cytidine triphosphate
Cu-Zn-SOD	Copper-and-zinc-containing SOD
CVD	Cardiovascular disease
Cys	Cysteine
CySS	Cystine
DAG	Diacylglycerol
DC	Dendritic cell
DES	Diethylstilbestrol
DGC	Dystroglycan complex
DHA	Docosahexaenoic acid
DIO	Deiodinase
DISC	Death-induced signalling complex
DLMO	Dim-light melatonin onset
DM	Diabetes mellitus
DMARD	Disease-modifying anti-rheumatic drug
DMD	Duchenne muscular dystrophy
DNA	Deoxyribonucleic acid
mtDNA	Mitochondrial DNA
DNIC	Dinitrosyl iron complex
DR	Death receptor
DUOX	Dual oxidase
ΔE	Energy
ECG	Electrocardiogram
ECM	Extracellular matrix
EC-SOD	Extracellular SOD
ED	Erectile dysfunction
EDC	Endocrine disrupting chemical
EDRF	Endothelium-derived relaxing factor
EGCG	Epigallo-catechin-3-gallate
E_h	REDOX Potential

ER	Endoplasmic reticulum
ERK	Extracellular signal-regulated kinase
ETC	Electron transport chain
FAD	Flavin adenine dinucleotide
FADD	Fas-associated death domain
FADH2	Reduced FAD
Fas	Fibroblast-associated cell surface
FFA	Free fatty acid
FGF	Fibroblast growth factor
FMN	Flavin mononucleotide
FOXO	Forkhead transcription factor, class O
FSH	Follicular stimulating hormone
G6PDH	Glucose-6-phosphate dehydrogenase
ΔG	Gibbs free energy
GABA	γ-aminobutyrate
GAG	Glycosaminoglycan
GCL	Glutamate cysteine ligase
GDP	Guanosine diphosphate
GI	Gastrointestinal
GIP	Gastric inhibitory peptide
Gln	L-glutamine
Glu	L-glutamic acid
GPx	GSH-Peroxidase
eGPx	Extra-cellular GPx
GRx	Glutaredoxin
GSH	Glutathione
GSH-R	GSH-Reductase
GSNO	S-nitrosoglutathione
GSSG	Oxidized glutathione (Glutathione disulfide)
GST	Glutathione-S-transferase
GTP	Guanosine triphosphate
GULO	Gulonolactone oxidase
HAART	Highly active antiretroviral therapy
HBOT	Hyperbaric oxygen therapy
HDL	High-density lipoproteins
HF	Heart failure
HIV	Human immunodeficiency virus
HMG CoA	3-hydroxyl-3-methylglutaryl-coenzyme A
HNE	Human neutrophil elastase
4-HNE	4-hydroxynonenal
HO	Heme oxygenase
HRT	Hormone replacement therapy
IBBM	Intestinal brush-border membranes

IBD	Inflammatory bowel disease
ICAM	Intercellular adhesion molecule
ICP	Intra-cavernosal pressure
IDDM	Insulin-dependent diabetes mellitus
IFN	Interferon
IgA	Immunoglobulin A
IgE	Immunoglobulin E
IgG	Immunoglobulin G
IGF	Insulin-like growth factor
IGT	Impaired glucose tolerance
IKK	Iκ-B kinase
IL	Interleukin
IMS	Inter-membrane space
IMT	Intima-media thickness
IVF	*In vitro* fertilization
J / kJ	Joule / kilo-Joule
JNK	c-jun N-terminal kinase
K	Kelvin
Keap-1	Kelch-like (erythroid cell-derived)-associated protein-1
LDL	Low-density lipoproteins
L-DOPA	L-3,4-dihydroxyphenylalanine
LF	Lactoferrin
LH	Luteal hormone
LOx	Lipoxygenase
LPO	Lactoperoxidase
LPS	Lipopolysaccharide
LTP	Long term potentiation
M/mM/μM/nM	Molar / milli-Molar / micro-Molar / nano-Molar
MAP	Mitogen activated protein
MAPK	MAP kinase
MAT	Methionine-adenosyl-transferase
mcg RAE	Micro-gram of retinol activity equivalent
MD	Mitochondrial dysfunction
MDA	Malondialdehyde
Met	Methionine
Met-Hb	Methemoglobin
Met-R-SO	Methionine-R-sulfoxide
Met-S-SO	Methionine-S-sulfoxide
MHC	Major histocompatibility complex
MI	Myocardial infarction
MMP	Matrix metalloproteinase
Mn-SOD	Manganese-containing SOD
MPO	Myeloperoxidase

MS	Multiple sclerosis
MSC	Mesenchymal stem cells
MSG	Monosodium glutamate
Msr	Methionine sulforeductase
MT	Metallothionein
MTA	Methylthioadenosine
MUFA	Monounsaturated fatty acid
NAC	N-acetylcysteine
NAD	Nicotinamide adenine dinucleotide
NADH	Reduced NAD
NADP	Nicotinamide adenine dinucleotide phosphate
NADPH	Reduced NADP
NET	Neutrophil extra-cellular trap
NF-κB	Nuclear factor-κB
NFT	Neurofibrillary tangle
NIDDM	Non insulin-dependent diabetes mellitus
Ni-SOD	Nickel-containing SOD
NK	Natural killer (cell)
NMDA	N-methyl-D-aspartate
NO	Nitric oxide
NOS	NO synthase
cNOS	Constitutive NOS
eNOS	Endothelial NOS
iNOS	Inducible NOS
mtNOS	Mitochondrial NOS
nNOS	Neuronal NOS
NOx	NADPH oxidase
Nrf2	Nuclear factor-erythroid-2-related factor 2
NRTI	Nucleoside reverse transcriptase inhibitor
NSAID	Non-steroidal anti-inflammatory drug
3-NT	3-nitrotyrosine
OA	Osteoarthritis
8-OHdG	8-oxodeoxyguanosine
OPG	Osteoprotegerin
ORAC	Oxygen radical absorbancc capacity
OXPHOS	Oxidative phosphorylation
PAI	Plasminogen activator inhibitor
PAMP	Pathogen-associated molecular pattern
PARP	Poly-(ADP-ribose) polymerase
PCOS	Polycystic ovary syndrome
PCR	Polymerase chain reaction
PD	Parkinson's disease
PDGF	Platelet-derived growth factor

PEPCK	PEP carboxykinase
PG	8-epi-prostaglandin
PHG	Portal hypertensive gastropathy
Phox	Phagocyte oxidase
PKC	Protein kinase C
PMF	Proton motive force
PMN	Polymorphonuclear leukocyte
PNS	Peripheral nervous system
PPAR	Peroxisome proliferator-activated receptor
Prx	Peroxiredoxin
PSI	Photosystem I
PSII	Photosystem II
PSSG	Protein S-glutathionylation
PTH	Parathyroid hormone
PUFA	Polyunsaturated fatty acid
ΔQ	Heat flow
RA	Rheumatoid arthritis
RANK	Receptor for activator of NF-KB
RANKL	Receptor for activator of NF-KB, ligand
RBC	Red blood cells
RDA	Recommended dietary allowance
REDOX	Reduction-Oxidation
Ref-1	REDOX effector factor 1
R-ELF	RNase-enriched lactoferrin
RNA	Ribonucleic acid
mRNA	Messenger RNA
rRNA	Ribosomal RNA
tRNA	Transfer RNA
RNase	Ribonuclease
RNS	Reactive nitrogen species
ROP	Retinopathy of prematurity
ROS	Reactive oxygen species
RPE	Retinal pigment epithelium
RS	Reactive species
RSNO	S-nitrosothiol
RSH/RSSR	Thiol/disulfide REDOX couple
RSS	Reactive sulfur species
ΔS	Entropy
SAH	S-adenosyl-homocysteine
SAID	Steroidal anti-inflammatory drug
SAMe	S-adenosyl-methionine
SAPK	Stress-activated protein kinase
Sec	Seleno-cysteine

SeMet	Selenomethionine
SFA	Saturated fatty acids
SIRS	Systemic inflammatory response syndrome
SIRT	Sirtuin
Sirtuin	Silent information regulator-like protein
SN	Substantia nigra
SNO	S-nitrosylation
SOD	Superoxide dismutase
SR	Sarcoplasmic reticulum
S-S	Disulfide
T2DM	Type 2 diabetes mellitus
TAC	Total antioxidant capacity
TBARS	Thiobarbituric acid-reactive substances
TGF-β	Transforming growth factor-β
THF	Tetrahydrofolate
TLR	Toll-like receptors
TNF	Tumor necrosis factor
α-TP	α-tocopherol
α-TTP	α-tocopherol transfer protein
TPO	Thyroid peroxidase
Trx	Thioredoxin
TrxR	Trx-Reductase
TrxSS	Trx oxidized
UDP	Uridine diphosphate
UTP	Uridine triphosphate
UV	Ultraviolet
V / mV	Volt / Milli-volt
VCAM	Vascular cell adhesion molecule
VEGF	Vascular endothelial growth factor
VLDL	Very low-density lipoproteins
VSMC	Vascular smooth muscle cell
ΔW	Work
WAT	White adipose tissue
XDH	Xanthine dehydrogenase
XO	Xanthine oxidase
XOR	Xanthine oxidoreductase

Glossary

Active Ingredient is the chemical in a drug, nutraceuticals or supplement that exerts desired action.

Adenosine Triphosphate (ATP) is a nucleotide (part of DNA, the genetic material) that occurs in living cells; and a major source of cellular energy.

AIDS (Acquired Immune Deficiency Syndrome) is the final stage of HIV disease, which causes severe damage to the immune system.

Alcohol-induced Gastritis is an inflammation, irritation, or erosion of the lining of the stomach caused by excessive alcohol consumption. It can occur suddenly (acute) or develop gradually (chronic).

Alzheimer's Disease (AD) is a progressive dementia that affects a large group the aging population.

Amino Acids are the 20 building blocks of protein: alanine, arginine, aspara-gine, aspartic acid, cysteine, glutamic acid, glutamine, glycine, histidine, iso-leucine, leucine, lysine, methionine, phenylalanine, proline, serine, threonine, tryptophan, tyrosine, and valine. Eleven amino acids are produced by the body and the remaining 9 (known as essential amino acids) come from the diet.

Anabolism is the phase of metabolism in which simple substances are synthesized into the complex biochemical molecules of living tissue.

Angiogenesis is the physiological process of new blood vessel formation from pre-existing vessels.

Antagonist is an agent that acts against and blocks an action.

Antioxidant is a molecule that can safely interact with free radicals and terminate cell-damaging chain reactions. Antioxidants protect cells from free radicals.

Apoptosis is a distinct natural process of programmed cell death that affects single cells without releasing harmful substances to the surrounding area.

Arthritis is a group of clinical conditions that affect the health of bone joints.

Assimilation is conversion of nutrient into the fluid or solid substance of the body, by the processes of digestion and absorption.

Asthma is a chronic lung disease that inflames and narrows the respiratory airways.

Atherosclerosis is a vascular condition of hardening of the arteries. It occurs when fat, cholesterol, and other substances build up in the walls of arteries and form hard structures called 'plaques'. Over time, these plaques can block the arteries and cause problems throughout the body.

ATP Synthase is an enzyme that is directly responsible for the production of ATP via chemi-osmotic phosphorolation. It uses the proton gradient created by several of the other carriers in the ETC to drive a mechanical rotor. Energy from the rotor is utilized in ATP generation (from ADP + phosphate).

Atrophy is partial or complete wasting away of a body part. Causes of atrophy include poor circulation, malnutrition, hormone deficiency, reduced nerve impulse, or lack of exercise.

Autism is a disorder of neural development characterized by impaired social interaction and verbal and non-verbal communication, and by restricted, repetitive or stereo-typed behavior.

Autopoiesis is the process whereby a system, organization, or organism produces and replaces its own components and distinguishes itself from its environment.

Basal Metabolic Rate (BMR) is the body's resting level of energy expenditure; the number of calories needed to sustain basic life processes during a 24-hour period.

Bioavailability refers to the amount of an active compound in food, medication, or supplement that is absorbed in the intestines and ultimately available for biological activity in the body.

Bio-Energetics is the study of the flow and transformation of energy through biological systems.

Bio-Informatics is the study of information processes in biological systems. It involves storing, retrieving, organizing and analyzing biological data. Within the context of this book, this phrase refers to nucleic acids and molecular genetics.

Bio-Replenishment is the innate ability of a living organism to continuously refill its expended (depleted) chemicals that are vital to maintain homeostasis and negative entropy, while aging.

Biosphere refers to the part of earth's crust, waters, and atmosphere in which living organisms exist or that is capable of supporting life. It can be also defined as the global sum of all ecosystems.

Body Mass Index (BMI) is a standardized ratio of weight to height, and is often used as a general indicator of health. BMI can be calculated by dividing body weight (in kilograms) by the square of body height (in meters). A BMI between 18.5 and 24.9 is considered normal for most adults.

Bone Mineral Density (BMD) is the density of minerals (such as calcium) in bones, which is measured by using special X-ray, computed tomography (CT) scan, or ultrasound. This information is used to estimate the strength of bones.

Bone Modeling is a process in which bone is sculpted during growth to ultimately achieve its proper shape. During modeling, the bone resorption and bone formation occur on separate surfaces (formation and resorption are not coupled).

Bone Remodeling is a continuous process throughout life, in which damaged bone is repaired, mineral homeostasis is maintained, and bone is reinforced for increased stress. It involves replacement of old bone tissue by new bone tissue (formation and resorption are coupled).

Bone Resorption is the process by which osteoclasts break down bone and release the minerals from bone fluid to the blood.

Buffer is a chemical solution that resists change in pH by neutralizing both acids and bases to maintain the original acidity or alkalinity.

Calorie is a unit of measurement for the amount of energy that is released from food upon oxidation by the body. Carbohydrate, protein, fat and alcohol provide calories in the diet. Carbohydrate and protein have 4 calories per gram, fat has 9 calories per gram, and alcohol has 7 calories per gram.

Calpains are a family of cellular proteases that are activated by elevated cytosolic levels of free calcium.

Cancer is the uncontrolled growth of abnormal cells (also called malignant cells) in the body.

Carbon Cycle is one of the major biogeochemical cycles describing the flow of essential elements from the environment to living organisms and back to the environment again. This process is required for the building of all organic compounds.

Cardiovascular Disease (CVD) refers to a class of diseases that involve the heart, the blood vessels (arteries, capillaries, and veins) or both.

Carotenoids are natural red and yellow pigments synthesized by plants and other photosynthetic organisms. These compounds are responsible for the bright colors of various fruits and vegetables.

Cartilage is a gel-like, porous, elastic, slippery tissue that coats the ends of the bone. Normal cartilage provides a durable, low-friction, load-bearing surface for joints.

Catabolism is breakdown of more complex substances into simpler ones together with release of energy.

Catalase is a common enzyme found in nearly all living organisms that are exposed to oxygen, where it functions to catalyze the decomposition of hydrogen peroxide to water and oxygen. It has one of the highest turnover numbers of all enzymes; one molecule of catalase can convert 40 million molecules of hydrogen peroxide to water and oxygen each second.

Celiac Disease is a condition that damages the lining of the small intestine and prevents it from absorbing parts of food that are important for health. The damage is due to a chemical reaction from gluten in the food. This chemical is found in wheat, barley, rye, and possibly oats.

Cell Signaling is part of a complex system of communication that governs basic cellular activities and coordinates cell function.

Chemiosmotic Theory explains the movement of ions across a selectively permeable membrane, down their electro-chemical gradient. More specifically, it relates to the generation of ATP by the movement of protons (H+ ions) across membrane during cellular respiration.

Cholecalciferol is a form of vitamin-D, also called vitamin-D3. It is structurally similar to steroids such as testosterone, cholesterol, and cortisol.

Chondrocytes are basic cartilage cells, critical for balance and function of bone and joints.

Chronic Obstructive Pulmonary Disease (COPD) is a progressive dysfunction of the small airways causing poor airflow to the lungs and makes it hard to breathe. Such breakdown of the lung tissue breakdown is known as emphysema.

Circulatory System is a physiological system that circulates blood and lymph throughout the body.

Clinical Trial is a pre-planned study to evaluate the safety, efficacy, or optimum dosage schedule (if appropriate) of one or more diagnostic, therapeutic, or prophylactic drugs, devices, or techniques selected according to predetermined criteria of eligibility and observed for predefined evidence of favorable and adverse effects.

Collagen is the main protein found in all connective tissues of the body, including muscles, ligaments, cartilage and tendons. It provides the bone with a framework on which minerals can be laid down, most importantly calcium phosphate and hydroxyapatite.

Cortex is a tissue that forms the outer layer of an organ or structure. For example, cerebral cortex layer of neurons located in the cerebrum of the skull.

Cortisol is a hormone released by the cortex of the adrenal gland when a person is under stress. Cortisol helps regulate blood pressure and cardiovascular function, as well as the body's use of proteins, carbohydrates and fats.

Cystic Fibrosis (CF) is an inherited disease characterized by the buildup of thick, sticky mucus that can damage many of the body's organs.

Cytokines are chemicals that are secreted by various types of cells and act on other cells to stimulate or inhibit their function.

Daily Value (DV) is a reference value used on labels for dietary supplements and are based on a 2000 Calorie intake for adults and children. DV is based on Daily Reference Values (that apply to fat, saturated fat, cholesterol, carbohydrate, protein, fiber, sodium, and potassium) and Reference Daily Intakes (of essential vitamins and minerals).

Dehydrogenases are enzymes that catalyze oxidation reactions.

Dehydrogenation is a chemical process whereby hydrogen is removed from molecules. Often times is considered synonymous to oxidation.

Diabetes is a group of metabolic dysfunctions in which a person has high blood sugar, either because the pancreas does not produce enough insulin, or because cells do not respond to the insulin.

Dietary Supplement is any product taken by mouth that has a 'dietary ingredient', which may include vitamins, minerals, herbs, and amino acids, as well as substances such as enzymes, organ tissues, metabolites, extracts or concentrates. Dietary supplements can be found in many forms such as pills, tablets, capsules, liquids or powders.

Digestive system is a biological system which chemically breaks down food and nutrients to be used as energy by cells in the body.

DNA (or deoxyribonucleic acid) is a nucleic acid that contains the genetic instructions used in the development and functioning of all known living organisms and some viruses.

Dosage is the total amount of a dietary supplement given to, taken or absorbed by an individual.

Double-Blind Study is when neither the patient nor the doctor knows whether the patient is receiving the study drug or a placebo. The code is revealed at the end of the study.

Duchenne Muscular Dystrophy (DMD) is an inherited disorder that involves muscle weakness, which quickly gets worse.

DXA (previously DEXA), is a dual energy X-ray absorptiometry which measures bone mineral density (BMD) to provide evidence of osteoporosis (or loss of bone mass).

Efficacy is the capacity or power of a drug or dietary supplement to produce a desired effect or to affect a claimed health condition.

Electron Transport Chain is a process in which the NADH and [FADH2] produced during glycolysis, B-oxidation, and other catabolic processes are oxidized thus releasing energy in the form of ATP.

Electrophiles functional groups which are electron deficient and seek out electrons.

Emphysema is a type of chronic obstructive pulmonary disease (COPD) involving damage to the air sacs (alveoli) in the lungs. As a result, the body does not get the oxygen it needs and makes it hard to catch ones breath.

Endocrine System is a system of glands that produces hormones, which regulate body's growth, metabolism, and sexual development and function. The endocrine system includes the pituitary gland, thyroid, parathyroids, adrenal glands, pancreas, ovaries (in women) and testes (in men).

Endothelial Dysfunction (ED) is a systemic pathological state where there is an imbalance between vasodilating and vasoconstricting substances produced by (or acting on) the endothelium.

Enthalpy (H) – the heat content used or released in of a reacting system at constant pressure.

Entropy (S) – the quantitative measure of disorder [or randomness] in a system. The concept comes out of thermodynamics, which deals with the transfer of heat energy within a system.

Enzymes are complex proteins that cause a specific chemical change in all parts of the body. They are biological catalysts with three characteristics: (i) their basic function is to increase the rate of a reaction; (ii) they act specifically with only one reactant (called a substrate) to produce products; (iii) the are regulated from a state of low activity to high activity and vice versa. Most cellular reactions occur about a million times faster than they would in the absence of an enzyme.

Erectile Dysfunction (ED) is the inability to attain and/or maintain penile erection.

Estrogen is a female sex hormone essential for the development and functioning of the female reproductive system, including menstrual cycles and pregnancy. It has a key role in maintaining strong and healthy bones.

Excretory System is a biological system that regulates the chemical composition of body fluids by removing metabolic wastes and retaining the proper amounts of water, salts, and nutrients.

FOX (Forkhead box) Proteins – a family of transcription factors that play important roles in regulating the expression of genes involved in cell growth, proliferation, differentiation, and longevity.

Free Radicals are atoms or groups of atoms with an odd (unpaired) number of electrons. They can be formed when oxygen interacts with certain molecules. Due to unstable nature and high reactivity, free radicals can damage cells.

Gastric (peptic) Ulcer refers to open sores that develop on the inside lining of the stomach. The most common symptom of a peptic ulcer is abdominal pain.

Gibbs Free Energy (ΔG) – the amount of energy available for doing work in a chemical reaction at constant pressure and temperature. It combines thermodynamic concepts like enthalpy and entropy in order to determine the spontaneity of a reaction.

Glycolysis is the metabolic pathway that converts glucose ($C_6H_{12}O_6$), into pyruvate ($CH_3COCOO^- + H^+$). The free energy released in this process is used in the formation of high-energy compounds such as ATP and NADH. It is an anaerobic reaction.

G-Proteins – also known as guanosine nucleotide-binding proteins, are a family of proteins involved in transmitting signals from a variety of different stimuli outside a cell into the inside of the cell. They function as molecular switches.

Growth Hormone (or somatotropin) is a powerful anabolic hormone that affects all systems of the body and plays an important role in muscle growth.

Haloperoxidases are enzymes that are able to mediate the oxidation of halides by hydrogen peroxide.

Heart Failure (HF) – when the heart is unable to provide sufficient pump action to maintain blood flow (circulation) to meet body' demand.

Helicobacter pylori (H. pylori) is a micro-aerophilic bacterium found in the stomach. It is a common cause of peptic ulcers and *H. pylori* infection is present in about half the world population.

Hematopoiesis is the formation of blood cells in the body. All blood components derive from the hematopoietic stem cells located in the bone marrow.

Highly Active Antiretroviral Therapy (HAART) is the use of multiple drugs that act on different viral targets to suppress HIV replication and reduce the likelihood of the virus developing resistance.

Homeostasis is the physiological capacity of an organism to regulate itself by rapidly restoring internal conditions in response to sudden changes in the external environment.

Hormone Replacement Therapy (HRT) is the administration of exogenous hormones to replace those the body is unable to produce; HRT most often refers to the replacement of estrogen and progesterone following menopause.

Hormones are chemical messengers that transport signals from one cell to another in different parts of the body. Only a small amount of hormone is required to alter cell metabolism.

Human Immunodeficiency Virus (HIV) is a pathogen that attacks the immune system and leaves the body vulnerable to a variety of life-threatening infections and cancers. It causes AIDS.

Hydrolysis is a reaction in which water is used in breaking of a bond in a molecule. The reaction mainly occurs between an ion and water molecules and often changes the pH of a solution. In chemistry, there are three main types of hydrolysis: salt hydrolysis, acid hydrolysis, and base hydrolysis.

Hypertension is a chronic medical condition in which the blood pressure in the arteries is elevated.

Hypoxia is a pathological condition whereby the partial pressure of oxygen (PO_2) in the tissues is lower than normal.

Immune System is a biological system that protects against disease and foreign substances the body encounters.

***In Vitro* Fertilization (IVF)** is the joining of a woman's egg and a man's sperm outside the body. Usually it is done in a laboratory.

Inflammatory bowel disease (IBD) is a broad term that describes conditions with chronic or recurring immune response and inflammation of the gastrointestinal tract. The two most common inflammatory bowel diseases are ulcerative colitis and Crohn's disease.

Integumentary System protects the body from various kinds of damage from outside the body. It may serve to waterproof, cushion, and protect the deeper tissues, excrete wastes, and regulate temperature, and is the attachment site for sensory receptors to detect pain, sensation, pressure, and temperature.

IU (International Unit) is a unit of measurement of vitamin activity determined by biological methods rather than by direct chemical analysis.

Kinases are a family of enzymes that catalyze phosphoryl group transfers with ATP. The word 'kinase' is derived from the Greek word 'kinein' which means 'to move'.

Lactoferrin (LF) is a glyco-protein found naturally in milk, saliva, tears, blood, synovial fluid in bone joints, sperm, various secretions, tissues and other parts of body. LF is credited with numerous functions including metal regulation (iron, copper, zinc, manganese, cobalt, vanadium, etc.), antimicrobial, antioxidant, immune modulation, and anti-inflammatory effects. LF is a quintessential bio-replenishment for mammalian development and growth.

Lipids are a group of molecules which includes fats and oils, waxes, phospholipids, steroids, fat soluble vitamins (A,D,E,K) and some other related compounds. One common property they all have in common that they are hydrophobic. The main biological functions of lipids include energy storage, signaling, and acting as structural components of cell membranes.

Lymph is a clear, watery, sometimes faintly yellowish fluid derived from body tissues that contains white blood cells and circulates throughout the lymphatic system, returning to the venous bloodstream through the thoracic duct. Lymph acts to remove bacteria and certain proteins from the tissues, transport fat from the small intestine, and supply mature lymphocytes to the blood.

Lymphatic System is a biological system that oversees the absorption of excess fluid and its return to the blood stream, fat absorption and the immune system.

Mcg (Microgram) is equal to one millionth of a gram.

Menopause is the time in a women's life when menstruation permanently stops. Sometimes referred to as the 'change of life', occurs due to decrease in production of estrogen and progesterone by the ovaries.

Metabolism is a set of chemical reactions that occur in organisms to sustain life. These processes allow organisms to grow and reproduce, maintain their structures, and respond to their environment.

Mg (Milligram) is equal to one thousandth of a gram.

Mineral is a naturally occurring, inorganic substance with a definite chemical composition and a crystalline structure. Minerals are generally classified into the following chemical classes: silicates, carbonates, sulfates, halides, oxides, sulfides, phosphates and metals.

Mitochondria are structures found in most eukaryotic cells (the cells that make up plants, animals).They are sometimes described as 'cellular power house' because they generate most of the cell's supply of adenosine triphosphate (ATP) and are used as a source of chemical energy. In addition to supplying cellular energy, mitochondria are involved in other tasks such as signaling, cellular differentiation, cell death, as well as the control of the cell cycle and cell growth.

Muscle Atrophy is the wasting or loss of muscle tissue.

Muscular System is a biological system that permits movement of the body and maintains posture.

Myocardial Reperfusion Injury is the tissue damage caused when blood supply returns to the tissue after a period of ischemia or lack of oxygen.

Nervous System is a biological system that coordinates the voluntary and involuntary actions by transmitting signals between different parts of its body.

Nitrogen Cycle is the process by which nitrogen is converted between its various chemical forms from the environment to living organisms and back to the environment again. The important processes in the nitrogen cycle include fixation, ammonification, nitrification, and denitrification.

Non-steroidal anti-inflammatory drugs (NSAID)s are a class of drugs that provide analgesic (pain killing) and antipyretic (fever-reducing) effects, and, in higher doses, anti-inflammatory effects.

Nucleic Acids are a group of complex compounds made up of nucleotides. They are found in all living cells and viruses and are composed of purines, pyrimidines, carbohydrates, and phosphoric acid. In the form of DNA and RNA they control cellular function and heredity.

Nucleophiles are functional groups that are capable of donating electrons.

Osteoarthritis (OA) is a non-inflammatory degenerative joint disease that occurs mostly among the elderly. The cartilage in joints breaks down; as a result, the joint becomes irritated and swollen (inflamed).

Osteoblasts are cells responsible for making new bone, maintaining the balance of calcium in the blood and bone. They form new bone tissue in response to increasing demands of the bone turnover.

Osteoclasts are cells that break down the bone, release minerals into the blood stream, to serve various functions in the body. They dissolve bone tissue by secreting enzymes and regulate calcium and phosphate levels in body fluids.

Osteoporosis (porous bone) is a skeletal condition that develops when bone is no longer replaced as quickly as it is removed. Abnormal loss of bony tissue results in fragile bones due to low BMD; most common in post-menopausal women.

Oxidation is the loss of electrons or an increase in oxidation state by a molecule, atom, or ion.

Oxidative Phosphorylation (Respiration) is an apparently more complex process where a proton-motive force across a membrane powers an ATP synthase enzyme complex (a molecular machine) to make ATP from ADP and inorganic phosphate. Oxidative phosphorylation is coupled to REDOX pathways that release large amounts of free energy from food molecules, and makes the vast majority of the ATP in respiring cells.

Oxidative Stress is an imbalance in free radical formation within a cell or organism, most commonly in the form of reactive oxygen or nitrogen species (ROS/RNS), and a biological system's ability to readily detoxify the reactive intermediates or to repair the resulting damage.

Parkinson's Disease (PD) is a progressive, age-related mobility disorder with neuropathology characterized by degeneration of neurons in the substantia nigra.

Peroxidase belongs to a group of enzymes that catalyze the oxidation of a substance by a peroxide.

Peroxiredoxins (PRx) are a ubiquitous family of antioxidant enzymes that also control cytokine-induced peroxide levels which mediate signal transduction in mammalian cells.

pH (potential of hydrogen) is the measure of the acidity or alkalinity of a solution. The pH scale ranges between 0 and 14. Water with a pH of 7 is neutral; a pH of less than 7 is acidic; a pH greater than 7 is alkaline.

Phagocyte is a cell that engulfs and absorbs waste material, harmful microorganisms, or other foreign bodies in the bloodstream and tissues. They play an important role in biological immunity as they protect organisms from diseases and are essential for fighting infections.

Photo-Phosphorylation (Photosynthesis) is a process used by plants and some microorganisms such as algae and some types of bacteria, to convert light energy, normally from the sun, into chemical energy to form organic compounds. Some compounds such as sugars that are formed from CO2 and water can be used to fuel the organisms' activities.

Portal Hypertensive Gastropathy involves changes in the mucosa of the stomach in patients with portal hypertension; by far the most common cause of this is cirrhosis of the liver.

Prooxidants are chemicals that induce oxidative stress, either by generating reactive oxygen species or by inhibiting antioxidant systems. The oxidative stress produced by these chemicals can damage cells and tissues partly through its production of reactive oxygen species.

Proteins are large molecules consisting of one or more chains of amino acids. They perform a vast array of functions within living organisms such as catalyzing metabolic reactions, replicating DNA, responding to stimuli, and transporting molecules from one location to another,transport oxygen and other substances within an organism.

Recommended Daily Allowance (RDA) is the average daily dietary intake level that is sufficient to meet the nutrient requirement of nearly all (97 to 98 percent) healthy individuals in a particular life-stage and gender group.

REDOX (reduction-oxidation) reactions include all chemical reactions in which atoms have their oxidation state changed. REDOX reactions involve the transfer of electrons between chemical species.

REDOX Couple is an oxidizing and reducing agent which appear on opposite sides of a half-equation. REDOX couples are analogous to conjugate acid-base pairs and behave in much the same way. The stronger an oxidizing agent, the weaker the corresponding reducing agent, and the stronger a reducing agent, the weaker the corresponding oxidizing agent.

REDOX Homeostasis is the maintenance of constant electro-chemical potential and ionic-concentration gradient across a cellular boundary, despite fluctuations in the electro-chemical potential of the external environment and changing identities and activities of electron donors and acceptors.

REDOX Medicine is the field which deals with the metabolic states of health and disease, based on the movement of reactive chemical species (ROS, RNS, RSS, etc) through cellular pathways.

REDOX potential (Eh) is a measure of the affinity of a substance for electrons (its electro-negativity). This is also known as standard electrode potential.

REDOX Signaling is cellular communication using short-lived ROS, RNS or RSS, or through changes in the electrochemical potential of REDOX couples within biological systems.

Reduction is the gain of electrons or a decrease in oxidation state by a molecule, atom, or ion.

Reproductive System is a biological system that contains the means for reproduction.

Respiratory (or Ventilatory) System is a biological system which supplies the blood with oxygen in order to deliver oxygen to all parts of the body.

Rheumatoid Arthritis (RA) is an autoimmune disorder where the body attacks itself; the inflamed lining of the joint (synovium) progresses to severe damage and in few cases, deformity. Inflamed joints are painful and stiff.

Ribonuclease (RNase) is a protein molecule involved in the formation of new blood vessels. Milk contains RNase types 2 and 4.

Ribonucleic acid (RNA) belongs to a family of large bio-functional molecules with vital roles in the coding, decoding, regulation and expression of genes.

Senescence is the organic process of growing older and showing effects of increasing age.

Sepsis is a complex inflammatory host response to microbial infection, frequently occurs after trauma, hemorrhage, burn or surgery and manifested as 'Systemic Inflammatory Response Syndrome (SIRS)'.

Sex Hormones are a class of steroid hormones that regulate the growth and function of the reproductive organs or stimulate the development of the secondary sexual characteristics.

Skeletal System is a biological system which provides the frame and stability to the body.

Substrate-level Phosphorylation (Fermentation) is a type of metabolic pathway that results in the formation of adenosine triphosphate (ATP) or guanosine triphosphate (GTP) by the direct transfer and donation of a phosphoryl (PO3) group to adenosine diphosphate (ADP) or guanosine diphosphate (GDP) from a phosphorylated reactive intermediate.

Sulfur Cycle is the collection of processes by which sulfur moves to and from minerals and biological systems. It involves Assimilative sulfate reduction, Desulfurization, Dissimilative sulfur reduction, Oxidation of hydrogen sulfide, Oxidation of elemental sulfur.

Synovium is the fluid-filled lining of a bone joint. It secretes synovial fluid that lubricates tissues where friction would otherwise occur and also supplies nutrients and oxygen to the cartilage.

T-Cells are a type of white blood cells called lymphocytes that play an important role in the immune system such as helping the body fight diseases or harmful substances.

Testosterone is a sex hormone secreted by the adrenal glands in both males and females. This anabolic steroid plays an important role in immune function and libido.

Thermodynamics 1st Law: The increase in internal energy of a closed system is equal to the difference of heat supplied to the system and the work done by it.

Thermodynamics 2nd Law: Heat cannot spontaneously flow from a colder location to a hotter location.

Thermodynamics 3rd Law: As a system approaches absolute zero the entropy of the system approaches a minimum value.

Transition metals refer to 40 elements that belong to groups 3–12 of the periodic table. In the atomic ground state, the highest-energy electron of those elements is in the 'd-orbita'l. Most of these elements have an incomplete 'd' sub-shell, or can give rise to cations with an incomplete 'd' sub-shell.

Vitamin is an organic substance essential in small amounts to support growth and activity of the body. Vitamins can be obtained from plant or animal sources.

Water Cycle is the evaporation and condensation cycle that controls the distribution of the earth's water as it evaporates from bodies of water, condenses, precipitates, and returns to those bodies of water. It is also called the hydrologic cycle.

Image Credits

Note: The artist, photographer or copyright holder name is followed by a parenthesis (which refers to the page number in the book with the corresponding image).

Juan Carlos Torres (1, 7, 19, 27-29, 32, 33, 36, 37, 47, 48, 51, 57, 83, 86, 90, 103, 105, 119, 120, 123, 142, 136, 148, 153, 159, 163, 165, 167, 173, 184, 189, 205, 206, 208, 215, 217, 218, 227, 234, 242, 245, 246, 258-260, 266, 269, 278, 285, 288, 289, 294, 301, 305, 309, 310, 317, 318, 332, 348, 351, 356, 389, 390, 430, 446, 448, 477, 479, 483, 485, 487, 489); Jennifer Walker Dickson (Dr. Naidu's Image).

The following copyrighted images were licensed from Shutterstock:

Front pages: Mopic (ii); abstract (v); Aprilphoto (vi); and Wieteke de Kogel (xiii).

Chapter-1: 89studio (2); Eric Isselee (4); Bruce Rolff (14); Mara008 (16); Leonid Andronov (18); Lorelyn Medina (20); Leonid Andronov (21, 22); Triff (23); Mopic (24); petarg (30); BlueRingMedia (34); Mopic, yuriy kulik (35); molekuul.be (39); and Ashray Shah (41).

Chapter-2: Luka Skywalker (50); Andrei Marincas (52, 55-58, 70, 71); Designua (54); mart, snapgalleria (56); iurii (59); molekuul.be (61, 80, 82, 88, 95-97); gashgeron (75); Sergey Dubrov (77); 3DDock (79); Poznyakov (81); LSaloni (85); Sebastian Kaulitzki (87); O2creationz (89); Leonid Andronov (92, 93); Vladimir Arndt (99); and Gayvoronskaya_Yana (100).

Chapter-3: Alila Medical Images (127); ancroft (137); Alex Luengo (143); Sebastian Kaulitzki (146); Natalia Karpova (149); stanislaff (150); Photosani (151); Mopic (161); Jan Kaliciak (165); Pan Xunbin (171); Juan Gaertner (181); Sergey Nivens (183); and andreiuc88 (203).

Chapter-4: Alila Medical Images (229, 236, 237, 240, 244, 247, 248, 255, 261, 263, 265, 267, 273, 275, 280, 282, 283, 304, 306, 307,317, 322, 324, 336, 342, 343, 352); Sebastian Kaulitzki (219, 277, 299, 308, 334); VALIK-NOVIK (221); Blamb (222, 286, 339, 345, 349); udaix (224); Creations (225); Juan Gaertner (229); molekuul.be (230); bikeriderlondon (231); BlueRingMedia (235, 338, 359); yumofoto (239); koosen (252); stihii (262, 313, 314); Sergii Korolko (271); BioMedical (276); Alex Mit (279); petarg (281); Designua (282, 283, 311); Heiti Paves (291); Fedorov Oleksiy (292); Marcin Pawinski (293); Alex Luengo (296, 304, 311, 333); Anatomical Design (297); Sofia Santos (313); Giovanni Cancemi (321); arka38 (322); petarg; Bochkarev Photography (327); Leonid Andronov (328); ermess (331); GRei (333); hkannn (338); Leptospira (340); jovan vitanovski (345); nobeastsofierce (347); GRei (353); Levent Konuk (355); and ramcreations (364).

Chapter-5: agsandrew (395); Best3d (396, 399, 403, 407, 410, 418, 422, 426); bitt24, HLPhoto, iampuay, Gayvoronskaya_Yana (398); GGRIGOROV, florin opera, Immagy, silver-john (402); Mopic (405); JIANG HONGYAN, Stephen Mcsweeny, O.Bellini, phomphan (406); byggarn.se, Ambient Ideas, Valentyn Volkov, wjarek (408); Andrea Danti (409); Shawn Hempel, Bairachnyi Dmitry, Silberkorn, vsl (414); Elena Schweitzer (416); Chayapol Plairaharn, Boris15, Curioso, JeniFoto (421); Iakiv Pekarskyi, Elena Itsenko, Christian Fischer, kaband (425); joannawnuk, bitt24, nui7711 (430); molekuul.be (432, 460); Joerg Beuge, Africa Studio, Volosina, Smit (438); petarg (434, 439, 442, 443); Hayati Kayhan, chien321, Africa Studio, Paul Cowan (441); Peeradach Rattanakoses (444); Milosz Bartoszczuk, Monticello, Madlen, Sergio Martinez (445); Maksim Chaikou (451); Sebastian Kaulitzki (459); Heiti Paves (461); and svry, Nataliia Pyzhova, Volosina, Little_Desire (463).

Synopsis: 89studio (481)

The following copyrighted images were from Wikipedia: "File:Glycolysis2.svg" © 2012 JohnyAbb, used under Creative Commons Attribution-Share Alike 3.0 license: HYPERLINK "http://creativecommons.org/licenses/by-sa/3.0/deed.en" http://creativecommons.org/licenses/by-sa/3.0/deed.en (40)

Notes

Notes